LOWER PALAEOZOIC ROCKS OF THE WORLD

Volume 1 · Cambrian of the New World

Cambrian
of the
New World

Edited by
C. H. HOLLAND

WILEY–INTERSCIENCE
a division of John Wiley & Sons Ltd
LONDON   NEW YORK   SYDNEY   TORONTO

Library of Congress Catalog Card No. 70-122342
ISBN 0 471 40624 4

Frontispiece: *Low-level oblique aerial photograph looking
north-westwards along the axis of the major syncline in the
Main Range of the central Rocky Mountains. Mount Robson
(3,954 m), shown in the centre, is the highest mountain in the
Canadian Rockies and consists entirely of Cambrian strata.
It represents one of the finest continuous Cambrian sections
in the New World. Virtually all the rocks shown in the photograph are Middle and Upper Cambrian carbonates over
3,000 metres thick.*

Photograph by K. T. Hyde of Calgary.

*Printed in Great Britain by
J. W. Arrowsmith Ltd., Winterstoke Road,
Bristol BS3 2NT*

PREFACE

This volume on the Cambrian System in the New World is the first to be published of a projected series intended eventually to cover the Lower Palaeozoic rocks of the world. The intention is to provide a survey which is at once authoritative and up to date, comprehensive and yet reasonably circumscribed. The wide scope of the individual volumes and the relatively small number of authors responsible for them are intended to result in true syntheses, illuminated by the authors' own views on treatment and emphasis, but retaining some measure of common organization.

The six parts of the present volume range across much of the length of the earth from arctic North America to Chile. The beautifully exposed bare rock sections of Arctic Canada and Greenland, tantalizingly difficult of access, provide but part of a story to be pursued in terms of Spitsbergen and Scotland in a subsequent volume. The Cambrian rocks of the remainder of Canada range widely from the Newfoundland development, which fits so convincingly with the Cambrian picture of the British Isles, to the splendid limestone sections of the Rocky Mountains and to the Burgess Shale of British Columbia: a connoisseur's rock to all palaeontologists. The Cambrian of the United States with its grand pattern of central incomplete cratonic development and flanking areas of the Appalachian and western regions provides incidentally many examples of the intricate evolution of modern stratigraphical nomenclature. Finally, the scattered evidence of the Cambrian Period in the whole sub-continent of South America is brought together in the last part of the volume, which treats not only the comparatively well known sections of Argentina but also the more ambiguous records from other South American countries.

Later volumes of the series will not only describe the three Lower Palaeozoic systems throughout the world in a regional sense, but will also include parts on faunas and other more general features. Thus, a subsequent Cambrian volume will include an essay on the Pre-Cambrian–Cambrian boundary and an introduction to the Lower Palaeozoic as a whole.

Charles Hepworth Holland

Trinity College,
Dublin.

CONTRIBUTING AUTHORS

A. V. Borrello   *Department of Geology, Faculty of Natural Sciences, National University of La Plata, Argentina*

J. W. Cowie   *Department of Geology, University of Bristol, England*

Christina Lochman-Balk   *New Mexico Institute of Mining and Technology, Socorro, New Mexico*

F. K. North   *Carleton University, Ottawa, Canada*

Allison R. Palmer   *Department of Earth and Space Sciences, State University of New York at Stony Brook, U.S.A.*

CONTENTS

THE CAMBRIAN OF THE GREAT BASIN AND ADJACENT AREAS, WESTERN UNITED STATES

Allison R. Palmer

Department of Earth and Space Sciences, State University of New York at Stony Brook

Contents

1. Introduction

The Great Basin of western United States is a broad semi-arid region that includes parts of western Utah, south-eastern Idaho, south-eastern California, and most of Nevada. It is bounded on the north by Tertiary and Quaternary volcanic rocks of the Snake River Plain, on the west and south-west by the Sierra Nevada Mountains and the San Andreas fault, and on the east by the Wasatch Front, Colorado Plateau, and the lower reaches of the Colorado River. The area is characterized topographically by numerous isolated elongated north-south mountain ranges usually less than 20 kilometres wide and from 80 to 160 kilometres long. These rise about 1,600 metres above intervening broad, flat-bottomed basins averaging 10 to 20 kilometres in width. Cambrian rocks are exposed in many of these mountain ranges and also in the high country of the Wasatch Mountains along the eastern margin of the Great Basin in northern Utah and south-

1

eastern Idaho, as well as the lower part of the Grand Canyon which dissects the Colorado Plateau in north-western Arizona (Figs. 1 and 2).

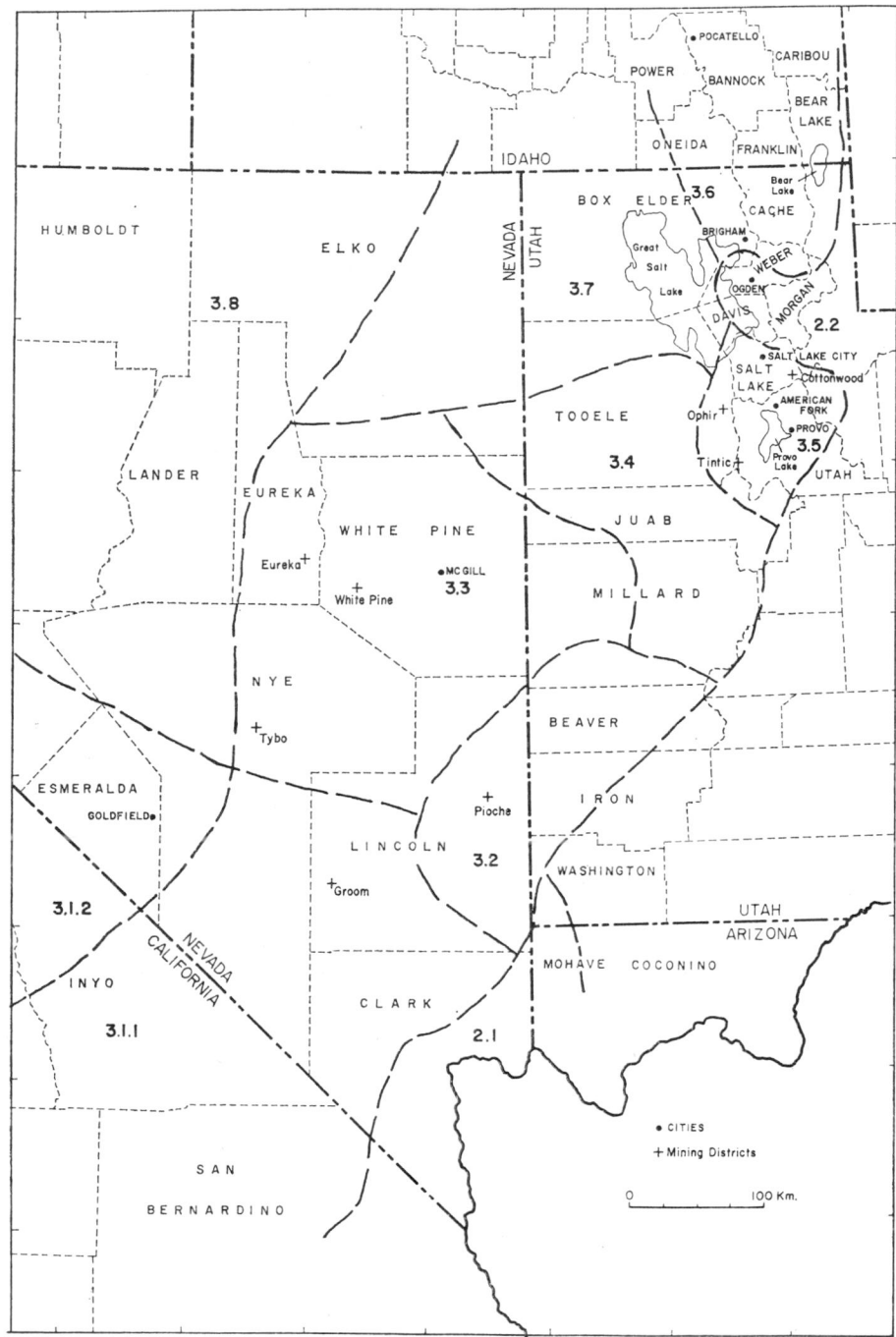

FIG. 1. Index map of the Great Basin region showing principal geographical features, exclusive of outcrop areas, mentioned in the text and outlines of the areas described in Chapters 2 and 3. Numbers refer to chapter headings.

The thickness of the blanket of Cambrian sediments that covered the Great Basin and adjacent areas on the east generally exceeded 2,700 metres. During the late Palaeozoic, this blanket was disrupted along the north–south Antler orogenic belt in central Nevada by eastward overthrusts with tens of kilometres of displacement (ROBERTS and others, 1958). During the Late Mesozoic, a parallel overthrust belt, the Sevier orogenic belt (ARMSTRONG, 1968a), with similar large horizontal displacements disrupted the Cambrian blanket near the eastern margin of the region. Lesser disruptions took place along numerous low-angle normal faults of pre-Tertiary age, with displacements up to a few kilometres, as well as generally younger high-angle normal faults. Intrusions of Mesozoic and Tertiary ages have metamorphosed some Cambrian sections.

The numerous partial to complete Cambrian sections now exposed in many of the mountain ranges provide information about events that influenced sedimentation and faunal evolution during most of the Cambrian period over an area of about 400,000 square kilometres. The Cambrian environments at different times ranged from near-shore in the eastern part of the region across a complex of carbonate banks to deeper water sediments in the western areas. In order to develop the full story of the Cambrian history of the whole region, it will be necessary first to consider sub-regions where there is some degree of stratigraphical homogeneity.

1.1. History of Cambrian studies in the Great Basin region

The earliest Cambrian fossils reported from the Great Basin region were archaeocyathids in a small collection from the Silver Peak area of the White–Inyo Mountains region (Chapter 3.1.1) (MEEK, 1868). Descriptions of other small collections, mostly of trilobites and brachiopods, made by geologists of the federally supported western reconnaissance surveys appeared during the following 15 years, but the major early studies of the Cambrian System are all included in the works of C. D. Walcott. His monograph on the palaeontology of the Eureka District (WALCOTT, 1884) was the first of a series of publications, spanning a period of 40 years, which established much of our present knowledge of the Cambrian System of the region. WALCOTT's extensive researches on the Cambrian rocks of all of North America (YOCHELSON, 1967) are still a source of much valuable information, even though they have been superseded in many instances by more detailed reports of the areas which he studied.

Because WALCOTT's work was not confined to the Great Basin region, many of his larger publications include data from this region along with data from other parts of the continent. These include general stratigraphical summaries (WALCOTT, 1891a, 1891b, 1908b) and descriptive reports on various fossil groups (WALCOTT, 1886, 1908a, 1910, 1912, 1916a, 1916b, 1925). Following WALCOTT's death in 1927, C. E. RESSER, who had been his assistant, continued to publish descriptions and nomenclatural modifications of Cambrian fossils from the extensive collections in the United States National Museum. Many of his papers involve species and specimens from the Great Basin region (RESSER, 1928, 1935, 1936, 1937, 1938, 1939a, 1939b, 1942). Contributions to the regional aspects of the stratigraphy of the Cambrian System in the period before 1950 were made by DEISS (1938) and WHEELER (1943, 1948). DEISS re-examined the Great Basin sections described by WALCOTT (1908b) and published revised descriptions. WHEELER prepared a series of regional cross-sections and synthesized most of the stratigraphical data existing to that time. A substantial increase in interest in this region by the United States Geological Survey in the years following the Second World War, as well as an influx of students from many universities, has produced a steady flow of new information about

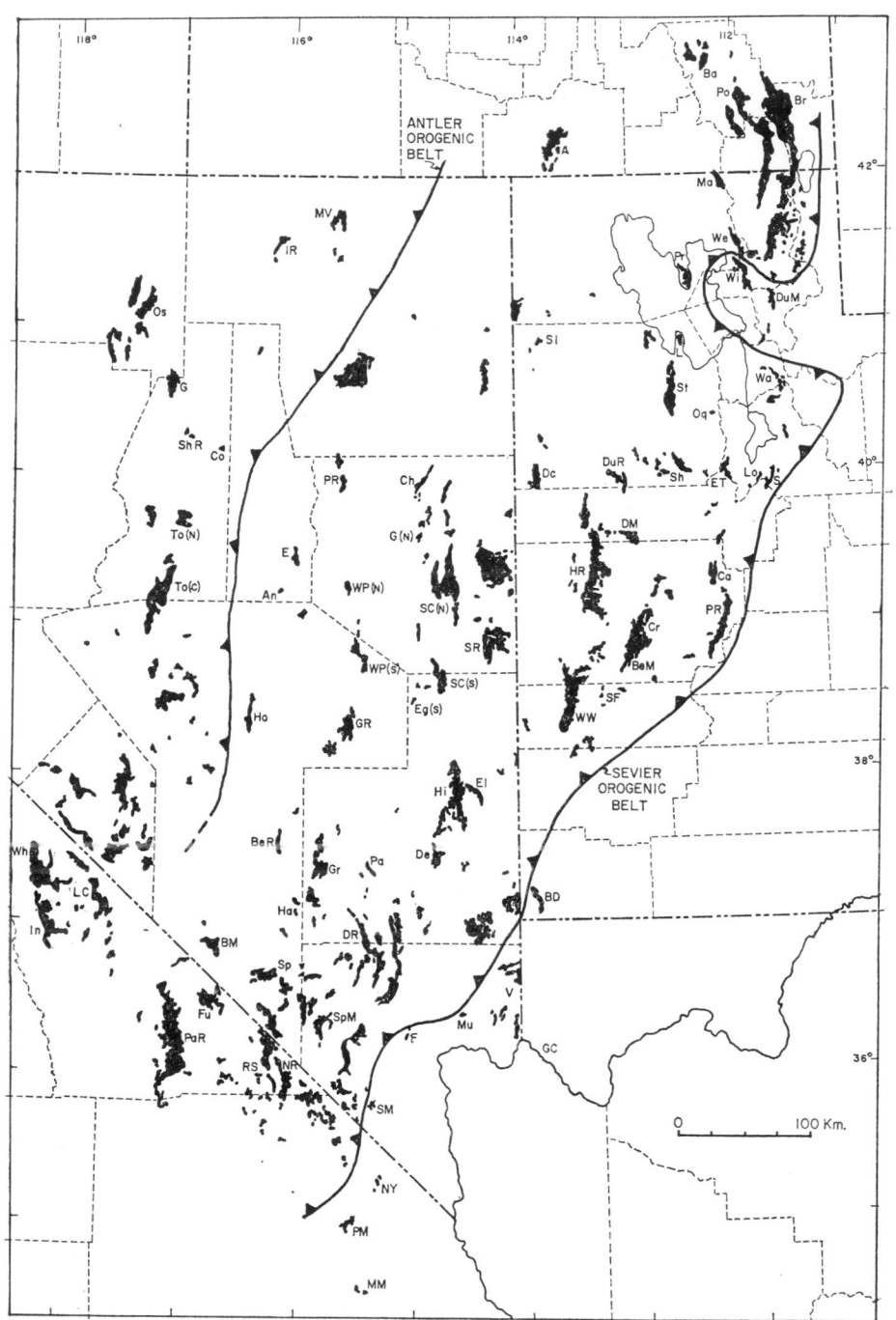

A—Albion Range
An—Antelope Range
Ba—Bannock Range
BM—Bare Mountain
BR—Bear River Range
BeM—Beaver Mountains
BD—Beaver Dam Mountains
BeR—Belted Range
Ca—Canyon Range
Ch—Cherry Creek Range
Co—Cortez Range
Cr—Cricket Range
DC—Deep Creek Range
De—Delamar Range
DR—Desert Range
DM—Drum Mountains
DuR—Dugway Range
DuM—Durst Mountain
E—Eureka area
ET—East Tintic Mountains
G(N), Eg(S)—northern and southern Egan Range
El—Ely Range
F—Frenchman Mountain
Fu—Funeral Mountains
G—Galena Range
GC—Grand Canyon
GR—Grant Range
Gr—Groom Range
Ha—Halfpint Range
Hi—Highland Range
Ho—Hot Creek Range
HR—House Range
IR—Independence Range
In—Inyo Mountains
L—Lakeside Mountains
LC—Last Chance Range
Lo—Long Ridge
Ma—Malad Range

MM—Marble Mountains
Mo—Morman Mountains
MV—Mt. Velma area
Mu—Muddy Mountains
NY—New York Mountains
NR—Nopah Range
Oq—Oquirrh Mountains
Os—Osgood Mountains
Pa—Pahranagat Range
PaR—Panamint Range
PR—Pavant Range
Po—Portneuf Range
Pr—Promontory Range
PM—Providence Mountains
RS—Resting Springs Range
RR—Ruby Range
SF—San Francisco Mountains
S—Santaquin area
SC(N)(S)—northern and southern Schell Creek Range
SM—Sheep Mountain
Sh—Sheeprock Range
ShR—Shoshone Range
SI—Silver Island Range
SR—Snake Range
Sp—Specter Range
SpM—Spring Mountains
St—Stansbury Range
To(N)(C)—northern and southern Toiyabe Range
V—Virgin Mountains
WW—Wah Wah Mountains
Wa—Wasatch Mountains
We—Wellsville Mountains
Wh—White Mountains
WP(N)(S)—northern and southern White Pine Range
Wi—Willard Peak

FIG. 2. Index map of the Great Basin region showing distribution of known areas of outcrop of Cambrian rocks, with identifications for the areas mentioned in the text.

Cambrian rocks and faunas since that time. Most of this newer information, as well as older references dealing with local areas, is cited in the appropriate chapters that follow.

A previous synthesis of existing knowledge about the Cambrian System in the Great Basin was prepared for the Twentieth International Geological Congress in Mexico (PALMER, 1956). Since then, stratigraphical summaries of parts of the Cambrian System for areas within the Great Basin have been presented by MAXEY (1958), ROBERTS and others (1958), BENTLEY (1958), ROBISON (1960), PALMER (1960a), STEWART (1966), and BARNES and CHRISTIANSEN (1967). Regional faunal studies have been published only for parts of the Late Cambrian by ROBISON (1960) and PALMER (1965).

This chapter could not have been prepared without the experience gained by the author during the period 1950–1966 when he was Cambrian geologist for the United States Geological Survey. He takes sole responsibility for all interpretations presented here and for all undocumented information. In addition to published material, the following persons permitted the author to use their unpublished stratigraphical data or their maps of areas of Cambrian outcrops needed for the preparation of Figure 2: H. J. BISSELL, H. R. CORNWALL, L. F. HINTZE, ROGER HOPE, R. K. HOSE, F. J. KLEINHAMPL, S. S. ORIEL, and J. S. SHELTON.

1.2. Classification of the Cambrian System

Division of the Cambrian System into lower, middle and upper series was first suggested by WALCOTT (1891) and has been followed by all subsequent workers. At least two names have been proposed for each of the series (Lower: Waucoban, Georgian; Middle: Albertan, Acadian; Upper: Saratogan, Croixan). These names were derived from regions where the particular series were believed to be best developed. The merits of use of particular names have been discussed at various times but, because of the relatively simple three-fold division of the system, they are not required for clear communication. Most workers consider these names to be unnecessary jargon in a science already cluttered with such terms, and prefer to use only lower, middle and upper for the Cambrian series.

A stage classification has been proposed only for the Upper Cambrian (HOWELL and others, 1944), based on the formational succession of the Minnesota–Wisconsin region (LOCHMAN-BALK, this volume pages 79–167), and the names Dresbachian, Franconian, and Trempealeauan are currently used to identify the lower, middle and upper parts of the Upper Cambrian Series. Only the top of the Trempealeauan Stage, however, which is probably a biomere boundary (Chapter 4) has any biostratigraphical significance outside of the type region. In southern and western United States where the faunal successions are nearly complete, the upper and lower boundaries of the Franconian Stage and the lower boundary of the Dresbachian Stage are within evolutionary complexes of the Conaspid, Pterocephaliid and Crepicephalid biomeres. In these areas, the stage boundaries have been arbitrarily placed at interzonal boundaries.

The problem of an adequate zonal classification is discussed in the sections of this book dealing with the Great Basin and Mid-continent regions of the United States (see pages 55–59 and Lochman chapter). There is still a considerable amount of work to be done in the Cordilleran region of both the United States and Canada before a stable zonation of the Cambrian System for North America can be established.

2. Cambrian rocks east of the Sevier orogenic Belt

Two areas of autochthonous Cambrian sequences east of the Sevier orogenic belt are known. One in the south-western part of Utah, southern Nevada, and extreme south-eastern California and the other in the area between Salt Lake City and Ogden in northern Utah (Fig. 1). Both of these areas are characterized by relatively thin Cambrian sequences that rest unconformably on Pre-Cambrian metamorphic rocks and are unconformably overlain by Devonian or younger Palaeozoic rocks.

2.1. The Marble Mountains–Virgin Mountains region

In the Marble Mountains–Virgin Mountains region, autochthonous sequences are known in the Beaver Dam Mountains of south-western Utah (REBER, 1952); the Mormon Mountains (WHEELER, 1943), Virgin Mountains (McNAIR, 1951), Muddy Mountains LONGWELL, 1928), southern Spring Mountains (HEWETT, 1931, 1956), Frenchman Mountain (LONGWELL, unpublished), and Sheep Mountain (HAZZARD and MASON, 1953) in south-eastern Nevada; and in the New York Mountains (HEWETT, 1956), and Providence and Marble Mountains (HAZZARD, 1933; HAZZARD and MASON, 1936) of south-eastern California (Fig. 3).

All of the sequences consist of a basal sandstone or quartzite formation resting unconformably on Pre-Cambrian igneous or metamorphic basement. This formation is overlain by shales and siltstones with minor limestone interbeds that are, in turn, overlain by a thick sequence of limestones and dolomites. At least the upper half of all sections is characterized by dolomites. In all of the areas, the Cambrian rocks are unconformably overlain by rocks of Devonian or younger Palaeozoic ages.

The general stratigraphical succession and relation to younger and older rocks is a westward continuation of the Cambrian stratigraphy described in the Grand Canyon by McKEE (1945), which includes the basal Tapeats Sandstone, the Bright Angel Shale, the Muav Limestone, and an upper unnamed dolomite unit. However, the stratigraphical nomenclature applied to these rocks includes locally derived names and names derived from either the Grand Canyon sequence or from Cambrian sequences of the allochthonous Pioche area to the north (Chapter 3.2).

The basal formation is a cross-bedded, generally coarse-grained sandstone or quartzite described as either the Tapeats Sandstone or the Prospect Mountain Quartzite. Its varying thickness shows no regional trends and seems largely to reflect original topographical irregularities on the surface of the Pre-Cambrian gneissic terrain over which it transgressed. The thicknesses range from about 40 metres on Sheep Mountain and Frenchman Mountain to about 325 metres in the Marble Mountains. Locally a basal conglomerate is developed and conglomeratic beds or lenses within the sandstone sequence are also locally present.

Although the formation has yielded no fossils, the lower beds of the overlying shale and siltstone sequence have yielded late Early Cambrian olenellids at many localities so the sandstone is almost certainly of Early Cambrian age.

In the Providence and Marble Mountains of California, the generally shaly or silty sequence lying between the older quartzites and younger carbonates has been divided into three formational units by HAZZARD (1933, 1954). The lowest of these is the Latham Shale, a greenish-grey platy shale 15 to 20 metres thick with some thin interbeds of sandy limestone. This unit has yielded a rich fauna of olenellids described by RESSER (1928), CRICKMAY (1933), and RICCIO (1952), including species of *Bristolia, Paedeumias*, and *Olenellus*.

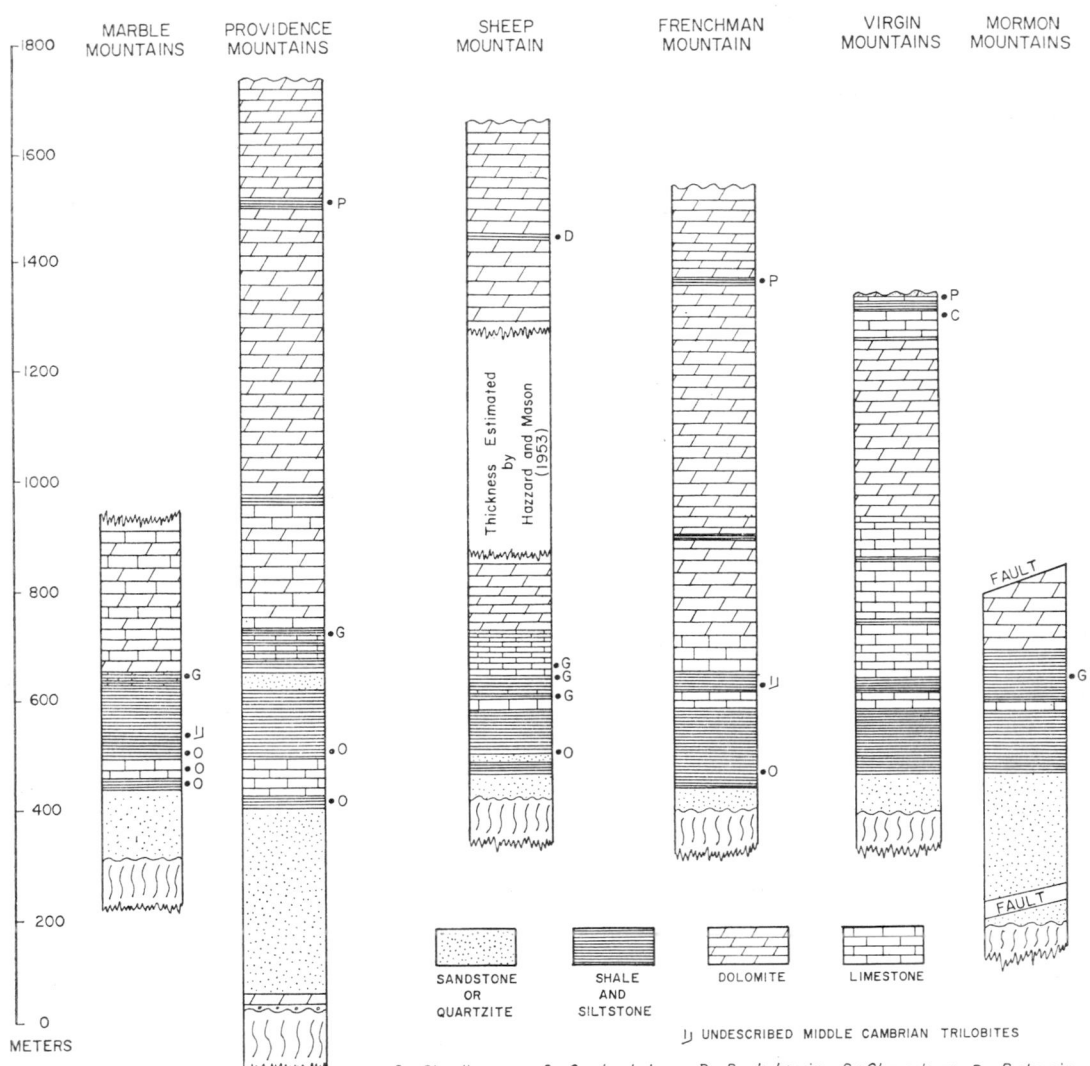

FIG. 3. Comparison of the Cambrian sections of the Marble Mountains–Virgin Mountains area.

The middle unit, the Chambless Limestone, is a massive weathering dark to light grey limestone that is 60 metres thick in the Providence Mountains and 30 metres thick in the Marble Mountains. This unit is characterized by abundant oncolites which are commonly identified as *Girvanella* in American Cambrian sections. Olenellid trilobites are also reported from this unit but have not been described.

The upper unit, the Cadiz Formation, is a heterogeneous succession of shale, siltstone and thin limestone or dolomite beds ranging in thickness from 155 metres in the Marble Mountains to 210 metres in the Providence Mountains. The lower 10 metres of the Cadiz Formation has yielded olenellid trilobites and the upper beds have yielded the early Middle Cambrian genus *Glossopleura* and some associated forms. Older undescribed Middle Cambrian trilobites have been collected from beds about 40 metres above the base of the formation. The formation is poorly fossiliferous, and the only descriptions of trilobites from it are by RESSER (1928), MASON (1935), and STOYANOW and SUZUKI (1955).

To the north, there is no equivalent of the Chambless Limestone and the entire interval of shales and siltstones between the quartzites, and the lowest distinctly cliff-forming carbonate has been referred to either the Pioche Shale or the Bright Angel Shale. This interval averages about 110 metres in thickness between Sheep Mountain and the Mormon Mountains and is about 60 metres thick in the Beaver Dam Mountains (REBER, 1952).

Overlying the Bright Angel (Pioche) in all areas north of the Providence Mountains is a limestone unit ranging from 16 to 60 metres in thickness that has been identified as the Lyndon Limestone (WHEELER, 1943; McNAIR, 1951; REBER, 1952). This unit is usually separated from the higher main mass of carbonate rocks by a shaly interval ranging from 30 to 80 metres in thickness which has been identified by WHEELER, McNAIR, and REBER (*op. cit.*) as the Chisholm Shale. The Chisholm interval has yielded trilobites of the genus *Glossopleura* at many localities, indicating a correlation with the upper beds of the Cadiz Formation to the south.

The remainder of the Cambrian section above the Chisholm Shale or the Cadiz Formation is predominantly dolomite. In the Providence Mountains and at Frenchman Mountain the dolomite section is over 700 metres thick. Elsewhere it is thinner because of greater pre-Devonian erosion. In the Mormon and Virgin Mountains, WHEELER (1943) identified this carbonate unit as the Peasley Limestone and correlated it with rocks of similar stratigraphical position in the Pioche district (see Chapter 3.2). McNAIR (1951) identified several of the Pioche units in the Virgin Mountains. The validity of these identifications which extend minor carbonate units of the thick allochthonous sequences into the thinner autochthonous sequences is still to be determined.

In the autochthonous carbonate sequence, from the Marble Mountains to the Virgin Mountains, there are two thin shaly marker units: one is about 215 metres above the base of the sequence, and the other is 415 to 450 metres higher. The upper shaly unit, which has yielded Upper Cambrian trilobites of Late Dresbachian age (PALMER and HAZZARD, 1956) is correlated with a similar shaly unit at the base of the Nopah Formation in the allochthonous sequences of the Death Valley region (page 15). In the Providence Mountains, the dolomites below this shaly marker unit, including the lower shaly marker unit, were described as the Bonanza King Formation (HAZZARD and MASON, 1936) and the upper shaly marker unit and overlying dolomites were identified as the Cornfield Springs Formation. Subsequently PALMER and HAZZARD (*op. cit.*) considered the Cornfield Springs Formation in the Providence and Marble Mountains and the Nopah Formation in the Spring Mountains–Death Valley region (Chapter 3.1.1) to be correlative units and the name Cornfield Springs could be replaced by Nopah for the uppermost carbonate sequence of the autochthonous sequences.

2.2. The Salt Lake City–Ogden region

Between Salt Lake City and Ogden, Utah (Figs. 1 and 2) all of the Cambrian sections rest unconformably on Pre-Cambrian gneisses of the Farmington Canyon complex. Very little detailed information about these sections is available.* EARDLEY (1944) described the exposures in the Durst Mountain area and MAXEY (1958) described a section at Willard Peak north of Ogden but neither author provided any faunal data. The basal

* Since the completion of the manuscript an excellent study of the Cambrian rocks of this area has been published by Rigo (1968) which includes section descriptions, regional correlations, and a discussion of stratigraphical evidence for magnitude of movement of the eastern edge of the Sevier Orogenic Belt.

quartzite and overlying shale were identified as Prospect Mountain and Pioche by
MAXEY and as Tintic and Ophir by EARDLEY. The overlying limestones were referred to
the Langston Formation by MAXEY and were undesignated by EARDLEY. MAXEY's
nomenclature seems inappropriate, but the stratigraphy of this whole area needs
re-examination. The basal quartzite is between 300 and 360 metres thick and has an
arkosic basal conglomerate in contact with the underlying gneisses. The overlying shaly
unit includes 45 to 60 metres of silty and micaceous shales and thin interbedded quartzites. The carbonate sequence consists of interbedded units of thin-bedded silty limestone,
massive dolomite, and shales truncated at various distances above the base by either thrust
faults or unconformably overlying Devonian or younger Palaeozoic rocks.

In the canyon east of Ogden, dark thin-bedded limestones at the top of the lowest
prominent cliff-forming limestone have yielded a rich trilobite and brachiopod fauna
including species of *Bathyuriscus*, *Ptychagnostus*, and *Linnarssonia*. About 180 to 270
metres higher in this section, a shaly interval above a sequence of banded white and grey
dolomites and below a sequence of massive grey dolomites has yielded species of
Eldoradia, indicating that most of the Middle Cambrian section is preserved in the area.
Details of this section are not yet published.

3. Cambrian rocks within and west of the Sevier orogenic belt

Most of the Cambrian rocks of south-eastern Idaho, western Utah, eastern and southern Nevada, and south-eastern California can be integrated into a single regional
stratigraphical framework in which the principal facies components are a complex of
carbonate banks generally flanked to both east and west by regions of contrasting and
generally terrigenous sedimentation. Shifting of the margins of the carbonate banks and
of environments within the banks has produced many local differences in stratigraphical
sequences which are reflected in locally developed stratigraphical terminologies.

In central Nevada, the scattered outcrop areas included in the Antler orogenic belt
are still too poorly known in detail to integrate into the stratigraphical framework to the
east. They are largely composed of silty thin-bedded limestones, shales, and siltstones,
above thick quartzitic sequences. All have been badly deformed and few of the units
except for the quartzites are resistant, so that even in areas where the rocks are present
the outcrops are generally poor.

Sub-regions of the Great Basin having some degree of stratigraphical homogeneity are
discussed in the following pages, beginning with the California-Nevada boundary region
in the south and continuing north-eastwards to Idaho. The central Nevada areas are
discussed in a separate chapter.

3.1. Southern Nevada and adjacent parts of California

Cambrian sections south of the 38th parallel, exclusive of the Pioche and Delamar
mining districts, represent two contrasting formational sequences designated by STEWART
(1966) as the Spring Mountains–Death Valley and White–Inyo Mountains facies.
Although these facies designations were applied primarily to the dominantly terrigenous
Early Cambrian and early Middle Cambrian rocks, they are also appropriate for the
dominantly carbonate later Middle Cambrian and Late Cambrian rocks, and the areas in
which they occur are designated here as the Spring Mountains–Death Valley and White–
Inyo Mountains regions.

Plate 1, Fig. 1. Complete Cambrian section of the Nopah range, California. The foreground is of Wood Canyon Formation. The background is of carbonates of the Bonanza King and Nopah Formations with Carrara Formation at the base. (See page 11).

(To face page 10)

Plate 1, Fig. 2. Zabriskie Quartzite and overlying Carrara and Bonanza King Formations, Bare Mountain, Nevada. (See page 13).

The Spring Mountains–Death Valley region is located south of a line between the southern Last Chance Range and the Pahranagat Range. North of this line, and principally to the west is the White–Inyo Mountains region (Figs. 1 and 2).

3.1.1. The Spring Mountains–Death Valley region

The sequence of formations and members described for the Spring Mountains–Death Valley region has evolved from a number of local studies conducted in the region since 1929.* This formational sequence can be recognized over an area of about 15,000 square kilometres and the regional relationships of all units have been fully clarified by the work of STEWART (1966), CHRISTIANSEN and BARNES (1966), and BARNES and CHRISTIANSEN (1967). Figure 4 compares two of the best known sections through this facies and shows the historical development of the stratigraphical nomenclature for the region which is necessary for understanding many of the earlier publications.

The Stirling Quartzite is the oldest unit that might possibly include beds of Cambrian age. It consists of greyish-red and pinkish-grey quartzite and quartzitic conglomerate; common greyish-red and greenish-grey phyllitic sandstone; and rare dolomitic sandstone, dolomite, and limestone distributed among five informally designated members (STEWART 1966, 1967; STEWART and BARNES, 1966). The uppermost member has been correlated by STEWART (1966) with the upper part of the Reed Dolomite of the White–Inyo Mountains region which was considered by CLOUD and NELSON (1966) to be of Early Cambrian age (Chapter 3.1.3). However, STEWART (1966) considered the Lower Cambrian–Pre-Cambrian boundary to be within the middle member of the overlying Wood Canyon Formation based on correlation with the oldest trilobite bearing beds of the White–Inyo Mountains region. In the Spring Mountains–Death Valley region the uppermost member of the Stirling Quartzite ranges from 180 to 360 metres in thickness and is conformably underlain by at least 2,100 metres of probably Pre-Cambrian sediments.

The Wood Canyon Formation includes three regionally distributed members (STEWART, 1966; STEWART and BARNES, 1966; BARNES and CHRISTIANSEN, 1967): relatively thin lower and upper members composed of siltstone, fine-grained quartzite, and minor dolomite and limestone; and a thick middle member composed of coarse-grained cross-bedded quartzite, quartzitic conglomerate, and minor siltstone. In the Nopah Range (Plate 1), Desert Range, and Groom Range, the Wood Canyon Formation is about 630 to 675 metres thick. At Bare Mountain to the west, the formation is about 1,125 metres thick and the gross lithology has changed sufficiently so that the local name Daylight Formation (CORNWALL and KLEINHAMPL, 1964) was proposed for the unit before the regional stratigraphy was worked out.

The oldest fossils in the Spring Mountains–Death Valley region are trilobite, echinoderm, brachiopod, and archaeocyathid fragments from the upper member of the Wood Canyon Formation. The trilobites represent the olenellid genera *Nevadella* and *Judomia*?, and the echinoderm fragments include helicoplacoid plates. Together with the archaeocyathids, these are key elements for correlation of the upper member of the Wood Canyon Formation with the Poleta Formation of the White–Inyo Mountains region.

* NOLAN (1929), Spring Mountains; HAZZARD (1937), Nopah–Resting Springs Ranges; HUMPHREY (1945), and BARNES and CHRISTIANSEN (1967), Groom Range; McALLISTER (1952), northern Panamint Range; JOHNSON and HIBBARD (1957), BARNES and PALMER (1961), BARNES and BYERS (1961), BARNES and others (1962), Halfpint Range; CORNWALL and KLEINHAMPL (1961, 1964), Bare Mountain; RESO (1963), Pahranagat Range; BURCHFIEL (1964), Specter Range; STEWART (1965), Last Chance Range; HUNT and MABEY (1966), Panamint Range and Funeral Mountains; and STEWART and BARNES (1966), Desert Range.

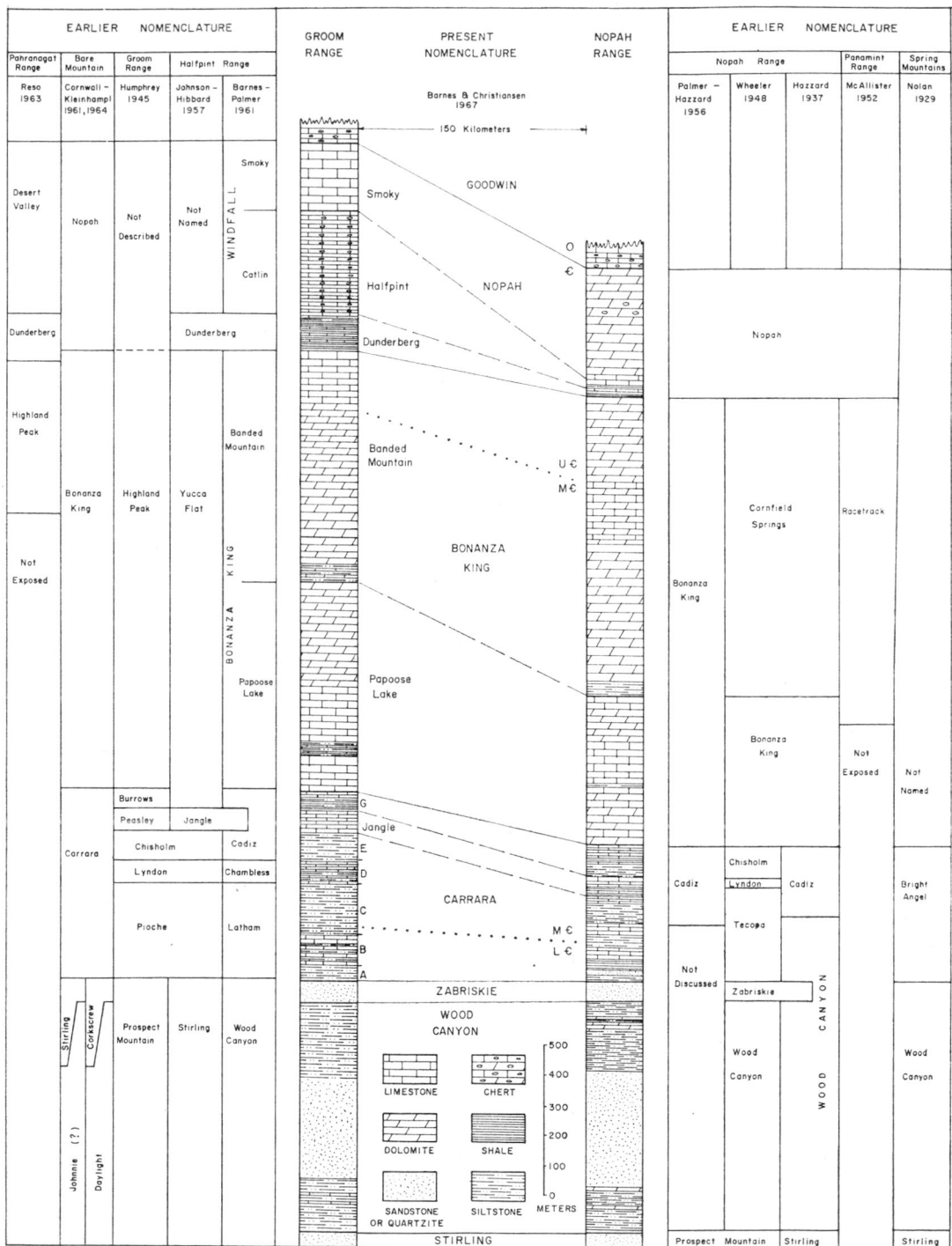

FIG. 4. Representative stratigraphical sections for the Spring Mountains–Death Valley region with a correlation chart of the nomenclature that has been applied to the rocks of the region.

The Zabriskie Quartzite is a homogeneous sequence of pale-red, fine- to coarse-grained quartzite often with well developed *Scolithus* tubes. In the Desert Range, beds identified as Zabriskie are only 2 metres thick. In its type area in the Nopah Range this formation is 50 metres thick and at Bare Mountain (Plate 1) it is over 300 metres thick and has been locally named the Corkscrew Quartzite (CORNWALL and KLEINHAMPL, 1964). Farther north-west the Zabriskie intertongues with the Harkless and Saline Valley Formation of the White–Inyo Mountains facies (Chapter 3.1.2).

All of the terrigenous units from the Zabriskie downwards thicken westward and north-westward and show a marked decrease in average grain size. In the same direction the amount of limestone and dolomite increases. This evidence for an eastward source for the terrigenous materials and a more marine basin to the west is supported by data on transport direction of the terrigenous materials obtained from studies of the dip of cross-strata in all of the formations (STEWART, 1966).

Overlying the Zabriskie Quartzite is the Carrara Formation which reflects a major transition from the earlier times of dominantly terrigenous sedimentation to later times of dominantly carbonate sedimentation. The formation consists of three complete sedimentary cycles and the lower part of a fourth cycle analogous to the Cambrian Sullivan-type Grand Cycles described by AITKEN (1966) from the Rocky Mountains in Alberta, Canada. Each cycle has a characteristic silty or shaly lower part and a generally limestone upper part. The contact between the limestone upper part of one cycle and the silty or shaly lower part of the overlying cycle is invariably abrupt. However, there is a gradual transition from siltstone and shale to limestone beds between the lower and upper parts of a single cycle. In most outcrops of the Carrara Formation the limestone upper parts of the Grand Cycles form distinct cliffs or groups of ledges separating the slope-forming siltstone and shale units. In the south-eastern part of Death Valley, the limestone parts of the Grand Cycles are noticeably more silty than to the north-west. Conversely, the lower parts of the Grand Cycles in the north-west include a considerable amount of thin-bedded limestone, particularly in the upper part of the formation.

The thickness of the Carrara formation ranges from 405 metres in the Nopah Range to about 540 metres at Bare Mountain and in the Groom Range. Individual Grand Cycles range from 100 to 150 metres in thickness. The first Grand Cycle (Members A and B, Fig. 4) and the lower 10 metres of the second Grand Cycle (Members C and D, Fig. 4) are of Early Cambrian age. Silty beds and some thin limestones in the lower part of the first Grand Cycle contain species of *Olenellus*, *Paedeumias*, *Bristolia*, *Peachella*, and many undescribed olenellids. The limestone upper part of this Grand Cycle, which is probably the south-eastern continuation of the Mule Springs Limestone of the White–Inyo Mountains region, has yielded *Olenellus*, *Bonnia*, and antagmid trilobites. The lower 10 metres of the second Grand Cycle has yielded species of *Olenellus*, *Paedeumias*, and some undescribed olenellids. Higher in the second Grand Cycle, in the lower beds of the limestone part, *Kochaspis*, *Schistometopus*, and other ptychoparioid trilobites typical of North American earliest Middle Cambrian beds have been found. The uppermost beds of the limestone part of this cycle have yielded a rich fauna of trilobites and molluscs of the *Albertella* zone (Chapter 4) that includes species of *Albertella*, *Albertelloides*, *Dolichometopsis*, *Ptarmigania*, '*Prozacanthoides*', and *Oryctocephalites*. Scattered limestones within the lower and middle parts of the third Grand Cycle (Member D and the Jangle Member, Fig. 4) have also yielded specimens of *Albertella*. The lower part of the fourth Grand Cycle (Member F, Fig. 4), which is the only part that is included within the Carrara Formation, has yielded species of *Glossopleura*. The limestone part of the third Grand Cycle, bracketed by *Albertella* below and *Glossopleura* above, has been named the Jangle Limestone (JOHNSON and HIBBARD, 1957). This is correlative, and perhaps coextensive, with the Lyndon Limestone of Pioche area (Chapter 3.2). The shaly half-cycle at the top of the Carrara Formation is the correlative of the Chisholm Shale of the Pioche area.

Earlier misidentifications of the Chisholm Shale and Lyndon Limestone in the northern regions of the Spring Mountains–Death Valley facies (HUMPHREY, 1945; JOHNSON and HIBBARD, 1957; Fig. 4) resulted from reliance on physical and superpositional similarity for correlation of the prominent upper Grand Cycles of the Carrara Formation with units in the Pioche district. Palaeontological data obtained since those reports were written show the correct correlation and point to the danger of long-distance physical correlation of units within transgressive sequences without such data.

A similar problem developed in the Nopah Range where the limestone upper part of the second Grand Cycle (unit 4N of HAZZARD, 1937), which is entirely of Middle Cambrian age, forms a prominent cliff and is lithologically and topographically like the Early Cambrian Chambless Limestone of the autochthonous sequences east of the Sevier orogenic belt (Chapter 2.1) with which it has been miscorrelated. The resulting usage of Cadiz for the beds overlying unit 4N (HAZZARD, 1937; PALMER and HAZZARD, 1956) is thus also incorrect.

The usage of Lyndon and Chisholm in the Nopah Range and the proposal of Tecopa Shale for the underlying beds (WHEELER, 1948) have not been accepted in the most recent work in this region (STEWART, 1966; BARNES and CHRISTIANSEN, 1967).

The contact between the Carrara Formation and the overlying Bonanza King Formation is conformable and gradational between the lower and upper parts of a Grand Cycle. Above this contact, the remaining parts of all Cambrian sections in the region are predominantly composed of limestones or dolomites of the Bonanza King and overlying Nopah Formations.

Within the Bonanza King Formation, details of the carbonate sequences vary from locality to locality but are not yet reflected in a complex member nomenclature. In general, the south-eastern sections are almost entirely dolomite. To the north and northwest, significant intervals of limestone are present. The total thickness of the formation ranges from about 900 to about 1,350 metres with the thickest sections in the eastern part of the region.

Two thin, slightly to moderately silty, limestone units are recognizable in most well exposed sequences of the Bonanza King Formation. The lower unit, about 40 to 60 metres thick, is found 125 to 135 metres above the base of the formation and has yielded fragmentary specimens of *Alokistocare* and *Glossopleura*. The higher unit is also about 40 to 60 metres thick. It is found from 270 to 470 metres above the lower silty interval and has yielded trilobites referable to *Ehmania*? (PALMER and HAZZARD, 1956). The base of this unit has been used as the datum for dividing the formation into a lower, Papoose Lake, member and an upper, Banded Mountain, member (BARNES and PALMER, 1961). The upper 100 metres of the Banded Mountain Member has yielded trilobites of the Late Cambrian *Crepicephalus*, *Aphelaspis*, and *Dicanthopyge* zones (PALMER, 1965; Chapter 4) The Middle-Upper Cambrian boundary thus lies somewhere within the Banded Mountain Member.

In the Nopah Range, HAZZARD (1937) identified two formations having the same limits as the Papoose Lake and Banded Mountain Members in the interval that is now recognized as the Bonanza King Formation. Because of faulty palaeontological data, the

Middle Cambrian silty unit separating the members was correlated with the Upper Cambrian silty unit of the formation then called Cornfield Springs in the Providence Mountains (Chapter 2.1). Thus, only the Banded Mountain Member was identified in the Nopah Range as Bonanza King Formation. The Papoose Lake Member in the Nopah Range was misidentified as Cornfield Springs. The correlation error was corrected by PALMER and HAZZARD (1956), and these units are now recognized as the lower and upper members of the Bonanza King Formation.

Overlying the Bonanza King Formation is a silty or shaly unit ranging in thickness from 33 metres in the Nopah Range and Bare Mountain to about 100 metres in the Groom Range. Above this is a sequence of limestones or dolomites 380 to 540 metres thick which extends to the base of the Ordovician Pogonip Group. This carbonate sequence and the underlying shaly or silty unit constitute the Nopah Formation. The shaly or silty unit has been correlated with the Dunderberg Shale of the Eureka district (JOHNSON and HIBBARD, 1957) and was designated as the Dunderberg Member of the Nopah Formation by CHRISTIANSEN and BARNES (1967). In the Groom Range and in the Halfpint Range, the carbonate sequence has been divided into a lower, Halfpint, member consisting of 315 metres of cherty, thin-bedded, grey limestone; and an upper, Smoky, member consisting of 200 metres of massive, partly stromatolitic, generally chert-free limestone (CHRISTIANSEN and BARNES, 1966; BARNES and BYERS, 1961). The Halfpint Member thins to the south by change of facies into the more massive Smoky Member and the Nopah Formation above the basal Dunderberg Member in the Nopah Range is almost entirely composed of the Smoky Member (CHRISTIANSEN and BARNES, 1966).

The Dunderberg Member has yielded trilobites of the *Dunderbergia* and *Elvinia* zones of late Dresbachian and early Franconian age (PALMER and HAZZARD, 1956; PALMER, 1965b). The Halfpint Member has yielded trilobites of the *Ptychaspis-Prosaukia* Zone of late Franconian age (Chapter 4), and the upper part of the Smoky Member has yielded Trempealeauan trilobites and also the problematical mollusc(?) *Matthevia* (BARNES and BYERS, 1961; YOCHELSON and others, 1965).

The upper contact of the Nopah Formation marks a change from generally massive relatively chert-free beds of the Smoky Member to thin-bedded, partly silty and cherty beds assigned to the Goodwin Limestone which locally yield early Ordovician fossils (CHRISTIANSEN and BARNES, 1966). This contact is also the approximate Cambrian–Ordovician boundary.

3.1.2. The White–Inyo Mountains region

Cambrian sequences in this region (Fig. 5) are much less homogeneous than those of the Spring Mountains–Death Valley region. They are characterized by a significant development of Early Cambrian carbonate units and, in Nevada, a thin Middle and Late Cambrian succession of thin-bedded cherty limestones. The Middle and Late Cambrian sequences in California are predominantly composed of dolomite units comparable to those of the Spring Mountains–Death Valley region.

Knowledge of the present regional stratigraphy of the White–Inyo Mountains region is largely the result of work by NELSON (1962, 1965), ROSS (1963), and STEWART (1965) in California; and by ALBERS and STEWART (1962) and MCKEE and MOIOLA (1962) in Nevada. These works superseded or clarified the older work in this area by TURNER (1902), WALCOTT (1908a), and Kirk (*in* KNOPF, 1918). Historical development of the present stratigraphical terminology is discussed by NELSON (1962).

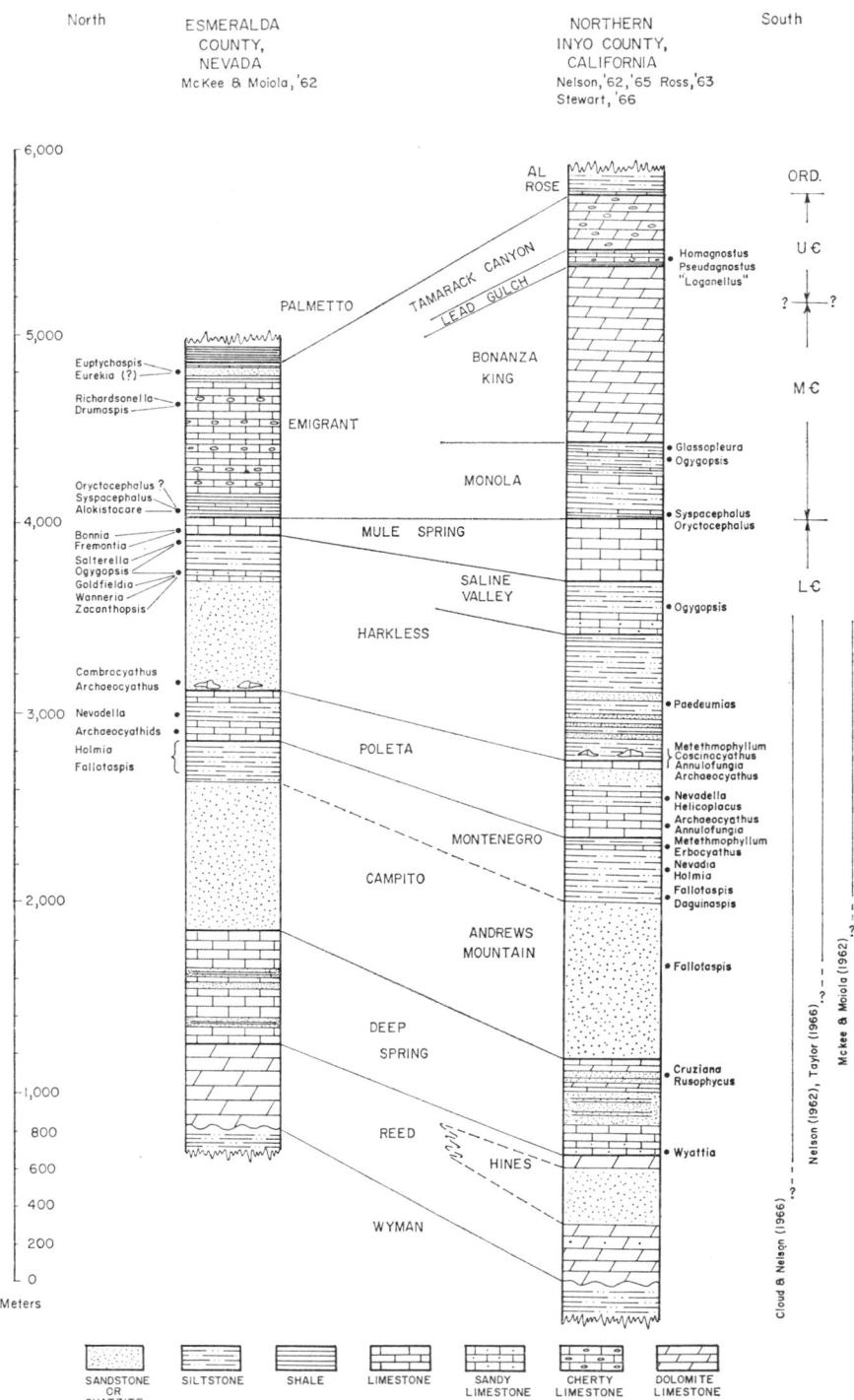

FIG. 5. Representative columnar sections of the thin and thick sequences of the White–Inyo Mountains region, and locations of various suggested positions for the base of the Cambrian System.

The oldest unit that might possibly include beds of Cambrian age is the Reed Formation. In its type area in the White Mountains of California, and in Nevada, this formation is composed of massive, locally oolitic, light-coloured dolomite. Its thickness ranges from about 300 metres in Nevada to about 600 metres in California. In the Inyo Mountains, south of the type area, the Hines tongue, a middle member of quartzite, sandy dolomite, and calcareous sandstone increases in thickness southwards to nearly 245 metres. The lower contact of the Reed Formation is a subtle regional unconformity on the older phyllitic argillites of the Wyman Formation (NELSON, 1962). The uppermost three metres of the Reed Formation have yielded the problematical mollusc (?) *Wyattia* (TAYLOR, 1966). This was considered to be a Pre-Cambrian fossil by TAYLOR, but CLOUD and NELSON (1966) used it as suggestive evidence for a Cambrian age (Chapter 3.1.3).

Conformably above the Reed Formation is a succession of five alternating carbonate and quartzite units totalling about 540 metres in thickness which is referred to the Deep Spring Formation. The carbonate units are massive to thin-bedded, blue to grey, locally sandy, limestones or dolomitized limestones and constitute the upper, lower, and middle members of the formation. The intervening quartzites are usually black and fine-grained and are locally interbedded with siliceous shales. Several localities within this formation have yielded tracks and trails including forms identified as *Rusophycus* and *Cruziana* (CLOUD and NELSON, 1966). No skeletal fossils are known from these beds.

The Deep Spring Formation is conformably overlain by a sequence of massive, dark, locally cross-stratified quartzitic sandstone and interbedded grey siltstone and shale about 800 metres thick. This quartzitic unit is the Andrews Mountain Member of the Campito Formation. It is overlain by the Montenegro Member, about 180 metres of grey shale and interbedded fine-grained quartzitic siltstone and sandstone with local thin lenticular archaeocyathid limestone beds.

The oldest trilobite fossils from western United States are specimens of *Fallotaspis* found about 450 metres below the top of the Andrews Mountain Member (NELSON and HUPÉ, 1964). The Montenegro Member has yielded species of *Fallotaspis* and *Daguinaspis* from its lower beds, and species of *Nevadia, Holmia, Laudonia,* and *Judomia*? from its upper beds (WALCOTT, 1910; NELSON and DURHAM, 1966). In addition, limestone bioherms in the upper part of the member have yielded abundant solitary species of *Ethmophyllum, Metethmophyllum, Ajacicyathus, Erbocyathus, Pycnoidocyathus,* and *Syringocyathus* (GANGLOFF, *in* NELSON and DURHAM, 1966).

The Montenegro Member is conformably overlain by the Poleta Formation. This formation is about 360 metres thick in the Inyo Mountains and thins northwards into the White Mountains and into Nevada where it is about half this thickness. Throughout its extent, it has a tripartite division, and consists of lower and upper massive grey-blue limestone members and a middle member of grey-green shale and siltstone and thin quartzites. The lower limestone member contains abundant archaeocyathids including species of *Archaeocyathus, Ajacicyathus, Annulofungia, Ethmophyllum, Pycnoidocyathus,* and *Syringocyathus* (OKULITCH, 1943; McKEE, 1963; GANGLOFF, *op. cit.*)

The middle member has yielded trilobites referable to *Nevadella, Olenellus, Laudonia, Fremontia,* and unnamed ptychoparioids; abundant *Scolithus* tubes in some quartzite beds; echinoderms referable to the remarkable echinozoan *Helicoplacus* and to undescribed edrioasteroids and eocrinoids; and undescribed inarticulate and articulate brachiopods (NELSON and DURHAM, 1966; DURHAM and CASTER, 1963; DURHAM and others, 1966).

The upper limestone member and limestone bioherms in the lower part of the over-lying Harkless Formation have yielded archaeocyathids referable to *Coscinocyathus*, *Archaeocyathus*, *Pycnoidocyathus*, *Annulofungia*, and *Metethmophyllum* (GANGLOFF, *op. cit.*).

The Harkless Formation is a unit of fine-grained quartzites, siltstones, and shales in varying proportions which is about 540 to 630 metres thick. In the Inyo Mountains, quartzite is the dominant rock type. Northwards into the White Mountains and into Nevada, siltstones and shales become more prominent. Locally, limestone bioherms with archaeocyathids are present in the silty lower part of the formation.

In the White–Inyo Mountains area of California, the Harkless Formation is conformably overlain by about 250 metres of interbedded medium- to coarse-grained quartzitic sand-stone and blue-grey arenaceous limestone of the Saline Valley Formation. Absence of the distinctive limestones in Nevada has resulted in inclusion of rocks equivalent to the Saline Valley in an expanded Harkless Formation (MCKEE and MOIOLA, 1962; ALBERS and STEWART, 1962).

The Harkless–Saline Valley interval has yielded scattered olenellids throughout its extent. The upper beds have yielded local concentrations of *Salterella* and trilobites referable to *Bonnia*, *Ogygopsis*, *Wanneria*, and *Paedeumias* (NELSON, 1963). At one locality near Goldfield, Nevada, sandy limestone beds believed to be equivalent to beds of the lower Saline Valley Formation have yielded a rich association of olenellid, corynexo-choid, oryctocephalid, and ptychoparioid trilobites (PALMER, 1964).

The uppermost Early Cambrian unit is the Mule Spring Limestone, a distinctly bedded, commonly mottled, blue-grey limestone characterized by abundant oncolites generally referred to *Girvanella*. This unit is about 300 metres thick in California, but thins to about 70 metres in Nevada.

The Mule Spring Limestone seems to be a thick western equivalent of the limestone at the top of the first Grand Cycle of the Carrara Formation (Chapter 3.1.1). However, it includes both older and younger beds. In the Carrara Formation, a *Bristolia-Peachella* fauna is consistently found in thin limestones a few metres below the main limestone unit of the first Grand Cycle. Other olenellids are found up to 10 metres above the lime-stone in the overlying siltstones. In contrast, siltstones immediately above the Mule Spring Limestone in Esmeralda County, Nevada, have yielded an early Middle Cambrian fauna (ALBERS and STEWART, 1962) and the *Bristolia-Peachella* fauna is found within the lower beds of the formation (STEWART, 1965).

The Cambrian section above the Mule Spring Limestone in the White–Inyo Mountains area of California is distinctly different from the section in Esmeralda County, Nevada, and the nearby northern Last Chance Range in California (Fig. 5). In the White–Inyo Mountains area, the Mule Spring Limestone is overlain by the Monola Formation (NELSON, 1965). This formation consists of three members: a lower member of inter-bedded silty limestone and siltstone about 200 metres thick; a middle member of well-bedded blue-grey cliff-forming limestone with a few interbeds of siltstone about 40 metres thick; and an upper member of dark-brown weathering, platy siltstone capped by grey and blue-grey limestone and interbedded with limy shale, about 125 metres thick. The lower member has yielded species of *Syspacephalus* and *Oryctocephalus*, and the upper member has yielded species of *Ogygopsis* and *Glossopleura* (STOYANOW, 1958*; NELSON, 1965). This indicates a correlation of the formation with the lower part of the Emigrant

---

* Examination of the pygidium identified by STOYANOW as *Albertella* aff. *stenorachis* Rasetti shows that it is a broken pygidium of a species of *Glossopleura*.

Formation of Esmeralda County, Nevada, and the upper part of the Carrara Formation of the Spring Mountains–Death Valley region.

The Monola Formation is conformably overlain by about 800 metres of laminated to thick-bedded dolomite characterized by alternating light-grey and dark-grey layers several centimetres to a few tens of metres thick, which has been identified as the Bonanza King Formation (Ross, 1963).

Conformably above the Bonanza King Formation, about 100 metres of thin-bedded limestone, siltstone, dolomite, chert, and shale has been described as the Lead Gulch Formation (Ross, 1963). The lower part of the formation is predominantly shale. Thin-bedded limestones in the upper part of the formation have yielded species of *Pseudagnostus* and '*Loganellus*' that characterize the lower part of the Halfpint Member of the Nopah Formation of the Spring Mountains–Death Valley region. The shales of the lower part of the Lead Gulch Formation are probably the equivalent of the shales of the Dunderberg Member of the Nopah Formation.

The Lead Gulch Formation is conformably overlain by about 270 metres of thin- to thick-bedded grey dolomite with some chert, which has been named the Tamarack Canyon Formation (Ross, 1963). This unit has not yielded any fossils, but the overlying limestone, siltstone, and shale of the Al Rose Formation have yielded fossils of early Ordovician age. The Tamarack Canyon Formation is very probably the lithic equivalent of the Smoky Member of the Nopah Formation (Chapter 3.1.1).

In Esmeralda County, Nevada, and in the northern part of the Last Chance Range, California, the Mule Spring Limestone is overlain by the Emigrant Formation. This formation includes a basal member of siliceous shales and flaggy mudstone with some interbeds of thin-bedded chert and platy limestone about 145 metres thick, and an upper member of thin-bedded blue to grey limestone alternating with chert that is at least 630 metres thick and may be considerably thicker. Locally, intraformational limestone breccias up to 3 metres thick are found. The Lower member has yielded early Middle Cambrian trilobites referable to *Alokistocare* and *Syspacephalus*. The upper member has yielded Late Cambrian trilobites including species of *Richardsonella* and *Drumaspis* of Franconian age, and species of *Euptychaspis* and *Eurekia*(?) of Trempealeauan age (McKee and Moiola, 1962).

The Emigrant Formation represents a thin-bedded cherty limestone facies that contrasts with the thicker-bedded more massive limestone and dolomite facies of the time-equivalent Bonanza King and Nopah Formations to the south and east. This is a typical facies contrast between sequences assigned to the carbonate belt and the outer detrital belt (Palmer, 1960a; Chapter 5) in the southern part of the Great Basin.

The boundary between these two regional facies belts in the Middle and Late Cambrian can be fairly precisely defined in two areas. In the Last Chance Range, sections in the northern part include beds assigned to the Emigrant Formation, and sections in the southern part include beds assigned to the Carrara and Bonanza King Formations (Stewart, 1965). In the Belted Range (Fig. 2), a partial section of beds above the Zabriskie Quartzite includes about 450 metres of beds equivalent to the Carrara Formation. This section lacks the distinctive development of all of the limestone members and is overlain by a sequence of thin-bedded limestones about 420 metres thick resembling the Emigrant Formation (Cornwall, 1967). It contrasts strongly with the section in the Groom Range about 30 kilometres to the east (Fig. 4).

3.1.3. The problem of the base of the Cambrian System

The Pre-Cambrian–Early Cambrian succession of sedimentary rocks in the White–Inyo Mountains region includes the most fossiliferous and complete Early Cambrian succession in the United States. Thus, any discussion of the criteria for distinguishing Early Cambrian from Pre-Cambrian rocks is necessarily limited by the quality and quantity of the factual data on sedimentary relations or faunal distribution that can be clearly related to this succession.

Several of the more recently suggested positions for the earliest Cambrian rocks are shown on Fig. 5. They illustrate two different classes of solutions to the problem of determining the oldest Cambrian rocks. One is philosophical and involves attempts to utilize criteria that may be of time significance on a global scale. The other is practical and involves the desire to reach an empirical solution to the problem.

The practical considerations that recognize the oldest formation (or member) containing Cambrian fossils as wholly Cambrian (McKEE and MOIOLA, 1962), or that identify as Cambrian only those rocks that contain Cambrian fossils (NELSON, 1962) beg the philosophical question of what, in fact, identifies a fossil as Cambrian rather than Pre-Cambrian. Older solutions that recognize the base of the oldest sedimentary cycle with Cambrian fossils or the first 'major' unconformity below beds with Cambrian fossils as representing the 'boundary' between Cambrian and Pre-Cambrian are catastrophist views that reflect stratigraphical teaching in the United States largely prior to 1940.

The philosophical arguments are based on faunal data from the White and Inyo Mountains in California. The oldest fossils are indifferently preserved dolomitized conical objects from the upper dolomite beds of the Reed Formation that were described in detail by TAYLOR (1966). They range from 4 to 6 millimetres in length and 2 to 3 millimetres in diameter and seem to have a thickened margin near the adapical end and a slightly bulbous apex. Three layers of shell are suggested, as well as an internal solid structure at the apical end. TAYLOR named these objects *Wyattia reedensis* and suggested possible affinities to the molluscan order Globorilida.

About 360 metres above the *Wyattia*-bearing beds in the terrigenous lower part of the upper member of the Deep Spring Formation CLOUD and NELSON (1966) recorded fragmentary trace fossils identified as *Cruziana* and *Rusophycus* for which there is considerable evidence of an arthropod, and probably trilobite, origin. No shelly remains of trilobites or other fossils have been found in the Deep Spring Formation. The next younger fossils are olenellids identified as *Fallotaspis* that are found about 630 metres above the *Cruziana-Rusophycus* bearing-beds in the middle of the Andrews Mountain Member of the Campito Formation. There is agreement by all workers that these are of Cambrian age.

TAYLOR (1966) described *Wyattia* as an unequivocal Pre-Cambrian fossil. He noted the occurrence of *Fallotaspis* as the oldest trilobite in the Campito Formation as well as in the North African Cambrian succession and suggested that the base of the *Fallotaspis* biozone is an 'excellent biostratigraphic basis for recognition and correlation of the Lower Cambrian–Pre-Cambrian boundary.' CLOUD and NELSON (1966), however, develop the argument that the base of the Palaeozoic and probably the Cambrian is marked by an evolutionary burst of metozoans and therefore this boundary must be below the beds with *Wyattia*. They leave open the possibility of recognizing a Pre-Cambrian Palaeozoic System but imply that the bottom of the Cambrian 'in the usual sense' is at least below the beds with *Cruziana* and *Rusophycus* because of their probable trilobite origin.

There is no consensus among palaeontologists as to which of the several philosophical positions is most acceptable for distinguishing Cambrian from Pre-Cambrian beds and thus no attempt is made to indicate a boundary in the White–Inyo Mountains region (Fig. 5).

3.2. The Pioche region

West of the easternmost thrusts of the Sevier orogenic belt, Cambrian sections from the Delamar Range, Nevada, to the Wah Wah and Beaver Mountains in Utah (Fig. 2) are characterized by an Early Cambrian dominantly quartzitic sequence, a transitional sequence of alternating finer terrigenous materials and carbonates ranging from late Early Cambrian to early Middle Cambrian in age, and a thicker younger sequence of carbonate rocks of Middle and Late Cambrian age. These sequences are grossly similar to the sequences of the Spring Mountains–Death Valley region but differ in details that have resulted in a totally different stratigraphical terminology. The Pioche mining district, which includes the Ely and Highland Ranges, is the source for most of the stratigraphical nomenclature for these rocks, and the region is therefore referred to here as the Pioche region.

The gross aspects of the stratigraphy in the Ely and Highland Ranges were first described by WALCOTT (1886). Later work by PACK (1906a) and WALCOTT (1908a, 1916b) added some detail that was incorporated in a general study by WESTGATE and KNOPF (1927, 1932) in which the present formational sequence was established. This includes, from base to top, the Prospect Mountain Quartzite, Pioche Shale, Lyndon Limestone, Chisholm Shale, Highland Peak Formation and Mendha Formation. Restriction of the boundaries of the Pioche Shale because of a desire to recognize separately the Early Cambrian and Middle Cambrian parts was suggested by MASON (in GRABAU, 1936) and DEISS (1938) but was rejected (WHEELER and LEMMON, 1939), and the name Pioche has since been consistently applied to all of the beds between the Prospect Mountain Quartzite and the Lyndon Limestone. WHEELER and LEMMON (1939) divided the Highland Peak Formation into 17 members designated only by the letters A to Q. Subsequently (WHEELER, 1940, 1948) some of these received names [Peasley (A), Burrows (B), Burnt Canyon (C), and Condor (F)] and the name Highland Peak was restricted to the younger carbonates. MERRIAM (1964) reinstituted the use of Highland Peak for the entire carbonate interval between the Chisholm Shale and the Mendha Formation, thus recognizing the regional utility of this more inclusive designation, and recommended that all previously named subdivisions by WHEELER be recognized as members. He also showed that units in the Ely Range that WHEELER (1948) identified by the House Range names Dome and Swasey were miscorrelated and he named new members for these units: Step Ridge (D and E, 'Dome') and Meadow Valley (G, 'Swasey'). The same misidentification was also made by WHEELER (op. cit.) in the Wah Wah Mountains.

In addition to the study of the lower part of the Highland Peak Formation, MERRIAM presented a detailed description of the Pioche Shale in which he recognized four shale members designated by letters and two named limestone members: the Combined Metals Limestone Member and the Susan Duster Limestone Member. The present sequence of formations and members in the Ely and Highland Ranges is shown on Fig. 6.

Elsewhere in the Pioche region, CALLAGHAN (1937) and WHEELER (1943) have applied the Ely–Highland Range nomenclature to the section in the Delamar Range. BENTLEY (1958) described some of the Late Cambrian units in the Wah Wah Mountains. MILLER (1966) and EAST (1966) have provided some lithological data on the older carbonate

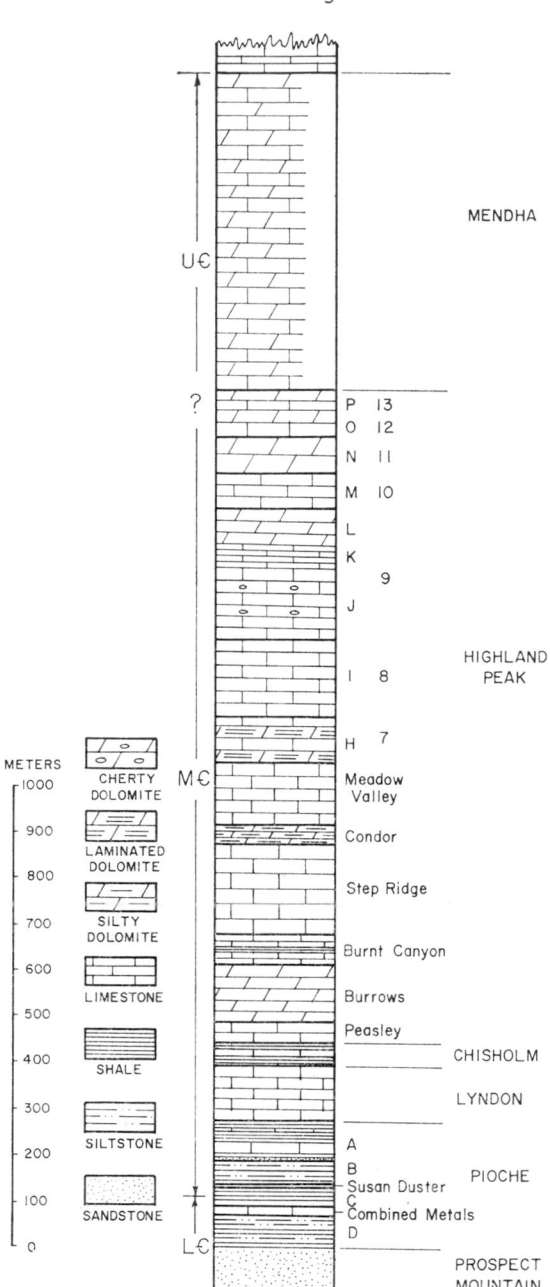

FIG. 6. Columnar section of the rocks of the Pioche Mining district.

sequences of the Wah Wah and San Francisco Mountains. WOODWARD (1968) recognized only the Prospect Mountain Quartzite and a small amount of Pioche Shale lying above a thick conformable sequence of older terrigenous sediments in the Beaver Mountains of Utah.

The lithological characteristics of the formations of the Pioche region, described below, are primarily derived from studies in the Ely and Highland Ranges.

The Prospect Mountain Quartzite is the oldest unit of probable Cambrian age in the Pioche region. It is characteristically thick- to massive-bedded, coarse-grained to conglomeratic quartzite coloured various shades of red or brown and often with well-developed cross-laminations. Locally thin-bedded or shaly intervals, finer-grained quartzite, and zones of grey quartzite are present. The maximum thickness of rocks assigned to this formation in the region is about 2,000 metres in the Wah Wah Mountains (MILLER, 1966) where the base is buried beneath valley alluvium. The basal part has been seen only in the Beaver Mountains (WOODWARD, 1968) where it is a feldspathic conglomerate and pebbly feldspathic sandstone about 70 metres thick, resting with possible unconformity on about 1,500 metres of older quartzite and siltstone units. In this area, the thickness of the Prospect Mountain Quartzite was estimated by WOODWARD to be in excess of 1,100 metres. Southward, only the upper 270 to 720 metres of the formation are present above the valley alluvium. The only fossils from the Prospect Mountain Quartzite are trace fossils represented by unidentified crawl marks and vertical tubes of *Scolithus* (MERRIAM, 1964). Because the immediately overlying beds of the Pioche Shale contain the late Early Cambrian olenellid *Bristolia*, which is also found in the basal beds of the Carrara Formation immediately above the Zabriskie Quartzite in the Spring Mountains–Death Valley region, the Prospect Mountain Quartzite is at least partly of Early Cambrian age and correlates with the Zabriskie Quartzite and possibly parts of the underlying Wood Canyon and Stirling Formations of that region.

The Pioche Shale is actually a unit of heterogeneous lithologies including micaceous and non-micaceous shale, siltstone, sandstone, and limestone. The proportion of limestone increases upwards in the formation and geographically from the Wah Wah Mountains south-westwards to the Delamar Range.

In the Wah Wah Mountains, MILLER (1966) identified as the Pioche Shale 155 metres of olive-green, red, and brown micaceous shale containing a few thin quartzite interbeds. Overlying this is 7 metres of very silty limestone followed by about 55 metres of oolitic and oncolitic grey limestone and micaceous red, green, and brown shale which is included by MILLER with the overlying undifferentiated Middle and Late Cambrian carbonates. In the regional context, these beds would more likely be included in the Pioche Shale, thus increasing the formational thickness in the Wah Wah Mountains to about 210 metres.

In the Delamar Range (CALLAGHAN, 1937), the Pioche interval is about 270 metres thick and contains scattered thin beds or units of limestone.

In the Ely–Highland Range area, where the only detailed studies of the formation have been made, six members have been recognized by mining geologists for many years and were formally described by MERRIAM (1964). These seem to have validity only in that area, and even there the boundary between the upper two members is not everywhere recognizable.

The oldest member, the D-Shale Member, is characteristically micaceous, greenish to yellowish, unevenly-bedded shale with some interbedded siltstone. Its thickness increases westwards from about 55 metres in the Ely Range to about 90 metres in the Highland Range. Throughout the Ely–Highland Range area it is overlain conformably by the Combined Metals Member, a unit of sandy and argillaceous, thin-bedded, grey to black limestone varying from 12 to 24 metres in thickness. This is the principal ore-bearing horizon of the Pioche mining district and has been extensively mined for lead and zinc (Chapter 6).

The Combined Metals Member and the underlying D-Shale Member are both of Early Cambrian age. The D-Shale Member contains a *Fremontia fremonti-Bristolia bristolensis* faunule comparable to that found in the basal beds of the Pioche Shale in the Eureka district (Chapter 3.3) and the lower part of the Carrara Formation (Chapter 3.1.1) (PALMER, *in* MERRIAM, 1964). The Combined Metals Member has yielded a silicified faunule, including ontogenetic stages, of *Olenellus*, *Paedeumias*, *Crassifimbra*, and rare individuals of *Zacanthopsis*. Parts of this faunule were described by PALMER (1957, 1958).

The C-Shale Member is a distinctive light-brown non-micaceous shale without significant siltstone interbeds, which maintains a thickness of about 35 metres throughout the Ely–Highland Range area. Only indeterminate ptychoparioid trilobites have been found in this member and it may be of either Early or Middle Cambrian age.

Above the C-Shale Member is a thin limestone unit, rarely more than 5 metres thick, which has been designated the Susan Duster Member. It is a well bedded, very fine-grained, medium- to light-grey limestone with argillaceous partings which is locally mineralized. In areas where it is not mineralized, it has yielded a rich assemblage of trilobites and brachiopods characteristic of the earliest Middle Cambrian. These include species of *Kochaspis*, *Mexicella*, *Onchocephalus*, *Poliella*, *Schistometopus*, *Strotocephalus*, and *Acrothele* (PALMER, *in* MERRIAM, 1964).

Above the Susan Duster Member, the A-Shale and B-Shale Members can be recognized only in the Ely Range. The B-Shale Member consists of unevenly-bedded greenish-brown micaceous shales, silty shales, and fine sandstones. It is distinguished from the overlying A-Shale Member in lacking any limestone interbeds. The combined thickness of the A-Shale and B-Shale Members ranges from 95 to about 160 metres in the Ely–Highland Range area. Some of the apparent thickness variation may be due to minor faulting which is pervasive in the area.

The B-Shale Member is unfossiliferous. The A-Shale Member has yielded species of *Albertella* and *Mexicella* which characterize the *Albertella* Zone of the Middle Cambrian throughout the Cordilleran region (Chapter 4). This fauna has also been found in the lower part of the third Grand Cycle of the Carrara Formation (Chapter 3.1.1).

The Lyndon Limestone, a cliff-forming unit of dark- and light-grey thin- to thick-bedded limestones, lies conformably above the Pioche Shale throughout the Pioche region. Individual beds may be oolitic and cross-bedded, aphanitic or fine-grained. Locally, dark- and light-coloured members can be recognized. The formation is about 100 metres thick in the Ely–Highland Range area and thins both northwards and southwards. In the Delamar Range it is only 35 metres thick, and in the Wah Wah Mountains a correlative cliff-forming unit is about 45 metres thick. No fossils are known from the Lyndon Limestone. However, its position between beds bearing the *Albertella* and *Glossopleura* faunas indicates a correlation with the Jangle Limestone Member of the Carrara Formation in the Spring Mountains–Death Valley region (Chapter 3.1.1) and the Howell Limestone in the House Range (Chapter 3.3). The pattern of regional thickness changes suggests that the Jangle, Lyndon, and Howell may represent tongues reflecting more or less simultaneous eastward spreading of carbonate sedimentation from major carbonate bank complexes that lay to the west (Chapter 5).

The Chisholm Shale overlies the Lyndon Limestone conformably and consists of very fine-textured, smooth, non-calcareous shales, calcareous shales, and limestones that vary in proportion in different areas. It is richly fossiliferous and has yielded trilobites, brachiopods, and echinoderms including species of *Glossopleura*, *Zacanthoides*, *Alok-*

istocare, *Athabaskia*, *Diraphora*, and *Gogia* (WALCOTT, 1886, 1912, 1916b, 1925; PACK, 1906b; RESSER, 1935; PALMER, 1954; ROBISON, 1965). The shale increases in thickness southward from 20 metres in the Wah Wah Mountains (MILLER, 1966) to about 45 metres in the Ely–Highland Range area (MERRIAM, 1964) and about 100 metres in the Delamar Range (CALLAGHAN, 1937). Farther to the south-west, the silty thin-bedded limestone interval which characterizes the upper member of the Carrara Formation is a correlative unit (Chapter 3.1.1). In the House Range, to the north-east, a shale correlated with the Chisholm is present between the Howell and Dome Limestones (Chapter 3.3).

The Highland Peak Formation, which conformably overlies the Chisholm Shale, includes about 1,350 metres of limestones and dolomites that have been divided in the Ely–Highland Range area into locally identified members (WHEELER and LEMMON, 1939; MERRIAM, 1964) based on composition, colour differences, or differences in terrigenous content. The lower six members have been named and the upper seven members have been informally designated by numbers (MERRIAM, 1964).

The Peasley Member, at the base of the formation, is a dark-grey massive limestone about 50 metres thick which is characterized by an abundance of oolites, algal oncolites, and pelletal calcarenites.

The Burrows Member, about 110 metres thick, is generally massive, dolomitic light-grey rock, characterized by irregularly shaped white calcitic branching structures generally described as 'Bluebird' structures from the dolomite formation bearing that name in the East Tintic Mountains (Chapter 3.5).

The Burnt Canyon Member consists of about 65 metres of thin- to medium-bedded, dark-grey, fine-grained limestone with local concentrations of argillaceous matter, some thin shale units, and edgewise conglomerates in a 25-metre thick zone about 8 metres above the base. This shaly zone is the first fossiliferous zone above the Chisholm Shale. It contains species of *Kootenia* and abundant undescribed ptychoparioid trilobites similar to those that characterize the Whirlwind Member of the Swasey Formation in the House Range (Chapter 3.3).

The Step Ridge Member is a unit of locally varying thickness, generally more than 180 metres thick, which includes a variety of limestones ranging in colour from white to dark-grey and in lithology from lithographic to mottled and oolitic. The oolitic phases are characteristically cross-bedded with alternating dark- and light-coloured patches.

The Condor Member is a distinctive relatively non-resistant unit of brown weathering, generally thin-bedded to laminated, silty dolomites and limestones about 35 metres thick that contrasts strongly with the underlying Step Ridge Member in both colour and topographical expression. From its lithology and its general position in the carbonate sequence of the Highland Peak Formation, this is probably correlative with the silty unit that characterizes the basal unit of the Banded Mountain Member of the Bonanza King Formation (Chapter 3.1.1). It was miscorrelated by WHEELER (1948) with the unit in the House Range now called the Whirlwind Formation (ROBISON, 1960b).

The Meadow Valley Member, which is the youngest named member of the Highland Peak Formation, consists of about 125 metres of thick-bedded to massive, dark-grey, medium- to fine-grained mottled limestones. This is also the youngest unit cropping out in the main part of the Pioche mining district at the north end of the Ely Range.

The younger units of the Highland Peak Formation, numbered seven to thirteen by MERRIAM (1964) and lettered H to Q by WHEELER and LEMMON (1939), are found only in much faulted areas on the east side of the Highland Range and in a well-exposed section near Panaca at the south end of the Ely Range. They constitute about 810 metres of dark- to light-grey massive to laminated limestones and dolomites which have only local stratigraphical significance. The distinctive late Middle Cambrian trilobite *Eldoradia* has been found about 180 metres above the top of the underlying Meadow Valley Member. No other fossils are known from this part of the Highland Peak Formation.

The Mendha Formation, which overlies the Highland Peak Formation, is one of the most poorly known Cambrian units in the Great Basin. No complete section is known in the Pioche region and its thickness has only been estimated to exceed 630 metres. Its type area in the Highland Range is an area of much-faulted poorly-exposed carbonate rocks on the east side of the range. The basal beds are described by WESTGATE and KNOPF (1932) as thin-bedded, grey, oolitic, coarsely-crystalline fossiliferous limestones. Rocks of this lithology have yielded the Late Cambrian trilobites *Aphelaspis* and *Dicanthopyge* typical of the lower part of the Dunderberg Formation (PALMER, 1965b). Younger shales with *Housia*, and still younger limestones with the latest Cambrian trilobites *Eurekia* and *Dikelocephalus* and the earliest Ordovician trilobites *Symphysurina* and *Kainella*, are also known (MERRIAM, 1964). Older limestones with trilobites identified as *Crepicephalus* and *Tricrepicephalus* (WESTGATE and KNOPF, 1932; PALMER, unpublished notes, 1952) have also been included in the Mendha Formation, so the exact nature of its lower contact needs to be re-examined.

Because of the variety of included rocks, the uncertainty about the contact between the Highland Peak and Mendha Formations, inadequateness of paleontological data, and lack of clear regional persistence of the carbonate units, the correlative sequences in the Wah Wah Mountains have been cited only as undifferentiated Middle and Late Cambrian carbonates (MILLER, 1966).

BENTLEY (1958) described a Late Cambrian section in the northern Wah Wah Mountains and identified a unit of thin-bedded limestone and shale as the 'Dunderberg' shale. The over- and underlying units were identified with names Notch Peak and Orr from the House Range section to the north (Chapter 3.3). Re-examination of the well-exposed 'Dunderberg' unit by PALMER (1963, unpublished data) showed that the lower thin-bedded limestones of this unit contain species of *Aphelaspis* and *Prehousia* typical of the lower part of the Dunderberg Formation (the middle part of the Orr Formation) of the House Range and that the shales about 13 metres higher contain trilobites of the *Elvinia* Zone, typical of the Corset Spring Shale found at the top of the Orr Formation as originally described in the House Range (Chapter 3.3). These faunas are normally separated by about 100 metres of beds in the House Range. The thin sequence in the Wah Wah Mountains may reflect the widespread disconformity between beds of Dresbachian (*Prehousia*) and Franconian (*Elvinia*) age which is well known from the mid-continent region and has been previously reported in the Great Basin only along the eastern margin in the Nopah Range in California and the Tintic region of Utah (PALMER, 1965b). The appropriateness of usage of Dunderberg and Orr for Late Cambrian units in the Wah Wah Mountains is questionable. Alternative suggestions must await new work in the area.

3.3. The Eureka–House Range region

The House Range in western Utah is at the eastern apex of a triangular region that broadens westward to include all Cambrian sequences in eastern Nevada east of the Antler orogenic belt and between the latitudes of 38° N. and 40° N. (Fig. 2). Areas for which some information on the Cambrian sequences is available include the Eureka mining district and the Hot Creek and Ruby Ranges in the west and the White Pine,

Grant, Cherry Creek, Egan, Schell Creek, Snake, and House Ranges progressively to the east.

All Cambrian sections in this region contrast strongly with those of other regions of eastern Nevada and western Utah in having thick sequences of Middle and Late Cambrian shales and thin-bedded silty limestones alternating with more massive sequences of clean carbonate rocks, instead of rather continuous sequences of relatively clean carbonate rocks. However, the sections in most mountain ranges within the region differ sufficiently among themselves so that local stratigraphical terminologies have been applied to part or all of the Cambrian stratigraphical succession in the Eureka mining district (NOLAN and others, 1956), Hot Creek Range (FERGUSON, 1933), Cherry Creek and northern Schell Creek and Egan Ranges (YOUNG, 1960), southern Egan Range and Schell Creek Range (KELLOGG, 1963), Snake Range (DREWES and PALMER, 1957), and House Range (DEISS, 1938) (Fig. 7).

Throughout the region the oldest rocks of probable Cambrian age are medium- to coarse-grained, light-coloured, medium- to thick-bedded and generally cross-bedded quartzites with rare thin siltstone and conglomerate units. In the Schell Creek Range, thin units of extrusive volcanic rocks are reported within the quartzite sequence (KELLOGG, 1963; DREWES, 1967), and a unit of micaceous argillite about 115 metres thick is described by KELLOGG below the uppermost 30-metre quartzite unit. These rocks are all included within the Prospect Mountain Quartzite. YOUNG (1960) and DREWES (1967) have recorded additional quartzites and argillites below rocks they identified as Prospect Mountain Quartzite. These were assigned by YOUNG to the Piermont Group of Pre-Cambrian age. DREWES has reported that the 30-metre thick ledge-forming shaly unit separating the upper 450-metre quartzite of the Piermont Group from similar quartzites of the Prospect Mountain is not recognizable in his area, about 25 kilometres to the south, and he includes quartzites probably equivalent to the upper part of the Piermont Group in his Prospect Mountain Quartzite. Elsewhere in the Eureka–House Range region, a base for the Prospect Mountain Quartzite has not been recorded but up to 1,300 metres of quartzites have been included in the formation. Trilobites from sandy beds in the lower part of the overlying Pioche Shale in the Eureka district include species of *Olenellus*, *Paedeumias*, and *Peachella*. These indicate correlation with the lower beds of the Pioche Shale of the Pioche region (Chapter 3.2) and the lower part of the Carrara Formation of the Spring Mountains–Death Valley region (Chapter 3.1.1). Thus the top of the mass of Lower Cambrian quartzitic rocks throughout eastern and southern Nevada and adjacent parts of California is approximately synchronous.

A unit of sandstones, siltstones, shales, and thin limestones which overlies the Prospect Mountain Quartzite has been consistently identified as the Pioche Shale throughout the Eureka–House Range region. In the northern Egan Range, FRITZ (1968) has recognized a lower siltstone member 125 metres thick and an upper member of interbedded limestone and siltstone 255 metres thick. In all other areas, the Pioche Shale has not been subdivided, and reported thicknesses range from 195 metres in the southern Schell Creek Range (KELLOGG, 1963) to generally about 90 to 135 metres in other areas. Some of the differences in thickness may be caused by lack of consistency in choosing the upper limit for the formation. In the House Range, above 90 metres of sandstones and sandy shales of the Pioche Shale, DEISS (1938) has included about 55 metres of interbedded sandstones and limestones in a separate formation, the Tatow Limestone. In the southern part of the White Pine Range, above about 100 metres of micaceous shale with thin sandstone and black limestone of the Pioche Shale, MOORES and others (1968) include about 100 metres

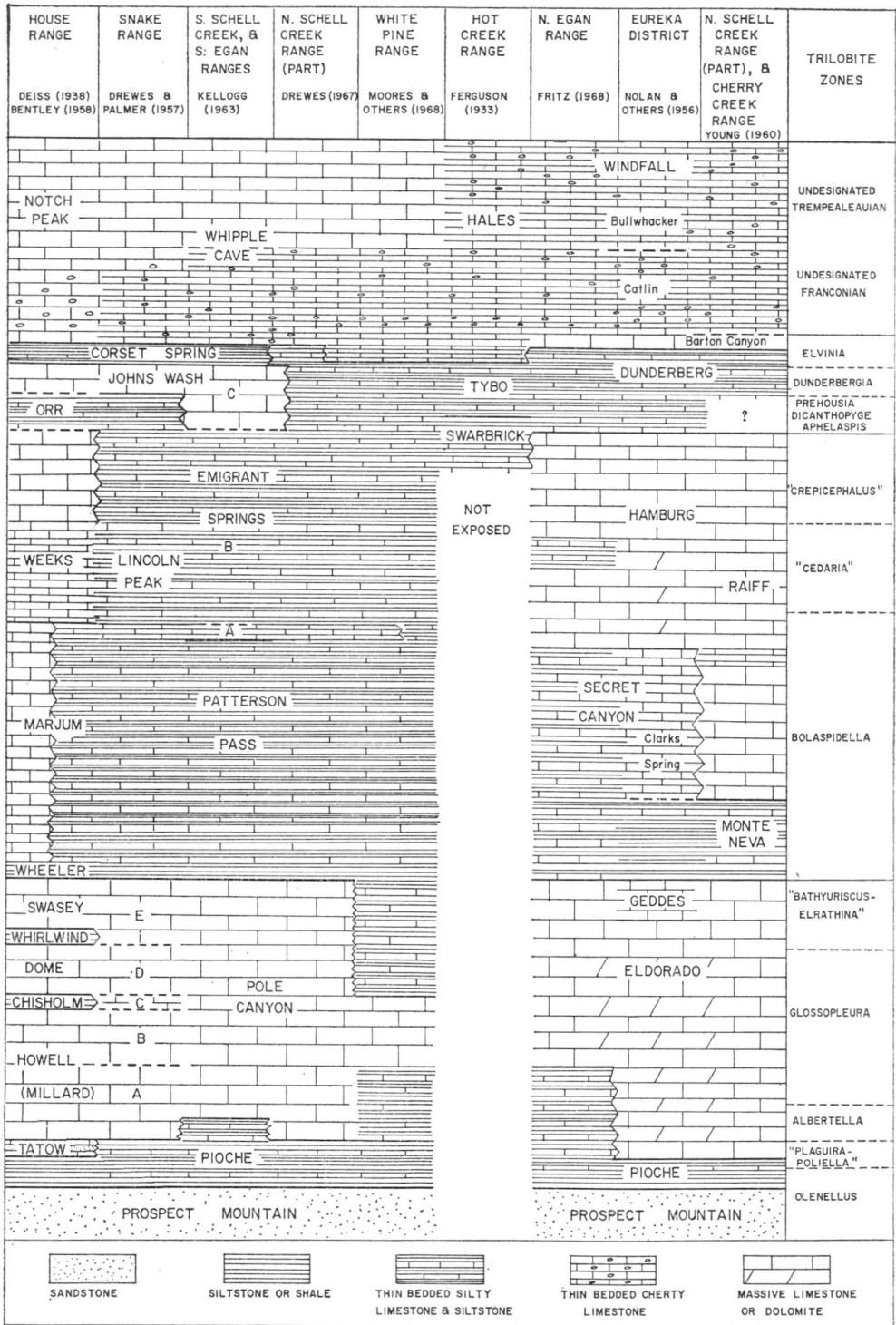

Fig. 7. Correlation chart of general lithologies and formational nomenclature in the Eureka–House Range region.

of limestone and interbedded shale in the lower part of the Pole Canyon Limestone. Elsewhere the top of the Pioche Shale has been placed at the base of a thick sequence of massive cliff-forming limestones. Unpublished faunal data indicate that the Tatow Limestone and a part of the beds included in the lower Pole Canyon Limestone by MOORES and others contain early Middle Cambrian trilobites of *Albertella* Zone or older age and are correlative with beds included in the Pioche Shale of the Pioche region.

Although some trilobites from the Pioche Shale of the Eureka district were described by WALCOTT (1884, 1910), the only recent study of Pioche Shale faunas in the Eureka–House Range region is by FRITZ (1968). He records a small fauna of Early Cambrian trilobites, including species of *Olenellus, Bonnia, Zacanthopsis, Crassifimbra*, and *Antagmus* from a thin limestone unit 50 metres above the base of the lower siltstone member in the northern Egan Range. The overlying siltstone and limestone member has yielded a rich early Middle Cambrian *Albertella* Zone fauna, including species of *Albertella, Ptarmiganoides, Poliella, Pagetia, Oryctocephalus, Oryctocephalites*, and assorted ptychoparioid trilobites from the lower 110 metres, and a small fauna including species of *Glossopleura* from the top of a thin-bedded limestone unit separated from the massive cliff-forming carbonates overlying the Pioche Shale by about 40 metres of nearly limestone-free shale. This is the only locality where shaly beds as young as the *Glossopleura* Zone have been included in the Pioche Shale and it is thus important for regional facies considerations (Chapter 5). In the Schell Creek Range, KELLOGG (1963) and DREWES (1967) report unidentified early Middle Cambrian trilobites from the upper part of the Pioche Shale, and DREWES also lists species of *Olenellus, Bonnia*, and *Crassifimbra* from the lower part of the formation.

Above the rocks assigned to the Pioche Shale and to the partly terrigenous sequences of the Tatow Limestone and the lower member of the Pole Canyon Limestone in the southern White Pine Range, all Cambrian sections throughout the Eureka–House Range region are characterized by the appearance of a thick cliff-forming unit of limestones with some interbedded dolomite.

The thickness of this unit ranges from 430 metres in the northern Egan Range (FRITZ, 1968) to an estimated 810 metres in the Eureka district (NOLAN and others, 1956) and varies between 450 and 630 metres in most other parts of the region. In the southern White Pine Range only 105 metres of massive limestone are present (MOORES and others, 1968), and a comparably thin sequence has been observed to the south in the Grant Range and to the north in the Ruby Range (SHARP, 1942). In the Eureka district, this unit was named the Prospect Mountain Limestone by HAGUE (1883). Later WALCOTT (1908b) renamed it the Eldorado Limestone to avoid confusion with the Prospect Mountain Quartzite. WHEELER and LEMON (1939) changed the designation to Eldorado Dolomite and separated off the uppermost 100 metres, consisting of thin- to medium-bedded dark-grey limestone as the Geddes Limestone. The Eldorado Dolomite in its type area consists of an estimated 720 metres of badly fractured, often dolomitized rocks lacking any comprehensible subdivisions. YOUNG (1960) and FRITZ (1968) identified the Eldorado Limestone in the northern Egan Range.

In the Snake Range, the unit correlative with the Eldorado Dolomite and Geddes Limestone consists of five members: lower, middle and upper members of dark-coloured limestone and intervening members of light-coloured limestone. These constitute the Pole Canyon Limestone (DREWES and PALMER, 1957). In addition to the Snake Range, this formation has been identified in the Schell Creek Range, southern Egan Range, and White Pine Range.

In the House Range, the sequence comparable to the Pole Canyon Limestone is interrupted by two shaly units and has been divided into several formations (WALCOTT, 1908b; DEISS, 1938; ROBISON, 1960b). The lowest formation is the Howell Formation which is about 250 metres thick and includes dark-coloured limestones in its lower half, light-coloured limestones in its upper half, and interbedded shales in its uppermost 15 metres. The dark-coloured limestones were separately named the Millard Limestone by WHEELER (1948). The shales contain species of *Glossopleura* and were correlated with the Chisholm Shale of the Pioche region and the middle dark-coloured limestone member of the Pole Canyon Limestone by ROBISON (1960b). The overlying 95 metres of massive dark-coloured limestone represents the Dome Limestone. The remaining part of the sequence was originally described as the Swasey Formation (WALCOTT, and DEISS, *op. cit.*). It includes 90 metres of dark-coloured limestone at its top and 40 metres of shales and interbedded limestones at its base. WHEELER (1948) correlated the shaly unit with the Condor Member of the Highland Peak Formation (Chapter 3.2); however, ROBISON (1960) showed that this was incorrect, proposed the name Whirlwind Formation for this shaly interval, and restricted the name Swasey to the overlying dark limestones. Both the Whirlwind and restricted Swasey are fossiliferous but the faunas are undescribed.

The dark limestones referred to the Swasey, upper Pole Canyon, and Geddes Limestones all contain trilobites that have been referred to the Middle Cambrian *Bathyuriscus-Elrathina* Zone (Chapter 4) (NOLAN and other, 1956; DREWES and PALMER, 1957; ROBISON, 1964b). Except for descriptions of a few species from the Eureka district (PALMER, 1954) this fauna is undescribed from the Great Basin. It includes species of *Ptychagnostus*, *Peronopsis*, *Tonkinella*, *Pagetia*, *Elrathina*, and *Parkaspis*.

Throughout the Eureka–House Range region, the beds assigned to the Eldorado, Geddes, Pole Canyon, or Swasey Limestones are overlain by thick sequences of siltstones or shales with some thin limestone interbeds commonly alternating with massive, generally clean carbonates. Wherever palaeontological data are available, the lower beds contain trilobites assignable to the lower part of the late Middle Cambrian *Bolaspidella* Zone or the underlying *Bathyuriscus-Elrathina* Zone. The interval between these Middle Cambrian beds and regionally widespread Franconian and younger Late Cambrian limestones and dolomites ranges in thickness from about 630 to about 1,350 metres, is often richly fossiliferous, and constitutes the most complex part of the Cambrian System of the Great Basin.

In the Eureka district (NOLAN and others, 1956) the interval begins with 75 metres of shale that constitutes a lower unnamed member of the Secret Canyon Shale. Above this is 135 metres of thin-bedded silty limestone of the Clarks Spring Member. Toward the top of this member, a few units of thick-bedded limestone appear below about 300 metres of massive dolomites and limestones which constitute the Hamburg Dolomite. Throughout the district, the Hamburg is in faulted contact with a unit of shales and thin limestone interbeds at least 90 metres thick which constitutes the Dunderberg Shale and marks the top of the heterogeneous Middle-Late Cambrian interval. The upper beds of the Secret Canyon Formation have yielded species of *Bolaspidella*, *Eldoradia*, *Asaphiscus*, and *Modocia*, and the lower part of the overlying Hamburg Dolomite has yielded species of *Bolaspidella*, *Modocia*, *Asaphiscus*, and *Holteria*. A few of these were described by PALMER (1954) but much of the fauna is undescribed. The Dunderberg Shale has yielded a rich assemblage of trilobites of the *Dunderbergia* and lower *Elvinia* Zones (RESSER, 1942; PALMER, 1960d, 1965b).

Plate 2. Upper Cambrian Orr Formation, east side of House Range, Utah. (See page 31).

(To face page 30)

In the northern Egan Range, the Cherry Creek Range, and the northern Schell Creek Range north of the latitude of McGill, Nevada, the Middle-Late Cambrian heterogeneous interval begins with the Monte Neva Formation, an interval of platy silty limestone, siltstone, and shale 165 metres thick (YOUNG, 1960). This is overlain by the Raiff Limestone which consists of three members: a lower member, 435 metres thick, of massive light- to dark-grey limestone; a middle member, 90 metres thick, of thin-bedded limestone and subordinate shale; and an upper member, 315 metres thick, of massive cliff-forming light- to dark-grey limestone. Above the Raiff Limestone is another unit of siltstones, shales, and thin limestones about 145 metres thick which is correlated with the Dunderberg Shale. The Monte Neva Formation has yielded undescribed species of *Marjumia* and *Trymataspis* typical of the middle part of the *Bolaspidella* Zone. The upper beds of the Raiff Limestone have yielded a rich fauna of *Crepicephalus*-Zone trilobites including species of *Coosia*, *Komaspidella*, *Cedaria*, *Carinamala*, and *Deiracephalus*. The Dunderberg Formation has yielded over 100 species of trilobites assigned to the *Aphelaspis*, *Dicanthopyge*, *Prehousia*, *Dunderbergia*, and lower *Elvinia* Zones of Late Cambrian age (PALMER, 1962a, 1962b, 1965b).

In the House Range (WALCOTT, 1908b; DEISS, 1938; BENTLEY, 1958; ROBISON, 1964b), the Middle-Late Cambrian heterogeneous interval begins with 145 metres of shaly limestones and calcareous shales of the Wheeler Formation which have yielded the well known complete weathered-out specimens of *Elrathia kingii* (MEEK) (BRIGHT, 1959) found in most North American university trilobite collections, as well as less common specimens of species of *Bolaspidella*, *Asaphiscus*, and *Olenoides*. Above this, about 405 to 420 metres of thin-bedded silty limestones and shales constitute the Marjum Formation. Within the House Range, this formation changes facies from north to south. In the northern part of the range, the formation is predominantly thin-bedded limestone with silty mottling. The amount of shale and silt increases southwards over a distance of about 20 kilometres until the formation is about 38% shale (ROBISON, 1964b). The thin-bedded limestones have yielded a rich fauna of unusual eocriniods (ROBISON, 1965) and sponges (WALCOTT, 1920; RIGBY, 1966), as well as silicified trilobites of the *Bolaspidella* Zone including species of *Modocia*, *Marjumia*, *Bathyuriscus*, *Bolaspidella*, *Ptychagnostus*, *Cotalagnostus*, and *Olenoides* and, toward the top, species of *Lejopyge* and *Holteria* (ROBISON, 1964a, 1967). The uppermost beds of the Marjum Formation in the shaly sequences are less shaly than the lower beds and form a distinctive cliff. Approximately 360 metres of thin-bedded slope-forming silty limestones constitute the overlying Weeks Formation, which contains abundant Late Cambrian trilobites of the *Cedaria* and early *Crepicephalus* Zones. Some of the trilobites, including species of *Densonella*, *Crepicephalus*, *Tricrepicephalus*, and *Cedaria*, have been described by WALCOTT (1916a, 1916b) and ROBISON (1960a). The Orr Formation (Plate 2) conformably overlies the Weeks Formation. As restricted by BENTLY (1958), it consists of three members: a lower limestone member about 215 metres thick which includes zones of algal stromatolites and abundant biocalcarenites; a middle member of siltstones and thin-bedded silty limestones about 135 metres thick; and an upper cliff-forming limestone member about 35 metres thick. The lower member has yielded a *Crepicephalus*-Zone fauna including species of *Kingstonia*, *Crepicephalus*, *Tricrepicephalus*, and *Terranovella*, some of which have been described by ROBISON (1960a). The middle member includes trilobites of the *Aphelaspis*, *Dicanthopyge*, *Prehousia*, and lower *Dundergergia* Zones (PALMER, 1965b). The upper unit is not fossiliferous. Above this unit is about 45 metres of shales with some interbedded limestones which are referable to the Corset Spring Shale and constitute the upper unit of the heterogeneous Middle and Late Cambrian interval. Trilobites from this unit are referable to the *Elvinia* Zone and include species of *Housia, Irvingella,* and *Kindbladia* (PALMER, 1965b).

In the Snake Range (DREWES and PALMER, 1957) and the northern Schell Creek Range south of the latitude of McGill (DREWES, 1967), the Middle-Late Cambrian heterogeneous interval begins with a thick sequence of shales, siltstones, and thin-bedded limestones in excess of 630 metres in thickness which has been named the Lincoln Peak Formation. Its lower beds have yielded species of *Ptychagnostus*; a cliff-forming limestone unit in its middle part has yielded species of *Lejopyge;* slightly younger limestones have yielded species of *Tricrepicephalus, Cedaria,* and *Carinamala;* and the upper 115 metres of the formation have yielded abundant trilobites of the *Aphelaspis, Dicanthopyge, Prehousia,* and lower *Dunderbergia* Zones (PALMER, 1965b). Above the Lincoln Peak Formation is a cliff-forming, oolitic, relatively clean limestone unit about 80 metres thick which constitutes the Johns Wash Limestone. This formation is the exact equivalent of the upper member of the Orr Formation. Except for a small collection in its uppermost beds which has yielded species of *Parahousia* and *Elviniella* typical of the lower part of the *Elvinia* Zone, the formation is unfossiliferous. About 23 metres of shales with a thin middle limestone member constitutes the Corset Spring Shale at the top of the heterogeneous Middle-Late Cambrian interval. This formation has yielded a rich fauna from the lower *Elvinia* Zone including species of *Pterocephalia, Kindbladia, Iddingsia,* and *Housia* (PALMER, 1965b).

In the southern Schell Creek Range and the southern Egan Range (KELLOGG, 1963), the Middle-Late Cambrian heterogeneous interval is very similar to that of the Snake Range. It begins with about 630 metres of shale, siltstone, and thin-bedded silty limestone assigned to the Patterson Pass Shale. Overlying this is a cliff-forming unit of thin-bedded silty limestone and intraformational limestone conglomerate ranging from 35 to 85 metres in thickness which includes a major contorted zone indicative of soft sediment slumping. The overlying 450 metres consists of mudstone and thin-bedded silty limestone in the lower half and thin-bedded limestone in the upper half. This unit and the thin cliff-forming underlying limestone unit are Members A and B of the Emigrant Springs Limestone. Together with the Patterson Pass Shale, they constitute a unit equivalent to the Lincoln Peak Formation of the Snake Range. The Patterson Pass Shale contains abundant but generally poorly preserved agnostids including species of *Hypagnostus*. The cliff-forming limestones of Member A of the Emigrant Springs Limestone contain abundant agnostids including species of *Lejopyge* and seem to be the equivalent of the *Lejopyge*-bearing limestones in the middle of the Lincoln Peak Formation of the Snake Range and the cliff-forming limestones at the top of the Marjum Formation in the House Range. The lower beds of Member B contain species of *Cedaria* and *Tricrepicephalus* and the upper beds contain species of *Coosina*, a species typical of the upper part of the *Crepicephalus* Zone in the mid-continent region.

Member C of the Emigrant Springs Limestone, the uppermost member, is characterized by 205 metres of oolitic, medium- to fine-grained, massive- to medium-bedded cliff-forming limestone containing trilobite faunas of the *Aphelaspis, Dicanthopyge,* and *Prehousia* Zones (PALMER, 1965b). This is overlain by about 40 metres of shales and inter-bedded limestones containing trilobite faunas of the uppermost *Dunderbergia* and lower *Elvinia* Zones (PALMER, 1965b). These units are equivalent to the Johns Wash Limestone and Corset Spring Shale of the Snake Range. The principal difference is that the Johns Wash Limestone in the southern Schell Creek Range and southern Egan Range is significantly thicker and its base is about two trilobite zones older than in the Snake Range. Comparison of sections from the two areas (PALMER, 1965b) shows that these changes are due to a lateral facies change from thin-bedded silty limestones of the upper Lincoln Peak Formation to more massive cleaner limestones of the Johns Wash Limestone.

In the White Pine and Grant Ranges (HUMPHREY, 1960; MOORES and others, 1968), the Middle-Late Cambrian heterogeneous interval is characterized by about 1,080 metres of thin-bedded silty limestones and shales assigned to the Lincoln Peak and Dunderberg Formations. HUMPHREY (1960) recognized seven units in the upper 765 metres of this sequence in the White Pine mining district at the north end of the White Pine Range. The lowest exposed unit is about 30 metres of dark-grey to black platy limestone which has yielded the late Middle Cambrian trilobite *Holteria*. Overlying this are about 240 metres of thin-bedded, brownish, locally micaceous shale lacking fossils; 300 metres of thin-bedded, shale-parted limestone which has yielded species of the Late Cambrian *Aphelaspis* Zone from its upper beds; 90 metres of greenish shale; 30 metres of interbedded shale and limestone; 40 metres of greenish shale; 70 metres of interbedded shale and thin-bedded limestone which have yielded trilobites of the *Dunderbergia* Zone; and 10 metres of nodular shaly limestone. There are no massive limestone units within the Middle-Late Cambrian heterogeneous sequence of this area.

In the Tybo mining district of the Hot Creek Range, FERGUSON (1933) described an incomplete Cambrian section which represents the upper part of the Middle-Late Cambrian heterogeneous interval. The lowest exposed unit consists of thin-bedded silty limestone described as the Swarbrick Limestone. This limestone has yielded trilobites of the *Aphelaspis* Zone from its upper beds and is probably the same as the 300 metres unit of thin-bedded limestone included within the Lincoln Peak Formation of the White Pine Range. The Swarbrick Limestone is overlain by a shale formation, the Tybo Shale, which has a few thin limestone beds containing trilobites of the *Dunderbergia* Zone. This formation was estimated by FERGUSON (1933) to be about 450 metres thick, but the figure is probably excessive. Immediately overlying limestone beds contain trilobites of the *Elvinia* Zone, and units correlative with the Tybo Shale elsewhere in the Great Basin are only about 180 to 225 metres thick. The Tybo Shale correlates with the upper 215 metres of the beds assigned to the Lincoln Peak and Dunderberg Formations in the White Pine Range and to the complete Dunderberg Formation of the northern Schell Creek, northern Egan, and Cherry Creek Ranges.

In the Ruby Range, the Cambrian sequence described by SHARP (1942) has many similarities with the Middle-Late Cambrian heterogeneous sequence of the White Pine and Grant Ranges. He briefly described about 1,800 metres of Cambrian beds, without any formal subdivisions, overlying the Prospect Mountain Quartzite. Except for about 90 metres of dark-grey, coarsely-crystalline, massive limestone with some quartzite interbeds and lenses, the entire sequence is composed of units of thin-bedded, arenaceous and argillaceous limestone and limy argillite with only rare thicker-bedded limestones. More recent observations in the Ruby Range by PALMER (1960a, 1965b) show that one of the distinctive units in the southern part of the Ruby Range is composed of thin-bedded limestone and interbedded argillite capped by a few metres of thicker-bedded limestone. It is more resistant than the adjacent units and contains late Middle Cambrian trilobites including species of *Holteria*. This unit is overlain by several hundred metres of argillite without noticeable limestone beds which is transitional upwards into a sequence of alternating limestone and argillite units with trilobites of the *Dicanthopyge*, *Prehousia*, and *Dunderbergia* Zones. These beds are lithologically and faunally equivalent to the beds of the Dunderberg Shale in the Cherry Creek and northern Egan Ranges to the east and in the Eureka district to the south. However, the massive clean carbonate interval of the Hamburg or Raiff Limestones is entirely replaced by silty argillites.

Throughout the Eureka–House Range region, the remainder of the Late Cambrian rocks consist of cherty and non-cherty limestones and dolomites which have been assigned to different local formational units or have been included in undifferentiated units of Cambrian and Ordovician age.

In the Eureka district (NOLAN and others, 1956) this interval was described as the Windfall Formation and was divided into two members. The lower, Catlin, member consists of a basal unit of massive chert-free limestone about 9 metres thick overlain by about 75 metres of thin-bedded, dark-grey cherty limestones. The chert-free basal limestone contains trilobites of the lower Franconian *Elvinia* Zone in its lower part (PALMER, 1965b) and has yielded species of *Drumaspis* of Middle Franconian age in its upper part. The overlying cherty limestones include species of *Lotagnostus*, *Bienvillia*, and *Richardsonella* also of Franconian age, and the uppermost beds include species of *Eurekia*, *Tatonaspis*, and *Saukiella* of Trempealeauan age.

The upper, Bullwhacker, member consists of about 115 metres of thin-bedded, silty, generally chert-free limestones and some massive limestone interbeds. It has yielded a rich fauna of Trempealeauan trilobites and brachiopods including species of *Eurekia*, *Euptychaspis*, *Rasettia*, *Richardsonella*, *Elkania*, and *Conodiscus*, some of which have been described by WALCOTT (1925), KOBAYASHI (1935), and ULRICH and COOPER (1938).

The Windfall Formation has been recognized in several other areas. In the Antelope Range south-west of Eureka, it consists almost entirely of rocks of Catlin lithology (NOLAN and others, 1956). In the northern Schell Creek and Egan Ranges, YOUNG (1960) has recognized a basal massive non-cherty unit as the Barton Canyon Limestone Member. In the White Pine and Grant Ranges, MOORES and others (1968) have included in the lower part of the Windfall Formation about 135 metres of thin-bedded non-cherty limestone. At least the lower 50 metres of this unit contain species of the upper *Elvinia* Zone. Trilobites of this age are found no higher than 7 metres above the base of the Windfall Formation in the Eureka district. The Windfall Formation and overlying Ordovician limestones in the White Pine Range were included by HAGUE (1883) in the original designation of the Pogonip Limestone. This name has been restricted in almost all subsequent publications to Ordovician rocks only.

In the House Range, the beds above the Corset Spring Shale are all included in the Notch Peak Formation. This formation consists of thin-bedded to massive, cherty limestones and dolomites aggregating about 565 metres in thickness (WALCOTT, 1908b; BENTLEY, 1958). Locally, the upper part of the formation contains algal stromatolites through as much as 70 metres of beds. Some of these algal stromatolites show evidence of wave erosion (HOSE, 1961).

In the Snake Range and part of the Schell Creek Range (DREWES and PALMER, 1957; DREWES, 1967), the complicated geological structure has precluded subdivision of the generally massive Cambrian and Ordovician limestones and dolomites, and no formal nomenclature has been applied to these rocks which are estimated to aggregate about 900 metres in thickness.

In the southern Schell Creek and Egan Ranges, KELLOGG (1963) included the cliff-forming sequence of Late Cambrian limestones and dolomites in the Whipple Cave Formation. This is composed of a lower member of cherty thin-bedded limestones about

180 metres thick and an upper member of more massive limestones and dolomites about 360 metres thick. The lower member has trilobites of the uppermost *Elvinia* Zone in its basal beds (PALMER, 1965b) and the distinctive brachiopods *Billingsella* and *Ceratreta* about 50 metres above its base. The upper member has yielded species of the distinctive Trempealeauan trilobites *Eurekia* and *Bowmania* about 135 metres below its top (KELLOGG, 1963).

In the Hot Creek Range, thin-bedded, black cherty limestones similar to those of the Catlin Member of the Windfall Formation include trilobites of the *Elvinia* Zone in their basal part and early Ordovician trilobites of the *Symphysurina* Zone in their upper part. These limestones are badly deformed and no accurate thickness can be obtained, but they are included within the unit described as the Hales Limestone by FERGUSON (1933).

The boundary between Cambrian and Ordovician rocks in the Eureka–House Range region is within a continuous succession of carbonate rocks. For most mapping purposes it has been placed at the contact of the beds of the predominantly Ordovician Pogonip Group with the Windfall, Notch Peak, or Whipple Cave Formations. However, in the Eureka district, the lowest few metres of the Pogonip Group contain specimens of the Trempealeauan trilobite *Eurekia* (NOLAN and others, 1956). In the southern Egan and Schell Creek Ranges, there is no faunal control for the upper 135 metres of the Whipple Cave Formation or the lower 60 metres of the overlying House Limestone above which are species of *Symphysurina*. Thus an unknown part of this 190 metre interval may be of Cambrian age and the system boundary cannot be precisely placed relative to the formation boundary. In the House Range, at least the upper 70 metres of the Notch Peak Formation lack any faunal control to identify these beds as Cambrian. In other areas of the Eureka–House Range region, the position of the Cambrian–Ordovician boundary is even less precisely known.

3.4. The Central Utah region

The Cambrian sections of the Deep Creek, Dugway, Stansbury, and Sheeprock Ranges north of the House Range, and the Drum Mountains, Cricket, Canyon, and Pavant Ranges east of the House Range (Fig. 2) present a striking contrast to the slightly thicker sections of the Eureka–House Range region to the south and west. This contrast is primarily due to changes in Middle Cambrian and early Late Cambrian carbonate and shale-siltstone relationships. The thick, massive, generally cliff-forming early Middle Cambrian carbonate units of the Eureka–House Range region are replaced eastward and northward by an interval of thinner limestone and thicker shale units, and the thick shaly and silty units of the late Middle Cambrian and early Late Cambrian of the Eureka–House Range region are replaced eastward by nearly continuous sequences of limestones and dolomites (Fig. 8).

The Tintic–southern Wasatch region to the east (Chapter 3.5) has a gross stratigraphy similar to that of the central Utah region, but the sections are only slightly more than half as thick.

Within the central Utah region, published information is available for the Deep Creek Range (NOLAN, 1935; BICK, 1966), Dugway Range (STAATZ and CARR, 1964), Stansbury Range (RIGBY, 1958), Sheeprock Range (COHENOUR, 1959), Drum Mountains (CRITTENDEN and others, 1961), and Canyon Range (CHRISTIANSEN, 1952). However, a unified stratigraphical nomenclature has not been established. NOLAN (1935) and STAATZ and CARR (1964) proposed many new local stratigraphical names; CRITTENDEN and others

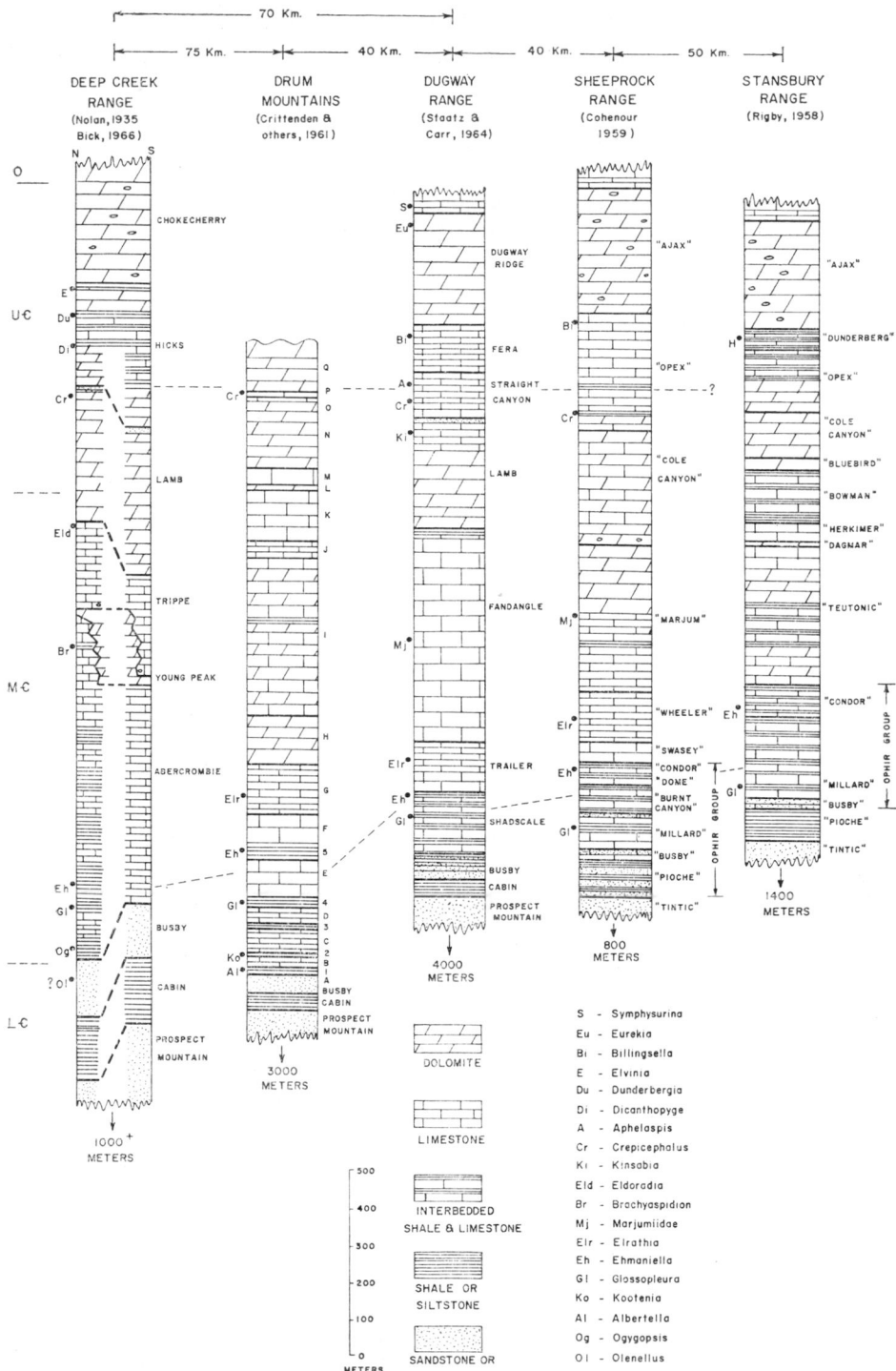

FIG. 8. Comparison of columnar sections and stratigraphical terminology for the rocks of the central Utah region.

(1961) used only informal letter or number designations for all stratigraphical units; and RIGBY (1958), COHENOUR (1959), and CHRISTIANSEN (1952) used existing stratigraphical names derived primarily from the House Range to the south-west and the east Tintic Mountains or Oquirrh Mountains to the north-east. The result is a confused welter of names which is difficult to summarize.

In the Deep Creek Range, the Cambrian section begins within a sequence of quartzite and shale units, The upper 765 metres of this sequence are predominantly quartzite and the lower 630 metres consist of alternating units of quartzite and shale with individual units ranging up to 180 metres in thickness. All of these units were assigned to the Prospect Mountain Quartzite by NOLAN (1935), but only the upper 300 metres of the upper quartzite sequence was designated as Prospect Mountain by BICK (1966). The Prospect Mountain Quartzite is overlain by about 150 metres of shale described by NOLAN as the Cabin Shale. This unit has yielded a few scraps of indeterminate brachiopods and contains the oldest fossils in the Deep Creek Range. The Cabin Shale is overlain by about 135 metres of fine- to coarse-grained quartzite with some shaly interbeds that was named the Busby Quartzite by NOLAN. Possible olenellid scraps have been recorded from the Busby Quartzite by BICK. BICK used the name Pioche Shale for NOLAN's Cabin Shale. However, if the olenellid scraps in the Busby Quartzite can be confirmed, then the Busby is more nearly the correlative of quartzite beds which lie below the Pioche Shale in areas to the south; and the underlying 'Pioche' Shale of BICK is an older shale unit, perhaps reflecting the same marine incursion into the area of dominant quartz sand deposition which is represented by the upper Wood Canyon Formation in the Spring Mountains–Death Valley region to the south (Chapter 3.1.1) and the argillite in the upper part of the Brigham Quartzite in south-eastern Idaho (Chapter 3.6).

Above the Busby Quartzite in the northern part of the Deep Creek Range is a sequence of alternating thin-bedded limestone units and shale units totalling about 800 metres in thickness, with individual units of shale and limestone ranging from a few metres to several tens of metres in thickness. This sequence was described as the Abercrombie Formation by NOLAN. Less than 10 kilometres to the south, the shale units of the Abercrombie Formation have nearly all disappeared and the Abercrombie is reduced to less than 540 metres of thin-bedded limestones.

The formation has yielded a few fossils indicating an age range through the lower two-thirds of the Middle Cambrian. The oldest fossils are specimens of *Ogygopsis* from silty beds at the base of the formation. This genus is found at many localities in northern Utah in beds immediately above the basal quartzites. Slightly higher shale beds have yielded the *Glossopleura* and overlying *Ehmaniella* faunas which indicate correlation with the shales of the uppermost Howell and Whirlwind Formations of the House Range to the south. The uppermost shales of the Abercrombie Formation have yielded specimens of *Brachyaspidion*, a small trilobite known from the lower *Bolaspidella* Zone (ROBISON, 1964a). Thus at least part of the Abercrombie is as young as the Wheeler Shale of the House Range.

The top of the Abercrombie was placed at the base of a massive dolomite described by NOLAN as the Young Peak Dolomite. He showed that this unit thins northward within his area by lateral gradation with thin-bedded limestones of the upper part of the Abercrombie Formation. Within a distance of 7 kilometres, the Young Peak Dolomite thins from about 180 metres to only a few metres which barely serve to separate the thin-bedded limestones of the Abercrombie Formation from similar thin-bedded limestones in the overlying Trippe Limestone. Southwards, within a distance of 10 kilometres,

BICK shows that the lower beds of the Young Peak Dolomite persist and the upper beds, are gradational along strike into the thin-bedded limestones of the overlying Trippe Limestone.

Because of these relationships, the Trippe Limestone, which has a maximum thickness of about 235 metres, also shows considerable thickness variation. The upper beds of the Trippe have yielded specimens of *Eldoradia*, a distinctive latest Middle Cambrian trilobite.

The Trippe Limestone is overlain by about 335 metres of dolomites of the Lamb Dolomite. An interval of thin-bedded limestones and some included sandstones about 50 metres thick marks the top of this formation. BICK (1966), BENTLEY (1958), and ROBISON (1960a) have reported the early Dresbachian trilobites *Crepicephalus* and *Tricrepicephalus* from this interval as well as the problematical genus *Kinsabia*.

The overlying Hicks Formation is composed of alternating units of dolomite and thin-bedded silty limestone or siltstone, which separate the older predominantly dolomitic sequences from the latest Cambrian cherty dolomites of the Chokecherry Dolomite. The Hicks also shows significant lateral facies changes within the Deep Creek Range. In the northern exposures described by NOLAN, the Hicks is predominantly composed of thin-bedded sandy and oolitic dolomite with only a small proportion of thin-bedded silty limestone. Southwards, the limestone intervals thicken and the whole formation increases in thickness from 180 to almost 360 metres. A shaly interval about 10 metres thick near the top of the Hicks Formation is persistent throughout the area and contains trilobites of the *Elvinia* Zone (BENTLEY, 1958), which indicates its probable equivalence to the Corset Spring Shale of the Eureka–House Range region. The thin-bedded limestones lower in the formation have yielded trilobites of the *Dicanthopyge*, *Prehousia* and lower *Dunderbergia* Zones (PALMER, 1965b) and can be correlated with the middle part of the Orr Formation of the House Range.

The cherty Chokecherry Dolomite which overlies the Hicks Formation is about 360 metres in maximum thickness. Nolan believed all of the formation to be Ordovician and to lie unconformably on the Hicks. The work of BICK and a better knowledge of regional stratigraphy now strongly support the idea that the formation is conformable on the Hicks and is at least in part of Late Cambrian age. It is thus one of several locally named Late Cambiran cherty dolomite units within the Great Basin.

In the Drum Mountains, about 75 kilometres south-east of the Deep Creek Range, CRITTENDEN and others (1961) have described a section which is in part transitional between the nearby House Range to the West and the other sections of the central Utah region. Basal quartzites, totalling about 2,700 metres in thickness, were all assigned to the Prospect Mountain Quartzite. These are overlain by about 40 metres of shale and siltstone referred to the Cabin Shale and about 60 metres of quartzite referred to the Busby Quartzite. The remaining stratigraphical units were unnamed. All carbonate units were designated with letters and all shale units with numbers. At the base of the sequence, a 7-metre unit of dolomite and dolomitic sandstone (Dolomite A) is overlain by an equal thickness of shales (Shale 1) which contains the trilobite *Albertella*. If the identification of the underlying Busby Quartzite is correct, this is additional evidence that the Cabin Shale which underlies the Busby, and the Pioche Shale which has the *Albertella* fauna in its upper part, are probably not, as claimed by BICK (1959), the same unit. The thin-bedded Limestones B, C, and D and the included Shales 2 and 3 which total about 145 metres are the equivalents of the massive Howell Limestone of the House Range (Fig. 7), and they illustrate clearly the facies change from massive limestone to thin-bedded limestone and shale in a general north-easterly direction. Shale 4, about 20 metres thick, has yielded *Glossopleura* and is the approximate correlative of the shales at the top of the Howell Limestone of the House Range. Limestone 3, a massive unit 100 metres thick, is overlain by Shale 5, 40 metres thick, which contains species of *Ehmaniella* and *Solenopleurella* typical of the Whirlwind Formation. Thus, Limestone E is the correlative of the Dome Limestone of the House Range and shows that the Dome has maintained its character farther to the north-east than the older limestone units of the House Range. Limestone F is another massive limestone, about 76 metres thick, which is the probable equivalent of the Swasey Limestone of the House Range. This is followed by 115 metres of thin-bedded limestone of Limestone G. This unit has yielded species of *Elrathia* and *Asaphiscus* which indicate a correlation with the Wheeler Shale. The correlation reflects a significant facies change to less shaly rocks to the north-east and supports the idea that the shaly rocks of the Wheeler, Lincoln Peak, Patterson Pass, and Secret Canyon (Fig. 7) were deposited to the west of the carbonate sequences, in contrast to the older shales which increase in abundance in the sections to the east.

Overlying Limestone G is a succession of 10 limestone or dolomite units aggregating almost 1,000 metres in thickness. Thin-bedded limestones of Limestone P, about 950 metres above Limestone G, have yielded an early Dresbachian fauna including species of *Tricrepicephalus* and *Coosella* indicating a correlation with the thin-bedded limestones at the top of the Lamb Dolomite of the Deep Creek Range to the west and the Dugway Range to the north. The underlying 320 metres of massive carbonates comprising Limestones or Dolomites K to O could be the equivalent of the Lamb Dolomite and, at least in part, also of the lower carbonate unit of the Orr Formation of the House Range. In the area described by CRITTENDEN and others, the upper part of the Cambrian section is not exposed. Further north in the Drum Mountains, a full Cambrian section is present but has not been described.

In the Dugway Range (STAATZ and CARR, 1964), about 60 kilometres east of the Deep Creek Range and 40 kilometres north of the Drum Mountains, the stratigraphy is sufficiently distinct from known areas elsewhere in the region to require that, except for the basal terrigenous units and dolomites identified as the Lamb Dolomite, all of the formational units be locally named. At the base of the section, the Prospect Mountain Quartzite includes about 3,600 metres of beds. The overlying Cabin Shale includes 45 metres of interbedded quartzite and siltstone. The unit identified as Busby Quartzite includes a lower quartzite member 25 metres thick and an upper member composed of about 30 metres of siltstone.

The overlying Shadscale Formation is composed of units of silty limestone, shale, or dolomite aggregating 155 metres in thickness. The occurence of the *Ehmaniella* fauna in the upper 45 metres and *Glossopleura*, about 20 metres lower, indicates that this formation includes beds equivalent to the Howell, Dome, and Whirlwind Formations of the House Range and that the massive Dome Limestone like the Howell Limestone has now changed facies to thin-bedded limestones and interbedded shales.

The next younger Trailer Formation includes a lower massive limestone member about 50 metres thick overlain by 78 metres of thin-bedded limestones with *Elrathia*. These members are thus equivalent to Limestones F and G of the Drum Mountains, and the lower member is probably the north-eastern continuation of the Swasey limestone.

Above the Trailer Formation is a unit of 505 metres of massive alternating units of dark and light coloured limestone assigned to the Fandangle Limestone, which is overlain by 190 metres of massive dolomite and 80 metres of thin-bedded limestone and

sandstone of the Lamb Dolomite. Rare marjumiid trilobites in the Fandangle Limestone
indicate a correlation with the Marjum Formation of the House Range, and the occurence
of the problematical Dresbachian fossil *Kinsabia* in the upper thin-bedded limestones
and sandstones of the Lamb Dolomite confirms its correlation with Lamb Dolomite in
the Deep Creek Range. The overlying 110 metres of thin-bedded limestone and more
massive dolomite units of the Straight Canyon Formation has yielded faunas of the
Crepicephalus and *Aphelaspis* Zones of Dresbachian age indicating a correlation with
the Hicks Formation in the Deep Creek Range and in part with the Opex Formation
of the Tintic–Southern Wasatch region (Chapter 3.5). The next overlying 120 metres of
mottled thin- to thick-bedded limestone and dolomite have been assigned to the Fera
Limestone and have yielded the brachiopod *Billingsella* and conaspid trilobites of middle
Franconian age about 30 metres below the top. The remainder of the Cambrian section
is composed of 265 metres of massive coarse dolomite of the Dugway Ridge Dolomite
which has yielded the Trempealeauan trilobite *Eurekia* about 35 metres below the top.
The Dugway Ridge Dolomite and Fera Limestone occupy the stratigraphical position
of the Notch Peak Limestone of the House Range and the Chokecherry Dolomite of
the Deep Creek Range.

The Stansbury, Sheeprock, Canyon, and Pavant Ranges lie in a north–south line
that extends south for almost 200 kilometres from the Great Salt Lake and lies about
50 kilometres to the east of the Dugway Range and Drum Mountains (Fig. 2). The
published work on these areas has all been by Utah geologists who have attempted,
with varying degrees of success, to use existing formational names for the Cambrian
sequences.

In the Stansbury Range (RIGBY, 1958), the lowest exposures are quartzites, about
1,260 metres thick, assigned to the Tintic Quartzite. These are overlain by about 80 metres
of shale and quartzite identified as Pioche Shale. The next 270 metres of beds, including
a basal unit of sandstone and sandy limestone (25 metres) and overlying units of limestone or interbedded limestone and shale, were assigned to the Ophir Group. The shales
in the middle of the sequence have yielded species of *Glossopleura* and in the upper part
species of *Ehmaniella*. The basal unit was correlated with the Busby Quartzite, and the
faunas indicate a correlation of the remainder of the interval with the Shadscale
Formation in the Dugway Range to the southwest. The probably improper usage of
Pioche for the unit below the Busby has already been commented on (p. 37), and RIGBY's
usages of Millard (i.e. Howell) and Condor for intervals within the Ophir Group now
seem inappropriate. The Howell lost its identity before even reaching the intervening
Dugway Range and Drum Mountains, and the name Condor, for the *Ehmaniella*
bearing beds, perpetuated an earlier misuse of that name in the House Range by
WHEELER (p. 30). The next 810 metres of the section is composed of units of limestone,
dolomite, and interbedded limestone and shale. The formational nomenclature used for
these rocks represents an interleaving of names derived from the thin sequences of the
Oquirrh and East Tintic Mountains to the east, with the additional usage of Dunderberg
Shale for a shaly interval at the top. The names for these rocks are shown in quotation
marks on Figure 8 because the Hartmann, Bowman, and Lynch of the Oquirrh Range
and the Teutonic, Dagmar, Herkimer, Bluebird, and Cole Canyon of the East Tintic
Mountains are correlative sequences (Fig. 9) and the Dunderberg Shale is not recognizable in central Utah (PALMER, 1960c). New work is needed in the Stansbury Range to
determine more appropriate designations for the parts of the section where these names
have been used. The uppermost 270 metres of the section consists of massive cherty

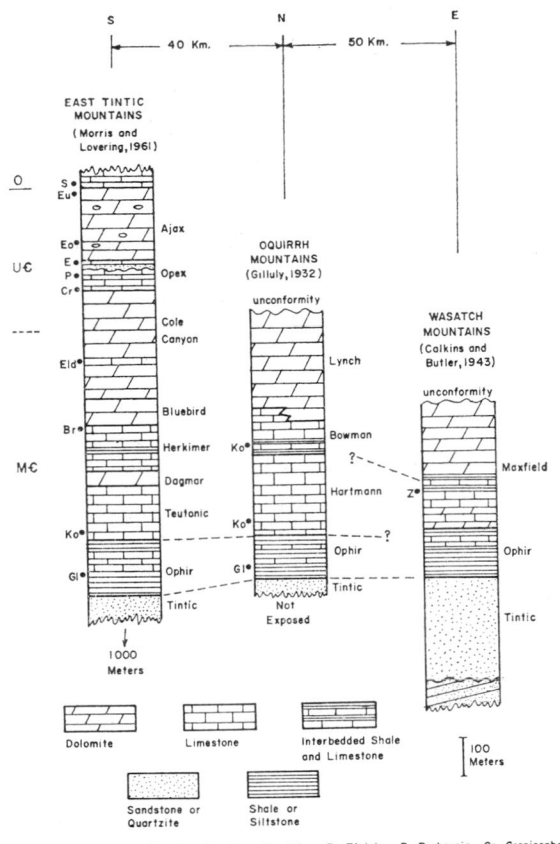

FIG. 9. Comparison of columnar sections and stratigraphical terminology for the Tintic–southern Wasatch region.

dolomites which have been identified as the Ajax Dolomite after a similar unit is the same stratigraphical position in the East Tintic Mountains (Chapter 3.5).

In the Sheeprock Range (COHENOUR, 1959) the lowest exposures are 720 metres of medium-grained quartzite assigned to the Tintic Quartzite. Above this is a sequence of about 325 metres of beds assigned to the Ophir Group. This includes 100 metres of shaly and cross-bedded quartzites at the base assigned to the Pioche shale; 20 metres of dolomite with an *Albertella* Zone fauna and 20 to 30 metres of quartzite collectively identified as the Busby Quartzite; 60 metres of medium-bedded limestone, 30 metres of shale with *Glossopleura*, and 15 metres of pisolitic limestone all assigned to the Millard Formation; 60 to 70 metres of interbedded shale and thin-bedded limestone assigned to the Burnt Canyon Formation; 12 to 25 metres of massive limestone assigned to the Dome Limestone; and about 40 metres of shale with *Ehmaniella* assigned to the Condor Formation.

Above the rocks assigned to the Ophir Group are about 40 to 50 metres of massive dark limestone identified as the Swasey Limestone. These are overlain by 120 to 130 metres of thin-bedded dark-coloured limestone assigned to the Wheeler Formation which has yielded species of *Elrathia* and *Peronopsis*. The next younger unit includes, in ascending order; 110 metres of thin-bedded limestone, 16 metres of shale, 70 metres of

limestone with marjumiid trilobites, and 165 metres of dolomite—all assigned to the
Marjum Formation. Above the Marjum about 270 metres of alternating dark- and
light-coloured limestone and dolomite are assigned to the Cole Canyon Dolomite. The
basal bed of this unit has black chert concretions in a white laminated dolomite.
The next younger unit begins with 40 metres of limestone and dolomite followed by 3
metres of shale, 70 metres of limestone, 10 metres of shale, and 170 metres of thin-bedded
limestone, all assigned to the Opex Formation. The shales and overlying limestones
have yielded faunas ranging in age from the Dresbachian *Crepicephalus* Zone to the
mid-Franconian *Conaspis* Zone. The lithology and faunas indicate that this interval is
the equivalent of the Straight Canyon and Fera Limestones of the Dugway Range. A
fault of unknown, but probably small, displacement separates the beds assigned to the
Opex formation from predominantly dark cherty dolomites assigned to the Ajax
Dolomite which marks the top of the section.

As with the Stansbury Range, increases in knowledge of the regional stratigraphy
of western Utah now require reassessment of the stratigraphical assignments of many of
the units in the Sheeprock Range, particularly in the Ophir Group, in order to establish
a consistent stratigraphical nomenclature for the rocks of the central Utah region.

The Cambrian sequences of the Canyon, Pavant, and Cricket Ranges are not well
known. The Canyon Range section includes at its base about 450 metres of quartzite,
identified as the Tintic Quartzite, which overlies about 1,800 metres of older quartzitic
rocks assigned to the Pre-Cambrian (CHRISTIANSEN, 1952). The 295 metres of shales and
limestones above the quartzites were assigned to the Ophir Formation, and the remaining
1,420 metres of limestones and dolomites were unidentified. The only reported trilobites
are species of an '*Ehmania* fauna' from the uppermost shales of the Ophir Formation,
which suggests a correlation with the uppermost shaly beds assigned to the Ophir Group
or the Shadscale Formation in the Sheeprock and Dugway Ranges to the north. Thin
shale and sandy dolomite units about 855 metres above the base of the undifferentiated
carbonate sequence are at about the stratigraphical position of the Dresbachian shaly
or sandy units in areas to the north. According to the geological map of Utah, the uppermost part of this carbonate sequence in the area of CHRISTIANSEN's section is now
considered to be of early Ordovician age, so that a complete Cambrian section is present
in the area.

Detailed sections of the rocks in the Cricket and Pavant Ranges have not been published,
but summary columnar sections prepared by HINTZ (1969) indicate a stratigraphy
similar to that described for other sections of the central Utah region. In the Cricket
Range, about 50 kilometres south-east of the House Range, greatly thinned massive
equivalents of the Howell, Dome, and Swasey limestones can be recognized, separated by
thick shale units bearing the *Glossopleura* and *Ehmaniella* faunas. There is no apparent
equivalent to the Wheeler Shale, and not enough detail is available for comparison
with younger units elsewhere in the region. The late Dresbachian trilobite *Aphelaspis*
is recorded from shales towards the top of the exposed section which indicates that the
younger parts of the Cambrian section are not exposed. In the Pavant Range, above the
basal quartzite and an overlying interval of about 110 metres of shale, sandstone, and
limestone, the remainder of the section is composed of limestone and dolomite units
totalling about 1,170 metres in thickness. The only faunal information is a report of the
late Middle Cambrian trilobite *Eldoradia* from the limestones about 630 metres above
the base of the carbonate sequence. Ordovician rocks are reported in normal sequence
above the Cambrian carbonates in this range.

3.5. The Tintic-southern Wasatch region

A number of small outcrop areas from the East Tintic and Oquirrh Mountains, about 30 kilometres east of the Stansbury and the Sheeprock Ranges, to the Wasatch Mountains as far north as Salt Lake City (Fig. 2) contain Cambrian sections differing from those of the central Utah region primarily in aggregate thickness of the section (Fig. 9). The sections of the central Utah region above the basal quartzites range from 1,350 to about 1,700 metres in thickness. The section in the East Tintic Mountains, which is the only one in the area which is not truncated by unconformably overlying Devonian rocks, is only slightly more than 900 metres thick. In addition, none of the sections of this region have any significant limestone or dolomite beds lying below *Glossopleura*-bearing shales. This demonstrates the gradual rise in time from west to east of the earliest carbonates which reflects the general eastwards shift of facies throughout the Cambrian of the Great Basin region.

Local stratigraphical nomenclatures have been proposed for the East Tintic Mountains, Oquirrh Mountains, and the Wasatch Mountains south of Salt Lake City. Most of the formational units in the East Tintic Mountains have been identified with reasonable confidence eastwards into the Wasatch Mountains in the Santaquin region. North along the Wasatch Mountains between Provo and Salt Lake City, the carbonate part of the sequence has all been included within the Maxfield Limestone.

In the East Tintic Mountains (MORRIS and LOVERING, 1961), the Cambrian section begins with about 900 metres of medium- to coarse-grained Tintic Quartzite that rests unconformably with a locally developed basal conglomerate, on quartzitic rocks assigned to the Pre-Cambrian Big Cottonwood Group. About 335 metres above the base of the Tintic Quartzite, a lava flow (?) ranging up to 12 metres in thickness has been recorded. This has also been reported in the Long Ridge area to the east (ABBOTT, 1951) and suggests a local volcanic event of probable Early Cambrian age.

Above the Tintic Quartzite, the Ophir Formation consists of a basal shale member about 60 metres thick that has yielded species of *Glossopleura* from its upper beds; a middle mottled and banded limestone member about 50 metres thick; and an upper shale member about 25 metres thick. The remainder of the section above the Ophir is almost entirely composed of sparsely fossiliferous carbonate rocks. These are well described and illustrated by MORRIS and LOVERING, and many of the descriptive terms applied to distinctive limestone and dolomite lithologies in this region and the central Utah region are derived from examples in the East Tintic Mountains.

The lowest of the carbonate units is the Teutonic Limestone, which consists of about 125 metres of mottled, oolitic or pisolitic, cross-bedded limestone with units of *Girvanella* oncolites in its basal part. This unit has yielded species of *Kootenia* and *Alokistocare*. It is separated from the lithologically similar Herkimer Limestone by the distinctive white-weathering, laminated Dagmar Dolomite, about 27 metres thick.

The Herkimer Limestone, about 115 metres thick, has a medial shale member about 10 metres thick, and the upper limestones are characterized by local intraformational limestone pebble conglomerates. The uppermost bed has yielded a few specimens of *Brachyaspidion* which indicates a correlation with the lower *Bolaspidella* Zone of late Middle Cambrian age. The Herkimer is overlain by about 65 metres of massive dark-coloured dolomite or limestone of the Bluebird Dolomite which includes many small white twig-like dolomite bodies that give it a distinctive character.

Above the Bluebird Dolomite is a succession of alternating units of white-weathering laminated dolomite and dark massive dolomite or limestone totalling about 250 metres in thickness which is described as the Cole Canyon Dolomite. A limestone unit abut 90 metres above the base of this formation yielded the latest Middle Cambrian trilobite *Eldoradia*.

The top of the Cole Canyon Dolomite is marked by the first shales of a unit of shales, oolitic and conglomeratic limestones, and sandstones about 50 to 80 metres thick which comprises the Opex Formation. The lower beds of the Opex contain species of the early Dresbachian genus *Coosella*. Towards the top of the formation, below a thin sandstone unit, species of *Prehousia* have been collected, and beds above the sandstone have yielded a large assemblage of trilobites of the *Elvinia* Zone. The absence of faunas of the *Dunderbergia* Zone and the short stratigraphical distance between beds with the *Prehousia* and *Elvinia* faunas demonstrate that the Late Cambrian disconformity so widespread in the mid-continent region of the United States is reflected at least this far to the west.

Overlying the Opex Formation is 170 metres of predominantly dark cherty dolomite of the Ajax Formation. The lower 60 metres, below a 10 metre white-weathering dolomite designated as the Emerald Member, has yielded silicified brachiopods of the early Franconian genus *Eoorthis*. The upper beds of the formation have yielded a few specimens of the Trempealeauan genus *Eurekia*.

Similar sequences of units of comparable thickness but lacking the upper dolomite, which is lost by pre-Devonian erosion, are present in the Long Ridge and Santaquin areas about 25 and 35 kilometres, respectively, to the east (BRADY, 1965).

About 35 kilometres north of the East Tintic Mountains, a small exposure of Cambrian rocks in the Oquirrh Range has been described by GILLULY (1932). The lowest beds are about 100 metres of quartzite assigned to the Tintic Quartzite. The overlying 100 metres of shales with a few thin limestones were referred to the Ophir Shale. *Glossopleura* was collected about 17 metres above the base of this formation.

Above the Ophir are 210 metres of alternating massive and thin-bedded limestone units described as the Hartmann Formation. The thin-bedded limestones have yielded specimens of *Kootenia*. This formation is overlain by the Bowman Limestone, a unit of shales and limestones which also yielded species of *Kootenia*. The upper 25 metres of the Bowman Limestone is laterally gradational with massive grey dolomites in the lower part of the overlying Lynch Dolomite. This formation is at least 245 metres thick and is overlain uncomformably by Devonian rocks.

In the Wasatch Mountains between Provo and Salt Lake City, the Cambrian sections include only three formations: the Tintic Quartzite, Ophir Shale, and Maxfield Limestone; and are everywhere unconformably overlain by Devonian or younger Palaeozoic rocks. The Tintic Quartzite throughout the area rests unconformably on Pre-Cambrian sedimentary rocks, usually with a basal conglomerate, and ranges between 240 and 385 metres in thickness. It is gradational upward into the silty, micaceous shales of the Ophir Formation.

In the Provo area (BAKER, 1964), the Ophir Shale is about 80 metres thick and is overlain by about 270 metres of massive- to thin-bedded limestone and massive dolomite and one 9-metre interval of shale, assigned to the Maxfield Limestone.

In the vicinity of American Fork, 25 kilometres to the north (BAKER and CRITTENDEN, 1961), and in the area immediately south-east of Salt Lake City (CALKINS and BUTLER, 1943; CRITTENDEN and others, 1952), the Ophir Shale has been divided into three members. The lower member consists of about 80 metres of silty micaceous shale with abundant tracks and trails but very few shelly fossils. Fragmentary trilobites near the base are of Middle Cambrian age. The middle member consists of about 30 metres of grey limestone with wavy siliceous bands. The upper member consists of brown-weathering calcareous shales about 55 metres thick. The overlying Maxfield Limestone ranges in thickness from 70 to over 270 metres due to variations in depth of erosion in pre-late Devonian or early Mississipian time. In its maximum development south-east of Salt Lake City (CALKINS and BUTLER, 1943), three members can be recognized. The lower member includes about 55 metres of oolitic or mottled limestone and dolomite and is capped by a white-weathering laminated dolomite. The middle member includes about 90 metres of thin-bedded silty or mottled limestones, limestone pebble conglomerate, and shale. The upper member consists of oolitic or mottled dolomite lacking significant shales. Rare fossils in the middle member are of Middle Cambrian age.

3.6. The northern Utah–south-eastern Idaho region

North of the latitude of Brigham City, Utah, Cambrian sections in the Wellsville Mountains and the Malad, Bannock, Portneuf, and Bear River Ranges (Fig. 2) have a reasonably uniform stratigraphy, but changes in some early Middle Cambrian and Late Cambrian units are of considerable regional significance.

The basic stratigraphical sequence for the region (Fig. 10) was first described by WALCOTT (1908b) in ascending order as the Brigham, Langston, Ute, Blacksmith, Bloomington, Nounan, and St. Charles Formations. Subsequent work by RICHARDSON (1913), MANSFIELD (1927), ANDERSON (1928), DEISS (1938), RESSER (1939a, 1939b), WILLIAMS and MAXEY (1941), WILLIAMS (1948), COULTER (1956), and MAXEY (1958) has in some cases added refinement in the form of member names and in clarification of some stratigraphical relationships but has not substantially modified the formational succession proposed by Walcott. Only in the Portneuf Range (ORIEL, 1964, 1965) and in the Pocatello area (CARR and TRIMBLE, 1962), both in the north-western part of the area, are significant major changes in the stratigraphy noted, and detailed work in those areas is not yet fully published.

The name Brigham has been applied to parts of a thick sequence of quartzitic rocks that lie beneath the Cambrian shale and carbonate sequences throughout the northern Utah–south-eastern Idaho region. The exact amount of the several thousand metres of quartzites to be included in the Brigham is a subject of unresolved discussion among geologists working in the area. The discrimination of Cambrian and Pre-Cambrian parts of this sequence is also unresolved.

In the Portneuf Range in Idaho, an indeterminate species of *Olenellus*, reported by Oriel (1964) from the middle part of a unit of 100 metres of quartzite overlying an argillite unit about 225 metres thick at the top of the main quartzite sequence, is the only record of fossils from within the quartzite sequence. It establishes an Early Cambrian age for at least some of the quartzites in the north-western part of the northern Utah–south-eastern Idaho region. However, in some areas to the east and south, the basal limestone member of the overlying Langston Formation contains a Middle Cambrian *Albertella*-Zone fauna. The contact between this limestone and the Brigham Quartzite is gradational. Because there is at least one early Middle Cambrian fauna in the Great Basin that is older than *Albertella* (Chapter 4), the uppermost quartzite beds are probably of early Middle Cambrian age in areas east and south of the Portneuf Range.

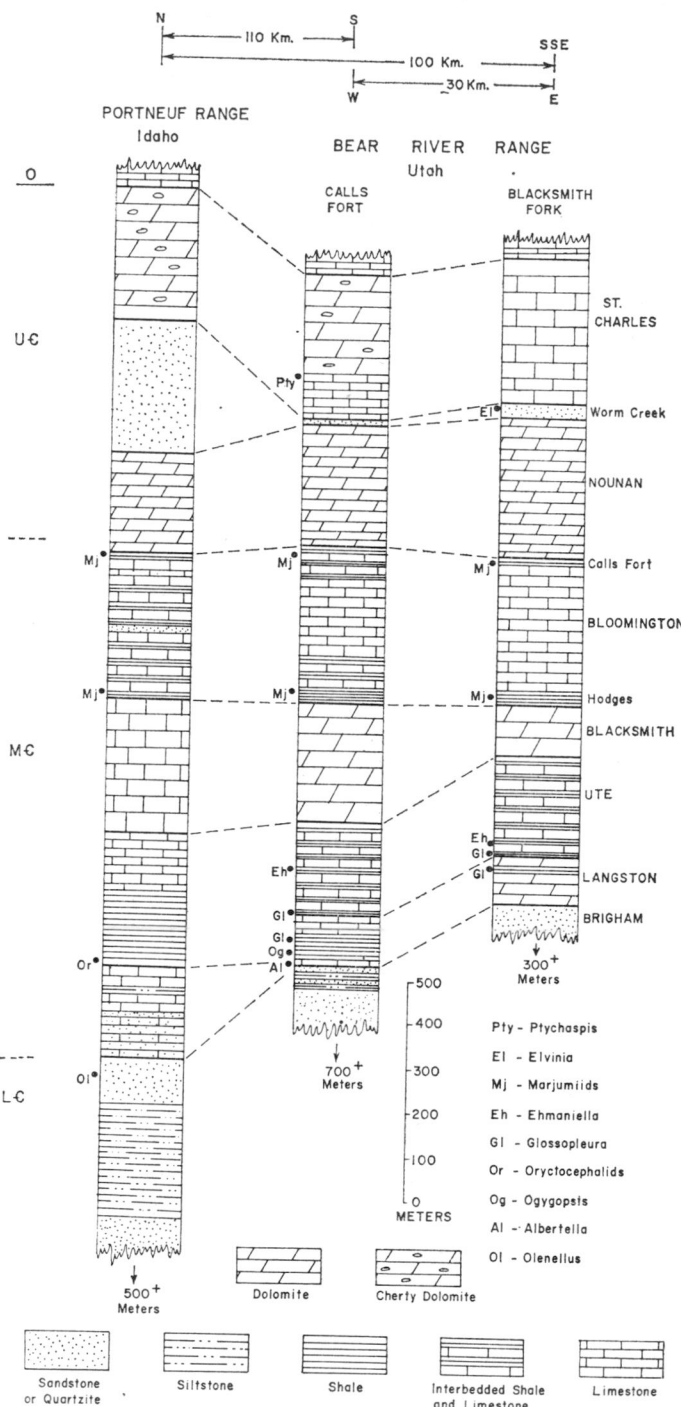

FIG. 10. Representative columnar sections and stratigraphical terminology for the northern Utah–south-eastern Idaho region.

Rock units immediately above the Brigham Quartzite in different parts of the northern Utah–south-eastern Idaho region represent two contrasting facies. In the eastern part of the Bear River Range from the vicinity of Bear Lake southwards, the Brigham Quartzite is overlain by a cliff-forming unit of brownish-weathering dolomites about 100 metres thick which usually has a thin *Glossopleura*-bearing shale or thin-bedded limestone unit about 15 metres below its top. This is the unit to which Walcott first applied the name Langston.

Throughout the Bear River Range north of the latitude of Bear Lake and in the western part of the Bear River Range to the south, the Malad Range, and the Wellsville Mountains, the Brigham Quartzite is overlain by a limestone unit usually less than 10 metres thick followed by an interval between 95 and 135 metres thick which consists of shales grading upwards into thin-bedded limestones and finally into an upper cliff-forming unit of limestone or dolomite (MAXEY, 1958). The cliff-forming units of both facies are overlain by a sequence of shales and limestones bearing correlative trilobite faunas which are assigned to the Ute Formation.

WALCOTT (1908b) believed that richly fossiliferous shales, which he called the Spence Shales, that cropped out in an area of poor exposures on the east side of the Bear River Range north of Bear Lake, were the same as the basal shales of the Ute Formation. However, WILLIAMS and MAXEY (1941) and MAXEY (1958) showed that these shales were the same as the shales above the basal limestone of the western facies and thus represented a completely older shale unit than the shales at the base of the Ute Formation. MAXEY applied the name Langston to western facies rocks that are correlative with the type Langston and recognized the basal Naomi Peak Limestone Member and overlying Spence Shale Member. Locally, the Naomi Peak Member may be absent and the Spence Member may rest directly on the Brigham Quartzite. At one locality in northern Utah that is intermediate between eastern and western facies areas, limestone containing an *Albertella*-zone fauna lies between dolomites of the eastern facies and the Brigham Quartzite (WILLIAMS and MAXEY, 1941).

The Naomi Peak and Spence Members have yielded rich faunas of trilobites and other fossils described principally by WALCOTT (1916b) and RESSER (1939a, 1939b). However, neither author had adequate stratigraphical data for many of the faunal elements and a re-evaluation of the actual faunal associations is needed. The Naomi Peak Limestone Member contains a varied fauna of the *Albertella* Zone comparable to faunas from the northern Egan Range (Chapter 3.3) and the Halfpint Range (Chapter 3.1.1) in Nevada. The Spence Shale Member includes a variety of oryctocephalid and ptychoparioid trilobites in its lower part, as well as species of the corynexochoid *Glossopleura* and *Zacanthoides*. Exact faunal criteria for the boundary between the *Albertella* and *Glossopleura* Zones of the early Middle Cambrian can be defined by future careful work in this area.

In the Portneuf Range about 25 kilometres west of the north end of the Bear River Range, about 115 metres of green and black shale and mudstone with faunas typical of the Spence Shale overlie nearly 180 metres of limestones, with some interbedded siltstone in the upper part and interbedded sandstone and quartzite in the lower part, which are gradational downwards into the Brigham Quartzite. ORIEL (1964) referred the limestones to the Langston Formation and the overlying shales to the Ute Formation. However, it is more probable that the limestones represent a greatly thickened equivalent of the Naomi Peak Limestone and the shales are the Spence Shales.

In the Bannock Range, about 50 kilometres to the north-west of the Portneuf Range, there is no limestone equivalent to the Naomi Peak Limestone (CARR and TRIMBLE, 1962). Thus, the Portneuf Range seems to be located near an area of maximum carbonate sedimentation during early Middle Cambrian time in the northern Utah–south-eastern Idaho region (Chapter 5).

The regional stratigraphy above the rocks assigned to the Langston Formation is somewhat more uniform. The Ute Formation consists of alternating units of shales and thin-bedded limestones. It is gradational upwards into a massive unit of cliff-forming limestones or dolomites referred to the Blacksmith Formation. Because of the gradational nature of the contact between these formations, consistent criteria for placement of the boundary have not been established and their thicknesses vary reciprocally over the region. The Blacksmith and Ute Formations both have thickness ranges from 145 to 205 metres, but their combined thickness in any one area is usually about 340 metres. In the Portneuf and Bannock Ranges, although equivalents of the Ute and Blacksmith are present, they have not yet been separately identified. The Ute Formation is moderately fossiliferous but its faunas have not been carefully studied. Locally, the uppermost beds of the Blacksmith Formation have yielded specimens of *Bolaspidella* indicating a late Middle Cambrian age.

The massive carbonates of the Blacksmith Formation or its equivalents are overlain throughout the northern Utah–south-eastern Idaho region by an interval of shales and thin-bedded limestones, averaging about 360 metres in thickness, which is referred to the Bloomington Formation. In the Bear River Range and the Wellsville Mountains, this formation consists of three members: lower and upper shale members and a middle member of thin-bedded limestones. Rare fossils scattered thoughout the formation, but principally in the shale members, are all representative of the late Middle Cambrian *Bolaspidella* Zone. The lower member has been named the Hodges Shale Member (RICHARDSON, 1913), and the upper member is the Calls Fort Shale Member (MAXEY, 1958). The middle member is unnamed. In the Bannock Range near Pocatello, Idaho, the Bloomington Formation is not clearly divisible into three members, and the middle part of the formation includes, in addition to thin-bedded limestones, several units of sandstone up to 25 metres thick. Absence of sands in correlative horizons to the east and south is strong evidence for a northern or western source for these terrigenous materials during late Middle Cambrian time.

The Bloomington Formation is overlain by 270 to 360 metres of thin-bedded dolomites and some limestones referred to the Nounan Formation. Fossils from limestones in the upper and lower parts of the formation are of Dresbachian age but are largely unstudied.

The St. Charles Formation which overlies the Nounan Formation consists of three members. The lowest member is locally quartzitic sandstone with some sandy limestone interbeds which was named the Worm Creek Quartzite (RICHARDSON, 1913). This unit thickens greatly north-westwards from Utah into Idaho. In the Wellsville Mountains it is only 2 metres thick; in the Bear River Range in northern Utah it is 25 metres thick; and in the northern part of the Bear River Range in Idaho it is 115 metres thick (WILLIAMS, 1948). In the Portneuf Range to the west, it is 270 metres thick (ORIEL, 1965). Fossils from limestones in the Worm Creek Member include species of *Elvinia*, *Dokimocephalus*, and other typical *Elvinia* Zone trilobites. The Worm Creek Quartzite is the exact correlative of the Corset Spring Shale of the Eureka-House Range region and reflects the widespread regressive event that resulted in the Dresbachian–Franconian disconformity in the mid-continent region and in the Tintic and Wah Wah Ranges within the Great Basin (Chapters 3.2., 3.5). However, the north-westward thickening suggests that the primary source for the sand may have been the same source that provided late Middle Cambrian sand in the Pocatello area.

The upper two members of the St. Charles Formation are unnamed. Above the Worm Creek Member is an interval of about 100 metres of thin-bedded limestones which has yielded rich *Ptychaspis*-Zone fauna in the northern Bear River Range (LOCHMAN and HU, 1959). The remainder of the St. Charles Formation is composed of massive cherty dolomites which are overlain by thin-bedded fossiliferous early Ordovician limestones. The total thickness of the St. Charles Formation averages between 225 and 315 metres.

3.7. The north-west Utah region

Several localities in north-western Utah and adjacent Idaho have Cambrian sequences about which too little information is available to include them with confidence in the better known regions to the east and south. They do, however, represent areas in which new or additional work can provide information valuable for understanding of the regional relationships of the Cambrian rocks.

The Promontory Range along the north margin of Great Salt Lake has a full Cambrian section briefly described by OLSON (1956), which has many of the nomenclatural problems of the Cambrian sections of the central Utah region (Chapter 3.4). Names for the Middle Cambrian units were derived from the House Range about 200 kilometres to the south (including some Pioche district names misused in the House Range by WHEELER, 1948), and names for the Late Cambrian units were derived from the north-eastern Utah–south-eastern Idaho region about 50 kilometres to the east. At least 360 metres of quartzites at the base of the section rest conformably on older quartzitic rocks and were referred to the Prospect Mountain Quartzite. The overlying transition zone of thin-bedded quartzites and shales, about 100 metres thick, was referred to the Pioche Shale. The lower part of the carbonate-shale sequence begins with a thin massive limestone 3 metres thick overlain by 16 metres of platy siltstone which has yielded specimens of *Ogygopsis*. Above this a sequence of about 340 metres of thin-bedded silty limestones, shales, and occasional units of massive limestone was divided into six mappable units to which House Range nomenclature from 'Millard' to 'Swasey' was applied. Several shale intervals about 100 metres above the base have yielded species of *Glossopleura*. Slightly higher shales have yielded specimens of *Ehmaniella* and a fauna with abundant specimens of *Kootenia* has been collected from the top of the 'Swasey'. Above the 'Swasey' is about 60 metres of black fissile shale and limestone lithologically much like the Wheeler Shale with which it was identified. However, the next 560 metres of thin-bedded limestones and shales which were assigned to the Marjum Formation have many minor lithological features characteristic of the middle member of the Bloomington Formation of northern Utah and an upper shaly member has yielded late Middle Cambrian trilobites. The 'Marjum' is overlain by 630 metres of limestone and dolomite identified as Nounan Dolomite; 27 metres of Worm Creek Quartzite; and 325 metres of limestone and dolomite referred to the St. Charles Formation. The stratigraphy of this area must be re-evaluated to establish a more appropriate nomenclature for at least the lower series of mappable units.

The Lakeside Mountains, on the south shore of Great Salt Lake about 50 kilometres south-west of the Promontory Range, also have a complete Cambrian section exposed. This was very briefly described by YOUNG (1955) who assigned with question names for the carbonate units derived from the thin sequence in the Oquirrh Mountains (Chapter 3.5). The thickness of the section is much more comparable to sections in the central Utah region (Chapter 3.4). About 225 metres of quartzite at the base of the section are separated by an unexposed interval from about 1,350 metres of limestones and dolomites, with some interbedded shales, which were questionably divided into the Hartmann, Bowman, and Lynch Formations. No faunal data were given, and new work on this area must be published before it can be related to nearby areas.

In the Silver Island Range 80 kilometres to the west, a full Cambrian section was reported by SCHAEFFER (1960) and the formational sequence of the House Range was identified. However, ROBISON and PALMER (1968) examined the carbonate sequence and found that the lowest exposures in a continuous section of about 630 metres of carbonates which reportedly included House Range Formations from 'Millard' to 'Swasey' contained Late Cambrian trilobites. Trilobites of this age were also found higher in this sequence and in an interval of about 90 metres of shales and silty thin-bedded limestones which had been identified as the Wheeler Shale. This unit is faunally and lithologically correlative with the lower part of the Dunderberg Shale and the middle shaly member of the Orr Formation of the Eureka–House Range region (Chapter 3.3). Above this is about 135 metres of limestones overlain by a thin shaly unit with an *Elvinia* Zone fauna. These had been included in the Marjum and Weeks formations which are completely older, unrelated units, and the units are properly correlated with the Johns Wash Limestone and Corset Spring Shale of the Eureka–House Range region. Thus, what was reported as a thin Late Cambrian sequence and has influenced isopach maps of the northern Great Basin is actually a very thick Late Cambrian sequence, and new work is need to re-evaluate all of the Cambrian stratigraphy of this area.

In the Albion Range of southern Idaho about 120 kilometres south-west of Pocatello, ARMSTRONG (1968) has demonstrated the Palaeozoic age of a deformed and meta-morphosed sequence of rocks, formerly thought to be Pre-Cambrian, by the discovery of echinoderm columnals in a marble unit. He considers the lower two formations of this sequence to be of probable Cambrian age. The Elba Quartzite at the base lies unconformably on older crystalline schists and gneisses. It ranges from 3 to 270 metres in thickness and is laterally gradational with schists of the lower part of the Conner Creek Formation. This formation is estimated to be at least 1,260 metres thick and its upper part includes limestone and dolomite marble and the columnal bearing beds. There is no clear evidence about how much, if any, of the Elba–Conner Creek sequence is Cambrian. If it is Cambrian, it has considerable palaeogeographical significance. Only east of the Sevier orogenic belt have thin Cambrian quartzites been found resting unconformably on gneissic basement (Chapter 2). In the northern Utah–south-eastern Idaho region to the east (Chapter 3.6), several thousand metres of conformable quartzitic sediments are present below the Cambrian carbonate-shale sequence. Thus, if the Albion Range section is Cambrian, it could define the western margin of a trough that subsided rapidly during the Late Pre-Cambrian and Early Cambrian. Additional work is needed in this part of the Great Basin to clarify the relations of the Palaeozoic rocks.

3.8. Cambrian rocks of the Antler orogenic belt in northern and central Nevada

The Antler orogenic belt includes a number of small areas of generally highly deformed and sparsely fossiliferous rocks assigned to the Cambrian System. None of these areas has an unbroken succession of Cambrian rocks, and in all of them reported thicknesses are only rough estimates because of complex structure, lack of lithological variety, and insufficient faunal control. Nevertheless, this region is important for an understanding of Cambrian relationships in the Great Basin because of the facies contrasts between sequences within the region, and between this region and the regions east of the Antler orogenic belt.

The two most nearly complete Cambrian sequences in the Antler orogenic belt are in the Osgood Mountains (HOTZ and WILLDEN, 1964) and in the northern Toiyabe Range (STEWART and PALMER, 1967). Partial sequences are known in the Galena Range (ROBERTS, 1964), the northern Shoshone Range (GILLULY and GATES, 1965), the central Toiyabe Range (FERGUSON and CATHCART, 1954; MEANS, 1962), the Independence Range (DECKER, 1962), and the Cortez Range (GILLULY and MASURSKY, 1965) (Fig. 2).

In the Independence Range near the northern border of Nevada, DECKER (1962) has described a sequence of strongly deformed rocks including three formations of probable Cambrian age. The oldest unit consists of about 1,260 metres of medium- to fine-grained quartzite referred to the Prospect Mountain Quartzite. This is overlain by the Edgemont Formation, about 205 metres of slate and phyllite, in part calcareous, which has yielded fragments of *Olenellus*. The overlying Porter Peak Formation includes a lower member of thin-bedded sandy limestone and limestone conglomerate estimated to be about 360 metres thick and an upper member of massive limestone estimated to be over 600 metres in thickness. The report of a 'Streptelasma' from the upper part of this formation indicates that it is at least in part younger than Cambrian.

In other areas of northern Nevada, isolated unfossiliferous exposures of quartzite have been referred to the Prospect Mountain Quartzite. In addition, COASH (1967) has described a thick unfossiliferous sequence of interbedded limestones and argillites of possible Cambrian age above the Prospect Mountain Quartzite in the Mt. Velma area as the Tennessee Mountain Formation.

In the Osgood Mountains, the north-westernmost Cambrian area in the Great Basin, the oldest formation of probable Cambrian age is the Osgood Mountain Quartzite. This formation is a thick unit of medium- to thick-bedded, frequently cross-bedded, fine- to medium-grained unfossiliferous quartzite which is locally pebbly and has a locally developed upper member of interbedded shales and quartzites named the Twin Canyon Member (HOTZ and WILLDEN, 1964). The lower contact of the formation is not exposed. Estimated thickness of the exposed part of the formation is about 1,080 metres and the Twin Canyon Member may be as much as 720 metres thick.

The Preble Formation, which rests conformably on either the Twin Canyon shaly member or the main body of the Osgood Mountain Quartzite, consists of phyllitic shales in its lower part and interbedded limestones and shales in its middle and upper parts. Its thickness is estimated to be between 2,500 and 3,600 metres, but complex structure and lack of distinctive units within the formation prevent accurate thickness measurement and also accurate placement of fossiliferous units. Faunas of at least three different ages are known. The oldest fauna of probable early Middle Cambrian age contains *Kootenia;* the next younger fauna contains late Middle Cambrian conchostracans referable to *Svealuta primordialis* (LINNARSSON); and the youngest fauna contains species of the Dresbachian trilobites *Tricrepicephalus*, *Kingstonia*, and *Meteoraspis*.

The contact between the Preble Formation and the next younger Paradise Valley Chert is not known, and outcrops of the two formations are several kilometres apart. The Paradise Valley Chert is predominantly composed of dark-brown to black bedded chert with some thin intervals of siliceous shale and thin-bedded silty limestone. Its estimated thickness is about 100 to 180 metres. Faunas of two slightly different Dresbachian ages have been collected from the formation. The older fauna contains species of *Crepicephalus* and *Deiracephalus* and is not much younger than the youngest fauna of the Preble, thus suggesting that not much stratigraphical interval is missing between the formations. The younger fauna contains species of *Aphelaspis*, *Olenaspella*, and *Glyptagnostus* (PALMER, 1962b).

The Paradise Valley Chert is overlain conformably (?) by the Harmony Formation, a distinctive unit of thick-bedded to massive, generally graded arkosic sandstones more than 1,440 metres thick. A basal shale unit about 25 to 50 metres thick and lenses of limestone between 16 and 50 metres thick which occur about 900 metres above the base of the formation are the principal deviations from the typical arkosic lithology. The limestones have yielded faunas of two different Late Cambrian ages. An older fauna of Franconian age has yielded species of *Dartonaspis*, and a younger fauna of Trempealeauan age has yielded species of *Eurekia* and *Idiomesus*.

In the Galena Range about 60 kilometres to the south-east, Roberts (1964) has described the Scott Canyon Formation: a unit predominantly composed of chert, argillite, and greenstone which is more than 1,440 metres thick. Locally, limestone lenses up to 70 metres thick and 145 metres long are associated with the greenstones and have yielded archaeocyathids, algae, and undetermined trilobites. The archaeocyathids include species of *Protopharetra*, *Eucyathus*, *Ajacicyathus*, *Sajanocyathus*, and *Pycnoidocyathus* and the algae represent *Epiphyton* and *Renalcis*. The greenstones are locally pillowed. Other exposures in the area, entirely isolated by faults, include over 900 metres, of unfossiliferous, pebbly, graded-bedded feldspathic sandstone, micaceous sandstone, arkose, and shale which are assigned to the Harmony Formation.

In the northern part of the Shoshone Range about 25 kilometres farther to the south-east (GILLULY and GATES, 1965), a structural window has exposed a complexly folded unit of chloritic phyllite, metadolerite, greenstone, black limy slate, and mottled shaly limestone without top or bottom that aggregates about 450 to 630 metres in thickness. This unit, described as the Schwin Formation, has yielded deformed agnostids referable to *Peronopsis*, *Ptychagnostus*, and *Hypagnostus* which establish its age as at least in part Middle Cambrian. Isolated fault pieces of unfossiliferous dolomite, quartzite, and arkosic sandstone referred to the Eldorado Dolomite, Prospect Mountain Quartzite, and Harmony Formation are also present in the area.

In the Cortez Range another 35 kilometres to the south-east, unfossiliferous dolomite unconformable below the Ordovician Eureka Quartzite in a structural window has been referred to the Hamburg Dolomite (GILLULY and MASURSKY, 1965).

In the northern Toiyabe Range about 160 kilometres south of the Osgood Mountains, a nearly complete Cambrian section is present (STEWART and PALMER, 1967) (Fig. 11), but none of the units has been named. The oldest exposed rocks are quartzites, siltstones, and sandstones totalling about 360 metres in thickness which have yielded indeterminate trilobite scraps. These are overlain by a three-part unit about 100 metres thick, consisting of lower and upper cliff-forming limestones and a middle interval of limestones and siltstones. The lower limestones have yielded species of *Bristolia* and the upper limestones have yielded species of *Olenellus* and *Paedeumias*. These beds are overlain by about 1,170 metres of thin-bedded unfossiliferous rocks including a lower siltstone and silty limestone unit 205 metres thick; 630 metres of laminated aphanitic limestone; 205 metres of interbedded laminated limestone, siltstone, and shale; and about

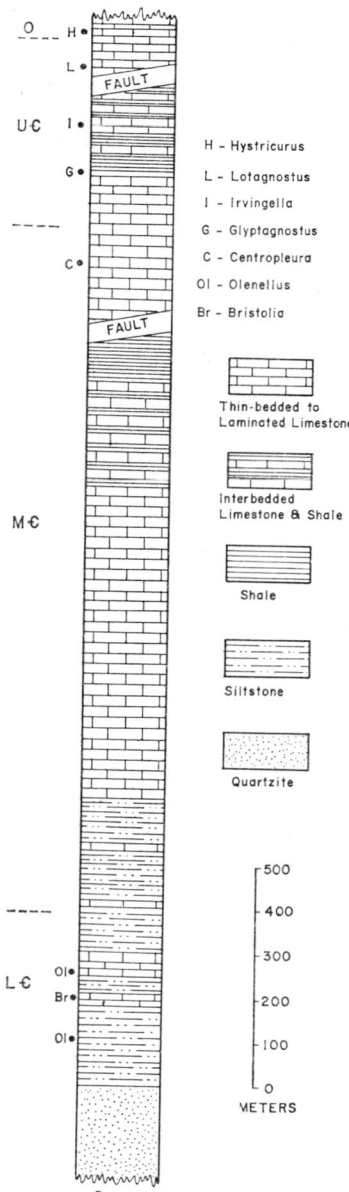

NORTHERN TOIYABE RANGE

(Stewart & Palmer, 1967)

H – Hystricurus

L – Lotagnostus

I – Irvingella

G – Glyptagnostus

C – Centropleura

Ol – Olenellus

Br – Bristolia

Thin-bedded to
Laminated Limestone

Interbedded
Limestone & Shale

Shale

Siltstone

Quartzite

500

400

300

200

100

0
METERS

FIG. 11. Columnar section in the Callaghan window in the
northern Toiyabe Range.

100 metres of dark-grey shale. These rocks are separated by faults from a younger sequence which begins with over 300 metres of thin-bedded silty limestone containing rare specimens of the late Middle Cambrian trilobite *Centropleura* (PALMER and STEWART, 1968). Above this is at least 180 metres of shale and silty limestone containing species of the Late Cambrian trilobites *Glyptagnostus* in its lower part and *Irvingella*, *Elvinia*, *Housia*, and *Litocephalus* in its upper part. This sequence is largely correlative with the Dunderberg Formation in the Eureka–House Range region to the east (Chapter 3.3). The thin-bedded and laminated limestone units below these beds may be the facies equivalents of the more massive largely dolomitic units of the Eldorado and Hamburg Dolomites.

Above an intervening fault, a unit of still younger thin-bedded limestones over 100 metres thick contains Late Cambrian trilobites referable to *Bienvillia* (?) and *Lotagnostus* in its lower part and early Ordovician trilobites in its upper beds.

In the central part of the Toiyabe Range over 900 metres of quartzite, graywacke, and some limestone at least in part of Cambrian age have been referred to the Gold Hill Formation (FERGUSON and CATHCART, 1954). Limestones at the top of this quartzitic sequence have yielded species of *Archaeocyathus* and *Ethmophyllum* (WASHBURN, 1966). About 630 metres of intensely deformed unfossiliferous thin-bedded limestones, argillaceous siltstones, and phyllites referred to the Broad Canyon sequence (MEANS, 1962) separate the Gold Hill Formation from younger rocks with Ordovician fossils. In a separate fault slice within the central Toiyabe Range, the Gold Hill Formation is overlain by a thick sequence of strongly deformed, laminated to thin-bedded limestone referred to the Crane Canyon sequence (MEANS, 1962) which has yielded Late Cambrian trilobites.

No detailed information is available about other areas of Cambrian rocks in the Antler orogenic belt. However, because of the generally poorly fossiliferous character of much of the lower and middle Palaeozoic succession of the region, discovery of new areas of Cambrian rocks and more information about known areas can be expected as work progresses on details of the geology of this complex region.

3.9. Cambrian rocks from north-eastern Washington related to the Great Basin region

Two slightly differing thick successions of poorly exposed, structurally disturbed, and sparsely fossiliferous Cambrian rocks crop out in the north-eastern part of Stevens County and northern Pend Oreille County, Washington. These rocks are southern continuations of similar successions in adjacent parts of Canada and contrast in stratigraphy and faunal content with the rocks of the Pend Oreille Lake area in Idaho to the east. Very little substantial new information about the general Cambrian stratigraphy in this area has been added since an earlier summary by LOCHMAN (1956).

The oldest rocks are quartzites and argillites assigned to the Addy Quartzite near Colville (BENNETT, 1941; BECRAFT and WEIS, 1963) and the Gypsy Quartzite near Metaline Falls (PARK and CANNON, 1943; DINGS and WHITEBREAD, 1965) and Northport (CAMPBELL, 1947). The Addy Quartzite rests with angular unconformity on Pre-Cambrian sediments and consists of alternating units of quartzite and argillite totalling about 1,170 metres in thickness. The Gypsy Quartzite is a unit of quartzite, grit, and phyllite which is between 1,530 and 2,520 metres thick. It conformably overlies a heterogenous unit of limestones, phyllites, grits, and quartzites of probable Pre-Cambrian age assigned to the Monk Formation. In nearby Canada, the Reno Formation is the equivalent of the Gypsy Quartzite.

In the Colville area, the Addy Quartzite is overlain by about 1,350 metres of thin-bedded, fine- to coarse-grained limestone described as the Old Dominion Limestone. In the Metaline Falls area, the Gypsy Quartzite is overlain by the Maitlen Phyllite, a grey-green banded phyllite with units of quartzite, sericite schist, and limestone which is at least 1,170 metres thick. In Canada, this unit is the Laib Formation, and a distinctive lower limestone member about 70 metres thick has been named the Reeves Member. YATES (1964) recognized the Reeves Member in the Maitlen Phyllite in Washington.

The Maitlen Phyllite is overlain by the Metaline Limestone, about 1,620 metres of lime-stones and dolomites, known in Canada as the Nelway Limestone. The Metaline Limestone consists of lower and upper limestone units and a middle dolomite unit. The lower limestone unit is thin-bedded and includes some shale units. Its thickness is the greatest of the three units of the formation, but cannot be exactly determined because of unsatis-factory exposures and large-scale deformation. The middle dolomite unit is characterized by fine-grained, banded dolomites and is 180 to 360 metres thick. The upper limestone unit is massive limestone at least 180 metres thick. In parts of northern Stevens County, the upper unit is composed of thin-bedded dolomite and dolomite breccia instead of massive limestone (YATES and ROBERTSON, 1958). Throughout the Metaline Falls area, the Metaline Limestone is overlain by black graptolitic units of the Ledbetter Slate.

Evidence for dating the Cambrian rocks is sparse and does not permit any detailed correlations to better known regions. In the Colville area, only Early Cambrian rocks have been recognized. The upper part of the Addy quartzite has yielded the distinctive olenellid *Nevadella addyensis* (OKULITCH, 1951) which indicates a correlation with the Poleta Formation of the Nevada–California area of the Great Basin region and a position in the lower part of the Lower Cambrian. Limestones in the lower part of the overlying Old Dominion Limestone have yielded a rich archaeocyathid fauna (GREGGS, 1959) which is also present in the Reeves Member of the Maitlen Phyllite. Thus, the Addy and Gypsy quartzites seem to be correlative units, and the Old Dominion Limestone is a correlative of at least the Maitlen Phyllite in the Metaline Falls area.

The lower thin-bedded limestone and shale unit of the Metaline Limestone in the Metaline Falls area has yielded a rich trilobite fauna including species of *Ogygopsis*, *Elrathina*, *Pagetia*, and *Olenoides* in its upper part (MCLAUGHLIN and ENBYSK, 1950; DINGS, 1965). This is a medial Middle Cambrian fauna that indicates an approxi-mate correlation of the fossiliferous interval with the Stephen Formation in British Columbia. YATES and ROBERTSON (1958) have recorded phosphatic brachiopods of Middle or Late Cambrian age in the thin-bedded dolomites of the upper unit of the Metaline Limestone. A complete Cambrian section is possibly present in the Metaline Falls area but its details are obscured by the lack of sufficient outcrops and adequate faunal control.

4. Cambrian biostratigraphy of the Great Basin region

The biostratigraphy of the Cambrian deposits of the Great Basin region, which is based entirely on trilobites, is still evolving and no definitive synthesis can yet be presented. The reason for this is that many of the fossiliferous successions in the Great Basin are as yet unstudied, and the zonal nomenclature currently in use includes names that have been uncritically applied to unstudied parts of the faunal succession. All faunal studies in the Great Basin since 1950 have used zones in the sense of faunizones (LOCHMAN-BALK and WILSON, 1958) or assemblage zones (ROBISON, 1964a, 1964b), in which stratigraphically significant assemblages of fossils are designated by the names of one or two of the typical included genera. Older studies designated both assemblages and fossiliferous intervals as zones without making a clear distinction between them.

The most recent attempt at a complete biostratigraphical synthesis in this region (LOCHMAN-BALK and WILSON, 1958) actually preceded any of the definitive studies upon which a lasting biostratigraphy could be based. Therefore, this chapter will indicate some of the possiblities availiable for establishing a modified biostratigraphy based on Great Basin data.

The Early Cambrian faunal succession is at present rather loosely divided into upper and lower 'Olenellus' subzones (LOCHMAN-BALK and WILSON, 1958). Criteria for the upper subzone have been the presence of non-olenellid trilobites associated with olenellids, whereas the lower subzone was believed to include only olenellids. However, at no one place in the Great Basin region has the Early Cambrian trilobite sequence been described. The only places in the Great Basin where the Early Cambrian biostratigraphy can be worked out are in the Spring Mountains–Death Valley and White–Inyo Mountains regions (Chapter 3.1). Although the rock succession is not richly fossiliferous, enough detail can be obtained to provide a useful framework into which less fossiliferous intervals elsewhere in the Great Basin can be correlated. The best published summary of the Early Cambrian faunas is a series of plates with very little accompanying discussion in a mimeographed guidebook to the Cambrian succession of the White–Inyo Mountains, California, prepared by NELSON and DURHAM (1966). The Campito Formation includes at least two olenellid faunas: an older fauna characterized by fallotaspids which is found in the quartzites of the Andrews Mountain Member and in the lower part of the Montenegro Member; and a younger fauna characterized by *Nevadia* and *Holmia* which is found in the upper part of the Montenegro Member. The olenellids in the middle part of the overlying Poleta Formation are generally quite distinct from most of those in the Montenegro Member. Species of *Laudonia* and *Nevadella* and some new forms are typical elements, but trilobites possibly related to this fauna have also been reported from the uppermost beds of the Montenegro Member of the Campito Formation, and their relationships to the *Nevadia–Holmia* fauna, with which they are not associated, remain to be worked out. Trilobites are rare in the Harkless Formation and are mostly undescribed olenellids related to younger forms. The overlying Saline Valley Formation has yielded a rich Early Cambrian trilobite fauna (PALMER, 1964), including species of *Wanneria*, which seems to be slightly older than the olenellid faunas of the Carrara Formation and is certainly older than the *Bonnia–Olenellus–Paedeumias* faunas of the overlying Mule Springs Limestone. The lower Carrara Formation contains at least two distinct olenellid faunas. The shales and the lower part of the limestone of the first Grand Cycle (see p. 13) are characterized by species of *Bristolia* and *Peachella* and a number of new olenellids, as well as the longer ranging genera *Olenellus* and *Paedeumias*. The lower shales at the base of the second Grand Cycle contain an association of species of *Olenellus*, *Paedeumias*, and undescribed genera that constitute the youngest olenellid fauna of the Great Basin region. Thus, potentially as many as seven different biostratigraphical entities can be recognized in the Early Cambrian sequences of the Great Basin.

The Middle Cambrian faunal successions are somewhat better known, but even in this part of the Cambrian, the only detailed descriptive studies that involve adequate stratigraphical data are by ROBISON (1964a) and FRITZ (1968) and these deal only with restricted parts of the section. The oldest Middle Cambrian faunas are found in the basal beds of the Emigrant Formation, the second Grand Cycle of the Carrara Formation, and the middle part of the Pioche Formation. They are unstudied and have been included uncritically in the *Plagiura–Poliella* or *Wenchemnia–Stephenaspis* Zones because they have some elements of these faunas as described by RASETTI (1951) from Canada, and in the Death Valley and Pioche regions they are clearly older than the *Albertella* faunas.

Faunas assigned to the *Albertella* Zone, which seems to be an acceptable biostratigraphical unit for the Great Basin, have been described by RESSER (1939b) from the Naomi Peak Limestone Member of the Langston Formation in the northern Utah–south-eastern Idaho region (Chapter 3.6) and by FRITZ (1968) from the Pioche Formation in the northern Egan Range of the Eureka–House Range region (Chapter 3.3). Undescribed *Albertella* Zone faunas are well developed in the upper part of the Pioche Shale of the Pioche region (Chapter 3.2) and in the upper part of the second Grand Cycle and most of the third Grand Cycle of the Carrara Formation in the Spring Mountains–Death Valley region (Chapter 3.1.1).

The next younger biostratigraphical unit is the *Glossopleura* Zone which also seems to be an acceptable unit for the Great Basin but which lacks any modern biostratigraphical study. Faunas of the Chisholm Shale in the Pioche region (Chapter 3.2), described principally by PACK (1906b) and WALCOTT (1916b); from the upper part of the Cadiz Formation in the Marble Mountains (Chapter 2.1), described by MASON (1935); and from the Spence Shale Member of the 'Langston' Formation in the northern Utah–south-eastern Idaho region (Chapter 3.6), described by RESSER (1939a), are all referable to this zone. Beds with *Glossopleura* and very little else are known from many localities in the Great Basin, and an interval of at least 135 metres between the upper beds of the Carrara Formation and the first silty zone in the overlying Bonanza King Formation, both of which have yielded species of *Glossopleura*, gives an indication of the vertical extent of this zone. Perhaps the best area to study details of the *Glossopleura* Zone and its boundaries within the Great Basin is the northern Utah–south-eastern Idaho area (Chapter 3.6) where a reasonably continuous fossiliferous sequence is present from the Naomi Peak Limestone with an *Albertella* fauna, up into the Spence Shale, and where the lower few metres of the Ute Formation yield assemblages containing *Glossopleura* and also assemblages assigned to the younger '*Bathyuriscus–Elrathina*' Zone.

The interval between the *Glossopleura* Zone and the *Bolaspidella* Zone includes a variety of faunas that are presently included arbitrarily in the *Bathyuriscus–Elrathina* Zone. None of these faunas has been described in detail, and neither *Bathyuriscus* nor *Elrathina* seems to be a characteristic element. A careful biostratigraphical study of the Ute Formation in the northern Utah–south-eastern Idaho region (Chapter 3.6) and of the Whirlwind and Swasey Formations in the House Range (Chapter 3.3) would be a valuable contribution to our knowledge of the faunas of this interval. Usually, the lowest beds in this interval are crowded with several small ptychopariod trilobites, including species of *Ehmaniella*. Its uppermost beds locally contain a rich fauna including species of *Tonkinella* and *Pagetia* (DREWES and PALMER, 1957; ROBISON, 1964b).

The *Bolaspidella* Zone, which is the youngest Middle Cambrian Zone, has been studied in detail only in the House Range and Drum Mountains (Chapters 3.3, 3.4) by ROBISON (1964a, 1964b) who recognized, in ascending order, three subzones: the *Bathyuriscus fimbriatus*, *Bolaspidella contracta*, and *Lejopyge calva* subzones. His careful work for the first time established a precise correlation between the late Middle Cambrian rocks of western United States and Scandinavia (Fig. 12). In fact, it is much easier to correlate details of the *Bolaspidella* Zone sequence from the Wheeler, Marjum, and Weeks Formations of the House Range to Sweden than it is to correlate these

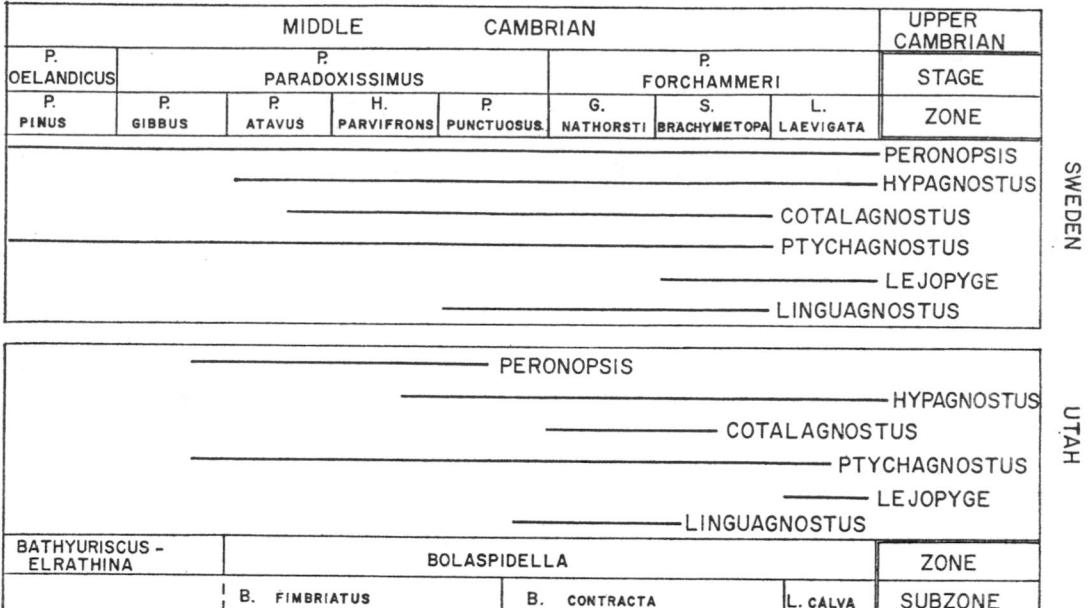

FIG. 12. Comparison of the ranges of the agnostids of the late Middle Cambrian Wheeler and Marjum Formations with those of the late Middle Cambrian of Sweden (after ROBISON, 1964b).

details from the House Range to the Bloomington Formation in the northern Utah–south-eastern Idaho region and the Meagher Limestone and Park Shale in the northern Rocky Mountain region. The reason for this seems to be due to lithofacies and biofacies differences. The fossiliferous rocks of the House Range represent the outer detrital belt and the correlative fossiliferous rocks to the north-east represent either the carbonate belt or the inner detrital belt (Chapter 5).

Failure to recognize that faunal sequences in the inner and outer detrital belts, particularly, can be quite different may have contributed to some biostratigraphical problems in the Cordilleran region which are only now becoming apparent. The sequence of faunas up through the lower part of the 'Bathyuriscus–Elrathina' Zone in the southern half of the Great Basin reflects almost completely the fauna of the inner detrital belt and inner parts of the carbonate belt which are not particularly rich in species or genera; whereas all younger faunas in this region are generally much richer in genera and species and represent the outer detrital belt or outer parts of the carbonate belt. In the northern part of the Great Basin, the relations of the Spence Shale with its rich fauna to the dolomites of the Langston Formation suggest that this shale may be representative of the outer detrital belt; but this relationship needs to be carefully examined. The present biostratigraphical scheme for the Cordilleran region makes no distinction between faunal assemblages from the inner and outer detrital belts and thus has placed in vertical sequence faunas that may have evolved in different environments. An analysis of the contrasts in faunal evolution on the inside and outside of the carbonate belt is needed before a totally acceptable biostratigraphy of the Cambrian of western United States can be prepared.

The Late Cambrian faunal successions in the Great Basin include many rich faunas. However, only a part of the sequence has been described in any detail (PALMER, 1962b,

1965b). Early Upper Cambrian faunas correlative with the *Cedaria* and *Crepicephalus* Zones of the mid-continent region are well developed in the shale and siltstone sequences of the House, Schell Creek, and Snake Ranges (Chapter 3.3); but many of the species assemblages are quite distinct from contemporaneous mid-continent assemblages. No meaningful zonation of this interval can be made until these faunas have been studied.

Everywhere in the Great Basin as well as elsewhere in North America, the *Crepicephalus–Cedaria* faunas are succeeded abruptly, without evidence of evolutionary gradation or stratigraphical hiatus, by faunas of the *Aphelaspis* Zone. The abrupt faunal change is reported to be slightly diachronous (PALMER, 1962b). The rich faunas of this zone and the succeeding *Dicanthopyge, Prehousia, Dunderbergia,* and *Elvinia* Zones are found in the Dunderberg and Corset Spring Shales and their correlatives and the lower part of the overlying carbonate sequences (PALMER, 1960a, 1962b, 1965a,b). These zones all contain related trilobite assemblages and represent a single evolutionary complex. The *Elvinia* Zone is succeeded abruptly, also without evidence for evolutionary gradation or stratigraphical hiatus, by largely undescribed younger Cambrian faunas. The stratigraphical interval represented by the trilobite complex contained between the *Aphelaspis* and *Elvinia* Zones, and bounded by non-evolutionary, non-diastrophic, but perhaps diachronous, boundaries has been defined as the Pterocephaliid biomere (PALMER, 1965a,b). Similar abrupt changes in faunas seem to be present at the lower boundaries of the '*Bathyuriscus–Elrathina*' and *Bolaspidella* Zones in the Middle Cambrian and at the Cambrian–Ordovician boundary, so that perhaps four or more biomeres are represented in the faunal succession of the Great Basin region. The biomere boundaries seen to be unaffected by lithofacies differences that seem to affect some of the faunal assemblages, and they may provide a basic biostratigraphical framework within which regional facies distinctions can be recognized by a varied zonal nomenclature.

The Cambrian faunas of the Great Basin region above the *Elvinia* Zone are not described, and no appropriate zonal nomenclature can be applied to them. They are well developed in the upper carbonate sequences of the Eureka–House Range and Pioche regions (Chapters 3.2, 3.3) where they include an outer detrital belt fauna which is correlative with the Franconian *Conaspis* and *Ptychaspis–Prosaukia* Zones of the mid-continent region. This, however, is almost totally distinct from the faunas of those zones and is closely similar to faunas from similar facies in Alaska (PALMER, 1967) and from boulders in the Levis conglomerates of Quebec (RASETTI, 1944). Only in the St. Charles Formation in the Bear River Range at the eastern margin of the Great Basin is a late Franconian assemblage typical of the mid-continent region known (LOCHMAN-BALK and HU, 1959). Some Trempealeauan trilobite assemblages include elements of the faunas of the carbonate belt in the mid-continent region, but the details of these faunas and their relationships are not yet described.

The top of the Cambrian throughout the Great Basin region and elsewhere in the United States is marked by an abrupt change to the faunas of the early Ordovician *Symphysurina* Zone. This change is very probably a biomere boundary as noted above.

In addition to trilobites, many of the Cambrian rocks of the Great Basin, particularly those adjacent to the margins of the carbonate belt, have rich faunas of inarticulate brachiopods which were partly described by WALCOTT (1912b) before the advent of modern acid-preparing techniques. These are in need of modern study, and afford an excellent opportunity to establish a brachiopod biostratigraphy comparable to the trilobite biostratigraphy and to determine whether the events that produced the biomere boundaries in the trilobite successions are also reflected in associated organisms.

5. Cambrian history of the Great Basin region

The Cambrian period was a quiet time in the Great Basin region. No major orogenic events are indicated in the sedimentary sequences. Instead, the stratigraphical record shows a gradual marine inundation of the region during the early part of the Cambrian and continued gradual subsidence in a marine environment, with only one short but significant reversal, until well into early Ordovician time.

In detail, the Cambrian stratigraphy is very complex and not all relationships can be resolved on a regional scale. Throughout Middle and Late Cambrian time, carbonate sediments form the greater part of most sedimentary sequences and at least two major environmental regimes are indicated. The presence of units of cross-bedded oolites, algal oncolites, and stromatolites, and massive limestones and dolomites, all relatively free of significant terrigenous detritus, is strong suggestive evidence for their genesis in relatively shallow water on carbonate banks. Rocks such as these are the dominant components of most of the medium- to thick-bedded carbonate sequences of the Great Basin east of the Antler orogenic belt.

In contrast, thin-bedded, silty, dark-coloured, frequently laminated limestones form thick, monotonous limestone or interbedded limestone-siltstone sequences in the Antler orogenic belt and many parts of Nevada and western Utah east of this belt. These sediments reflect quieter and probably deeper water environments than those of the carbonate banks.

Another important factor affecting regional sedimentary patterns is cyclicity on several scales. Large scale cyclic patterns are reflected in Grand Cycles (AITKEN, 1966) which are most clearly developed in the Carrara Formation (Chapter 3.1.1). In each of these, a basal terrigenous unit grades into an overlying limestone sequence which increases in purity upwards and then is abruptly replaced by the terrigenous unit of the next Grand Cycle. Similar cycles, but on the scale of metres instead of tens of metres, are well developed in the upper part of the Dunderberg Formation in the northern Egan and Schell Creek Ranges in the Eureka–House Range region (Chapter 3.3) and in the Ute Formation in the southern part of the Bear River Range and the Wellsville Mountains in northern Utah (Chapter 3.6). Repetitions of different types of carbonate sediments are present in many areas and are strikingly displayed in the Cole Canyon Dolomite in the East Tintic Mountains (Chapter 3.5), where units of dark structureless dolomite a few metres thick alternate with units of laminated white dolomite of similar thickness.

The causes for these cyclic fluctuations are not known. Those of the scale of Grand Cycles affect large areas and are very important for purposes of correlation. Reflection of the smaller cycles in regional stratigraphy has not been studied.

A genetic study of the Cambrian carbonate rocks and the spatial relations of the environments which can be identified, as well as an analysis of the regional effect of cyclic sedimentation at all scales, will be necessary for further refinement of much of the Cambrian stratigraphy of the Great Basin region.

The terrigenous component of the Cambrian sections is also in need of evaluation in a regional context. The Late Cambrian Worm Creek Quartzite thickens north-westwards in the northern Utah–south-eastern Idaho region (Chapter 3.6), and thick sand bodies are present in the western occurrences of the late Middle Cambrian Bloomington Formation of the same region. The lower and upper shale members of the Bloomington Formation also thicken to the north-west. In the Antler orogenic belt (Chapter 3.8) thick arkosic sequences constitute the Late Cambrian Harmony Formation in north-western Nevada, and terrigenous silt generally increases northwards in the thin-bedded Middle and Late Cambrian limestone sequences. All of these indicate a major sediment source to the north of the Great Basin, probably in western Idaho, during much of Cambrian time. The name Montania suggested by DEISS (1941) is an appropriate name for this feature.

Another major source for terrigenous materials, at least in early Middle Cambrian time, seems to have been located to the east of the Great Basin in the vicinity of the present 40th parallel. This is reflected in the relatively large proportion of terrigenous material in the lower parts of the Middle Cambrian sequences of the central Utah region (Chapter 3.4) as compared to the Eureka–House Range region (Chapter 3.3), and the fact that in the Tintic–southern Wasatch region (Chapter 3.5) carbonate sedimentation was not really significant until the later part of *Bathyuriscus–Elrathina* Zone time.

Early Middle Cambrian shale units in the Delamar Range in south-eastern Nevada are significantly thicker than corresponding units in the Pioche region to the north (Chapter 3.2) and the Spring Mountains–Death Valley region to the south (Chapter 3.1.1), suggesting the possibility of a third source area to the east in the vicinity of the present 38th parallel.

Sources for the shales in the lower parts of the Middle Cambrian sections in the northern part of the Deep Creek Range in western Utah (Chapter 3.4), in the northern Egan Range in eastern Nevada (Chapter 3.3), and in the Promontory Range north of Great Salt Lake (Chapter 3.7) are not clear. Sandy units of Late Cambrian *Crepicephalus*-Zone age in the Deep Creek and Dugway Ranges in western Utah (Chapter 3.4) also do not have a clear source. Resolution of these problems is needed before the regional Cambrian stratigraphy can be fully understood.

Despite the limitations discussed above, a good many regional relationships are considerably clearer now than they were at the time of the last attempt at a regional synthesis (PALMER, 1956).

At the beginning of the Cambrian period, sedimentation over most of the Great Basin region was from terrigenous sources to the north and east (Fig. 13). Earliest Early Cambrian source areas which are now within the Sevier orogenic belt or its northern extension are recorded in the Tintic–southern Wasatch region (Chapter 3.5) and in western Montana, where gently deformed Pre-Cambrian sedimentary sequences are overlain by thin sequences of late Early or early Middle Cambrian quartzites. East of the Sevier orogenic belt, the early Early Cambrian source areas included both meta-morphic and sedimentary complexes now present beneath the thin basal quartzitic sequences. In the Salt Lake City–Ogden region (Chapter 2.2) and in south-eastern Nevada (Chapter 2.1), only the metamorphic complexes remain. Later Pre-Cambrian sediments which were deformed prior to the beginning of Cambrian sedimentation are now recorded only in the Uinta Mountains and the Grand Canyon east of the Great Basin region.

Marine conditions spread into the present Great Basin area from the south-west, as attested to by the presence of tracks and trails well below the lowest trilobites in the White–Inyo Mountains region (Chapter 3.1.2). The spread of marine conditions east-wards reached a maximum in the early part of the Early Cambrian as shown by the lime-stones of the Poleta Formation and the correlative fossiliferous shales and siltstones of the upper Wood Canyon Formation in south-eastern California and southern Nevada

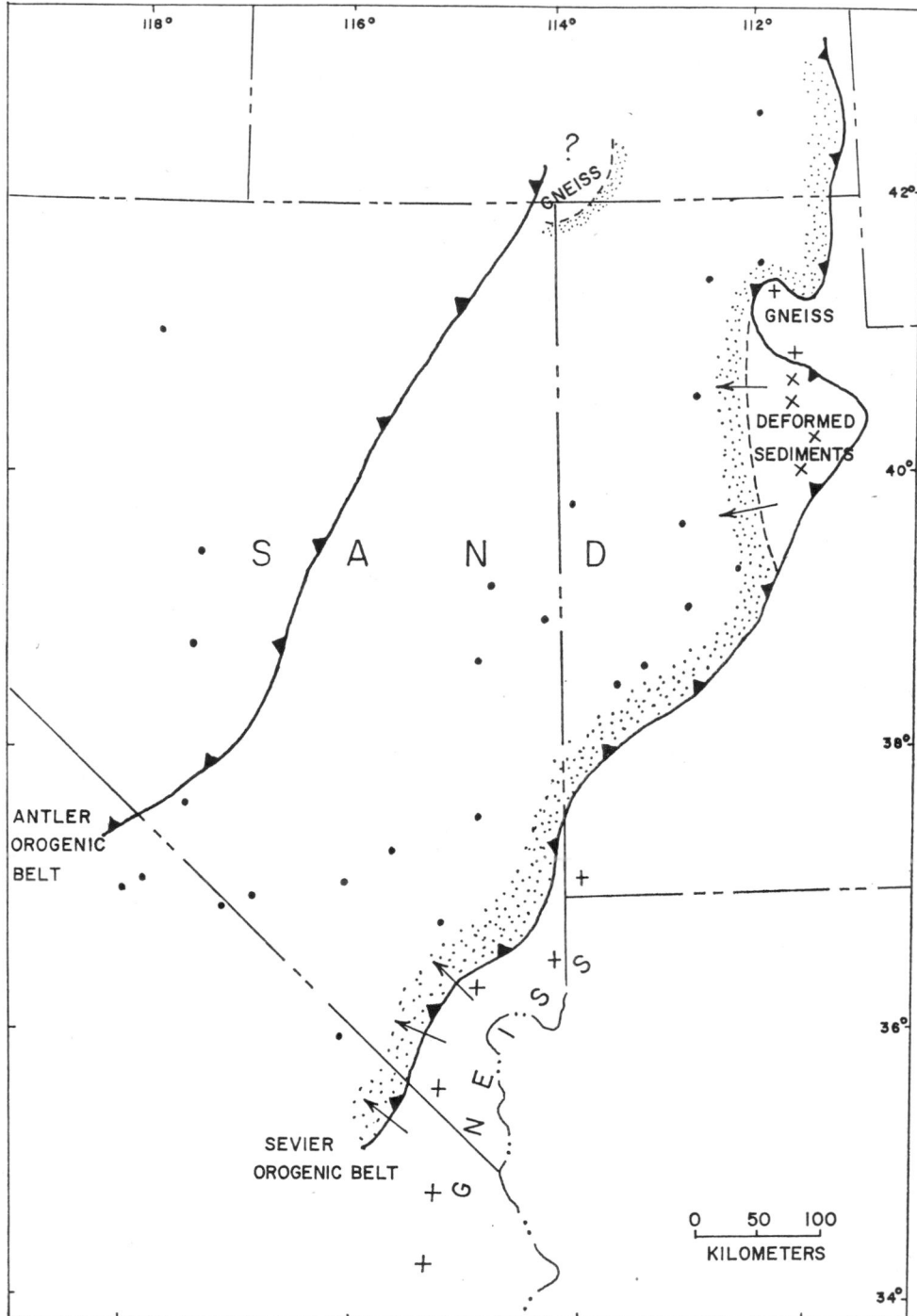

FIG. 13. Paleogeographical map of the Great Basin region for the early part of the Early Cambrian, showing the location and composition of nearby source areas.

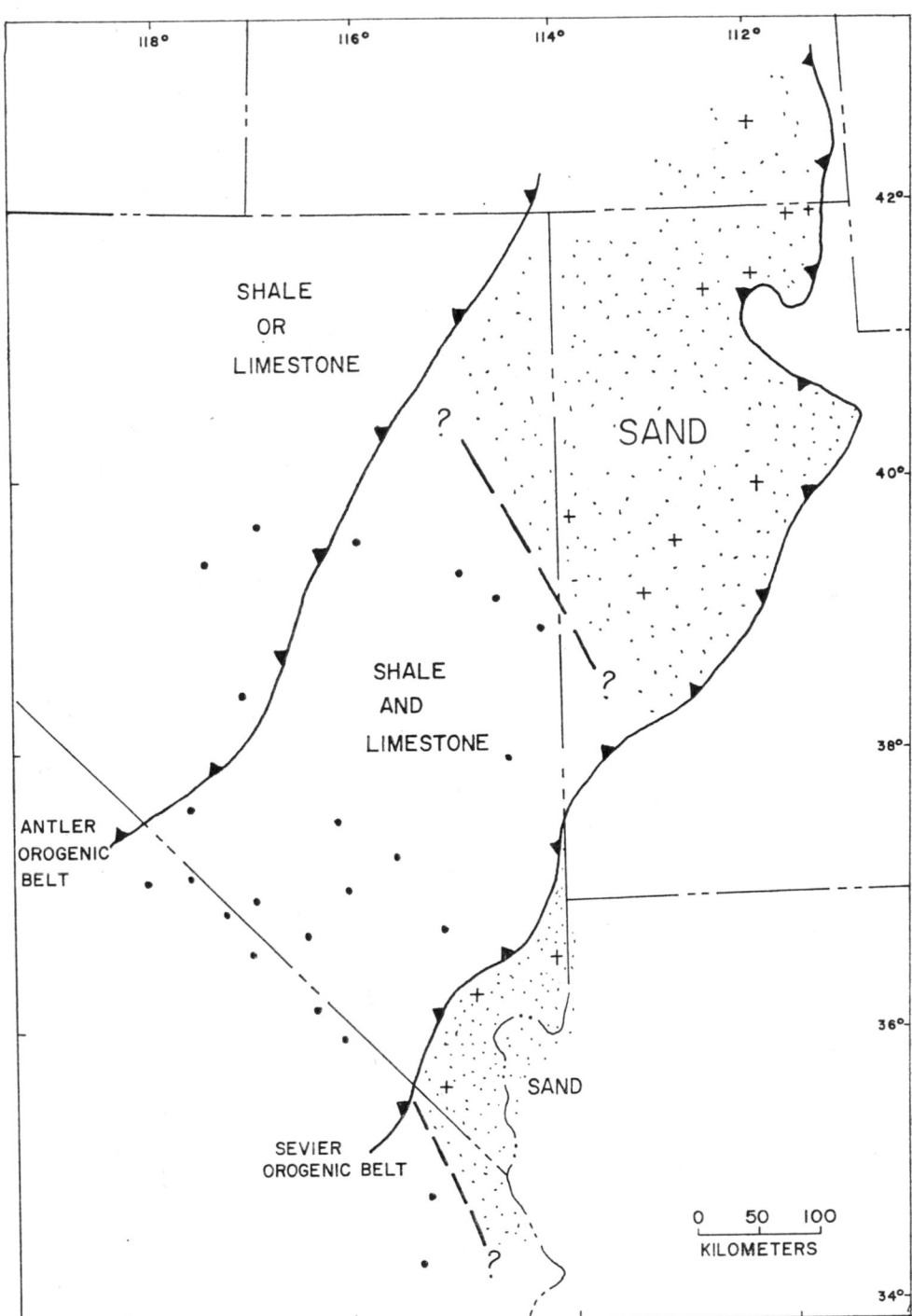

FIG. 14. Paleogeographical map of the Great Basin region during the late Early Cambrian showing facies distributions apparently offset by post-Cambrian tectonic activity.

(Chapter 3.1). An unnamed shaly sequence towards the top of the Prospect Mountain Quartzite in the southern Schell Creek Range in eastern Nevada (Chapter 3.3), the Cabin Shale of the Deep Creek Range in western Utah (Chapter 3.4), and the argillite unit near the top of the Brigham Quartzite in the Portneuf Range in south-eastern Idaho (Chapter 3.6) may also reflect this Early Cambrian spread of marine conditions. Later, but still within the Early Cambrian, a return to less marine conditions in these areas is indicated by the Zabriskie Quartzite and Harkless Formation, and perhaps the upper Brigham, Prospect Mountain, and Busby Quartzites.

A major transgressive event near the end of the Early Cambrian is indicated by the spread of quartz sand sedimentation eastwards across the Grand Canyon region and the development of marine shales and westward-thickening carbonates of latest Early Cambrian age in the southern and western parts of the Great Basin (Chapters 2.1, 3.1, 3.2, 3.3, 3.8). North-east of the Eureka–House Range region between the Antler and Sevier orogenic belts (Chapters 3.4, 3.5, 3.6, 3.7) and in the Salt Lake City–Ogden area east of the Sevier orogenic belt (Chapter 2.2), there is no evidence for Early Cambrian sediments above the quartzites, and fossils from the earliest post-quartzite beds in the north-eastern Great Basin are of *Albertella* Zone or younger Middle Cambrian ages. The north-westerly trend of the boundary between areas of sand and shale-limestone sedimentation during the late Early Cambrian seems to have been offset by eastward movements of both the Antler and Sevier orogenic belts (Fig. 14).

A major centre for carbonate bank sedimentation during late Early Cambrian time was located in the northern Last Chance Range area of the Spring Mountains–Death Valley region and in the adjacent White–Inyo Mountains region (Chapter 3.1). This is represented by the Mule Spring Limestone and the limestone at the top of the first Grand Cycle of the Carrara Formation. South-eastwards in the Nopah Range and at Eagle Mountain and northward in the Groom Range, the limestones from this centre become increasingly silty and rubbly, indicating that the carbonate bank was losing its identity in those directions. Because of this, the approximately correlative Chambless Limestone of the Providence and Marble Mountains south-east of the Nopah Range and east of the Sevier orogenic belt (Chapter 2.1) may represent a separate centre for carbonate sedimentation. The relation of the Mule Spring Limestone to correlative carbonates in the northern Toiyabe Range (Chapter 3.8) is not known. All of these carbonate bank areas were short-lived, and there were no carbonate banks anywhere in the Great Basin region at the end of the Early Cambrian epoch.

Evidence for Early Cambrian volcanic activity is present only in parts of the central and northern Great Basin region. Thin basaltic flows (?) are reported in the basal quartzite sequences in the Schell Creek Range (KELLOGG, 1963; DREWES, 1967), the East Tintic Mountains (MORRIS and LOVERING, 1961), and the Promontory Range (EARDLEY and HATCH, 1940). Volcanic events, at least in part submarine, are represented by pillow lavas and greenstones associated with archaeocyathid limestones in the Galena Range (ROBERTS, 1964). The present geographical relations of this volcanic sequence to the outcrop area of the Osgood Mountains Quartzite, about 60 kilometres to the north-west, have been used by ROBERTS and others (1958) as evidence that the volcanic sequence was transported eastwards relative to the clean quartzite sequence during the Antler orogeny.

The Middle Cambrian epoch records the establishment of several major areas of carbonate bank sedimentation which coalesced by the middle of the epoch to form a north-east trending belt of carbonate banks (Fig. 15). This belt separated eastern and western sedimentary sequences with significant terrigenous components that were described by PALMER (1960a,b) as the inner and outer detrital belts (Figs. 15 and 16).

Three different centres seem to be represented by the earliest Middle Cambrian carbonate bank sediments. The carbonate banks which developed from each of these centres expanded and contracted several times during early Middle Cambrian time. The expansions were predominantly eastwards and appear in each local area as the upper elements of Grand Cycles. On a regional scale they appear as eastward thinning limestone tongues. Available faunal evidence, though meagre, indicates that the eastward expansion was gradual so that the lower parts of each of these tongues are diachronous. The termination of a Grand Cycle was abrupt and was reflected across most of each carbonate bank, so that the sediments at the beginning of each new Grand Cycle seem to be approximately synchronous.

The most south-westerly centre for carbonate bank sedimentation was in the southern Last Chance Range-Bare Mountain area of the Spring Mountains–Death Valley region (Chapter 3.1.1), only slightly east of the carbonate bank centre which formed the Early Cambrian Mule Spring Limestone. The earliest sediments from this centre are represented by the limestones at the top of the second Grand Cycle of the Carrara Formation. These become increasingly silty towards the north and east indicating a gradual transition to a silty facies in a manner similar to their Early Cambrian counterparts. North-westwards in Esmeralda County, Nevada, and in the northern Last Chance Range, California, they are abruptly replaced by thin-bedded, silty, deeper water limestones of the lower part of the Emigrant Formation. Thus, along the trend of Death Valley, a transect across the earliest Middle Cambrian carbonate bank can be studied.

The second expansion of this carbonate bank is represented by the Jangle Limestone Member of the Carrara Formation. This extended eastwards into the present autochthonous areas of southern Nevada, where it is represented by a thin limestone identified as the Lyndon Limestone (Chapter 2.1).

The third expansion of this carbonate bank is represented by the lower beds of the Papoose Lake Member of the Bonanza King Formation in the Great Basin and the Muav Limestone of the Grand Canyon. Except for brief influxes of silt within the Papoose Lake Member and at the base of the overlying Banded Mountain Member, which probably represent the basal parts of two more Grand Cycles, the south-western carbonate bank remained as a major palaeographical feature over 250 kilometres wide until almost the end of early Late Cambrian (Dresbachian) time.

In the central part of the Great Basin between Eureka and the Snake Range (Chapter 3.3), a second centre for carbonate bank sedimentation was established (Fig. 15A). In the Antler orogenic belt to the west, except for undated dolomites tentatively identified as the Eldorado or Hamburg Dolomites in structural windows in the Shoshone and Cortez Ranges (Chapter 3.8), all carbonate sequences are thin-bedded and relatively silty and are considered to represent the outer detrital belt.

The oldest deposits from this bank are the basal beds of the Eldorado Dolomite and Pole Canyon Limestone (Fig. 7) in the Eureka district and in the Snake and northern Schell Creek Ranges. By *Albertella* Zone time, the bank had spread eastwards to the House Range region, where it is represented by the Howell Limestone, and southward to the Pioche region (Chapter 3.2), where it is represented by the Lyndon Limestone. It is possible that at this time the central and south-western banks briefly coalesced, but the Lyndon Limestone thins rapidly southwards from Pioche into the Delamar Range and there is no information in the critical interval between this area and the Desert and

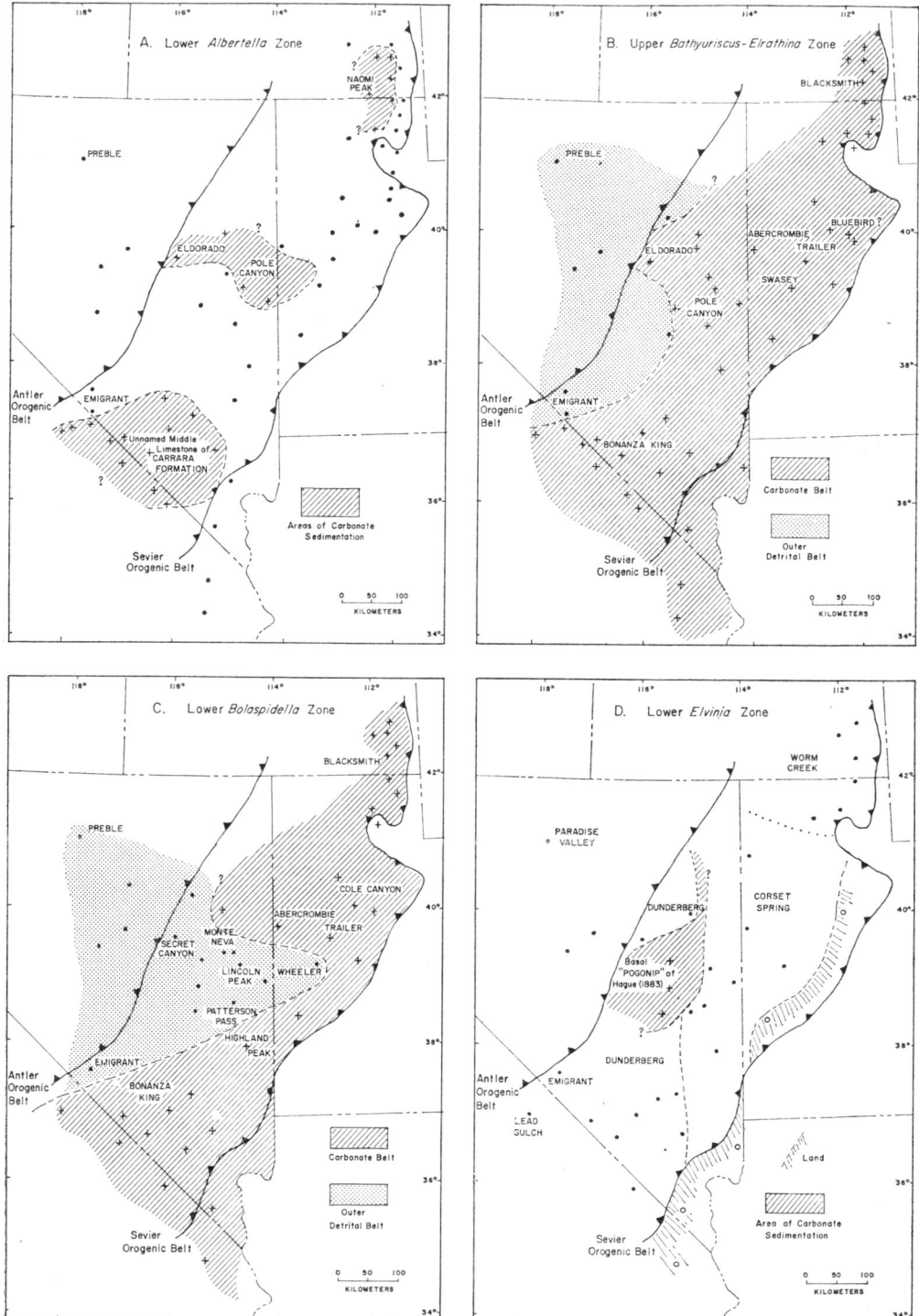

FIG. 15. Paleogeographical maps of the Great Basin region showing the principal contrasts in facies distribution for the Cambrian Period.

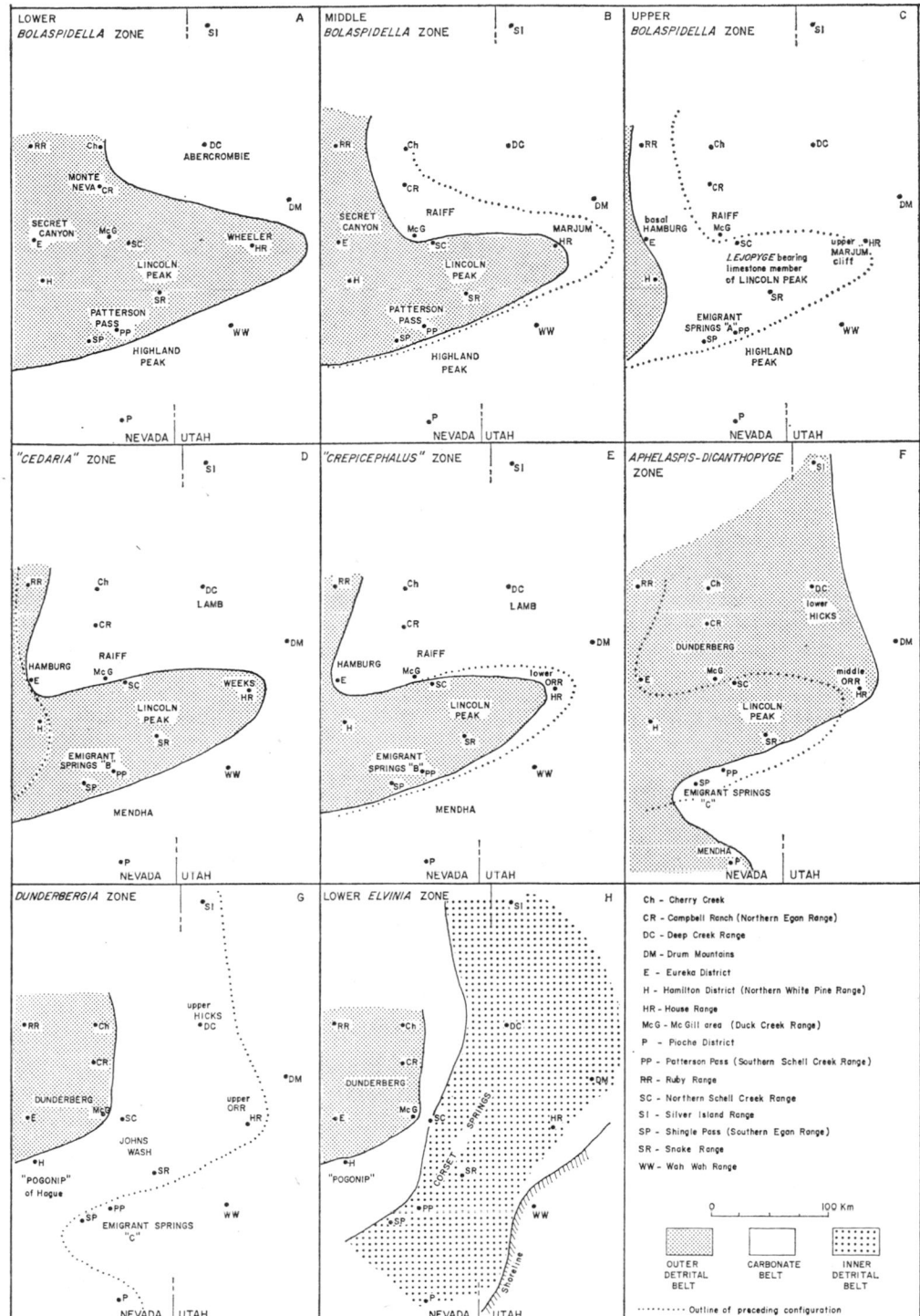

FIG. 16. Paleogeographical maps showing changes in configuration of the western margin of the carbonate belt from Late Middle Cambrian to earliest Franconian time (between the times represented by Fig. 15C and D), and their effect on stratigraphical nomenclature. Compare with Fig. 7.

Groom Ranges 75 kilometres to the south-west where the correlative Jangle Limestone is present. A major reduction in area of the carbonate bank in the region of the House Range and in the Pioche region is represented by the *Glossopleura*-bearing Chisholm Shale and equivalent shaly beds at the top of the Howell Limestone. However, carbonate sedimentation persisted without significant interruption in the areas to the north-west (Fig. 7). A second expansion of the bank produced the Dome Limestone of the House Range region and the Peasley and Burrows Limestones of the Pioche region. It probably resulted in the coalescence of the central and south-western bank into a single grand bank having an area of over 125,000 square kilometres. The second reduction in area of this bank produced the Whirlwind Shale in the House Range and the Burnt Canyon Member of the Highland Peak Formation in the Pioche region. This shaly unit may correlate with the silty unit within the Papoose Lake Member of the Bonanza King Formation to the south-west.

The next expansion of the grand bank during '*Bathyuriscus-Elrathina*' Zone time resulted in the spread of carbonate sediments northwards into south-eastern Idaho and eastwards across the present autochthonous regions, as well as eastwards into the House Range and Pioche regions (Fig. 15B). The units representing this great expansion of the carbonate bank are the Step Ridge Member of the Highland Peak Formation in the Pioche Region (Chapter 3.2), the Swasey Formation of the House Range (Chapter 3.3), the Teutonic Limestone in the East Tintic Mountains, the Maxfield Limestone in the Wasatch Range (Chapter 3.5), and the Ute and Blacksmith Formations of northern Utah and south-eastern Idaho (Chapter 3.6).

Earlier, the northern Utah–south-eastern Idaho region was the site for a third carbonate bank (Fig. 15A) centred in the vicinity of the Portneuf Range and represented initially by the Naomi Peak Limestone. The full extent of this carbonate bank has not been worked out but there is no clear connection across northern Utah to the carbonate bank of the central region. It seems to have been eliminated towards the end of *Albertella* Zone time and to have been buried by the Spence Shales. The remaining details of pre-Ute stratigraphy in this region are very uncertain. At about the time of elimination of the Naomi Peak bank, a second carbonate bank seems to have become established a short distance to the east and to have produced the Langston Dolomite. During the later part of *Glossopleura* Zone time, this bank was shifted westwards and its eastern margin was flooded by terrigenous sediments now represented by a thin unit of *Glossopleura*-bearing shales in the upper part of the Langston Dolomite (Chapter 3.6). The bank spread again over the eastern areas by the end of *Glossopleura* Zone time and then was eliminated and buried by shales of the basal Ute Formation.

The spread of carbonate sediments across the whole of the Great Basin east of the Antler orogenic belt at the end of *Bathyuriscus–Elrathina* Zone time marks the end of a major phase of development of the Great Basin stratigraphical framework (Fig. 15B). Beginning with *Bolaspidella* Zone time, the western margin of the grand carbonate bank was apparently depressed in the Eureka–House Range region (Chapter 3.3), forming a broad embayment terminating just east of the House Range (Figs. 15C and 16A). This embayment persisted through early Late Cambrian time.

In the Eureka–House Range embayment, the outer detrital belt shales of the Lincoln Peak, Secret Canyon, Monte Neva, Patterson Pass, and Wheeler Formations (Fig. 7) mark the cessation of carbonate sedimentation. The margin between the carbonate banks and the Eureka–House Range embayment during the remainder of Middle Cambrian time and through early Late Cambrian (Dresbachian) time shifted locally

(Fig. 16) so that carbonate bodies of several varied ages are intercalated with the outer detrital belt shales. In late Dresbachian time, the outer detrital belt shales spread eastwards into some areas of former carbonate deposition (Fig. 16F). On the northern margin of the embayment, the carbonate bank spread southwards near the end of the Middle Cambrian to form the Hamburg and Raiff Formations of the Eureka district and the northern Egan and Schell Creek Ranges (Figs. 16B and C). In the House Range area, shift of bank margin westwards during *Crepicephalus* Zone time produced the lower limestone member of the Orr Formation (Fig. 16E). The northern margin of the embayment shifted to the north in *Aphelaspis* Zone time to produce the lower beds of the Dunderberg Formation in the Eureka district, and the northern Schell Creek, Cherry Creek, and Egan Ranges; the lower beds of the Hicks Formation in the Deep Creek Range; and the middle member of the Orr Formation in the House Range (Fig. 16F). At the same time, the carbonate bank at the southern margin of the embayment spread northwards to form the upper member of the Emigrant Springs Limestone in the southern Schell Creek and Egan Ranges. The bank also spread westwards, eliminating the eastern tip of the embayment during *Dunderbergia* Zone time, and producing the Johns Wash Limestone and equivalent units in the Snake, House, and Deep Creek Ranges, and the lower limestones of HAGUE's (1883) Pogonip Limestone in the Grant and White Pine Ranges (Chapter 3.3).

Meanwhile, with a few exceptions noted below, in most areas east of the embayment carbonate bank sedimentation persisted through most of late Middle Cambrian and Dresbachian time. During early *Bolaspidella* Zone time, the bank area was briefly, but extensively, reduced. This is reflected by the Hodges Shale Member of the Bloomington Formation in northern Utah (Chapter 3.6), siltstones in the lower part of the Banded Mountain Member of the Bonanza King Formation in the southern Great Basin (Chapter 2.1, 3.1), and by the silty Condor Member of the Highland Peak Formation in the Pioche region (Chapter 3.2). In the later Middle Cambrian, sands from Montania to the north-west, and perhaps also shales, flooded parts of the northern bank. The shales form the Calls Fort Member of the Bloomington Formation. Shales and sands of uncertain source which bear *Crepicephalus* and *Aphelaspis* Zone fossils are reported from the top of the Lamb Dolomite and correlative units in the Deep Creek, Dugway, and Sheeprock Ranges in the central Utah region (Chapter 3.4); from the Opex Formation in the East Tintic Mountains (Chapter 3.5); and from unnamed units in the Wah Wah Range in south-western Utah (Chapter 3.2). They are also reported from the Virgin Mountains in southern Nevada east of the Sevier orogenic belt (Chapter 2.1). Elsewhere in southern Nevada and adjacent California, the bank was briefly eliminated during *Dunderbergia* and *Elvinia* Zone time by an eastward spread of outer detrital belt sediments which formed the Dunderberg Shale of most of the southern region (Chapter 3.1.1) and the shales of the Lead Gulch Formation of the Inyo Mountains (Chapter 3.1.2).

Just prior to *Elvinia* Zone time a brief, continent wide lowering of sea level produced an erosional surface as far west as the East Tintic Mountains and the Wah Wah Range (Chapter 3.5, 3.2) and a major reorientation of the palaeogeography (Fig. 15D). A flood of terrigenous sediment, perhaps in a large part from Montania, produced the Worm Creek Quartzite and Corset Spring Shale of northern and central Utah (Chapters 3.3, 3.4, 3.5, 3.6, 3.7). This was followed, still within *Elvinia* Zone time, by elimination of the Eureka–House Range embayment and a return to carbonate sedimentation over the entire Great Basin region east of the Antler orogenic belt. This palaeogeography persisted without major change into Ordovician time. In the western part of this region, some units in the Windfall Formation may be classed as outer detrital belt carbonate units, but the details of their relations to the carbonate banks have not yet been worked out; and they contrast strongly with the shalier underlying units.

In the Antler orogenic belt (Chapter 3.8), detailed relationships of the Middle and Late Cambrian rocks have not been worked out. With the exception of a few massive carbonate rocks exposed in structural windows, they are all representative of the outer detrital belt. At about the time of formation of the Eureka–House Range embayment, a period of Middle Cambrian volcanism is recorded in the Schwin Formation in the Shoshone Range. During the later part of the Cambrian, while carbonate sedimentation was re-established over the Great Basin region east of the Antler orogenic belt, a flood of coarse arkoses reached the north-western part of Nevada to form the Harmony Formation. These seem to reflect a period of uplift and rapid erosion of Montania to the north.

In summary, Early Cambrian time is dominated by terrigenous and predominantly quartz sand sedimentation throughout the Great Basin. Carbonates appear at this time only in the westernmost areas.

In early Middle Cambrian time, a series of carbonate banks developed and eventually coalesced into a north-east trending carbonate belt. Most of the stratigraphy of the region during this time reflects the interaction between these banks and sources of terrigenous sediments to the east. A fairly stable boundary between the banks and areas of deeper water sediments of the outer detrital belt to the west persisted throughout this time.

During late Middle Cambrian and early Late Cambrian time, an embayment from the west which received outer detrital belt sediments indented the margin of the carbonate belt in the central Great Basin region. To the east, carbonate bank sedimentation persisted with only a few minor incursions of terrigenous materials from the north or east, and deeper water sedimentation persisted to the west.

In lower *Elvinia* Zone time, a major short-lived regressive event eliminated the embayment in the carbonate belt. Following this event, carbonate bank sedimentation returned to nearly all of the areas which had had earlier carbonate banks. Deeper water sedimentation continued in the western areas. This palaeogeographical configuration continued without significant modification into the Ordovician.

6. Mineral deposits in the Cambrian of the Great Basin region

The Great Basin and areas immediately adjacent to it have been major areas of mineral production in western United States since the latter part of the 19th century. Several significant mining districts, some of which are discussed below, and many small prospects and briefly exploited mineralized areas have been developed in the Cambrian sequences. Mining activity in most of these areas has been minimal since the 1930's.

Almost all of the mineral deposits in Cambrian rocks are related to fracture zones that carried mineralizing solutions from nearby Mesozoic or Tertiary igneous intrusions. This intrusive activity largely post-dates the regional thrust and low-angle normal faulting. The most productive ore bodies have been replacement deposits of silver–lead ores with some zinc, gold, or copper. Vein deposits within the lower quartzites have produced copper, gold, or silver. Contact metamorphic ore deposits are not common but have been a source for tungsten.

The replacement deposits are either bedded or massive. The principal deposits have been reported from the Cottonwood, Tintic, and Ophir districts in Utah (Chapter 3.5) and the Eureka, Pioche, and Groom districts in Nevada (Chapters 3.1, 3.2, 3.3).

In Utah, limestones within the Ophir Shale and adjacent to the white laminated dolomite of the Maxfield Formation of the Cottonwood district (CALKINS and BUTLER, 1943) have produced rich ores of lead and silver with associated zinc and gold. Limestones within the Ophir Shale in the Ophir district have produced lead, zinc, copper, and silver ores (GILLULY, 1932). In the Tintic district (LINDGREN and LOUGHLIN, 1919; MORRIS and LOVERING, 1961), limestones within the Ophir Shale have been a rich source of lead–silver ores and some horizons within the Teutonic, Herkimer, and Opex Formations have also been productive of these metals. In addition, the Ajax Dolomite has been a host for copper ores as well as for a large replacement body of halloysite.

In Nevada, limestones within the Pioche shale, the Lyndon Limestone, and some units of the Highland Peak Formation in the Pioche district (WESTGATE and KNOPE, 1932) have yielded silver–lead–zinc ores with some associated copper and tungsten. In the Eureka district (NOLAN, 1962), large massive replacement bodies of lead–silver ore which are relatively rich in gold are present in the Eldorado and Hamburg Dolomites. Bedded ores are present within the Windfall Formation. In the Groom district (HUMPHREY, 1945), lead–silver ores have been mined from limestone units within the Carrara Formation.

In the Tintic and Pioche districts, veins within the Tintic and Prospect Mountain Quartzites have been sources for lead–silver and copper–gold ores. In addition, the Tintic Quartzite in the Tintic district has been quarried for high-grade silica.

In the Osgood Mountains (Chapter 3.8), limestones of the Preble Formation in contact with late Cretaceous intrusives have yielded tungsten ores (HOTZ and WILLDEN, 1964). Other tungsten deposits are known but are not fully described in the Pole Canyon Limestone in the Snake Range.

Almost all of the bedded replacement deposits except those in the Windfall Formation of the Eureka district are in areas where carbonate bodies are tonguing eastwards with shaly rocks of the inner detrital belt. Study of the mineralization of the rocks of the Cambrian System in the Great Basin as a function of their regional stratigraphy has never been attempted, but it should produce interesting and perhaps significant information of value in prospecting for new ore deposits.

Bibliography

ABBOTT, W. O. (1951). Cambrian diabase flow in central Utah. *Compass* 29, 51.

AITKEN, J. D. (1966). Middle Cambrian to Middle Ordovician cyclic sedimentation, southern Rocky Mountains of Alberta. *Bull. Can. Pet. Geol.* 14, 405.

ANDERSON, A. L. (1928). Portland cement materials near Pocatello, Idaho. *Pamph. Idaho Bur. Mines and Geol.* 28.

ARMSTRONG, R. L. (1968a). The Sevier orogenic belt in Nevada and Utah. *Bull. Geol. Soc. Am.* 79, 249.

ARMSTRONG, R. L. (1968b). Mantled gneiss domes in the Albion Range, southern Idaho. *Bull. Geol. Soc. Am.* 75, 1295.

ALBERS, J. P. and STEWART, J. H. (1962). Precambrian (?) and Cambrian stratigraphy in Esmeralda County, Nevada. *Profess. Paper U.S. Geol. Surv.* 450-D, 24.

BAKER, A. A. (1964). Geologic map and sections of the Orem quadrangle, Utah. *U.S. Geol. Survey Geol. Quad. Map* GQ-241.

BAKER, A. A. and CRITTENDEN, M. F., Jr. (1961). Geology of the Timpanogos Cave quadrangle. *U.S. Geol. Surv. Geol. Quad. Map* GQ-132.

BARNES, HARLEY and BYERS, F. M., Jr. (1961). Windfall formation (Upper Cambrian) of Nevada Test site and vicinity, Nevada. *Profess. Paper U.S. Geol. Surv.* 424-C, 103.

BARNES, HARLEY and CHRISTIANSEN, R. L. (1967). Cambrian and Precambrian rocks of the Groom District, Nevada, southern Great Basin. *Bull. U.S. Geol. Surv.* 1244-G.

BARNES, HARLEY and PALMER, A. R. (1961). Revison of stratigraphic nomenclature of Cambrian rocks, Nevada Test Site, and vicinity, Nevada. *Profess. Paper U.S. Geol. Surv.* **424-C**, 100.

BARNES, HARLEY and others (1962). Cambrian Carrara formation, Bonanza–King Formation, and Dunderberg shale east of Yucca Flat, Nye County, Nevada. *Profess. Paper U.S. Geol. Surv.* **450-D**, 27.

BECRAFT, G. E. and WEIS, P. L. (1963). Geology and mineral deposits of the Turtle Lake quadrangle, Washington. *Bull. U.S. Geol. Surv.* **1131**.

BENNETT, W. A. G. (1941). Preliminary report on Magnesite deposits of Stevens County, Washington. *Rept. Inv., Div. Geol. Wash. Dept. Cons. and Devel.* **5**.

BENTLEY, C. B. (1958). Upper Cambrian stratigraphy of western Utah. *Res. Stud. Brigham Young Univ. Geol. Ser.* **5**.

BICK, K. F. (1959). Stratigraphy of Deep Creek Mountains, Utah. *Bull. Am. Assoc. Pet. Geol.* **43**, 1064.

BICK, K. F. (1966). Geology of the Deep Creek Mountains, Tooele and Juab Counties, Utah. *Bull Utah Geol. Min. Surv.* **77**.

BRADY, M. J. (1965). Thrusting in the southern Wasatch Mountains, Utah. *Brigham Young Univ. Res. Stud. Geol.* **12**, 3.

BRIGHT, R. C. (1959). A paleoecologic and biometric study of the Middle Cambrian trilobite *Elrathia kingii* (Meek). *J. Paleont.* **35**, 83.

BURCHFIEL, B. C. (1964). Precambrian and Paleozoic stratigraphy of the Specter Range quadrangle, Nye County, Nevada. *Bull. Am. Assoc. Pet. Geol.* **48**, 40.

CALKINS, F. C. and BUTLER, B. S. (1943). Geology and ore deposits of the Cottonwood–American Fork area, Utah. *Profess. Paper U.S. Geol. Surv.* **201**.

CALLAGHAN, EUGENE (1937). Geology of the Delmar district, Lincoln County, Nevada. *Bull. Univ. Nev.* **31**.

CAMPBELL, C. D. (1947). Cambrian rocks of north-eastern Stevens County, Washington. *Bull. Geol. Soc. Am.* **58**, 597.

CARR, W. J. and TRIMBLE, D. E. (1962). Paleozoic rocks measured southwest of Pocatello, Idaho. *U.S. Geol. Surv. Open File Rept.*

CHRISTIANSEN, F. W. (1952). Structure and stratigraphy of the Canyon Range, central Utah. *Bull. Geol. Soc. Am.* **63**, 717.

CHRISTIANSEN, R. L. and BARNES, HARLEY (1966). Three members of the Upper Cambrian Nopah formation in the southern Great Basin. *Bull. U.S. Geol. Surv.* **1244-A**, 49.

CLOUD, P. E., Jr. and NELSON, C. A. (1966). Phanerozoic–Cryptozoic and related transitions: New evidence. *Science* **154**, 3750.

COASH, J. R. (1967). Geology of the Mt. Velma quadrangle, Elko County, Nevada. *Bull. Nev. Bur. Mines* **68**.

COHENOUR, R. E. (1959). Sheeprock Mountains, Tooele and Juab Counties. *Bull. Utah Geol. and Mineralog. Surv.* **63**.

CORNWALL, H. R. (1967). Geology of southern Nye County, Nevada. *U.S. Geol. Surv. Open File Rept.*

CORNWALL, H. R. and KLEINHAMPL, F. J. (1961). Geology of the Bare Mountain quadrangle, Nevada. *U.S. Geol. Surv. Geol. Quad. Map* **GQ-157**.

CORNWALL, H. R. and KLEINHAMPL, F. J. (1964). Geology of the Bullfrog quadrangle, and ore deposits related to Bullfrog Hills caldera, Nye County, Nevada, and Inyo County, California. *Profess. Paper U.S. Geol. Surv.* **454-J**.

COULTER, H. W. (1956). Geology of the southeast portion of the Preston quadrangle, Idaho. *Pamph. Idaho Bur. Mines and Geol.* **107**.

CRICKMAY, C. H. (1933). *in* Hazzard, (1933).

CRITTENDEN, M.D., Jr., SHARP, B. J., and CALKINS, F. C. (1952). Geology of the Wasatch Mountains east of Salt Lake City, Parleys Canyon to Traverse Ridge, *in* Geology of the central Wasatch Mountains, Utah. *Utah Geol. Soc. Guidebook* **8**, 1.

CRITTENDEN, M. D., Jr., STRAEZEK, J. A., and ROBERTS, R. J. (1961). Manganese deposits in the Drum Mountains, Juab County, Utah. *Bull. U.S. Geol. Surv.* **1082-H**, 493.

DECKER, R. W. (1962). Geology of the Bull Run Quadrangle, Elko County, Nevada. *Bull. Nev. Bur. Mines.* **60**.

DEISS, C. F. (1938). Cambrian formations and sections in part of the Cordilleran trough. *Bull. Geol. Soc. Am.* **49**, 1067.

DEISS, C. F. (1941). Cambrian geography and sedimentation in the central Cordilleran region. *Bull. Geol. Soc. Am.* **52**, 1085.

DINGS, McC. G. and WHITEBREAD, D. H. (1965). Geology and ore deposits of the Metaline zinc–lead district, Pend Oreille County, Washington. *Profess. Paper U.S. Geol. Surv.* **489**.

DREWES, HARALD (1967). Geology of the Connors Pass quadrangle, Schell Creek Range, east-central Nevada. *Profess. Paper U.S. Geol. Surv.* **557**.

DREWES, HARALD and PALMER, A. R. (1957), Cambrian rocks of the southern Snake Range. *Bull. Am. Assoc. Pet. Geol.* **41**, 104.

DURHAM, J. W. and CASTER, K. E. (1963). Helicoplacoidea—A new class of echinoderms. *Science* **140**, 820.

DURHAM, J. W. and others (1966). *Treatise on invertebrate paleontology*, Part. U. Echinodermata 3, Asterozoa, Echinozoa, V. 1–2. Lawrence, Kansas.

EARDLEY, A. J. (1944). Geology of the north-central Wasatch Mountains, Utah. *Bull. Geol. Soc. Am.* **55**, 819.

EARDLEY, A. J. and HATCH, R. A. (1940). Proterozoic (?) rocks in Utah. *Bull. Geol. Soc. Am.* **55**, 795.

EAST, E. H. (1966). Structure und stratigraphy of San Francisco mountains, western Utah. *Bull. Am. Assoc. Pet. Geol.* **50**, 901.

FERGUSON, H. G. (1933). Geology of the Tybo district, Nevada. *Bull. Nevada Univ.* **27**.

FERGUSON, H. G. and CATHCART, S. H. (1954). Geology of the Round Mountain quadrangle, Nevada. *U.S. Geol. Surv. Geol. Quad. Map* **GQ-40**.

FRITZ, JAMES (1968). Lower and early Middle Cambrian trilobites from the Pioche shale, east central Nevada, U.S.A. *Palaeontogy*, **11**, 183.

GILLULY, JAMES (1932). Geology and ore deposits of the Stockton and Fairfield quadrangles, Utah. *Profess. Paper U.S. Geol. Surv.* **173**.

GILLULY, JAMES and GATES, OLCOTT (1965). Tectonic and Igneous geology of the northern Shoshone Range, Nevada. *Profess. Paper U.S. Geol. Surv.* **465**.

GILLULY, JAMES and MASURSKY, HAROLD (1965). Geology and ore deposits of the Cortez quadrangle, Nevada. *Bull. U.S. Geol. Surv.* **1175**.

GRABAU, A. W. (1936). Paleozoic formations in the light of the pulsation theory; vol. 1, Lower and Middle Cambrian pulsations.

GREGGS, R. G. (1959). Archaeocyatha from the Colville and Salmo areas of Washington and British Columbia. *J. Paleont.* **33**, 63.

HAGUE, ARNOLD (1883). Abstract of report on the geology of the Eureka district, Nevada. *U.S. Geol. Surv. 3rd Ann. Rept.* 237.

HAZZARD, J. C. (1933). Notes on the Cambrian rocks of the eastern Mohave desert, California, with a paleontological report by Colin H. Crickmay. *Bull. Dept. Geol. Sci., Univ. Calif. Publ.* **23**.

HAZZARD, J. C. (1937). Paleozoic section in the Nopah and Resting Springs mountains, Inyo County, California. *Calif. Jour. Mines and Geol.* **33**, 273.

HAZZARD, J. C. (1954). Rocks and structure of the northern Providence Mountains, San Bernardino County, California. *Bull. Calif. Dept. Nat. Res.* **170**, 27.

HAZZARD, J. C. and MASON, J. F. (1936). Middle Cambrian formations of the Providence and Marble Mountains, California. *Bull. Geol. Soc. Am.* **47**, 229.

HAZZARD, J. C. and MASON, J. F. (1953). The Goodsprings dolomite at Goodsprings, Nevada. *Am. Jour. Sci., 5th ser.* **251**, 643.

HEWETT, D. F. (1931). Geology and ore deposits of the Goodsprings quadrangle, Nevada. *Profess. Paper U.S. Geol. Surv.* **162**.

HEWETT, D. F. (1956). Geology and mineral resources of the Ivanpah quadrangle, California and Nevada. *Profess. Paper U.S. Geol. Surv.* **275**.

HINTZE, L. F. (1969). Stratigraphic rock columns for Utah and adjacent areas. *Utah Geological Survey*. In press.

HOSE, R. K. (1961). Physical characteristics of Upper Cambrian Stromatolites in western Utah. *Profess. Paper U.S. Geol. Surv.* **424-D**, 245.

HOTZ, P. E. and WILLDEN, C. R. (1964). Geology and Mineral Deposits of the Osgood Mountains Quadrangle, Humboldt County, Nevada. *Profess. Paper U.S. Geol. Surv.* **431**.

HOWELL, B. F. and others (1944). Correlations of the Cambrian formations of North America. *Bull. Geol. Soc. Am.* **55**, 993.

HUMPHREY, F. L. (1945). Geology of the Groom district, Lincoln County, Nevada. *Bull. Nev. Univ.* **39**.

HUMPHREY, F. L. (1960). Geology of the White Pine mining district, White Pine County, Nevada. *Bull. Nev. Bur. Mines* **57**.

HUNT, C. B. and MABEY, D. R. (1966). Stratigraphy and structure, Death Valley, California. *Profess. Paper U.S. Geol. Surv.* **494-A.**

JOHNSON, M. S. and HIBBARD, D. E. (1957). Geology of the Atomic Energy Commission Nevada proving grounds area, Nevada. *Bull. U.S. Geol. Surv.* **1021-K,** 333.

KELLOGG, H. E. (1963). Paleozoic stratigraphy of the southern Egan Range, Nevada. *Bull. Geol. Soc. Am.* **74,** 685.

KIRK, EDWIN (1918), *in* KNOPF, ADOLPH (1918). A geologic reconnaissance of the Inyo Range and the eastern slope of the southern Sierra Nevada, California, with a section on the stratigraphy of the Inyo Range by Edwin Kirk. *Profess. Paper U.S. Geol. Surv.* **110.**

KOBAYASHI, TEIICHI (1935). The *Briscoia* fauna of the late Upper Cambrian in Alaska, with descriptions of a few Upper Cambrian trilobites from Montana and Nevada. *Jap. J. Geol. Geog.* **12,** 39.

LINDGREN, WALDEMAR and LOUGHLIN, G. F. (1919). Geology and ore deposits of the Tintic mining district. *Profess. Paper U.S. Geol. Surv.* **107.**

LOCHMAN, CHRISTINA (1956). The Cambrian of the Rocky Mountains and southwest deserts of the United States and adjoining Sonora province, Mexico, *in Symposium, El Sistema Cambrico, su paleogeografia y el problema de su base, XX Int. Geol. Congress, Mexico,* 529.

LOCHMAN-BALK, CHRISTINA and HU, C. H. (1959). A *Ptychaspis* faunule from the Bear River Range, south-eastern Idaho. *J. Paleont.* **33,** 404.

LOCHMAN-BALK, CHRISTINA and WILSON, J. L. (1958). Cambrian biostratigraphy in North America. *J. Paleont.* **32,** 312.

LONGWELL, C. R. (1928). Geology of the Muddy Mountains, Nevada. *Bull. U.S. Geol. Surv.* **798.**

McALLISTER, J. F. (1952). Rocks and structure of the Quartz Spring area, northern Panamint Range, California. *Spec. Rept. Calif. Dept. Nat. Res.* **25.**

McKEE, E. D. (1945). Cambrian history of the Grand Canyon region, Part I, Stratigraphy and ecology of the Grand Canyon Cambrian. *Carnegie Inst. Wash. Pub.* **563.**

McKEE, E. H. (1963). Ontogenetic stages of the archaeocyathid *Ethmophyllum whitneyi* Meek. *J. Paleont.* **37,** 287.

McKEE, E. H. and MOIOLA, R. J. (1962). Precambrian and Cambrian rocks of south-central Esmeralda County, Nevada. *Am. J. Sci.* **260,** 530.

McNAIR, A. H. (1951). Paleozoic stratigraphy of part of north-western Arizona. *Bull. Am. Assoc. Pet. Geol.* **35,** 503.

MANSFIELD, G. R. (1927). Geography, geology and mineral resources of south-eastern Idaho. *Profess. Paper U.S. Geol. Surv.* **152.**

MASON, J. F. (1935). Fauna of the Cambrian Cadiz formation, Marble Mountains, California. *Bull. So. Calif. Acad. Sci.* **34,** 97.

MAXEY, G. B. (1958). Lower and Middle Cambrian stratigraphy in northern Utah and southeastern Idaho. *Bull. Geol. Soc. Am.* **69,** 647.

McLAUGHLIN, K. P. and ENBYSK, B. B. (1950). Middle Cambrian trilobites from Pend Oreille County, Washington. *J. Paleont.* **24,** 466.

MEANS, W. D. (1962). Structure and stratigraphy in the central Toiyabe Range, Nevada. *Calif. Univ. Pubs. Geol. Sci.* **42,** 71.

MEEK, F. B. (1868). Note on *Ethmophyllum* and *Archaecyathus. Am. J. Sci., 2nd ser.* **46,** 62.

MERRIAM, C. W. (1964). Cambrian rocks of the Pioche Mining district, Nevada. *Profess. Paper U.S. Geol. Surv.* **469.**

MILLER, G. M. (1966). Structure and stratigraphy of southern part of Wah Wah Mountains, southwest Utah. *Bull. Am. Assoc. Pet. Geol.* **50,** 858.

MOORES, E. M. and others (1968). Tertiary tectonics of the White Pine–Grant Range region, east-central Nevada and some regional implications. *Bull. Geol. Soc. Am.* **79,** 1703.

MORRIS, H. T. and LOVERING, T. S. (1961). Stratigraphy of the East Tintic Mountains, Utah. *Profess. Paper U.S. Geol. Surv.* **361.**

NELSON, C. A. (1962). Lower Cambrian–Precambrian succession, White–Inyo Mountains, California. *Bull. Geol. Soc. Am.* **73,** 139.

NELSON, C. A. (1963). Stratigraphic range of *Ogygopsis. J. Paleont.* **37,** 244.

NELSON, C. A. (1965). Monola formation *in* COHEE, G. V. and WEST, W. S., Changes in stratigraphic nomenclature by the U.S. Geological Survey, 1963. *Bull. U.S. Geol. Surv.* **1194-A,** 29.

NELSON, C. A. and HUPÉ, PIERRE (1964). Sur l'existence de *Fallotaspis* et *Daguinaspis,* trilobites marocains, dans le Cambrien inferieur de Californie, et ses consequences. *C. R. Acad. Sci. Paris* **258,** 621.

NELSON, C. A. and DURHAM, J. W. (1966). Guidebook for field trip to Precambrian–Cambrian succession, White–Inyo Mountains, California. Mimeographed.

NOLAN, T. B. (1929). Notes on the stratigraphy and structure of the northwest portion of the Spring Mountains. *Am. J. Sci.* **17**, 461.

NOLAN, T. B. (1935). The Gold Hill mining district, Utah. *Profess. Paper U.S. Geol. Surv.* **177**.

NOLAN, T. B. and others (1956). The stratigraphic section in the vicinity of Eureka, Nevada. *Profess. Paper U.S. Geol. Surv.* **276**.

NOLAN, T. B. (1962). The Eureka mining district, Nevada. *Profess. Paper U.S. Geol. Surv.* **406**.

OKULITCH, V. J. (1943). North American Pleospongia. *Spec. Paper, Geol. Soc. Am.* **48**.

OKULITCH, V. J. (1951). A Lower Cambrian fossil locality near Addy, Washington. *J. Paleont.* **25**, 405.

OKULITCH, V. J. (1954). Archaeocyatha from the Lower Cambrian of Inyo County, California. *J. Paleont.* **28**, 293.

OLSON, R. H. (1956). Geology of the Promontory Range. *Utah Geol. Soc. Guidebook* **11**, 41.

ORIEL, S. S. (1964). Brigham, Langston, and Ute formations in Portneuf Range, southeastern Idaho (Abs.). *Program, Rocky Mountain section, Geol. Soc. Am. Mtng.* **35**.

ORIEL, S. S. (1965). Geology of the southwestern 1/4 Bancroft, quadrangle, Idaho. *U.S. Geol. Surv. Min. Fuels Inv. Map* **MF-299**.

PACK, F. J. (1906a). Geology of Pioche, Nevada and vicinity. *Columbia Univ. Sch. Mines. Quart.* **27**, 285.

PACK, F. J. (1906b). Cambrian fossils from the Pioche Mountains, Nevada. *J. Geol.* **14**, 290.

PALMER, A. R. (1954). An appraisal of the Great Basin Middle Cambrian trilobites described before 1900. *Profess. Paper U.S. Geol. Surv.* **264-D**, 55.

PALMER, A. R. (1956). The Cambrian system of the Great Basin in western United States, *in* RODGERS, JOHN, ed. *El. Sistema Cambrico, su paleogeografia y el problema de su base—symposium, Pt.* **2**: *Int. Geol. Cong. 20th*, Mexico, 663.

PALMER, A. R. (1957). Ontogenetic development of two olenellid trilobites. *J. Paleont.* **31**, 105.

PALMER, A. R. (1958). Morphology and ontogeny of a Lower Cambrian ptychopariond trilobite from Nevada. *J. Paleont.* **32**, 154.

PALMER, A. R. (1960a). Some aspects of the early Upper Cambrian stratigraphy of White Pine County, Nevada, and vicinity. *Intermountain Assoc. Pet. Geol. Guidebook to Geol. of east central Nev.* 53.

PALMER, A. R. (1960b). Early Late Cambrian stratigraphy of the United States (Abs.). *J. Wash. Acad. Sci.* **50**, 8.

PALMER, A. R. (1960c). Identification of the Dunderberg shale of Late Cambrian age in the eastern Great Basin. *Profess. Paper U.S. Geol. Surv.* **400-B**, 289.

PALMER, A. R. (1960d). Trilobites of the Upper Cambrian Dunderberg shale in the Eureka district, Nevada. *Profess. Paper U.S. Geol. Surv.* **334-C**, 53.

PALMER, A. R. (1962a). Comparative ontogeny of some opisthoparian, gonatoparian, and proparian trilobites. *J. Paleont.* **36**, 87.

PALMER, A. R. (1962b). *Glyptagnostus* and associated trilobites in the United States. *Profess. Paper U.S. Geol. Surv.* **374-F**.

PALMER, A. R. (1964). An unusual Lower Cambrian trilobite fauna from Nevada. *Profess. Paper U.S. Geol. Surv.* **483-F**.

PALMER, A. R. (1965a). The Biomere—a new kind of biostratigraphic unit. *J. Paleont.* **39**, 149.

PALMER, A. R. (1965b). Trilobites of the Late Cambrian Pterocephaliid Biomere in the Great Basin, United States. *Profess. Paper U.S. Geol. Surv.* **493**.

PALMER, A. R. (1967). Cambrian trilobites of east-central Alaska. *Profess. Paper U.S. Geol. Surv.* **559-B**.

PALMER, A. R. and HAZZARD, J. C. (1956). Age and correlation of Cornfield Springs and Bonanza King formations in southeastern California and southern Nevada. *Bull. Am. Assoc. Pet. Geol.* **40**, 2494.

PALMER, A. R. and STEWART, J. H. (1968). A Paradoxidid trilobite from Nevada. *J. Paleont.* **42**, 177.

PARK, C. F., Jr. and CANNON, R. S. Jr. (1943). Geology and ore deposits of the Metaline quadrangle, Washington. *Profess. Paper U.S. Geol. Surv.* **202**.

RASETTI, FRANCO (1944). Upper Cambrian trilobites from the Levis conglomerate (Quebec). *J. Paleont.* **18**, 229.

RASETTI, FRANCO (1951). Middle Cambrian stratigraphy and faunas of the Canadian Rocky Mountains (Alberta–British Columbia). *Smiths. Misc. Colln.* **116**, no. 5.

REBER, S. J. (1952). Stratigraphy and Structure of the south-central and northern Beaver Dam mountains, Utah. *Guidebook to Geol. of Utah* 7, 101.

RESO, ANTHONY (1963). Composite columnar section of exposed Paleozoic and Cenozoic rocks in the Pahranagat Range, Lincoln County, Nevada. *Bull. Geol. Soc. Am.* 74, 901.

RESSER, C. E. (1928). Cambrian fossils from the Mojave desert. *Smiths. Misc. Colln.* 81, no. 2.

RESSER, C. E. (1935). Nomenclature of some Cambrian trilobites. *Smiths. Misc. Colln.* 93, no. 5.

RESSER, C. E. (1936). Second contribution to nomenclature of Cambrian trilobites. *Smiths. Misc. Colln.* 95, no. 4.

RESSER, C. E. (1937). Third contribution to nomenclature of Cambrian trilobites. *Smiths. Misc. Colln.* 95, no. 22.

RESSER, C. E. (1938). Fourth contribution to nomenclature of Cambrian fossils. *Smiths. Misc. Colln.* 97, no. 10.

RESSER, C. E. (1939a). The Spence shale and its fauna. *Smiths. Misc. Colln.* 97, no. 12.

RESSER, C. E. (1939b). The *Ptarmigania* strata of the northern Wasatch Mountains. *Smiths. Misc. Colln.* 98, no. 24.

RESSER, C. E. (1942). New Upper Cambrian trilobites. *Smiths. Misc. Colln.* 103, no. 5.

RICCIO, J. F. (1952). The Lower Cambrian Olenellidae of the southern Marble Mountains, California. *Bull. So. Calif. Acad. Sci.* 51, 25.

RICHARDSON, G. B. (1913). Paleozoic section in northern Utah. *Am. J. Sci., 4th ser.* 36, 406.

RIGBY, J. K. (1958). Geology of the Stansbury Mountains, eastern Tooele County, Utah. Utah Geol. Soc. *Guidebook to Geology of Utah* 13.

RIGBY, J. K. (1966). *Protospongia hicksi* Hinde from the Middle Cambrian of western Utah. *J. Paleont.* 40, 549.

RIGO, R. J. (1968). Middle and Upper Cambrian stratigraphy in the autochthone and allochthon of northern Utah. *Brigham Young Univ. Res. Stud. Geol.* 15, 31.

ROBERTS, R. J. (1964). Stratigraphy and structure of the Antler Peak quadrangle, Humboldt and Lander Counties, Nevada. *Profess. Paper U.S. Geol. Surv.* 459-A.

ROBERTS, R. J. HOTZ, P. E., GILLULY, JAMES, and FERGUSON, H. G. (1958). Paleozoic rocks of north-central Nevada *Bull. Am. Assoc. Pet. Geol.* 42, 2813.

ROBISON, R. A. (1960a). Some Dresbachian and Franconian trilobites of western Utah. *Brigham Young Univ. Res. Stud.* 7, no. 3.

ROBISON, R. A. (1960b). Lower and Middle Cambrian stratigraphy of the eastern Great Basin. *Intermountain Assoc. Pet. Geol. Guidebook to Geol. of east central Nev.*, 43.

ROBISON, R. A. (1964a). Late Middle Cambrian faunas from western Utah. *J. Paleont.* 38, 510.

ROBISON, R. A. (1964b). Middle-Upper Cambrian boundary in North America. *Bull. Geol. Soc. Am.* 75, 987.

ROBISON, R. A. (1964c). Upper Middle Cambrian stratigraphy of western Utah. *Bull. Geol. Soc. Am.* 75, 995.

ROBISON, R. A. (1965). Middle Cambrian eocrionoids from western North America. *J. Paleont.* 39, 355.

ROBISON, R. A. (1967). Ontogeny of *Bathyuriscus fimbriatus* and its bearing on affinites of corynexochoid trilobites. *J. Paleont.* 41, 213.

ROBISON, R. A. and PALMER, A. R. (1968). Revision of Cambrian stratigraphy, Silver Island Mountains, Utah. *Bull. Am. Assoc. Pet. Geol.* 52, 167.

ROSS, D. C. (1963). New Cambrian, Ordovician and Silurian formations in the Independence quadrangle, Inyo County, California. *Profess. Paper U.S. Geol. Surv.* 475-B, 74.

SCHAFFER, F. E. (1960). Geology of the Silver Island Mountains, Box Elder and Tooele Counties, Utah, and Elko County, Nevada. *Utah Geol. Soc. Guidebook to the Geology of Utah* 15.

SHARP, R. P. (1942). Stratigraphy and structure of the southern Ruby Mountains, Nevada. *Bull. Geol. Soc. Am.* 53, 647.

STAATZ, M. H. and CARR, W. J. (1964). Geology and mineral deposits of the Thomas and Dugway Ranges, Juab and Tooele Counties, Utah. *Profess. Paper U.S. Geol. Surv.* 415.

STEWART, J. H. (1965). Precambrian and Lower Cambrian strata in the Last Chance Range area, Inyo County, California, *in* COHEE, G. V. and WEST, W. S., Changes in stratigraphic nomenclature by the U.S. Geological Survey, 1964. *Bull. U.S. Geol. Surv.* 1224-A, 60.

STEWART, J. H. (1966). Correlation of Lower Cambrian and some Precambrian strata in the southern Great Basin, California and Nevada. *Profess. Paper U.S. Geol. Surv.* 550-C, 66.

STEWART, J. H. (1967). Possible large right-lateral displacement along fault and shear zones in the Death Valley–Las Vegas area, California and Nevada. *Bull. Geol. Soc. Am.* 78, 131.

STEWART, J. H. and BARNES, HARLEY (1966). Precambrian and Lower Cambrian formations in the Desert Range, Clark County, Nevada, *in* COHEE, G. V. and WEST, W. S., Changes in stratigraphic nomenclature by the U.S. Geological Survey, 1965. *Bull. U.S. Geol. Surv.* **1244-A**, 35.

STEWART, J. H. and PALMER, A. R. (1967). Callaghan window—a newly discovered part of the Roberts Thrust, Toiyabe Range, Lander County, Nevada. *Profess. Paper U.S. Geol. Surv.* **575-D**, 56.

STOYANOW, ALEXANDER (1958). *Sonoraspis* and *Albertella* in the Inyo Mountains, California. *Bull. Geol. Soc. Am.* **69**, 347.

STOYANOW, ALEXANDER and SUSUKI, TAKAEO (1955). Discovery of *Sonoraspis* in southern California. *Bull. Geol. Soc. Am.* **66**, 467.

TAYLOR, M. E. (1966). Precambrian Mollusc-like fossils from Inyo County, California. *Science* **153**, 3732.

TURNER, H. W. (1902). A sketch of the historical geology of Esmeralda County, Nevada. *Am. Geol.* **29**, 261.

ULRICH, E. O. and COOPER, G. A. (1938). Ozarkian and Canadian Brachiopoda. *Spec. Paper, Geol. Soc. Am.* **13**.

WALCOTT, C. D. (1884). Paleontology of the Eureka District. *Mon. U.S. Geol. Surv.* **8**.

WALCOTT, C. D. (1886). Second contribution to the studies on the Cambrian faunas of North America. *Bull. U.S. Geol. Surv.* **30**.

WALCOTT, C. D. (1891a). Correlation papers—Cambrian. *Bull. U.S. Geol. Surv.* **81**.

WALCOTT, C. D. (1891b). The North American Continent during Cambrian time. *U.S. Geol. Surv., 12th Ann. Rept.* **523**.

WALCOTT, C. D. (1908a). Cambrian trilobites. Cambrian Geology and Paleontology I. *Smiths. Misc. Colln.* **53**, 13.

WALCOTT, C. D. (1908b). Cambrian sections of the Cordilleran area, Cambrian Geology and Paleontology I, *Smiths. Misc. Colln.* **53**, 167.

WALCOTT, C. D. (1910). *Olenellus* and other genera of the Mesonacidae. Cambrian Geology and Paleontology I. *Smiths. Misc. Colln.* **53**, 231.

WALCOTT, C. D. (1912). Cambrian Brachiopoda. *Mon. U.S. Geol. Surv.* **51**.

WALCOTT, C. D. (1916a). Cambrian trilobites. Cambrian Geology and Paleontology III. *Smith. Misc. Colln.* **64**, 157.

WALCOTT, C. D. (1916b). Cambrian trilobites. Cambrian Geology and Paleontology III. *Smiths. Misc. Colln.* **64**, 303.

WALCOTT, C. D. (1920). Middle Cambrian spongiae. *Smiths. Misc. Colln.* **67**, no. 6.

WALCOTT, C. D. (1925). Cambrian and Ozarkian trilobites. Cambrian Geology and Paleontology V., *Smiths. Misc. Colln.* **75**, 61.

WASHBURN, R. H. (1966). Paleozoic stratigraphy in Toiyabe Range, southern Lander County, Nevada (Abs.). *Program, Rocky Mtn. Sec. Geol. Soc. Am. Mtng.*, 57.

WESTGATE, L. G. and KNOPF, ADOLPH (1927). Geology of Pioche, Nevada and vicinity. *Trans. Am. Inst. Min. and Met. Eng.* **75**, 816.

WESTGATE, L. G. and KNOPF, ADOLPH (1932). Geology and ore deposits of the Pioche district, Nevada. *Profess. Paper U.S. Geol. Surv.* **171**.

WHEELER, H. E. (1940). Revisions of the Cambrian stratigraphy of the Pioche District, Nevada. *Bull. Nev. Univ.* **34**.

WHEELER, H. E. (1943). Lower and Middle Cambrian stratigraphy of the Great Basin area. *Bull. Geol. Soc. Am.* **54**, 1781.

WHEELER, H. E. (1948). Late Pre-Cambrian–Cambrian stratigraphic cross-section through southern Nevada. *Bull. Nev. Univ.* **42**.

WHEELER, H. E. and Lemmon, D. M. (1939). Cambrian formations of the Eureka and Pioche districts, Nevada. *Bull. Nev. Univ.* **33**.

WILLIAMS, J. S. (1948). Geology of the Paleozoic rocks, Logan quadrangle, Utah. *Bull. Geol. Soc. Am.* **59**, 1121.

WILLIAMS, J. S. and MAXEY, G. B. (1941). Cambrian section in the Logan quadrangle, Utah and vicinity. *Am. J. Sci.* **239**, 276.

WOODWARD, L. A. (1968). Lower Cambrian and Upper Precambrian strata of Beaver Mountains, Utah. *Bull. Am. Assoc. Pet. Geol.* **52**, 1279.

YATES, R. G. (1964). Geologic map and sections of the Deep Creek area, Stevens and Pend Oreille Counties, Washington. *U.S. Geol. Surv. Misc. Geol. Inv. Map.* **I-412**.

YATES, R. G. and ROBERTSON, J. F. (1958). Preliminary geologic map of the Leadpoint quadrangle, Stevens County, Washington. *U.S. Geol. Surv. Min. Res. Map* **MF-137**.

YOCHELSON, E. L. (1967). Charles Doolittle Walcott. *Bib. Mem.* 39.

YOCHELSON, E. L. and others (1965). Stratigraphic distribution of the Late Cambrian mollusk *Matthevia* Walcott, 1885. *Profess. Paper U.S. Geol. Surv.* **525-B**, 73.

YOUNG, J. C. (1955). Geology of the southern Lakeside Mountains, Utah. *Bull. Utah Geol. Min. Surv.* **56**.

YOUNG, J. C. (1960). Structure and stratigraphy in north central Schell Creek Range. *Intermountain Assoc. Pet. Geol. Guidebook to geology of east central Nevada*, 158.

THE CAMBRIAN OF THE CRATON OF THE UNITED STATES

Christina Lochman-Balk

New Mexico Institute of Mining and Technology, Socorro, New Mexico

Contents

1. Introduction

Cambrian deposits on the craton consist largely of Upper Cambrian marine sediments lying upon Pre-Cambrian basement rocks. Two exceptions are (1) the eastern and western edges of the craton where marine Middle Cambrian deposits underlie the Upper Cambrian, and (2) two areas within the craton where marine and non-marine sediments and volcanics of Early? and/or Middle? Cambrian age intervene between the basement and the Upper Cambrian (Fig. 1).

The Upper Cambrian sections of the craton are thin (average 300 metres or less), contain many sedimentary breaks (disconformities) recognized by lithological and by faunal criteria, and show internal evidence of slow deposition and much reworking. Three major marine transgressions and regressions occurred across the craton during the Late Cambrian and many sections show numerous cyclic oscillations of sea level (OSTROM, 1965; TASCH, 1951a, 1951b). The Cambrian section, during the entire Phanerozoic, has not undergone deep burial, severe structural deformation, or metamorphism

79

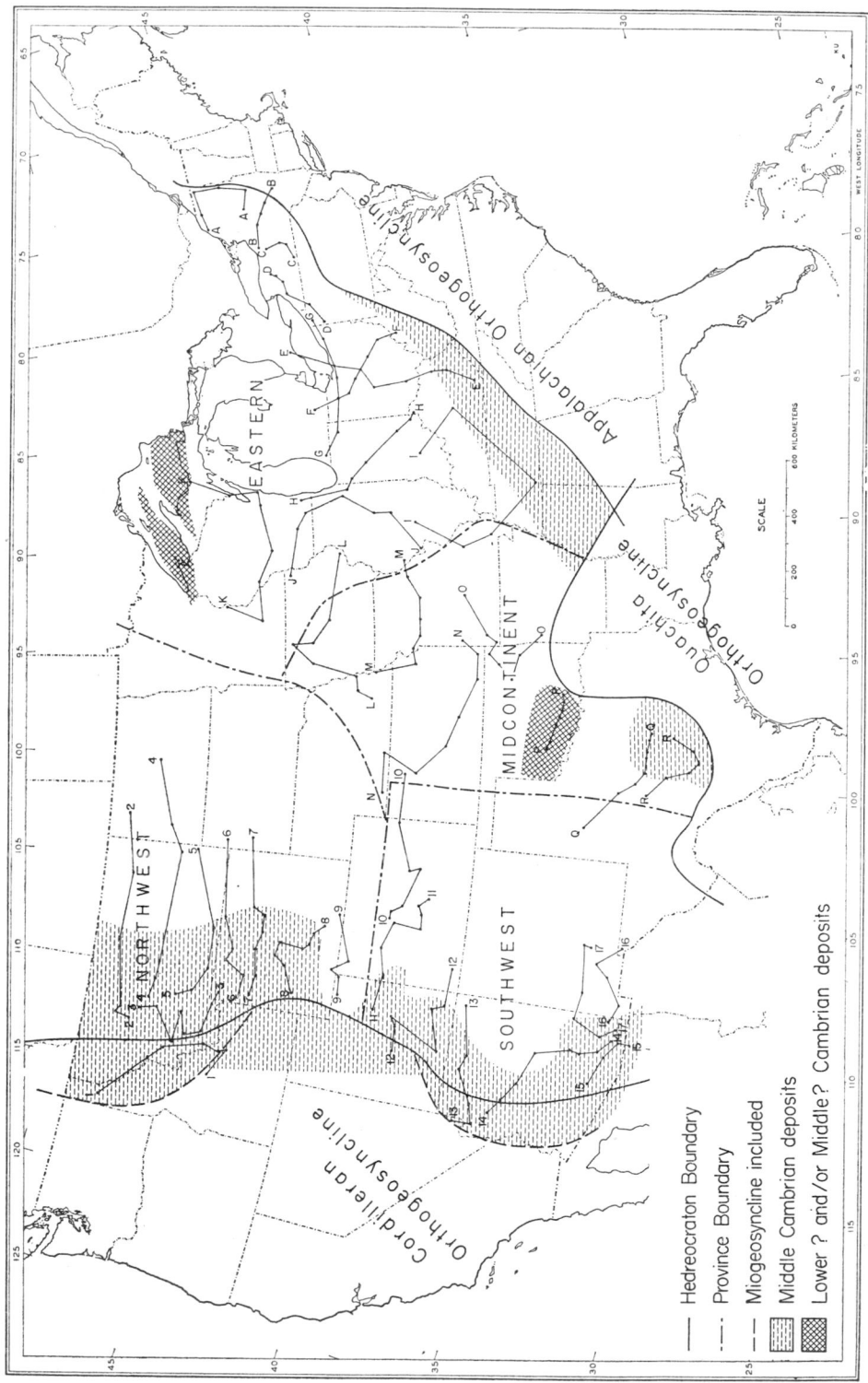

Fig. 1. Boundaries of Craton and its subdivisions; location of sections.

on a regional scale. The deposits were eroded during several periods of the Palaeozoic, and their degradation continues to the present. At widely scattered localities hydrothermal emanations have introduced valuable metallic ore deposits into the sediments.

Coarse to fine sandstones, siltstones, and sandy and pure carbonates are the common rock types and intergrade both vertically and laterally. Shale rarely occurs in significant thicknesses but is an argillaceous impurity in the siltstones and carbonates. Glauconite is common and widespread. Intraformational pebble conglomerates and oolites are abundant. Many carbonate units are dolostones and recent studies consider the dolomitization to be early diagenetic (HOWE, 1966, 1968).

The boundaries of the hedreocraton were defined by KAY (1947, 1951) (and see Fig. 1) and lie generally on the present 2,000 foot (approximately 650 metre) total Cambrian isopach. The western boundary coincides with the Wasatch Line, the tectonic boundary between strongly deformed sediments of the miogeosyncline and weakly deformed sediments. Rapid increase in thickness of Cambrian sediments west of this line suggests that it actually is close to the position of the Cambrian hingeline. The eastern border coincides with several major thrust fault fronts—Logans Line in the north to the Cumberland Block in the south. Only in Pennsylvania may the present boundary lie close to the position of the Cambrian hingeline. Miogeosynclinal deposits have been moved north and west over the hingeline and onto the craton. In south-central United States the border lies along the Ouachita Front (FLAWN and others, 1962), a zone of regional faulting and metamorphism. Along this zone orthogeosynclinal deposits overlie and conceal the craton deposits.

The stratigraphical sections are grouped in four regional provinces for presentation but all provincial boundaries are arbitrary (Fig. 1).

Concepts used in recognizing lithostratigraphical and biostratigraphical units and in interpreting the environmental and paleogeographical significance of their distribution currently differ among geologists. The recommendations embodied in the American Code of Stratigraphic Nomenclature (1961) are followed. The observable stratal unit is the lithostratigraphical unit—*the formation*. It may contain hiatuses representing intervals when sedimentation ceased upon the sea floor, either because materials failed or the interface had reached the profile of equilibrium. The boundaries of the formation have no time significance. In Upper Cambrian sections on the craton the base of a formation is often time-transgressive and the top time-regressive.

The faunal zones used are *assemblage zones* based on a number of trilobite genera which are associated at different localities. The succession of trilobite assemblages listed on page 88 can be recognized throughout the craton and into the adjoining miogeosynclines. The contemporaneous existence of two zonal assemblages, postulated by PALMER (1965) in the biomere concept, cannot be demonstrated upon the craton. Each assemblage occurs in the same stratigraphical position relative to other assemblages. Certain genera of each assemblage are cosmopolitan throughout the United States, but others show geographical restrictions which appear to have been caused by ecological controls. Studies have not yet designated facies faunules. No faunal zone is characterized by a particular species (as in European practice). This situation reflects an early 20th century concept that trilobite species had very restricted geographical ranges, which rendered them useless for cratonic correlation. The concept is no longer accepted and the earlier described faunas must be restudied.

Boundaries of the faunal zones are diachronous across the craton from the edge of the outer shelf to interior cratonic sites. The faunules demonstrate that in interior cratonic sites significant lacunae occur between zones whereas near the borders of the craton shorter lacunae occur or a transition faunule may be found. The standard Upper Cambrian section of the United States (Fig. 2) is in the Upper Mississippi Valley, an interior cratonic position. The faunal assemblage zones were established from this section (HOWELL and others, 1944; RASCH, 1952). Their boundaries are sharp and are depositional breaks usually demonstrable by physical criteria. Toward the borders of the craton a more complete sedimentary sequence occurs with older and younger faunules of each assemblage zone. Zonal boundaries must then be arbitrarily designated. Both exact coincidence and marked non-coincidence of lithostratigraphical and biostratigraphical boundaries may occur. The miogeosynclinal sections of eastern Nevada (PALMER, 1965) have furnished the most complete Upper Cambrian faunal succession known to date, especially for the Dresbachian.

Detailed petrological examination of Cambrian sections for interpreting the environments of deposition has just begun (OSTROM, 1965; KULICK, 1965; LEBAUER, 1965; HOWE, 1966, 1968). These studies indicate widespread shallow submarine platforms subjected to sea level fluctuations, a situation encouraging cosmopolitan faunal assemblages, but affording local ecological variations favouring the development of facies faunules.

2. Tectonic framework of the craton

The sites of deposition, their ecologies, and the thickness and lithologies of the sections were controlled by the degree of tectonism undergone by the tectonic elements of the craton. The tectonic pattern of the Cambrian persisted into and through the Lower Ordovician. This major tectonic cycle, termed the Sauk Sequence (SLOSS, 1950), began with the initial Middle Cambrian transgression upon the craton and ended with the Early Ordovician regression, the upwarping of the craton, and the forming of most of the important Palaeozoic tectonic elements.

Broad stable shelves which were subjected to only sporadic minor downwarping occupied most of the craton. On the western border a narrow linear unstable shelf merged into several tectonically negative miogeosynclinal basins, and was interrupted in the north by the weak positive dome, Montania. On the eastern border the unstable shelf is abnormally narrow in the north, because much of it was concealed or destroyed by later thrusting (DENNISON and WOODWARD, 1963). In the south it appears wider south of the central Kentucky fault scarps. The rapid increase in thickness of the Cambrian sections on the unstable shelf (Sections E-E, I-I*) led WOODWARD (1961) to postulate an Early Cambrian steep continental slope (coastal declivity) or a fault scarp, from Pennsylvania through West Virginia.

Deep drilling has revealed the presence of many large normal fault systems in the basement rocks (RUDMAN, SUMMERSON, and HINZE, 1965). The largest system is the east-west scarp from western Illinois to West Virginia: the Rough Creek Fault Zone (Fig. 3). Marked thickening of Late Cambrian sediments across the structure indicates movement during the Cambrian. RODGERS (1963) recognized the scarp as part of his Adirondack–Ozark Hingeline. The position of shorter scarps may show relationship to the later Palaeozoic basins.

* A series of comparative sections of the Cambrian of the Craton is given in Chapter 12 (pages 139–157). These are referred to in the text as Sections A-A to R-R and Sections 1-1 to 17-17. All lines of section are indicated on Fig. 1 (page 80).

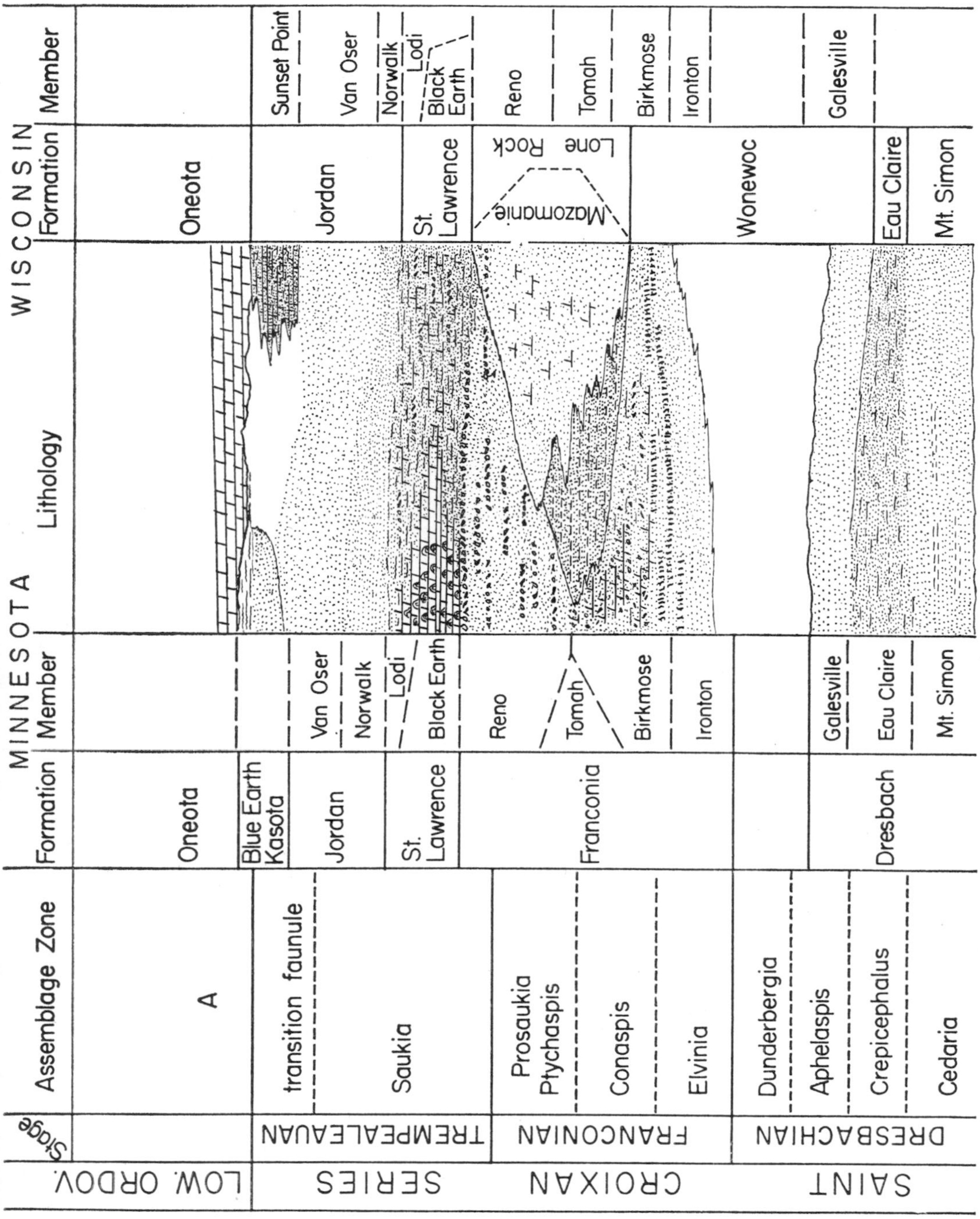

FIG. 2. Saint Croixan type area—modern nomenclature of lithic units. For key to lithological symbols see page 140.

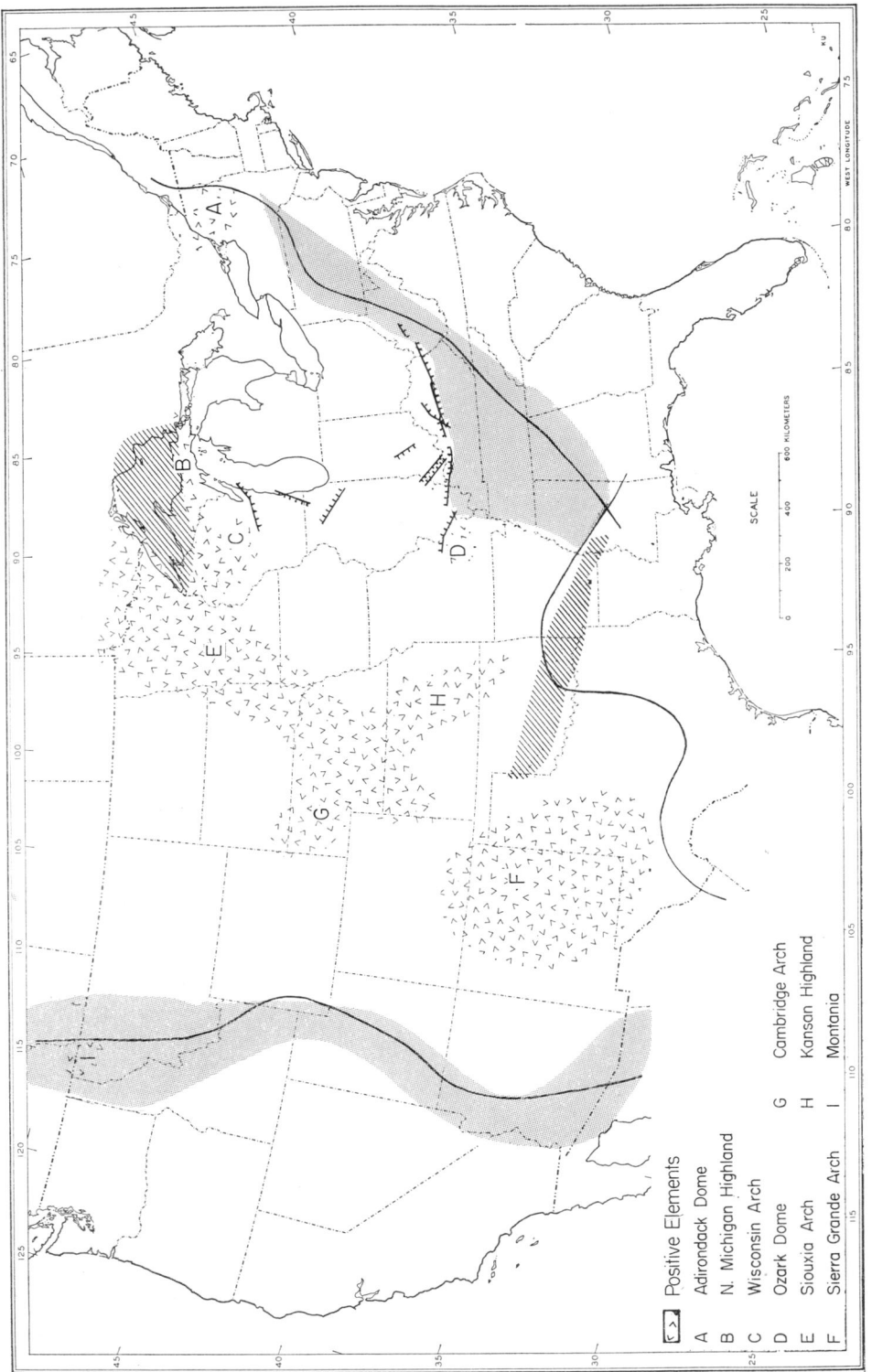

FIG. 3. Tectonic Framework of Craton during the Cambrian.

Positive Elements

A Adirondack Dome
B N. Michigan Highland
C Wisconsin Arch
D Ozark Dome
E Siouxia Arch
F Sierra Grande Arch
G Cambridge Arch
H Kansan Highland
I Montania

Stable Shelf
Unstable Shelf
Early Cambrian Cratonic Basins
Fault Scarp

There were no intracratonic basins during the Upper Cambrian. The two basins recently recognized (Fig. 1) originated in the late Pre-Cambrian and persisted into the Early Cambrian or the early Late Cambrian (Lake Superior Basin). Both had become tectonically neutral and functioned as part of the shelf area during the Late Cambrian.

Most of the positive elements upon the craton were inherited from Late Pre-Cambrian elements, and many have remained positive to the present. The Adirondack Arch, the Ozark Dome, the Wisconsin Highlands, and the Kansas Highlands were inactive during the Upper Cambrian. They were much reduced by subaerial erosion and mostly covered by Early Ordovician seas. The Northern Michigan Highlands were deeply eroded during the Cambrian and by the beginning of the Franconian had become part of the stable shelf (HAMBLIN, 1957). The Transcontinental Arch was a major tectonic element trending south-west across the craton from the Canadian border. It was separated into Siouxia and Sierra Grande by a sag developing in the Trempealeauan. A north-west trending branch was the Cambridge Arch. These three elements were moderately positive during the Upper Cambrian, especially the Franconian.

3. Lower? and/or Middle? Cambrian deposits

In two separate basins, apparently inherited from the late Pre-Cambrian tectonic pattern, sediments of Early Cambrian ? age occur (Fig. 1). In both areas the section is known from surface exposures as well as in subsurface.

Jacobsville–Bayfield Group: The Bayfield and Hinckley Formations are clastic units frequently encountered in subsurface in Wisconsin and Minnesota. The Jacobsville Formation, exposed along the south shore of Lake Superior, is a clastic unit (0–610 metres thick) of quartz sand grains (75%), feldspar (15%), pyroxene, amphibole, and rock fragments (10%). The red to red-brown colour is universally mottled by white spots, blotches, and streaks. HAMBLIN (1958) recognized four lithofacies: (1) lenses of conglomerate, especially common at the base; (2) a thick medium-grained sandstone with lenticular bedding—the dominant facies; (3) a massive sandstone in 2 to 6 metre beds; (4) a thin, less common, siltstone and shale. Hamblin concluded that the Jacobsville represents non-marine deposition during the Early Cambrian in a downwarping basin which persisted from Late Keweenawan to Dresbachian time. The Jacobsville clastics are not genetically related either in their composition or their dispersal drainage pattern to the Late Keweenawan clastics. The Jacobsville appears at several localities to be unconformable upon the Late Keweenawan Freda Sandstones, and the Dresbachian Chapel Rock Member lies upon it with an angular unconformity.

Southern Oklahoma Geosyncline: In the region of the Wichita Mountains HAM, DENISON, and MERRITT (1964) recognized four groups of igneous and sedimentary rocks for which radiometric ages of 550 m.y. to 525 m.y. have been obtained. Stratigraphical and structural relationships of the four groups suggest deposition in an intracratonic basin extending across southern Oklahoma during Early and Middle Cambrian time. The two youngest groups are exposed at the surface and also found in subsurface. The older groups are known only from the subsurface.

The Carlton Group (525 m.y.) consists of rhyolite flows and tuffs over 1,370 metres thick and represents an extensive volcanic field intruded by the Wichita Granite Group of the same age. The underlying Navajoe Mountain Group (550 m.y.) has a drilled thickness of 320 metres and consists of interbedded basalt, spilite, andesite, and altered palagonite tuffs, partly of marine deposition. It is considered the extrusive equivalent of the Raggedy Mountain Gabbro Group (535 m.y.), which consists of olivine gabbro, anorthosite, and diorite in a layered intrusion (possibly 3,050 metres thick) injected as an elongate lens into the Tillman Group, of metagreywacke, argillite, quartzite, and bedded cherts in a sequence at least 4,570 metres thick. This sedimentary sequence is identified as of marine eugeosynclinal deposits of probable Early Cambrian age.

4. Middle Cambrian faunal zones

In the Middle Cambrian five faunal zones are recognized, ascending: *Plagiura-Poliella*, *Albertella*, *Glossopleura*, *Bathyuriscus-Elrathina*, and *Bolaspidella* (LOCHMAN-BALK and WILSON, 1958). The faunal assemblages of these zones are known primarily through the studies of RASETTI (1957) and ROBISON (1964) on the miogeosynclinal assemblages of the western United States and Canada. Diagnostic genera of the *Albertella Glossopleura*, *Bathyuriscus-Elrathina*, and *Bolaspidella* Zones occur in the outer shelf sections of Montana, Wyoming, and Arizona. Fossils are quite abundant in some of the Montana and Wyoming sections, but the material is largely undescribed. DEISS (1939) described a few of the Montana faunules and RESSER (1945) described some of the species obtained from the Grand Canyon sections.

In the southern portion of the Eastern Region late Middle Cambrian fossils, *Coelaspis* and *Micromitra* (identified by C. LOCHMAN, 1947), have been obtained from one well (E. W. BEELER #1) in south-central Tennessee. It is probable that the basal sands and overlying shales and siltstones of the deep wells of south-central and south-eastern Kentucky are also latest Middle Cambrian in age, but no fossils have been obtained from them.

5. Saint Croixan series: the standard Upper Cambrian section

E. EMMONS (1838) described the rocks of Saint Lawrence County, New York, as (descending) calciferous sand rock, transition limestone, sandstone of Potsdam, and primary strata (see Section A-A). By 1847 the New York Geological Survey had adopted the Potsdam Sandstone (Potsdam Epoch) for the lowest stratified fossiliferous rock, and its meagre fauna was regarded as older than any fauna in the Silurian System of Great Britain. From 1847 to 1863 HALL (1863) investigated these strata and their fossils in New York, Canada, Pennsylvania, Virginia, Iowa, Wisconsin, and Minnesota. In the latter states in 1850 he recognized the correlation of the Lower Magnesian Limestone with the Calciferous Sand Rock and the Potsdam Sandstone with the underlying Lower Sandstone. These strata had been carefully mapped and described during the D. D. OWEN survey from 1839 to 1850 along the bluffs of the upper Mississippi valley. OWEN (1852) accepted HALL's correlation and described in detail the lithology, thickness, and stratigraphical succession of these strata and a few of the fossils characteristic of each horizon. HALL (1863) under the title of 'Preliminary notice of the fauna of the Potsdam Sandstone' described 42 species from the upper Mississippi valley and 2 from New York. The sequence of sandstones in Minnesota and Wisconsin thus became the best known Cambrian strata lithologically and faunally. The sands are locally very fossiliferous with sandstone moulds of disarticulated trilobites and the calcareous-phosphatic shells of the Lingulacea. But the section is thin and contains numerous physical and faunal hiatuses.

WINCHELL (1873, 1888) named all the sandstones and shales under the Lower Magnesian series and above his Potsdam Sandstone (the Bayfield Group, see page 85) the Saint Croix Sandstone from the section in Saint Croix Valley. HALL and SARDESON (1895) established the Lower Magnesian Series to extend from the Shakopee dolomite to the base of the Saint Lawrence dolomites and shales, and referred the underlying non-dolomitic sandstones to the Potsdam Sandstone. BERKEY (1897) in the Saint Croix Dalles region noted that the Saint Croix Formation of WINCHELL extended from the top of the Jordan Sandstone to at least the base of the Dresbach Sandstone. WINCHELL (1900) included the Jordan and Saint Lawrence Formations at the top and the Dresbach at the base of his Saint Croix Series which overlies the Hinkley Sandstone (Bayfield Group, page 85). ULRICH and SCHUCHERT (1902), and SCHUCHERT (1910) used the term 'Saint Croix transgression', emphasizing the diastrophic significance of this first widespread Palaeozoic transgression upon the craton. ULRICH (1911) placed the term 'Saint Croixan or Upper Cambrian' on his correlation chart which may be considered the original designation. Later the upper part of the section was involved in the 'Ozarkian problem' and geologists were slow to recognize the numerous lateral facies changes and the varied physical environments they reflected. The large fossil collections have not yet been thoroughly studied. STAUFFER, SWARTZ, and THIEL (1939) gave an historical survey of the nomenclature of the various formational units recognized in Minnesota and Wisconsin and attempted to co-ordinate the nomenclature of the two states. BERG (1953, 1954); NELSON (1951, 1956); BELL, FENIAK, and KURTZ (1952); and BELL, BERG, and NELSON (1956) described some of the faunal assemblages and the modern stratigraphy of the type area. The term 'Croixan' or 'Croixian' is often used loosely for the series (KEROHER, 1965).

Figure 2 shows the modern nomenclature. The base rests unconformably on pre-Late Cambrian red clastics (Bayfield Group) or Pre-Cambrian basalts, metamorphics, or granite. The upper boundary is placed at the disconformity at the base of the overlying Oneota or at the appearance of *Symphysurina* in a conformable sequence of sandstones.

6. Upper Cambrian faunal zones

The Upper Cambrian begins with the appearance of the *Cedaria* trilobite assemblage and ends with the disappearance of the Saukid genera. The beginning of the Ordovician occurs at the appearance of the *Symphysurina* assemblage. The Upper Cambrian–Lower Ordovician boundary is arbitrary at many sites on the craton. The regression at the end of the Trempealeauan affected mainly the borders of the positive elements, and shallow seas remained over the outer shelves. A transitional trilobite assemblage occurs at this position, but the graptolite and brachiopod species cross the arbitrary boundary without change (LOCHMAN-BALK and WILSON, 1967).

Radiometric age dates of 570 m.y., 515 m.y., and 495 m.y. (COWIE, 1964) have been accepted for the base of the Lower Cambrian, the base of the Upper Cambrian, and the top of the Upper Cambrian respectively.

Eight faunal assemblage zones are recognized in Upper Cambrian strata. Seven are represented faunally on the craton and the position of the *Dunderbergia* Zone is occupied, except in Texas, by a disconformity. The *Dicanthopyge* Zone and *Prehousia* Zone named by PALMER in Great Basin sections were correlated by him (PALMER, 1965b) with the *Aphelaspis* Zone in Texas. The faunal zones were assigned to three stages

(HOWELL and others, 1944) based upon and named from the three lithostratigraphical units recognized in the type area. At the base of each there is a marine transgression, and at the top of each a regression. The statement of the British Mesozoic Committee that 'the base of each stage should be regarded as fixed for all time' is followed in placing the *Dunderbergia* Zone in the Dresbachian Stage (PALMER, 1965b). The base of the Dresbachian Stage is the transgression which brings the *Cedaria* Zone assemblage into a region, the base of the Franconian the transgression bringing the *Elvinia* Zone assemblage, and the base of the Trempealeauan is the appearance of the Saukid Zone assemblage.

Range charts, Figs. 4, 5, 6, and 7, show the total range of each genus upon the craton by combining the teilzones of each from all described localities. For many genera this combined teilzone represents the biozone of the genus. However, there are some genera which appear in association with the immediately earlier faunal assemblage in extra-cratonic (miogeosynclinal) sites. *Modocia* appears first in the *Bolaspidella* Zone, then invades the craton in the *Cedaria* Zone, and finally reaches the interior states in the *Crepicephalus* Zone. *Lonchocephalus* and *Welleraspis* appear in the late *Cedaria* Zone and follow a quiet water, inner sublittoral facies onto the craton during the *Crepicephalus* Zone (ROBISON, 1964b). *Apachia* appears in the Great Basin sections with the *Dunderbergia* assemblage (PALMER, 1965b). During the Trempealeauan *Acheilops*, *Leiocoryphe*, *Leiobienvillia*, *Theodenisia*, *Triarthropsis*, *Apatokephaloides*, *Idiomesus*, and *Heterocaryon*, all genera common in the late Franconian and Trempealeauan *Hungaia* assemblage of extracratonic sites, invade the outer shelf sites (Llano region, Texas: WINSTON and NICHOLLS, 1967); but only *Acheilops* and *Triarthropsis* penetrate to the upper Mississippi valley where the cratonic Saukids dominate the assemblages.

The range charts reveal the abrupt faunal changes which occur on the craton at the base of the *Aphelaspis* Zone, the *Elvinia* Zone, the *Conaspis* Zone, the *Ptychaspis-Prosaukia* Zone, and the *Saukia* Zone. Several different phenomena produced these changes in generic composition.

The change from the *Crepicephalus* to the *Aphelaspis* assemblage saw the almost complete extinction of the older assemblage, with only *Blountia* and *Glaphyraspis* persisting briefly. This change is not reflected in the sedimentation at any cratonic site, and must be attributed to waters invading the shelf areas. The large and varied *Crepicephalus* assemblage is replaced abruptly by the small *Aphelaspis* assemblage, of one or two species each of only six genera. Most of the species are represented by a large

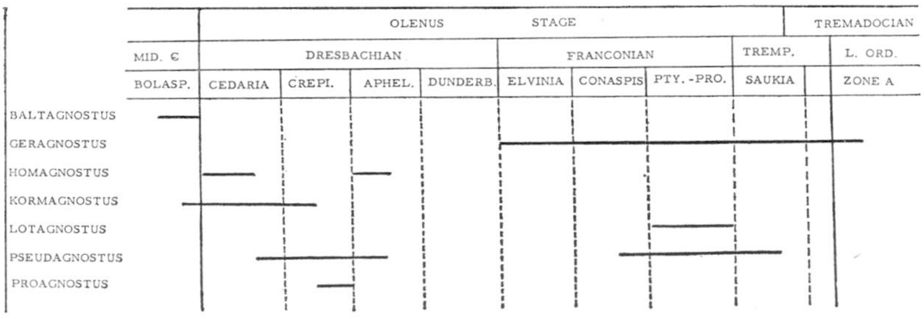

FIG. 4. Range of Upper Cambrian Agnostids on the Craton and correlation of North American and European stages.

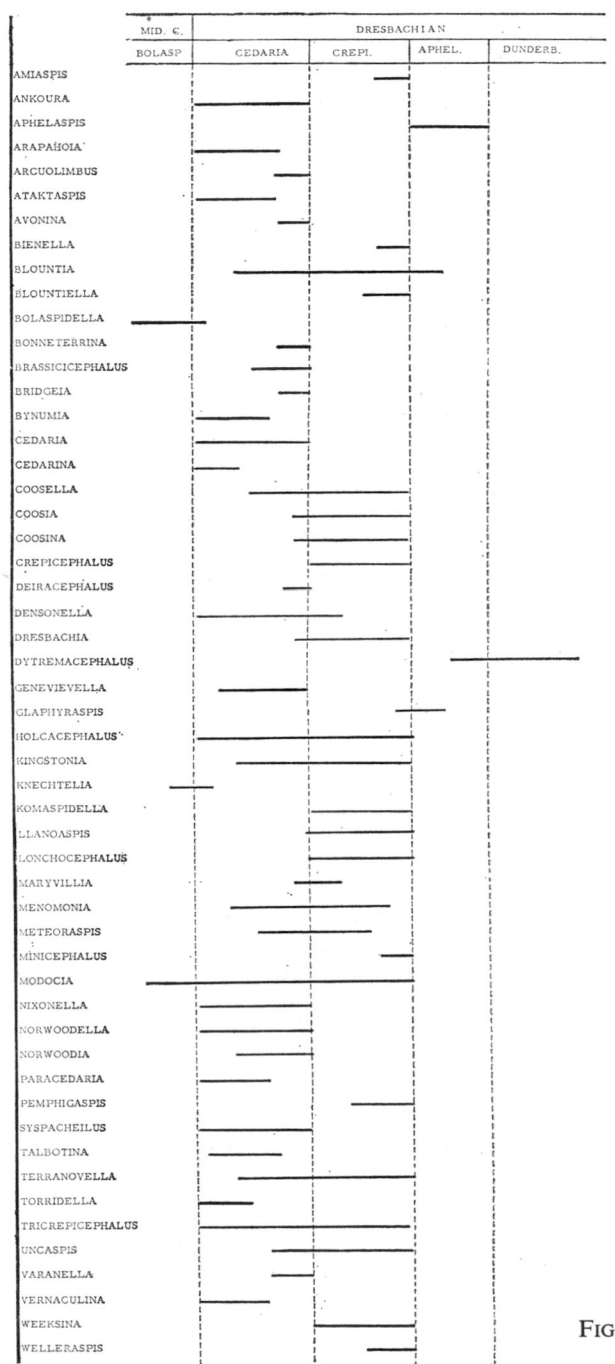

FIG. 5. Range of early Upper Cambrian trilobite genera on the Craton.

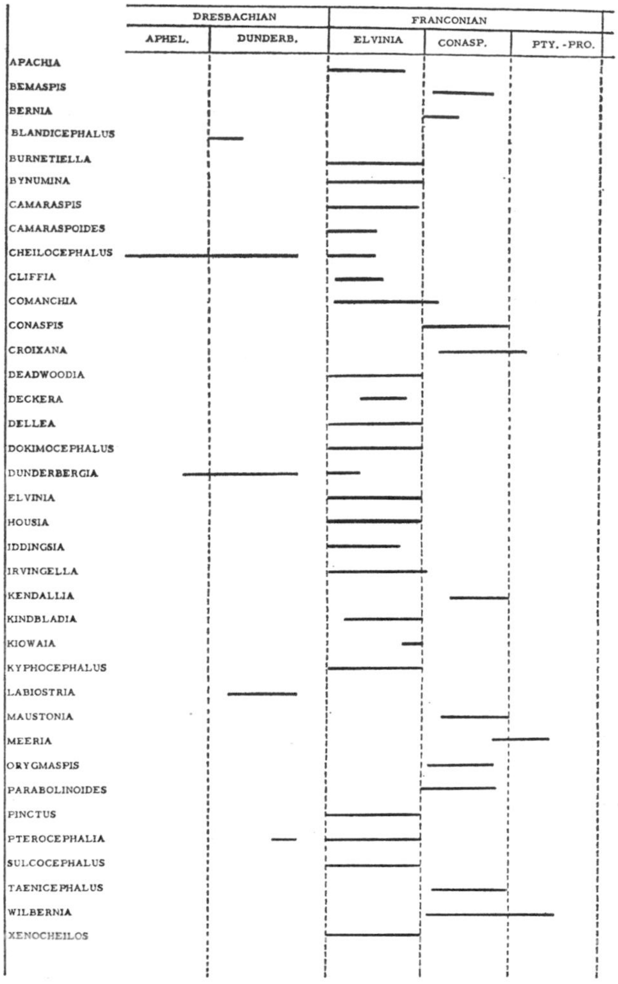

FIG. 6. Range of middle Upper Cambrian trilobite genera on the Craton.

number of individuals. Limestone coquina lenses of *Aphelaspis* cranidia and pygidia occur from Tennessee to Montana. The environmental change was so abrupt that only in the Llano district, Texas and north-eastern Tennessee miogeosynclinal sections is the suggestion of a transitional faunule known. Shortly after the appearance of the *Aphelaspis* assemblage upon the craton, a significant cratonic regression is reflected lithologically in the sediments. The marine regression moved progressively outward from the north central states, and it is only along the borders of the craton (Texas) that the *Dunderbergia* assemblage is recognized. The entire craton to its borders was finally emergent.

The abrupt appearance of the *Elvinia* Zone assemblage marks the Franconian marine trangression and the cratonward migration of Pterocephalid biofacies genera which had developed in extracratonic sites during *Aphelaspis* and *Dunderbergia* Zones time.

The sudden disappearance of the *Elvinia* Zone assemblage must also be attributed to relatively rapid environmental changes in the marine waters which did not affect the local sedimentation. In Wisconsin the two faunas may be separated by only a bedding plane in the sandstone sequence; and in the Williston basin Montana species of the two assemblages occur on each side of a 1 mm shale lamina (LOCHMAN, 1964). The early *Conaspis* faunules contain only three trilobite genera, represented by numerous individuals of one or two species, together with an abundance of several species of the articulate brachiopod, *Eoorthis*.

By late *Conaspis* Zone time the migrant stocks had become adapted to shelf conditions and a diversification into genera and species to fill the many unoccupied habitats began.

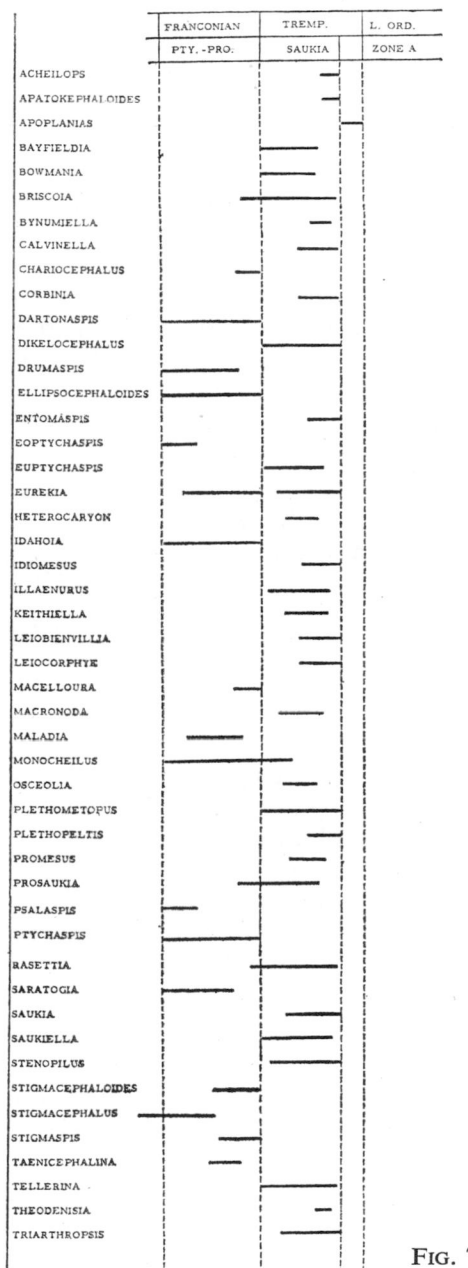

FIG. 7. Range of late Upper Cambrian trilobite genera on the Craton.

The base of the *Ptychaspis-Prosaukia* Zone lies near the start of this evolutionary burst (LOCHMAN-BALK and WILSON, 1958).

The boundary between the *Ptychaspis-Prosaukia* and *Saukia* Zones is arbitrary at most sites on the craton. The fauna is characterized by the continued diversification of the Conaspid stock in the appearance of genera of Dikelocephalidae, Saukidae, Eurekidae, and Ptychaspidae and the extinction of the Parabolinoididae. In later *Saukia* Zone time the assemblage becomes more distinctive by the migration into the outer shelf sites of numerous small genera from the extracratonic assemblages. This migration was related to the continued expansion of carbonate shoals across the shelf sites as the seas transgressed the remnants of the positive elements.

Facies faunules in the cratonic faunal assemblages have not yet been recognized. LOCHMAN (1968) finds evidence that *Lonchocephalus, Glaphyraspis, Welleraspis*, and *Komaspidella* are members of a faunule adapted to the warm quiet waters of shallow inner sublittoral lagoons. *Pemphigaspis* has been regarded as a planktonic or pseudo-planktonic genus (TASCH, 1952). The uncommon and sporadic occurrence of the Agnostid genera in the cratonic assemblages as compared to their abundance throughout the extracratonic faunules suggests a planktonic mode of life for them also. *Blountia* is found frequently in pockets in algal bioherms (LOCHMAN and HU, 1961) and is believed to have adapted to the high energy environment of the columnar (digitate) algal reefs (HOWE, 1966).

Intercontinental correlation has been attempted for many years especially between North America and Europe but rendered difficult and uncertain because of the marked dissimilarity between the craton (Pacific) and the extracraton (Atlantic) assemblages (HOWELL and others, 1944; LOCHMAN-BALK and WILSON, 1958). Discovery of faunas occupying intermediate environment sites and study of nektonic forms are allowing more accurate correlations (ROBISON, 1964a; D. L. CLARK, pers. comm. 1967 on conodonts; PANTOJA-ALOR and ROBISON, 1967).

7. Environments of deposition

The palaeogeography and palaeoecology (environments of deposition) have been combined on a series of maps (Figs. 8 to 24, pages 95–103) for each significant age. All environments of deposition on the craton are characterized by relatively shallow water depths. On the inner shelf littoral, intertidal, and shallow sublittoral (infralittoral) are the only environments indicated. Sections from outer shelf sites are thicker but the sediments indicate that deposition was also predominately in the intertidal and shallow sublittoral zones, and the thickening was caused by slight but persistent negative tectonism of the outer shelf. Swales with water depths in the circalittoral zone were present but uncommon. The outer shelf was dominated by widespread tidal shoals maintained by calcareous algae whose position with respect to the shoreline was determined by the extent of clear, normal marine water.

Ten categories have been chosen for mapping in order to combine and relate: (1) the position of the environment with reference to the shore, (2) the dominant lithotope of the sea floor, (3) the water depth, and (4) the energy of the environment. Details of these are as follows:

(1A) *Littoral—transgressive sand lithotope.* A clean, well-sorted quartz sand, commonly with ripple marks and cross-bedding, is interpreted as a high energy beach environment. The thickness of the unit is highly variable and a quartz pebble conglomerate may be present at the base. In some of the thicker sections members of coarse granules and pebbles may alternate with sand members and occasionally red shales may appear (TEMPLETON, 1950). The beds are unfossiliferous except for sporadic patches of innumerable comminuted inarticulate brachiopod valves.

(1B) *Littoral—regressive sand lithotope*. A clean, well-sorted, coarse-grained quartz sand with common cross-bedding is interpreted as a high energy beach environment. The unit is never more than about 60 metres thick and thins toward the outer shelf. The lower beds may contain patches of fossil fragments but otherwise the unit is entirely unfossiliferous.

(2A) *Tidal (intertidal to infratidal)—sand lithotope*. These sand flat deposits are characterized by thinly bedded, poorly sorted, glauconitic quartz sands grading vertically and laterally into lenses of silt, shale, granules, and dirty limes. Some areas are extensively burrowed and unfossiliferous, others have many lenses of fossil fragments. Intraformational pebble conglomerates may be few or numerous. The lithotope extended from the supratidal zone to the infratidal. The energy varied from low to high in tidal channels. Actual deposition was very sporadic and slight. Tides and currents reworked the material at the interface and slowly moved it towards the sublittoral zone.

(2B) *Tidal (intertidal to infratidal)—mud lithotope*. This environment of mud flats is characterized by silt and clay, thinly bedded and with fine glauconite grains, grading laterally into dirty limes and fine sand. The deposits are unfossiliferous except for rare inarticulate brachiopod valves. Intraformational pebble conglomerates may be few or numerous and very coarse. The lithotope extends from the supratidal zone to the infratidal and the energy is predominantly low with rare high-energy episodes. Actual deposition was sporadic and slow.

(2C) *Tidal (lagoon)—sand/mud lithotope*. Limited areal sites of fine-grained clastics trapped behind a barrier are located adjacent to the shore and characterized by few fossils, occasional carbonate debris, and thinly laminated beds. The energy is predominately low.

(2D) *Tidal (algal shoals)—carbonate lithotope*. Extensive shoals developed widely over the outer shelf and extended across the inner shelf nearly to the shore, depending upon the presence of clear normal sea water. Stromatolites occupied sites extending from the supratidal zone to the infratidal zone and ranging from low-energy lagoons to high-energy tidal channels. Fossil fragments were originally common, but have been destroyed by the prevailing early diagenetic dolomitization.

(2E) *Tidal (intertidal to infratidal)—carbonate ooze lithotope*. Lime mud flats with minor amounts of argillaceous clay and/or fine quartz sand or silt, as impurities or thin beds or lenses. Medium to fine glauconite grains are locally abundant. Rounded, medium-sized, flat-pebble conglomerates are usually common, as well as the thin micrite beds. Fossils are scattered, but of moderate abundance, often concentrated as a coquina matrix of the pebble conglomerates. The lithotope usually grades laterally into 2B with the increase in fine clastic material; and on the infratidal side may grade into 3C.

(3A) *Sublittoral shelf—argillaceous sand lithotope*. A moderately sorted, argillaceous or silty, quartz sand occurs as the seaward continuation of 2A. Glauconite may be abundant and calcareous cement is common. The clay may occur as matrix, as thin shale partings in thin-bedded sands, or uncommonly as shale sequences or lenses up to 3 or 4 metres thick. Scattered fossil debris is abundant. Ripple marks and current cross-bedding are present as well as thin layers of rounded pebble conglomerates. The water depth is within the infralittoral range and the environment passes seawards by gradual increase of carbonate beds and lenses into one of the carbonate environments, 2D or 3C.

(3B) *Sublittoral shelf—sandy mud lithotope*. This is an environment of relatively limited areal extent on the shelf which has a substrate of fine sand, silt, fine glauconite grains, clay, and occasional rounded frosted sand grains. It may develop seawards of 2B and represents areas where the water was quiet enough most of the time to accumulate predominately the winnowed fines. Trails, burrows, and scattered shells of inarticulate brachiopods may occur, but fossils are not abundant. It appears usually as local thin beds or lenses passing laterally by increase of lime into the dirty limes of 2D or 3C.

(3C) *Sublittoral shelf—carbonate lithotope*. Thin to medium-bedded limes, with small amounts of sand, silt, or clay, developed most frequently on outer shelf sites, but can extend close to the shore. Infralittoral water depths of medium to high energy are distinguished from 3A and 3B by the absence of most terrigenous-derived material and from 2D by greater water depths. Fossils may be absent, scattered, or locally very abundant, and oolite beds and lenses may be locally abundant. Size sorting of the organic debris may occur. The unit occurs now as crystalline limestones in which any dolomitization can be demonstrated to be late.

(3C$_1$) *Sublittoral shelf—a variant of the carbonate lithotope*. Brachiopod biostromes composed of numerous valves usually of *Eoorthis* and *Billingsella* occur most commonly during the Franconian. They appear to represent in-situ accumulations which developed in hard-bottom, clear, medium-energy areas, probably at depths close to 45 metres at which stromatolites did not or could not maintain themselves. The biostromes vary from 15 cms to as much as 3 metres in thickness.

(3D) *Sublittoral shelf—swales—mud lithotope*. These are sea bottom areas of variable size and shape which lie appreciably lower than the surrounding shelf floor. They may occur in either inner or outer shelf sites but appear to be more abundant in the latter. The low-energy, often stagnant waters of the bottoms permitted settling of the finest clastics in thin laminae, or the slow accumulation of organic debris as lag concentrates in 1–2 mm laminae. But slumping from the surrounding shelf shoals introduced pebble conglomerates, broken exoskeletons, and sand grains of varying sizes, often well rounded and mixed with well-rounded glauconite grains.

ENVIRONMENTS OF DEPOSITION

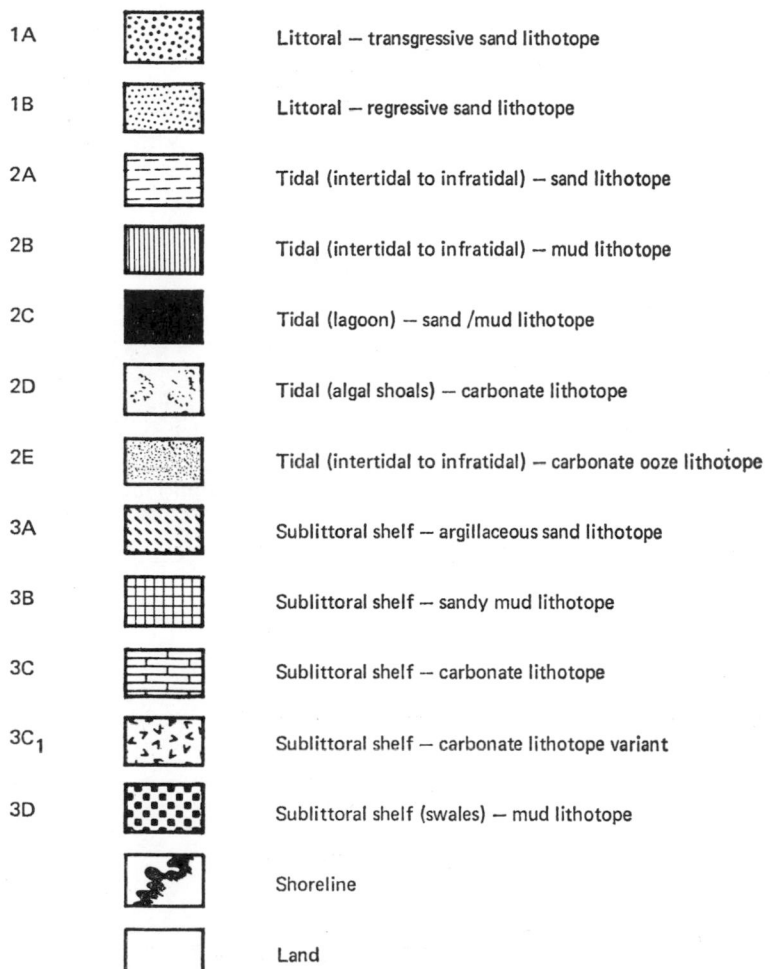

1A		Littoral — transgressive sand lithotope
1B		Littoral — regressive sand lithotope
2A		Tidal (intertidal to infratidal) — sand lithotope
2B		Tidal (intertidal to infratidal) — mud lithotope
2C		Tidal (lagoon) — sand /mud lithotope
2D		Tidal (algal shoals) — carbonate lithotope
2E		Tidal (intertidal to infratidal) — carbonate ooze lithotope
3A		Sublittoral shelf — argillaceous sand lithotope
3B		Sublittoral shelf — sandy mud lithotope
3C		Sublittoral shelf — carbonate lithotope
3C$_1$		Sublittoral shelf — carbonate lithotope variant
3D		Sublittoral shelf (swales) — mud lithotope
		Shoreline
		Land

FIG. 8. Key to symbols used to indicate environments of deposition in the series of palaeogeographical maps (Figs. 9–24) and referred to in the text (pages 92–94).

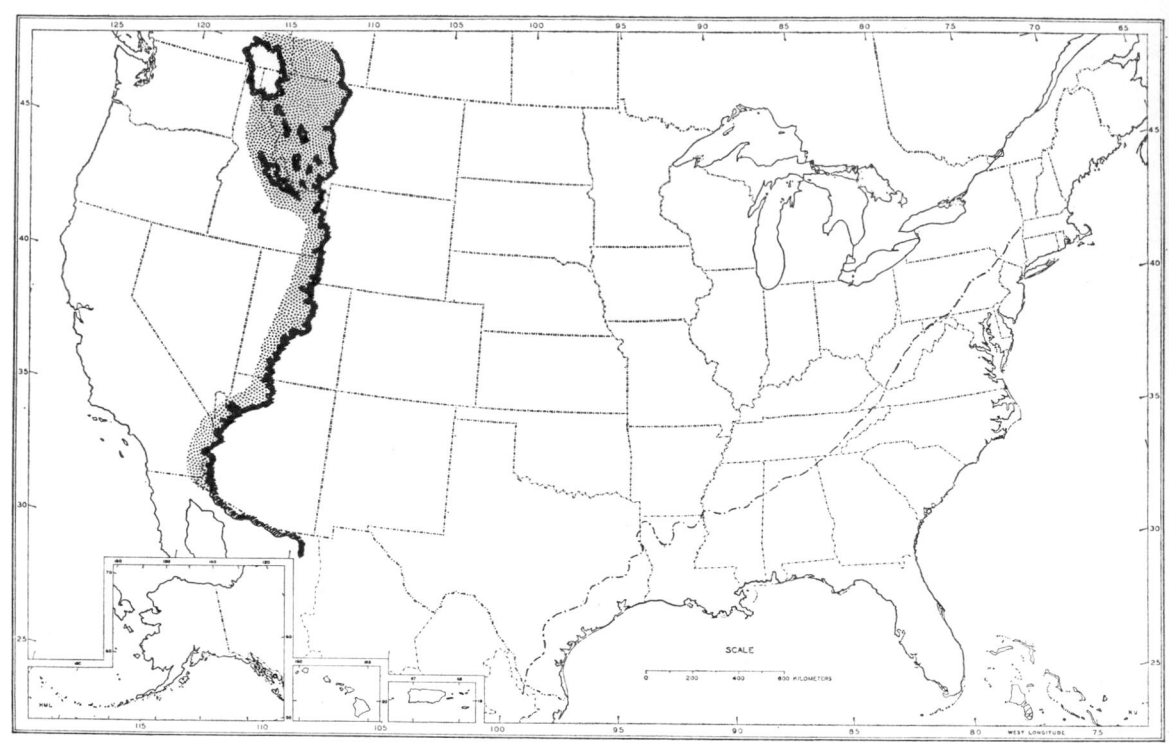

Fig. 9. Palaeogeography and palaeoecology—Middle Cambrian, *Albertella* Zone.

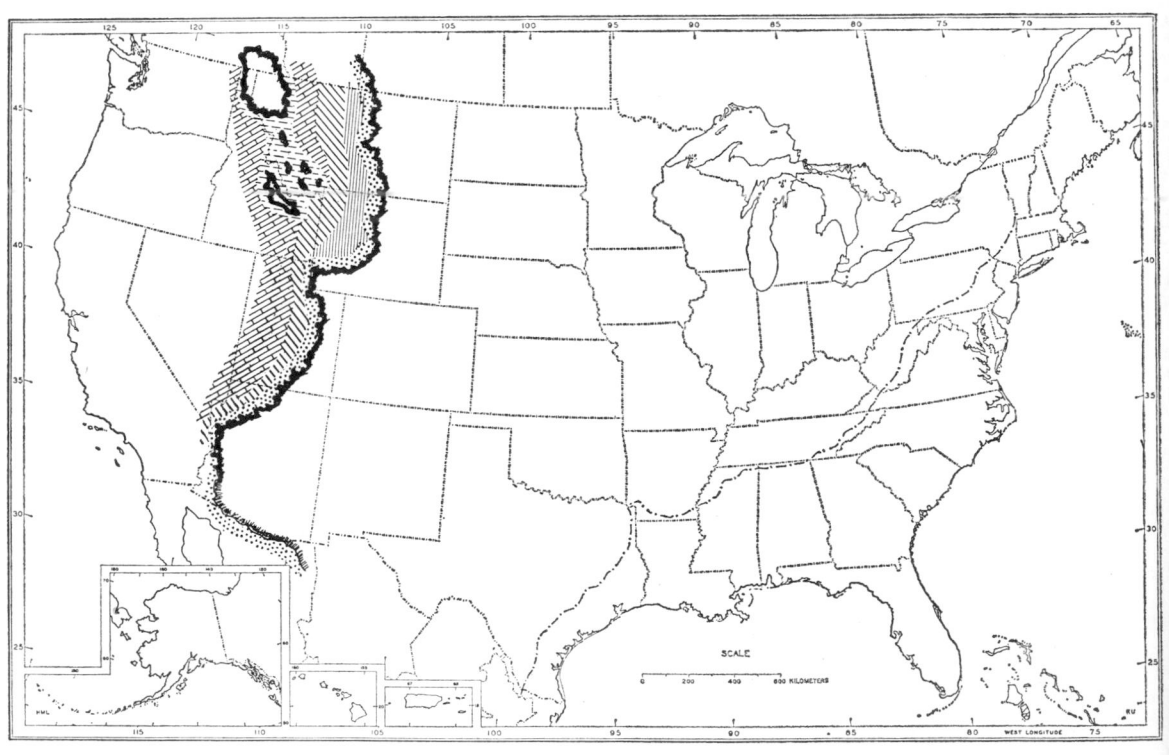

Fig. 10. Palaeogeography and palaeoecology—Middle Cambrian, *Glossopleura* Zone.

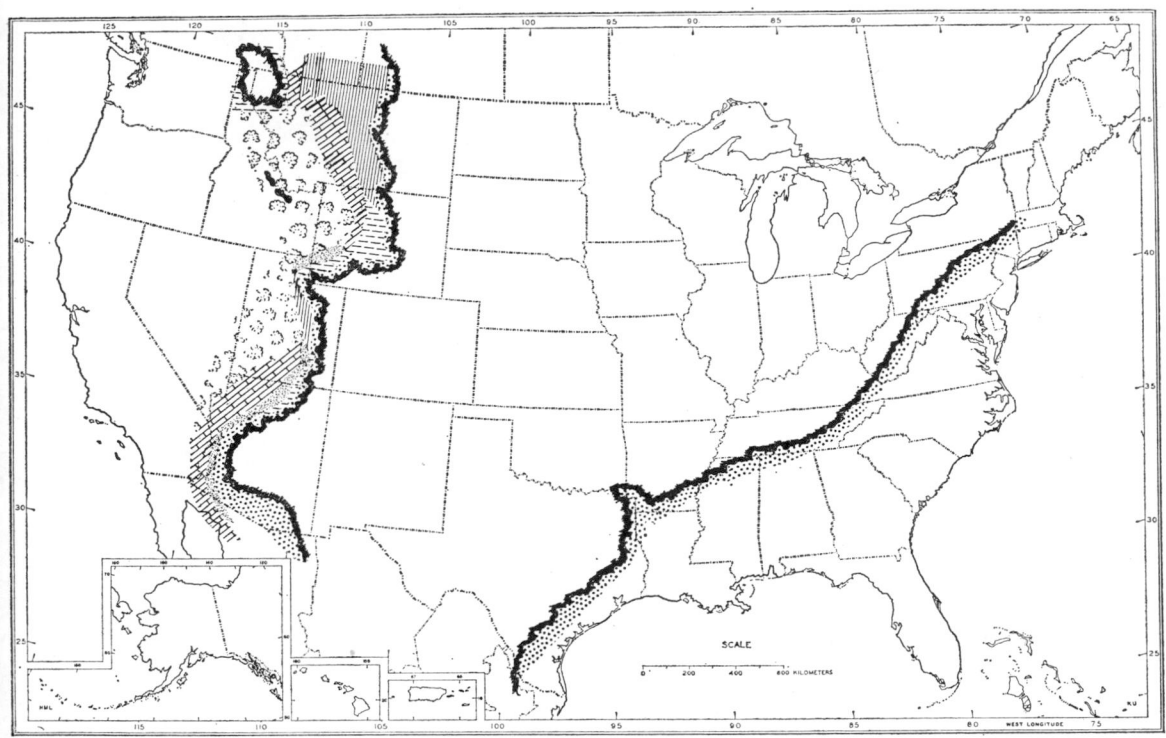

Fig. 11. Palaeogeography and palaeoecology—Middle Cambrian, *Bathyuriscus-Elrathina* Zone.

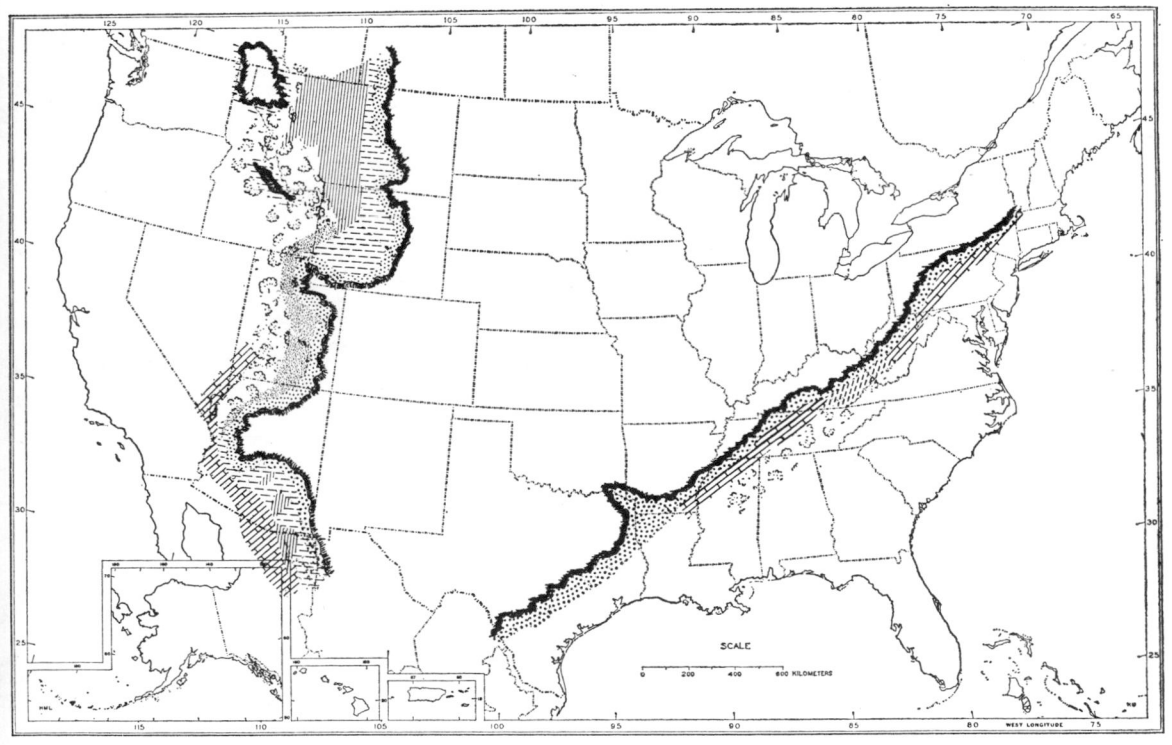

Fig. 12. Palaeogeography and palaeoecology—Middle Cambrian, *Bolaspidella* Zone.

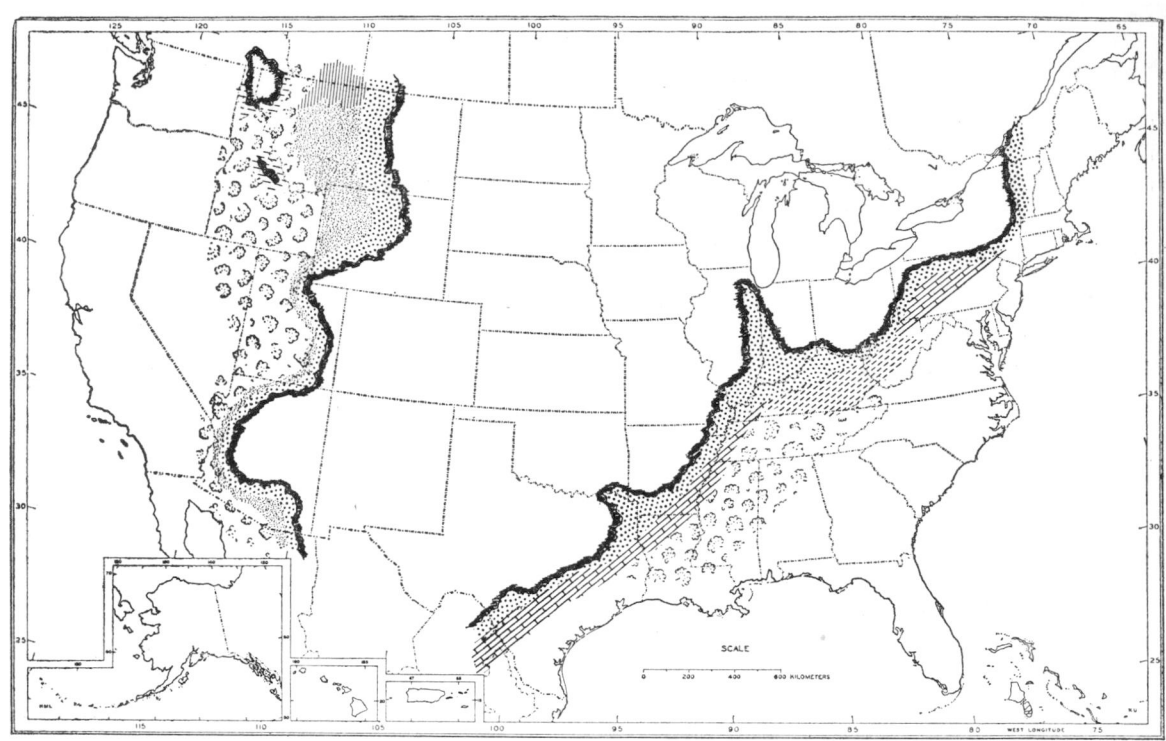

FIG. 13. Palaeogeography and palaeoecology—Upper Cambrian, early *Cedaria* Zone.

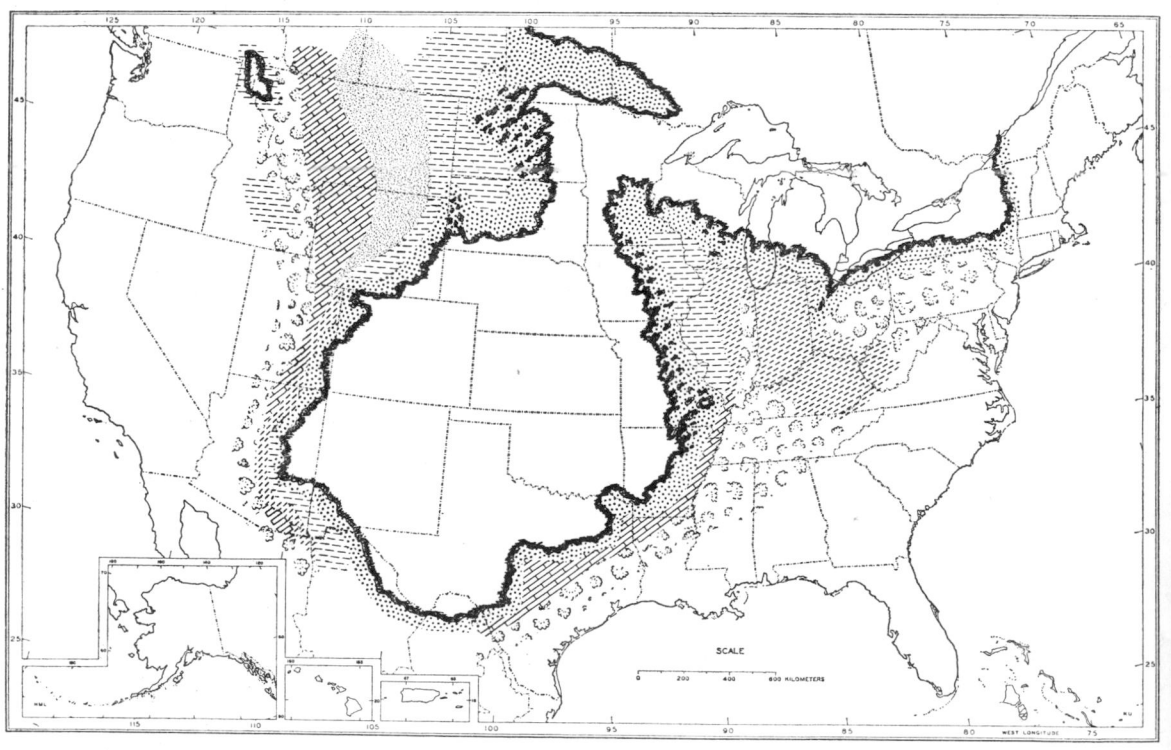

FIG. 14. Palaeogeography and palaeoecology—Upper Cambrian, late *Cedaria* Zone.

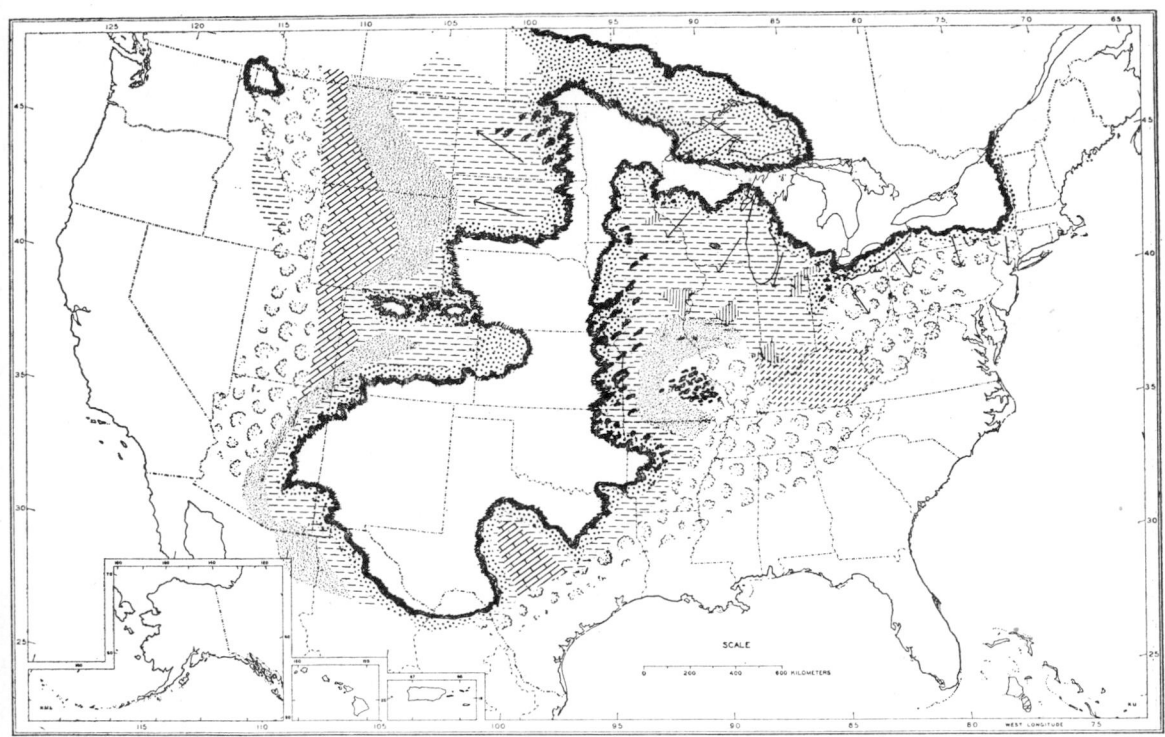

FIG. 15. Palaeogeography and palaeoecology—Upper Cambrian, *Crepicephalus* Zone.

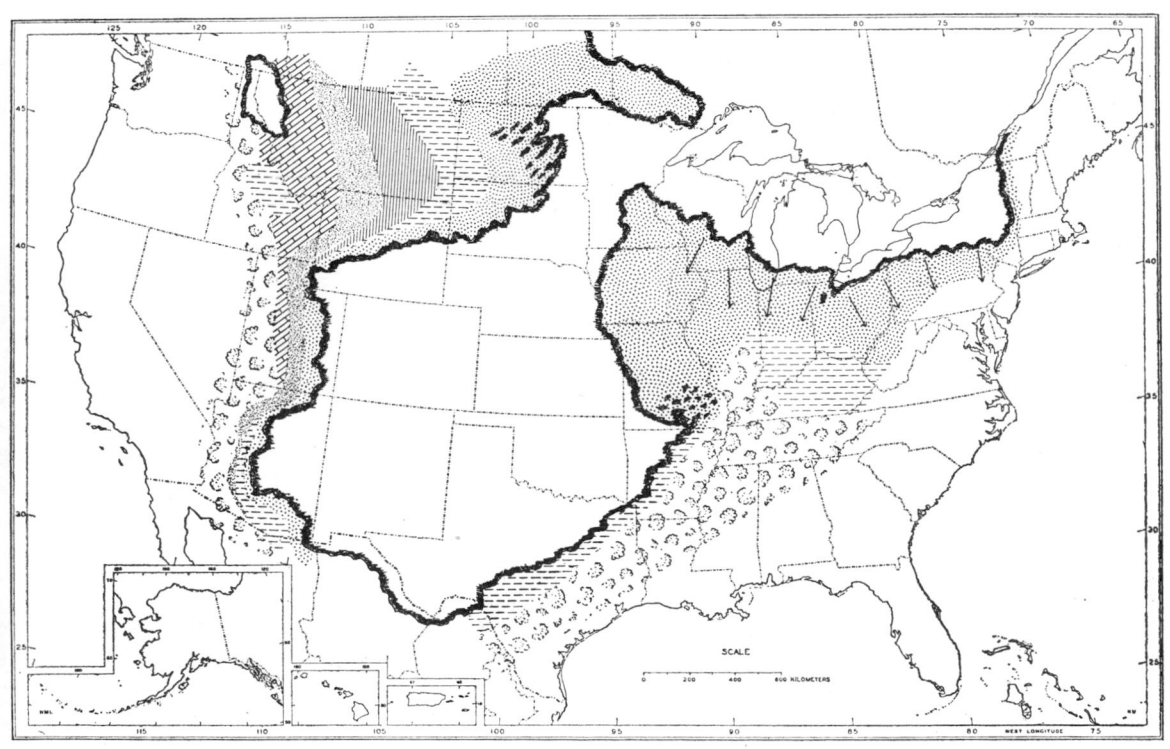

FIG. 16. Palaeogeography and palaeoecology—Upper Cambrian, early *Aphelaspis* Zone.

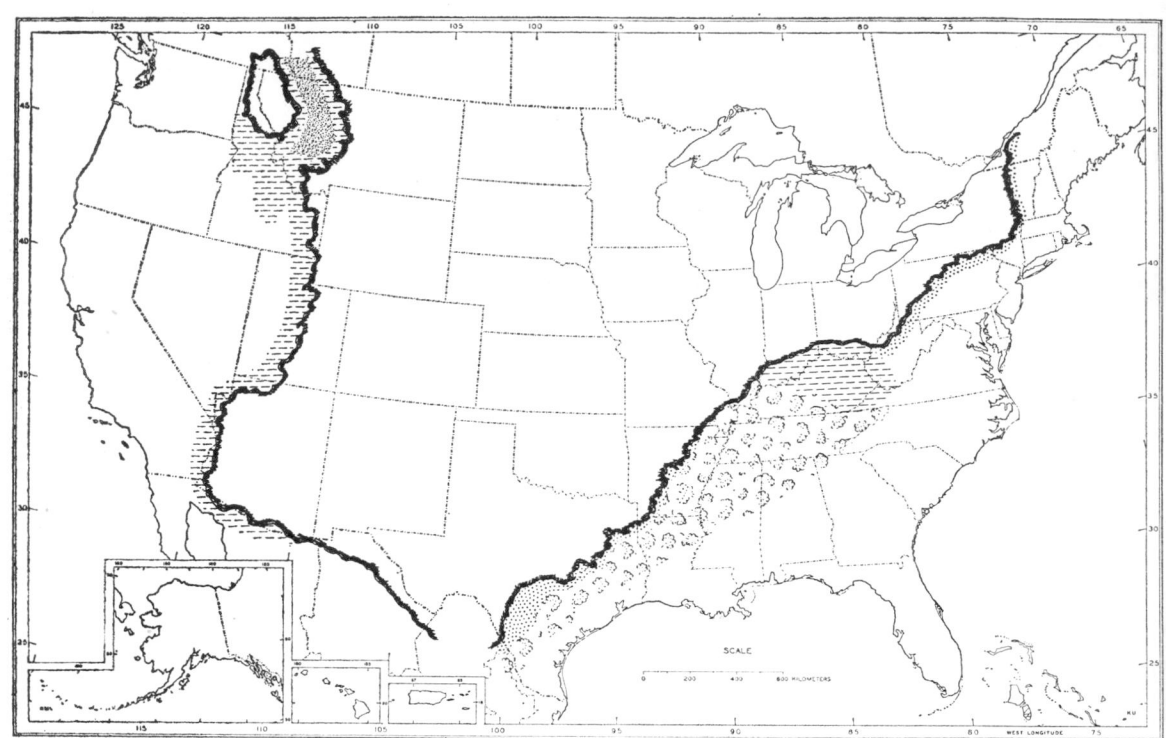

FIG. 17. Palaeogeography and palaeoecology—Upper Cambrian, late *Aphelaspis* and *Dunderbergia* Zones.

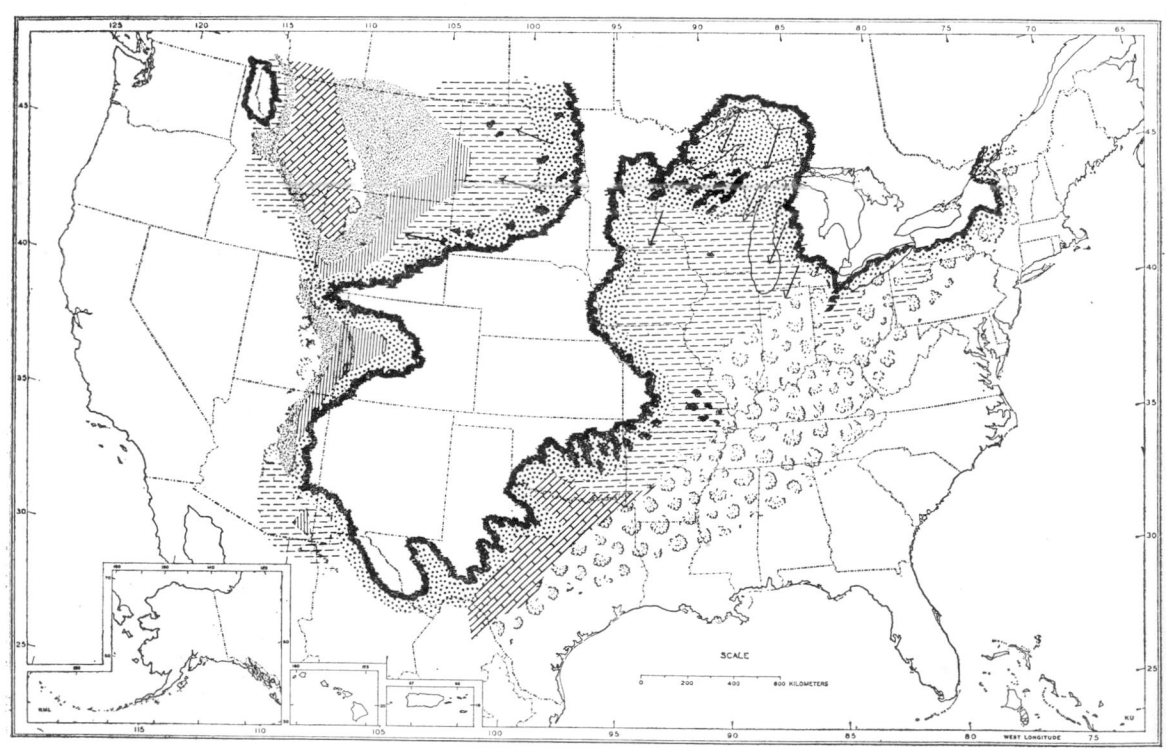

FIG. 18. Palaeogeography and palaeoecology—Upper Cambrian, *Elvinia* Zone.

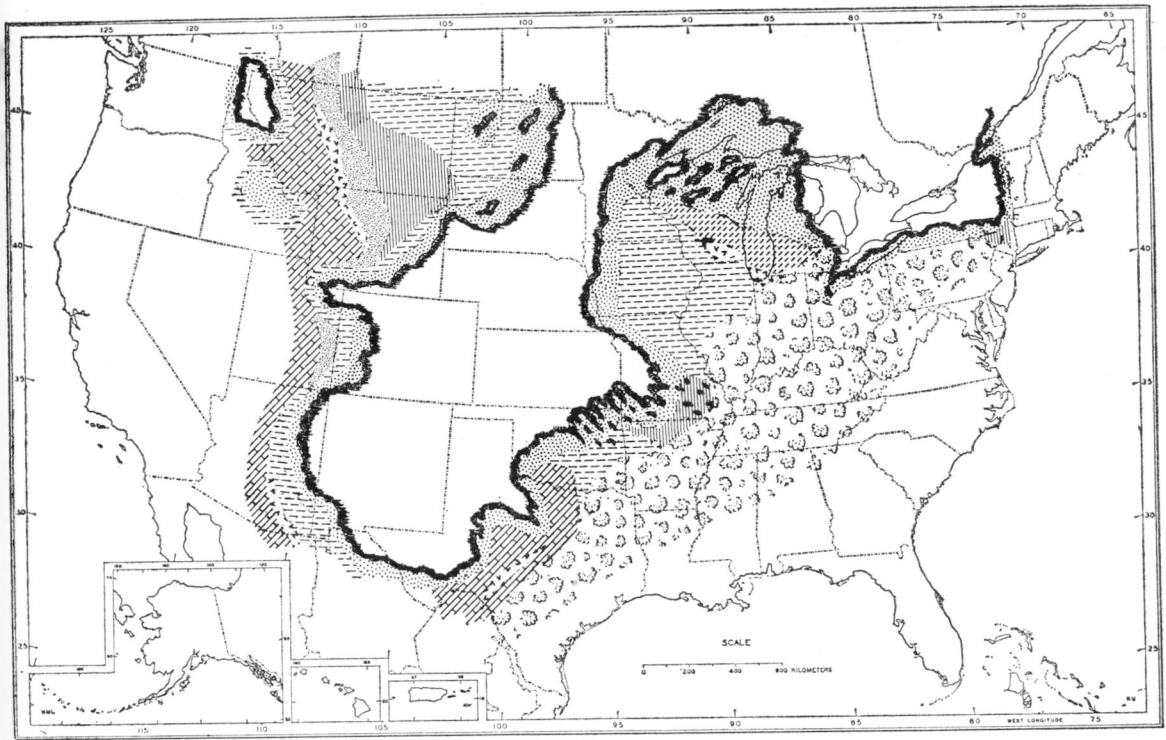

FIG. 19. Palaeogeography and palaeoecology—Upper Cambrian, early *Conaspis* Zone.

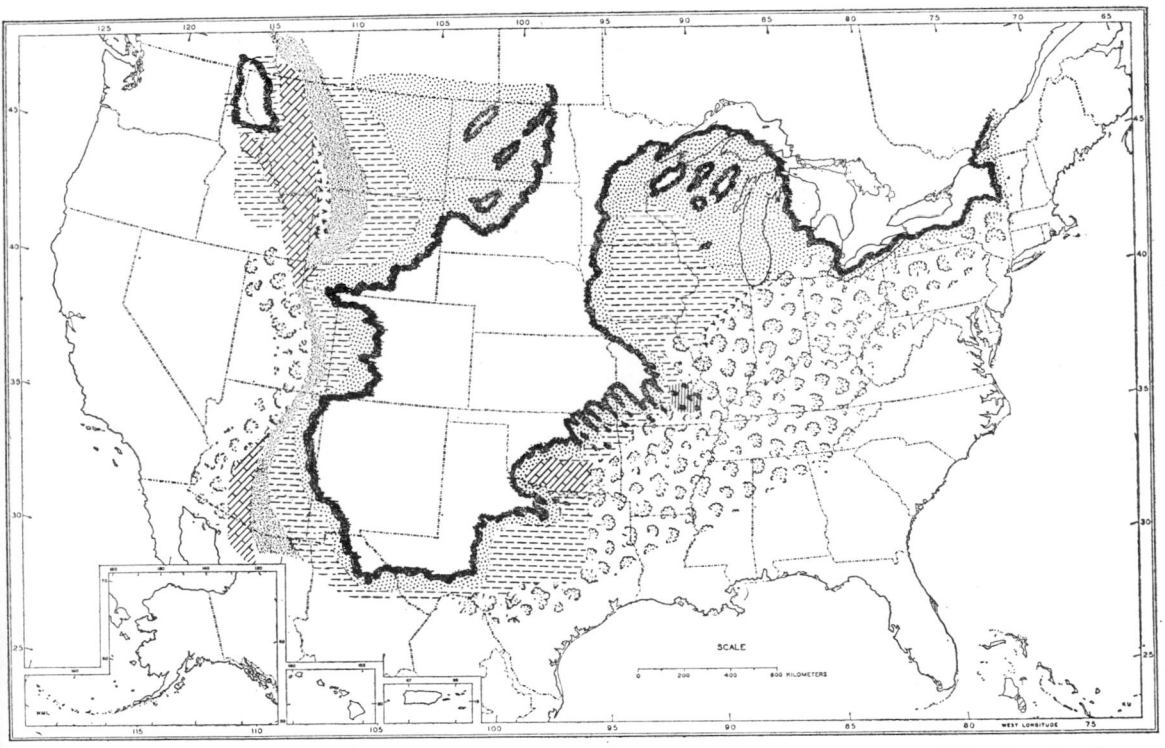

FIG. 20. Palaeogeography and palaeoecology—Upper Cambrian, late *Conaspis* Zone and *Ptychaspis* subzone.

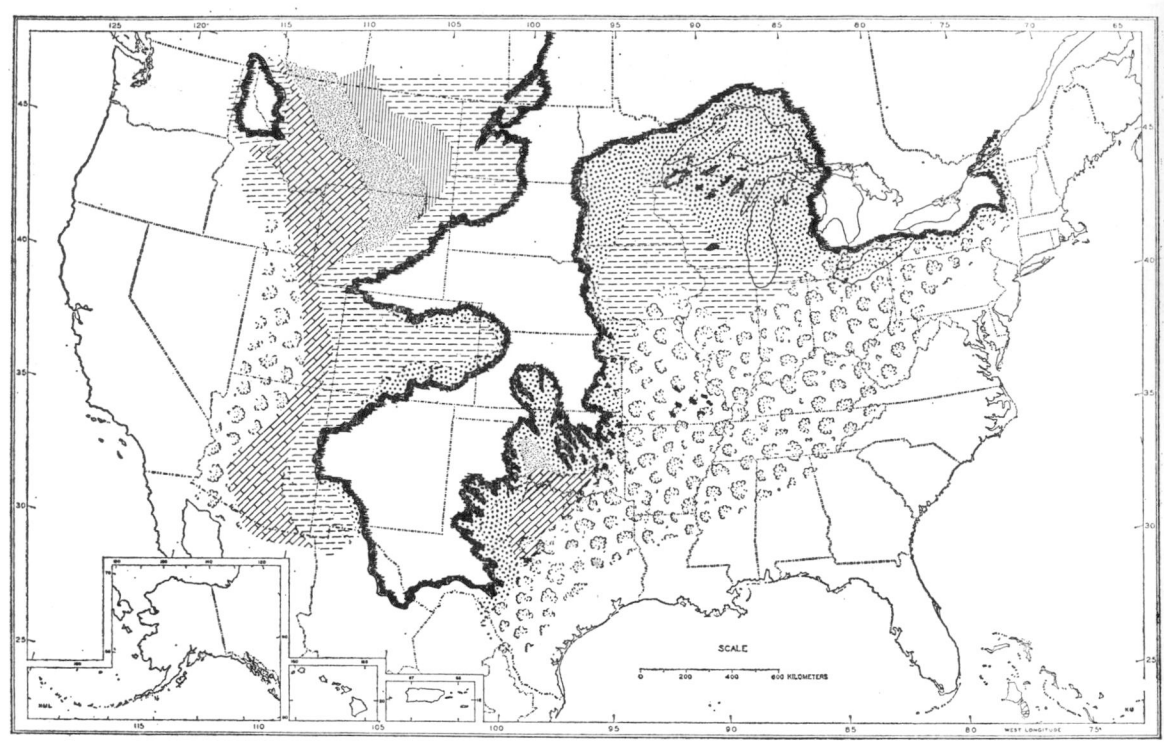

FIG. 21. Palaeogeography and palaeoecology—Upper Cambrian, *Prosaukia* subzone and earliest *Saukia* Zone.

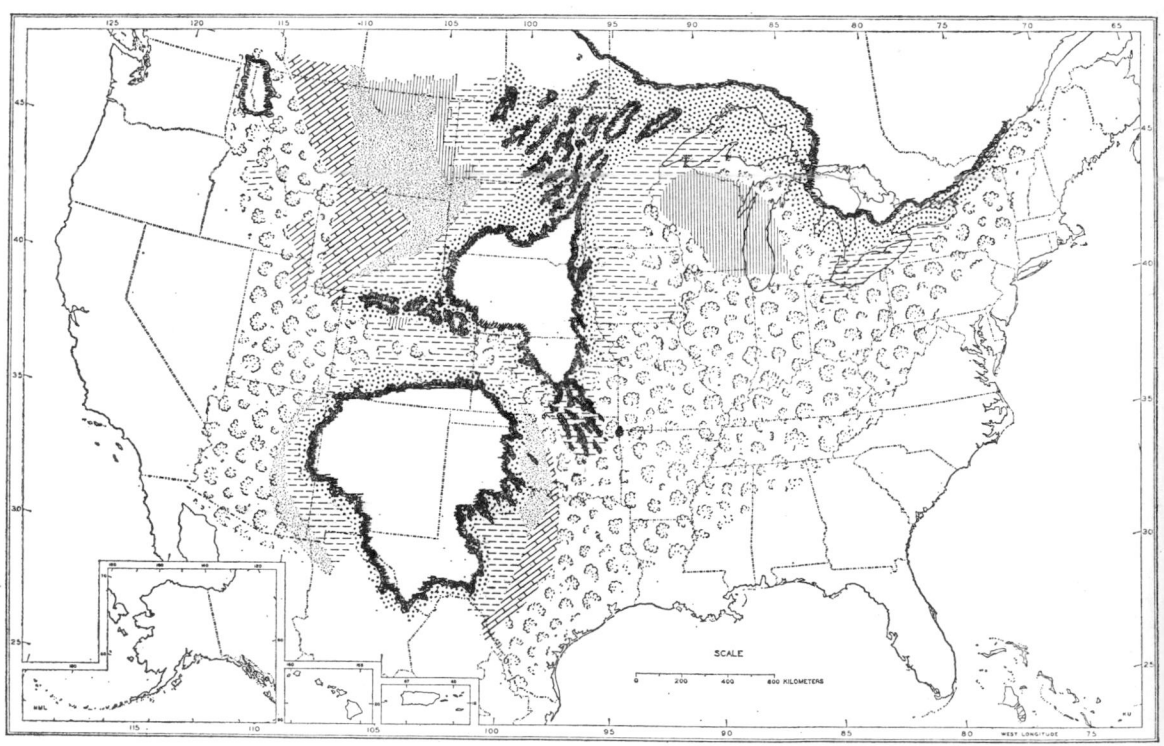

FIG. 22. Palaeogeography and palaeoecology—Upper Cambrian, early middle *Saukia* Zone.

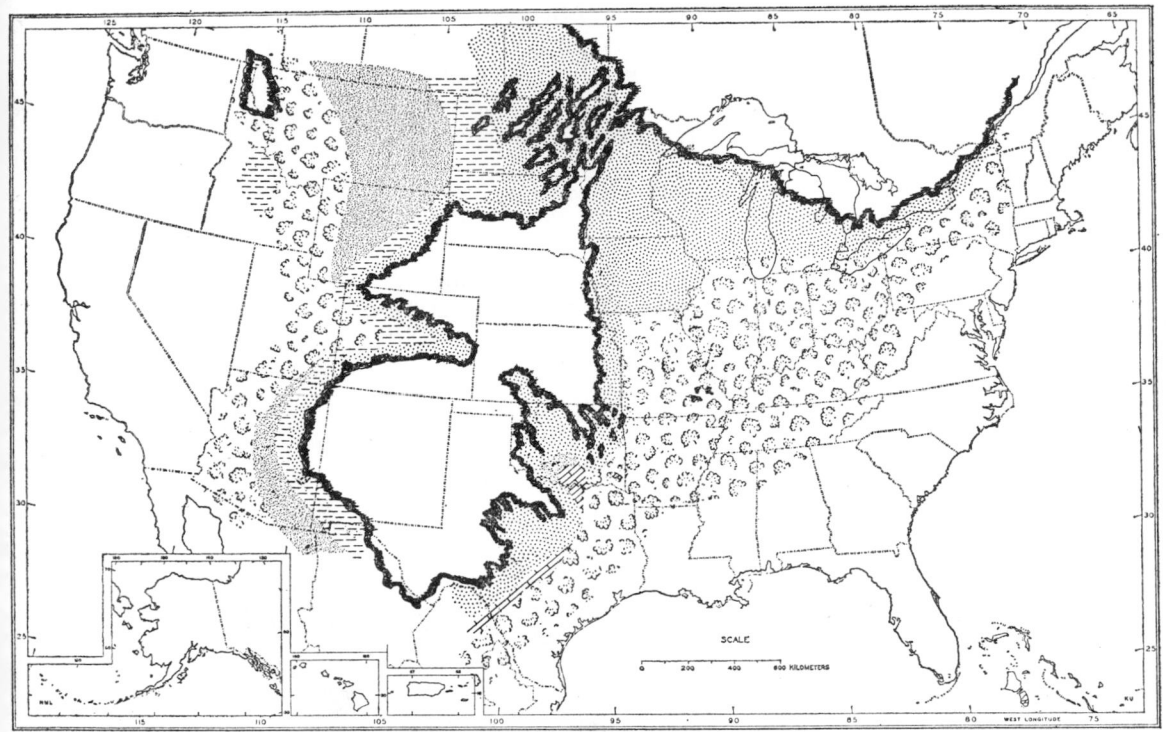

FIG. 23. Palaeogeography and palaeoecology—Upper Cambrian, late *Saukia* Zone.

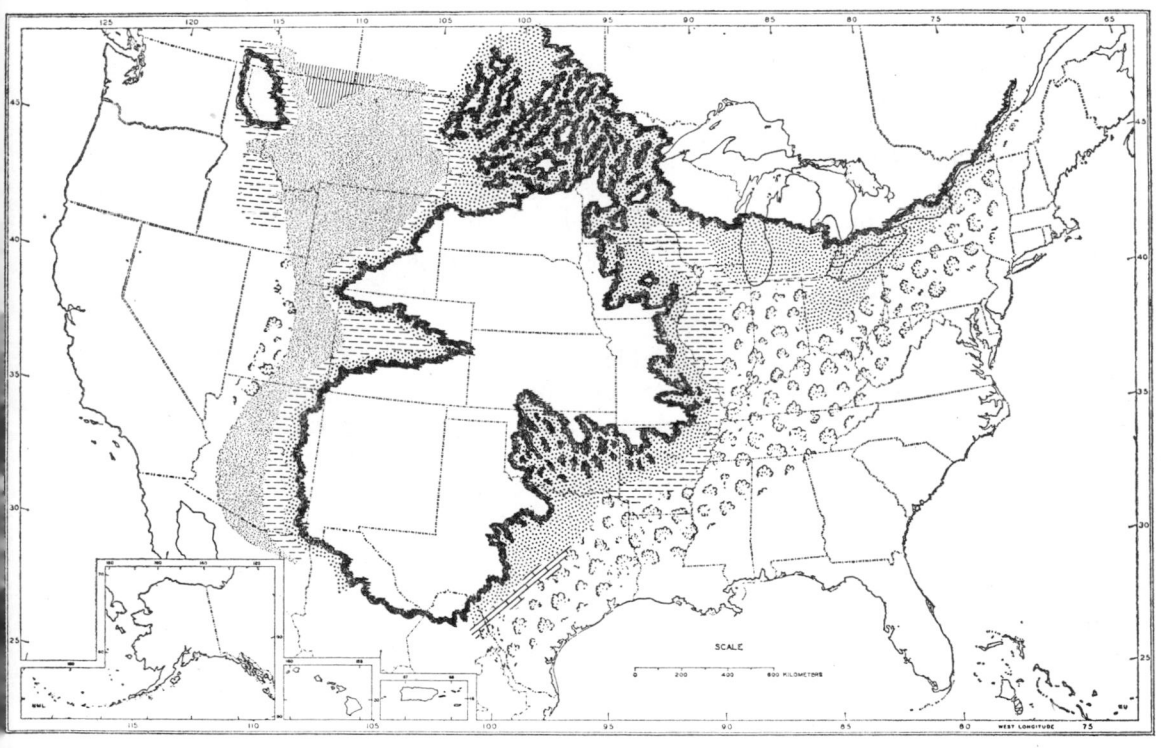

FIG. 24. Palaeogeography and palaeoecology—Upper Cambrian, transition faunule.

8. The Eastern Region

8.1. Lithostratigraphy

In the Eastern Region of the craton, roughly 2,590 square kilometres, the Cambrian now crops out in only two areas, a narrow band around the Adirondack Mountains in the north-east and a broader band around the Wisconsin Highlands in the north-west. The latter area is the type area of the Saint Croixan Series. Subsurface lithology is moderately well known from drilling in the central interior states primarily to develop water resources in the sands, but in the north-east deep test wells have explored the gas and oil possibilities (CALVERT, 1964b, 1965; FLAGLER, 1966).

In the subsurface adjoining the outcrop areas lithic units are readily recognized and close control across Illinois permits demonstration of a southward transition from the sand facies into a carbonate facies. The lithic units from the Adirondacks can be carried into western New York, Pennsylvania, and eastern Ohio in the subsurface. But in central Kentucky south of the Rough Creek Fault Zone the sections thicken rapidly and the lithic units are different.

The Upper Mississippi valley section first described by Owen (1852) from exposures in the bluffs along the Mississippi River is a sequence of coarse to fine-grained quartz arenites, often glauconitic, and sometimes calcareous and filled with fossils (calcareous-phosphatic shells of inarticulate brachiopods and sandstone moulds of trilobite fragments). Shale is very rare and dolomites appear only as thin beds near the top of the section. But the sands have proven to be highly variable when traced laterally; the fossil occurrences are abundant but sporadic; and the exact definition of the lithic units and their boundaries has not yet been agreed upon. A generalized and brief characterization of the units, in ascending order, is given: (see Fig. 2).

Mount Simon Sandstone: a fine- to coarse-grained, poorly sorted, quartz arenite with a local basal arkosic conglomerate. Thin beds of red and green micaceous shale may appear in the upper and lower part, but carbonates are absent. The only fossils are rare tracks and trails (*Climactichnites*) in the upper part. The thickness is from less than 30 metres to over 760 metres and reflects both relief on the Pre-Cambrian surface and movement on the Sandwich Fault Zone during deposition. TEMPLETON (1950) described seven members from surface studies. The boundary between the Mount Simon and the overlying Eau Claire is gradational and may show an interbedding of the two lithologies through 6 or more metres.

Eau Claire Formation: highly variable lithically of interbedded and intergrading glauconitic, micaceous siltstone; shale; dolomite; sandy dolomite; and fine grained, well sorted, quartz arenite. Shales and carbonates thicken southward in the subsurface. Glauconite and fossil occurrences are common (*Cedaria* and *Crepicephalus* Zone faunules). Thickness ranges from 15 metres in outcrop to 175 metres in southern Illinois. BUSCHBACH (1964) recognizes three members in northern Illinois. The contact with the overlying Galesville is sharp and may be disconformable.

Galesville Formation: a fine- to medium-grained quartz arenite, locally coarse-grained, with no mica and little very fine-grained glauconite. At the base in north-east Illinois is a dark brown, sandy dolomite. Otherwise, carbonate, silt, and shale are conspicuously absent. The sand is well-sorted and often cross-bedded. EMRICH (1966) restudied the Galesville at the type section and on the basis of characteristic lithology recognized 27 metres of Galesville, including 11 metres in the lower part which carries an *Aphelaspis*

Zone faunule. He also lowers the top 8·5 metres, placing the higher beds in the overlying Ironton. The Galesville thins to a featheredge on the Wisconsin Arch. The contact with the Eau Claire is a sharp lithic change and northwards is an erosional boundary as the Galesville rests on the middle or lower part of the Eau Claire.

Ironton Sandstone: a medium- to coarse-grained, slightly glauconitic, poorly sorted, clean to partly silty quartz arenite. Thickness varies from 5·8 metres in outcrop to over 30 metres in north-east Illinois. BUSCHBACH (1964) differentiated four members in the subsurface, all with varying amounts of dolomite. Southwards the dolomite increases and the unit passes into the Knox Dolomite in central Illinois and Indiana. In outcrop many beds have distinctive lithologies: a widespread, brown, iron-stained, fossiliferous coarse sandstone may be the original Ironton. Dark grey-green, glauconitic wormstones are conspicuous and may pass laterally into cross-bedded greensands or silty, shaly sandstones. Fossils of the *Elvinia* Zone assemblage (the late *Camaraspis convexa* faunule) are found only in the uppermost metre or so of the formation as lithically defined. In exposures the contact between the Ironton and the Galesville is usually abrupt and locally a relief of several centimetres is seen. Indication of minor reworking is observed at a few localities. In the subsurface differentiation of the Galesville from the overlying Ironton is on the basis of the finer and more uniform grain size of the Galesville. Such differentiation suggests a considerable relief at the contact and a southward thickening of the Ironton at the expense of the Galesville (EMRICH, 1966, figs. 4, 5). Locally the Galesville appears absent and the Ironton rests on the Eau Claire. EMRICH cautions that some of the apparent relief may result from facies changes in the Galesville, the lower Ironton, or both. But the indicated relationship of the Galesville and Ironton is most consistent with the interpretation of the Galesville as a regressive sandstone, and the known absence of most of the *Aphelaspis* Zone, all of the *Dunderbergia* Zone, and much of the *Elvinia* Zone.

Franconia Formation: a very glauconitic, fossiliferous, fine-grained quartz arenite and siltstone. Shales and silty, argillaceous dolomites become more common southwards, and the unit passes into the Knox Dolomite in central Illinois (BUSCHBACH, 1964). Shades of green and grey-green are common from the abundant, fine-grained glauconite. Wormstones and intraformational pebble conglomerates are common in the upper part. Thickness varies from 15 metres to about 600 metres in subsurface. Several members were recognized by BERG (1954) on the basis of glauconite content. In Wisconsin (OSTROM, 1967) the name Franconia has been dropped and the upper three members—ascending: Birkmose, Tomah, and Reno—constitute the *Lone Rock Formation*. This unit passes north and east into a non-glauconitic, cross-bedded, fine-grained dolomitic quartz arenite, the *Mazomanie Sandstone*, which almost completely replaces the Lone Rock facies.

St. Lawrence Formation: consists of the basal Black Earth Member; a massive, buff, sandy dolomite with thin interbedded dolomitic siltstones; and the Lodi Member, a dolomitic siltstone and very fine quartz arenite. The Black Earth, called the Nicollet Creek Member, is 10·7 metres thick in Minnesota and shows algal reef structures. It thins eastwards to 15 cms in central Wisconsin, and as it thins the Lodi lithology appears above and below it and eventually entirely replaces it. Southwards both units become dolomite in northern Illinois and thicken to 60 metres, and the lithology is identified as the Potosi Dolomite. The contact with the underlying Reno Member appears to be conformable. In early studies the highest intraformational conglomerate of the Reno was interpreted as a basal conglomerate and included in the St. Lawrence. The upper contact is conformable with the overlying Jordan Sandstone and is placed at the usually sharp lithic change.

Jordan Sandstone: a well-bedded, fine- to coarse-grained, quartz arenite with local sandstone pebble conglomerates. Three members are recognized: the basal Norwalk Sandstone, a fine quartz arenite with minor carbonate and silt; the Van Oser Sandstone, a medium-to coarse-grained, massive, cross-bedded quartz arenite which overlaps the Norwalk from west to east; and the upper Sunset Point, a medium-bedded sandy dolomite or dolomitic sandstone with worm burrowing locally. The Sunset Point dolomites may be distinguished from the overlying Oneota by the absence of chert, cryptozoons, and oolites, but the contact appears even and conformable.

Southwards in the subsurface across Illinois all units except the Mount Simon pass laterally into carbonates, and these lithic units can be correlated with the Upper Cambrian sequence of south-eastern Missouri. The Beeler well of south-central Tennessee contains 1,067 metres of carbonates below the Lower Ordovician boundary and late Middle Cambrian fossils were obtained from thin shale beds 122 metres above the basement. The Spears #1 well of central Kentucky is south of the Rough Creek fault zone and shows also a marked increase in thickness, but in this case, in shales and siltstones in the lower part of the section. Fossils of late Dresbachian age were obtained from these shales and siltstones 381 to 396 metres above the basement. Thus, while to the west the thickening of the carbonates appears to have occurred during both Dresbachian and Franconian time, to the east the thicker clastic section belongs mainly to the Dresbachian. None of these thick subsurface units should be given names of surface lithic units, though they may seem to resemble them superficially. CALVERT (1962, 1963, 1964) attempted to apply the nomenclature from the Appalachian miogeosyncline to the central interior units. Such usage does not recognize the southward facies changes in the subsurface which can be demonstrated in Illinois and Indiana. The coarse sands of the Mount Simon pass laterally into calcareous siltstones and shales in the lower part of the Lawrence County, Indiana deep wells; farther south-eastwards into calcareous shales and sandstones of the Williams #1 and Knuckles #1; and finally into the calcareous shales and limestones of the Nolichucky Formation of the miogeosyncline.

When stage correlation is indicated, lithic units of known Upper Cambrian age are shown as correlatives of the Appalachian miogeosyncline lithic units of known Lower and Middle Cambrian age (Spindletop Research, 1965). A clearer and more accurate picture is obtained if the observed facies changes are handled as *lateral variations of a defined stratigraphical unit* (in this instance the time-stratigraphical units of the Upper Cambrian stages). It cannot be demonstrated that the Lower Cambrian Erwin Sandstone is the earliest and easternmost deposit of a continuously transgressive sea. Deposition of this sand did not extend beyond the geosyncline.

The New York Adirondack section was the earliest Cambrian outcrop area known, but exposures are limited and the fossils are rare. The lateral changes in lithology away from the Adirondacks have now been resolved (FISHER, 1962) and the lithic units can be traced southwards in the subsurface. A generalized and brief characterization of the units, in ascending order, is given:

Potsdam Sandstone: a fine to coarse-grained quartz arenite, with no clay size particles and minor amounts of quartz silt, conglomerate, and breccia. Frosted and well-rounded quartz grains are abundant. In St. Lawrence County a poorly-sorted, red quartzite breccia, 8 to 10 metres thick, occurs at the base and is considered a talus-alluvial fan breccia by OTOVOS (1966). Ripple marks, cross-bedding, and channel cut-and-fill structures characterize much of the unit; but locally fine sands and silts occur in sub-parallel and parallel thin laminations with mud-cracked bedding planes. The scattered fossils demonstrate that with increasing amounts of calcareous cement the Potsdam sandstone grades both vertically and laterally into the Theresa Formation.

Theresa Formation: a dolomitic sandstone and sandy dolomite. The two lithic units form a single transgressive sequence rising in the section from south to north (Section A-A). The Theresa Formation in its type area is largely Lower Ordovician, and FISHER and HANSON (1951) revived the *Galway Formation* for the lithic equivalent known to be Upper Cambrian in age. The Galway Formation is unnecessary since it is the lithic equivalent of the Theresa and was deposited during a single transgressive cycle.

Hoyt Limestone: dark grey dolomitic limestone, oolitic limestone, algal biostromal limestone, and dolomites. Quartz sand fillings in the algal reefs are common, but chert is absent. The Cryptozoon reefs contain abundant gastropod and trilobite fossils, but this facies occurs only along the south-eastern side of the outcrop area.

Ritchie Formation: a white-weathering, slightly dolomitized, blue-grey micrite which may be a lateral facies of the Hoyt, but is isolated by faults.

Little Falls Dolomite: a sandy and cherty dolomite with large well-rounded quartz sand grains most abundant in the lower part and increasing geographically northwards. A few *Elvinia* Zone fossils occur in the lower beds in the south-eastern outcrop area and a few scattered Cryptozoon reefs occur higher in the section, but dolomitization and silicification have been extensive. Cavities lined with quartz crystals—'the Herkimer diamonds'—and deposits of anthraxolite, believed derived from the Cryptozoon reefs, are well-known features of the formation.

The New York lithic units can be traced in subsurface through south and western New York (FLAGLER, 1966) and into north-western Pennsylvania (WAGNER, 1961) and eastern Ohio. The basal sandstone thins away from the Adirondacks and the sandy carbonates and cherty dolomites increase steadily in thickness. Fine clastics appear in the southernmost wells (R. OLIN #1 of Section C-C). Sections E-E, F-F, and G-G show the westward transition of the New York lithic units into the upper Mississippi valley sequence. The continuity is lost over the later site of the Findlay Arch in west-central Ohio, where the post-Canadian pre-St. Peter erosion (which produced the *Knox Unconformity*) has cut deepest into the older sequence because of the inception of the Findlay Arch during this time. The Hope Natural Gas well 9634, of West Virginia, lies close to the eastern boundary of the craton. It was studied and correlated by WOODWARD (1959) who used the upper Mississippi valley terminology in both a lithic and a time-stratigraphical sense. The prominent sand in the middle of the sequence was recognized as the Franconia–Dresbach (the Galesville Sandstone) which is an important marker. In Section F-F the stage boundaries are placed following WOODWARD and the lithic terminology of CALVERT is shown on the right-hand side.

The Cambrian section throughout the Eastern Region contains no mineral deposits of economic importance other than a small production of gas and oil in Ohio and Kentucky (CALVERT, 1964). Five deep test wells in 1966–1967 went to basement in Tennessee but all information on them is confidential.

8.2. Palaeogeography and palaeoecology

In the Eastern Region the site of deposition was a broad continental slope which was exposed to the border of the orthogeosyncline until late in the Middle Cambrian. From then until late in the Lower Ordovician the area was gradually inundated by shallow epeiric seas, interrupted by several periods of marine regression. The palaeoslope in the eastern half was south, changing slowly to south-east into eastern Ohio. In the western half the slope was south to south-west, changing to south-easterly off the Transcontinental Arch (POTTER and PRYOR, 1961; EMRICH, 1966). A north-west slope existed in the Lake Superior area during the Dresbachian (HAMBLIN, 1959) but in early Franconian a south to south-west palaeoslope extended northwards onto the Canadian Shield.

The Pre-Cambrian surface was highly variable. South of the Rough Creek fault system, subaerial erosion had reduced the surface to a low, rolling coastal plain with no more than a hundred or so metres relief. Low escarpments may have followed the Rough Creek fault zone. To the north the surface became increasingly rougher with a relief of 300 to 460 metres common. Ridges of gneiss and quartzite, revealed by recent deep drilling (SUMMERSON, 1962; RUDMAN, SUMMERSON, and HINZE, 1965), trended south–south-westwards from the Canadian Shield. The Adirondack dome, the Northern Michigan Highlands, the Wisconsin Highlands, and the Ozark dome were positive elements forming the backbone of the upland regions. During the Upper Cambrian they were tectonically neutral as part of the wide shelf area and by early Ordovician subaerial erosion had reduced them to near sea level. Wide river valleys formed lowland areas and in the western half several graben valleys are known (RUDMAN, SUMMERSON, and HINZE, 1965).

Cambrian sedimentation suggests that the craton was covered by a thick soil cover; that there were no mountain ranges on the Canadian Shield near the shoreline, and that no large drainage system reached the continental shelf in this region. The supply of clastics, predominantly of sand and silt size, is derived from the immediate shores; is reworked often; and is limited to the northern (shoreward) half of the shelf. The supply of clay-size particles is conspicuously small and was scattered in local lenses over the inner shelf. From the beginning of the Franconian the deficiency of clastics was emphasized by the steady shoreward encroachment of carbonates.

During the Upper Cambrian the shelf was subjected to one major and three minor marine regressions, yet the sediments indicate no appreciable upwarping of the craton surface or the positive domes at any time. This feature suggests that the observed sea level changes were eustatic and that the craton experienced only very slow downwarping during the entire Upper Cambrian. The major regression during the later half of the Dresbachian, followed by the major transgression of the Franconian, may have been caused by the last glacial phase of the late Pre-Cambrian (Eocambrian) glacial period. The minor regressions of the beginning of the *Conaspis* Zone, the end of the Franconian, and the end of the Trempealeauan, which caused only a moderate expansion of shore areas, could also be explained by minor post-glacial climatic pulses. A significant tectonic movement (upwarp) of the craton did not occur until late in the Canadian (Lower Ordovician).

In the late Middle Cambrian as the sea transgressed the edge of the outer shelf, the transgressive littoral environment developed briefly. Continued slow downwarping brought the area into the sublittoral zone and as the water cleared a carbonate facies developed. In south-eastern Kentucky intertidal algal shoals appeared briefly while south-westwards in Tennessee a normal marine sublittoral environment developed where the exoskeletons of animals mingled with some clastics to form limestones. To the north-east, as shown in the deep wells of south-eastern Kentucky, the algal shoal could not maintain itself against a steady supply of silt reaching the region. Continued downwarping, attributable probably to movement on the Rough Creek fault zone, may have brought the sea floor into the lower energy circalittoral environment. Farther south the cleaner waters permitted the calcareous algae, invading the shelf from the miogeosyncline, to become established on the infralittoral sea floor. Intertidal shoals were built up on the slowly sinking shelf and spread laterally along the shelf as depth and turbidity of the water permitted.

Continued minor downwarping brought the earliest Dresbachian seas northwards across the shelf following graben valleys (Fig. 3) across south-west Indiana and Illinois, but to the east the Indiana–Ohio Platform and the Adirondacks remained above the sea. In the graben valleys fluvial sands apparently merge upwards into marine littoral sands and form abnormally thick sections. The east-west and south-west trend of the shoreline in the eastern half of the region may have influenced the development of a south-west flowing longshore current carrying fine clastics winnowed from the shore sites to settle over the outer shelf in south-eastern Kentucky.

By late *Cedaria* Zone time the littoral sand environment had reached the upper Mississippi valley, crossed Indiana and Ohio and the southern edge of the Adirondacks. Downwarping was so slow that a strip of intertidal sand flats up to 240 kilometres wide appeared off shore and passed gradually seawards into argillaceous sands of the shallow sublittoral zone. The latter zone extended seawards as much as 160 kilometres before water conditions permitted the calcareous algae to build up intertidal–infratidal shoals. During *Crepicephalus* Zone time all environment zones moved shorewards. The algal shoals of south-central Tennessee expanded north and west into Illinois and south-eastern Missouri. To the north-east a current spread the winnowed fines in a belt across Illinois and Indiana into eastern Kentucky. A shallowly submerged ridge, following the trend of the later Findlay Arch, deflected the clastics southwards and in the eastern half the algal shoals encroached northwards nearly to the shore.

The beginning of *Aphelaspis* Zone time is marked by nearly complete disappearance of the varied, warm-water, *Crepicephalus* assemblage and the sudden appearance on the same lithotopes of a small, cool-water, *Aphelaspis* assemblage. After a relatively short time span in the shoreward sites and a longer time in the outer shelf sites, evidence of a major marine regression appears in the sediments. A regressive sand appears in the upper Mississippi valley and moves southwards onto the algal shoal lithotope. A zone of oolite development appears in this lithotope and is interpreted as marking the time of regression, when high-energy tidal channels were more abundant. Shorewards some sections show a thin unit of sands and silts, believed to represent the clastic residue left from a time of subaerial solution. During the later Dresbachian algal tidal shoals persisted on the outer shelf.

As the sea moved across the shelf of western Canada it entered the drainage system of the interior Lake Superior Basin with a north-west palaeoslope. By *Crepicephalus* Zone time the shore had spread across the basin and intertidal sand flats developed along shore. Seawards a sublittoral sand environment probably appeared but has been removed by later erosion.

No tectonic upwarp can be noted over the craton shelf during the late Dresbachian subaerial exposure. As the early Franconian transgression moves northwards across the upper Mississippi valley, the eroded Wisconsin Highlands and North Michigan Highlands are crossed and the sea enters the Lake Superior Basin from the south-west. Cross-bedding shows a south-west palaeoslope during the Franconian, a slope which could have been produced by slight tilting or simply by drainage changes. In the upper Mississippi valley the reworking of Galesville regressive sands produced a lithically similar transgressive sand, the basal Ironton. To the south the mud and silt soils of the reliefless coastal plain settled to the sea floor as the water rose and the calcareous algae encroached northwards again over the inner shelf.

Franconian sedimentation is characterized by a steady decrease in the already small supply of clastics available from the shore sites. Glauconite is a very common mineral in the early Franconian sand flats where the addition of sand is so infrequent that belts of thoroughly burrowed sands can be traced for nearly 160 kilometres. The low-energy water cannot move the small amount of fines off the flat and they are trapped in local pools. Seaward the region is protected from the ocean by a belt of intertidal algal shoals up to 480 kilometres wide. In the north-west littoral sands continue as the waves slowly erode the Transcontinental Arch and to the north-east the waves slowly reduce the resistant rocks of the Adirondacks and move northwards across the Arch.

The beginning of the *Conaspis* Zone marks a second sudden faunal change. Again a large varied assemblage is replaced by a small assemblage of entirely different genera. Most sections show no lithic change at this position (Sections B-B, H-H, F-F, I-I). Only near-shore sites suggest a minor regression by the appearance of a littoral regressive sand environment (Mazomanie Formation) and subaerial exposure of some shores.

During the later half of the Franconian the tidal sand flats transgressed slowly northwards and descendants of the small *Conaspis* assemblage evolved into the genera of the diversified *Ptychaspis–Prosaukia* assemblages. In latest Franconian the tidal flats were frequently exposed and the dried surface broken into fragments. This extensive shallowing may have been caused by a minor marine regression of very minor upwarping of the Wisconsin Highlands. The highest of these coarse pebble conglomerates is often a convenient top of the Reno Member. On the outer shelf the environment of the broad intertidal algal shoals remains unchanged, while on the inner shelf, in Indiana and Ohio, locally derived clastics are no longer available and the algal shoals expand over the area.

The base of the Trempealeauan is placed at the appearance of *Saukiella* and *Calvinella* of the Saukidae and *Dikelocephalus* of the Dikelocephalidae. A succession of subzones in the Trempealeauan has been recognized in Wisconsin (RAASCH, 1952) and Texas (WINSTON and NICHOLLS, 1967), which appear to reflect both local biofacies and a time sequence. There is an early, rather provincial assemblage; a middle assemblage in which genera of extracratonic derivation appear; and a late impoverished assemblage of the last Saukid and Dikelocephalid species. In most sections no significant lithic change occurred. In the eastern half of the region a renewed, but slight, marine transgression carried the littoral zone across the Adirondacks into Canada. In the upper Mississippi valley the marine transgression submerged the Wisconsin Highlands completely producing sand shoals from which the finer silts were carried down the palaeoslope to settle out over southern Wisconsin and eastern Minnesota. The algal shoals moved northwards from Missouri and Illinois and maintained a precarious position on the tidal flats of Wisconsin, Minnesota, and northern Michigan during the early Trempealeauan. But by middle *Saukia* Zone time the tidal sand flats were spreading south and south-west across the upper Mississippi valley, and the gradual upward change from fine-grained, through medium, to coarse-grained sands in the upper part of the Jordan indicates an advancing regressive littoral zone. At the end of the Trempealeauan subaerial surfaces were more widespread than earlier in the stage in the upper

Mississippi valley and the Transcontinental Arch; but all the southern shelf persisted as broad, continuous, intertidal algal shoals on which the north-western regression is recorded only by the appearance of well-rounded sand grains, sand lenses, or thin sand beds. Even in the upper Mississippi valley thin beds and lenses of sandstone and siltstone occur, indicative of continued local deposition (STAUFFER and THIEL, 1941) until the marine trangression which brought *Symphysurina* and *Bellefontia* into the region.

9. The Mid-Continent Region

9.1. Lithostratigraphy

In the Mid-Continent Region, roughly 960 kilometres east–west by 1,600 kilometres north–south, the Cambrian strata crop out in four widely spaced, areally restricted locations—the north-east corner of Iowa, the Saint Francois Mountains of south-eastern Missouri, the Arbuckle and Wichita Mountains of southern Oklahoma, and the Llano uplift of central Texas. In the northern half of the region the subsurface Cambrian contains important aquifers and is known from water well logs. In the southern half the strata are known from a number of oil wells which have been completed to basement (BARNES and others, 1959; McCRACKEN, 1965).

The lithostratigraphy of Iowa and Nebraska relates directly to that of the outcrop area of the Upper Mississippi Valley of the Eastern Region, and the terminology of the lithic units is retained. In north-east Allamakee County (SCHULDT, 1943) outcrops along the western bluffs of the Mississippi consist entirely of sandstones and siltstones with a single thin dolomite and much dolomitic cement near the top of the section. At but one locality the top 60 cms of the Galesville is exposed, so most of the Dresbachian is in subsurface. Wells reaching basement rocks reveal a relief of some thousand metres on the Pre-Cambrian surface. The Mount Simon Sandstone at the base of the section is the unit most affected by this irregularity. The Cambrian section may reach a thickness of nearly 300 metres comparable to that in Wisconsin. The section thins both to the west and the south-west, and the lower units disappear as the Transcontinental Arch is approached.

The lithic units recognized are (Sections J-J, L-L):

Mount Simon Sandstone: in north-east Iowa is a coarse- to medium-grained quartz arenite, well to poorly sorted, white to brown in colour, with angular to subrounded grains. Glauconite and calcareous cement are absent, but argillaceous impurities occur either as clay-coated sand grains or thin, grey-green, micaceous shale beds. The thickness varies from 183 to 46 metres. Westwards the unit thins to 20 metres and thin calcareous siltstones appear in the sands. Eighty kilometres farther north-west only 6 metres of sandstone overlying 3 metres of green, non-calcareous shale can be assigned to the Mount Simon, which is not recognized any farther west.

Eau Claire Formation: in north-east Iowa varies from a fine-grained quartz arenite to a siltstone to silty shales, white to buff in colour with a thickness of 43 to 37 metres. Mica, glauconite, and dolomite cement are present. South-westwards to Webster County the unit thins to 29 metres and consists of glauconitic, silty limestones with interbeds of green calcareous shale. To the north-west the unit has become a grey-green, glauconitic, slightly sandy shale with 15 metres of argillaceous calcareous sand at the top.

Galesville Sandstone: can be recognized as a distinct lithic unit throughout eastern and central Iowa but is not continuous. It is a regressive sand and its sporadic occurrence may be attributed either to non-deposition or subsequent erosion. It is a medium to fine-grained, clean, well sorted, white quartz arenite, usually unconsolidated and unfossiliferous; 15 to 27 metres thick in the east, to 9 metres thick in the north-west.

Franconia Sandstone: The clastics of the Franconian stage have not been subdivided in this region although the subsurface lithology suggests that east–west and north–south facies are present. The unit is distinguished by its high glauconite content. In the north-east (SCHULDT, 1943) it is 49 metres thick and consists of a basal unit of medium to very coarse-grained, poorly sorted quartz arenite with layers of glauconite, overlain by a fine-grained, glauconitic, well-sorted angular quartz arenite, with local carbonate cement or thin dolomite beds. South and south-westwards dolomitic siltstones and shales are interbedded with the fine grained sandstones. In Webster County a lower 47 metres of very glauconitic, fine-grained, light-grey limestone interbedded with grey-green, glauconitic shales is overlain by 32 metres of grey, very glauconitic, dolomitic siltstone. In Kossuth County the unit consists of 30 metres of glauconitic, dolomitic, silty green shale overlain by 9 metres of very glauconitic dolomite. Farther west in Ida County the unit lies directly on the basement and consists of a basal 9 metres of calcareous sandy shale overlain by 15 metres of intercalated glauconitic, sandy, thin dolomites and green sandy shales, and an upper 46 metres of interbedded green silty shales and dolomitic glauconitic sands. The Franconia is not recognized farther west.

St. Lawrence Dolomite: in the north-east consists of a thin (5 to 60 cms) basal glauconitic, sandy, pink dolomite, overlain by 1·5 to 6 metres of thin-bedded, slightly dolomitic, glauconitic siltstone, lacking the greensand conglomerates of the Wisconsin section. These beds are overlain by 6 to 9 metres of buff, slabby, dolomitic siltstones identified as the Lodi facies. Southwards and westwards the Lodi grades rapidly into different lithologies. Southwards in Clayton County the St. Lawrence comprises 52 metres of coarsely crystalline, grey, silty dolomite with sparse glauconite, and in Scott County it consists of a similar thickness of blue-grey dolomite. Westwards the unit is a silty, blue-grey, pyritic dolomite with interbedded dolomitic sands near the top, thinning from 61 metres in Story County to 15 metres in Kossuth County.

Jordan Sandstone: a regressive sand unit. In the north-east it is 30 to 37 metres thick, and comprises a lower, very fine-grained, buff to white, dolomitic sandstone and an upper, coarse-grained, well to poorly sorted, unconsolidated quartz arenite either massive or cross bedded. Locally 1 to 7 metres of thin, flat-pebble conglomerates and thin-bedded fine sandstones immediately underlying the Oneota are assigned to the *Sunset Point Member.* Southwards the thickness varies from 12 to 37 metres and several silty or sandy dolomite beds occur as far west as Kossuth County. In eastern Nebraska 7·5 to 12 metres of slightly dolomitic, clean quartz arenites lying on the Pre-Cambrian may be the Jordan.

The exposures in the Saint Francois Mountains furnish the type sections of the lithic units in Missouri, Kansas, and north-eastern Oklahoma. These units are traced successfully in the subsurface if a westward increase in clastics is recognized (MCCRAKEN, 1965; KOENIG, 1966) (Sections M-M, N-N, O-O). The lithic units are:

Lamotte Sandstone: a basal transgressive sandstone, rising in the section westwards and thinning. It consists of white or red to brown quartz arenite, with arkose and conglomerate locally near the base and red to purple silty shales, lenses of sandy limestone, and dolomite near the top. In the east it varies from 152 metres in deep valleys to zero against the persistent Pre-Cambrian islands of the Saint Francois Mountains.

Bonneterre Dolomite: dominantly a light-grey, medium-to fine-grained, medium-bedded dolomite, of stromatolites and calcarenites (HOWE, 1966; 1968) with sandy, glauconitic fossiliferous limestones, blue-grey shales, thin-bedded micrites and/or reef breccias in the lower part (OHLE and BROWN, 1954). The formation is the host rock for the south-eastern Missouri lead and zinc deposits, and sedimentary structures control the ore bodies (SNYDER and EMERY, 1956; SNYDER and ODELL, 1958). The unit averages 122 metres in the Saint Francois Mountains, thickens rapidly to 457 metres to the south-east and thins steadily west and south-west to zero at the Missouri border (KOENIG, 1966) as the Lamotte climbs in the section. It carries a late *Cedaria* Zone fauna at the base (LOCHMAN, 1940) and a late *Crepicephalus* Zone faunule near the top (LOCHMAN, 1968).

Elvins Group: a sequence of very clastic carbonates, conformably overlying the Bonneterre, originally named the Elvins Formation. In the outcrop area a lower Davis Formation and an upper Derby-Doerun Formation are now recognized. But the lithic distinctions between the two units cannot be traced in the subsurface and the earlier term Elvins is used.

Davis Formation: consists of intercalated beds and lenses of fine-grained sandstone, siltstone, shale, dolomite, sandy limestone and limestone pebble conglomerates. Glauconite is common, especially in the clastic beds. The lower 15 metres are predominately clastics. Eighteen metres below the top the 'Marble Boulder Bed' is an horizon of mound stromatolites buried and preserved intact by an influx of mud. Fossils of the *Conaspis* and *Elvinia* Zones are moderately abundant and include a 30 to 60 cms bed of *Eoorthis*. The unit averages 60 metres in thickness and thins steadily westwards to zero just beyond the eastern border of Kansas. Through western Arkansas and eastern Oklahoma it is continuous in the subsurface into the Honey Creek Formation of the Arbuckle Mountains.

Derby-Doerun Formation: consists of thin-bedded, fine-grained, buff, argillaceous and silty dolomites intercalated with thin beds of siltstone and shale. Glauconite is common in the lower part. Beds of hexactinellid and other sponge spicules and echinoderm plates provide a marker horizon 15 metres below the top. The presence of less than 10% of chert is in contrast to the overlying Potosi. The unit averages 45 metres in the east, thins westwards to zero at the west Missouri border, but continues into north-eastern Oklahoma as a thin unit.

Potosi Dolomite: a massive, thick-bedded, medium to fine-grained, brown-grey dolomite, characterized by an abundance of drusy quartz cherts, conformably overlying the Derby-Doerun. The fresh rock gives off a petroliferous odour. Commercial barite occurs in the dolomite and in the deep red residual clays. Stromatolite reefs are common (HOWE, 1966). The thickness varies from some 90 metres to zero around the persistent Pre-Cambrian islands in south-eastern Missouri. In the subsurface the unit thins to 9 metres in western Missouri and cannot be traced into Kansas and Oklahoma.

Eminence Dolomite: a medium to massive-bedded, light grey dolomite with several distinctive chert types: nodular chert, angular chert fragments, white oolitic chert, and massive chert boulders. Silicified moulds and casts of gastropods are locally abundant (ULRICH and BRIDGE, 1931), and near the top masses of *Cryptozoon*. The fauna belongs to the *Saukia* Zone. Most of the major caves and large springs of Missouri are developed in the Eminence. The formation averages 69 metres in south-eastern Missouri and increases to 107 metres in south-central districts. It is easily traced by the chert residues. In north-central Missouri thin sand beds appear near the top, and the unit thins to about 52 metres in eastern Kansas. Northwards (in Nebraska) it grades into the Jordan Sandstone. Across southern Kansas and north-east Oklahoma silt, sand, shale, and glauconite are present in the dolomite. McCRACKEN (1965), tracing the Potosi-Eminence as a single unit, notes a rapid thinning south-westwards from 198 metres to zero in Creek County, Oklahoma (Section O-O). The top of the Eminence is easily picked out by the appearance of sandy dolomite or the Gunter Sandstone. Across south-central Kansas (Section N-N) the Eminence rests upon a basal transgressive sand which grades upward into grey sandy, silty, cherty dolomite. The unit averages 38 metres in eastern Kansas. In Western Kansas dolomitic sands are common in the upper and lower beds. The unit is traced northwards into the Jordan dolomitic sandstone of south-western Nebraska. The sections of KOENIG (1966) and McCRAKEN (1965) and the fossils reported by McELROY (1965) indicate that the basal sand and overlying dolomite of the Kansas sections is a facies of the Eminence of Missouri, and not the Lamotte and Bonneterre as previously designated by KIRBY and KEROHER (1932).

The Arbuckle and Wichita Mountains in south-central and south-western Oklahoma are the third outcrop area, about 965 kilometres south-west of the Saint Francois Mountains. The thick lower carbonate section of the Arbuckle Mountains, a sequence spanning the Upper Cambrian–Lower Ordovician boundary, comprises the Arbuckle Group. The lateral transition of a western limestone sequence into an eastern dolomite sequence may be traced in the exposures (Section P-P). The lithic units are:

Timbered Hills Group: consists of the basal transgressive Reagan Sandstone and the overlying Honey Creek Formation. As initial deposits on a Pre-Cambrian surface of over 300 metres relief their thickness may vary from zero to nearly 245 metres and the boundary between the two units is arbitrary.

Reagan Sandstone: is a buff-brown, poorly sorted, poorly cemented quartz arenite, arkosic and conglomeratic near the base. Near the top, glauconite may appear, and thick-bedded, ferruginous sandstones with hematite oolites reach thicknesses of some 10 metres locally. Bedding is absent, irregular, cross, or evenly laminated. The contact with the Honey Creek may be sharp, and is placed where calcareous, fossiliferous glauconitic sands appear.

Honey Creek Formation: in the west consists of intercalated lower beds of fine-grained, calcareous, glauconitic quartz arenite and coarsely crystalline, glauconitic, fossiliferous limestone; and upper beds of limestone only. In the east the lower beds are a sandy, silty, glauconitic dolomite and the upper beds very slightly sandy dolomite. The abundant fossils in the limestones belong to the *Elvinia* and *Conaspis* Zones. In many sections the Honey Creek and overlying Fort Sill contact is conformable and gradational, but local channels in the Honey Creek dolomite occur in the eastern Arbuckles (FREDRICKSON, 1956).

Arbuckle Group: begins with the Fort Sill Formation, the most persistent limestone unit in the area and includes all the overlying Upper Cambrian and Lower Ordovician units.

Fort Sill Formation: a thin-bedded, yellow-grey to purple, dense limestone (micrite), locally sandy, shaly, or oolitic. Beds of flat-pebble limestone conglomerate are scattered through the unit and a conspicuous development of conglomerates is used as the base of the overlying Signal Mountain. Five faunal subzones belonging to the *Ptychaspis–Prosaukia* Zone and the lower part of the *Saukia* Zone are present (FREDERICKSON, 1956). The formation is 180 to 45 metres thick in the west and thins eastwards to 24 metres with intertonguing slabby, fine-grained, yellow dolomites. In the central Arbuckle Mountains the Fort Sill shows abrupt gradation of limestone into dolomite accompanying rapid thinning in the vicinity of buried hills (HAM, 1955).

Signal Mountain Formation: a thin- to thick-bedded, dark grey, medium crystalline limestone with glauconite and lenses of dolomite common in the lower part. At the type locality thick beds of trilobite coquinas and flat pebble limestone conglomerates are conspicuous, especially in the lower half. The unit ranges from 75 metres to over 120 metres where the Royer Dolomite is absent. The contact with the overlying McKenzie Hill Formation is conformable and gradational, but there is a sharp lithic change to the overlying Butterly Dolomite. HAM used two brachiopod horizons, a lower *Finkelnburgia* and an upper *Apheoorthis*, as marker beds in the Arbuckle Mountains. The trilobite assemblages belong to the *Saukia* Zone.

Royer Dolomite: a coarsely crystalline, white, grey, pink dolomite with brown, very rough, irregular weathering. It is the facies equivalent of the upper Fort Sill and the lower Signal Mountain Formations and ranges from zero to 224 metres in thickness. It is unfossiliferous except for the silicified shell layer of *Plectotrophia* and *Mesonomia* which HAM traced from the Fort Sill into the Royer (FREDERICKSON, 1956).

Butterly Dolomite: a light to dark grey, fine to coarse crystalline, sandy dolomite with coarse arkosic sands in the upper beds. It is the facies equivalent of the Signal Mountain and ranges from 45 to 133 metres. The unit continues eastwards in the subsurface into Arkansas where it is called the Eminence (McCRACKEN, 1965). Laminated thin beds of fine crystalline dolomite distinguish the Butterly from the Royer. The only fossils are the silicified shell horizons of *Finkelnburgia* and *Apheoorthis* which can be traced eastwards into the lower part of the Butterly. Dolomite deposition continued uninterrupted across the Cambro-Ordovician boundary in the eastern Arbuckles.

The Llano district of central Texas, the fourth outcrop area, 400 kilometres south of the Arbuckle Mountains, contains the type sections of the lithic units of Texas. The *Riley Formation* is a marine transgressive—regressive sequence of three members. It contains fossils of a late *Bolaspidella* faunule and the *Cedaria*, *Crepicephalus*, *Aphelaspis*, and lower *Dunderbergia* Zones.

The *Hickory Sandstone Member* is a non-calcareous, non-glauconitic, quartz arenite with a basal quartz pebble conglomerate locally, the quartz pebbles being often wind-facetted. Feldspar grains and pebbles occur and cross-lamination is present through the unit. The unit is usually yellow-grey to olive-grey but a distinctive red zone appears near the top in the north-west caused by iron oxide coating on well polished sand grains. The zone can be traced south-east into the red zone of the Cap Mountain Limestone. The Hickory ranges from 84 to 152 metres in outcrops, but thins westwards in the subsurface. The boundary is gradational both vertically and laterally into the Cap Mountain Limestone.

The *Cap Mountain Limestone Member* is a sandy, silty, glauconitic, bioclastic limestone with terrigenous material decreasing from base to top. Colour varies from brown, yellow, olive-grey to reddish grey. Oolites are common in the upper half. North and west the limestones pass into the Hickory Sandstone from the base upwards and the thickness ranges from 198 metres in the south-east to zero in the north and west. The upper boundary is arbitrarily placed where the Lion Mountain Sandstone lithology becomes dominant.

The *Lion Mountain Sandstone Member* consists of quartzose greensand; glauconitic quartz arenite; thin beds of sandy limestone, shale, and siltstone; and lenses of trilobite and brachiopod coquinas. The dusky-green colour of the sands contrasts strikingly with the white to cream, cross-bedded coquinites. The thickness is constant within a range of 8·8 to 20·7 metres. The top is an erosional unconformity and 80 kilometres to the west-north-west the unit is gone. Southwards in the subsurface the limestones become thicker and siltstone and sandstone disappear. It is possible that deposition may have been continuous into the base of the Wilberns (Section R-R).

The *Wilberns Formation* is the transgressive unit of the middle and late Upper Cambrian, and Earliest Ordovician. It includes four members:

The *Welge Sandstone Member* at the base consists of yellow brown, massive bedded, non-glauconitic, coarse to medium-grained, well-sorted, quartz arenite. The basal metre or so may contain earthy, reworked Lion Mountain sands or poorly sorted granules. It carries faunas of the *Elvinia* Zone. To the south-east the Welge becomes glauconitic and in the subsurface passes into a greensand. The unit varies from 3·4 to 10·7 metres and its boundary with the overlying Morgan Creek is gradational through a few beds.

The *Morgan Creek Limestone Member* is a glauconitic, bioclastic, oolitic limestone of shades of grey-green. The base is sandy and interbeds of fine-grained, darker green, silty limestones and stromatolite biostromes appear in the upper half. The algal biostromes up to a metre thick frequently overlie or pass laterally into oolitic limestone. To the south-east patches of yellow-orange dolomite become common and in the subsurface the unit passes laterally into a coarse-grained, silty, sandy dolomite with only 7·6 metres of limestone at the base. In the subsurface to the west the characteristic lithology persists for 120 kilometres before sands and silts become predominant; but to the north the member passes rapidly into calcareous sandstone, from which fossils indicating its age equivalence to the Morgan Creek were obtained. Between 12 and 18 metres above the base of the member there occurs an example of the '*Irvingella* coquinite', a sheet several cms thick of the disarticulated exoskeletons of a species of *Irvingella*. Immediately overlying are several cms of an *Eoorthis* coquinite, composed mainly of large *Eoorthis remnicha* valves, but also containing some specimens of *Billingsella*, *Ceratreta*, *Irvingella*, *Berkeia*, and *Comanchia* in the basal 2·5 cms. *Parabolinoides* appears in the top 2·5 cms. *Elvinia* and *Conaspis* Zone faunas occur in the lower half of the member and *Ptychaspis* subzone faunules appear in the upper half.

The *Point Peak Member* consists of intercalated light-grey, calcareous siltstone, olive-grey, silty limestone; varicoloured limestone pebble conglomerates; thin grey shales; and grey stromatolite biostromes. Locally, calcareous siltstone forms a basal unit. The stromatolites occur as thin lenses scattered along a single horizon, as continuous beds several kilometres square, or as masses 12 metres square or 90 metres thick and traceable for kilometres. The large biostromes in the west pass eastwards into dolomite, and the gradational boundary into the San Saba Dolomite is difficult to place. The thickness ranges from 57 to 7·6 metres, and as the Point Peak thins to the north-east and the south the San Saba thickens (BARNES and others, 1959). The top of the *Prosaukia* subzone lies below the middle of the Point Peak in the west, near its upper boundary in the north-east, and in the San Saba Dolomite in the south-east (BELL and BARNES, 1961). To the west and north-west the Point Peak siltstones merge into sands and the age equivalent units shown in Sections Q-Q and R-R are based on electric log characteristics.

The *San Saba Member* consists of a limestone facies and a dolomite facies, which vary considerably laterally and vertically. The limestone in the west is grey-green, medium to fine-grained, glauconitic, thin to thickly-bedded, and becomes finer-grained, very thin-bedded, and non-glauconitic eastwards as it passes into fine-grained, medium-bedded, pink to grey mottled dolomites. In central Llano district the limestone overlies the dolomite, but to the south-west the stromatolite dolomite is predominant and is interbedded with and overlain by clean calcareous sandstone. The fossils belong to the middle and upper *Saukia* Zone and Zones A and B of the Early Ordovician (WINSTON and NICHOLLS, 1967).

9.2. Palaeogeography and palaeoecology

The Mid-Continent Region occupies the eastern portion of the Transcontinental Arch and the adjoining continental shelf sloping south and south-east to the Appalachian miogeosyncline. The region is the site of numerous scattered mountain masses of Pre-Cambrian intrusives, resistant ridges of quartzite, and intervening valleys and plains developed on schists or the soft shales of the Red Clastics. A relief of some thousand metres is known from Iowa to southern Oklahoma. Inundation of this land formed an irregular rugged coast line, characterized by deep embayments, long headlands, and numerous islands. (SCHULDT, 1943; HOWE, 1966; McCRACKEN, 1965; HAM, 1955). Initial dips averaging 10° are common in the basal sandstones, and many of the larger islands were not entirely reduced and covered until Early Ordovician. The westward encroachment of the sea is significantly impeded by the elevation of the coast and the resistance of the rocks. McCRACKEN (1965) noted a few faults in the subsurface of north-eastern Oklahoma and south-west Missouri which may have been active, and there is apparently some basinal downwarping during Upper Cambrian in the Arbuckle geosyncline. The continental shelf experienced only slow minor downwarping during the Upper Cambrian, and the position of the shoreline reflects a balance between the transgressive and regressive phases of the ocean and the character of the coast.

The sediment pattern throughout the Upper Cambrian shows a predominance of clastics in the north (this portion may actually be regarded as the western side of the Eastern Region). A steady increase in carbonates southwards occurs through northern Missouri, and in the southern half carbonates become predominant. In Texas the presence of wind-facetted and pitted quartz sands and pebbles in the basal sandstones suggests (BELL and BARNES, 1961) that the climate of the land was semi-arid. But these features may reflect high-energy conditions along an open oceanic coast. HOWE (1968) commented on the marked differences in carbonate sedimentation between the exposed northern coasts and the protected southern coasts of the Saint Francois islands.

Late in *Bolaspidella* Zone time the westwards transgressing sea reaches the central Texas basin. A *Bolaspidella* faunule appears in the basal Hickory Sandstone and in the overlying lower sandy carbonates.

During the early *Cedaria* Zone continued transgression defines the central Texas basin, and a broad headland persists through northern Texas. The shore trends north-eastwards and the sea just reaches the south-eastern Missouri basin. Supratidal algal reefs build westwards and protect a series of long narrow lagoons laced by shallow tidal channels. Glauconite and lime muds accumulate in the quiet waters while oolites form near the channels and shore. The abundant trilobite and brachiopod fauna concentrate in the sublittoral lagoons and near the tidal channels, separated from the open ocean environment by the supratidal algal reefs.

By late *Cedaria* Zone time the upper Mississippi Valley is inundated and clean littoral sands deposited along the western shore. Eastward narrow sand flats develop as fine sands accumulate. The argillaceous sand lithotope of the sublittoral zone merges into the lime mud lagoons of the southern half of the region. Among the islands of the rugged coast a variety of bottom-conditions and habitats develop which vary both vertically and laterally, but are too small in area to be shown on the maps (OHLE and BROWN, 1954; HOWE, 1968).

The Dresbachian transgression culminates during *Crepicephalus* Zone time. The shore line reaches its farthest westward position for this stage. In the north the sand flats widen and locally pass into mud flats as terrigenous material becomes finer. Southwards in Missouri tidal flats of lime mud and argillaceous mud develop shoreward of the encroaching algal reefs. Stromatolite biostromes build over reduced islands or close to the shore of still exposed land, and numerous small varied habitats persist among the islands. Southwards the stromatolite reefs border the tidal flats to central Texas where lime muds, oolites, glauconite, and abundant organic debris accumulate in wide sublittoral lagoons and tidal channels. The abundance of oolites in the top beds of the Cap Mountain Limestone suggests the lagoon waters became shallower about the time the *Aphelaspis* fauna made its appearance. On the craton the *Aphelaspis* assemblage appears suddenly and the *Crepicephalus* assemblage disappears equally abruptly. Only in Texas have a few genera of the older fauna been reported with *Aphelaspis* genera (PALMER, 1954). However, the local lithofacies does not change. Only after an appreciable time lapse does evidence appear in the sections of a marine regression. The faunal change is attributed to a lowered water temperature, which could be related to a glacial cycle causing the widespread regression observed at the end of the Dresbachian.

During the *Aphelaspis* Zone clean, well-sorted regressive sands spread along the shore in the north. Often they are removed by later erosion, and the occurrence and thickness of the Galesville is now sporadic. Across Missouri as the lime mud flats and algal reefs are laid bare there is no sand available for deposition. By *Dunderbergia* Zone time the shore has retreated to southern Illinois and the south-eastern corner of Missouri. To the south its position is not certainly known except in Texas. Here numerous stagnant swales form on the lagoon floor where an abundance of glauconite accumulates. Occasionally storms introduce sand and masses of organic debris into the sites. South-east of the Llano district the Lion Mountain sands appear to pass in the subsurface into a sandy glauconitic limestone. This well probably records the position of a tidal channel across the exposed algal reef. However, during *Dunderbergia* Zone time most of the central Texas basin was exposed, and a disconformity is present in the Llano district sections (BELL and BARNES, 1961).

The Franconian transgression during *Elvinia* Zone time differs from the earlier transgression. The sea now crosses a low coastal plain covered with sand or clay loam soils. In the north broad sand flats reappear with numerous shallow swales where glauconite forms. Shifting currents disperse the glauconite on the flats. Although a regression of the sea occurs during the *Conaspis* Zone and Early *Ptychaspis* subzone, the environmental conditions and the position of the western shoreline remain static throughout the Franconian. Southwards the small sand supply gradually fails and in south-eastern Missouri carbonates appear. The deposits of the *Elvinia* Zone record the encroachment of the stromatolite biostromes upon the sand flats as the water level rises. Local accumulations of lime and clay muds settle among the islands and along sheltered southern shores. The quiet waters of the regressive sea during the *Conaspis* Zone permit a large mud flat to develop in southern Missouri. By the *Ptychaspis* subzone terrigenous debris fails and the stromatolites from the barrier reefs build shoreward in the clearing water as minor downwarping of the coastal plain permits. In north-eastern Oklahoma supratidal algal flats often extend to the island shores, but minor amounts of silt and clay still drift into the region.

The sea enters the southern Oklahoma area for the first time in the early Franconian and encounters a hilly Pre-Cambrian surface. The palaeogeography and sedimentation in this area indicate the beginning of downwarping of a narrow basin trending north-west to south-east through the western Arbuckles. Immediately to the south the northern Texas headland partially separates the Oklahoma basin from the central Texas basin. In the Oklahoma basin the coarse, cross-bedded, poorly sorted Reagan sands represent the first, relatively rapid, transgression of the sea. Soon sand bars close the re-entrant valleys and form sandy-bottomed, quiet lagoons, the sites of oolitic hematite deposits. Farther east in sublittoral lagoons glauconite forms, lime muds and fine glauconite settle out in the quiet waters, and piles of organic debris and sand are stranded along the sides of tidal channels. To the east the algal reefs build shoreward as the tidal channels carry only small amounts of sand into the area.

In central Texas the *Elvinia* Zone transgression oversteps the Dresbachian sediments, and clean basal sands extend into western and south-western Texas. The fossils in sands from a well in Runnels County demonstrate the equivalence of these sands to the Morgan Creek limestones in the Llano district. The sublittoral zone is probably continuous northwards to Oklahoma. Sandy glauconitic biosparites are abundant initially, but gradually these alternate with, and then become subordinate to, silty lime muds as the algal barrier reefs expand in length and breadth and broad protected lagoons form.

The abrupt appearance of the *Conaspis* assemblage and disappearance of the *Elvinia* assemblage is unmarked lithologically. The amount of terrigenous material falls nearly to zero in the early *Conaspis* Zone and regressive conditions set in. Faunal evidence suggests some significant environmental change at this time; a drop in water temperature, increased storminess, or changed current direction. Near the top of the *Elvinia* Zone there appears a widespread coquinite several centimetres thick of *Irvingella major* debris. *Irvingella* is present but not common in the faunules of the lower beds. Overlying the *Irvingella* coquinite within several centimetres is another widespread coquinite of *Eoorthis reminicha* and *Billingsella*, also some centimetres thick and with occasional specimens of *Irvingella*, *Comanchia*, and *Berkia* in the base. This situation is comparable to the association of a few specimens of *Elvinia* Zone genera in the basal bed of the *Conaspis* assemblage at such far separated localities as Wisconsin and Montana (LOCHMAN, 1964). WILSON and FREDERICKSON (1950) note the geographical distribution of the *Irvingella* and *Eoorthis* coquinites along the eastern side of the Transcontinental Arch. This distribution suggests that the coquinites develop as a result of catastrophic storms or currents which destroy, and wash shoreward onto the tidal flats, biota that normally live in the deeper channels crossing the reef or on its oceanward side. The coquinites are valuable marker beds for mapping as throughout the shelf they occur in the same stratigraphical sequence and mark the relative position of the *Elvinia-Conaspis* faunal change.

In southern Oklahoma during the *Conaspis* Zone and the *Ptychaspis* subzone continued failure of clastic material permits the algal reefs to extend shoreward over the silty tidal flats to the north and north-west as well as build up to supratidal position and form a nearly continuous barrier across the Arbuckle basin. Organic debris, sand, silt, and oolites wash onto the flats along the western and south-western shores and fine lime muds accumulate in the quiet waters of the central and south-eastern lagoons. As the water level continues to drop, the tidal flats and shallow lagoon floors are frequently exposed and broken up into flat pebbles. By *Ptychaspis* subzone time the drop in sea level stops the shoreward growth of the algal reefs. Fine lime muds from the reefs form the lower beds of the Fort Sill which is the only limestone extending into the eastern Arbuckles.

In Texas, unlike Oklahoma, the slow marine regression is accompanied by the reappearance of fine clastics in steadily increasing amounts. The silty lime muds of the *Ptychapsis* subzone in the Llano district grade north-westwards into the Point Peak siltstone. Point Peak sedimentation marks low sea level in the Texas area and is characterized by facies of siltstone, shale, and silty limestone, representing fines settling in the quiet tidal lagoons; intraformational conglomerates from the frequently exposed tidal flats; and stromatolite biostromes penetrating the lagoons wherever fines reach zero. Near the top of the Point Peak a 30 cms to 1 metre zone of fossiliferous limestone with abundant silicified valves of *Plectotrophia* and *Billingsella* is a marker bed which also occurs in the Arbuckle Mountain section FREDERICKSON, 1956). Eastwards in Texas the *Plectotrophia* specimens spread over 12 metres or more of beds which pass laterally into the dolomite facies (BARNES and others, 1959). This observation suggests the brachiopod habitat as the clear protected lagoon waters along the inner reef edge.

The last Upper Cambrian transgression began during the *Prosaukia* subzone and reached maximum extent by the middle Trempealeauan. In the northern part of the region the shoreline was slowly pushed westwards, but the trend of the Nemaha Ridge appears to have formed a persistent barrier in South Dakota, eastern Nebraska, and north-eastern Kansas. Tidal flats of glauconitic fine sands persist over most of the area. In Missouri and southern Iowa the stromatolite reefs follow the deepening sea northwards and westwards. Hexactinellid and other sponges are a significant faunal element of the deeper waters in and around the reefs. By middle *Saukia* Zone time the northern end of the Transcontinental Arch is shallowly submergent as an island archipelago. The stromatolite reefs and biostromes build northwards over the tidal flats across central Iowa into Minnesota, and across Missouri into eastern Kansas. The tidal flats and beaches are narrow, but erosion of the hilly shore of Siouxia prevents stromatolite growth up to the shore line. In south-eastern Missouri nearly all the islands of the Saint Francois Mountains are now below sea level and overgrown by the stromatolites. In north-eastern Oklahoma, islands still rise above the sea but they are small and low and their shores are covered with algal mats.

In the southern half of the region the rising sea penetrates a broad north-westward trending valley across northern Oklahoma and western Kansas and clean transgressive sands are deposited. The algal reefs move inland, north and west, across the lime-mud lagoons. By *Saukia* Zone time the stromatolites occupy the shallow clean waters in the centre of the strait and extend to the western border of Kansas. Fluctuating currents control the distribution of sand, tidal flats, and stromatolite reefs within the strait. The sublittoral waters of the lagoons in southern Oklahoma gradually fill with lime mud until during the *Saukia* Zone the area becomes carbonate tidal flats with scattered tidal pools and ponds. FREDERICKSON (1959) reports the presence of detrital dolomite clasts in the Signal Mountain beds of the Arbuckles, but notes that nowhere can evidence of a disconformity be found. The clasts are contemporaneous debris torn from the algal barrier reef. In Texas the presence of a north-east to south-west trending barrier of stromatolite reefs is suggested by BARNES and others, (1959) and the west to north-west encroachment of the stromatolites with the transgressing sea of the *Prosaukia* Zone is shown in the outcrops of the Llano district. The algal masses become larger and more numerous at the top of the Point Peak and pass without interruption into the San Saba lithology. By the middle Trempealeauan broad sand flats line the western shore and slope gradually into the deeper water of sublittoral lagoons behind the reefs. The glauconitic, coarse-grained biosparites of the west, accumulated by waves and tidal currents, pass eastwards into nonglauconitic lime mud from the reefs.

In the later half of the Trempealeauan a regression is indicated by the eastward movement and the seaward spread of a regressive sandstone. Shoaling conditions appear throughout the northern half as the Jordan sands spread across Iowa. A band of tidal sand flats with stromatolite biostromes scattered along major tidal channels remains in the upper Mississippi Valley. In Missouri the regression extends eastward to the Illinois border and exposes most of the coastal shelf. The regressive sands are later eroded or reworked by the transgressing seas of the Early Ordovician to form the Gunter Sandstone.

To the south the Kansas strait is exposed and a sheet of regressive sands spreads southwards. Shallow lagoons and lime mud tidal flats exist for a time but finally coarse arkosic sands reach the Arbuckle Mountains area. In the Wichita Mountains the similar lithologies of the upper Signal Mountain and basal McKenzie Hill Formations suggest that shallow tidal pools and tidal channels persist. In Texas the regressing sea brings fine sands into the Llano district and tidal sand flats build eastwards into the lagoon, which is narrowed but persists west of the barrier reef. In the lagoon fine lime muds settle or masses of organic debris are stranded. Stromatolite biostromes build up along the tidal channels until they are destroyed by an influx of sand (WINSTON and NICHOLS, 1967). The environment remains unchanged as the Earliest Ordovician trilobite assemblage enters the area with another marine transgression.

10. The South-West Region

10.1. Lithostratigraphy

In the South-West Region, roughly 970 kilometres square, Cambrian sections are relatively thin and exposures are found mainly in the river canyons of the Colorado Plateau and scattered mountain uplifts. Remnants of Upper Cambrian occur in the Rocky Mountains of Colorado, but the eastern half of the region was land throughout the Cambrian (Fig. 25). Several sections from the Cordilleran geosyncline in the west have been included to show their relationship to the sections of the outer shelf. Scattered deep wells in eastern Utah and wells in eastern Colorado and adjoining Kansas provide the only information in those areas.

Lithofacies changes occur across the shelf from west to east and from north to south. In eastern Utah and adjoining north-west Arizona the sections of the outer (unstable) shelf are more closely related to those of the adjoining miogeosynclinal basin of central Utah than to the stable shelf sections of Colorado. In south-eastern Arizona the shelf borders the Sonoran miogeosynclinal basin and most of the unstable shelf sites are located in northern Mexico and were destroyed during Mesozoic and Cenozoic orogenies. The southern Arizona and New Mexico sections are on the stable shelf. The western shelf of Sierra Grande was repeatedly exposed to subaerial erosion during the early Palaeozoic and an unknown thickness of Upper Cambrian sediments was removed.

The nomenclature of the lithic units usually changes at the state boundaries. In central Colorado (Sections 10-10, 11-11) the variable thickness (0 to 183 metres) of the basal Sawatch Sandstone or Quartzite reflects deposition on a Pre-Cambrian surface of moderate relief. In the Sawatch Mountains three members are recognized: (1) a lower

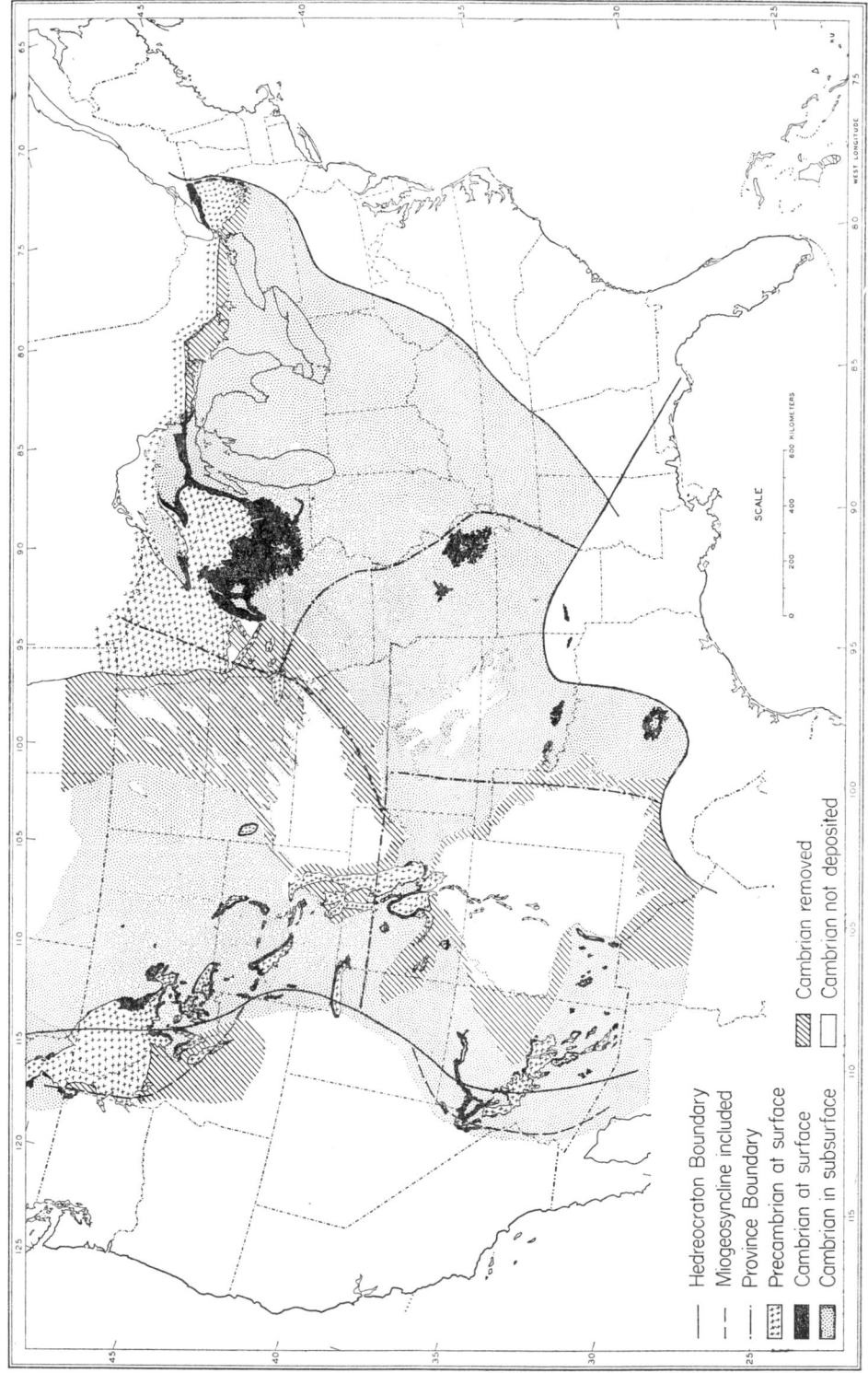

FIG. 25. Areal distribution of Cambrian on the Craton.

white quartzite with sporadic basal conglomerates and arkoses, (2) a middle glauconitic red-brown sandstone with some calcareous beds, and (3) an upper white quartzite. In Glenwood Canyon to the north-west only a single quartzite unit is recognized. Sporadic inarticulate brachiopod and trilobite remains indicate a Dresbachian and Franconian age for the formation. The overlying Peerless Formation consists of interbedded thin dolomites, sandy dolomites, dolomitic sandstones, and shales. Ripple marks, 'fucoid' markings, tracks, and trails are common on the irregular bedding planes. The variable thickness (1 to 30 metres) is due both to lateral thinning during deposition and post-Cambrian erosion. Late Franconian and also Trempealeauan fossils have been obtained from this lithic unit. The Peerless is recognized by BERG and ROSS (1959) on the east side of the Front Range, overlying the Sawatch and disconformably underlying the Lower Ordovician Manitou Formation. Approximately 160 to 240 kilometres to the east, along the Colorado–Kansas border, a similar section occurs in the subsurface, and in one well the basal Sawatch is dated by fossils as upper *Crepicephalus* Zone (MCELROY, 1965).

In north-western Colorado the Dotsero Formation overlies the Sawatch and consists of interbedded argillaceous sandy dolomites, shales, thin limestones, limestone pebble conglomerates, and a massive 1·5 metre algal reef at the top occurring as a marker bed throughout much of an area of over 1,000 square kilometres (BASS and NORTHROP, 1953). Trilobites and graptolites of Trempealeauan age are common in the beds.

To the west the thicker sections of the unstable shelf are known from well records. The Utah nomenclature from exposures in the southern Wasatch Mountains is used. The Tintic Quartzite consists of clean quartz arenite with quartz pebble lenses in the lower third. Calcareous, glauconitic, cross-bedded sands and thin shales appear in the upper half of the unit, and the uppermost beds grade vertically into the overlying Ophir Shale. The Ophir consists of brown to olive-green, micaceous shales with irregular bedding planes often covered with tracks and trails. Lenses and thin beds of dark grey shaly limestones become common in the upper half of the unit and have been traced westwards as tongues of a carbonate unit. Fossils of the *Glossopleura* fauna occur on the bedding planes.

The Maxfield Limestone is a light-grey, fine-grained limestone with many lithic variations: oolitic beds, limestone pebble conglomerate beds, dolomites and dolomitic limestones, and beds filled with white, twig-like, stromatolites. The lithologies intergrade vertically and laterally, and interbedded grey-green shales may become prominent in the upper part of the unit. The Maxfield and Ophir cannot be recognized in the anomalous section of the General Petroleum Company's Schulte #1 well, where sandy dolomites occupy the same interval. Westwards to the edge of the miogeosyncline the Maxfield is divided into two lithic units by the predominance of limestone in the lower portion—the Hartman Limestone—a dark, blue-grey, fine-grained limestone mottled and streaked with argillaceous laminae. Layers of edgewise conglomerate, oolitic beds, and horizons of filled worm burrows are present. The upper portion is the Bowman Formation of dolomite with thin interbeds of mottled limestone, oolites, edgewise conglomerates, and dark green calcareous shales. The shales increase rapidly eastwards in the upper half of this unit as seen in the Three States Natural Gas Company's Sinbad #1 well.

The Lynch Dolomite is the uppermost unit and may reach a total thickness of nearly 430 metres along the outer edge of the shelf, where it is overlain by Lower Ordovician beds. The thinner sections of the shelf are eroded remnants, The Lynch is a massive-bedded dolomite with local variations. The basal 90 metres may show rapid lateral and vertical gradations between light-grey mottled limestones and light to dark grey dolomites filled with twig-like stromatolites. Interbeds of grey dolomitic shale may be abundant (Section 12-12). The upper half of the unit is more persistently of dolomites finely laminated with shale or silt particles at recurrent horizons. Fossils have not been reported from most of the unit but a few rare specimens from the top beds are of Early Ordovician age and it is assumed that the Lynch extends from late Middle through all of Upper Cambrian time.

The nomenclature of the central Utah sections has usually been carried southwards in the subsurface to the Four Corners area. But as LOLEIT (1963) pointed out this area is much closer to the Grand Canyon geographically, the Cambrian sections are more closely related to the Grand Canyon sequence in their depositional history, and the Grand Canyon nomenclature should be used where possible. (Sections 12-12, 13-13). Thus the eastward 'shaling up' of the Bowman Formation and the lateral change into the Maxfield Formation is shown by comparing the Sinbad #1 well with the El Paso Natural Gas Company's Packsaddle #1 well (LOCHMAN-BALK, 1956b). But the eastward 'shaling up' in the Skelly Oil Company's Nokai #1A well relates to the shoreward lithological changes which are occurring in the Grand Canyon sequence exposed to the south-west along the Bass Trail.

The northern Arizona (Grand Canyon) sequence begins with the Tapeats Sandstone, a cliff-forming, coarse-grained, cross-bedded sandstone with a local basal conglomerate. The variable thickness reflects the uneven Pre-Cambrian surface. Near the top interbedded, fine-grained, sparsely glauconitic sands and shales form a transition zone into the overlying Bright Angel Shale. The Bright Angel is a series of laterally and vertically interbedded fine sandstones, siltstones, shales, brown dolomites, and thin hematite layers. The beds are thin to platy, laminated, grey-brown in colour, with mica and glauconite common, numerous worm burrows, and tracks and trails on the irregular bedding planes. Trilobite remains may be common, rare, or absent. McKEE (1945, fig. 1) noted the transgressive nature of both the Tapeats and the Bright Angel by the progressive eastward rise of the lithic units across the Late Lower Cambrian and Early Middle Cambrian faunal zones. East of the Little Colorado River fine sands are predominant in the Bright Angel.

The Muav Limestone is a massive-bedded, mottled, dense limestone (micrite) with local thin green shale partings. Small flat-pebble conglomerate layers were used as marker beds by McKEE (1945, fig. 1), who noted that the lower two-thirds of the limestone 'shales out' eastwards into the fine clastics of the Bright Angel and that algal limestone often occurs in these marginal positions. East of the Little Colorado the uppermost limestone becomes a calcareous siltstone and sandstone unit which is unnamed. From a maximum of 244 metres in the western Grand Canyon the unit thins to a featheredge near the Little Colorado River junction. Several small trilobite faunules indicate the *Bathyuriscus–Elrathina* and *Bolaspidella* Zones.

Throughout northern Arizona and southern Utah the Muav is overlain by a variable thickness of massive-bedded, white, granular, unfossiliferous dolomite, which was called undifferentiated Cambrian by McKEE (1945) of probable Upper Cambrian age, or Supra-Muav by WOOD (1956). In lithology and stratigraphical position the unit is similar to the Lynch Dolomite of central Utah, but as might be expected across a distance of 320 kilometres there is considerable difference in lithic details from the type Lynch. LOLEIT (1963) suggested that these post-Muav dolomites be tentatively referred to as 'Lynch' until they can be studied in detail and properly named. Post-Cambrian erosion has removed most of the dolomites from the eastern outcrops and the Four Corners area wells. A few reports of basal Ordovician conodonts from sandy dolomites and dolomitic sands in some wells suggest that remnants of the eastern clastic equivalents of the 'Lynch' dolomite may still be present in the subsurface of the San Juan Basin.

The lithic unit of south-western Colorado exposures and adjacent well sections is the Ignacio Sandstone or Quartzite. The unit consists of an extremely variable (vertically and laterally) series of indurated quartz arenites, siltstones, shales, thin beds of unfossiliferous sandy dolomite, and local basal conglomerates (BAARS and SEE, 1968). Tracks and trails are common on the irregular bedding planes and beds of worm-burrowed sands are conspicuous. Inarticulate brachiopod valves, locally abundant, indicate a Late Cambrian to basal Ordovocian age. BAARS and SEE (1968) described the disconformable relationship of the McCracken Sandstone Member (basal Upper Devonian) upon the Ignacio Sandstone in the exposures in the San Juan Basin.

Southwards from the Grand Canyon, exposures of the eroded remnants of this sequence can be traced to the vicinity of Jerome, Arizona (Section 14-14), beyond which Pre-Cambrian rocks surface the Mogollon Uplift. In the south-west corner of Arizona, Lower and Middle Cambrian boulders in fanglomerates of the New Water Mountains and a faulted fragment of a section in the Little Harquahala Mountains (LOCHMAN-BALK, 1956a) suggest the former existence of Cambrian deposits comparable to the western Grand Canyon sequence throughout this area.

In south-eastern Arizona and southern New Mexico the late Middle Cambrian and Upper Cambrian sequence is relatively thin. The basal Bolsa Quartzite or Sandstone consists of unfossiliferous clean, vitreous, white to yellow, cross-laminated, quartz arenites or quartzites, with thin shaly siltstone partings in the upper half and pink feldspar grains in the lower. A basal conglomerate may appear locally and pebbly grit lenses are common in the lower half. Regional variations in thickness reflect the irregular (relief up to about 150 metres) Pre-Cambrian surface. The Bolsa Quartzite was confused with the subjacent late Pre-Cambrian Troy Quartzite in early reports, but recent studies (KRIEGER, 1961 and SHRIDE, 1967) have clearly delineated the two units and demonstrated the unconformable relationship of the Bolsa to the Troy Quartzite. The Bolsa is transitional upward into the overlying Abrigo Formation by a steady increase in thin-bedded, sandy, micaceous shales intercalated with the upper fine-grained, thinly laminated sandstones. GILLULY (1956) gave a thickness of 15 to 30 metres for the transition zone and would place the arbitrary base of the Abrigo at the lowest thin limestone bed.

In the Clifton–Morenci mining district LINDGREN (1905) named a similar basal clean quartzite the Coronado Quartzite, but the section has not been restudied and the name is unused.

The Abrigo Formation is characterized in the type (Bisbee) area as a predominant carbonate unit, but the limestone beds grade rapidly to the north-west, north, and east into sandstones and shales. STOYANOW (1936) studied the lithology and faunules of the formation and defined five formations from portions of the Abrigo as he observed it in the area. His units were essentially palaeontological units and could not be followed in field mapping. Later authors have retained the Abrigo Formation and recognized several members of distinctive lithology within it, which can be traced throughout the area. At the type section on Mount Martin, HAYES and LANDES (1965) recognized: (1) a lower shaly member composed of interbedded grey to orange claystone and thin beds of grey micritic limestone alternating with argillaceous detrital limestone conglomerates (this member was called by STOYANOW the Cochise Formation in the Bisbee district and the Santa Catalina Formation in the northern Santa Catalina Mountains.); (2) the ribbed limestone member composed of interbedded micritic skeletal limestone and micritic limestone edgewise conglomerate, both interlaminated with crinkly laminae of silty lime (this member was locally called the restricted Abrigo by STOYANOW); (3) a sandy member composed of glauconitic, cross-laminated, quartz arenites, cemented by calcite or dolomite, and interbedded with some dolomitic marl (in the Santa Catalina Mountains this member is in the lower part of the Pepper-sauce Canyon Sandstone of STOYANOW); (4) An uppermost limestone member of thin-bedded, medium grey to pinkish, detrital-micritic limestones with argillaceous or sandy laminae (this is the Copper Queen Limestone of STOYANOW). The member is very sporadic in occurrence and has often been removed by erosion. The northward replacement of the carbonates by clastics apparently begins near the base of the ribbed limestone member and gradually rises in the section. In the Santa Catalina Mountains (Section 14-14) the entire section has an anomalously low carbonate content for its geographical position. The Southern Belle Quartzite of STOYANOW is a 7·5m sandstone bed at the base of the ribbed limestone member. At Holy Joe Peak to the north (KREIGER, 1961) fossils demonstrate that Abrigo deposition covers the same time span but sandy carbonates appear only in the uppermost beds. North of this locality the clastic Abrigo Formation has been thinned by erosion.

Fifty six km south of Bisbee a section of probable Cambrian rocks has been described from the Cananea mining district, Sonora, Mexico (MULCHAY and VELASCO, 1954). A basal clean quartzite with some feldspar fragments in the lower part is conformably overlain by a thin-bedded, highly altered limestone. No fossils have been found but the lithology, thicknesses, and stratigraphical position suggest that these units are lithic equivalents of the Bolsa and Abrigo.

In the Dos Cabezos and Chiricahua Mountains SABINS (1957) referred the basal Palaeozoic formation to the Bolsa Quartzite. The unit consists of a basal quartz pebble and boulder conglomerate overlain by a lower 90 to 120 metres of coarse to medium-grained, thick-bedded, cross-laminated, clean quartzite with some arkose in the lower part, and an upper 30 metres of slope-forming, thin-bedded siltstone and shaly sandstone with thin limestones and sandy dolomites at the top. This upper unit would now be called the Abrigo Formation. The upper unit has yielded fossils of the *Cedaria* and the *Aphelaspis* Zones. In the immediately overlying beds of sandy dolomite which Sabins placed in the El Paso Group, specimens of *Billingsella* were found, indicating a Franconian age and correlating these beds with the Copper Queen Limestone Member.

In south-western and south-central New Mexico the basal sandstone, the Bliss Formation, is extremely variable lithically, both laterally and vertically. Above the basal medium to coarse-grained quartz arenites are intercalated thin lenses and beds of buff to brown, glauconitic, calcareous fine sandstones; siltstones; dark brown to green glauconitic shales; sandy argillaceous glauconitic dolomites and limestones; and beds of oolitic arenaceous hematite. Near Silver City an *Elvinia* zone faunule has been found in the lower beds but eastwards along the Rio Grande Valley the lower beds carry a *Prosaukia* subzone faunule and the upper beds contain a few species of earliest Lower Ordovician age. All beds appear conformable and gradational but small scour channels have been noted across some of the bedding planes.

10.2. Palaeogeography and palaeoecology

The South-West Region was dominated throughout the Cambrian by the Sierra

Grande positive element, the southern half of the Transcontinental Arch. On the north-west border the Uinta Peninsula persisted as a minor positive element. Immediately to the south the minor negative element of the Central Colorado Embayment appeared sporadically during the Upper Cambrian, responding to the steady downwarping of the outer shelf bordering the central Utah miogeosyncline. The Uncompaghre Uplift of west central Colorado may have been a minor persistent positive element during the Cambrian (BAARS and SEE, 1968). Southward the Defiance positive element was a hilly land mass projecting westwards into and across Arizona. Fault blocks or resistant Pre-Cambrian rocks formed an irregular coast of peninsulas and bays along the north-west side (BAARS and SEE, 1968) and an embayed coastline has also been reported along the west-central side (WOOD, 1966). In south-eastern Arizona the south to south-west shelf was shallowly inundated from the Sonoran miogeosyncline.

During the early Middle Cambrian, *Albertella* Zone time, the shore line and littoral zone were located mainly within the miogeosynclinal regions to the west and only in the western Grand Canyon area did sands of the littoral zone reach the outer shelf. During *Glossopleura* Zone time downwarping of the outer shelf bordering the central Utah miogeosyncline became general. In eastern Utah the littoral sands pass seaward into fine clastics settling on a shallow sublittoral sea bottom. The silts and clays of the Ophir settle in a low-energy environment beyond the turbulence of the littoral zone. Sedimentation was sporadic and some areas may have built up nearly to tide level, with intervening deeper wide tide channels and pools. On the seaward side of the mud-bottom lithotope, carbonate oozes accumulated wherever the deposition of silt and mud particles ceased. Scattered tidal flats developed but broad shallow bays are more abundant. The lateral and vertical gradations between the Ophir and the overlying carbonate unit demonstrate the constantly shifting positions of the two lithotopes over the shallowly inundated coastal shelf.

The maximum Middle Cambrian transgression occurred during *Bathyuriscus–Elrathina* Zone time. In the Utah embayment a narrow sandy littoral zone passed seaward into narrow tidal flats on which mud, sand or carbonate ooze accumulated depending upon the local supply of detrital material and the current energy and direction. Worm-burrowed sand layers are common and all surfaces are exposed at low or neap tide and sporadically littered with locally derived clasts. The carbonate ooze flats pass seaward into the shallow sublittoral zone where bioclastic layers alternate with micrites. Farther seaward the intertidal banks of stromatolites (algae) encroach upon the quiet waters of the sublittoral zone embayments. The stromatolite platforms require the high energy and clear sunlit waters of the open ocean and are built shelfward from the miogeosyncline whenever a transgression drowns the outer edge of the tidal flats bordering the shore. Throughout the Cambrian the algae associated with the thick dolomite accumulations show a conspicuous intolerance of both quiet water conditions and any terrigenous material in the water.

In late Middle Cambrian, *Bolaspidella* Zone time, a regression occurred which is attributed to a eustatic sea level drop. The position of the shore line did not change significantly nor did an uplift of the positive element introduce significant amounts of clastics. The tidal flats of carbonate oozes and fine muds move seaward to the edge of the intertidal stromatolite banks. A shallow sublittoral zone occurs only on the seaward side of the algal reefs. The algae of these reefs have not received modern study and classification, but the abundance of twig-like bodies and ovoid masses in the carbonates of these horizons was commented upon in all the early mining reports.

In Arizona the hilly upland of resistant Pre-Cambrian rocks permitted penetration by the sea only along narrow river valleys (WOOD, 1968). It was not until *Bathyuriscus–Elrathina* Zone time that an appreciable inundation of the shelf occurred. The shore is paralleled by a wide belt of sand, reflecting the more abundant supply of resistant clastics as the waves pound the rugged coast. The water deepens more rapidly and sand is carried seaward into the sublittoral zone. The high-energy environment moves most of the fine clastics entirely out of the area and only a narrow carbonate and mud flat develops, which merges seaward into a normal clear water sublittoral zone. However, by *Bolaspidella* Zone time many resistant masses of Troy Quartzite had been eroded and broader valleys opened out on the less-resistant Apache Group rocks (SHRIDE, 1967, fig. 12). As sea level drops, the narrow sand littoral zone merges seaward into mud or sand tidal flats. Beds of worm-burrowed sands are common. The available terrigenous material settles on the flats and on the seaward side lime ooze, bioclastic debris, and small lime clastics from the adjacent flats accumulate in the shallow sublittoral zone. By the end of the Middle Cambrian the shore had built southward nearly 160 kilometres.

During the Dresbachian the eastward transgression of the sea was not nearly as widespread as in the North-West Region. The marine transgression beginning *Cedaria* Zone time floods the tidal flats of the Utah embayment and in the clear turbulent waters the algal reefs build shoreward over them (HOWE, 1967). The shore and beach sands remain in the same position and only a narrow carbonate tidal flat intervenes seaward. A eustatic rise in sea level, rather than any tectonic movement, is indicated. During Late *Cedaria* Zone time continued sea level rise and active shore line erosion result in an eastward movement of the shoreline and deepening of the water to a shallow sublittoral zone. Mud and silt from the eroded old soils settle out next to the mature beach sands and farther seaward the bottom accumulates lime ooze, clasts washed from the adjoining algal flats, and organic debris. In northern Arizona currents carry the sands and muds westwards as submarine bars over the late Middle Cambrian carbonate sites.

In *Crepicephalus* Zone time a minor downwarp or continued sea level rise produces a broad tidal embayment across Colorado and transforms the Uinta Peninsula into an island group. The sandy *Hyolithes*-bearing limestone fragments from the Ferris diatreme (CHRONIC and FERRIS, 1961) in the south-eastern corner of Wyoming may date from this inundation. Fossils in the basal sands of several wells indicate that the shoreline lay near the 100° W meridian in Kansas. The southern shore of the embayment trended nearly east–west to western Colorado where a Ria coast of headlands and steep islands was formed by fault blocks of resistant Pre-Cambrian rocks. Sand tidal flats occupied the entire embayment with local pools where sandy limes could accumulate. Sedimentation was slow and sporadic and channeling is frequent across the bedding planes of this unit (the Sawatch Sandstone). Clean littoral sands grade laterally into dirty worm-burrowed sands and siltstones. Only in western Colorado does the carbonate tidal flat lithotope appear and grade seaward into the shallow sublittoral zone, where carbonate oozes, clasts, and organic debris accumulate in the quiet waters sheltered by the seaward belt of algal reefs. *Aphelaspis* Zone faunules are unknown in Colorado. By this time the shoals of the Colorado embayment had built up above tide level and as sea level drops the shoreline returns to its earlier position in eastern Utah.

In southern Arizona during *Cedaria* Zone time resistant rocks along the south-western edge of the Defiance Uplift force the shoreline to curve off to the east forming the southern Arizona embayment. A band of clean transgressive sands extends as far south as the

Santa Catalina Mountains (SCHRIDE, 1967, fig. 12) (the Southern Belle Quartzite); but seaward only a narrow tidal flat remains as the intertidal algal reefs move shoreward with the sea level rise. During Late *Cedaria* Zone time rising sea level and shore erosion push the shore line into south-western New Mexico. A tidal sand flat develops near shore and the deepening water produces a shallow sublittoral zone into which mud and silt from the reworked soils settle. Seaward are areas of lime ooze and organic debris. By *Crepicephalus* Zone time broad carbonate flats border an inner belt of sand flats. The north–south facies change from sand to lime throughout the Dresbachian suggests that the curved northern shore of the embayment was maintained by a rugged and resistant headland of the Defiance positive element. By early *Aphelaspis* Zone time sea level falls and the shore moves westwards into Arizona. Littoral sands and tidal sand flats extend nearly to the Mexican border. A warm climate causes evaporation of the sea water in the shallow pools among the flats, and sandy dolomites rather than limestones appear in the sandy member of the Abrigo. Rare fossils of this zone have been found in these sandy dolomites and sandstones (GILLULY, 1956; COOPER and SILVER, 1964; SABINS, 1957).

No record of diagnostic species of the *Dunderbergia* Zone is known and it appears that the entire western shelf of Sierra Grande was emergent during the end of the Dresbachian.

The marine transgression marking the beginning of the Franconian is less noticeable in the South-Western Region than in the northern regions, a feature caused primarily by the steeper coastline of the region. The slow rise in sea level simply inundated the lowest areas of the coastal plain, revived the broad tidal flats, and restores the shoreline to the position it occupied about the end of *Crepicephalus* Zone time. In the Utah–Colorado embayment there is a moderate shoreward movement of the algal reefs in response to the transgression, but the reworking and shifting by the tides of silt and mud from the eroded soils kept the algal platform about 160 kilometres west of the low shoreline.

A marine regression began by the later part of the *Elvinia* Zone and characterizes the *Conaspis* Zone time. It causes a moderate westward growth of the shoreline, bordered by narrower tidal flats. Most important, however, was the seaward movement of the algal platforms, with the development of a continuous belt of shallow sublittoral lagoons in which lime oozes, small lime clastics, and organic debris slowly accumulated. No fossils have been found in the sands of this probable age in Colorado, and though the micrites are known from the wells of eastern Utah (LOCHMAN-BALK, 1956b) fossils have not yet been recovered. In north-east Arizona and south-west Colorado all Franconian and Trempealeauan deposits have been removed by post-Cambrian erosion. The palaeoecological patterns have to be restored from the remnants of the lithic units which occur to the west and to the east, wherever the inner shelf sites have been preserved.

By the end of *Conaspis* Zone time sea level had reached its lowest position. Many of the sublittoral lagoons had built up to carbonate ooze flats, and regressive conditions were stabilized. As the later Franconian began (*Ptychaspis* subzone time) a new marine transgression is first evidenced by the shoreward movement of the intertidal algal platforms over the earlier lagoon sites. The shore moves slightly eastwards and the tidal flats widen. Slow and continuous sea level rise reaches a maximum by the end of *Prosaukia* subzone time. The Central Colorado embayment is again crossed, placing the shore in western Kansas essentially at the earlier Dresbachian position. The entire embayment is the site of wide sand flats, shifting shallow tidal channels, and supersaline pools. The exoskeletal debris of trilobites was occasionally washed into pools or tossed on the flats. Sporadic fossils of the *Ptychaspis–Prosaukia* Zone are found in the sands and sandy dolomites as far east as the Front Ranges. Sedimentation was slow and sporadic and the fine sands and silts were often reworked. Tidal channeling on the bedding planes is common and lateral facies changes, reflecting the varying energy regimes on the flats, occur rapidly. At the entrance of the embayment the deepening waters behind the algal platforms form shallow sublittoral lagoons. On their landward side the lagoon waters receive small amounts of finest terrigenous material, usually sporadically in response to fluctuations in climate, and the micrites show grey-green clay or silt partings. Organic debris may be locally abundant. On the seaward side a supply of lime ooze and lime clasts is derived from the algal platforms.

The Dotsero Formation of north-western Colorado contains a characteristic middle *Saukia* Zone fauna and algal beds as well as thin sands, silts, and shales comprise the varied lithology. No *Saukia* Zone faunas are known farther east in Colorado, and in the Front Range sections (BERG and ROSS, 1959) the Early Ordovician beds lie disconformably upon the *Ptychaspis* Zone beds. But the persistent shoreward, eastward movement of the intertidal algal reefs during this time (the massive 1·5 metre Clinetop algal bed forms the top of the Dotsero) and the conspicuous failure of terrigenous material can best be explained by a continuing sea level rise, which reached its maximum height during middle *Saukia* Zone time and may have split the Transcontinental Arch into several large islands. The regressive condition at the end of the Trempealeauan is restored from the sections of eastern Utah where deposition apparently was continuous from latest Upper Cambrian into Early Ordovician. The easternmost known deposits are argillaceous limestone pebble conglomerates, representing the reappearance and seaward spread of wide carbonate ooze flats, which merge westward into the inner edge of the algal platforms.

To the south marine transgression of the early Franconian again inundates the southern Arizona embayment and moves eastwards into south-western New Mexico. Soil from the coastal plain settles over the broad, shallowly inundated, tidal flats. Fine sands are the usual lithotope, often thoroughly burrowed by worms, but locally silts and clays are abundant. Pools of supersaline waters are settling basins for scanty organic debris, and narrow tidal channels cross and recross the flats. All the deposits of this age are dominantly clastic and the position of the carbonate belts must have been at least 240 kilometres from shore.

During *Conaspis* Zone time a minor regression stabilizes the lithotopes along the entire western shore of Sierra Grande and similar lithotopes extend southwards from the Utah area. The configuration and position of the shore remains much the same. The carbonate lithotope of the shallow sublittoral lagoons appears in the western part of the region. Locally within these lagoons banks of *Billingsella* and *Eoorthis* build up where clear water and food are available. Occasional storm waves and high tides carry the valves landward and strand them over the sand flats, but the sites of the banks have not been located as in the North-West Region. By late *Conaspis* Zone time much of the sublittoral lagoon is built up into a carbonate ooze flat.

As sea level begins to rise during the *Ptychaspis* subzone, little or no change is made in the position of the shoreline or the bordering broad sand flats. The sublittoral lagoons gradually widen and deepen and probably the algal platforms encroach upon their seaward sides (but the sections containing these deposits have been removed by erosion).

As sea level reaches its maximum height during *Prosaukia* subzone time the shore is pushed eastwards to, or just beyond, the Rio Grande Valley in south-central New Mexico; but there is now no evidence of any farther northward spread. This is the farthest inland position of the shore during the rest of the Upper Cambrian. Broad sand flats are crossed by wide sublittoral tidal channels and seaward the deepening water forms wider sublittoral lagoons in which micrite and bioclastic limestones with abundant thin sand laminae accumulate. Currents and waves spread trilobite exoskeletal debris from the lagoon sites widely over the sand flats. In New Mexico the latest Upper Cambrian fauna known belongs to the *Prosaukia* subzone.

The sedimentation of the New Mexico sections suggests that the ecological conditions established at this time continued throughout the Trempealeauan until the marine regression near the end of the stage. Among the tidal sand flats, shallow ponds and pools formed in slight depressions on the flooded coastal plain. As time passed, silt and sand bars clogged the connecting tidal channels, and the water in these pools became super-saline. Primary dolomite was precipitated as thin laminae or beds among the sands. Near the end of the Trempealeauan conspicuous but thin layers of sandy oolitic hematite accumulated in many of these isolated pools. As all connection with marine waters was cut off by the regression, the ponds became stagnant freshwater bodies. Above the hematites, thin cross-bedded sands and silts with thin dolomite lenses and beds appear again. Scattered fragments of graptolites and brachiopods of earliest Ordovician age are present.

In Arizona the *Prosaukia* fauna is obtained from the few sandy dolomites at the top of each section which have been spared by post-Cambrian erosion (Sections 15-15, 16-16, 17-17). In the Trempealeauan the depositional history of this region was probably similar to that of the Utah–Colorado embayment to the north. During the early Trempealeauan, with the shoreline stabilized at or near its maximum eastward position of *Prosaukia* Zone time, and a minimum of terrigenous debris from the land, the lagoons were filled largely by carbonate ooze from the stromatolite platforms. As the floor built upward into the surf zone, the stromatolites extended shoreward over it until finally only a narrow flat of carbonate ooze and clay marked the line at which the stromatolites were halted by small amounts of mud. The sea regressed during late Trempealeauan and the littoral sands and the sand flats remain narrow—there is no material of this size being brought to the shore. As the waters shallow, the carbonate ooze flats widen at the expense of both the eastern sand flats and the stromatolite reefs, which die off shoreward as there is no longer a slowly downwarping (deepening) site of clear water in which they can build. The appearance of dolomite beds with clean quartz sands signalizes the beginning of the basal Ordovician transgression and not far above this horizon exo-skeletons of Early Ordovician trilobites are found.

11. The North-West Region

11.1. Lithostratigraphy

In the North-West Region, roughly 1,040 kilometres square, Cambrian exposures are abundant in the Rocky Mountains and in the numerous small uplifts on the High Plains east to the Black Hills. Also the subsurface Cambrian under the High Plains is often reached in oil company drillings, especially in the Williston Basin. The eastern third of the region is underlain by the Transcontinental Arch from which the originally thin Upper Cambrian sediments have been completely removed by Palaeozoic erosion. The deeply eroded scattered sections of Idaho are also included.

The Cordilleran miogeosyncline is characterized by a series of alternating negative marginal basins and mildly positive marginal uplifts (KRUMBEIN and SLOSS, 1963, p. 418). Montania, a small positive element, straddles the United States–Canadian border and is overstepped southwards by Middle Cambrian formations (FRITZ and NORRIS, 1966). In northern Idaho the sections around Pend Oreille Lake (Section 1-1) are the eroded remnants of the Middle Cambrian sections overstepping Montania from the south-west. Most of this marginal basin was destroyed by the later intrusion of the Idaho Batholith. The Gold Creek is a coarse-grained white, cross-bedded, unfossiliferous basal quartzite with pebble beds and lenses. The Rennie is of drab-green, argillaceous, micaceous shales with intercalated fossiliferous limestone lenses and nodules. The Lakeview is a shaly, thin-bedded fossiliferous limestone at the base, overlain by massive limestone and dolomites. This remnant sequence relates to the sections in the Philipsburg quadrangle, Montana, located on the eastern edge of the marginal basin. A comparable complete section of Middle and Upper Cambrian units was present in the north-central Idaho area. Predominantly quartzite sections of east-central Idaho have recently been assigned in part to Middle Cambrian and reveal the site of another marginal uplift which was tectonically active during the Middle Cambrian as well as later in the early Palaeozoic (BEUTNER and SCHOLTEN, 1967). The lithic units, unnamed, are a white to purple quartzite with pebble lenses; overlain by grey-green to tan slates and interbedded sandstone; succeeded by dolomitic sandstones and dolomites. Large *Skolithos* tubes are locally abundant in the quartzites. At Leatherman Pass, Middle Cambrian fossils occur about 150 metres above the base (KETNER, 1964). An overlying unit of 610 metres is undated but may be Middle and Upper Cambrian and/or Lower Ordovician, as the upper half (the Kinnikinic Quartzite) carries Middle Ordovician fossils.

Upon the shelf, lithofacies changes occur both in an east-west direction and in a north-south direction and a number of nomenclatural changes in lithic units across the region recognize these changes. The formations of north-western Montana were established by WALCOTT (1917) and DEISS (1936). Their cyclic development relates them to the units recently recognized by AITKEN (1968) in south-western Canada.

The basal Flathead Sandstone is a cross-bedded, thin to thick-bedded, clean quartz arenite with interbedded thin sandy micaceous shales near the top. The thickness can vary rapidly in response to the uneven Pre-Cambrian surface of deposition. The Flathead grades upward into the Gordon Shale, a sequence of drab-green and brown, fissile, micaceous shales with intercalated thin sandstones and dirty limestones. The overlying section consists of thick carbonate units cyclically interrupted by thinner intervals of calcareous grey-green shales. DEISS placed the shale interval as the basal part of each carbonate unit (with the exception of the Gordon Shale), because the lower contact of each shale interval is always abrupt upon the underlying carbonate, whereas the upper shale contact is gradational into the overlying carbonate beds. The available fossils suggest that the abrupt basal shale contact may be essentially synchronous within north-western Montana. AITKEN and GREGGS (1967) recorded the same phenomenon as it occurs in the Middle and Upper Cambrian sections of the southern Canadian Rockies.

The lowest carbonate, the Damnation Limestone, is a grey, fine-grained, thin to medium-bedded limestone with grey clay partings on which trilobite fragments are found. The Dearborn Limestone consists at the base of a 9 metre slope of drab sandy shales and thin sandy limestones, which grade upwards through intercalated limestones and thin shales into a 60 metre cliff of massive-bedded grey limestones. The lower shales are very fossiliferous. The overlying Pagoda Formation thins rapidly to the north-east as the shale content increases markedly, and it intertongues with the predominantly clastic unit, the Pentagon Shale. The basal shale interval of the Pagoda occupies about one-fifth the thickness and consists of green fissile shales interbedded with thin argillaceous limestone and several limestone pebble conglomerates. The upper cliff consists of cream to grey, fine-grained limestones with a thick-bedded oolitic limestone near the middle. Fossils were obtained both from the lower shales and from several limestones throughout the carbonate interval. The Pentagon Shale, as recognized by DEISS (1939), consists of inter-calated grey, thick-bedded, platy and lumpy shale; blue-grey, argillaceous, thin-bedded limestones; a few dark-grey fissile shales; and nodular to platy, white to brown, argil-laceous limestones. Trilobites and brachiopods are particularly abundant in the lower half of the formation. The Steamboat Limestone consists of a basal green shale interval overlain by several cliff-forming, fine-grained, thick-bedded limestones interrupted by slopes of interbedded green shales and thin-bedded grey limestones containing algal masses (DEISS, 1939, pl. 7). Late Middle Cambrian fossils occur in the shale units. The Switchback Shale rests conformably, but with abrupt lithic change, upon the Steam-boat. The formation consists of interbedded calcareous and arenaceous shales with some flaggy grey limestones and thin-bedded small limestone pebble conglomerates. Only a few fossils have been obtained from the limestones. The Devil's Glen Dolomite is the top unit of the sequence and, although the thickest, has been subjected to extensive post-Cambrian pre-Devonian erosion which must have removed most of the Upper Cambrian deposition. The Devil's Glen Dolomite consists of light grey, massive dolomite cliffs. Restudy is needed to determine whether the dolomite is early or late diagenetic in origin.

The central Montana lithic units (Sections 3-3, 4-4, 5-5) are the most widely recognized geographically and actually extend into northern Wyoming, although the names are usually changed at the state border. The basal Flathead Sandstone consists almost entirely of pure quartz arenite, but glauconite and hematite may appear in the upper beds which grade into the overlying Wolsey. Ripple marks and cross-bedding are common, but no fossils except a few broken inarticulate brachiopods are known. The Wolsey shale consists of grey-green to purple, fissile, micaceous shales interbedded with a few glauconitic shaly sandstones near the base and with lenses of silty, glauconitic limestone near the top. Worm castings and inarticulate brachiopod and trilobite fragments occur on the irregular bedding planes. The Meagher Limestone is a mottled micrite, thin-bedded, with conspicuous green shale partings. The mottling is of irregular yellow-tan areas in a grey to black matrix. Oolitic limestone, glauconitic bioclastic limestone, and limestone pebble conglomerates may appear throughout the unit. The carbonate beds are steadily replaced eastwards by shales and the entire unit 'shales out' near 110° W. The overlying Park Shale is largely the complement of the Meagher, ranging from zero in the west to over 150 metres in the north-eastern sections. It consists of grey-green, slightly micaceous shales interbedded with thin, brown, micaceous sandstones or grey limestones. In the middle and upper part fossiliferous pebble conglomerates become common. The Pilgrim Formation is a cliff-forming unit of thick- to thin-bedded grey limestone, oolitic mottled limestone, and limestone pebble conglomerates with thin partings or lenses of calcareous shale. Locally the top limestones contain much white quartz sand. Fossils may be abundant throughout the formation. Three members are recognized in the western sections (Radersburg Quadrangle) and can be traced eastwards to the Three Forks district (LOCHMAN-BALK, 1956). The Snowy Range Formation is the uppermost unit and contains three members. The Dry Creek Shale Member at the base, is of intercalated black-green fissile shale, sandy shale, fine-grained calcareous or pure quartz arenites, siltstones, and flaggy arenaceous limestones or thin limestone pebble conglomerates. This dominately clastic unit is often covered on the surface but forms a distinctive subsurface unit. The Sage Member consists of thick- to thin-bedded limestone pebble conglomerates intercalated with green calcareous shales. The shale increases steadily eastwards at the expense of the limes. In south-central Montana and northern Wyoming the base of the member contains a distinctive columnar algal reef (stromatolite type SH-V/LLH-C, LOGAN, REZAK, and GINSBURG, 1964). The Grove Creek member is deeply eroded in Montana but 38 metres occur in the northern Bighorn Mountains. The lower part contains thick-bedded limestone pebble conglomerates interbedded with shales; in the upper part fossiliferous limestones, limy siltstones, and thin arenaceous dolomites are common and the unit appears as a cliff or a series of ledges above the Sage Member.

In far western Montana (Philipsburg Quadrangle) the steady increase in carbonate near the edge of the shelf has produced a different succession (Sections 3-3, 1-1). The basal Flathead Quartzite is overlain by the Silver Hill Formation which consists of a lower interval of brown-green, chloritic, micaceous shales; a middle fine-grained, medium-bedded, mottled limestone; and an upper interval of interbedded shale and dark brown limestone. The overlying Hasmark Dolomite is a 300 to 430 metres cliff-forming carbonate unit. No trilobites or brachiopods are known but many horizons contain masses of twig-like algal bodies and pisolites (EMMONS and CALKINS, 1913, pl. 8). Locally tongues and lenses of grey-green shale (Park lithology) appear near the middle of the formation and several horizons of the dolomite are characterized by an increased argillaceous content. The overlying Red Lion Formation consists of a thin lower member of interbedded, black to dark-green shales, thin siltstones, and a few flaggy limestones; and a thick upper member of grey fine-grained limestones interbedded with distinctive irregular laminae of yellow siliceous siltstones and small flat pebble conglomerate layers (EMMONS and CALKINS, 1913, pl. 10).

The eastern Montana succession is known from the Little Rocky Mountains uplift and the Williston Basin well cores. The section is predominately clastic (shales and sandy shales) and the carbonates are nearly all edgewise limestone pebble conglomerates. KNECHTEL (1957) recognized only the basal Flathead Sandstone and the overlying Emerson Formation, a 300 metre unit of interbedded shales, a few thin grey bioclastic limestones, and thick lenses of edgewise conglomerate and trilobite coquina nodules. Fossils occur throughout the unit. Two hundred and forty kilometres farther east the subsurface sections of the Williston Basin contain no Middle Cambrian. Approximately one-half of the section consists of sandstone and siltstone, which, except for a basal sandstone unit, are intimately intercalated with thin green shales, thin crystalline limestones, and limestone pebble conglomerate lenses. The succession is the lithic equivalent of the Deadwood Formation, described from exposures at Deadwood, South Dakota, situated on the southern edge of the Williston Basin, and the name is used throughout the subsurface of the Basin. The Deadwood Formation thins eastwards to zero as a result of: (1) pinch out against the Transcontinental Arch, (2) off-lap during the Early Ordovician, and (3) pre-Chazyan erosion of the upper beds.

In Wyoming a similar succession of Middle and Upper Cambrian lithic units can be recognized but a different nomenclature is employed. In western Wyoming on Double

Top Peak (Section 7-7), BLACKWELDER (1918) recognized the basal Flathead Sandstone and the overlying Gros Ventre Formation, the latter consisting of a lower grey-green micaceous shale member; a middle member of fine-grained, dark grey, mottled limestone with green shale partings, and two to four interbedded green shale intervals; and an upper member of grey to purple fissile shales with an algal reef (LLH-S/LLH-C, LOGAN, REZAK and GINSBURG, 1964) (*Tetonophycus*) in the lower part and numerous thin limestones and limestone pebble conglomerates in the upper. B. M. MILLER (1936) named the middle carbonates the Death Canyon Member, and noted that the member becomes thinner eastwards by the progressive replacement of the carbonates by shales, from the lower part finally into the upper part. SHAW (1954) raised the Gros Ventre to group status and designated the lower Wolsey Shale, the middle Death Canyon Formation, and the upper Park Shale. The term 'Gallatin Formation' has been persistently and incorrectly used in Wyoming since 1896 when it was introduced from Montana by Weed in the Yellowstone region for all carbonates of presumed Upper Cambrian age above the Gros Ventre. B. M. MILLER (1936) designated the lower, cliff-forming, oolitic, and crystalline mottled limestone as the DuNoir Member. In 1955 SHAW and DeLAND recognized the overlying thin interbedded green shale and glauconitic limestone pebble conglomerates as the Dry Creek Shale. An upper, often cliff-forming unit of interbedded glauconitic limestone pebble conglomerates, thin-bedded grey crystalline limestone and green shales was named the Open Door Limestone. As in Montana all carbonate units 'shale out' eastwards and disappear near the 110° W. meridian. B. M. MILLER (1936) named as the Depass Formation the monotonous succession of intercalated thin sandstone, sandy shales, shaly sands, and dark green shales with a few thin dirty limestones and limestone pebble conglomerates in the upper part. It overlies the Flathead and underlies the Du Noir Limestone, or in the Northern Bighorn Mountains, the Snowy Range Formation (HANSEN, 1962). In the subsurface of the Powder River Basin sandstone and siltstone become prominent, the carbonates thin markedly, and the sequence is the Deadwood Formation.

The increase in clastics is also conspicuous in a southern and south-easterly direction across Wyoming (Section 8-8). The Flathead Quartzite and Depass Formation can be carried as far as the subsurface of the Lost Soldier oil field, but the sands of the Rawlins Uplift are distinctive and are lithically the same as the basal sands of the Uinta Mountains in north-eastern Utah and north-western Colorado (Section 9-9), named the Lodore Formation by UNTERMANN and UNTERMANN (1949). The Lodore consists of a basal clean quartz arenite (the Flathead equivalent) overlain by a cyclic sequence of two lithologies: a medium-bedded, cross-bedded, medium- to coarse-grained sandstone and a usually thinner, fine-grained, glauconitic sandstone with green shale partings, completely burrowed, i.e., wormstone. SHAW (1954) designated a top unit, the Buck Spring Formation, consisting of thin-bedded brown sandstone and siltstones with a few late Middle Cambrian fossils. Comparison with the subsurface section in well U.P. #4, North Baxter Basin, (Section 9-9) indicates that the Uinta Mountains and Rawlins Uplift were minor positive elements on the shelf probably during the Middle and Upper Cambrian as well as post-Cambrian. Their thin dominantly clastic sections were deeply eroded.

11.2. Palaeogeography and Palaeoecology

In the North-West Region the palaeoslope of the broad continental shelf is slightly north of west. The ancestral Cambridge Arch and Uinta Arch form minor positive elements on the inner shelf. The outer unstable shelf, bordering the miogeosynclinal basins, downwarps early in the Middle Cambrian and continues a slow, steady negative movement during the period. The thickest sections are located here and carbonate sedimentation similar to that of the miogeosynclinal basins is dominant. The edge of the shelf was a tectonically weakened zone and became complexly thrust faulted during the Laramide Revolution (LOCHMAN-BALK, 1955). In the north-western part of the shelf the marginal uplift on Montania (FRITZ and NORRIS, 1966) and the unnamed uplift of east-central Idaho (BEUTNER and SCHOLTEN, 1967) are minor positive elements which persist throughout the Cambrian and provide quartz sands and even sporadic arkose (in the Meagher Formation of the Bridger Range) seemingly anomalous to the deposits of the outer shelf.

All Middle and Upper Cambrian sections in Montana and North Dakota show a progressive increase in percent of clastics (shale and silt) in a north-easterly direction. AITKEN (1968) records the same phenomenon in the subsurface Plains sections of Alberta. One or more major drainage systems of the Canadian Shield must have reached the ocean along this western coast. The larger supplies of sand, silt, and argillaceous material contrast sharply with the much reworked thin sands of the Eastern Region.

The inundated coastal plain is of low relief only along the western margin. Eastwards to the Transcontinental Arch the surface becomes progressively higher and more hilly. In the Dakotas Pre-Cambrian quartzite ridges rise 300 metres or more above broad valleys developed on schist belts. The basement surface can be examined in a number of outcrops in the Black Hills. A deep soil covers the schist valleys and a thinner soil is developed on the ridges. The C zone of some of the old soil profiles is still present at the Cambrian–Pre-Cambrian contact in several localities.

The Cambrian sea entered north-western Montana in *Albertella* Zone time, and, following the palaeoslope, moved slowly east and south-east. The initial deposit is the clean quartz arenites of the littoral zone (the Flathead Sandstone) which may be locally conglomeratic or arkosic near granite. As the littoral zone moves eastwards the initial coarse clastics are followed by alternating thin layers of sandstone and micaceous shales as scattered tidal flats develop. Most of the sands are unfossiliferous, but rare inarticulate brachiopods occur in some of the upper shaly beds. Locally (Wind River Canyon section) *Skolithos*-burrowed sand flats may be common. The uppermost beds of the Flathead are characterized by the first appearance of glauconite and distinctive bands of hematite. Low sand barriers now cut off shallow stagnant lagoons from the open ocean, and little clastic material reaches these sites.

Seaward of the sand bars, clays and silt settle on the shallow bottom as thin undisturbed laminae. Currents occasionally introduce thin sand layers. Disarticulated trilobite exoskeletons are scattered on the irregular bedding planes, and small bioclastic lime lenses occur sporadically. This shale unit (Wolsey or Gordon) accumulated below surf level in the sublittoral zone. Thin limestone beds which may be intercalated with the shales represent local clearing of the sea bottom permitting a greater accumulation of carbonate. The marginal uplifts on the western edge of the shelf are too low to provide significant amounts of clastics but are the sites of persistant sand flats and bars. The clear water of the slowly sinking marginal basins favours the development of broad algal flats, which build eastward onto the outer shelf sites whenever and wherever the combination of shallow depth and clear water permit. The sections of the Philipsburg quadrangle show a development of algal flats during most of the middle and late Middle Cambrian, but only in the late *Bathyuriscus-Elrathina* zone time do the algal flats extend some distance eastwards onto the shelf. Their appearance is brief as in the case of the *Tetonophycus* reef in western Wyoming (BLACKWELDER, 1918), which is soon covered by the fine clastics of the Park Shale.

The marked cyclic deposition in the Middle Cambrian of units of carbonates and fine clastics noted by both DEISS (1939), for north-western Montana, and AITKEN (1968), for adjacent Alberta Rocky Mountain sections, could reflect a simple transgressive-regressive sequence, possibly in response to eustatic sea level change. But the progressive increase in shale content in the sections in a north-eastern direction, and the persistence of this sediment pattern throughout the Middle and Upper Cambrian, suggest an important source of clastics from the western shore of the Canadian Shield. The presence of one or more large river deltas to the north-east is indicated and the fluctuations in clastics could have been caused by climate control of the river drainage basins. In the late Middle Cambrian abundant oolites and pebble conglomerates indicate progressive shoaling over the carbonate sites.

To the south-east in Wyoming broad sand or mud flats cover most of the shelf. Mud cracks are common, indicating frequent subaerial exposure, and broad worm-burrowed sand flats border the southern shore.

Soon after the appearance of the *Cedaria* fauna on the western edge of the shelf, a slow steady eastward transgression of the sea begins, which by *Crepicephalus* Zone time has carried the shore to the eastern edge of the Dakotas. The basal sands are clean quartz arenites of the littoral zone with local conglomerates developed at the base of Pre-Cambrian quartzite ridges. There is no persistent downwarping of the shelf nor upwarping of the Transcontinental Arch, and wide tidal flats, on which sedimentation is slow and sporadic, extend as much as 800 kilometres westwards from the shore. In late *Cedaria* time the sea entered a major structural basin (also drainage basin) in southern Canada and at this time the abundant north-easterly supply of fine clastics ceases. The eastern sand flats grade laterally into a predominately lime ooze flat on which fine silt and clay particles settle sporadically. This lithotope produced the characteristic intercalated calcareous shales and limestone pebble conglomerates of the region. The pebble conglomerates are often coarse, angular to subangular, edgewise, and occur in thick lenses which can be traced into the original thin beds of limestone. The matrix may be a hash of broken trilobite exoskeletons. Westwards the pebbles become smaller, rounded, and arranged parallel to the bedding planes. The matrix is a fine-grained limestone similar to adjacent beds. These bedded limestones represent a sublittoral carbonate bottom, which receives lime clastics and minor amounts of silt and clay from the eastern flats and appreciable amount of lime ooze from the broad algal flats along the western edge of the shelf (STOCKMAN, GINSBURG, and SHINN, 1967). A constant fluctuation between tidal and sublittoral bottom across central Montana and western Wyoming is indicated by the erratic development of dolomite mottling in the Pilgrim Formation throughout this area. In western Montana stromatolites are common in the upper Hasmark Dolomite and are reported from the Dresbachian limestones and dolomites of south-western Alberta (AITKEN and GREGGS, 1967). The clear waters of the western edge of the shelf are occupied by a belt of intertidal algal flats.

A regression of the sea begins by early *Aphelaspis* Zone time, and is indicated by a westward shift of the carbonate facies and the establishment of a mud flat seaward of the sand flats. Continued regression during the *Aphelaspis* Zone places the shore in west-central Montana (GRANT, 1965, p. 81), and a low sandy coastal plain extends eastwards. In Wyoming and southern Montana the top beds of the carbonate unit become increasingly sandy and then carbonate deposition ceases abruptly. No section shows a disconformity, although most of the *Aphelaspis* Zone and all of the *Dunderbergia* Zone faunules are missing. The basal unit of the overlying Franconian formation suggests that the coastal plain was exposed long enough for a sandy loam soil to develop.

Franconian sedimentation begins with a widespread transgression of the sea across the shelf, accompanying very minor downwarping of the shelf and upwarping of the Transcontinental Arch. The initial deposit is a soft, fissile, grey-green, purple or reddish shale or siltstone and thin shaly sand. Locally angular quartz grains and feldspar appear in silt sizes. The beds are usually unfossiliferous but rarely inarticulate brachiopods occur. Most of the material is derived from the reworking and settling of the weathered soil in the sublittoral zone of the rapidly advancing sea. The basal clastic unit ranges from 6 to about 30 metres in thickness and is an important marker horizon throughout Montana, Wyoming, and the western Dakotas. At the top of this shale, limestone pebble conglomerates, thin grey limestones, or the 'columnar limestone'—an algal reef of *Collenia magna*—appear. The characteristic littoral and intertidal ecological zones are now re-established across the central and eastern part of the shelf and a clear water sublittoral zone in the west.

The Red Lion Formation of western Montana, deposited in the sublittoral zone, consists of thin-bedded micrites with wavy silt laminations and occasional beds of small, flat-pebble limestone conglomerates. Although the fine clastics could have bypassed the algal reefs to the east, a probable source is the persistent sand flats of the central Idaho uplift. Western currents could bring much fine carbonate into the area from miogeosynclinal sites. The importance of local currents is indicated by the basal Worm Creek Quartzite, a thick sand bar constructed south-easterly from the Central Idaho uplift during *Elvinia* Zone time. During *Conaspis* Zone time a minor regression occurs. The effect is most noticeable along the north-eastern shore sites of the Transcontinental Arch. Littoral sands and tidal sand flats move westwards and the mud flat zone disappears as the deltas of the Canadian rivers grow north-westwards. The *Collenia magna* reef disappears as the fine muds of the carbonate tidal flats move over it during the later part of the *Elvinia* Zone. But this site of the unstable shelf is occupied during the *Conaspis* Zone by *Eoorthis* and *Billingsella* 'beds'. Beds of these two brachiopods must have been common in many localities upon the craton during the Franconian, but their position is best known in the North-West Region and in Texas.

In the late Frasconian the *Prosaukia* Zone transgression brings the regressive phase to an end, but the eastern shore remains in the same position. A broad tidal flat, about 800 kilometres square, extends westwards to central Montana and Wyoming. The inner sand flat is probably supratidal much of the time, and only minor deposits accumulate on the adjoining mud flats. Limestone pebble conglomerates and thin shales of the carbonate tidal flat grade westwards into the micrites and small flat-pebble conglomerates accumulating on the shallow sublittoral bottom. The predominate carbonates form a low cliff (3–6 metres), which intertongues eastwards into the usual monotonous alternation of grey-green shale beds and 30–60 cms pebble conglomerate beds of the carbonate tidal flats (GRANT, 1965, p. 13; HANSEN, 1957, p. 48).

Post-Cambrian erosion has removed all deposits younger than Franconian in western and central Montana and Wyoming. But scattered patches in the eastern parts of the states show that originally Trempealeauan and Early Ordovician sedimentation were continuous across the shelf. Several cycles of subaerial erosion from pre-Middle Ordovician to pre-Lat Devonian have removed many of the early Palaeozoic deposits, including some and often all of the Upper Cambrian. Palaeoecological conditions must be reconstructed from the scattered remnants.

The earliest faunule of the *Saukia* Zone appears while physical ecological conditions remain the same as during the later Franconian. Downwarping of the shelf and widespread transgression of the Transcontinental Arch occurs by middle *Saukia* Zone time. The sea crosses the Arch in the north-east to enter the Upper Mississippi Valley. Tidal sand flat deposits occur in the Williston Basin wells of the western Dakotas, and this lithotope and/or littoral sands extend eastwards into the Eastern Region. Narrow mud flats appear primarily in the north-east and the carbonate intertidal flats expand eastwards. There is no positive movement of the Canadian Shield, or Transcontinental Arch. There is a noticeable deficiency of terrigenous material in the Trempealeauan sedimentation. Micrites and small flat-pebble conglomerates of the shallow sublittoral zone extend into central Montana and Wyoming. Most of this carbonate ooze was derived from the intertidal algal shoals encroaching from the western miogeosyncline sites (STOCKMAN, GINSBURG, and SHINN, 1967). The end of the Trempealeauan is characterized by continued shoaling as the carbonate tidal flats encroach westwards and the algal intertidal flats grow eastwards. Now only carbonate-producing tidal flats extend for over 800 kilometres across the shelf.

The shelf remains a vast intertidal flat from the end of the Trempealeauan into the Ordovician. Seaward building of deltas to the north continued and introduced enough fine silt and clay to discourage the continued growth of widespread algal flats. The carbonate tidal flat of pebble conglomerates becomes dominant, although here and there in favoured spots small intertidal algal reefs persist. The transition faunule of this time is characterized by an abundance of dendroid graptolites, some articulate and a few inarticulate brachiopods, and a single unique trilobite species, probably a last survivor of the *Saukia* assemblage. In the Black Hills, barrier beaches and sand bars built out from the low coast enclose a lagoon where layers of autochthonous glauconite and thin beds of trilobite coquinas accumulate (KULIK, 1965).

12. Comparative sections of the Cambrian of the craton

This chapter comprises a series of comparative sections of the Cambrian of the craton of the United States. Sections A-A to K-K refer to the Eastern Region, sections L-L to R-R to the Mid-Continent, sections 10-10 to 17-17 to the South-West Region, and sections 1-1 to 9-9 to the North-West Region. The locations for all the sections are given on Fig. 1 (page 80).

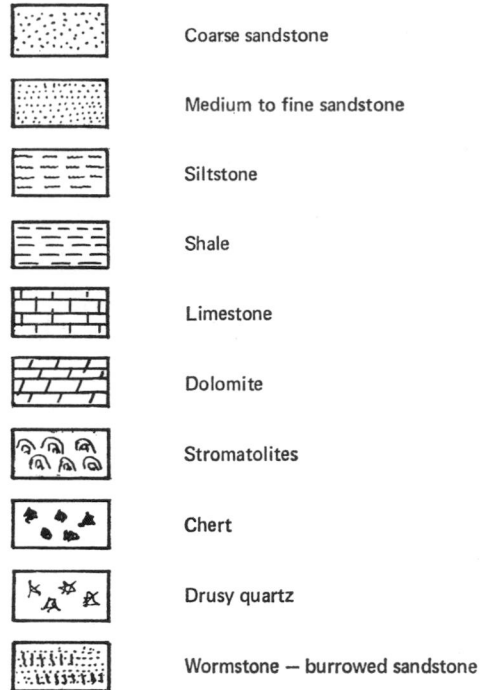

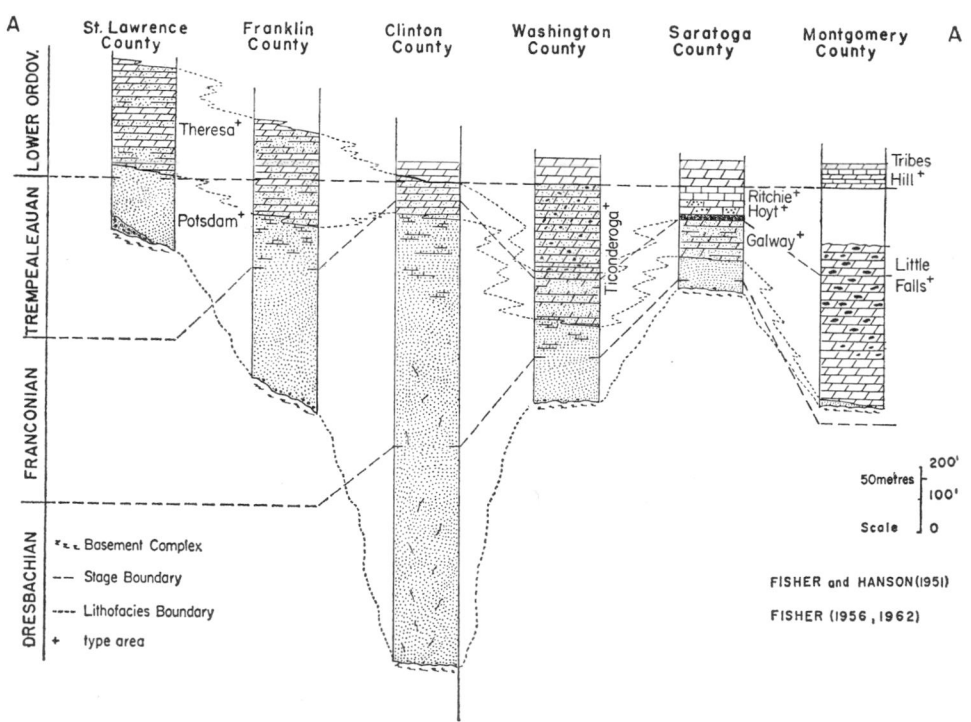

Section A–A of Eastern Region.

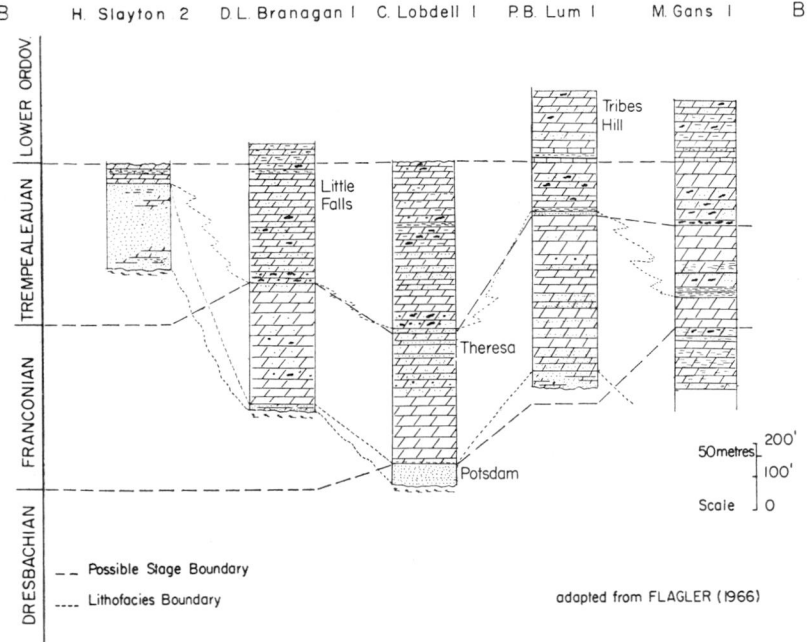

Section B–B of Eastern Region.

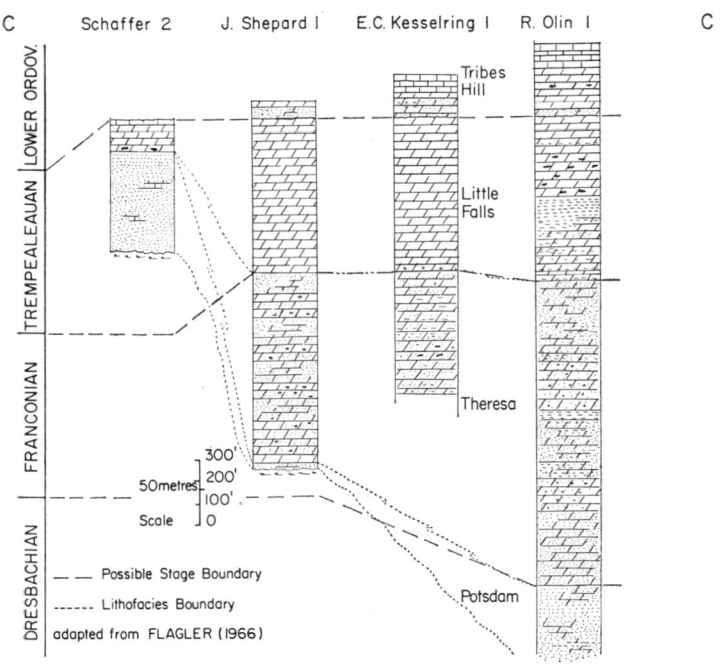

Section C–C of Eastern Region.

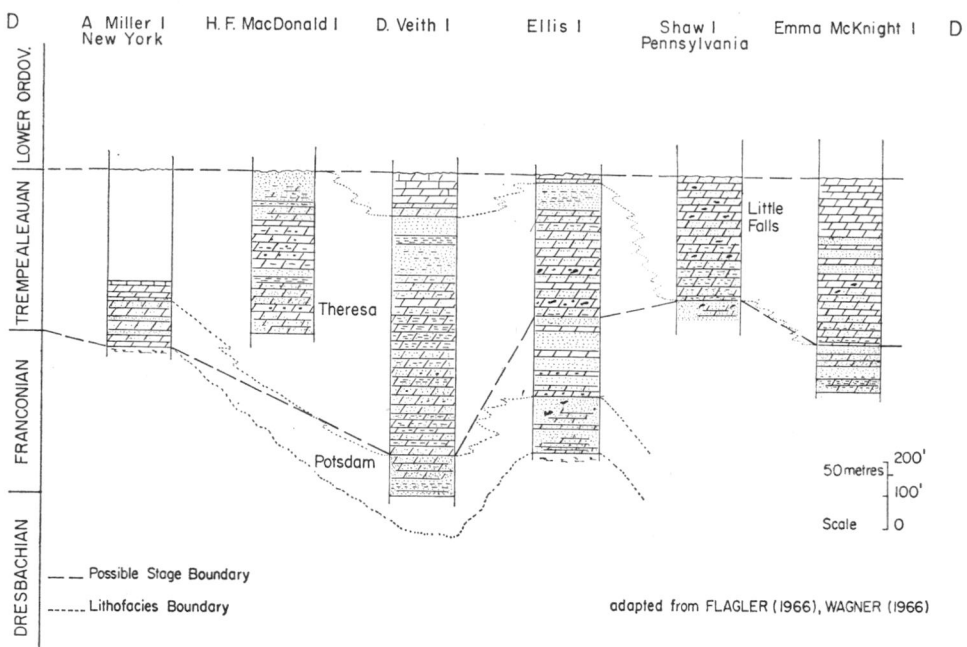

Section D–D of Eastern Region.

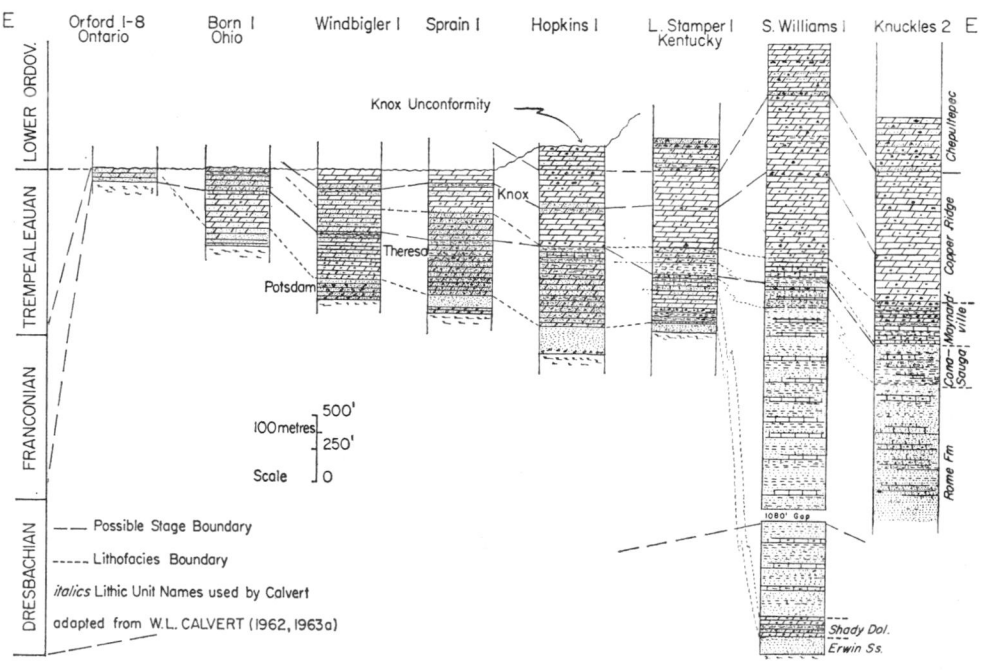

Section E–E of Eastern Region.

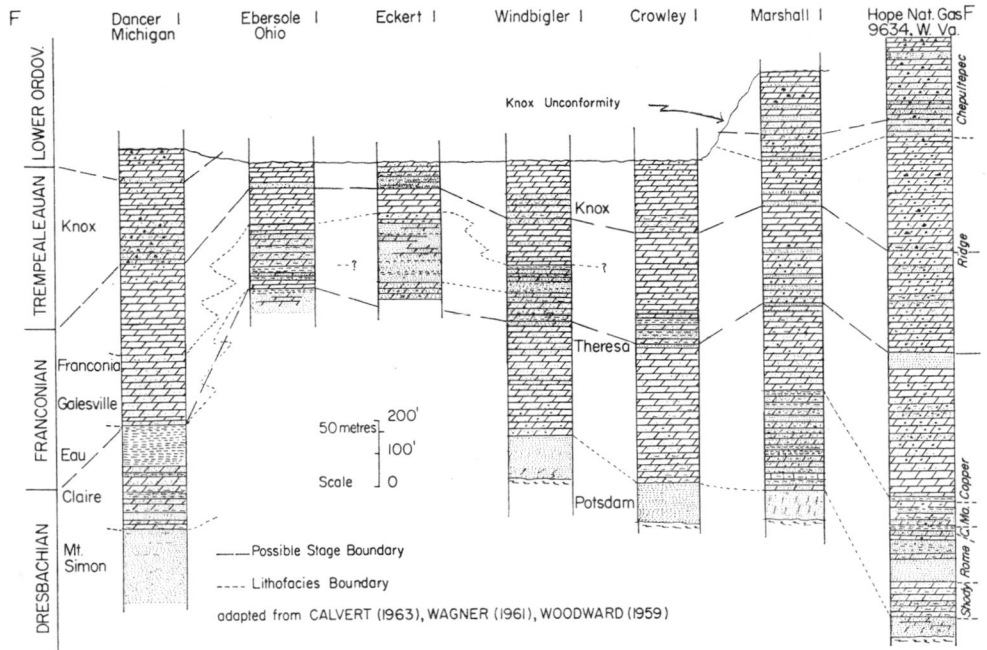

Section F–F of Eastern Region.

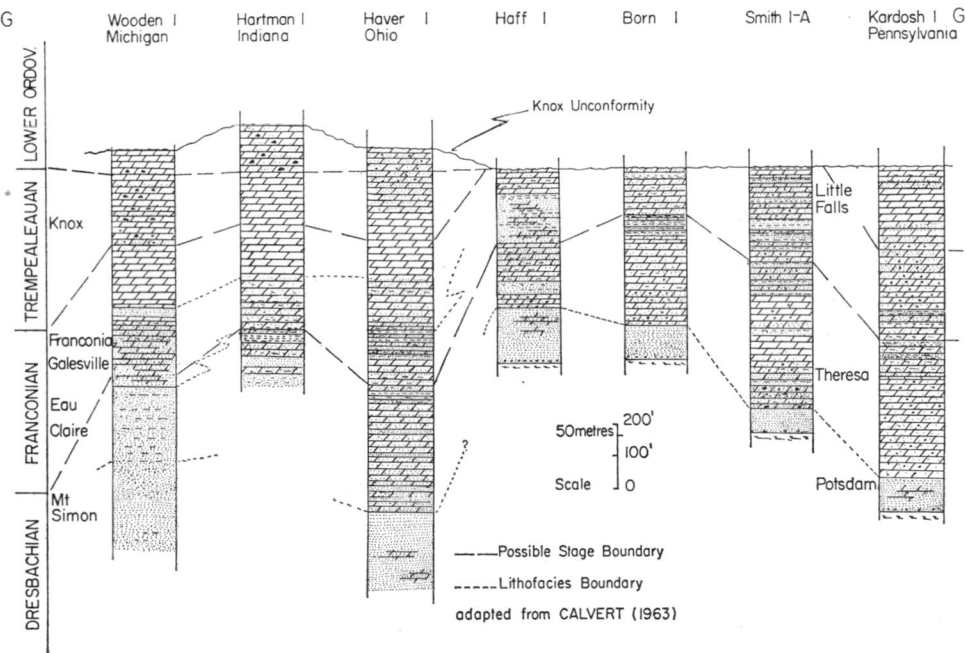

Section G–G of Eastern Region.

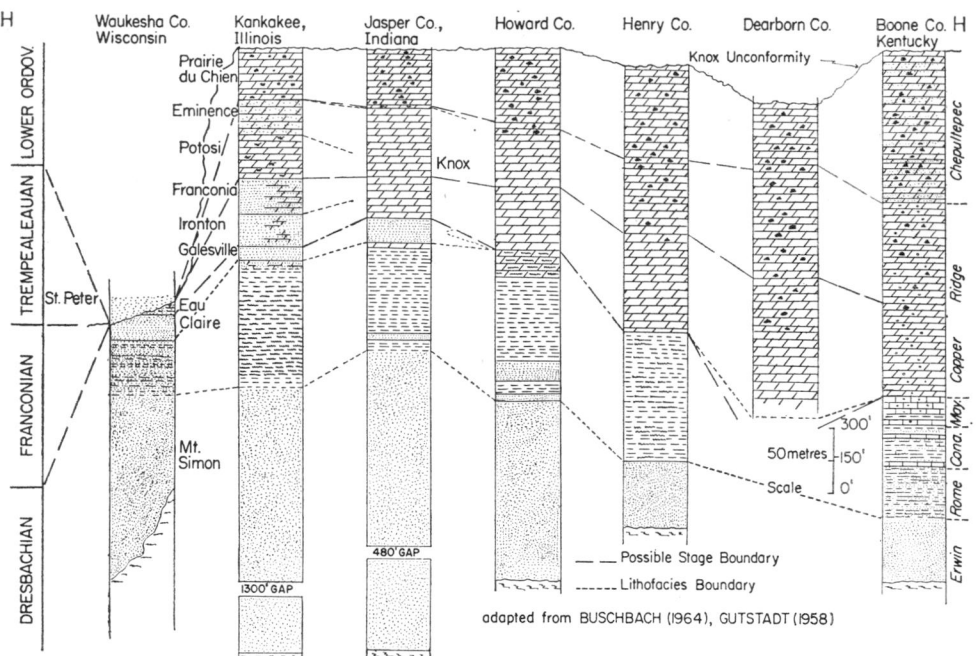

Section H–H of Eastern Region.

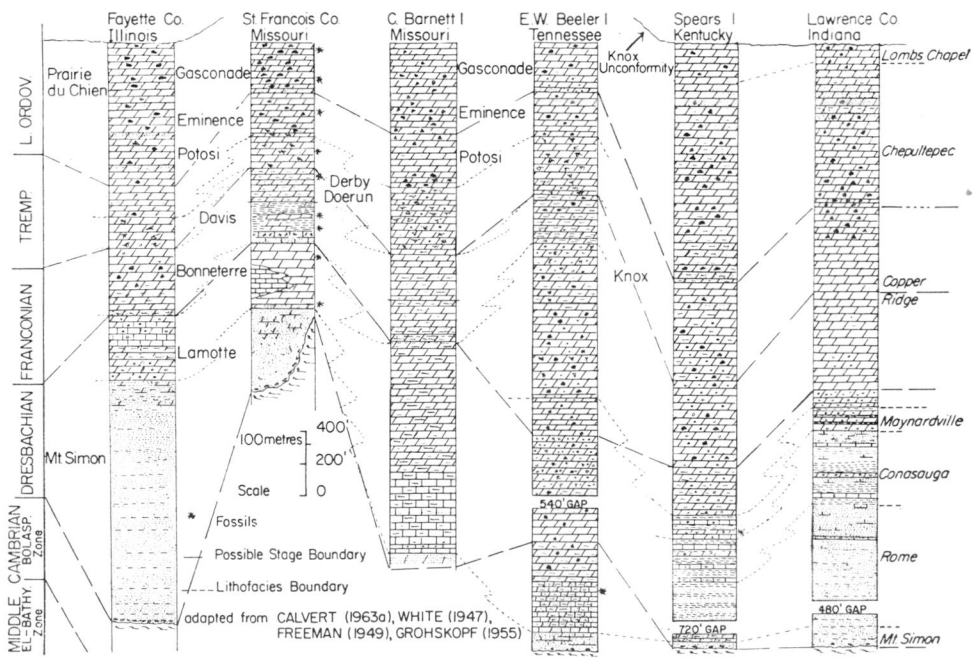

Section I–I of Eastern Region.

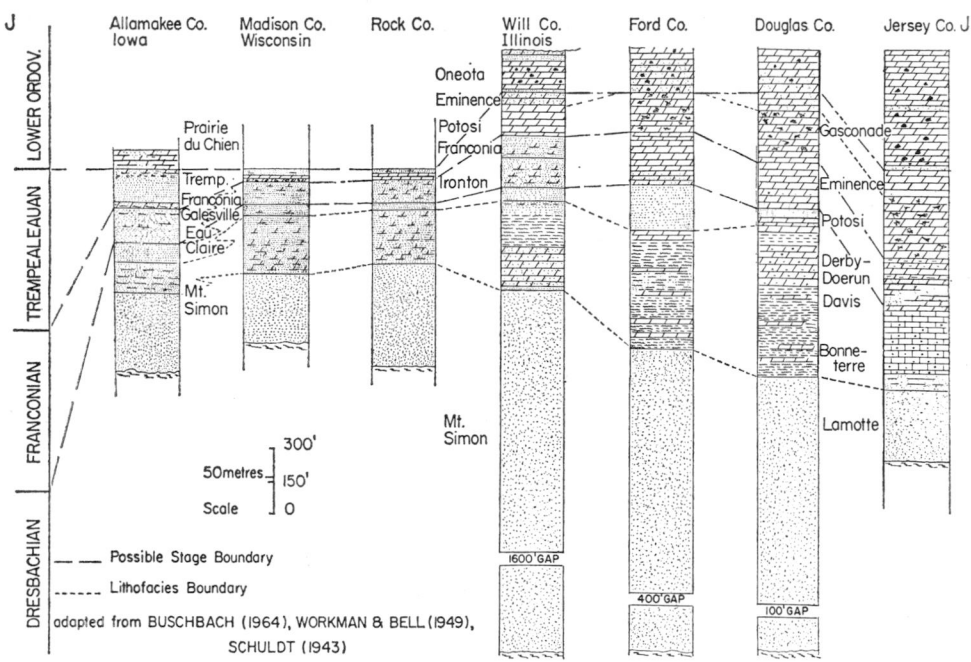

Section J–J of Eastern Region.

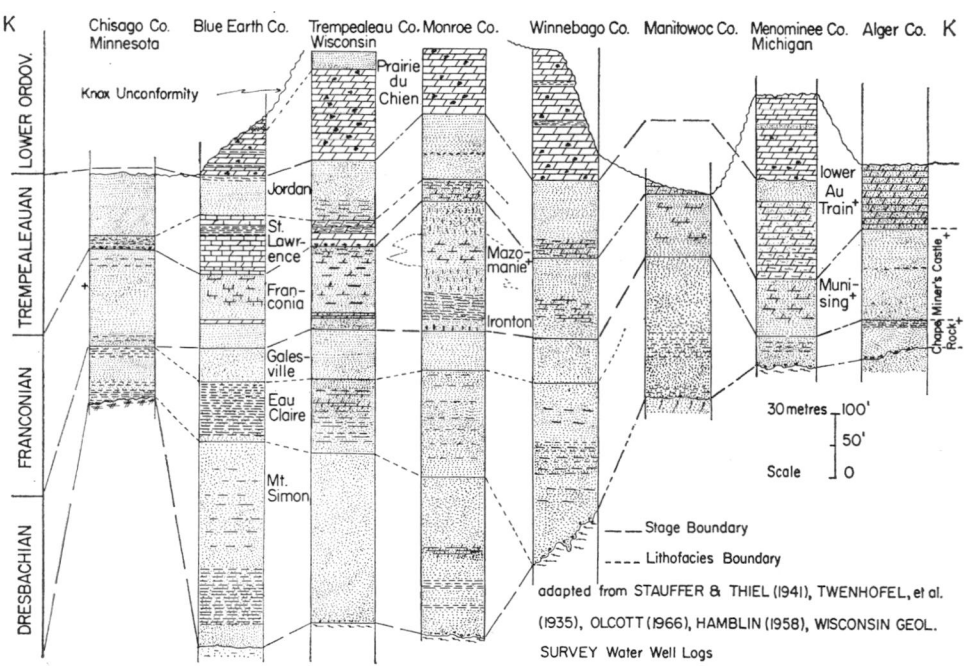

Section K–K of Eastern Region.

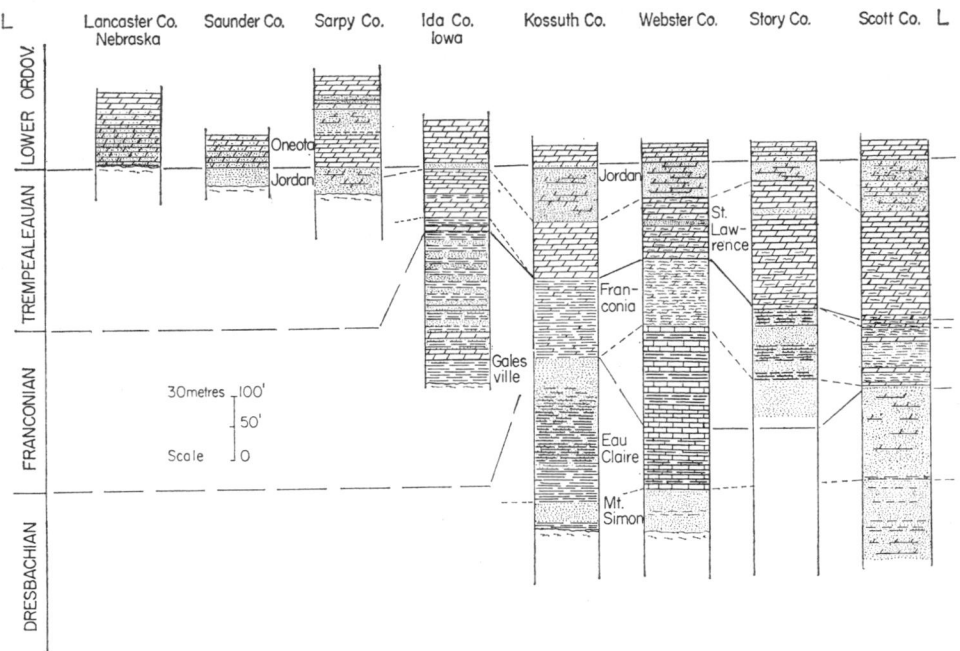

Section L–L of Mid-Continent Region.

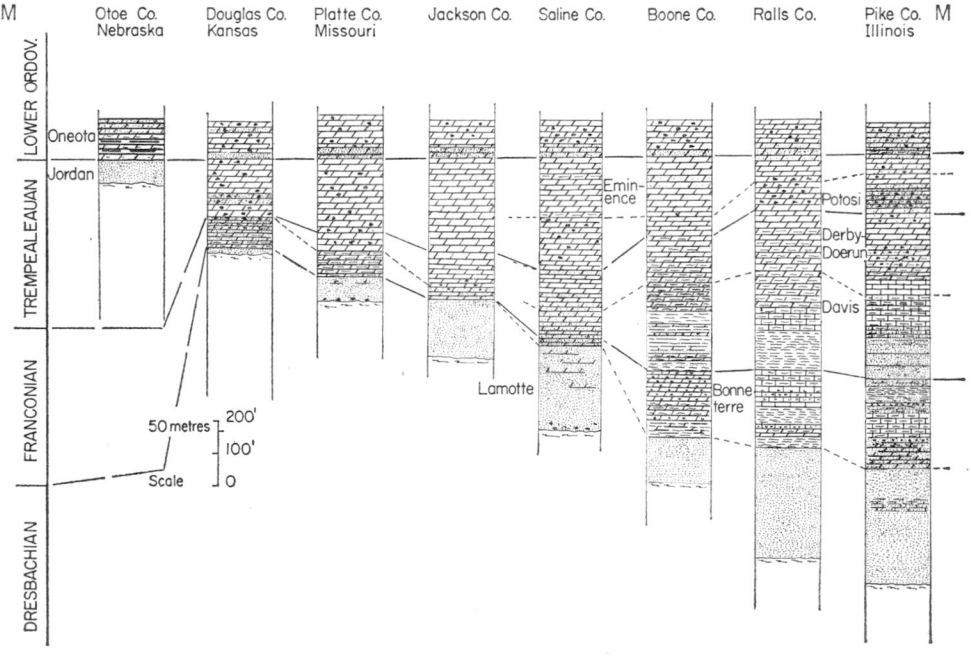

Section M–M of Mid-Continent Region.

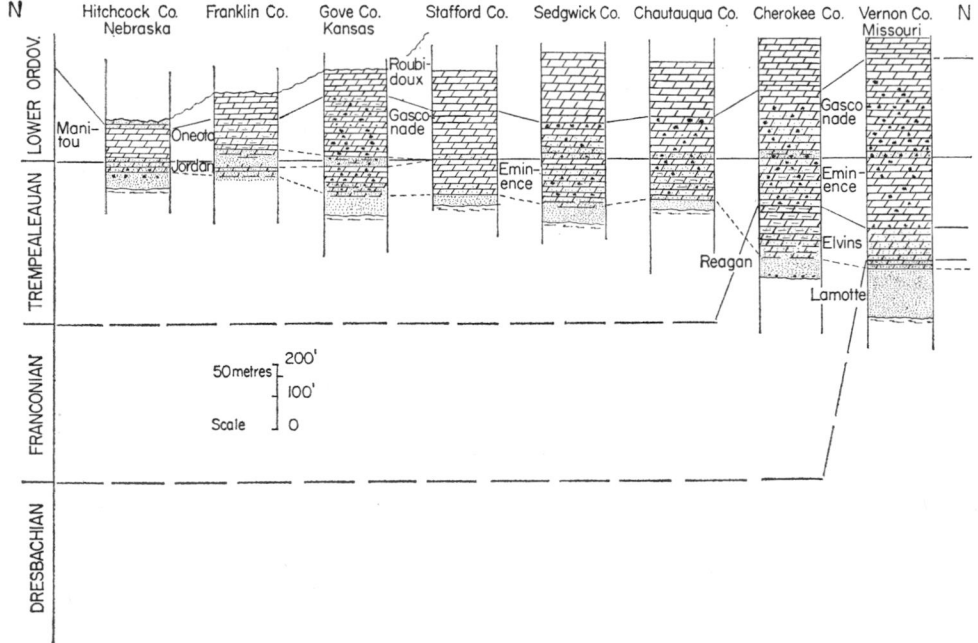

Section N–N of Mid-Continent Region.

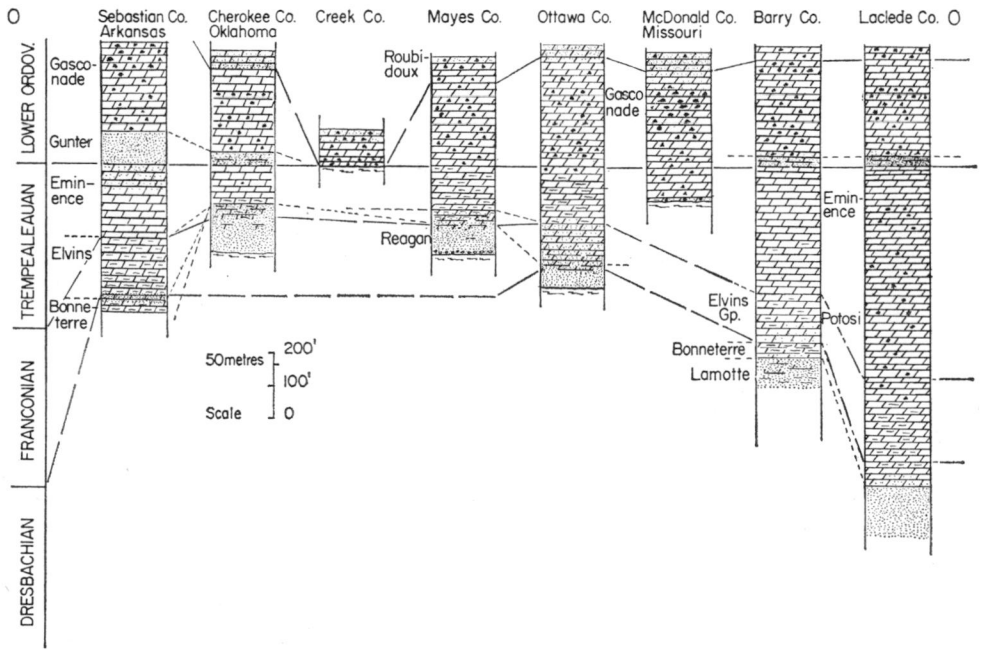

Section O–O of Mid-Continent Region.

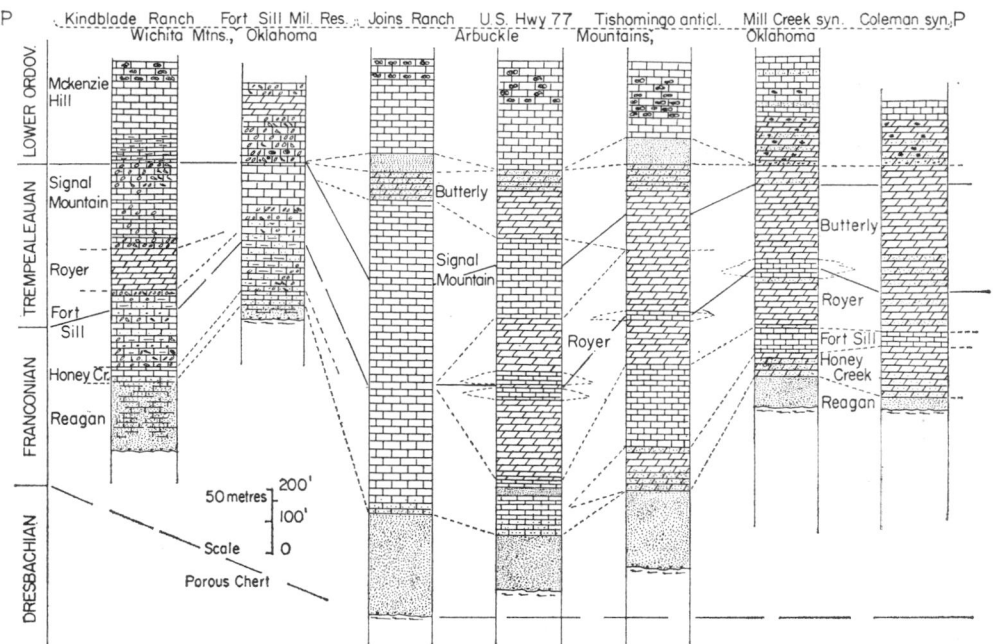

Section P–P of Mid-Continent Region.

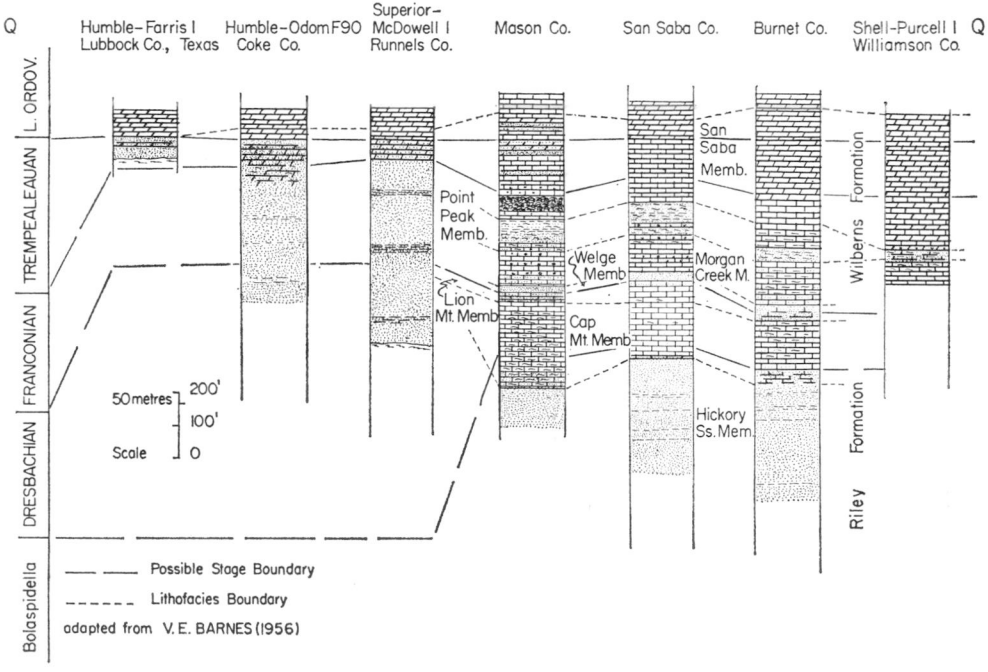

Section Q–Q of Mid-Continent Region.

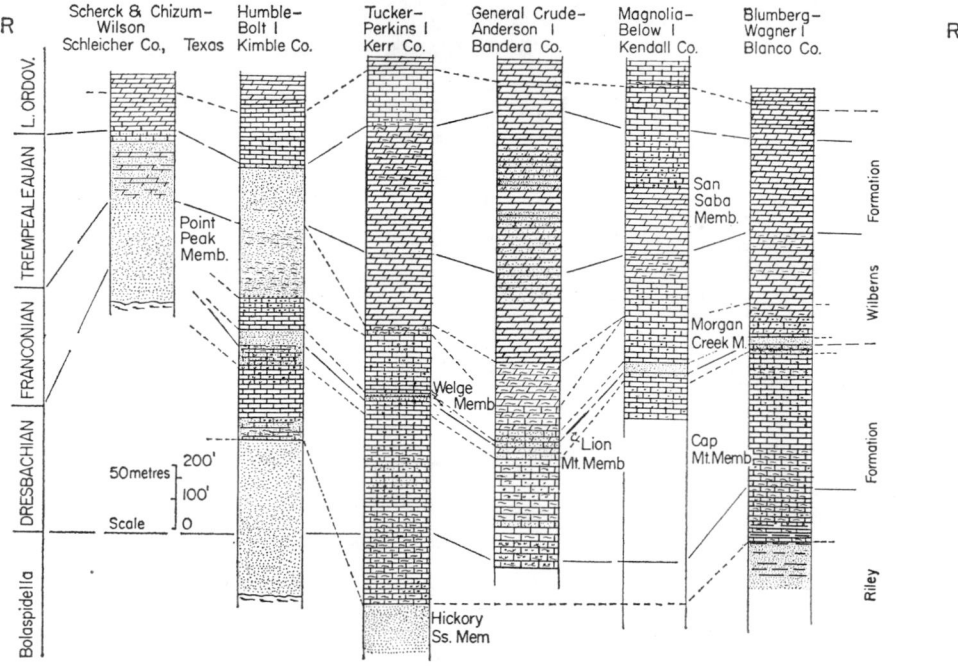

Section R–R of Mid-Continent Region.

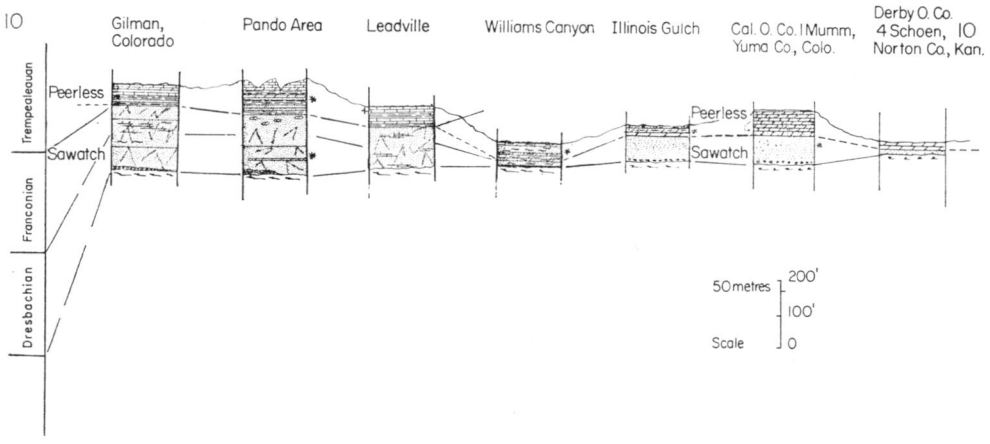

Section 10–10 of South-West Region.

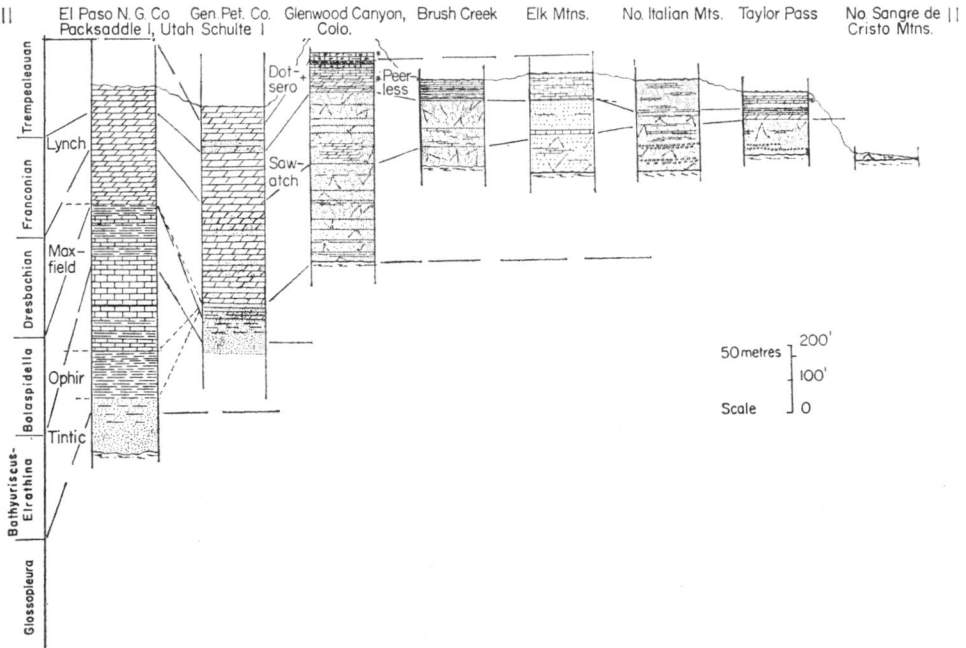

Section 11–11 of South-West Region.

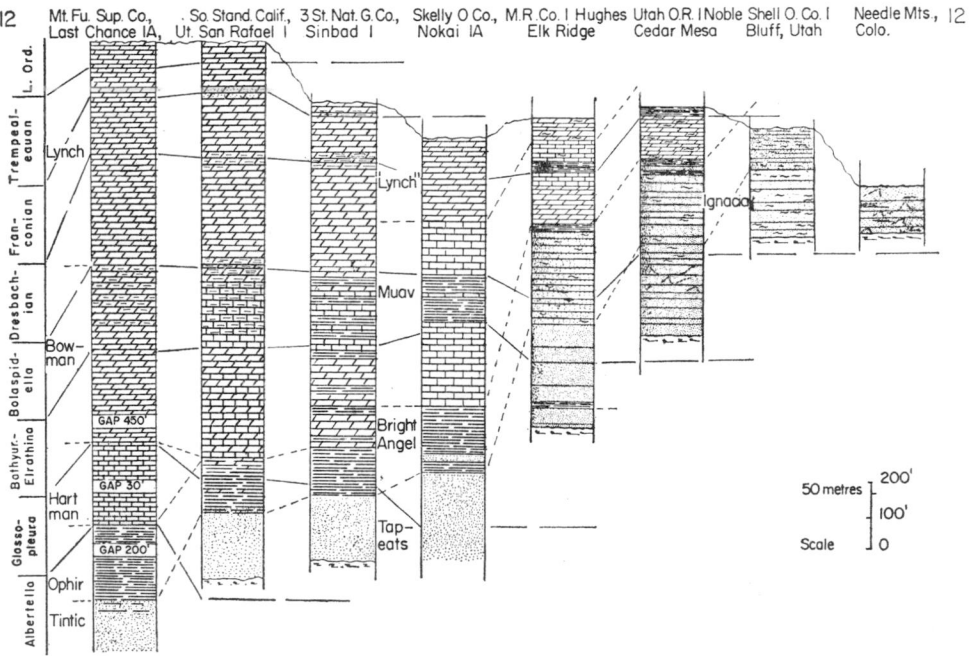

Section 12–12 of South-West Region.

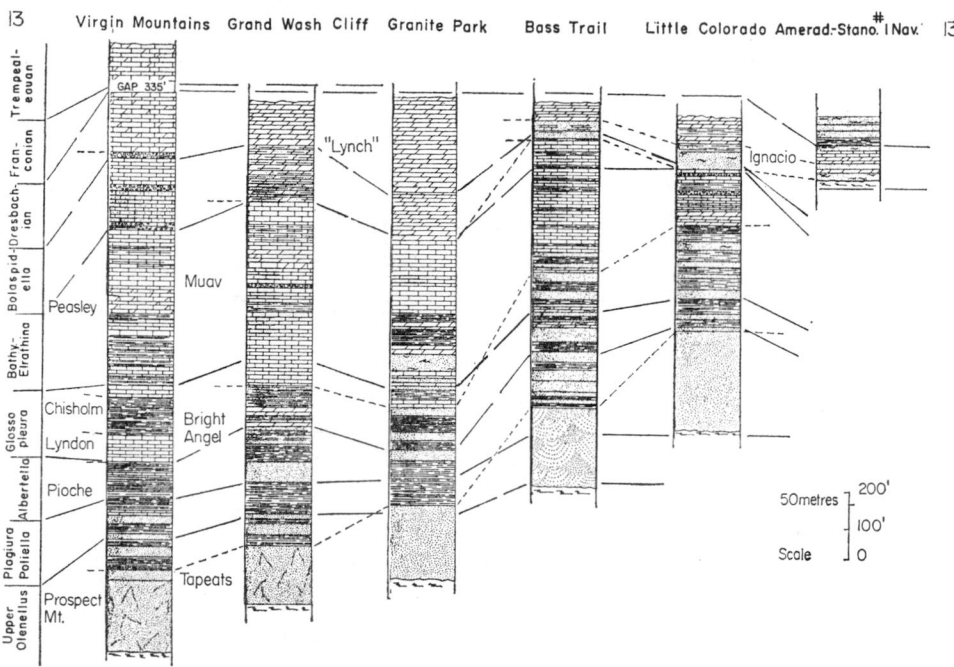

Section 13–13 of South-West Region.

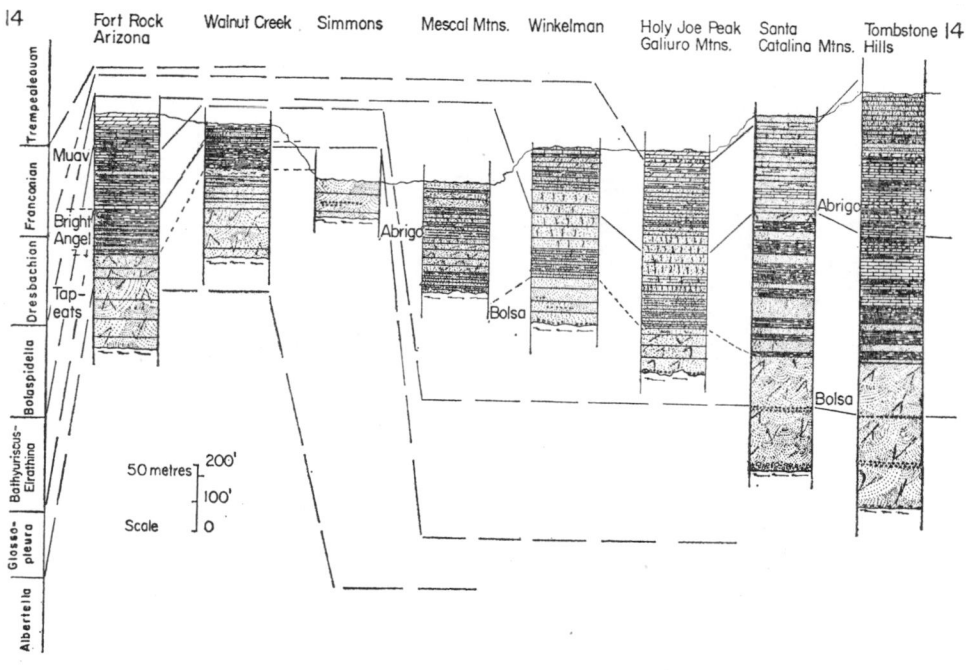

Section 14–14 of South-West Region.

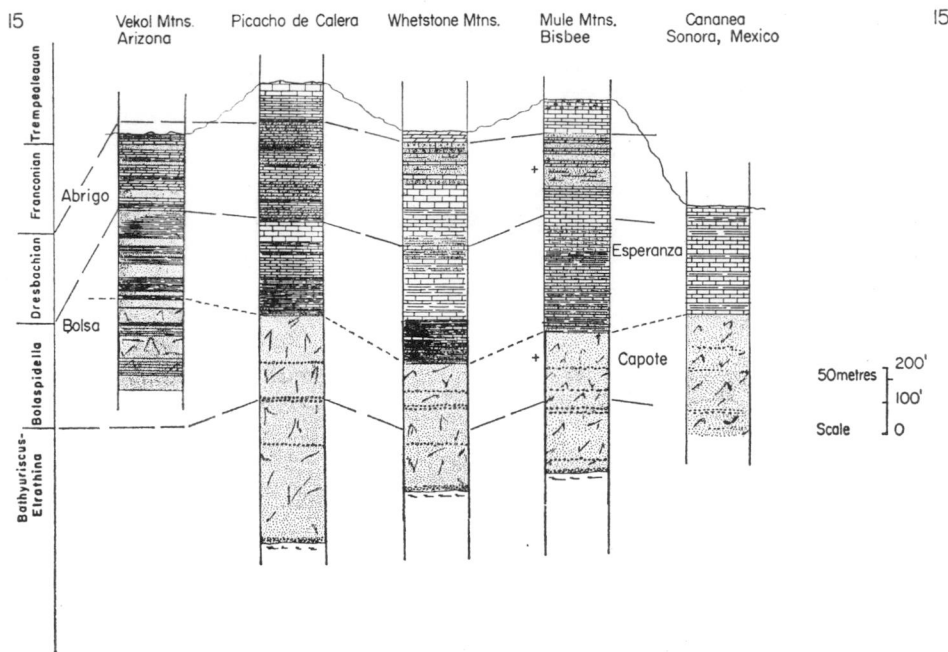

Section 15–15 of South-West Region

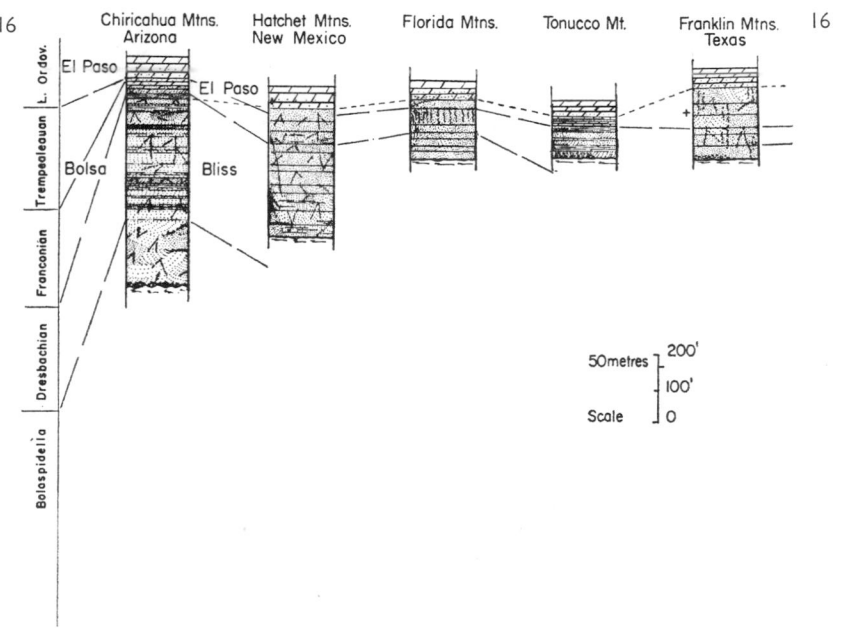

Section 16–16 of South-West Region.

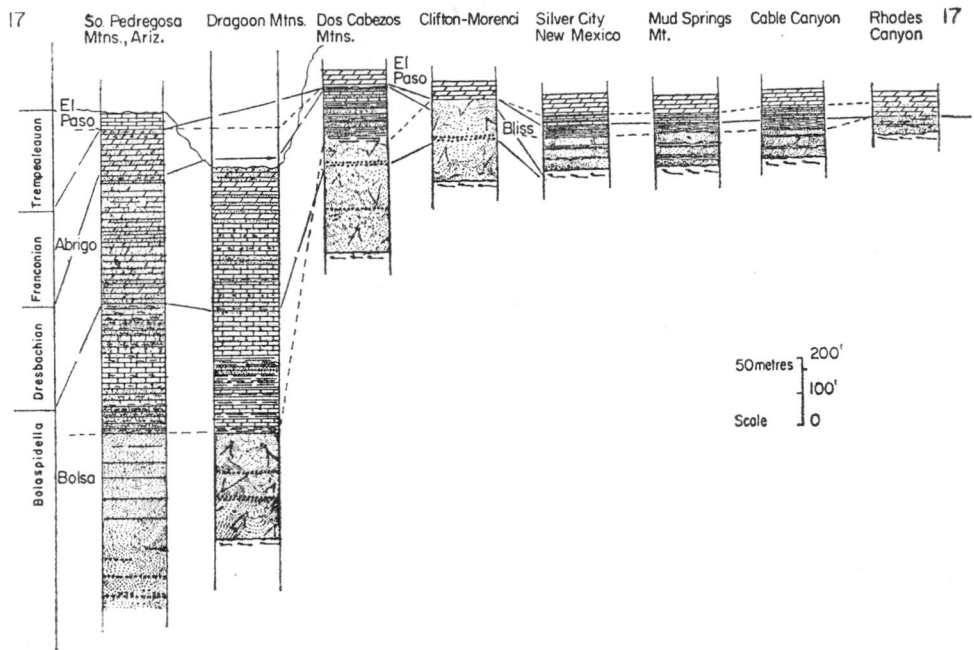

Section 17–17 of South-West Region.

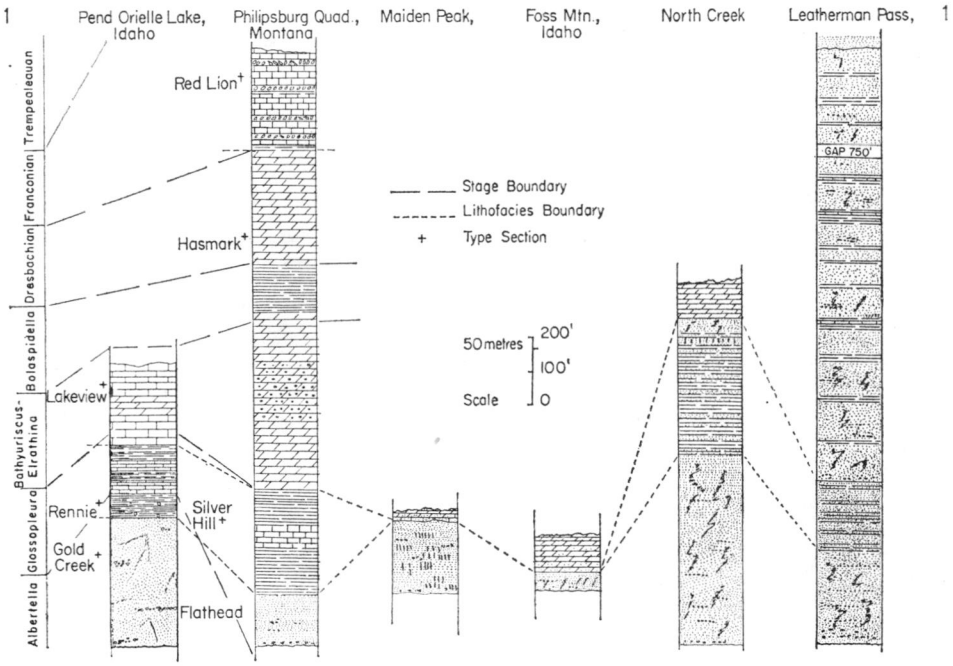

Section 1–1 of North-West Region.

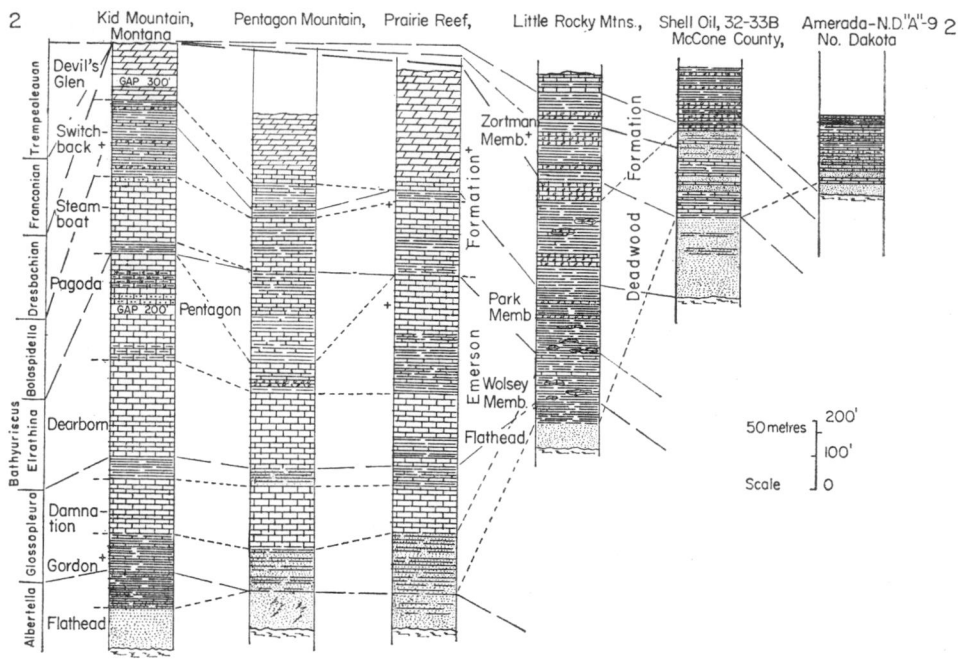

Section 2–2 of North-West Region.

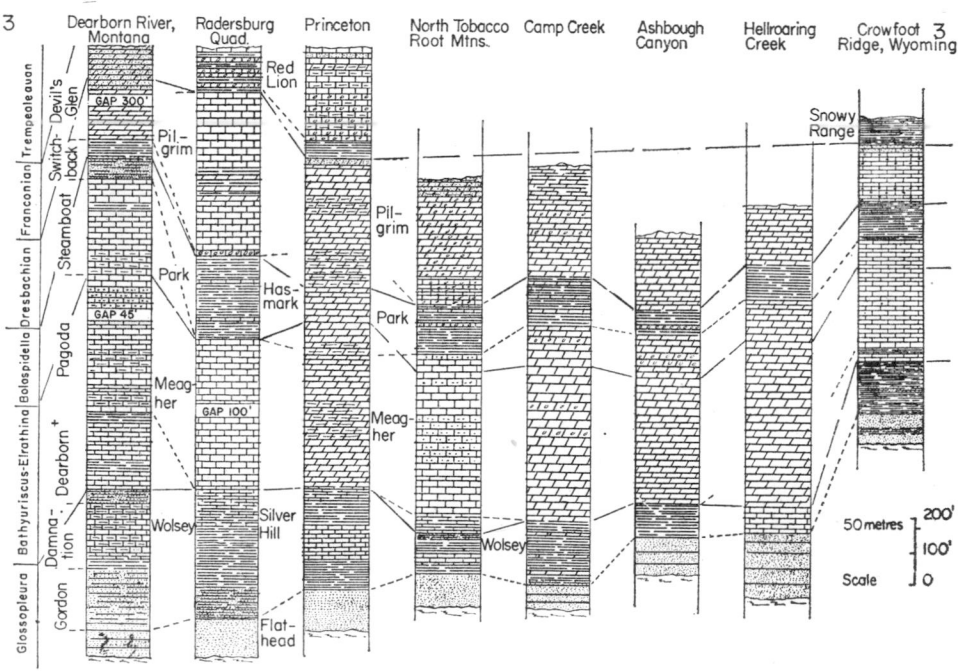

Section 3–3 of North-West Region.

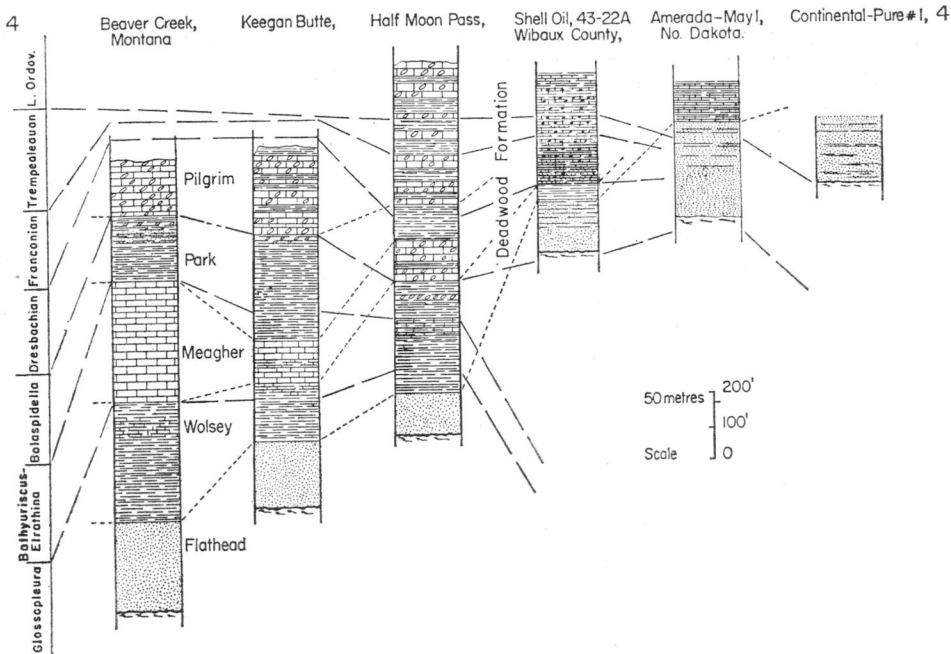

Section 4–4 of North-West Region.

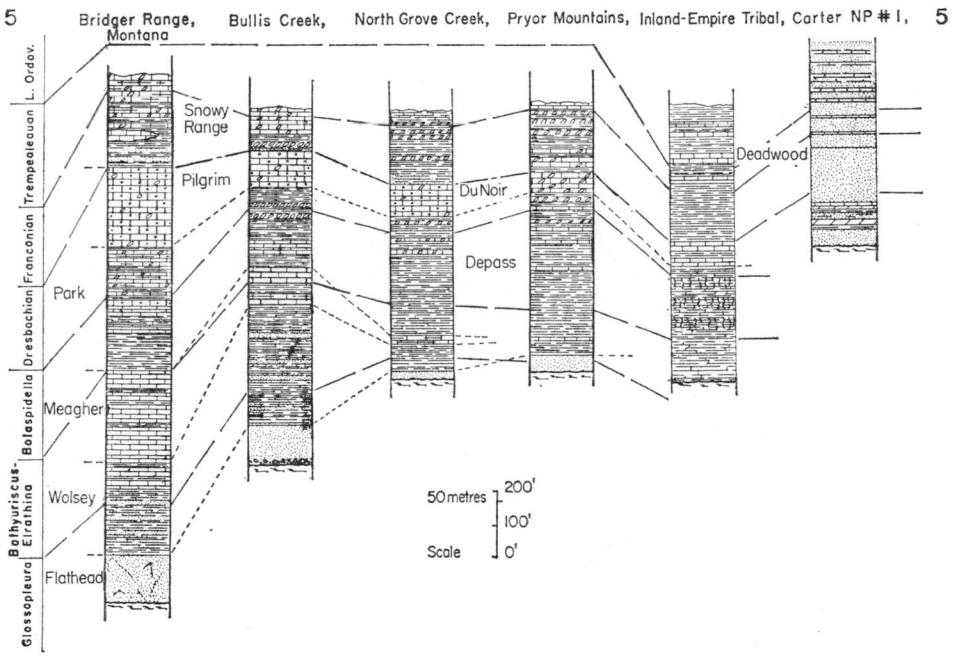

Section 5–5 of North-West Region.

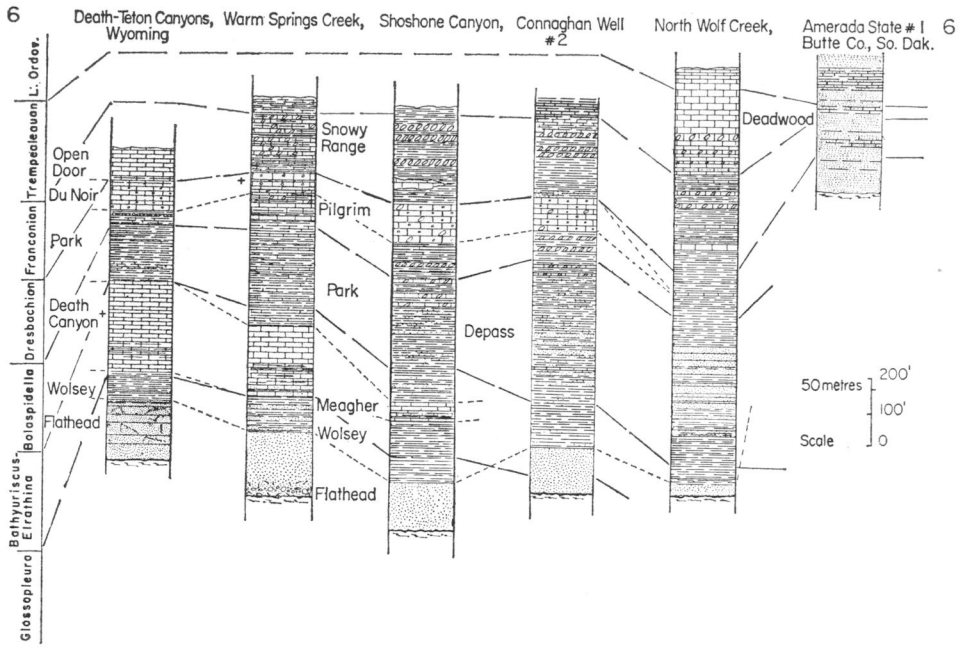

Section 6–6 of North-West Region.

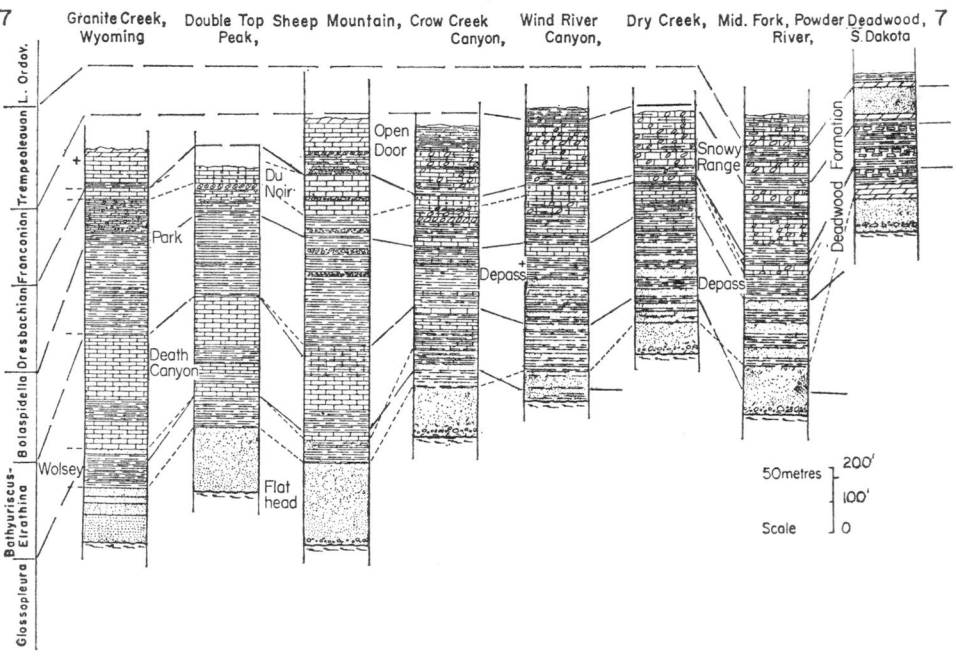

Section 7–7 of North-West Region.

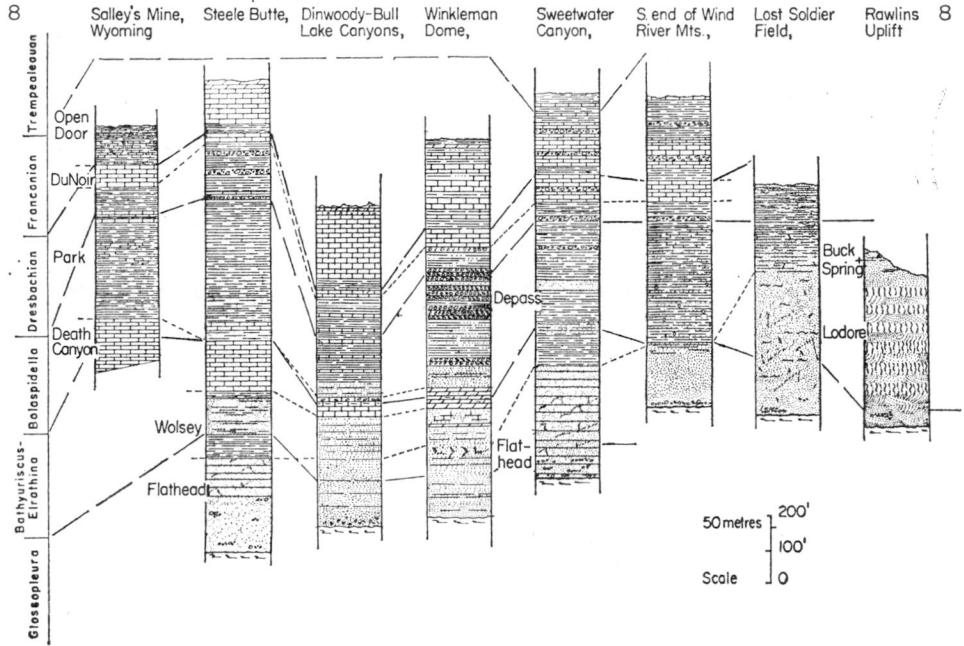

Section 8–8 of North-West Region.

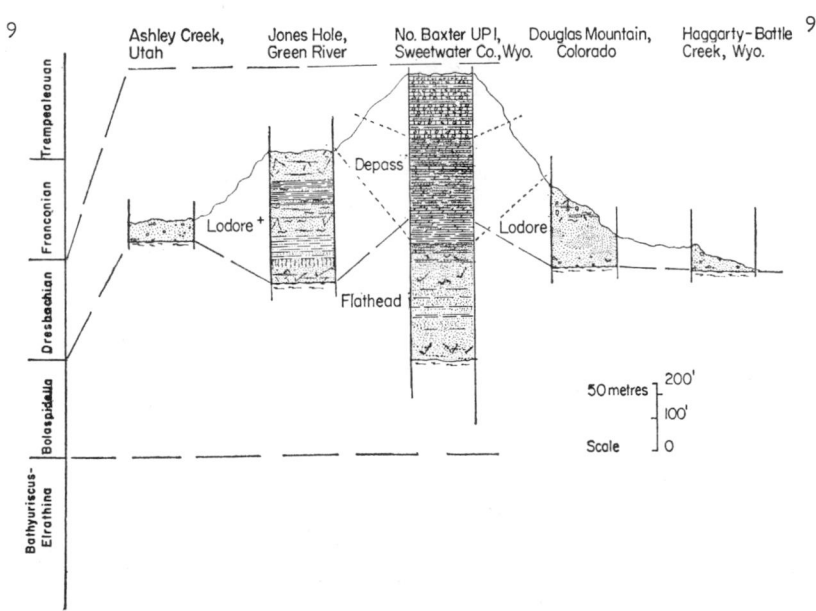

Section 9–9 of North-West Region.

13. Summary

The craton of the United States forms the southern half of the hedreocraton of the North American continent and remains emergent during the Lower Cambrian and most of the Middle Cambrian. The land surface is one of considerable relief and elevation with numerous mountain groups of Pre-Cambrian igneous and metamorphic rocks. The tectonic backbone of the craton is the Transcontinental Arch, a broad peninsula (2,250 kilometres long by 960 to 1,300 kilometres wide) trending north-east to south-west from south-west Ontario, Canada to Chihuahua, Mexico. The Arch is crossed midway by a north-west to south-east trending minor positive element, the Cambridge Arch–Central Kansas Highlands. In north-eastern United States a broad but short peninsula from eastern Canada extends from the Adirondack Dome to the Findlay Arch trend. These positive areas are inherited from Pre-Cambrian structures. Although not active tectonically during the Middle and Upper Cambrian and early Lower Ordovician, these areas are the last to be reduced by subaerial erosion and transgressed by the Cambrian and Early Ordovician sea. The Transcontinental Arch is bordered on the west by a coastal plain, broad (800 to 960 kilometres) on the north and sloping north-west, and narrow (320 to 480 kilometres) in the south with a west slope. The coastal plain on the east side of the Arch is narrow in the south half (because of the Ouachita Front overthrust) and slopes east-south-east; in the northern half it widens in the broad embayment of the Upper Mississippi Valley and the palaeoslope changes from south-east, through south, to south-west. The coastal plain continues eastwards along the southern border of the Canadian Shield with the palaeoslope changing from south-west to south-east at the trend of the Findlay Arch. Its true width in New York and Pennsylvania is unknown as it is concealed by Appalachian overthrusting. The coastal plain was an area of neutral to slightly negative tectonism during the Cambrian and Early Ordovician. Both coastal plains act as downwarping surfaces hinged along the edges of the continental slopes (the edge of the craton?) and only in three areas is evidence of basinal tectonism present: the inception of the Arbuckle geosyncline in southern Oklahoma; the beginnings of the southern Illinois Basin expressed in the marked thickening of the section in south-eastern Missouri and adjoining Kentucky, Tennessee, and southern Indiana; and probable downfaulting on the south side of the Rough Creek Fault Zone of Kentucky.

During the Lower and early Middle Cambrian negative tectonism was confined to the orthogeosynclinal sites and the marine waters of both the eastern and western oceans did not reach the borders of the craton. During later Middle Cambrian the seas reach the slowly down warping western coastal plain and inundate as much as 480 kilometres of it in the north-west. A comparable inundation of the south-eastern edge of the eastern coastal plain probably also occurs.

During the early Dresbachian the Upper Cambrian pattern of sedimentation upon the craton becomes apparent. Extremely slow but continuous downwarping of the coastal plains coupled with periodic eustatic sea level changes result in slow, sporadic sedimentation upon broad tidal flats. The transgressions and regressions upon the eastern and western coastal plains are synchronous when checked against the ordering of the faunal zones. Each transgression of the sea aids the subaerial denudation in reducing the relief and overall elevation of the coastal plains, and finally of the Transcontinental Arch. As a result each transgression covers more land area and during each regression broader tidal flats develop. During each regression the preceding deposits of the coastal plain are subaerially weathered, and the succeeding basal clastics are derived primarily from reworking of the soil.

The eastern and western coastal plains are separated by the Transcontinental Arch until the middle Trempealeauan, when a narrow strait appears across Kansas and a second strait across North Dakota and Minnesota. The Transcontinental Arch becomes two islands, Siouxia to the north and Sierra Grande in the south. During Early Ordovician times these islands are progressively reduced in size, reaching a minimum during the Roubidouxian transgression.

The cosmopolitan character of the Middle, and especially the Upper, Cambrian faunal assemblages is attributed to invasions by faunas which develop on the boreal shelves and move south as the coastal plains are inundated, or by faunas normally inhabiting the circalittoral depths of the eastern and western ocean regions. As the faunas are studied in detail more species will be found reflecting the varied habitats developing upon the coastal shelves. The Early Ordovician faunas reflect the rapid adaptations of various groups to the great variety of habitats opened up by the expansion of the seas over the entire craton.

After five transgressions: Dresbachian, early Franconian, latest Franconian to early Trempealeauan, Gasconadian, and Roubidouxian, separated by four characteristic regressions, a change in the pattern becomes apparent in the upper part of the Lower Ordovician. Slight but persistent upwarping of the Canadian Shield and Transcontinental Arch is initiated. By the end of the Early Ordovician these areas and all of the coastal plains are above sea level, completing the Sauk Sequence of deposition and returning the shore line of the hedreocraton to its position in the early Middle Cambrian.

The late Lower Ordovician tectonism exposes the entire craton but the amount of uplift is differential. On the western coastal plain south-western Montana and adjoining Wyoming are arched up higher than eastern Montana and adjoining North Dakota. On the eastern coastal plain the area of the Wisconsin Highlands and the trend of the Cincinnati–Findlay Domes are raised higher than the regions on either side. Subsequent subaerial erosion of these warped surfaces produces a widespread disconformity known as the Knox Unconformity in the east and the pre-Bighorn erosion surface in the west.

Bibliography

ADKINSON, W. L. (1963). Subsurface geologic cross section of Paleozoic rocks from Butler County to Stafford County, Kansas. *Kansas Geol. Survey, Oil and Gas Invest.* **28.**

ADKINSON, W. L. and VERODA, V. J. (1966). Paleozoic rocks in Kansas. *Cross Section Pub.* **4,** Amer. Assoc. Petroleum Geologists, Tulsa, Okla.

AITKEN, J. D. (1968). Cambrian sections in the easternmost southern Rocky Mountains and the adjacent subsurface, Alberta. *Geol. Survey Canada Paper* **66–23.**

AITKEN, J. D. and GREGGS, R. G. (1967). Upper Cambrian formations, southern Rocky Mountains of Alberta, an interim report. *Geol. Survey Canada Paper* **66–49.**

Amer. Comm. on Strat. Nomen. (1961). Code of stratigraphic nomenclature. *Bull. Amer. Assoc. Petroleum Geologists* **45,** 645.

BAARS, D. L. (1958). Cambrian stratigraphy of the Paradox Basin. *Inter. Mont. Assoc. Petrol. Geol. 9th Ann. Field Conf. Guidebook,* 93.

BAARS, D. L. and SEE, PAUL, D. (1968). Pre-Pennsylvanian stratigraphy and paleotectonics of the San Juan Mountains, south-western Colorado. *Bull. Geol. Society America* **79,** 333.

BARNES, V. E. (1956). Correlation of Pre-Simpson Paleozoic rocks *in* Barnes V. E. and others, (1959). *Univ. of Texas, Publ.* **5924.**

BARNES, V. E. and others (1959). Stratigraphy of the Pre-Simpson Paleozic subsurface rocks of Texas and south-east New Mexico. *Univ. of Texas, Bureau of Econ. Geol., Publ.* **5924,** 1.

BELL, W. C. and BARNES, V. E. (1961). Cambrian of central Texas. *El Sistema Cambrica, su paleogeografia y el problema de su base, Tomo III,* 484. XX Internat'l. Geol. Congress.

BELL, W. C., BERG, R. R., and NELSON, C. A. (1956). Croixan type area—upper Mississippi Valley. *El Sistema Cambrico, su paleogeografia y el problema de su base, Tome II*, 415. XX Internat'l. Geol. Congress.

BELL, W. C. and ELLINWOOD, H. L. (1962). Upper Franconian and Lower Trempealeauan Cambrian trilobites and brachiopods, Wilberns Formation, central Texas. *Jour. Paleontology* 36, 385.

BELL, W. C., FENIAK, O. W., and KURTZ, V. E. (1952). Trilobites of the Franconia Formation, southeast Minnesota. *Jour. Paleontology* 26, 175.

BERG, R. R. (1953). Franconian trilobites from Minnesota and Wisconsin. *Jour. Paleontology* 27, 553.

BERG, R. R. (1954). Franconian Formation of Minnesota and Wisconsin. *Bull. Geol. Society America* 65, 857.

BERG, R. R. and Ross, Jr., R. J. (1959). Trilobites from the Peerless and Manitou Formations, Colorado. *Jour. Paleontology* 33, 106.

BERKEY, C. P. (1897). Geology of the Saint Croix Dalles. *Amer. Geologist* 20, 345.

BEUTNER, E. C. and SCHOLTEN, ROBERT (1957). Probable Cambrian strata in east-central Idaho and their paleotectonic significance. *Bull. American Assoc. Petrol. Geologists* 51, 2305.

BLACKWELDER, ELLIOTT (1918). New geological formations in western Wyoming. *Washington Acad. Sci. Jour.* 8, 417.

BRANSON, C. C., HUFFMAN, G. G., and STRONG, D. M. (1965). Geology, oil and gas in Craig County. *Oklahoma Geol. Survey, Bull.* 99.

BURCHETT, R. R. and CARLSON, M. P. (1966). Twelve maps summarizing the geologic framework of south-eastern Nebraska. *Nebraska Geol. Survey, Rept. Invest.* 1, 12.

BURCHETT, R. R. and REED, E. C. (1937). *Centennial guidebook to the geology of southeastern Nebraska.* Nebraska Geol. Survey, Lincoln, Nebraska.

BUSCHBACH, T. C. (1964). Cambrian and Ordovician strata of north-eastern Illinois. *Illinois State Geol. Survey, Rept. Inv.* 218.

BUSCHBACH, T. C. (1965). Deep stratigraphic test well near Rock Island, Illinois. *Illinois State Geol. Survey Circ.* 394.

CALVERT, W. L. (1962). Sub-Trenton rocks from Lee County, Virginia to Fayette County, Ohio. *Ohio Geol. Survey, Rept. Inv.* 45.

CALVERT, W. L. (1963a). A cross section of Sub-Trenton rocks from Wood County, West Virginia to Fayette County, Illinois. *Ohio Geol. Survey, Rept. Inv.* 48.

CALVERT, W. L. (1963b). Sub-Trenton rocks of Ohio in cross sections from West Virginia and Pennsylvania to Michigan. *Ohio Geol. Survey, Rept. Inv.* 49.

CALVERT, W. L. (1964a). Pre-Trenton sedimentation and dolomitization, Cincinnati Arch province: theoretical considerations. *Bull. Amer. Assoc. Petroleum Geologists* 48, 166.

CALVERT, W. L. (1964b). Central Ohio yields oil from Cambrian erosional remnants. *World Oil* February and March issues.

CALVERT, W. L. (1964c). Sub-Trenton rocks from Fayette County, Ohio to Brant County, Ontario. *Ohio Geol. Survey, Rept. Inv.* 52.

CALVERT, W. L. (1965). Cambrian correlations. *Ontario Petrol. Inst., 4th Ann. Conf.* 4, Tech. Paper 2.

CARLSON, C. G. (1960). Stratigraphy of the Winnipeg and Deadwood Formations in North Dakota. *North Dakota Geol. Survey Bull.* 35.

CARLSON, M. P. (1967). Precambrian well data in Nebraska. *Nebraska Geol. Survey Bull.* 25.

CARLSON, M. P. (1968). Unpublished sample logs from Cambrian–Lower Ordovician of Nebraska: Personal communication.

CHASE, G. W., FREDERICKSON, E. A., and HAM, W. E. (1956). Resume of the geology of the Wichita Mountains, Oklahoma. *Petroleum Geology of southern Oklahoma* 1, 36. Amer. Assoc. Petroleum Geologists, Tulsa, Okla.

CHENOWETH, P. A. (1967a). Early Paleozoic overlap, southern Mid-Continent. *Oklahoma geology notes* 27, 170, (abst.).

CHENOWETH, P. A. (1967b). Southern Mid-Continent: past, present, future. *The Oil and Gas Journal* 65, Dec. 4, 130.

CHRISTIANSEN, F. W. (1963). Cambrian rocks of Utah. *Utah Geol. and Mineralog. Survey Bull.* 54, 45.

CHRONIC, JOHN and FERRIS, Jr., C. S. (1961). Early Paleozoic outlier in south-eastern Wyoming. *Rocky Mtn. Assoc. Geologists, 12 Ann. Field Cong.* 143.

COHEE, G. V. (1948). Cambrian and Ordovician rocks in Michigan Basin and adjoining areas. *Bull. Amer. Assoc. Petroleum Geologists* 32, 1417.

CONDRA, G. E., SCHRAMM, E. F., and LUGN, A. L. (1931). Deep wells of Nebraska. *Nebraska Geol. Survey, Bull.* **4**, 2nd ser.

CONDRA, G. E. and REED, E. C. (1939). Deep wells at Lincoln, Nebraska. *Nebraska Geol. Survey, Paper* **15**.

COOPER, JOHN R. and SILVER, LEON T. (1964). Geology and ore deposits of the Dragoon Quadrangle, Cochise County, Arizona. *U.S. Geol. Survey, Prof. Paper* **416**.

COWIE, J. W. (1964). The Cambrian Period. *Quart. Jour. Geol. Soc. London* **120**, 255.

CYGAN, N. E. and KOUCKY, F. L. (1963). The Cambrian and Ordovician rocks of the east flank of the Big Horn Mountains, Wyoming. *WGS-BGS Guidebook*, 26.

DARTON, N. H. (1925). A Resume of Arizona Geology. *Univ. of Arizona Bull.* **119**, Geol. Ser. 3.

DAWSON, T. A. (1960). Deep test well in Lawrence County, Indiana. *Indiana Geol. Survey, Rept. Prog.* **22**.

DECKER, C. E. (1939). Two Lower Paleozoic groups, Arbuckle and Wichita Mountains, Oklahoma. *Bull. Geol. Society America* **50**, 1311.

DEISS, CHARLES (1936). Revision of type Cambrian formations and sections of Montana and Yellowstone National Park. *Bull. Geol. Society America* **47**, 1257.

DEISS, CHARLES (1939). Cambrian stratigraphy and trilobites of north-western Montana. *Geol. Society America Sp. Paper* **18**.

DENNISON, John M. and WOODWARD, H. P. (1963). Palinspastic maps of central Appalachians. *Bull. Amer. Assoc. Petroleum Geologists* **47**, 666.

DORF, ERLING and LOCHMAN, CHRISTINA (1940). Upper Cambrian formations in southern Montana. *Bull. Geol. Society America* **51**, 541.

DRISCOLL, E. G. (1959). Evidence of transgressive–regressive Cambrian sandstones bordering Lake Superior. *Jour. Sed. Petrology* **29**, 5.

EMMONS, E. (1838). Report of the second geological district of the state of New York. *New York Geol. Survey, Ann. Rept.* **2**, 185.

EMMONS, W. H. and CALKINS, F. C. (1913). Geology and ore deposits of the Philipsburg Quadrangle, Montana. *U.S. Geol. Survey Prof. Paper* **78**.

EMRICH, G. H. (1966). Ironton and Galesville (Cambrian) Sandstones in Illinois and adjacent areas. *Illinois State Geol. Survey, Circ.* **403**.

EPIS, R. C. (1958). Early Paleozoic strata in southeastern Arizona. *Bull. Amer. Assoc. Petroleum Geologists* **42**, 2750.

EPIS, R. C. and GILBERT, C. M. (1957). Early Paleozoic strata in southeastern Arizona. *Bull. Amer. Assoc. Petroleum Geologists* **41**, 2223.

FETTKE, C. R. (1948). Subsurface Trenton and sub-Trenton rocks in Ohio, New York, Pennsylvania, and West Virginia. *Bull. Amer. Assoc. Petroleum Geologists* **32**, 1457.

FISHER, D. W. (1956). The Cambrian System of New York state. *El Sistema Cambrico, su paleogeografia y el problema de su base, Tomo II*, 321. XX Internat'l Geol. Congress.

FISHER, D. W. (1962). Correlation of the Cambrian rocks in New York state. *New York State Museum and Sci. Service Geol. Survey, Map & Chart.* Ser. 2.

FISHER, D. W. and HANSON, G. F. (1951). Revisions in the geology of Saratoga Springs, New York and vicinity. *Amer. Jour. Science* **249**, 795.

FLAGLER, C. W. (1966). Subsurface Cambrian and Ordovician stratigraphy of the Trenton Group-Pre-Cambrian interval in New York state. *New York State Museum and Sci. Service, Map and Chart.* Ser. 8.

FLAWN, P. T. and others (1961). The Ouachita System. *Univ. of Texas, Publ.* **6120**.

FOLEY, F. C., WALTON, W. C., and DRESCHER, W. J. (1953). Ground water conditions in the Milwaukee–Waukesha area, Wisconsin. *U.S. Geol. Survey Water-Supply Paper* **1229**.

FORRESTER, J. D. (1937). Structure of the Uinta Mountains. *Bull. Geol. Society America* **48**, 631.

FREDERICKSON, E. A. (1948a). Clarification of Upper Cambrian stratigraphy in Oklahoma. *Bull. Amer. Assoc. Petroleum Geologists* **32**, 1349.

FREDERICKSON, E. A. (1948b). Upper Cambrian trilobites from Oklahoma. *Jour. Paleontology* **22**, 798.

FREDERICKSON, E. A. (1949). Trilobite fauna of the Upper Cambrian Honey Creek Formation. *Jour. Paleontology* **23**, 341.

FREDERICKSON, E. A. (1956). Cambrian of Oklahoma. *El Sistem Cambrica, su paleogeografia y el problema de su base, Tomo II*, 483. XX Internat'l. Geol. Congress.

FREEMAN, LOUISE B. (1945). Geology and mineral resources of the Jackson purchase region, Kentucky-Paleozoic geology. *Kentucky Geol. Div.* Ser. 8, Bull. **8**, 12.

FREEMAN, LOUIS B. (1949). Regional aspects of Cambrian and Ordovician subsurface stratigraphy in Kentucky. *Bull. Amer. Assoc. Petroleum Geologists* **33**, 1655.

FRITZ, W. H. and NORRIS, D. K. (1966). Lower Middle Cambrian correlations in the east-central Cordillera. *Geol. Survey Canada Paper* **66-1**, 105.

GILLULY, JAMES and others (1956). General geology of central Cochise County, Arizona. *U.S. Geol. Survey Prof. Paper* **281**.

GOLDRING, WINIFRED (1938). Algal barrier reefs in the Lower Ozarkian of New York. *New York State Museum Bull.* **315**, 1.

GRANT, R. E. (1962). Trilobite distribution, Upper Franconia Formation (Upper Cambrian), southeastern Minnesota. *Jour. Paleontology* **36**, 965.

GROHSKOPF, J. G. (1955). Subsurface geology of the Mississippi embayment of southeast Missouri. *Missouri Geol. Survey and Water Res.* **37**.

GUSTADT, A. M. (1958). Cambrian and Ordovician stratigraphy and oil and gas possibilities in Indiana. *Indiana Geol. Survey, Bull.* **14**.

HALE, W. E. (1955). Geology and ground water resources of Webster County, Iowa. *Iowa Geol. Survey, Water Supply Bull.* **4**.

HALL, JAMES (1847). Description of the organic remains of the lower division of the New York System. *Paleontology of New York* **1**.

HALL, JAMES (1863). Preliminary notice of the fauna of the Potsdam Sandstone. *New York State Cab. Natur. Hist.* 16th Ann. Rept., 119.

HALL, C. W. and SARDESON, F. W. (1895). The Magnesian Series of the northwestern states. *Bull. Geol. Society America* **6**, 167.

HAM, W. E. (1949). Geology and dolomite resources, Mill Creek-Ravia area, Johnston County Oklahoma. *Oklahoma Geol. Survey Circ.* **26**.

HAM, W. E. (1955). Field conference on geology of the Arbuckle Mountain region. *Oklahoma Geol. Survey Guide Book* **111**.

HAM, W. E., DENISON, R. E., and MERRITT, C. A. (1964). Basement rocks and structural evolution, southern Oklahoma. *Oklahoma Geol. Survey Bull.* **95**.

HAM, W. E. and WILSON, J. L. (1967). Paleozoic epeirogeny and orogeny in the central United States. *Amer. Jour. Science* **265**, 332.

HAMBLIN, W. K. (1958). The Cambrian sandstones of northern Michigan. *Michigan Geol. Survey, Publ.* **51**.

HAMBLIN, W. K. (1961). Paleogeographic evolution of the Lake Superior region from Late Keweenawan to Late Cambrian time. *Bull. Geol. Society America* **72**, 1.

HAMBLIN, W. K. (1965). Basement control of Keweenawan and Cambrian sedimentation in Lake Superior region. *Bull. Amer. Assoc. Petroleum Geologists* **49**, 950.

HANSON, A. M. (1952). Cambrian stratigraphy in southwestern Montana. *Montana Bur. Mines and Geology, Memoir* **33**.

HANSON, A. M. (1957). Cambrian of Crazy Mountain Basin. *Billings Geol. Society, 8th Ann. Field Conf. Guidebook*, 48.

HANSON, A. M. (1962). Cambrian System in the Big Horn Canyon area. *Billings Geol. Soc., 12th Ann. Field Conf. 1961 Guidebook*, 51.

HAYES, P. T. and LANDIS, E. R. (1965). Paleozoic stratigraphy of the southern part of the Mule Mountains, Arizona. *U.S. Geol. Survey, Bull.* **1201-F**.

HEYLUM, E. B. (1961). Results of deep drilling in southwestern Utah. *Bull. Amer. Assoc. Petroleum Geologists* **45**, 252.

HOBBS, S. W., HAYS, W. H., and ROSS, R. J., Jr. (1968). The Kinnikinic Quartzite of central Idaho-redefinition and subdivision. *U.S. Geol. Survey Bull.* **1254-J**.

HOBBS, S. W., HAYS, W. H., ROSS, R. J., HUDDLE, J. W., and PALMER, A. R. (1967). Kinnikinic Quartzite separated into three distinctive units *in* Geological Survey Research for 1966. *U.S. Geol. Survey Prof. Paper* **550-A**, A75.

HOWE, W. B. (1966). Digitate algal stromatolite structures from the Cambrian and Ordovician of Missouri. *Jour. Paleontology* **40**, 64.

HOWE, W. B. (1968). Planar Stromatolites and burrowed carbonate mud facies in Cambrian strata of the Saint Francois Mountain area. *Missouri Geol. Survey and Water Res. Rept. of Invest.* **41**.

HOWE, W. B. and KOENIG, J. W. (1961). The stratigraphic succession in Missouri. *Missouri Geol. Survey and Water Res.* **40**, 2nd Ser.

HOWELL, B. F. and others (1944). Correlation of the Cambrian formations of North America. *Bull. Geol. Society America* **55**, 993.

IRELAND, H. A. (1955). Precambrian surface in northeastern Oklahoma and parts of adjacent states. *Bull. Amer. Assoc. Petroleum Geologists* **39**, 468.

JORDAN, LOUISE (1967). Geology of Oklahoma—A summary. *Oklahoma Geol. Notes* **27**, 215.

KAY, G. M. (1947). Geosynclinal nomenclature and the craton. *Bull. Amer. Assoc. Petroleum Geologists* **31**, 1289.

KAY, G. M. (1951). North American geosynclines. *Geol. Society America Memoir* **48**.

KEEFER, W. R. and VAN LIEU, J. A. (1966). Paleozoic formations in the Wind River Basin, Wyoming. *U.S. Geol. Survey Prof. Paper* **495-B**.

KEROHER, GRACE C. (1965). Lexicon of geologic names of United States. *U.S. Geol. Survey, Bull.* **1200**, 3386.

KEROHER, R. P. and KIRBY, J. J. (1948). Upper Cambrian and Lower Ordovician rocks in Kansas. *Kansas Geol. Survey, Bull.* **72**.

KNECHTEL, M. M. (1956). Geological note on Emerson Formation of Cambrian and probable Early Ordovician age in Little Rocky Mountains, Montana. *Bull. Amer. Assoc. Petroleum Geologists* **40**, 1994.

KOENIG, JOHN W. (1966). Stratigraphic cross section of Paleozoic rocks in northern Missouri. *Cross Section Pub.* **4**, 18. Amer. Assoc. Petroleum Geologists, Tulsa, Okla.

KRIEGER, MEDORA H. (1961). Troy Quartzite (Younger Precambrian) and Bolsa and Abrigo Formations (Cambrian), northern Galiuro Mountains, southeastern Arizona. *U.S. Geol. Survey Prof. Paper* **424 C**, 160.

KRUMBIEN, W. C. and SLOSS, L. L. (1963). *Stratigraphy and Sedimentation, 2nd edition.* W. H. Freeman and Co., San Francisco.

KULIK, JOSEPH W. (1965). Stratigraphy of the Deadwood Formation, South Dakota and Wyoming. *Unpub. Master's thesis, S. Dak. Sch. of Mines and Tech.*

LEBAUER, L. R. (1965). Genesis and environment of deposition of the Meagher Formation in southwestern Montana. *Jour. Sed. Petrology* **35**, 428.

LEE, WALLACE (1943). The stratigraphy and structural development of the Forest City Basin in Kansas. *Kansas Geol. Survey Bull.* **51**.

LEE, WALLACE (1956). Stratigraphy and structural development of the Salina Basin area. *Kansas Geol. Survey Bull.* **121**.

LIDIAK, E. G. (1963). Correlation of basement rocks with the Mid-Continent gravity anomaly in Nebraska and Kansas (abst.). *Soc. Explor. Geophysicists, 33rd. Internat. Meeting.*

LINDGREN, WALDEMAR (1905). The copper deposits of the Clifton–Morenci district, Arizona. *U.S. Geol. Survey Prof. Paper* **43**, 15.

LOCHMAN, CHRISTINA (1940). Fauna of the basal Bonneterre Dolomite (Upper Cambrian) of southeastern Missouri. *Jour. Paleontology* **14**, 1.

LOCHMAN, CHRISTINA (1947). Identification of fossils from Beeler deep test well, south-central Tennessee: letter to R. D. White.

LOCHMAN, CHRISTINA (1956). The Cambrian of the middle central interior states of the United States. *El Sistema Cambrico, su paleogeograpfia y el problema de su base, Tomo II.* pt. 2, 447. XX Internat'l. Geol. Congress.

LOCHMAN, CHRISTINA (1964). Upper Cambrian faunas from the subsurface Deadwood Formation, Williston Basin, Montana. *Jour. Paleontology* **38**, 33.

LOCHMAN, CHRISTINA (1968). *Crepicephalus* Faunule from the Bonneterre Dolomite (Upper Cambrian) of Missouri. *Jour. Paleontology* **42**, 1153.

LOCHMAN, CHRISTINA and DUNCAN, DONALD (1944). Early Upper Cambrian faunas of central Montana. *Geol. Society America Sp. Paper* **54**.

LOCHMAN-BALK, CHRISTINA (1955). Cambrian stratigraphy of the south and west margins of Green River Basin. *Wyo. Geol. Assoc. Guidebook, 10th Ann. Field Conf.*, 29.

LOCHMAN-BALK, CHRISTINA (1956a). The Cambrian of the Rocky Mountains and southwest deserts of the United States and adjoining Sonora Province, Mexico. *El Sistema Cambrico, su Paleogeografia y el problema de su base. Tomo II*, pt. 2, 529. XX Internat'l. Geol. Congress.

LOCHMAN-BALK, CHRISTINA (1956b). Cambrian stratigraphy of eastern Utah. *Intermont. Assoc. Petrol. Geol. 7th Ann. Field Conf. Guidebook*, 58.

LOCHMAN-BALK, CHRISTINA (1959). The Cambrian section in the central and southern Wasatch Mountains. *Intermont. Assoc. Petrol. Geol. 10th Ann. Field Conf. Guidebook*, 40.

LOCHMAN-BALK, CHRISTINA (1960). The Cambrian section of western Wyoming. *Wyoming Geol. Assoc. Guidebook*, 99.

LOCHMAN-BALK, C. and WILSON, J. L. (1958). Cambrian biostratigraphy in North America. *Jour. Paleontology* 32, 312.

LOCHMAN-BALK, C. and WILSON, J. L. (1967). Stratigraphy of Upper Cambrian–Lower Ordovician subsurface sequence in Williston Basin. *Bull. Amer. Petroleum Geologists* 51, 883.

LOGAN, B. W., REZAK, R., and GINSBURG, R. N. (1964). Classification and environmental significance of algal stromatolites. *Jour. Paleontology* 72, 68.

LOLIET, ALLAN J. (1963). Cambrian stratigraphic problems of the Four Corners area. *Four Corners Geol. Society—Shelf carbonates of the Paradox Basin symposium, 4th Field Conf.*, 21.

LUGN, A. L. (1934). Pre-Pennsylvanian stratigraphy of Nebraska. *Bull. Amer. Assoc. Petroleum Geologists* 18, 1597.

LYTLE, W. S. and others (1963). Oil and gas developments in Pennsylvania in 1962. *Penn. Topo. and Geol. Survey, Prog. Rept.* 165.

LYTLE, W. S. (1964). Oil and gas developments in Pennsylvania in 1963. *Penn. Topo. and Geol. Survey, Prog. Rept.* 166.

McCRACKEN, MARY H. (1965). The Cambro–Ordovician rocks of northeastern Oklahoma and adjacent areas. *Tulsa Geol. Society Digest* 32.

McELROY, M. N. (1965). Lithologic and stratigraphic relations between the Reagan Sandstone (Upper Cambrian) and sub-Reagan and supra-Reagan rocks in western Kansas. *Unpubl. Ph.D. thesis, Univ. of Kansas.*

McKEE, E. D. and RESSER, C. E. (1945). Cambrian stratigraphy of the Grand Canyon region. *Carnegie Inst. of Washington Publ.* 563.

McMANNIS, W. J. (1955). Geology of the Bridger Range, Montana. *Bull. Geol. Society America* 66, 1385.

McMANNIS, W. J. (1965). Resume of depositional and structural history of western Montana. *Bull. Amer. Assoc. Petroleum Geologists* 49, 1801.

MERRITT, C. A. (1967). Names and relative ages of granites and rhyolites in the Wichita Mountains, Oklahoma. *Oklahoma Geology Notes* 27, 45.

MILLER, B. M. (1936). Cambrian stratigraphy of northwestern Wyoming. *Jour. Geology* 44, 113.

Mountain Fuel Supply Company completion report (1962). North Baxter Basin, Union Pacific # 4.

MULCHAY, R. B. and VELASCO, J. R. (1954). Sedimentary rocks at Cananea, Sonora, Mexico *****. *Trans. Amer. Inst. Min. & Metal. Eng.* 199, 628.

MURRAY, R. C. (1955). Late Keweenawan or early Cambrian glaciation in upper Michigan. *Bull. Geol. Society, America* 66, 341.

NELSON, C. A. (1951). Cambrian trilobites from the Saint Croix valley. *Jour. Paleontology* 25, 765.

NELSON, C. A. (1956). Upper Croixan stratigraphy, upper Mississippi Valley. *Bull. Geol. Society America* 67, 165.

NORTON, W. H. (1928). Deep wells of Iowa. *Iowa Geol. Survey Annual Report*, 1927, 15.

OHLE, E. L. and BROWN, J. S. (1954). Geologic problems in the south-east Missouri lead district. *Bull. Geol. Society America* 65, 201.

OLCOTT, P. G. (1966). Geology and water resources of Winnebago County, Wisconsin. *U.S. Geol. Survey Water-Supply Paper* 1814.

OSTROM, M. E. (1964). Pre-Cincinnatian Paleozoic cyclic sediments in the upper Mississippi Valley. *Kansas Geol. Survey, Bull.* 169, 381.

OSTROM, M. E. (1965). Cambro–Ordovician stratigraphy of southwest Wisconsin. *Wisconsin Geol. and Nat. Hist. Survey, Infor. Circ.* 6.

OSTROM, M. E. (1967). Paleozoic stratigraphic nomenclature for Wisconsin. *Univ. of Wisc. Geol. and Nat. Hist. Survey, Inf. Circ.* 8.

OSTROM, M. E. and SLAUGHTER, A. E. (1967). Correlation problems of the Cambrian and Ordovician outcrop areas, northern peninsula of Michigan. *Mich. Basin Geol. Society Ann. Field Excursion.*

OTVOS, E. G., Jr. (1966). Sedimentary structures and depositional environments, Potsdam Formation, Upper Cambrian. *Bull. Amer. Assoc. Petroleum Geologists* 50, 159.

OWEN D. D. (1852). Report of a geological survey of Wisconsin, Iowa and Minnesota. Philadelphia.

PALMER, A. R. (1954). The faunas of the Riley Formation in central Texas. *Jour. Paleontology* **28**, 709.

PALMER, A. R. (1965a). Biomere—a new kind of biostratigraphic unit. *Jour. Paleontology* **39**, 149.

PALMER, A. R. (1965b). Trilobites of the Late Cambrian Ptericephaliid biomere in the Great Basin, United States. *U.S. Geol. Survey, Prof. Paper* **493**.

PONTOJA-ALOR, J. and ROBISON, R. A. (1967). Paleozoic sedimentary rocks in Ozxaca, Mexico. *Science* **157**, 1033.

PETERSON, EARL T. (1966). Paleozoic rocks in south-eastern Colorado. *Cross section Pub.* **4**. Amer. Assoc. Petroleum Geologists, Tulsa, Okla.

PETERSON, J. A. and others (1965). Rocky Mountain sedimentary basins-symposium. *Bull. Amer. Assoc. Petrol. Geologists* **49**, 1779.

POTTER, P. E. and PRYOR, W. A. (1961). Dispersal centers of Paleozoic and later clastics of the Upper Mississippi Valley and adjacent areas. *Bull. Geol. Society America* **72**, 1195.

RAASCH, G. O. (1935). Stratigraphy of the Cambrian System of the upper Mississippi valley. *Kansas Geol. Society, Ninth Annual Field Conf. Guidebook*, 302.

RASSCH, G. O. (1950). Current evaluation of the Cambrian-Keweenawan boundary. *Trans. Illinois Acad. Sci.* **43**, 137.

RAASCH, G. O. (1952). Revision of Croixan Dikelocephalidae. *Illinois Geol. Survey, Circ.* **179**, 137.

RAASCH, G. O. and UNFER, LOUIS (1964). Transgressive-regressive cycle in Croixan sediments (Upper Cambrian), Wisconsin. *Kansas Geol. Survey, Bull.* **169**, 427.

RANSOME, F. L. (1916). Some Paleozoic sections in Arizona and their correlation. *U.S. Geol. Survey, Prof. Paper* **98K**, 133.

RASETTI, FRANCO (1951). Middle Cambrian stratigraphy and faunas of the Canadian Rocky Mountains. *Smithson. Misc. Colls.* **116**, no. 3.

RASETTI, FRANCO (1956). The Middle and Upper Cambrian of western Canada. *El Sistema Cambrico su paleogeografia y el problema de su base* **2**, pt. 2, 735. XX Internat'l. Geol. Congress.

RASETTI, FRANCO (1965). Upper Cambrian trilobite faunas of northeastern Tennessee. *Smithson., Misc. Colls.* **148**, no. 3.

REED, E. C. (1938). Correlation of formations drilled in the Midland Forester well near Fremont Nebraska. *Nebraska Geol. Survey, Paper* **13**.

RESSER, C. E. (1945). Cambrian fossils of the Grand Canyon: Pt. 2 *in* Cambrian history of the Grand Canyon region. *Carneg. Inst. Wash. Publ.* **563**, 171.

ROBISON, R. A. (1964a). Late Middle Cambrian faunas from western Utah. *Jour. Paleontology* **38**, 510.

ROBISON, R. A. (1964b). Middle-Upper Cambrian boundary in North America. *Bull. Geol. Society of America* **75**, 987.

ROBISON, R. A. (1964c). Upper Middle Cambrian stratigraphy of western Utah. *Bull. Geol. Society America* **75**, 995.

RODGERS, JOHN (1956). The known Cambrian deposits of the southern and central Appalachian mountains. *El Sistema Cambrico, su paleogeografia y el problema de su base* **2**, pt. 2, 353. XX Internat'l. Geol. Congress.

RODGERS, JOHN (1963). Mechanics of Appalachian foreland folding in Pennsylvania and West Virginia. *Bull. Amer. Assoc. Petroleum Geologists* **47**, 1527.

ROSS, REUBEN J. (1951). Stratigraphy of the Garden City Formation in northeastern Utah and its trilobite faunas. *Peabody Mus. Nat. Hist., Yale Univ., Bull.* **6**.

RUDMAN, A. J., SUMMERSON, C. H., and HINZE, W. J. (1965). Geology of basement in midwestern United States. *Bull. Amer. Assoc. Petroleum Geologists* **49**, 894.

SABINS, Jr., F. F. (1957). Stratigraphic relations in Chiricahua and Dos Cabezos Mountains, Arizona. *Bull. Amer. Assoc. Petroleum Geologists* **41**, 466.

SCHOLTEN, ROBERT, KEENMON, K. A., and KUPSCH, W. O. (1955). Geology of the Lima region, southwestern Montana and adjacent Idaho. *Bull. Geol. Society America* **66**, 345.

SCHULDT, W. C. (1943). Cambrian strata of northeastern Iowa. *Iowa Geol. Survey, Annual Repts., 1940 and 1941* **38**, 383.

SCOTT, R. W. (1966). New Pre-Cambrian (?) formation in Kansas. *Bull. Amer. Assoc. Petroleum Geologists* **50**, 380.

SCOTT, R. W. and McELROY, M. N. (1964). Pre-Cambrian–Paleozoic contact in two wells in northwestern Kansas. *Kansas Geol. Survey Bull.* **170**.

SCHUCHERT, C. (1910). Paleogeography of North America. *Bull. Geol. Society America* **20**, 427.

SHAW, A. B. (1954). Correlation of the Paleozoic formations of Wyoming. *Wyo. Geol. Assoc. Guidebook*, *9th Ann. Field Conf., Chart II.*

SHAW, A. B. and DELAND, C. R. (1955). Cambrian of southwestern Wyoming. *Wyo. Geol. Assoc. Guidebook, 10th Ann. Field Conf.* 38.

SHRIDE, A. F. (1967). Younger Precambrian geology in southern Arizona. *U.S. Geol. Survey Prof. Paper* 566.

SLOSS, L. L. (1950). Paleozoic sedimentation in Montana area. *Bull. Amer. Assoc. Petroleum Geologists* 34, 423.

SLOSS, L. L. and MORITZ, C. A. (1951). Paleozoic stratigraphy of southwestern Montana. *Bull. Amer. Assoc. Petrol. Geologists* 35, 2135.

SNYDER, F. G. and EMERY, J. A. (1956). Geology in development and mining, southeast Missouri lead belt. *Min. Engineering* 8, 1216.

SNYDER, F. G. and ODELL, J. W. (1958). Sedimentary breccias in the southeast Missouri lead district. *Bull. Geol. Society America* 69, 899.

SPINDLETOP REASERCH (1965). Oil and gas possibilities of the Cambrian and Lower Ordovician in Kentucky. *Rept. for Kentucky Dept. Commerce, Louisville, Kentucky.*

STAATZ, M. H. and ALBEE, H. F. (1966). Geology of the Garns Mountain Quadrangle Bonneville, Madison and Teton Counties, Idaho. *U.S. Geol. Survey, Bull.* 1205.

STAUFFER, C. R. and THIEL, G. A. (1941). The Paleozoic and related rocks of southeastern Minnesota. *Minn. Geol. Survey, Univ. of Minn. Bull.* 29.

STAUFFER, C. R., SCHWARTZ, G. M. and THIEL, G. A. (1929). Saint Croixan classification of Minnesota. *Bull. Geol. Society America* 50, 1227.

STEINHILBER, W. L. and others (1961). Geology and ground water resources of Clayton County, Iowa. *Iowa Geol. Survey, Water Supply Bull.* 7.

STOCKMAN, K. M., GINSBURG, R. N. and SHINN, E. A. (1967). The production of lime mud by algae in south Florida. *Jour. Sed. Petrology* 37, 633.

STOYANOW, A. A. (1936). Correlation of Arizona Paleozoic formations. *Bull. Geol. Society America* 47, 459.

SUMMERSON, C. H. (1962). Precambrian in Ohio and adjoining areas. *Ohio Geol. Survey, Rept. of Invest.* 44.

TASCH, PAUL (1951a). Fauna and paleoecology of the Upper Cambrian Warrior Formation, central Pennsylvania. *Jour. Paleontology* 25, 275.

TASCH, PAUL (1951b). A cyclic occurrence of *Cryptozoon undulatum. Amer. Midland Natur.* 46, 151.

TASCH, PAUL (1952). The taxonomy and paleoecological significance of Pemphigaspid trilobites. *Jour. Paleontology* 25, 529.

TEMPLETON, J. S. (1950). The Mount Simon Sandstone in northern Illinois. *Trans. Illinois Acad. Science* 43, 151.

THOMAS, G. R. (1960). Geology of recent deep drilling in eastern Kentucky. *Kentucky Geol. Survey, Ser. 10, Sp. Publ.* 3, 10.

TWENHOFEL, W. H., RAASCH, G. O., and THWAITES, F. T. (1935). Cambrian strata of Wisconsin. *Bull. Geol. Society America* 46, 1687.

ULRICH, E. O. (1911). Revisions of the Paleozoic systems. *Bull. Geol. Society America* 22, 281.

ULRICH, E. O. (1933). Preliminary description of the Honey Creek, Fort Sill, Royer and Signal Mountain Formations of Oklahoma. *Bull. Geol. Society America* 43, 742.

ULRICH, E. O. and SCHUCHERT, C. (1902). Paleozoic seas and barriers in eastern North America. *New York State Mus. Bull.* 52, 633.

ULRICH, E. O., FOERSTE, A. F., and BRIDGE, J. (1931 [1930]). Paleontology of Late Cambrian and Early Ordovician formations in Missouri. *Missouri Bur. Geol. and Mines, 2nd ser.* 24, 186.

UNTERMANN, G. E. and UNTERMANN, B. R. (1949). Geology of Green and Yampa River canyons and vicinity, Dinosaur National Monument, Utah and Colorado. *Bull. Amer. Assoc. Petroleum Geologists* 3, 683.

WAGNER, W. R. (1961). Subsurface Cambro-Ordovician stratigraphy of northwestern Pennsylvania and bordering states. *Penn. Topo. and Geol. Survey, Prog. Rept.* 156.

WELLS, JACK and McCRACKEN, EARL (1964). Northwest Missouri oil and gas exploratory logs (1945–1963). *Missouri Geol. Survey & Water Res., Inf. Circ.* 17.

WILLIAMS, N. C. (1953). Late Precambrian and Early Paleozoic geology of western Uinta Mountains, Utah. *Bull. Amer. Assoc. Petroleum Geologists* 37, 2734.

WINCHELL, N. H. (1873). The geological and natural history survey of Minnesota. *Minn. Geol. and Nat. Hist. Survey, 1st Ann. Rept. for 1872*, 17.

WINCHELL, N. H. (1888). Geology of Minnesota. *Minn. Geol. and Nat. Hist. Survey, Final Rept.* 2.

WINCHELL, N. H. (1900). Geologic Map of Minnesota. *Minn. Geol. and Nat. Hist. Survey.*

WILSON, JAMES L. (1949). The Trilobite fauna of the *Elvinia* zone in the basal Wilberns Limestone of Texas. *Jour. Paleontology* **23**, 25.

WILSON, JAMES L. (1952). Upper Cambrian stratigraphy in the central Appalachians. *Bull. Geol. Society America* **63**, 275.

WINSTON, DON and NICHOLS, HARRY (1967). Late Cambrian and Early Ordovician faunas from the Wilberns Formation of central Texas. *Jour. Paleontology* **41**, 66.

WOOD, W. H. (1956). The Muav Limestone and the supra-Muav sequence at Yampai Cliffs, Arizona. *Plateau, Museum of No. Arizona* **29**, 25.

WOOD, W. H. (1966). Facies changes in the Cambrian Muav Limestone, Arizona. *Bull. Geol. Society America* **77**, 1235.

WOODWARD, H. P. and others (1959). A symposium on the Sandhill deep well, Wood County, West Virginia. *W. Va. Geol. Survey Rept. Invest.* **18**.

WOODWARD, H. P. (1961). Preliminary subsurface study of southeastern Appalachian Interior Plateau. *Bull. Amer. Assoc. Petroleum Geologists* **45**, 1634.

WORKMAN, L. E. and BELL, A. H. (1949). Deep drilling and deeper oil possiblities in Illinois. *Illinois State Geol. Survey, Rept. Inv.* **139**.

THE CAMBRIAN OF THE APPALACHIAN AND EASTERN NEW ENGLAND REGIONS, EASTERN UNITED STATES

Allison R. Palmer

Department of Earth and Space Sciences,
State University of New York at Stony Brook

Contents

1. Introduction

The principal Cambrian exposures of eastern United States are found in a sinuous belt generally less than 65 kilometres wide which extends for over 1,900 kilometres from the Canadian border to the northern edge of the coastal plain in central Alabama (Fig. 1). This is the Appalachian region. All of the rocks in this region have been subjected to middle or late Palaeozoic compressional deformation which has produced either asymmetrical folds with steep or overturned north-western limbs or east-dipping low-angle thrust faults of varying displacement, sometimes of the order of tens of kilometres. In addition, the rocks in the northern part of the region were involved in an earlier episode of westward gravity sliding during the Ordovician. The middle and late Palaeozoic deformational periods were accompanied by regional metamorphism which is most intense in the eastern parts of the Appalachian region particularly from Pennsylvania northwards.

Throughout the Appalachian region exposures are discontinuous and sections through the whole of the Cambrian system are dishearteningly few. Those parts of the region from New Jersey northwards were subjected to Pleistocene glaciation that removed the majority of the deeply weathered surface materials but left most of the area under an irregular blanket of glacial deposits and alluvium. In the central and southern Appalachians, deep weathering and soil cover limit outcrops. Thus, the stratigraphy of the Appalachian region has been pieced together from discontinuous exposures by detailed mapping and regional reconnaissance.

Although a few units have yielded rich invertebrate faunas locally, the bulk of the Cambrian deposits is not notably fossiliferous except for stromatolitic algae. As a result of the sparse palaeontological data and generally poor exposures, precise correlation of stratigraphical details has not been possible, and most stratigraphical units are either thick, somewhat heterogeneous in internal detail, and widespread; or thin and only locally recognized. Thicknesses are generally only approximate. Despite these problems, a number of regions of fairly homogeneous Cambrian stratigraphy can be identified within the Appalachian region, and their inter-relationships do provide the basis for a crude picture of the history of Cambrian sedimentation and the Cambrian palaeogeography for the region.

In eastern New England, east of the main Cambrian belt of the Appalachian region, only a few small scattered exposures of Cambrian rocks are known. They have considerable significance for the interpretation of the larger palaeotectonic framework of the Cambrian of eastern North America and will be separately discussed. Immediately west of the main Cambrian belt, Cambrian rocks are exposed at the surface in the Adirondack region of New York, and a few deep wells have penetrated these rocks in the areas to the south. These are discussed elsewhere in this volume.

The areas of more or less uniform stratigraphy are shown in Figs 2, 6, 9, and 12. In the northern Appalachians, these are: the predominantly slaty sequence of the St. Albans area of northern Vermont; the carbonate–sand sequences of the Milton-Bennington region of central and southern Vermont; the metamorphic sequences of eastern Vermont, western Massachusetts, and western Connecticut; the argillaceous and arenaceous sequences of the Taconic region of western Vermont and south-eastern New York; and the 'thin' dominantly carbonate sequences from north-western Massachusetts to eastern Pennsylvania. In the central Appalachians, these are: the main belt of older clastics and younger carbonates to the west of the South Mountain–Blue Ridge trend from Pennsylvania to southern Virginia; and the carbonate sequences south-east of the

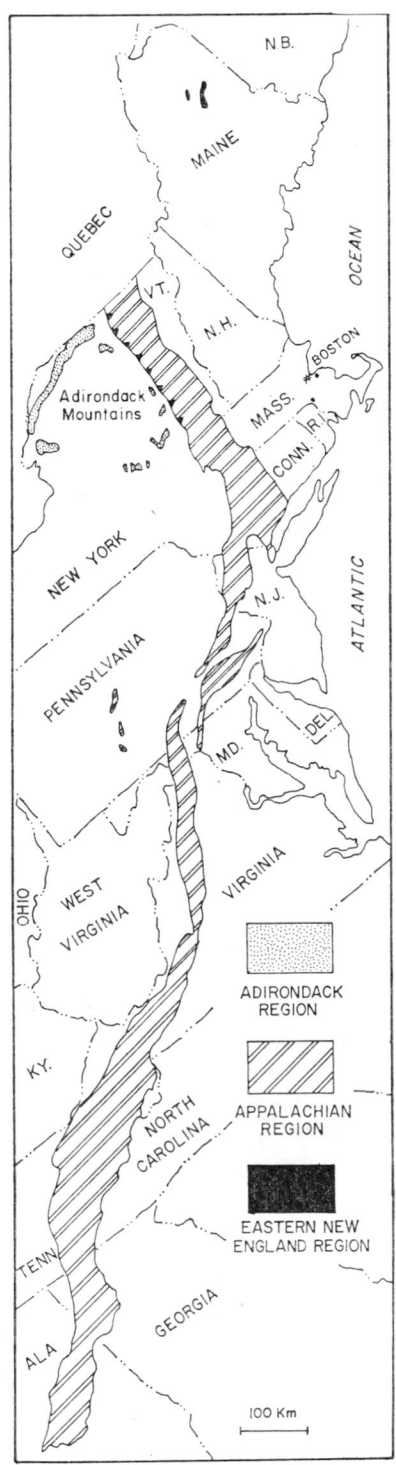

FIG. 1. The principal regions of Cambrian outcrops in eastern United States.

Triassic rocks in south-eastern Pennsylvania and Maryland. In the southern Appalachians, these are: the older clastic and younger interbedded shale and carbonate sequences of Tennessee and south-western Virginia; and the predominantly shale or thin-bedded limestone sequences, with overlying dolomites and underlying coarser clastics, of Georgia and Alabama. Each of these areas will be discussed separately below and then the regional relationships between them will be summarized. The formational names used in the text and their approximate correlations are shown on Figs 5, 8, 10, 13, and 14.

1.1. History of Cambrian studies in the Appalachian region

Cambrian studies in the Appalachian region before 1900 were largely involved in the complicated controversy about stratigraphical position of the rocks in the Taconic region of eastern New York and western Vermont (Chapter 2.3). However, information about local occurrences of Cambrian rocks and faunas began accumulating in the early 1840's. WALCOTT (1891) presented a detailed account of the earlier studies and clarified the age of the rocks of the Taconic region. As with western United States, WALCOTT dominated all phases of Cambrian activity in the Appalachian region during his lifetime, and much information is included in his general publications (WALCOTT, 1884, 1886, 1891, 1910, 1912, 1916a, 1916b). In subsequent years, most publications including Cambrian data have dealt with only parts of the region and they are cited in the appropriate chapters. A synthesis of data on Cambrian faunal distributions in the Appalachian region was presented by LOCHMAN-BALK and WILSON (1958) and syntheses of the Cambrian stratigraphy and faunas of the northern Appalachian and eastern New England regions (Chapters 2, 6) have been presented by SHAW (1961) and THEOKRITOFF (1968). RODGERS (1968) has made an important contribution to regional evaluation of the Cambrian facies in the central and northern Appalachians and earlier (RODGERS, 1956) presented useful summaries of the Cambrian stratigraphy in the central and southern Appalachians (Chapters 3, 4).

2. The northern Appalachians

This area includes all Cambrian exposures in Vermont, western Connecticut, western Massachusetts, south-eastern New York, New Jersey, and Pennsylvania west to the city of Reading (Figs 2 and 6). In the north, it is bounded on the west by Lake Champlain and the Hudson River and on the east by metamorphosed younger Palaeozoic rocks. In parts of south-eastern New York, New Jersey, and eastern Pennsylvania, it is bounded on the south-east by the Triassic lowlands or by Pre-Cambrian basement, and on the north-west by younger Palaeozoic rocks.

2.1. The St. Albans area

The northern 40 kilometres of the western New England Cambrian belt, west of the Hinesburg thrust, contains a Cambrian section that is in striking contrast to the sequence of Cambrian formations which extends to the south through the Milton–Bennington region. The only formations that this region shares with the Milton–Bennington region are the Lower Cambrian Cheshire Quartzite and the overlying Dunham Dolomite. Throughout the area, these formations are truncated at their bases by thrust faults and no Pre-Cambrian rocks are known.

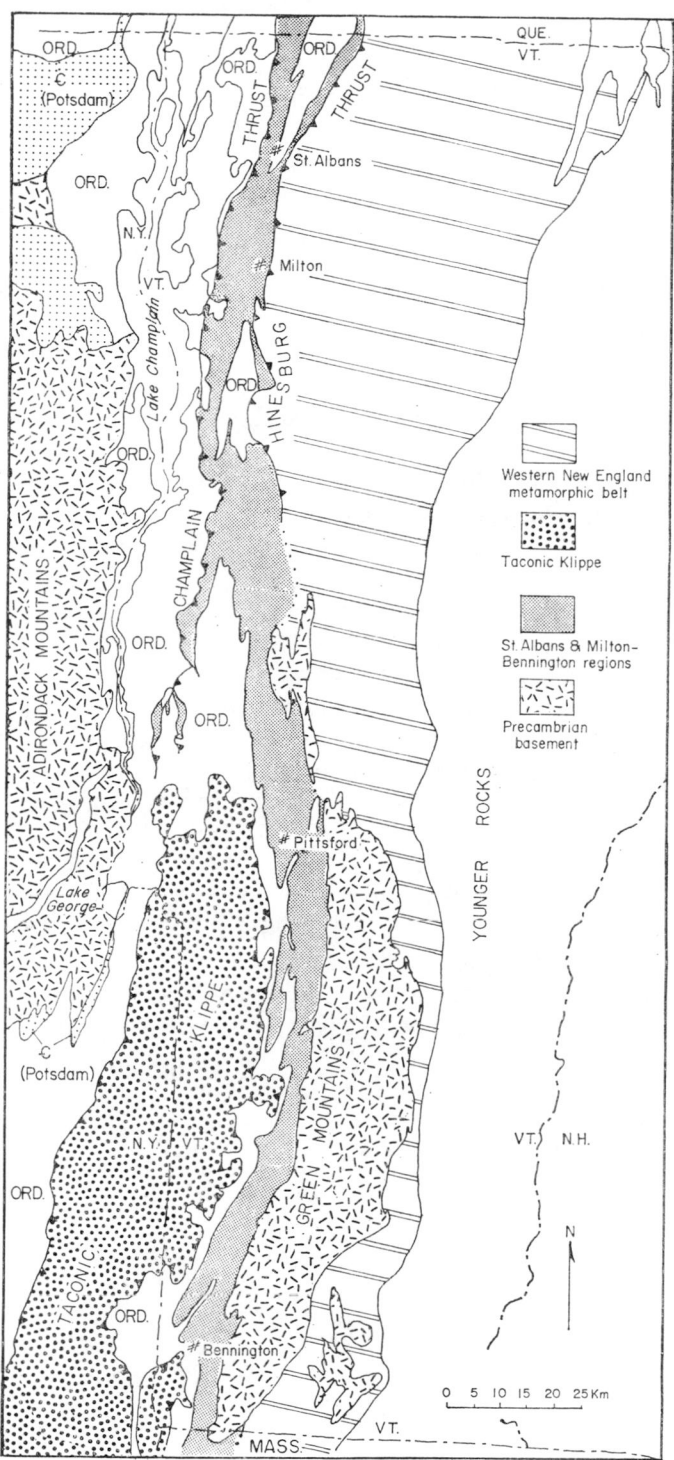

FIG. 2. Principal outcrop areas of Cambrian rocks in the northern part of the Northern Appalachian region.

The dominant structure of the St. Albans area is the St. Albans synclinorium. Most of the detailed information about the Cambrian rocks comes from beds exposed on the west limb of the synclinorium. SHAW (1958) has shown that much of the Cambrian System above the Lower Cambrian Dunham Dolomite in the west limb is represented by a sequence of shales (now slates) with a few thin interbeds of limestone or dolomite and several units of carbonate-clast conglomerate. This sequence ranges in age from late Lower Cambrian to Ordovician. Thicknesses are uncertain because of poor exposure and degree of deformation, but the shale-conglomerate sequence is of the order of 600 metres. Stratigraphical details within this sequence are complicated by a number of supposed unconformities (SHAW, 1958), some of which have been disputed by CADY (1960), and most of which were discounted by RODGERS (1968).

The dominantly shale (slate) sequence of the western part of the St. Albans region has been divided by SHAW (1958) and earlier workers into a series of alternating thick black shale and thin carbonate or conglomerate units. The lowest black shale unit is the Parker Shale which contains Early Cambrian fossils. This is overlain by a discontinous dolomite and dolomite conglomerate unit, the Rugg Brook Dolomite, which is unfossiliferous. The succeeding five units—three black shale units separated by two discontinous conglomerate units—are assigned by SHAW to the Woods Corners Group. The formations are, from oldest to youngest: the St. Albans Shale, Mill River Conglomerate, Skeels Corners Shale, Rockledge Conglomerate, and Hungerford Shale. The faunas from these formations represent a related complex of trilobites ranging in age from late Middle Cambrian to early Upper Cambrian (Dresbachian).

In the northern 10 kilometres of the St. Albans area, SHAW (1958) records the Saxe Brook Dolomite, estimated to be up to 180 metres thick, within the Woods Corners Group. This formation disappears within a few kilometres to the north and south. The Saxe Brook is separated by a narrow belt, mapped as Hungerford Shale (slate), from the overlying Gorge Formation, also known only in the northernmost part of the St. Albans area. The Gorge consists of a lower dolomite member about 165 metres thick, which is overlain by interbedded shales (slates) and limestones totalling about 120 metres in thickness. Faunas of late Dresbachian age from the lower part of the shale and limestone sequence in the Gorge Formation indicate that the lower dolomite member is probably no younger than early Dresbachian age. Re-evaluation of faunas from shales below the Saxe Brook Dolomite (p. 175) indicates that they are of late Middle Cambrian age. Thus the overlying Saxe Brook Dolomite may also be of early Dresbachian age, and the two dolomites may be the same unit repeated by faulting. Except for this unit, there is no significant development of relatively clean carbonate rocks above the Lower Cambrian Dunham Dolomite, either in the St. Albans region or in the Cambrian sequences along strike farther to the north in Quebec.

In the east limb of the St. Albans synclinorium, a white schistose quartzite is assigned to the Cheshire Quartzite by DENNIS (1964). Above the Cheshire is a thin sequence of rusty-weathering sandy dolomite, followed by a limy to dolomitic shale grading upwards into a shaly dolomite and black slate, and finally into an upper sandy dolomite or dolomitic sandstone. This sequence is assigned by DENNIS (1964) to the Bridgeman Hill Formation and includes in upward succession: the Dunham, Rice Hill, Oak Hill, and Rugg Brook members. The Bridgeman Hill Formation is overlain by the Sweetsburg Formation, a black slate with thin whitish silty layers. This is correlated with beds of the Woods Corners Group. According to DENNIS, this unit is overlain by the Middle Ordovician Morses Line black slate and there is no record of rocks equivalent to the

Gorge Formation. Because of intense deformation, no estimates of thickness have been given for the formations in this area.

2.1.1. Age and correlation

The Cambrian faunas from the St. Albans region have been described or revised in a series of papers by SHAW (1951–1968) and CLARK and SHAW (1968a, b). The most important elements of these faunas for regional correlation are the trilobites. Most of the other faunal elements are rare and stratigraphically non-diagnostic.

The Early Cambrian Dunham Dolomite has yielded a small fauna entirely of ptychoparioid trilobites totalling only four species from two localities (SHAW, 1955). These are typical North American trilobites referred by SHAW to: *Antagmus typicalis* RESSER, *Antagmus? simplex* RESSER, *Billingsaspis adamsii* (BILLINGS), and *Perimetopus arenosus* (BILLINGS).

The faunas of the overlying Parker Shale are considerably more diverse and SHAW (1955) recognized several species of olenellids all assigned to *Olenellus;* two species of corynexochoids assigned to *Bonnia;* single species of corynexochoids assigned to *Kootenia, Protypus, Prozacanthoides,* and *Zacanthopsis*; *Pagetides parkeri* (WALCOTT); and several species of nondescript North American types of ptychoparioids. No archaeocyathids have been recorded from any of the Early Cambrian rocks of this region.

A trilobite assemblage of supposedly early Middle Cambrian age was described by SHAW (1957) from a limestone conglomerate bed assigned to the upper Parker Shale (slate) below his Saxe Brook Dolomite. The fauna included as its most definitive elements: corynexochoids referred to *Orriella, Athabaskia, Kootenia* and *Zacanthoides;* agnostids referred to *Peronopsis;* and several ptychoparioids including forms referred to *Syspacephalus cadyi* SHAW and *Mexicella stator* (WALCOTT). On the basis of this last identification, the fauna was correlated with faunas of the early Middle Cambrian *Albertella* Zone of western United States, where *M. stator* is a distinctive element. However, re-examination of the fauna and re-collection from the critical locality, following a suggestion by RASETTI (in THEOKRITOFF, 1968), indicates that the trilobite identification as *M. stator* is most likely a species of *Semisphaerocephalus* and the forms identified as *Syspacephalus cadyi* are a species of *Spencella.* These trilobites are distinctive late Middle Cambrian forms, which is more in accordance with the presence of *Orriella* (= *Bathyuriscus*) in the assemblage, and both the age and stratigraphical assignment of this fauna should be re-evaluated.

Late Middle Cambrian assemblages are recorded by SHAW (1966b) from the St. Albans Shale which is lithologically identical to the Parker Shale but is usually separated from it by the unfossiliferous Rugg Brook Dolomite. The trilobites are not well preserved, but the presence of species assigned to *Meneviella* and *Centropleura* confirms the age assignment.

The trilobites assigned to the Skeels Corners Shale, which is lithologically much like the St. Albans Shale and is separated from it by discontinuous lenses of Mill River Conglomerate, were assigned an early Late Cambrian (Dresbachian) age by SHAW (1966a). However, the fauna contains ptychoparioids assigned to *Bolaspidella* and the distinctive corynexochoid *Hemirhodon* which are characteristic of beds of late Middle Cambrian age in western North America (ROBISON, 1964; PALMER, 1968). Although the fauna also includes trilobites referred to *Catillicephala,* and SHAW (1958) lists a fauna with this genus and several other genera of simple ptychoparioids usually considered to be of early Upper Cambrian age, as well as *Bolaspidella,* from the supposedly older Mill

River Conglomerate, a Late Cambrian age for the Skeels Corners faunas may be too young. In any event, the assemblages represented in the shales assigned to this formation are very close in age to beds on either side of the Middle-Upper Cambrian boundary.

The fauna of the Rockledge Conglomerate (SHAW, 1952) which separates the Skeels Corners Shale from the overlying Hungerford Shale represents a slightly younger part of the early Upper Cambrian and includes such typical early Dresbachian genera as *Blountia* and *Kingstonia*. The Hungerford Shale, a black shale lithologically like the Skeels Corners and St. Albans, contains a poorly preserved fauna of trilobites no younger than early Dresbachian, including specimens assigned to *Catillicephala*. Thus, the faunas recorded by SHAW from the St. Albans Shale, Mill River Conglomerate, Skeels Corners Shale, Rockledge Conglomerate, and Hungerford Shale contain a related complex of trilobites which ranges from late Middle Cambrian to early Upper Cambrian in age. Comparable faunal relationships are observed in western United States where no abrupt change in trilobite faunas takes place between Middle and Late Cambrian time.

Only a part of the fauna of the Gorge Formation has been described. CLARK and SHAW (1968b, c) have described two faunas from a 1·5 metre limestone bed in the lower part of the shaly middle member of the formation. The lower part of this bed has a large trilobite fauna including species of *Dunderbergia*, *Pterocephalops*, and *Quebecaspis*. This is a typical assemblage of the late Dresbachian *Dunderbergia* Zone. The upper part of this bed also has a large assemblage of trilobites including species of *Richardsonella*, *Theodenisia*, and *Prosaukia*. These trilobites indicate a late Franconian age for this part of the bed and the probable presence of a significant hiatus within the bed. Higher beds of the formation have yielded a considerable fauna of trilobites of Trempealeauan age described by RAYMOND (1924, 1937) and partly revised by RASETTI (1944). The trilobite faunas of this formation have many species and genera in common with faunas from limestone boulders in the Levis conglomerates of Quebec (RASETTI, 1944) and the Cow Head conglomerates of western Newfoundland (KINDLE and WHITTINGTON, 1958, 1959). Closely related trilobite faunas are also known from the Frederick Limestone in Maryland (p. 197), and from Cambrian sections in Nevada and Alaska.

2.2. The Milton–Bennington region

This region includes all Cambrian rocks in the area east of Lake Champlain and west of either the Hinesburg thrust or the Pre-Cambrian core of the Green Mountains (Fig. 2). Throughout the region, which extends for nearly 200 kilometres, the principal stratigraphical variability is in the older beds (Fig. 5, p. 183).

The oldest persistent stratigraphical unit in the region is the Cheshire Quartzite. This is a unit of massive white quartzite generally more than 270 metres thick. It is the oldest Cambrian unit exposed at the surface west of the Hinesburg thrust fault in the northern part of the region. South of Pittsford, Vermont, the Cheshire is underlain by a thin unit of schistose conglomeratic quartzite with local development of conglomeratic dolomite which is assigned to the Dalton Formation. However, from Pittsford northwards to the end of the Pre-Cambrian exposures, the pre-Cheshire stratigraphy is quite different. Generally schistose greywacke of the Pinnacle Formation, conglomeratic in its lower part, overlies the Pre-Cambrian basement. This formation is overlain by a thin unit of impure dolomite, the Forestdale or White Brook Dolomite which is, in turn, overlain by schists or phyllites of either the Fairfield Pond Member of the Underhill Formation or its equivalent, the Moosalamoo Phyllite. The thickness of these pre-Cheshire rocks is about 270 metres.

The Cheshire is everywhere overlain by the Dunham Dolomite. This is a sequence of dolomites which averages about 450 metres in thickness. It is generally massive and mottled in its lower part and becomes increasingly sandy upwards.

Above the Dunham, the section becomes increasingly quartzitic, and a unit of inter-bedded quartzites and impure dolomites which ranges from zero to 450 metres through-out the region is assigned to the Monkton Quartzite. In the thickest sections, south of Milton, Vermont, the Monkton is characterized by reddish quartzites and dolomites in its lower part and white or buff coloured quartzites, interbedded dolomites, and arenaceous shales in its upper part. Cross-beds, ripple marks, and mud cracks are common through-out the formation in this area. The Monkton thins rapidly to the north and east and more slowly southwards where it becomes more fine-grained and less reddish. CADY (1945) pointed out the probability of a western or north-western source for the Monkton terri-genous sediments. STONE and DENNIS (1964, p. 37) have considered this formation to be a deltaic deposit.

The Monkton grades upwards into the Winooski Dolomite, a generally thin-bedded sandy dolomite which ranges from 180 to 360 metres in thickness. Above the Winooski is the Danby Formation. This is generally characterized by white vitreous quartzites interbedded with sandy dolomites. In the northern part of the region, a few interbeds of black shale, one as much as 11 metres thick, have been reported (STONE and DENNIS, 1964, p. 44). The Danby Formation averages about 180 metres in thickness.

The youngest Cambrian formation in the Milton–Bennington region is the Clarendon Springs Formation. This is a generally massive limy dolomite at least 150 metres thick which has local development of black chert.

2.2.1. Age and correlation

Fossils are rare in the rocks of the Milton–Bennington region. Identifiable specimens have been found only in the Dunham, Monkton, and Clarendon Springs formations in the northern part of the region. The Winooski Dolomite, the Danby Formation, and all pre-Cheshire rocks are unfossiliferous. The Cheshire Quartzite has yielded scraps of olenellid trilobites at several localities. THEOKRITOFF (1968) reports an occurrence of *Olenellus* in the upper part of the Dunham Dolomite. The Monkton Quartzite has also yielded a few olenellids described by KINDLE and TASCH (1948). The Clarendon Springs Formation has yielded a rich trilobite fauna of Trempealeauan age at one locality, listed by STONE and DENNIS (1964). It includes *Hungaia magnifica* (BILLINGS) and species of *Levisella*, *Pseudosaukia*, *Onchonotus*, *Richardsonella*, and *Idiomesus*. This fauna is also known from the Gorge Formation of the St. Albans area and the Levis conglomerates in Quebec. STONE and DENNIS also report early Ordovician fossils from slightly higher beds in the Clarendon Springs Formation, so that the formation includes the Cambrian–Ordovician boundary in the Milton area. Further south, where the rocks have not yielded fossils, the system boundary has been placed between the dolomites of the Clarendon Springs Formation and overlying limestone assigned to the Shelburne Marble.

Detailed mapping of the Milton Quadrangle, Vermont, by STONE and DENNIS (1964) has demonstrated the existence of an important facies change which is responsible for the stratigraphical contrasts of the St. Albans and Milton–Bennington regions (Fig. 3). At the latitude of Malletts Bay, 8 kilometres south of Milton, the Cambrian succession includes the typical sequence of Dunham Dolomite, Monkton Quartzite, Winooski Dolomite, Danby Formation, and Clarendon Springs Formation. However, directly

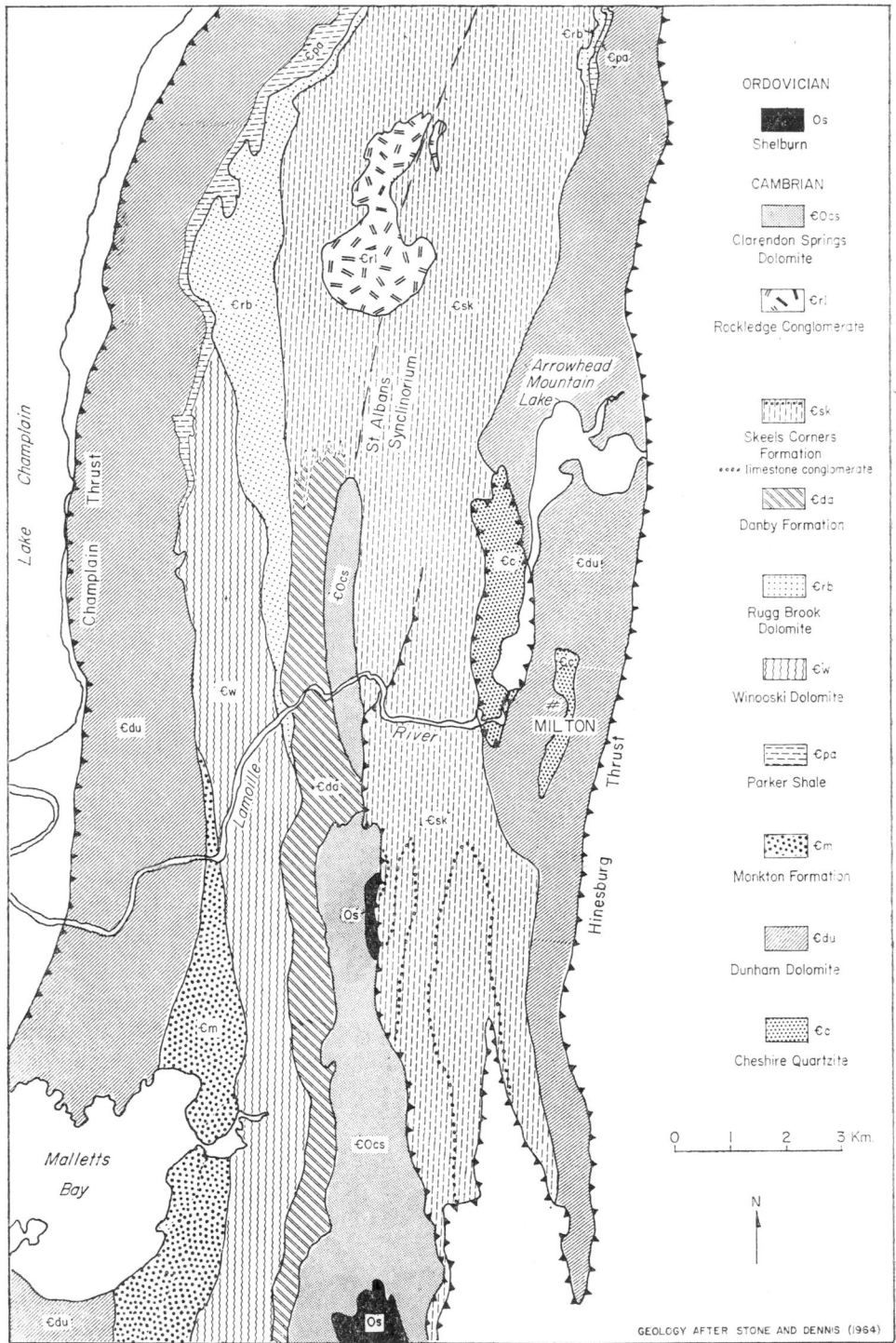

FIG. 3. Geological map of a part of the Milton area, Vermont, showing changes in the stratigraphy in the boundary area between the Milton-Bennington region and the St. Albans region.

west of Milton, the Monkton pinches out to the north, and the upper sandy dolomites of the Dunham are in direct contact with the cleaner dolomites of the Winooski. At the same latitude, the upper part of the Winooski is separated from the Danby by a unit of dolomites and dolomitic conglomerates. The conglomerates contain predominantly poorly rounded dolomite clasts as much as 0·5 metre in diameter. This unit is assigned to the Rugg Brook Dolomite.

At a latitude about 5 kilometres north of Milton, the Cambrian stratigraphy changes still further. A thin unit of black shales (slates), the Parker Shale, now appears between the Dunham and Winooski dolomites and the Winooski Dolomite is considerably thinner while the Rugg Brook Dolomite is correspondingly thicker. The overlying Danby Formation is abruptly replaced northwards by interfingering with a sequence of black shales which contains scattered horizons of limestone boulder conglomerates. These shales constitute the Woods Corners Group. Farther north in the St. Albans quadrangle the Rugg Brook Dolomite thins and becomes represented by discontinuous patches of dolomite conglomerate.

The Clarendon Springs Formation west of Milton forms the axial part of a large, southward plunging syncline, so that northwards facies changes in the Clarendon Springs Formation cannot be determined in the Milton area. On the east limb of the syncline, however, from the Milton area at least 12 kilometres southwards to the vicinity of Burlington, Vermont, the Clarendon Springs Formation overlies black slates of the Woods Corners Group. These, in turn, usually overlie conglomeratic dolomites of the Rugg Brook Dolomite. Thus, the lateral change from the quartzites and dolomites of the Danby Formation to the black shales (slates) of the Woods Corners Group, and from the bedded dolomites of the Winooski Dolomite to the dolomitic conglomerates of the Rugg Brook Dolomite, must also take place eastwards beneath the syncline. The fact that the Danby–Woods Corners change beneath the syncline must be as abrupt as the northward change which could be mapped is indicated at the north end of the syncline where the Clarendon Springs Formation in the synclinal axis has an outcrop width of less than a kilometre. Here, it lies on the Danby Formation on the west limb and on black shales of the Woods Corners Group on the east limb.

2.3. The Taconic region

The Cambrian rocks of this region are all included on a large klippe about 240 kilometres long and up to 25 kilometres wide, which extends southwards from southwestern Vermont into south-eastern New York (Figs. 2 and 6). ZEN (1967) has summarized the stratigraphy and complex history of stratigraphical nomenclature of the largely argillaceous and arenaceous rocks of this klippe. The clastic sequence of the klippe differs completely from the dominantly calcareous western New England sequence of Cambrian rocks which underlies the klippe.

The deceptively simple stratigraphy of the Taconic Klippe (Fig. 4) is the culmination of more than a century of work by many geologists. The rocks of the Klippe are included in several imbricated tectonic slices which have generally complex internal structures. All of the slices of the Klippe have been metamorphosed, with the degree of metamorphism increasing eastwards. Small-scale structural complexity, discontinuous exposures, and metamorphism have all contributed to the difficulty of working out detailed stratigraphical relationships. All palaeontological data come from the least metamorphosed Giddings Brook slice on the western side of the Klippe.

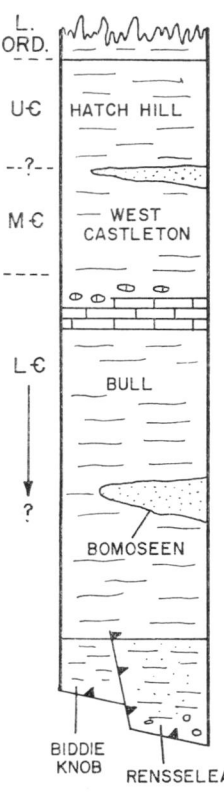

FIG. 4. Succession of Cambrian stratigraphical units in the Taconic region (after Zen, 1967).

The absence until recently of any record of faunas of post-Early Cambrian pre-Trempealeauan age in the Taconic region was considered to be due to a hiatus representing much of this time interval. New data reported by RASETTI (1967) and by BIRD and RASETTI (1968) indicate that much of the Cambrian System is represented in the Taconic region although the section seems to be thin and some evidence suggests abrupt lateral thickness changes of individual units. The thickness of the fossiliferous Cambrian deposits of the Taconic region probably does not exceed 330 metres. The conformable pre-fossiliferous beds are several times thicker.

The oldest rocks of the Klippe comprise a succession of dominantly purple and green slates. In the northern part of the Klippe, ZEN (1961, 1967) has recognized two formations within these slates. The lower formation, characterized by abundant chloritoid, is the Biddie Knob. The upper formation, which has subordinate amounts of greywacke, quartzite, and limestone pebble conglomerate, is the Bull Formation. Some of the lower beds of the Bull Formation are considered to be lateral equivalents of the Biddie Knob Formation. Slates, which constitute the bulk of the Bull Formation, are referred to as the Mettawee Slates. The lower half of the formation contains greywacke units separately named as the Rensselear Greywacke—a unit believed to be a facies equivalent of the lower part of the Mettawee Slates—and a younger, more areally persistent unit, the Bomoseen Greywacke. Thin quartzite members are present in the upper half of the formation and have received separate local names. Limestone pebble conglomerates are present only near the top of the formation and in the basal part of the overlying West Castleton Formation.

The purple and green slates of the Bull Formation are everywhere overlain by a unit of dominantly black and grey slates which constitutes the upper part of the Cambrian succession. The colour change between these two dominantly slate units is the most persistent stratigraphical boundary in the Cambrian sequences of the Taconic region. In the northern part of the region, the black and grey slate sequence is divided into two formations. The lower formation, characterized by dark slates ranging from hard, sandy, and cherty to fissile, graphitic, and pyritic, and rare thin-bedded limestones, is the West Castleton Formation (ZEN, 1961). The upper formation, which has thin quartzites and dolomitic sandstones interbedded with the dark slates is the Hatch Hill Formation (THEOKRITOFF, 1964). Southwards, the entire sequence of black and grey slates, thin-bedded carbonate rocks, and limestone pebble conglomerates lies above the Nassau Formation and is referred to the undifferentiated West Castleton–Hatch Hill Formations (BIRD and RASETTI, 1968).

2.3.1. Age and Correlation

Fossils representing all epochs of the Cambrian are now known from the interval that includes the uppermost Bull or Nassau, the West Castleton, and the Hatch Hill Formations (RASETTI, 1966, 1967; BIRD and RASETTI, 1968). The oldest trilobites, which are the primary elements for dating, are believed to be of late Early Cambrian age, so at least some of the lower unfossiliferous part of the Bull and Nassau formations may also represent the Cambrian Period.

The richest and most varied Cambrian faunas are in the Early Cambrian rocks which comprise the upper part of the Bull and Nassau formations and the lower part of the West Castleton Formation. The fauna, dominated by trilobites, is recorded mostly from thin-bedded limestones and from clasts in limestone pebble conglomerates. Rare occurrences of fossils in the matrix of the limestone conglomerates and in the slates are also known (BIRD and RASETTI, 1968). RASETTI (1967) has summarized the biostratigraphy of the Early and Middle Cambrian Taconic faunas and recognized three Early Cambrian faunal assemblages.

The oldest assemblage, characterized by the olenellid *Elliptocephala asaphoides* EMMONS, has been described by LOCHMAN (1956) with some additional illustration by THEOKRITOFF (1964) and RASETTI (1967). This fauna includes olenellid, corynexochoid, ptychoparioid, and eodiscid trilobites; non-trilobite arthropods; small archaeocyathids; molluscs; inarticulate brachiopods; annelids (?); and fossils of uncertain affinity. Most of the well preserved specimens are small. The large olenellids are almost always represented only by fragments.

The next younger assemblage, described by RASETTI (1966, 1967), has been designated as the *Acimetopus bilobatus* assemblage. It is characterized by a remarkably diverse fauna of unusual eodiscids, rare fragmentary olenellids, and corynexochoids. This assemblage is known at present only from a small area in the southern part of the Taconic region and from an outcrop near Elgin Station, L'Islet County, Quebec, about 64 kilometres north-east of Levis, Quebec (RASETTI, 1967, p. 19). Eodiscids closely related to the *Acimetopus bilobatus* assemblage have been described from the Purley Shales of central England by RUSHTON (1966).

The youngest Early Cambrian assemblage is the *Pagetides* assemblage, described by RASETTI and THEOKRITOFF (1967) and RASETTI (1967). The trilobites of this assemblage include early Cambrian agnostids, pagetiids, olenellids, corynexochoids, and ptychoparioids.

Middle Cambrian trilobite faunas are now known from the undifferentiated Hatch Hill–West Castleton rocks of the Taconic region (RASETTI, 1967). The oldest Middle Cambrian fauna is assigned to the *Bathyuriscus-Elrathina* Zone. It includes species of *Bathyuriscus, Oryctocephalus, Pagetia, Peronopsis,* and *Ptychagnostus* which are characteristic of this zone in many parts of North America, and the conocoryphid *Meneviella* known elsewhere in eastern North America, western Europe, and Siberia. Several faunules including trilobites referred to *Bathyuriscus, Hypagnostus, Ptychagnostus,* and *Centropleura* and the acrotretid *Pegmatreta* are assigned to the latest Middle Cambrian *Bolaspidella* Zone.

Late Cambrian faunules assigned to the *Aphelaspis* Zone of late Dresbachian age, and the *Hungaia magnifica* fauna of late Franconian or Trempealeauan age are recorded by BIRD and RASETTI (1968) but they have not yet been described. In addition, graptolites of Trempealeauan age have been recorded by BERRY (1962).

The Middle and Late Cambrian faunules share many genera with the Woods Corners Group and the Gorge Formation of the St. Albans region. Both areas are also characterized by the abundance of black shales and paucity or absence of carbonate rocks of Middle and Late Cambrian age. The deposits of both areas seem to represent typical accumulations of sediments immediately seaward of the carbonate banks which are represented by the rocks of the Milton–Bennington area.

2.4. The western New England metamorphic belt

East of the Hinesburg thrust and the axial Pre-Cambrian massifs of the Green Mountain anticlinorium, the Cambrian section contrasts strongly with the section of the Milton –Bennington belt and seems to represent a greatly thickened facies comparable to that of the St. Albans area and the Taconic region, but more strongly metamorphosed. The rocks represent metamorphic equivalents of coarse to fine-grained greywacke, siltstone, shale, and volcanic materials. Except for one or possibly more thin discontinuous intervals of Early Cambrian (?) age, carbonate sediments of any kind are conspicuously absent. The thickness of this succession of dominantly detrital rocks, which is now considered by most New England geologists to range from at least Early Cambrian into Middle Ordovician, has been estimated, perhaps excessively, to be from 9,000 to 15,000 metres (CADY, 1960, p. 541). Even if only half of this section is of Cambrian age, the thickness of the Cambrian rocks is more than twice the thickness of the rocks of the Milton–Bennington belt. The details of the regional stratigraphy have been worked out only in the Vermont region and are best summarized by DOLL and others (1961) (Fig. 5).

Above the Pre-Cambrian gneisses of the Mount Holly complex in the western part of the metamorphic belt in Vermont is a sequence of schists and greywackes with interbedded volcanics which is assigned to the Pinnacle, Underhill, Moosalamoo, and Fairfield Pond formations. These are overlain by quartzites assigned to the Cheshire Quartzite, which are the youngest rocks exposed east of the Hinesburg thrust fault and west of the axis of the Green Mountain anticlinorium. The Pinnacle Formation is composed of schistose greywackes which are conglomeratic near the contact with the Mount Holly complex and locally include pillowed and vesicular greenstones of the Tibbit Hill volcanic member. The Underhill, Moosalamoo, and Fairfield Pond include dominantly silvery to grey-green schists. In many areas a thin sandy dolomite member called the Forestdale or White Brook Dolomite, separates these formations from the Pinnacle Formation.

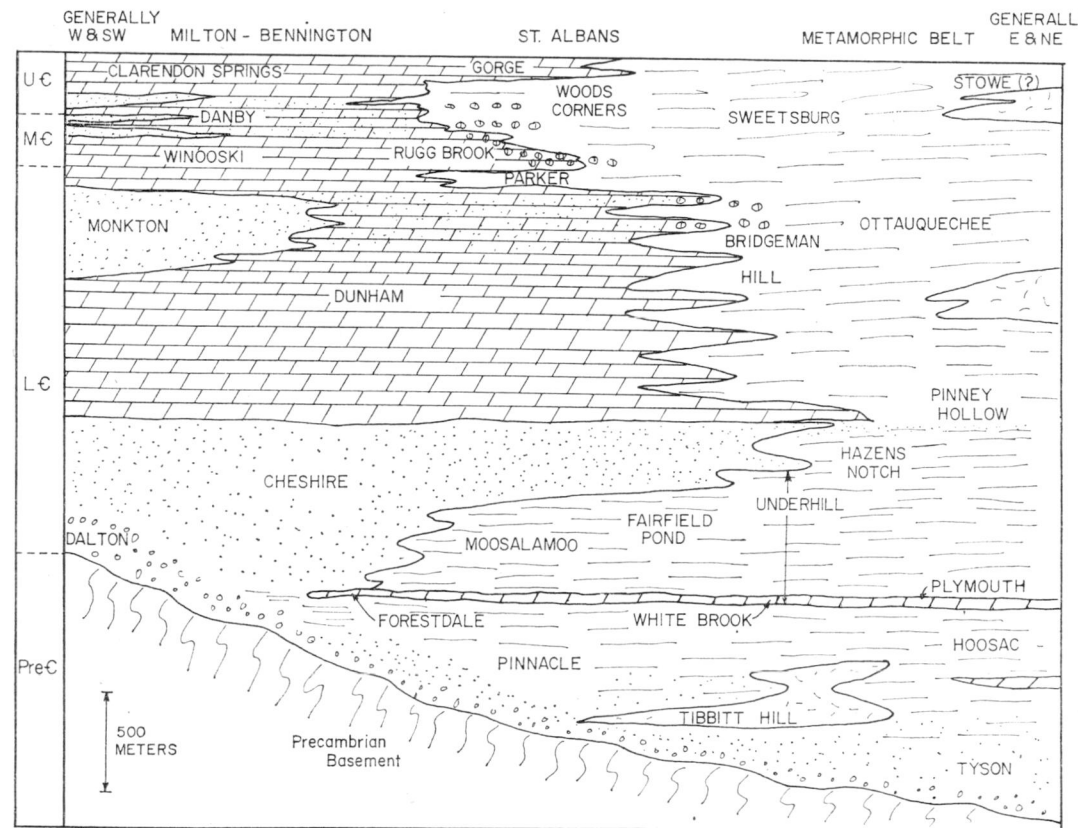

FIG. 5. Schematic reconstruction of the spatial relations of the Cambrian formations of Vermont. The scale is only valid for the western margin of the figure.

Eastwards, across the axis of the Green Mountain anticlinorium in northern Vermont, the upper part of the Pinnacle Formation seems to be replaced laterally by schists of the Underhill Formation and the upper part of the Underhill Formation seems to be replaced by carbonaceous and non-carbonaceous schist, quartzite, and gneiss of the Hazens Notch Formation.

On the east flank of the Green Mountain anticlinorium, the Hazens Notch Formation is overlain by black carbonaceous phyllite, quartzite, and schist of the Ottauquechee Formation. Above the Ottauquechee Formation is the Stowe Formation which consists of schists and phyllites with some interbedded greenstones and amphibolites.

In central and southern Vermont, a grossly similar stratigraphical succession is apparent in the metamorphics east of the Pre-Cambrian massifs at the axis of the Green Mountain anticlinorium. In this region, the basal unit of feldspathic quartz-mica schists and conglomerates is the Tyson Formation. Above the Tyson is the Hoosac Formation of quartz-mica schists with interbedded amphibolites and greenstones which also has a discontinuous thin sandy dolomite upper member, the Plymouth Member, comparable with the basal White Brook dolomitic member of the Underhill Formation to the north.

The Hoosac Formation is overlain by the Pinney Hollow Formation which is composed largely of pale green schist and phyllite and also includes some carbonaceous schist and interbedded amphibolite and greenstone. The sequence above the Pinney Hollow Formation is of Ottauquechee and Stowe formations, comparable with that in northern Vermont.

2.4.1. Age and correlation

Careful geological mapping in the Vermont–Quebec boundary area by OSBERG (1965) has shown that the black phyllites of the Ottauquechee and Stowe formations can be traced into the Sweetsburg Formation on the east limb of the St. Albans synclinorium. They are probably equivalent to the late Middle and early Upper Cambrian Woods Corners Group of the western part of the St. Albans region (p. 174).

This important correlation has stabilized the regional stratigraphical framework by demonstrating that the thick eugeosynclinal sequence of greywackes, black shales and argillites, and interbedded volcanics of the metamorphic belt of central Vermont is probably largely of Cambrian age and correlative with the thinner sequence of largely shallow water carbonates and quartzites of the Milton–Bennington region.

ZEN (1967) has presented strong arguments for the correlation of the units in the relatively thin section of argillaceous and arenaceous sediments of the Taconic klippe with individual units in the metamorphic belt. He has concluded that the original site of deposition of the rocks of the Taconic klippe was over the Pre-Cambrian axis of the Green Mountain anticlinorium and that the Cambrian succession of the klippe is intermediate in character between the miogeosynclinal deposits of the shelf and the eugeosynclinal deposits to the east.

The problem of the position of the base of the Cambrian System has not been resolved in the New England area. The oldest rock unit that can be certainly identified as Cambrian is the Cheshire Quartzite. However, lithic correlation of the Cheshire–Dunham–Monkton sequence of Early Cambrian units with the Hesse–Shady–Rome and Antietam–Toms-town–Waynesboro Early Cambrian sequences of the central and southern Appalachians (Chapters 3.1.1 and 4.1.1) seems reasonable. At least part of the section below the Hesse Quartzite is of Cambrian age. If the suggested correlation is correct, at least part of the pre-Cheshire section is probably also of Cambrian age.

2.5. The 'thin' sequences of Connecticut, Western Massachusetts, south-eastern New York, New Jersey, and eastern Pennsylvania

Between North Adams in north-western Massachusetts and the vicinity of Allentown and Reading in eastern Pennsylvania (Fig. 6) the Cambrian sequences share many common characteristics, and are significantly thinner than those in Vermont to the north and the remainder of Pennsylvania and the Appalachians to the west and south. Although the stratigraphical nomenclature is different for several areas within this region, the stratigraphy is fairly uniform.

At the base of the section is a relatively thin quartzitic unit variously named the 'Cheshire', Poughquag, or Hardyston, which may have locally developed basal conglomerates. This unit is usually less than 100 metres thick and rests on Pre-Cambrian schists and gneisses.

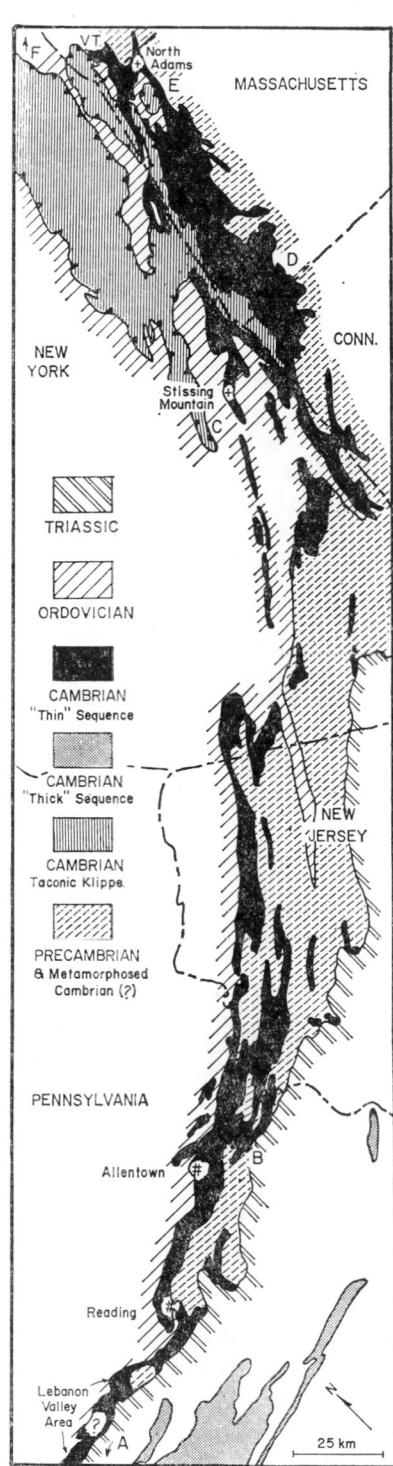

FIG. 6. Distribution of outcrops of the thin Cambrian sequences of the Northern Appalachian region. Letters A–F show the locations of the columnar sections of Figure 8.

Conformably above the quartzites is a sequence of limestones and dolomites about 540 to 810 metres thick which is unbroken by any regionally significant units of terrigenous rocks. This represents the rest of the Cambrian System and continues into the basal Ordovician. The sequence of carbonate rocks is usually poorly exposed and has yielded fossils at only a few localities. The poor exposure, moderate deformation, eastward metamorphism, and lack of internally persistent distinctive stratigraphical horizons has led to the development of a diverse stratigraphical nomenclature which has only local application.

In the North Adams area, (Figs 6E and 8E) the Cambrian part of the carbonate sequence is slightly metamorphosed and has been divided by HERZ (1961) into two formations, the Kitchen Brook Dolomite and the Clarendon Springs Dolomite. The Kitchen Brook Dolomite is about 300 metres thick. It includes much quartz sand and argillaceous material which increases in abundance upwards. The overlying Clarendon Springs Dolomite is about 240 metres thick. It consists of massive bedded dolomite and includes some thin lenticular layers of vitreous quartzites. The Clarendon Springs Dolomite is overlain by the Shelburne Marble of supposed Early Ordovician age.

In south-western Massachusetts and adjacent parts of Connecticut, (Figs 6D and 8D) ZEN and HARTSHORN (1966) have included all of the pre-Middle Ordovician carbonates within the Stockbridge Formation. They have divided the formation into seven lettered members of which the three lowest are considered to be Cambrian in age. The oldest unit, about 210 metres thick, is a massive mottled dolomite. Overlying this is about 180 metres of massive dolomite with interbeds of calcareous and feldspathic sandstone and calcareous siltstone. The upper unit, about 210 metres thick, includes massive dolomite characterized by floating rounded sand grains and chert.

In the Stissing Mountain area of south-eastern New York, (Figs 6C and 8C) the Cambrian part of the carbonate succession above the Poughquag Quartzite has been divided by KNOPF (1962) into three formations in ascending order: the Stissing Dolomite, Pine Plains Formation, and Briarcliff Dolomite. The Stissing Dolomite consists of about 145 metres of siliceous, shaly, generally aphanitic dolomite. This contrasts with the overlying Pine Plains Formation, which includes about 400 metres of extremely variable, rapidly alternating, more or less impure siliceous, oolitic, and stromatolitic dolomites and edgewise dolomite conglomerates. The Briarcliff Dolomite is a more uniform unit, consisting of about 200 metres of light coloured, limy dolomites.

In New Jersey, the 810 metres of Cambro-Ordovician dolomites and dolomitic limestones above the basal Hardystone Quartzite are assigned to the Kittatinny Limestone in most parts of the state except along the Delaware river where the more refined terminology of eastern Pennsylvania has been used (DRAKE and others, 1961; DRAKE, 65).

In eastern Pennsylvania, (Figs 6B and 8B) the carbonate rocks above the Hardyston Quartzite are considered to represent either two or three formations (compare HOWELL and others, 1950; DRAKE, 1965). Both papers agree on the lowest unit, the Leithsville Formation. This is a formation about 270 metres thick which contains interbedded units of thin-bedded impure dolomitic limestone, argillaceous sericitic shaly limestone, some sericitic shale, and massive dense dolomite in beds up to 3 metres thick. Abundant ripple marks and mudcracks attest to characteristic shallow-water conditions.

The overlying rocks are characterized by well developed cyclic sequences including algal stromatolites, oolitic dolomites, and edgewise conglomerates. A typical cycle, from bottom to top, consists of six units: 'textureless dololutite, dolarenite, oolitic dolarenite, dolorudite, cryptozoon dolomite, and dessication dolorudite' (DRAKE, 1965). These rocks have been divided into two formations, the Limeport Dolomite and Allentown Dolomite (restricted), by HOWELL and others (1950) and WILLARD (1955, 1961) largely on palaeontological grounds. However, DRAKE and others (1961) have not found the stratigraphical separation to be lithologically recognizable and they have returned to the original usage of Allentown Dolomite for the entire sequence above the Leithsville.

2.5.1. Age and correlation

The upper beds of the basal quartzite unit of the region have yielded fragmentary olenellid trilobites. The dating of the carbonate rocks is based on scattered occurrences of trilobites and other fossils in these largely unfossiliferous rocks. No fossils have been found in the Cambrian rocks of western Massachusetts and Connecticut. In the Stissing Mountain area of south-eastern New York (Fig. 7) the distinctive Early Cambrian phosphatic fossil *Discinella micans* (BILLINGS) has been found about 32 metres above the

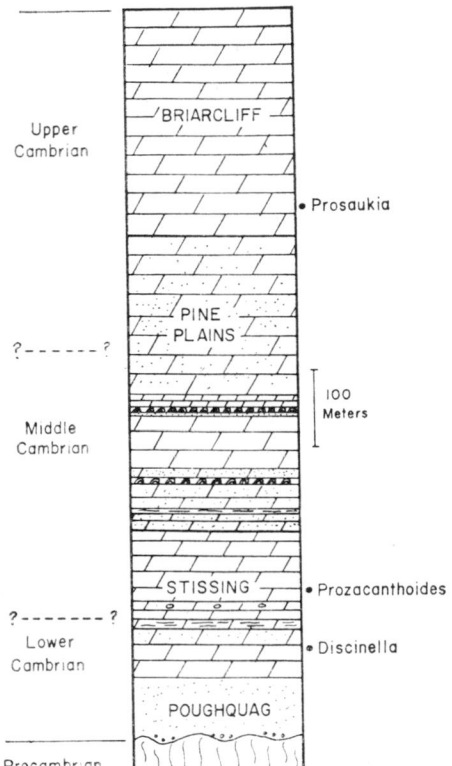

FIG. 7. The Cambrian rocks of the Stissing Mountain area showing the horizons that have yielded fossils.

base of the Stissing Dolomite. About 90 metres higher in the Stissing Dolomite, weathered dolomites with the trilobite *Prozacanthoides stissingensis* (DWIGHT) and the phosphatic brachiopod *Paterina stissingensis* (DWIGHT) are considered by KNOPF (1962) to represent an early Middle Cambrian age. The only other occurrence of datable fossils from the Stissing Mountain sequence is from 50 to 75 metres above the base of the Late Cambrian Briarcliff Dolomite (LOCHMAN, 1946). Here, the trilobites *Plethometopus knopfi* LOCHMAN and *Prosaukia briarcliffensis* LOCHMAN indicate a probable medial Late Cambrian (Franconian) age for the beds.

In New Jersey and eastern Pennsylvania, faunas representing Late Cambrian Dresbachian and Trempealeauan ages have been reported from the Kittatinny Limestone

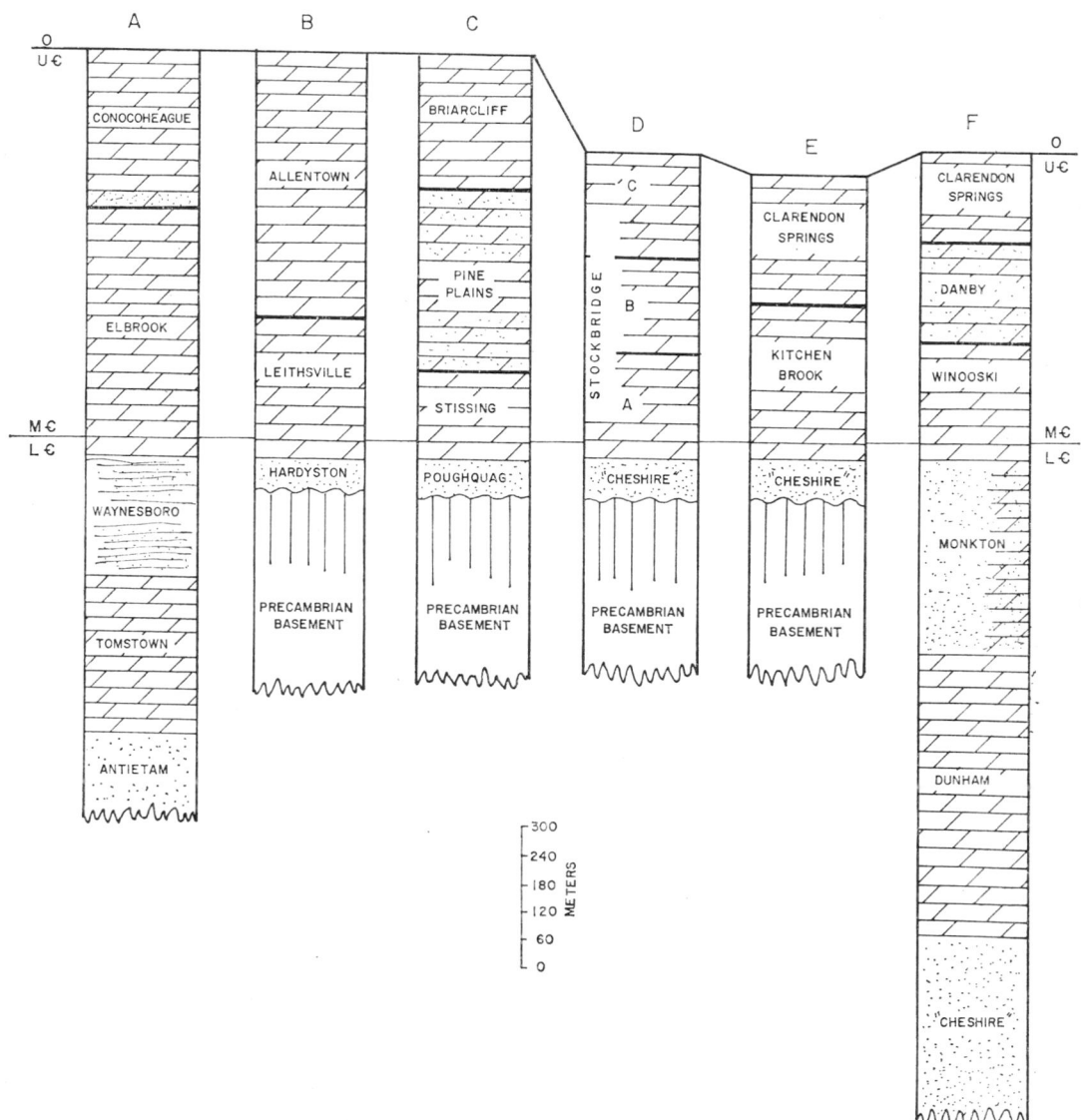

A—The Cumberland Valley area of the
Central Appalachians.
B—The Allentown area of eastern Pennsylvania.
C—The Stissing Mountain area of southeastern New York.

D—The Bashbish Falls area of southwestern Massachusetts.
E—The North Adams area of northwestern Massachusetts.
F—The Milton–Bennington area of western Vermont.

FIG. 8. Correlation of the formations of the thin Cambrian sequences of the Northern Appalachian region and the thick sequences to the north and south.

and the Allentown Dolomite by HOWELL (1945, 1955, 1957). Their precise stratigraphical occurrences have not been determined because of the poor exposures and internal deformation of the carbonate sequences.

The Cambrian sequences from eastern Pennsylvania to western Massachusetts provide important data for considerations of Early Cambrian palaeogeography. Traditionally, these sequences have been considered to be merely thinner versions of the much thicker Cambrian shelf sequences to the north-east and south-west (ZEN and HARTSHORN, 1966). However, available evidence suggests another interpretation (Fig. 8). The scanty faunal data indicate that the Lower-Middle Cambrian boundary lies within the lower part of the Stissing Dolomite and above the base of the Leithsville Formation. The units which contain this boundary to the north-east and south-west are the Winooski Dolomite (p. 177) and the Elbrook Dolomite (p. 193) respectively. In each case, these units are underlain by an Early Cambrian terrigenous unit (Monkton, Waynesboro) which must be the approximate correlatives of the basal Hardyston and Poughquag quartzites. The thickness of the western Massachusetts Cambrian sequences seems to relate them to the Stissing Mountain sequence rather than to the sequences of western Vermont. Thus the 'Cheshire' of HERZ (1961) and ZEN and HARTSHORN (1966) seems more likely to be the correlative of the Poughquag and Hardyston than of the Cheshire of western Vermont. Inasmuch as the Monkton and Waynesboro overlie thick sedimentary sequences which include older carbonate units wholly of Early Cambrian age, and still older conformable terrigenous units, the evidence presented above suggests that the area between eastern Pennsylvania and western Massachusetts was a positive area during much of the Early Cambrian and was only submerged just before the beginning of the Middle Cambrian.

3. The Central Appalachians

This region includes all Cambrian rocks south-westwards from the thin sequences of eastern Pennsylvania described in Chapter 2.5 to the transition into the southern Appalachian Cambrian sequences in the Virginia–Tennessee boundary area (Fig. 9). Except for a few small areas along the margins of the region, which are separately described, the stratigraphy is remarkably uniform.

3.1. The main Cambrian belt

From the Susquehanna River in central Pennsylvania south-westwards about 600 kilometres to the Virginia–Tennessee boundary area, the Cambrian sequences westwards from the South Mountain–Blue Ridge trend are all characterized by a lower clastic sequence of Early Cambrian and Pre-Cambrian age, and an upper carbonate sequence ranging from Early to Late Cambrian in age (Fig. 10). The most complete Cambrian sections are found along the eastern side of this belt. To the west and north-west, the lower parts of all sections have been faulted off, generally near the top of the Lower Cambrian succession. In south-western Virginia, the upper carbonate sequence includes several shale intervals which increase in thickness southwards at the expense of the intervening limestones. These sequences are all included in the southern Appalachian area (Chapter 4).

3.1.1. The lower clastic sequence

The formations of the lower clastic sequence are all assigned to the Chilhowee Group (RODGERS, 1956b). The thickness of the sediments included within the group ranges

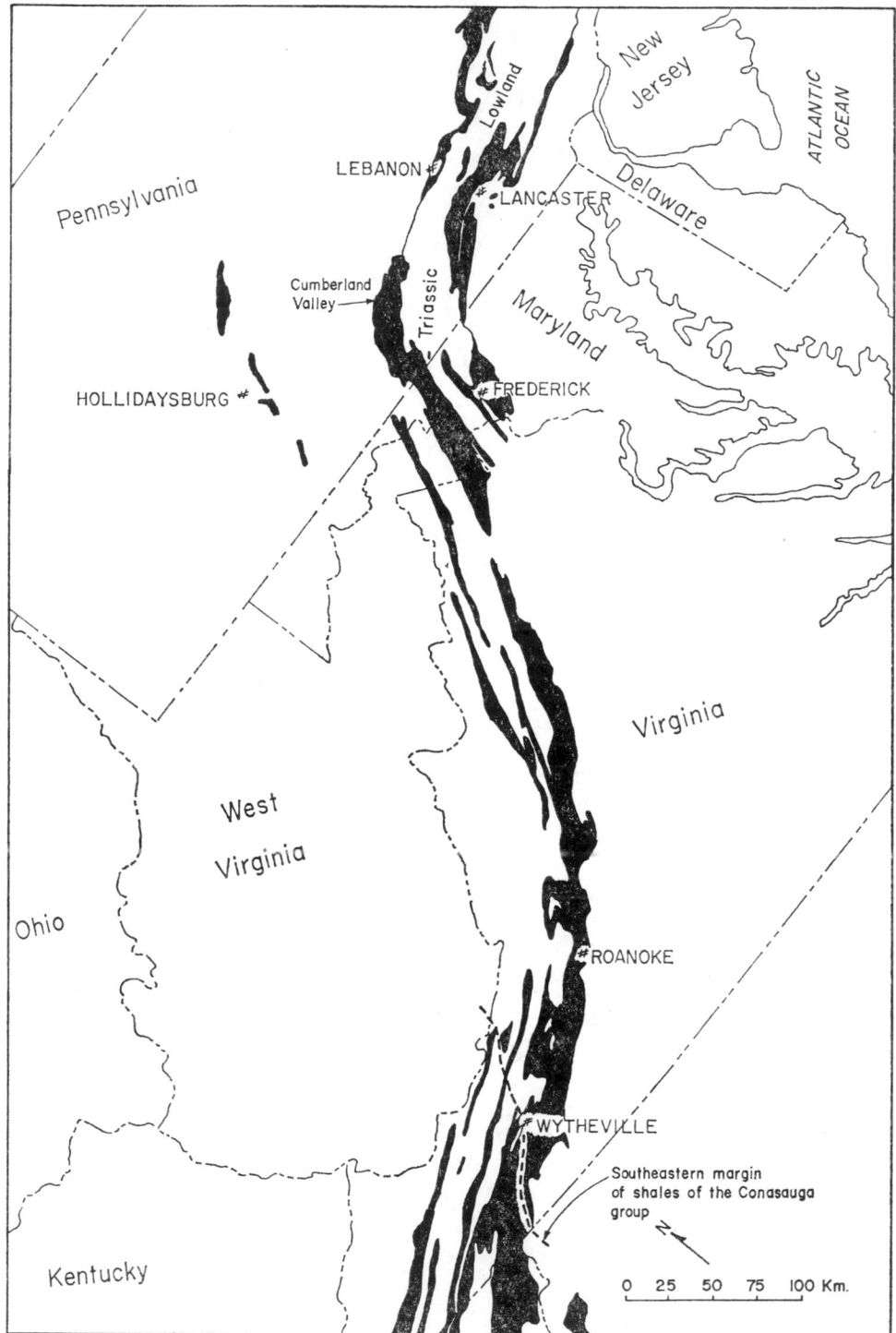

FIG. 9. Areas of outcrop of Cambrian rocks in the Central Appalachians.

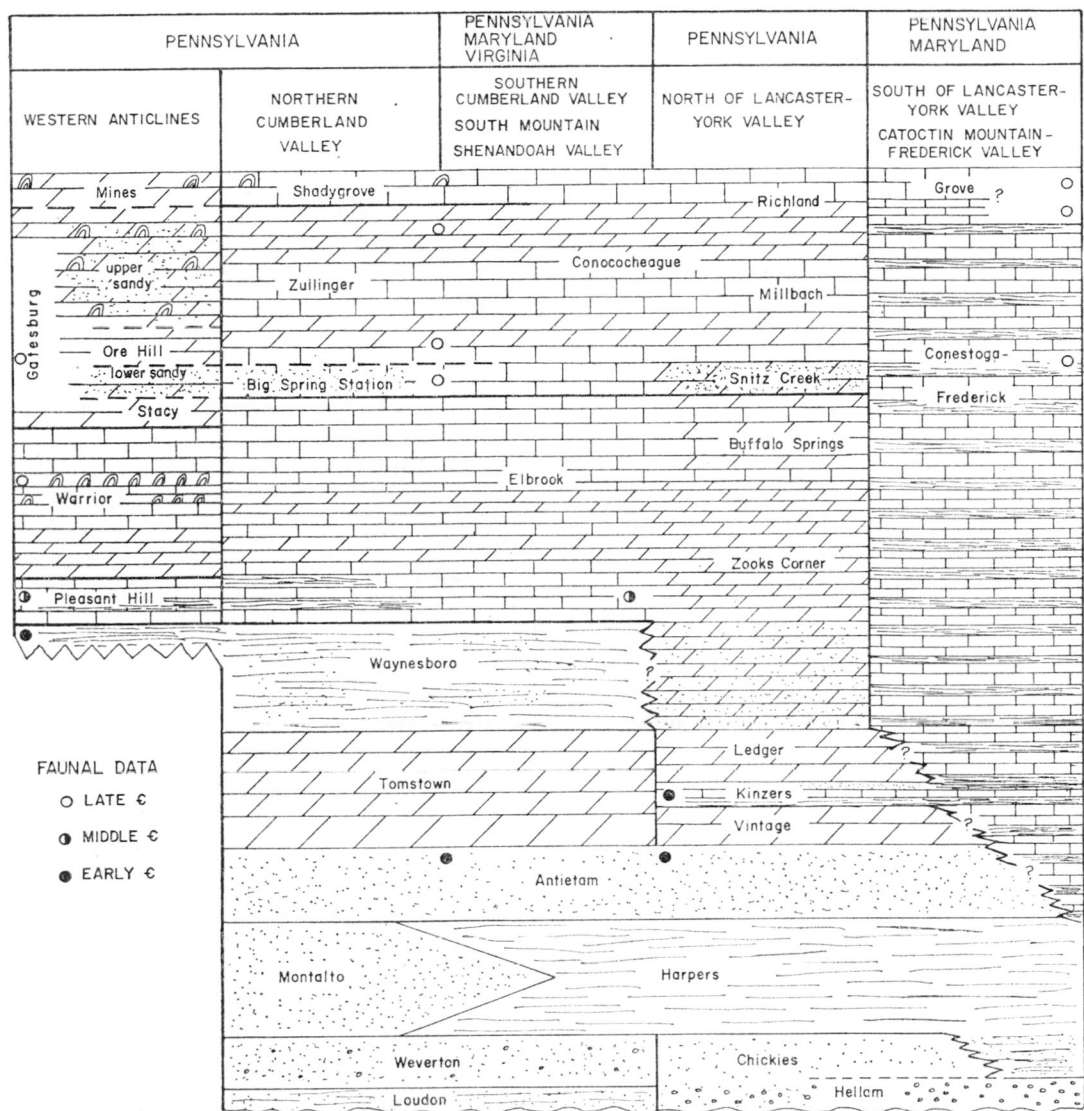

FIG. 10. Correlation of the Cambrian formations of the northern part of the Central Appalachians. Lithologies are schematically shown.

from 900 to about 3,000 metres. The lowest formation everywhere rests unconformably on either older sediments or on volcanics of Pre-Cambrian age. Detailed studies of several areas by STOSE and JONAS (1939, Pennsylvania), CLOOS (1951, Maryland), KING (1950, northern Virginia), and KING and FERGUSON (1960, Tennessee–Virginia boundary area) show the basic unity of the stratigraphy of the Chilhowee Group. Formational sequences that reflect local details of the Chilhowee group have been described for the Tennessee–Virginia boundary area, the northern Blue Ridge and South Mountains of Virginia and Maryland, and the northern part of the Cumberland Valley in Pennsylvania.

The Chilhowee Group of the Virginia–Tennessee boundary area includes, in ascending order: the Unicoi Formation—630 to 1,500 metres of feldspathic sandstone and conglomerate, with some beds of vitreous quartzite and shale, and rare intervals of amygdaloidal basalt; the Hampton Formation—160 to 415 metres of interbedded quartzite, sandstone, and shale with less resistant beds predominating; and the Erwin Formation—360 to 450 metres of interbedded quartzite, sandstone, and shale with quartzite predominating, which includes at its top about 30 metres of non-resistant shale and arkosic sandstone of the Helenmode Member. The Helenmode is overlain by the Shady Dolomite. *Scolithus* burrows are locally abundant in quartzites within the Erwin and Hampton formations.

In the northern Blue Ridge of Virginia and the South Mountain area of Maryland, the Chilhowee Group includes, in ascending order: the Loudon Formation—dull red and purple slate and altered volcanic rocks ranging from zero to 65 metres in thickness which seem to fill topographical irregularities on the surface of the older Catoctin greenstones, and which pinch out locally below the overlying quartzites; the Weverton Formation—300 to 315 metres of conglomerates, quartzites, and shales, with conglomerate and feldspathic quartzite dominating in its lower part and dark ferruginous quartzite interbedded with siltstone and sandstone in its upper part; the Harpers Formation—about 270 metres thick, consisting of a lower shale member about 70 metres thick and an upper member of siltstone and fine-grained sandstone; and the Antietam Quartzite—about 240 metres thick, composed of a lower vitreous quartzite member which contains *Scolithus* burrows and an upper member of interbedded white *Scolithus*-bearing quartzite and buff or brown sandstone. WHITAKER (1955b) has shown that the sands of the Weverton Formation came from the west.

In Pennsylvania, along the continuation of the strike of the Chilhowee Group of South Mountain, the Montalto Quartzite, a vitreous quartzite with *Scolithus* tubes, appears as a middle member of the Harpers Formation and thickens northwards so that at the northern end of the exposures, in north-western York County, the Harpers has been completely replaced by quartzite and the Chilhowee succession includes: the Loudon Formation—a heterogeneous assortment of argillaceous and arenaceous sediments about 100 metres thick; the Weverton Quartzite—conglomeratic feldspathic quartzite about 150 metres thick; the Montalto Quartzite—coarse granular to finely laminated white quartzite about 300 metres thick with abundant *Scolithus* tubes; and the Antietam Quartzite—lithologically similar to the Montalto Quartzite and separated from it by an interval of thin-bedded, less resistant quartzites, about 300 metres thick.

In Pennsylvania, Maryland, and northern Virginia, the Antietam of the main Cambrian belt is overlain by the Tomstown Dolomite.

The regional relationships of the units within the Chilhowee Group of the central Appalachians are summarized on Fig. 10. A basic tripartite sequence is apparent, consisting of a lower unit dominated by ferruginous and feldspathic quartzites and conglomerates; a middle unit dominated by shales or siltstones, except in Pennsylvania; and an upper unit dominated by white, vitreous, *Scolithus*-bearing quartzites. KING and FERGUSON (1960) showed that in the Tennessee–Virginia border area the lower formations thicken towards the east and become less coarsely clastic, suggesting a western source for the clastics and a deepening eastern basin. This is also indicated by the facies changes in the Harpers Formation between Pennsylvania and Maryland.

3.1.2. The upper carbonate sequence

Along the south-east side of the main Cambrian belt from southern Virginia to southern Pennsylvania, the stratigraphy of the upper carbonate sequence is remarkably uniform (Fig. 10). The uppermost unit of the Chilhowee Group is overlain by a generally massive dolomite unit that averages about 300 metres thick. In the southern areas, this is the Shady Dolomite; in Pennsylvania, Maryland, and northern Virginia it is the Tomstown Dolomite. Overlying the Shady–Tomstown is a clastic unit of comparable thickness which is characterized by interbedded red and green shale, reddish sandstone, and some limestone and dolomite units. In the southern areas, this is the Rome Formation and in Pennsylvania, Maryland, and northern Virginia it is the Waynesboro Formation.

Above the Rome–Waynesboro throughout the main Cambrian belt, the remaining part of the Cambrian System has generally been divided into two formations, the Elbrook and the overlying Conococheague. The Elbrook is characterized by thin-bedded or thinly laminated dolomite or limestone beds, generally with argillaceous partings or thin shaly interbeds. It ranges from 600 to about 900 metres in thickness throughout the region. The Conococheague contains less argillaceous material and generally thicker beds than the Elbrook. There is also a greater variety of rock types present, including sandstones, oolitic limestones and dolomites, algal stromatolites, and chert. Many individual beds are identical with beds in the Elbrook, however. The formation averages about 600 metres in thickness throughout the region.

In the Cumberland Valley of Pennsylvania, subdivisions of the Conococheague have been recognized. WILSON (1952) described the Big Spring Station Member at the base of the Conococheague Formation. This is a distinctive, largely clastic, unit characterized by argillaceous and arenaceous dolomites and siliceous orthoquartzites. MACLACHLAN and ROOT (1966) raised the Conococheague in the Cumberland Valley to group status and described two formations within it. The older formation, which includes the Big Spring Station Member, is the Zullinger Formation. This is characterized by cyclic sequences of laminated limestone and dolomite, calcarenite, and algal stromatolites. The younger formation, designated the Shadygrove, is characterized by pure, light-coloured stromatolitic and cherty limestones.

About 85 kilometres north-west of the Cumberland Valley in Pennsylvania, rocks of the upper carbonate sequence are exposed in the narrow axial parts of several anticlines. In one of these, about 70 metres of green and purple shale, sandstone, and fine conglomerate below the carbonates contains olenellid fragments and has been identified as the Waynesboro Formation. Thus, the carbonate sequence is comparable to the Elbrook–Conococheague sequence of the areas to the south and east. Three formations are recognized in this sequence, beginning with the oldest: the Pleasant Hill, Warrior, and Gatesburg.

The Pleasant Hill Formation consists of calcareous shales and silty or mottled, platy to massive, grey limestones totalling about 180 metres in thickness. The upper 70 metres are considerably less shaly than the lower part of the formation.

The Warrior Formation includes a heterogeneous assortment of limestones and dolomites with minor beds of sandstone, oolite, and algal stromatolites totalling about 400 metres in thickness. It lies between the more shaly Pleasant Hill Formation and the more sandy Gatesburg Formation. An interval at least 80 metres thick in the upper part of the formation has well developed sedimentary cycles. The ideal cycle has three parts (TASCH, 1951): the lower part, 1 to 2 metres thick, is characterized by thin-bedded limestone with some shaly or silty layers and some edgewise conglomerate; the middle part, generally 0·5 to 0·75 metres thick, contains massive black limestone which is frequently oolitic; and the upper part, usually less than 0·5 metres thick, is an algal stromatolite.

The Gatesburg Formation is a heterogeneous dolomitic unit about 540 metres thick. It has been divided into five members—three relatively clean carbonate units separated by two very sandy units. WILSON (1952) has reviewed their regional relationships. The lowest member is the Stacy Member, a unit of massive dolomite about 100 metres thick. Above this, an unnamed lower sandy member is characterized by alternations of dolomite and orthoquartzite through an interval of about 115 metres. This member, which is correlated with the Big Spring Station Member of the Conococheague Formation (Fig. 10), is separated from a similar unnamed sandy member about 200 metres thick by the relatively clean carbonates of the Ore Hill Member. The Ore Hill Member averages about 55 metres in thickness and grades from dominantly limestone in the southern anticlinal exposures to dominantly dolomite in the north. Eastwards and south-eastwards, it loses its identity when the upper sandy member is replaced by relatively clean dolomites. The youngest member is the Mines Dolomite Member, about 100 metres thick. This member is characterized by dark dolomite, siliceous oolite, and stromatolitic chert, and by the absence of orthoquartzite.

The lower and upper sandy members of the Gatesburg Formation are characterized by cyclic sedimentation analogous to that of the Warrior Formation. A typical cycle includes three elements. It begins with an orthoquartzite or shaly aphanitic dolomite usually 1 to 2 metres thick. This is overlain by a thin or medium bedded dolomite unit also about 1 to 2 metres thick. The upper unit consists of massive, coarsely crystalline, commonly oolitic or cherty dolomite topped by an algal stromatolite, totalling 1 to 3 metres in thickness. The thicknesses of the individual units vary so that complete cycles may range in thickness from 1 to 8 metres.

3.1.3. Age and correlation

The rocks throughout the main Cambrian belt of the central Appalachians are sparingly fossiliferous. Early Cambrian fossils are known only from Rome–Waynesboro and older formations. The oldest fossils, exclusive of *Scolithus*, are fragmentary olenellids and brachiopods reported at many localities from quartzites in the upper part of the Chilhowee Group. RESSER (1938) reports olenellids from the Rome Formation near Roanoke, Virginia, and BUTTS (1945) reports fragmentary olenellids from the Waynesboro Formation in isolated outcrops in Pennsylvania.

Middle Cambrian fossils are known only from the Elbrook and Pleasant Hill formations. A small collection from the lower part of the Elbrook Formation in southern Pennsylvania contains the early Middle Cambrian trilobite *Glossopleura bassleri* (RESSER, 1938). RASETTI (1965a) has described well preserved trilobites of late Middle Cambrian age from limestones in the upper 70 metres of the Pleasant Hill Limestone. These include species of *Olenoides*, *Asaphiscus*, and *Alokistocare*. Several of the species are similar or identical to species in the Maryville Limestone of the southern Appalachians in Tennessee.

Trilobites of middle Dresbachian age are known from the upper part of the Elbrook and Warrior formations and the Big Spring Station Member of the Conococheague Formation. The trilobites from the Elbrook Formation were reported by BUTTS (1940) from southern Virginia and by WILSON (1952) from southern Pennsylvania. One faunule containing the typical *Crepicephalus* zone genera *Komaspidella*, *Meteoraspis*, *Llanoaspis*,

Pemphigaspis, and *Terranovella* has been described by RASETTI (1961) from either uppermost Elbrook or lowermost Conococheague beds near Winchester, Virginia. The upper 300 metres of the Warrior Formation in central Pennsylvania have yielded *Crepicephalus* Zone trilobites, including species of *Tricrepicephalus*, *Kingstonia*, *Lonchocephalus*, and *Pemphigaspis*, described by TASCH (1951). WILSON (1952) listed the *Crepicephalus* Zone genera *Llanoaspis* and *Pemphigaspis* from the Big Spring Station Member of the Conococheague Formation in southern Pennsylvania.

Trilobites of early Franconian age are known only from the northern part of the central Appalachians in Pennsylvania, Maryland, and northern Virginia. They have been listed or described by WILSON (1951, 1952) from the Ore Hill Member of the Gatesburg Formation and from the Conococheague Formation above the Big Spring Station Member. He recognized three faunas, the *Pseudosaratogia magna* fauna, the *Elvinia* fauna, and the *Conaspis* fauna.

Younger Cambrian fossils are rare in the central Appalachian region. The Trempealeauan gastropod genus *Sinuopea* is listed by WILSON (1952) from the Mines Member of the Gatesburg Formation. Saukiid trilobites also indicative of Trempealeauan age have been described by WALCOTT (1914) and listed by WILSON (1952) from the upper part of the Conococheague Formation. The only significant Conococheague assemblages of Trempealeauan age include species of *Bowmania*, *Entomaspis*, *Idiomesus*, *Plethometopus*, and *Stenopilus*, which have been described by RASETTI (1959) from southern Pennsylvania and adjacent parts of Maryland. SANDO (1957) reported the Early Ordovician trilobites *Symphysurina* and *Clelandia* from the upper beds of the Conococheague in Maryland. Thus, the Cambrian–Ordovician boundary there does not conform to a formational contact. The apparent conformity of this system boundary and formational boundaries throughout the Appalachian region is more an expression of convenience for mapping than of any really precise control on its position.

3.2. Cambrian of south-eastern Pennsylvania and adjacent Maryland

The Cambrian rocks that crop out south-east of the Triassic lowlands in Pennsylvania, and east of Catoctin Mountain on both sides of the Triassic lowlands in Maryland, include a lower sequence of clastics referable to the Chilhowee Group of the main carbonate belt, and an upper sequence of carbonates which contrasts with the carbonate sequences of the main Cambrian belt and reflects important regional facies changes (Fig. 10).

The rocks of the Chilhowee Group have a general tripartite division comparable to that in the main Cambrian belt. The medial Harpers Formation is composed largely of argillaceous rocks comparable to those in the Harpers of the main Cambrian belt. However, there are notable changes in the upper and lower quartzitic units. The upper unit, the Antietam Quartzite, is only about 100 metres thick—about one-third of its thickness in the northern part of the main Cambrian belt. The lower unit includes a distinctive coarse conglomerate member at its base and the upper part grades southwards into the argillaceous rocks of the generally overlying Harpers Formation. In Maryland, the lower unit includes the conglomerate member of the Loudon Formation overlain by the Weverton Quartzite (WHITAKER, 1955a). In Pennsylvania, the basal Hellam Conglomerate is overlain by Chickies quartzite north of York, and the Chickies slate to the south (STOSE and JONAS, 1939). Because of lateral gradation of the upper parts of the Weverton and Chickies quartzites into argillaceous rocks southwards, the thickness of the lower quartzitic units decreases southwards from a maximum of about 180 metres.

WHITAKER (1955b) has indicated a western source for sands of the Weverton Quartzite.

The carbonate sequence changes character in this region from north to south. North of Lancaster, JONAS and STOSE (1930) described the sequence above the Antietam as: Vintage Dolomite, Kinzers Formation, Ledger Dolomite, Elbrook Limestone, and Conococheague Limestone.

The Vintage Dolomite, about 180 metres thick, is characterized by massive bedded to nodular and argillaceous dolomite. The Kinzers Formation is divided into three members: a lower member, about 30 metres thick, of black to dark grey shale; a middle member, about 25 metres thick, of limestones ranging from pure to argillaceous; and an upper member, about 15 metres thick, of argillaceous limestone and sandy limestone or dolomite. The Ledger Dolomite, estimated to be about 300 metres thick, is composed of massive light-grey dolomite. The Elbrook Limestone, also estimated to be about 300 metres thick, is composed of thin-bedded, fine-grained, dolomitic limestone with argillaceous partings. The Conococheague Limestone, also estimated to be about 300 metres thick, is composed of interbedded thick-bedded limestone and dolomite, chert layers, and algal stromatolites.

MEISLER and BECHER (1968) have refined the units assigned to the Elbrook and Conococheague limestones. They describe the sequence above the Ledger Dolomite as: Zooks Corner Formation, Buffalo Springs Formation, Snitz Creek Formation, Millbach Formation, and Richland Formation. All units above the Zooks Corner Formation are assigned to the Conococheague Group. The Zooks Corner Formation, characterized by silty and sandy dolomite, is estimated to be about 500 metres thick. The Buffalo Springs Formation is characterized by interbedded limestone and silty or sandy dolomite estimated to be between 450 and 1,125 metres thick. The Snitz Creek Formation is a silty or sandy dolomite about 115 metres thick. The Millbach Formation, characterized by a predominance of limestone and only a few beds of dolomite, is estimated to be 350 to 600 metres thick. The Richland Formation, characterized by interbedded limestone and dolomite, ranges from zero to about 160 metres in thickness. No explanation is given by MEISLER and BECHER for the considerable increase in estimated thickness of the post-Ledger rocks over that given for the same area by JONAS and STOSE. The formational nomenclature used for the units of the Conococheague Group was first proposed for the rocks of the Lebanon Valley 25 kilometres to the north (see p. 198). In that area, the entire section is overturned and has unusual thickness variations that seem anomalous on a broad regional scale. The differences in interpretation of formational thicknesses in both the Lebanon Valley and Lancaster areas reflect the difficulties of working with poorly exposed, moderately to strongly folded and faulted, essentially unfossiliferous rocks.

To the south and south-west, the units above the Ledger Dolomite are not present, and in some areas units down as far as the Antietam are missing. In their place is a thin-bedded dark blue argillaceous limestone interbedded with black shale and locally including beds of coarse, poorly sorted limestone conglomerate. This unit is the Conestoga Limestone in south-eastern Pennsylvania, and the Frederick Limestone in Maryland. In Pennsylvania, JONAS and STOSE (1930) and STOSE and JONAS (1939) believed that the Conestoga was an Ordovician unit unconformable on the older formations. However, RODGERS (1968) has suggested that this is a lateral facies of the carbonate units of the main Cambrian belt and the area north of Lancaster. A similar interpretation for the Frederick Limestone in Maryland was suggested as one of several possiblities to explain its relations to older units by WHITAKER (1955a). This interpretation is supported by the regional relationships of the trilobites of the Frederick Limestone and is incorporated in Fig. 10. However more work in south-eastern Pennsylvania is needed to verify the interpretation.

In Maryland, the Frederick Limestone is overlain by the Grove Limestone, a massive limestone about 180 metres thick.

3.2.1. Age and correlation

Fossils are known from this region principally from the Kinzers Formation in Pennsylvania and the Frederick and Grove limestones in Maryland. The lower member of the Kinzers Formation has yielded the richest fauna of Early Cambrian fossils known from eastern United States. This includes species of brachiopods, echinoderms, molluscs, trilobites, and other arthropods and has been described by RESSER and HOWELL (1938) with some additions to the echinoderms by DURHAM (1966). The trilobites and molluscs, which are the most definitive and wide-ranging elements, include species of *Olenellus*, *Paedeumias*, *Wanneria*, *Bonnia*, and *Salterella*. Early Cambrian fossils are also reported from the upper members of the Kinzers Formation by STOSE and JONAS (1939). These include archaeocyathid reefs from the middle member, and *Bonnia* and *Salterella* from the upper member. Because the underlying Vintage Dolomite and overlying Ledger Dolomite resemble the lower and upper parts of the Tomstown Dolomite, STOSE and JONAS (*op. cit.*) correlate the Vintage–Kinzers–Ledger sequence with the Tomstown Dolomite of the main Cambrian belt.* The overlying Elbrook Limestone is reported to have a reddish sandstone near its base east of Lancaster. Otherwise, there is no indication of the Waynesboro Formation which lies between the Tomstown and Elbrook in the main Cambrian belt, and the Elbrook in this area is considered to be in part a more limy facies of the Waynesboro. The poor exposure and strong deformation of the rocks south-east of the Triassic lowlands in Pennsylvania contribute to the difficulties of determining accurate thicknesses and lithological successions and to resolving many of the problems of correlation.

The Frederick Limestone has yeilded trilobite faunas of Dresbachian and Trempealeauan ages near Frederick, Maryland (RASETTI, 1959, 1961). The Dresbachian faunas include species of *Quebecaspis*, *Pterocephalops*, *Acmarhachis*, and *Dunderbergia*. The Trempealeauan faunas include species of *Richardsonella*, *Keithiella*, *Loganellus*, and *Apatokephaloides*. The lower part of the Grove Limestone has yielded Trempealeauan trilobites (RASETTI, 1959). These include species of *Onchonotus*, *Hungaia*, *Apatokephaloides*, *Richardsonella*, and *Bayfieldia*. Faunas similar to those of the Frederick and Grove limestones are known from the Woods Corners Group and Gorge Formation in the St. Albans area of Vermont; boulders in the Levis conglomerates of Quebec and the Cow Head conglomerates of Newfoundland; the Hillard Limestones of Alaska; and the Dunderberg and Windfall formations of Nevada. At all of these localities the deposits accumulated on the seaward side of carbonate banks, thus supporting the suggestion that the Frederick and perhaps the Conestoga limestones are south-eastern facies of the carbonate sequences of the main Cambrian belt.

3.3. Some problem areas

Two small areas, one in Pennsylvania and one in southern Virginia, have Cambrian sections which are not typical of the central Appalachian sequences, but nevertheless may be related to them.

* After completion of the manuscript for this section, KAUFFMAN and CAMPBELL (1969) have reported Middle Cambrian fossils from the upper part of the Kinzers and these have been confirmed by the writer. Thus, the stratigraphical position and regional relationships of the Kinzers must be re-evaluated.

3.3.1. The Lebanon Valley area

In the eastern part of the Lebanon valley, west of Reading, Pennsylvania, the Pre-Cambrian basement is overlain by about 180 metres of white quartzite referred to the Hardyston Formation by GEYER and others (1963). The Cambrian part of the overlying carbonate sequence has been divided into five formations, in ascending order (Fig. 11): the Leithsville (or Tomstown), Buffalo Springs, Snitz Creek, Millbach, and Richland

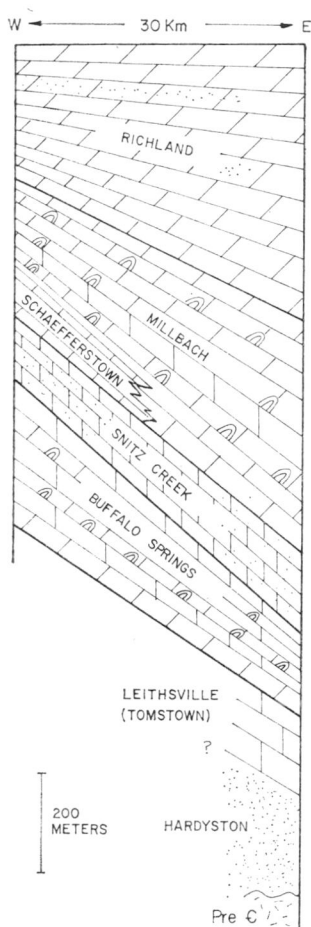

FIG. 11. The Cambrian succession of the Lebanon Valley area showing thickness changes from west to east.

formations. All are dolomitic and none has yielded any fossils other than algal stromatolites. The Leithsville (Tomstown) Formation is characterized here by massive dolomite with shaly interbeds. The Buffalo Springs and Millbach formations are both characterized by interbedded limestone and dolomite and by algal stromatolites. The Snitz Creek Formation, which separates the Buffalo Springs and Millbach formations, and the Richland Formation are both characterized by sandy and cherty dolomite. In the western part of the Lebanon valley, a limestone formation with shaly laminae, the Schaefferstown Formation, is present between the Snitz Creek and Millbach formations. Throughout the Lebanon valley, the Cambrian formations, with the exception of the

Hardyston and Leithsville (Tomstown) are strongly overturned, exposures are discontinuous, and thrust faults parallel to the regional strike are known. Only estimates of thickness for the formational sequence are possible. GEYER and others (1958, 1963) suggest that the upper formations nearly double in thickness in the 30 kilometres from the western part of the Lebanon Valley to the eastern part. In the eastern part, the succession of Cambrian carbonates above the Hardyston Quartzite is estimated to be nearly 1,400 metres thick. This is more nearly comparable with the thickness of the Cambrian succession above the Antietam Quartzite in the thick sequences of the Cumberland valley and the Lancaster area to the south and west than to the sequences above the Hardyston in the thin sequences of eastern Pennsylvania. The Buffalo Springs Formation has been correlated with the Elbrook Formation of the Cumberland Valley (MACLACHLAN, 1967). Both Tomstown and Leithsville have been used as names for the lowest carbonate unit. However, no unit comparable to the Waynesboro has been recognized in the Lebanon Valley area. The possibility remains that the thicknesses of the carbonate sequence in the eastern Lebanon Valley have been tectonically exaggerated and that the sequence really is related to the thin sequences of eastern Pennsylvania. Absence of any faunal control in the Lebanon Valley rocks precludes resolution of their regional stratigraphical relationships. MEISLER and BECHER (1968) have used these Lebanon Valley names in the thick carbonate sequence in the Lancaster Valley 25 kilometres to the south.

3.3.2. The Austinville-Wytheville area

In the area near Wytheville, Virginia, in the south-eastern part of the Main Cambrian belt (Fig. 9) the Early Cambrian part of the carbonate sequence changes dramatically. Because of real and inferred structural complications, the stratigraphical relations of the beds involved in these changes are still in dispute.

Near Wytheville, a typical succession of the carbonate sequence is developed consisting of the Shady Dolomite, Rome Formation, Elbrook Dolomite and Conococheague Limestone. The sequence below the red and green shales and sandstones of the Rome Formation consists of about 670 metres of limestones and dolomites that has been divided from top to bottom into the Ivanhoe Limestone (150 metres), Austinville Dolomite (300 metres), and Paterson Limestone (240 metres). BUTTS (1940) considered all three units to be members of the Shady but STOSE and JONAS (1938) and RODGERS (1956a) considered the Ivanhoe Limestone to be the lower member of the Rome Formation.

Near Austinville, 16 kilometres to the south-east, a series of fossiliferous limestones and associated dolomites is present which is quite unlike any Cambrian units to the north-west. According to CURRIER (1935) and BUTTS (1940) this sequence is estimated to exceed 1,080 metres in thickness and to overlie the Shady Dolomite. They consider the sequence to be most probably a south-eastern facies of the Rome and Elbrook formations. However, STOSE and JONAS (1938) reinterpreted the geology of the area and claimed that the section was duplicated by several faults. They recognize a series of three units: a lower dolomite formation correlated with the Vintage Dolomite of south-eastern Pennsylvania; a middle formation of fossiliferous limestones correlated with the Kinzers Formation; and an upper dolomite formation correlated with the Ledger Dolomite. The fossiliferous limestone sequence is considered to be only 270 metres thick, and the entire sequence of three formations is considered to be the south-eastern equivalent of the Shady Dolomite. Final resolution of these conflicting interpretations has not yet been made.

The lower part of the fossiliferous unit contains archaeocyathids, trilobites, brachiopods, and molluscs including species of *Olenellus*, *Bonnia*, *Austinvillia*, *Kutorgina*, and *Helcionella*. The upper part contains trilobites and brachiopods, including species of *Kootenia* and *Nisusia*. Most of the fossils have been described by RESSER (1938) with some additions by COOPER (1951). All of the fossils have been assigned an Early Cambrian age.

4. The Southern Appalachians

This region includes all Cambrian rocks in Tennessee, Georgia, Alabama, and parts of south-western Virginia (Fig. 12). Its northern boundary is placed at the northern limit of the shales of the Conasauga Group in the upper carbonate sequence, and arbitrarily at the Tennessee–Virginia boundary in the lower clastic sequence. It is divided into northern and southern sectors at the Tennessee–Georgia boundary.

4.1. The northern sector

Within this sector, the rocks of the lower clastic sequence together with the overlying Shady Dolomite of the carbonate sequence constitute a succession of units which differs only in local details from the correlative rocks in the central Appalachians (Chapter 3). The Middle Cambrian and early Upper Cambrian parts of the overlying succession, however, become increasingly shaly southwards, so that in southern Tennessee, the entire interval is occupied by shale (RODGERS, 1953) (Fig. 13).

4.1.1. The lower clastic sequence

Rocks of the lower sequence crop out only along the south-eastern margin of the Cambrian outcrop areas, as in the central Appalachians. For about 165 kilometres south-west of the Tennessee–Virginia boundary area, the clastic succession of Unicoi, Hampton, and Erwin formations described by KING and FERGUSON (1960) (see p. 192) and overlain by the Shady Dolomite is recognized. Farther south in Tennessee, although the Shady Dolomite is continuously recognized, the underlying clastics are assigned to a different sequence of formations. This area is the type area of the Chilhowee Group. Here, the group includes, in ascending order: the Cochran Formation—180 to 345 metres of unfossiliferous, coarse grained, largely conglomeratic, feldspathic sandstones which unconformably overlie the older sedimentary sequences of the Ocoee Formation; the Nichols Shale—about 210 metres of unfossiliferous black silty shale and siltstone; the Nebo Quartzite—60 to 120 metres of white vitreous quartzite with *Scolithus* burrows; the Murray Shale—110 to 165 metres of dark silty shale and siltstone with rare non-trilobite fossils; the Hesse Quartzite—165 to 180 metres of white vitreous quartzite with *Scolithus* burrows; and the Helenmode Formation—20 to 55 metres of silty shale, siltstone, and sandstone transitional to the Shady Dolomite (NEUMAN and NELSON, 1965).

4.1.2. The upper carbonate sequence

The upper sequence throughout the northern sector of the southern Appalachians begins with the heterogeneous assortment of generally reddish, but variegated, shales, sandstones, thin-dolomites, and rare conglomerates of the Rome Formation. Only in north-eastern Tennessee is a complete section across this formation known. There, it is about 360 metres thick. Throughout the rest of the region, the Rome is the oldest unit exposed in most thrust slices and rarely is more than the uppermost 180 metres preserved.

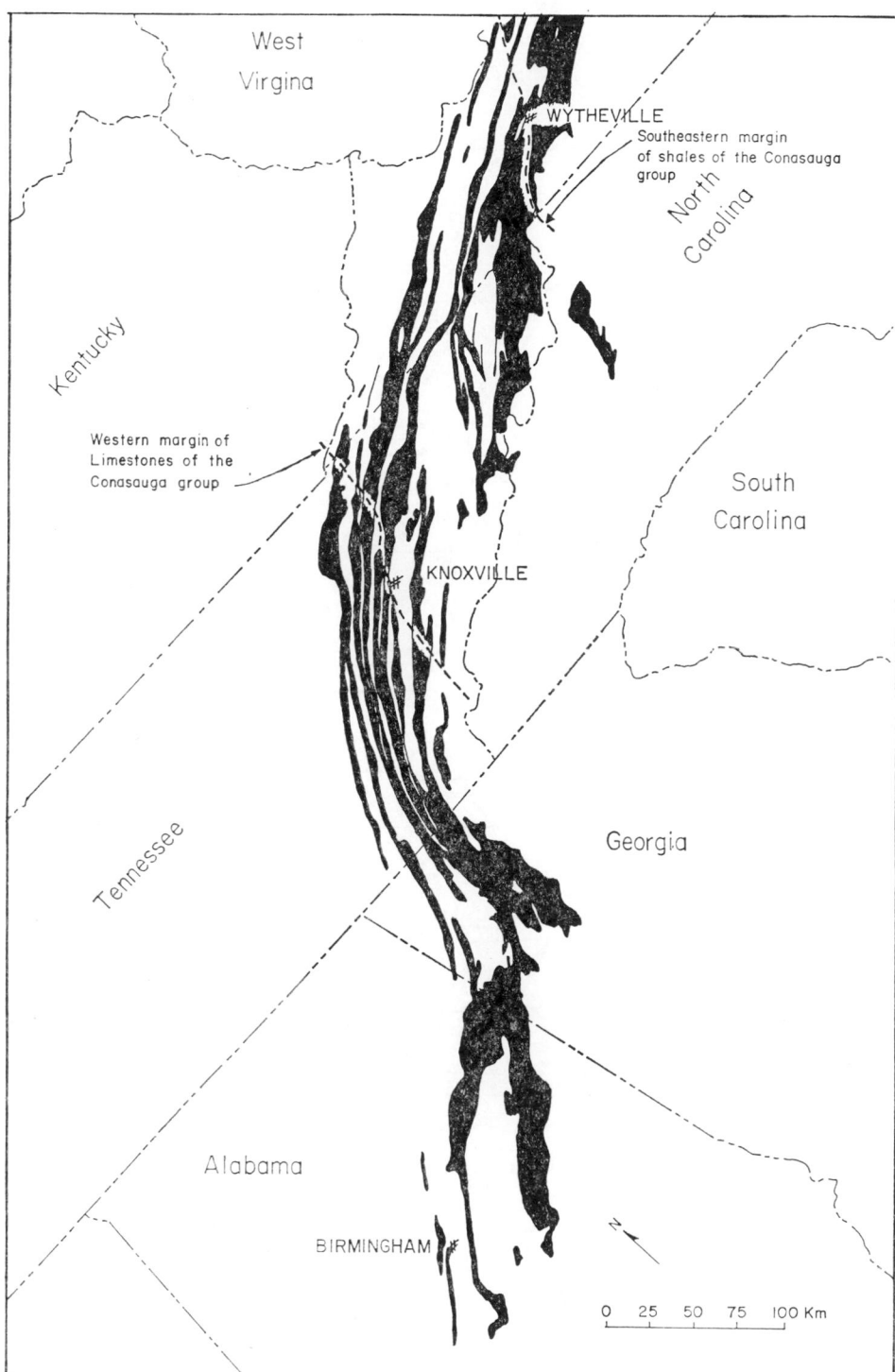

Fig. 12. Areas of outcrop of Cambrian rocks in the Southern Appalachians.

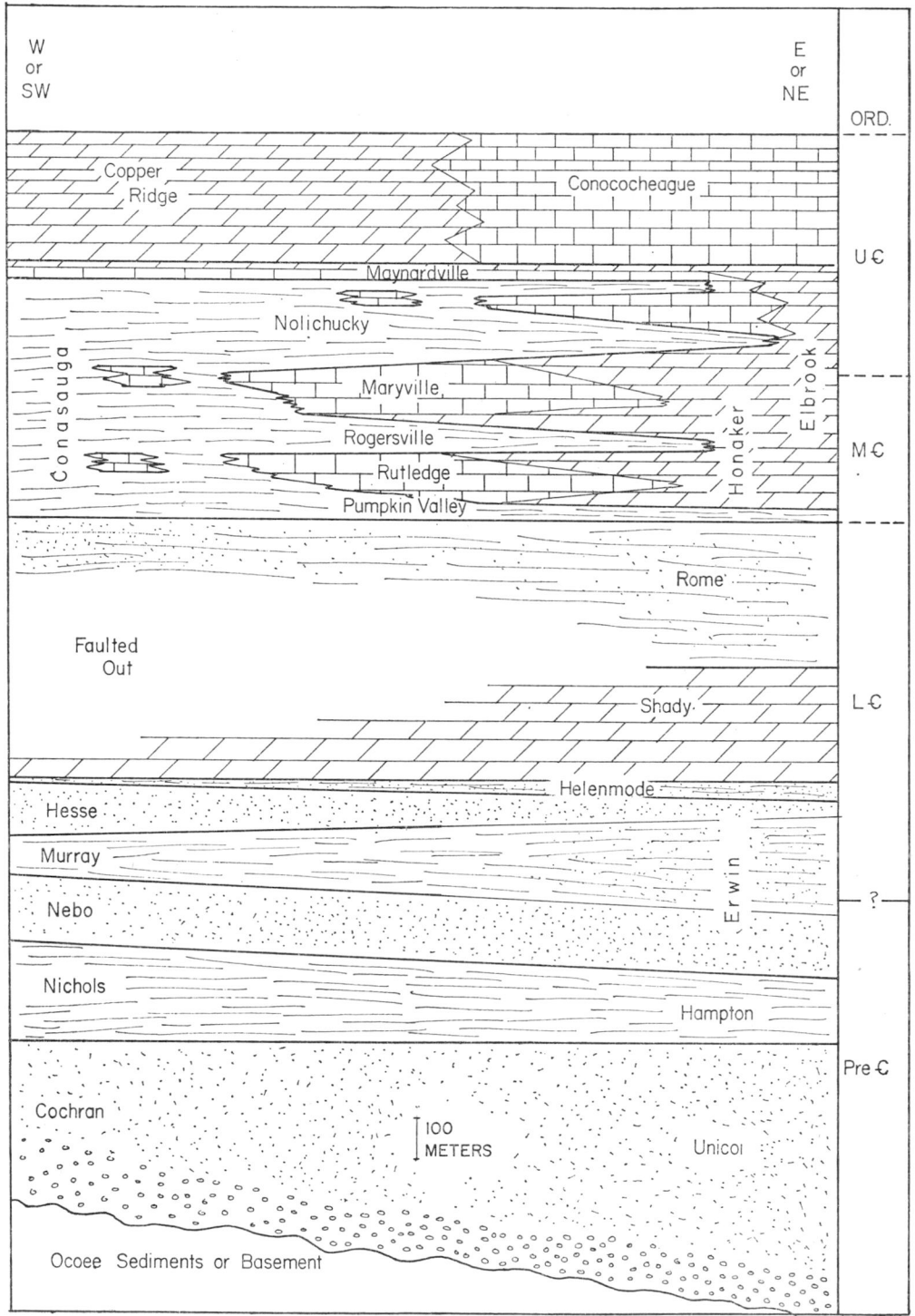

FIG. 13. Correlation and facies relationships of the Cambrian rocks of the northern sector of the Southern Appalachians.

The sequence above the Rome differs in the eastern and western parts of the northern sector of the southern Appalachians. East of a line that passes just west of Knoxville and crosses the entire strike belt of the southern Appalachians diagonally (Fig. 12), the sequence above the Rome Formation consists of alternating shale and limestone formations of the Conasauga Group, overlain by dolomites or limestones of the predominantly carbonate Knox Group. West of this line, the limestone formations of the Conasauga Group pinch out and the entire sequence between the Rome Formation and the carbonates of the Knox Group is included in the Conasauga Shale (RODGERS, 1953; HARRIS, 1964). The Conasauga Group (or Shale), throughout the region, is estimated to be about 600 metres thick. The Conasauga Shale contains a few scattered intervals of limestone but none is well enough exposed or persistent enough to be separately mapped. In north-eastern Tennessee, where the limestone units of the Conasauga can be mapped, the intervening shale units thin to the east and north-east as the adjacent limestones thicken (Fig. 13).

The top of the Rome Formation throughout the region is drawn at the highest sandstone bed. Between this and the Rutledge Limestone of the Conasauga Group is the Pumpkin Valley Shale, a unit of olive and purple shales ranging from 70 to 115 metres in thickness. The Rutledge Limestone consists of massive blue limestone up to 150 metres thick, with some units of oolitic limestone or limestone edgewise conglomerate. The Rogersville Shale, a unit of greenish shale up to 80 metres thick, overlies the Rutledge Limestone. This shale is overlain by the Maryville Limestone, a unit up to 200 metres thick containing most of the same types of limestone as the Rutledge Limestone. In extreme north-eastern Tennessee and parts of Virginia immediately to the north across the Cambrian outcrop belt, the Rogersville Shale is absent and the Rutledge and Maryville limestones merge into the Honaker Dolomite. Above the Honaker Dolomite or the Maryville Limestone are the green shales and interbedded limestones of the Nolichucky Formation ranging up to 225 metres in thickness. The Nolichucky, and the Conasauga Shale to the west and south-west, are overlain by the Maynardville Limestone, a unit of thin-bedded limestone ranging from 50 to 110 metres in thickness.

Above the Maynardville Limestone throughout most of the northern sector of the southern Appalachians, the remainder of the Cambrian section is included within a single formation, the Copper Ridge Dolomite. In north-eastern Tennessee and north-eastwards into Virginia, the Copper Ridge merges into the Conococheague Limestone. These units average between 270 metres and 360 metres in thickness. The Copper Ridge and Conococheague are characterized in this region by oolitic chert and dolomite, nodular chert, and abundant algal stromatolites. Sandy beds of the Chepultepec Dolomite that overlie the Copper Ridge and Conococheague throughout the region are considered to be basal beds of the Ordovician.

4.1.3. Age and correlation

The Early Cambrian rocks of this region include all formations from the Rome downwards to at least the Murray Shale of the Chilhowee Group. The oldest fossils in eastern United States are bivalved arthropods identified as *Indiana tennesseensis* (RESSER) from the lower part of the Murray Shale (LAURENCE and PALMER, 1963). Olenellid fragments have been reported from the Helenmode Formation at the top of the Chilhowee Group. The Shady Dolomite and Rome Formation have not yielded fossils in this region, but the Rome Formation, both to the north and south, yields olenellid trilobites.

The Middle Cambrian rocks of this region include the Pumpkin Valley Shale, Rutledge Limestone, Rogersville Shale, most of the Maryville Limestone and Honaker Dolomite, and the lower part of the Nolichucky Formation. Most of the faunas from these formations are described by RESSER (1938). The Pumpkin Valley Shale has yielded ptychoparioid trilobites referred to species of *Elrathiella*, and *Alokistocare*. The Rutledge Limestone has yielded several species of ptychoparioids, and corynexochoids referred to species of the early Middle Cambrian genera *Anoria* and *Glossopleura*. The Rogersville Shale has yielded principally ptychoparioid trilobites referred to species of *Ehmaniella*. The Maryville Limestone towards its south-western limit and the overlying Nolichucky Shale have yielded late Middle Cambrian trilobites referable to *Asaphiscus*, *Alokistocare*, and *Bolaspidella*.

Late Cambrian fossils are known primarily from the Nolichucky Shale and Maynardville Limestone in Tennessee, and parts of the underlying Maryville Limestone. The oldest fossils in the Nolichucky Shale, where it is underlain by the Honaker Dolomite, indicate that parts of the upper Honaker are also of Late Cambrian age. The faunas of the Nolichucky and adjacent formations have been described by RESSER (1938) and RASETTI (1965b). They are primarily trilobites and represent rich assemblages of species of the *Cedaria*, *Crepicephalus*, and *Aphelaspis* zones of Dresbachian age.

Fossils other than stromatolites are extremely rare in the Copper Ridge Dolomite and Conococheague Limestone. No identifiable fossils of Franconian age have been found in the southern Appalachians. Rare Trempealeauan fossils, primarily the gastropod *Scaevogyra*, have been reported from cherts in the Copper Ridge and Conococheague formations.

4.2. The southern sector

This sector includes the relatively poorly known Cambrian rocks of Georgia and Alabama at the extreme southern end of the Appalachian region. The topographical relief of the Appalachian mountains decreases in this region and, this, together with increasingly deep weathering, has resulted in extremely poor exposures. Most of the Cambrian rocks except for the Late Cambrian carbonates are relatively incompetent and have been intensely deformed. Despite this, the general stratigraphy has been worked out (Fig. 14) although some special problems still exist (p. 206).

4.2.1. The lower clastic sequence

The Chilhowee Group is represented in the southern sector by the Weisner Formation only. This formation includes about 540 metres of interbedded shale, quartzite, and conglomerate, and is cut off at its base by faults. Above this, rocks identified as Shady have been reported from a few localities. In the south-eastern outcrop belts of Georgia, these are coarse dolomites which are locally cherty. In Alabama, about 155 metres of thick-bedded limestone is reported to lie between the Weisner Formation and the Rome Formation and has been identified as Shady. The overlying Rome Formation has its type area in Georgia. It includes red and green shale, calcareous sandstone, and some thin limestone beds. Intense folding and poor exposure make accurate thickness determinations impossible and estimates for thickness of the Rome range between 270 metres and 540 metres.

4.2.2. The upper carbonate sequence

The upper sequence is generally poorly exposed throughout the southern sector. The Late Cambrian cherty dolomites of the Copper Ridge Dolomite persist to the southern limit of the exposures. Below these, and above the sandy Rome Formation, is a variety of carbonate and shale units, mostly assigned to the undifferentiated Conasauga Formation except in the south-western exposures in Alabama. In southern Tennessee, at the south end of the northern sector, the Conasauga in the central strike belts has several limestone units that can be locally mapped. In the south-eastern strike belts the Conasauga is almost entirely composed of shale. In northern Georgia, the Conasauga of both the central and south-eastern outcrop belts southwards becomes increasingly dominated by limestones. The north-western belts continue to have the Conasauga Formation dominated by shales.

In Alabama, rocks of the Conasauga and possibly correlative formations are exposed in several different areas with different characteristics (Fig. 14). The largest area is a broad outcrop belt just west of the Georgia boundary. Here, the Conasauga is composed

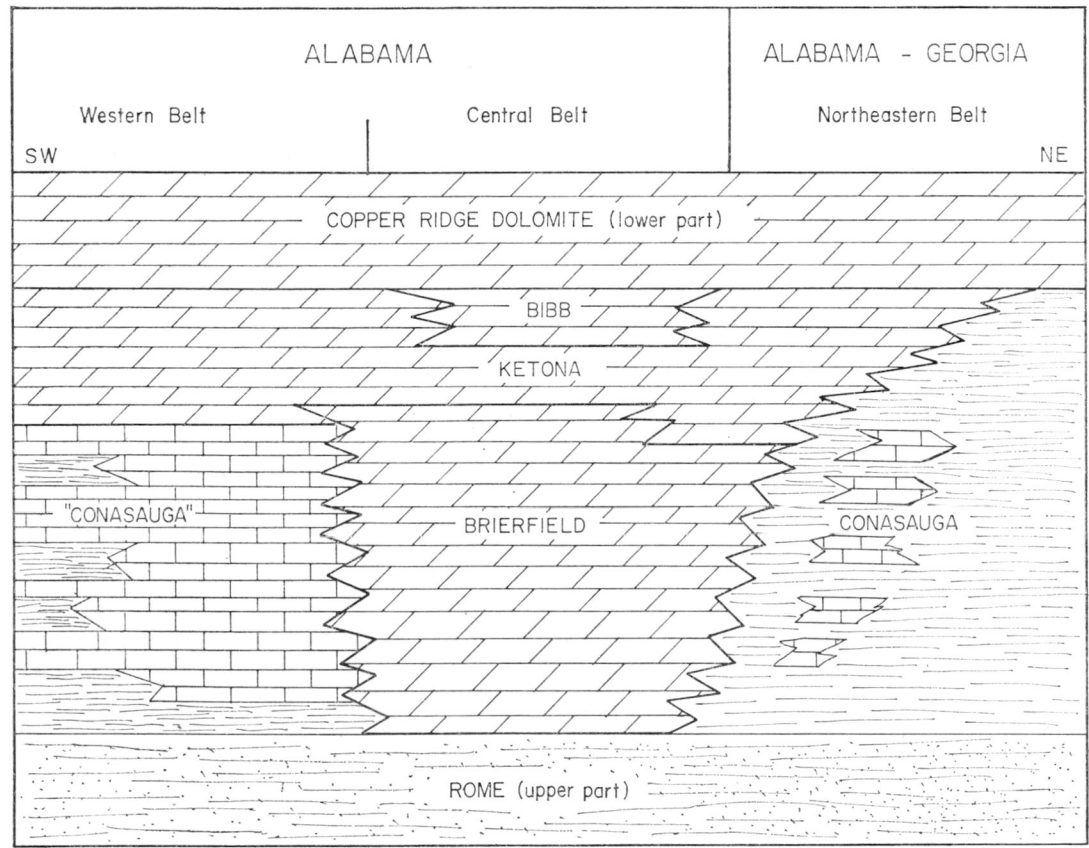

FIG. 14. Possible correlation and facies relationships of the Cambrian rocks of the southern sector of the Southern Appalachians.

predominantly of shales with occasional thin limestone interbeds and intervals of limestone nodules isolated within the shales. In some parts of this area the limestone nodules have been converted to chert. South-west of this area, particularly in the western thrust slices, the Conasauga Formation is composed dominantly of thin-bedded, argillaceous, dark blue, fine-grained limestone. Near the south-western limit of exposures in these western outcrop belts, some thick shale beds are again present within the Conasauga Formation.

From the latitude of Birmingham southwards, in the middle one of the three principal outcrop belts in that part of Alabama, a sequence of non-cherty dolomites is present between the Copper Ridge Dolomite and the Conasauga Formation. This has been divided into three units which total about 700 metres in thickness in their maximum development. The lowest unit, the Brierfield Dolomite, and the upper units, the Bibb Dolomite, are lithologically identical siliceous, coarse-grained, dark-coloured dolomites known only in this area. The Brierfield rests on limestones of the Conasauga Formation. The middle unit, the Ketona Dolomite, is a light-coloured, coarse-grained, pure dolomite of considerably greater areal extent. North of the latitude of Birmingham in the middle outcrop belt the Ketona immediately underlies the Copper Ridge and the Bibb Dolomite is believed to have been replaced by dolomite of the Ketona facies. In this area, the Ketona is reported to lie directly on the Rome Formation and the Conasauga Formation and Brierfield Dolomite are thought to have been removed prior to deposition of the Ketona. Farther north in the same strike belt, the Ketona also disappears and the Copper Ridge Dolomite is reported to lie unconformably on the Rome Formation (BUTTS, 1926). In the outcrop belt that includes Birmingham, the Ketona Dolomite is also present and rests everywhere with supposed unconformity on limestones of the Conasauga Formation.

The significance of the relationships of the Brierfield, Ketona, and Bibb dolomites, and the limestone phases of the Conasauga to one another, and to the other Cambrian rocks, is crucial in understanding the palaeogeography of the southern sector of the southern Appalachians. BUTTS considered that the dolomites were all a part of the Knox Group and younger than the early Upper Cambrian beds that constitute the upper part of the Conasauga Formation in its fullest development. Alternatively, the Brierfield and younger dolomites may represent a carbonate facies laterally equivalent to much of the Conasauga Shales and comparable with the Elbrook or Honaker formations in the southern part of the central Appalachians. This interpretation is indicated on Figs 14 and 15.

4.2.3. Age and correlation

Despite the poor exposures of rocks in the southern sector, a number of fossiliferous localities are known. Often, because of structural complications, the stratigraphical position of the faunas at these localities cannot be accurately determined.

Rocks of Early Cambrian age are represented by the Shady and Rome formations. The Shady Formation has yielded archaeocyathids and specimens of *Salterella*, and the overlying Rome Formation in its type area has yielded a fauna of well preserved olenellid trilobites.

Middle and Late Cambrian trilobites have been recorded from the Conasauga Formation in both Georgia and Alabama. The Middle Cambrian faunas include corynexochoids such as *Glossopleura*, *Olenoides*, *Kootenia*, and *Zacanthoides*, and a large assortment of ptychoparioids whose generic affiliations need re-evaluation. The Late Cambrian trilobites are all of Dresbachian age and include representatives of the *Cedaria*, *Crepicephalus*, and *Aphelaspis* zones.

The only Cambrian fossils from beds younger than the Conasauga Formation are specimens of the Trempealeauan gastropod *Scaevogyra* reported from the Copper

Ridge Dolomite. Many of the Cambrian fossils have been illustrated by BUTTS (1926) and are described by RESSER (1938). Additions to the Late Cambrian trilobite faunas have been made by PALMER (1962).

5. Sedimentation and palaeogeography of the Appalachian region

All Cambrian rocks of the Appalachian region were deposited in a subsiding marine basin along the east margin of the North American craton. Four more or less contemporaneous lithofacies, roughly paralleling the margin of the craton, can be recognized within the Cambrian sequences.

The lithofacies nearest the craton is largely composed of detritus derived from the craton. Rocks of this lithofacies comprise an inner detrital belt which includes the conglomeratic phases of the formations of the Chilhowee Group in the central and southern Appalachians and in the Cheshire and older sediments of the northern Appalachians; the quartz sands found at many levels in the Cambrian sequences throughout the region; and the red, green or brown shales of the Rome and Waynesboro formations and the Conasauga Group of the central and southern Appalachians. Evidence of a westward cratonic source for these sediments is abundant. In the northern Appalachians, the detritus of the Lower Cambrian Pinnacle and Monkton formations, and the Upper Cambrian Danby Formation, increases in grain size towards the west; and the quartzites of the Monkton and Danby formations become thicker and relatively more abundant in that direction. In the central Appalachians the currents depositing the sands of the Lower Cambrian Weverton Quartzite came from the west; the shales of the Lower Cambrian Harpers Formation are replaced to the north-west by quartzites of the Montalto Formation; and the sands of the Upper Cambrian Gatesburg Formation thicken to the north-west and lose their identity south-eastwards in the carbonate rocks. In the southern Appalachians, the detrital materials of the Lower Cambrian rocks of the Chilhowee Group and Rome Formation become coarse to the west, and the shales of the Middle and Upper Cambrian Conasauga Group pinch out eastwards into carbonate rocks.

Some units of the inner detrital belt, such as parts of the Monkton and Rome–Waynesboro formations, accumulated in very shallow waters, perhaps in the intertidal zone or even subaerially, as evidenced by mud cracks and red beds. Generally shallow subtidal conditions are suggested by glauconite in many units and by abundant *Scolithus* burrows in many of the Lower Cambrian orthoquartzite beds.

Seaward of the sediments of the inner detrital belt is a belt composed largely of carbonate sediments which are generally clean and have abundant oolites and algal stromatolites, indicative of accumulation in shallow water. Cyclic sequences of several kinds are also characteristic, and chert is locally abundant. These clean carbonate sediments are laterally gradational, often over distances of many kilometres, with the sediments of the inner detrital belt. Seaward transitions into sediments indicative of accumulation in deeper water are more abrupt. The belt of clean carbonate sediments is represented by all of the dolomites of the Appalachian region, the limestones of the Conococheague, Ore Hill, Warrior, and upper Pleasant Hill formations of the central Appalachians, and the Conococheague, Maynardville, Maryville, and Rutledge formations of the southern Appalachians.

Conditions of accumulation of the clean carbonate sediments are believed to be analogous to present-day conditions on the Bahama Banks. The thin-bedded argillaceous limestones or dolomites of the Elbrook, Conasauga, and lower Pleasant Hill of the central and southern Appalachians, and the Leithsville Formation of the southern part of the Northern Appalachians probably represent intervals transitional to the inner detrital belt.

The sediments that accumulated at the seaward margin of the carbonate banks are characterized by black shales, thin-bedded dark argillaceous limestones often interbedded with black shales, and zones of poorly sorted coarse carbonate clast conglomerate. Evidence for accumulation in water deeper than that of the adjacent banks is provided by graded beds of carbonate detritus in some areas and by the carbonate clast conglomerates which seem to be chaotic slumps into the black shale environment. In the Northern Appalachians, this lithofacies is represented by the Parker Shale, Woods Corners Group, and Gorge Formation in the St. Albans area; and the upper Bull, West Castleton, and Hatch Hill formations in the Taconic sequence. In the central Appalachians, the Frederick and Conestoga limestones and the Kinzers Formation in the Pennsylvania–Maryland area, and perhaps part of the problematical succession of the Austinville–Wytheville area, represent this lithofacies. In the southern Appalachians it is represented by some of the rocks assigned to the Conasauga Formation in Alabama.

A fourth lithofacies dominated by greywackes, shales (now slates), and interbedded volcanic rocks, and lacking significant units of carbonate sediment, occupied the sedimentary basin east of the transitional zone at the outer margin of the carbonate banks. In the northern Appalachians this is represented in the metamorphic sequence of central and eastern Vermont and in the lower part of the Bull Formation and much of the Rennsselear greywacke of the Taconic succession. In the central and southern Appalachians, it is probably represented in parts of the sequences of metamorphosed sediments of the piedmont region, but the relations of these rocks to the less metamorphosed sequences have not yet been satisfactorily worked out.

The preceding two lithofacies constitute an outer detrital belt of Cambrian sediments. The inner detrital belt and the carbonate belt represent sediments generally designated as miogeosynclinal and the outer detrital belt includes all sediments designated as eugeosynclinal. The shifting relations of the sediments of the inner and outer detrital belt and the carbonate belt during the Cambrian, and the relations of the sediments of the inner detrital belt to the craton provide the basis for understanding the Cambrian history of the Appalachian region.

The attempt to show the stratigraphical history of the Appalachian region by means of palaeofacies maps for selected times within the Cambrian period is complicated by post-Cambrian tectonic events. Imbricate westward thrusting of at least two ages has juxtaposed many formerly more widely separated Cambrian areas, and in some cases even reversed their geographical relations, so that the development of an understanding of the original spatial distribution of the Cambrian environments in this region is severely hampered. The palaeofacies maps of Fig. 15 have been constructed on the present geographical base to emphasize those areas where stratigraphical reconstructions indicate significant displacement of Cambrian facies patterns by post-Cambrian tectonic events. Except for the Taconic region, the magnitude of the displacements has not been resolved. In that region, ZEN (1967) has shown that the rocks were deposited over the site of the present Green Mountains and that they have been transported westwards about 30 kilometres.

Despite the difficulties of palaeogeographical reconstruction brought about by unresolved structural dislocations, the palaeofacies maps show the important features of

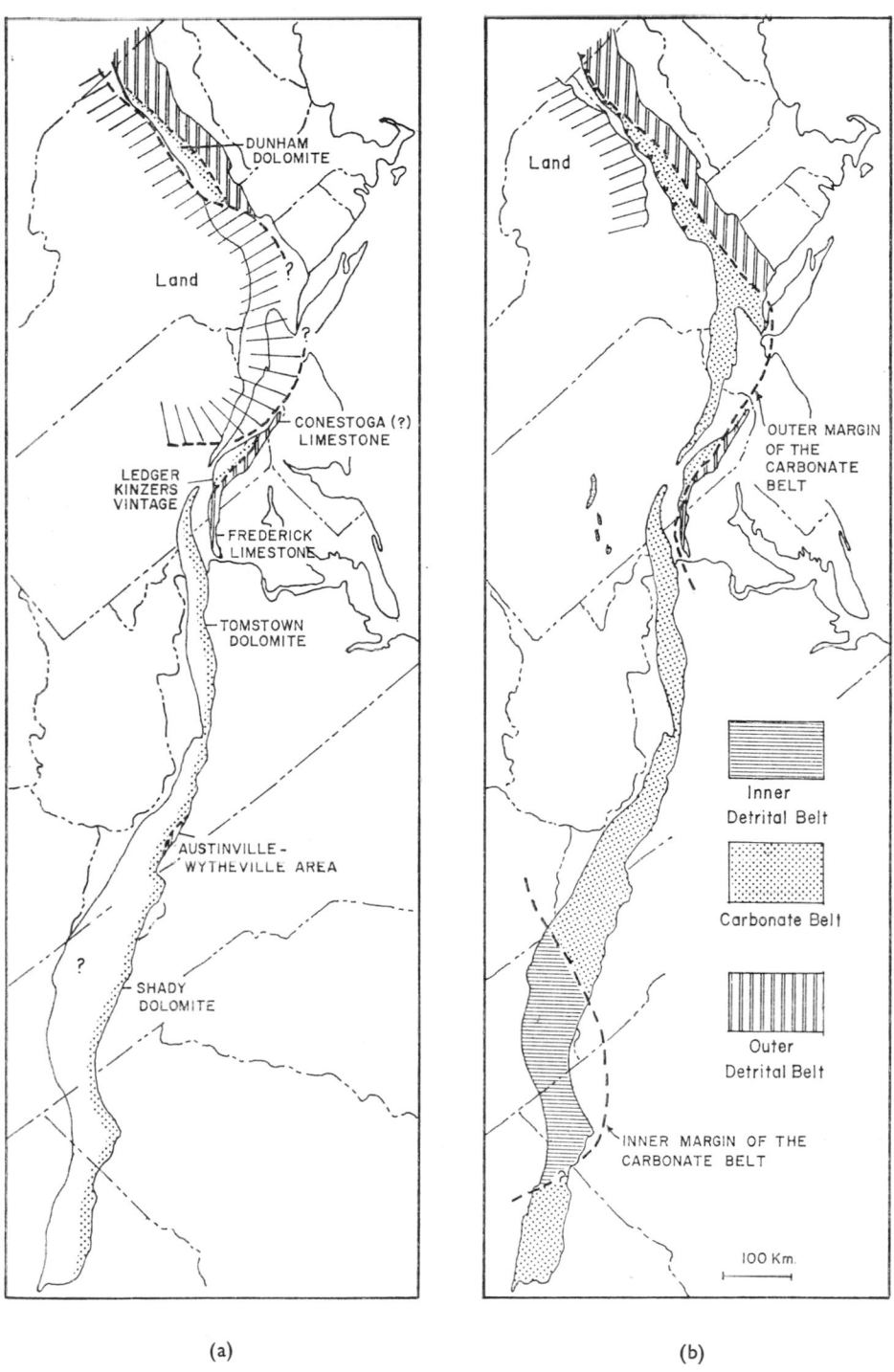

(a)   (b)

FIG. 15. Some aspects of the Cambrian palaeogeography of the Appalachian region:
(a) The Early Cambrian period of maximum transgression; (b) Late Middle Cambrian lithofacies distribution. Note the effect of the thrust fault along the western boundary in the northern part of the region.

Appalachian Cambrian palaeogeography. First, the major facies trends and the structural grain of the Appalachians are not strongly correlated. This is shown by the late Lower Cambrian peninsula extending south-eastwards across north-eastern Pennsylvania and south-eastern New York, the trend of the boundary between the carbonate belt and the outer detrital belt across northern Vermont during the Middle Cambrian and early Upper Cambrian, and the trend of the boundary between the inner detrital belt and the carbonate belt across the southern Appalachians during the same time.

Second, the generalization that the dolomite facies of the carbonate belt is usually north-west (on the cratonic side) of the limestone phase (RODGERS, 1968) does not seem to be supported by the facts. In the central Appalachians, the limestone phase is exposed only on the outer (eastern) margin of the carbonate belt. In the southern Appalachians, the limestone phase is exposed only in the inner (western) margin of the carbonate belt. An alternative generalization is that the dolomite phase of the carbonate belt is generally developed in the middle of the belt, and the margins of the belt more characteristically have limestones, this is supported by the fact that most of the fossiliferous limestones of the Appalachian region are found along either the inner or the outer margins of the carbonate belt.

6. The Eastern New England Cambrian

Three scattered areas in eastern New England with Cambrian fossils constitute the only certain record of Cambrian rocks east of the Appalachian region in eastern United States. Two of these areas are in eastern Massachusetts and one is in central Maine (Fig. 1).

The Cambrian rocks of eastern Massachusetts occur in small areas near Boston and one area adjacent to the north-eastern corner of Rhode Island. All of the outcrops are in areas of thick Pleistocene drift cover so that they provide very little detail of regional significance. The areas around Boston include dated rocks of both Early and Middle Cambrian age. The Early Cambrian rocks include green, grey, or red quartzitic slates all assigned to the Weymouth Formation. The Middle Cambrian rocks are light to dark grey, black or green, non-calcareous slate assigned to the Braintree Formation. The Weymouth Formation has an estimated thickness of between 100 metres and 180 metres. The Braintree Formation is about 300 metres thick.

The area of Cambrian rocks near the Rhode Island border is marked by three small outcrops in the vicinity of North Attleboro, Massachusetts. Here, a basal massive, white, partly cross-bedded quartzite 7 metres thick rests on granite and is overlain by 100 metres or more of red and green slates, fine-grained dark red or maroon silty limestone, and calcareous siltstone.

The principal area of Maine in which the probability of Cambrian rocks is reasonably well established is in northern Penobscot county and adjacent parts of Aroostook county to the east (NEUMAN, 1967; DOYLE and others, 1967). In this area, grey, green, and red slate and siltstone with about equal amounts of vitreous quartzite and lesser amounts of greywacke and tuff have been assigned to the Grand Pitch Formation (NEUMAN, 1967, p. 5). According to NEUMAN, it is no less than 1,485 metres in thickness and may possibly be much greater. It is unconformably overlain by rocks of Ordovician age.

6.1. Age and correlation

The most comprehensive study of the faunas of the rocks of the Boston area is by GRABAU (1900). Some additions and changes have been made by SHIMER (1907a, b), WALCOTT (1910), CLARK (1923), and SHAW (1950). The most significant elements of the Early Cambrian fauna are species of *Callavia* which clearly establish the correlation of the Early Cambrian beds of the Boston area with *Callavia*-bearing beds in eastern Newfoundland and in the Comley district of England. The additional presence of a smooth eodiscid, *Weymouthia nobilis* (Ford), and a species of *Strenuella*, both known elsewhere principally in Europe, emphasizes the European affinities of this fauna. The European affinities of the Boston area Cambrian are further emphasized by the Middle Cambrian fauna of the Braintree Formation which is dominated by *Paradoxides harlani* Green.

The fauna of the Cambrian rocks near Rhode Island was first described by SHALER and FOERSTE (1888) and verified by SHAW (1950). It is dominated by the ptychoparioid trilobite *Strenuella strenua* (Billings) and the conical fossils *Salterella* and *Hyolithes*. Some beds have abundant specimens of the tubular fossil *Coleoloides*. The lithology of some of the red, silty limestone beds with *Coleoloides* is identical to that of the Smith Point Limestone near the top of the Lower Cambrian succession of eastern Newfoundland, and to *Coleoloides*-bearing limestones in central England. The underlying granite has been dated at about 580 m.y. (FAIRBAIRN and others, 1967). A similar 'young' age for Pre-Cambrian granite is reported for the Holyrood granite that unconformably underlies the Lower Cambrian deposits of eastern Newfoundland (McCARTNEY, 1967).

In Maine, the Grand Pitch Formation contains the problematical fossil *Oldhamia* which has also been found associated with Early Cambrian trilobites in the Weymouth Formation of the Boston area. On the basis of this occurrence of *Oldhamia* and the relationships to the overlying Ordovician, the Grand Pitch has been assigned an Early Cambrian (?) age by NEUMAN (1967).

6.2. Palaeogeographical significance

Traditionally, the rocks and fossils of wholly western European aspect in the eastern New England area have been placed east of a hypothetical barrier to explain their absence from other parts of New England. However, some European elements are present in the faunas of the outer detrital belt rocks of Vermont and New York. Thus the 'barrier' was not totally effective. The palaeogeography of the northern Appalachian region shows that the area of central and western Vermont was a deeply subsiding marine basin with volcanic islands providing local sources for lavas and pyroclastics (Fig. 5 and see p. 182). Such volcanic islands would be more of a filter than a barrier. Therefore, it is most probable that a continuous marine environment existed between eastern and western New England. Because the fossiliferous Early Cambrian sediments of Massachusetts immediately overlie Pre-Cambrian basement, and because the oldest Cambrian beds in western New England are within a continuous marine sedimentary sequence, there must have been earlier marine sedimentation in western New England than in eastern New England. The intervening area would therefore include the axis of the Appalachian seaway. Although the evidence from eastern United States alone is not conclusive, the relations of the eastern New England rocks to other sequences of wholly European aspect in Nova Scotia and eastern Newfoundland, and the relations of the western New England sequences to sequences in western Newfoundland suggest that the axis of the Appalachian seaway continued northwards across central Newfoundland. The remarkable relations of sequences east of this seaway to those of central England and of the sequences west of this seaway to those of north-western Scotland strongly support the conclusion that central England and eastern New England were on the same side of the Appalachian seaway during the Cambrian (PALMER, 1969). Whether this seaway was of oceanic width as suggested by WILSON (1966) or whether it occupied an intracratonic geosyncline within the Eurasian–North American Pangea cannot yet be determined from evidence in the Cambrian rocks. There is little doubt, however, that the main Cambrian seaway lay between eastern and western New England and not in the position of the present Atlantic Ocean.

7. Cambrian History of eastern United States

A Cambrian history for eastern United States can describe in reasonable detail only those events affecting the western margin of the Appalachian seaway. Very little is known of events in eastern areas beyond the fact that marine sedimentation commenced in the Early Cambrian and continued at least into the Middle Cambrian. The Early Cambrian limestone in Massachusetts, judging from the comparable limestones in eastern Newfoundland, accumulated in very shallow turbid water in an area of primarily terrigenous sedimentation. During the Middle Cambrian, the eastern areas received only sediments of terrigenous or volcanic origin.

Along the western margin of the Appalachian seaway, a marine transgression was already in progress at the beginning of the Cambrian. The seaway was receiving terrigenous sediments from the craton to the west and volcanic materials from unknown sources to the east. The climax of this transgression is represented by the fine-grained terrigenous sediments of the middle formations of the Chilhowee Group in the central and southern Appalachians (Figs 10 and 13) and by the thin carbonates associated with the finer grained pre-Cheshire rocks of western New England (Fig. 5). The first record of Cambrian invertebrates in the Appalachian seaway came at this time.

A major return to terrigenous sedimentation is marked by the upper quartzites of the Chilhowee Group and the Cheshire Quartzite of the thick New England sequences. This was still marine sedimentation, however, as shown by scattered trilobite and brachiopod remains and abundant *Scolithus* burrows.

This phase was terminated abruptly by the development of a broad carbonate bank, represented by the Shady, Tomstown, and Dunham dolomites, along the entire Appalachian shelf. The shoreward limit of the bank is not precisely known. Its seaward margin was west of the sites of deposition of the Cambrian rocks of the Taconic region and the rocks of the New England metamorphic belt. It may be represented to the south by the Kinzers Formation and the Frederick and Conestoga limestones in south-eastern Pennsylvania and by the rocks of the Austinville area in Virginia (Fig. 15A). Archaeocyathids flourished locally on the bank and rich invertebrate faunas, including most of the phyla known from the Cambrian, existed along its seaward margins.

Towards the end of the Early Cambrian, the carbonate bank was almost completely eliminated by a flood of terrigenous sediments represented by the Rome, Waynesboro, and Monkton formations. Some of these sediments may have been deposited subaerially as suggested by local development of mud cracks and by widespread red colours. Others were certainly marine as indicated by trilobite remains and occurrences of glauconite.

Only in the Milton-Bennington region of Vermont and possibly in the Lancaster region of Pennsylvania is there evidence for continued carbonate sedimentation at this time.

A major transgressive phase began just before the end of the Early Cambrian. At this time, the broad Pre-Cambrian peninsula of northern New Jersey and adjacent areas was submerged and carbonate sedimentation spread into the northern and central Appalachians, where it was to remain dominant for the remainder of the Cambrian. Evidence that the spread of carbonates began before the end of the Early Cambrian is provided by the occurrence of Early Cambrian fossils in the Stissing Dolomite and Leithsville Formation, above the submerged peninsula (Figs 7, 8).

The generally westward spread of carbonates continued in the early Middle Cambrian and reached a temporary climax during the time of the *Glossopleura* Zone as reflected by the Rutledge Limestone of the southern Appalachians (Fig. 13). Through the remainder of Middle Cambrian time and into early Upper Cambrian (Dresbachian) time, the location of the main carbonate bank remained fairly stable. Its seaward margin did not extend as far east as that of the Early Cambrian bank, at least in the central and northern Appalachians, as shown by the presence above the earlier bank deposits of rocks of the marginal facies, represented by the Frederick and Conestoga (?) limestone in Pennsylvania and Maryland and by the formations of the Woods Corners Group in northern Vermont (Figs 5 and 10). In Alabama, at least some of the Late Cambrian rocks assigned to the Conasauga Group seem to represent the outer detrital belt as indicated by the regional relations of their trilobite faunas. Thus, a southern limit to the outer margin of the carbonate bank may be indicated in that area.

From the Middle Cambrian into Dresbachian time, the inner margin of the carbonate bank fluctuated back and forth across an interval of about 120 kilometres in the southern Appalachian region as shown by the intertonguing relationships of the limestones and shales of the Conasauga Group in Tennessee and Virginia. In late Dresbachian time, a particularly widespread regressive phase is represented by the Nolichucky Shale in the southern Appalachians, the lower sandy member of the Gatesburg Formation and the Big Spring Station sandy member of the Conococheague Limestone of the central Appalachians, and possibly by the Danby Formation of the northern Appalachians (Figs 5, 10, and 13). Following this, and throughout the remainder of the Cambrian, the carbonate bank was re-established over the entire Appalachian region and spread westwards across the continental interior. Its seaward margin also expanded slightly eastwards towards the close of the Cambrian as shown by the Grove Limestone above the Frederick Limestone in Maryland, and by the dolomites of the Gorge Formation in northern Vermont.

Throughout the period of development, spread, and fluctuation in dimensions of the carbonate banks, the marine region east of the banks received continued input of non-carbonate sediments including volcanogenic sediment and some extrusive volcanic rocks from a belt of offshelf volcanic islands.

In summary, the Cambrian history of the Appalachian Region records the interactions between a complex of carbonate banks and adjacent sources of terrigenous or volcanogenic sediment along the western part of an Appalachian seaway of uncertain breadth. Very little of significance happened at the end of the Cambrian in this region, and the Late Cambrian sedimentary patterns persist almost unchanged into the Early Ordovician.

Bibliography

BERRY, W. B. N. (1962). Stratigraphy, zonation and age of the Schagticoke, Deepkill, and Normanskill shales, eastern New York. *Bull. Geol. Soc. Am.* **73**, 695.

BIRD, J. M. and RASETTI, FRANCO (1968). Stratigraphic and biostratigraphic importance of the Lower, Middle and Upper Cambrian faunas in the Taconic sequence of eastern New York. *Sp. Paper, Geol. Soc. Am.* **113**.

BUTTS, CHARLES (1926). Geology of Alabama. *Spec. Rept. Geol. Surv. Ala.* **14**.

BUTTS, CHARLES (1940). Geology of the Appalachian Valley in Virginia—Pt. 1, Geologic text and illustrations. *Bull. Geol. Surv. Va.* **52**.

BUTTS, CHARLES (1945). Description of the Hollidaysburg and Huntington quadrangles, Pa. *Atl. folio U.S. Geol. Surv.* **227**.

CADY, W. M. (1945). Stratigraphy and structure of west-central Vermont. *Bull. Geol. Soc. Am.* **56**, 515.

CADY, W. M. (1960). Stratigraphic and geotectonic relationships in northern Vermont and southern Quebec. *Bull. Geol. Soc. Am.* **71**, 531.

CLARK, M. G. and SHAW, A. B. (1968a). Paleontology of northwestern Vermont. XIV. Type section of the Upper Cambrian Gorge formation. *J. Paleont.* **42**, 374.

CLARK, M. G. and SHAW, A. B. (1968b). Paleontology of northwestern Vermont. XV. Trilobites of the Upper Cambrian Gorge formation (lower part of bed 3). *J. Paleont.* **42**, 382.

CLARK, M. G. and SHAW, A. B. (1968c). Paleontology of northwestern Vermont. XVI. Trilobites of the Upper Cambrian Gorge formation (upper bed 3). *J. Paleont.* **42**, 1014.

CLARK, T. H. (1923). New fossils from the vicinity of Boston. *Proc. Boston Soc. Nat. Hist.* **36**, 473.

CLOOS, ERNST (1951). Stratigraphy of the sedimentary rocks *in* The physical features of Washington County. *Rept. Md. Dept. Geol. Mines, Water Res.* **14**, 17.

COOPER, G. A. (1951). New brachiopods from the Lower Cambrian of Virginia. *J. Wash. Acad. Sci.* **41**, 4.

CURRIER, L. W. (1935). Zinc and lead region of southwestern Virginia. *Bull. Geol. Surv. Va.* **43**.

DENNIS, J. G. (1964). The geology of the Enosburg area, Vermont. *Bull. Vt. Geol. Surv.* **23**.

DOLL, C. G., CADY, W. M., THOMPSON, J. B. Jr., and BILLINGS, M. P. (1961). Centennial geologic map of Vermont. *Geol. Surv. Vt.*

DOYLE, R. G., HUSSEY, A. M. II, CHAPMAN, C. A., OSBERG, P. H., PAVLIDES, LOUIS, and WARNER, JEFFREY (1967). Preliminary geologic map of Maine. *Geol. Surv. Me.*

DRAKE, A. A., McLAUGHLIN, D. B., and DAVIS, R. E. (1961). Geology of the Frenchtown quadrangle, New Jersey–Pennsylvania. *U.S. Geol. Surv. Geol. Quad. Map* **GQ-133**.

DRAKE, A. A. (1965). Carbonate rocks of Cambrian and Ordovician age, Northampton and Bucks Counties, eastern Pennsylvania, and Warren and Hunterdon Counties, western New Jersey. *Bull. U.S. Geol. Surv.* **1194-L**.

DURHAM, J. W. (1966). *Camptostroma*, an Early Cambrian supposed scyphozoan, referable to Echinodermata. *J. Paleont.* **40**, 1216.

FAIRBAIRN, H. W., MOORBATH, STEPHEN, RAMO, A. O., PINSON, W. H. Jr., and HURLEY, P. M. (1967). *Earth and Plan. Sci. Lett.* **2**, 321.

GEYER, A. R. and others (1958). Geology of Lebanon quadrangle. *Atl. Geol. Surv. Pa. 4th ser.* **167-C**.

GEYER, A. R., BUCKWALTER, T. V., McLAUGHLIN, D. B., and GRAY, CARLYLE (1963). Geology of the Womelsdorf quadrangle. *Atl. Geol. Surv. Pa. 4th ser.* **177-C**.

GRABAU, A. W. (1900). Paleontology of the Cambrian terranes of the Boston Basin. *Occ. Papers Boston Soc. Nat. Hist.* **4**, 601.

HARRIS, L. D. (1964). Facies relations of the exposed Rome formation and Conasauga group of northeastern Tennessee with equivalent rocks in the subsurface of Kentucky and Virginia. *Profess. Paper U.S. Geol. Surv.* **501-B**, 25.

HERZ, NORMAN (1961). Bedrock geology of the North Adams quadrangle, Massachusetts–Vermont. *U.S. Geol. Surv. Geol. Quad. Map* **GQ-139**.

HOWELL, B. F. (1945). Revision of the Upper Cambrian faunas of New Jersey. *Mem. Geol. Soc. Am.* **12**.

HOWELL, B. F. (1955). Upper Cambrian fossils of Northampton County, Pennsylvania. *Bull. Wagner Free Inst. Sci.* **30**, 45.

HOWELL, B. F. (1957). Upper Cambrian fossils from Bucks County, Pennsylvania. *Bull. Geol. Surv. Pa. 4th ser.* **G-28**.

HOWELL, B. F., ROBERTS, H. B. and WILLARD, BRADFORD (1958). Subdivision and dating of the Cambrian of eastern Pennsylvania. *Bull. Geol. Soc. Am.* **61**, 1366.

JONAS, A. I. and STOSE, C. W. (1930). Geology and mineral resources of the Lancaster quadrangle Pennsylvania. *Atl. Pa. Geol. Surv., 4th Ser.* **168**.

KAUFFMAN, M. E. and CAMPBELL, LYLE (1969). Revised interpretation of the Cambrian Kinzers formation in southeastern Pennsylvania. *Abst. NE sec. Geol. Soc. Am.*, 32.

KINDLE, C. H. and TASCH, PAUL (1948). Lower Cambrian fauna of the Monkton formation of Vermont. *Can. Field Nat.* **62**, 133.

KINDLE, C. H. and WHITTINGTON, H. B. (1958). Stratigraphy of the Cow Head region, western Newfoundland. *Bull. Geol. Soc. Am.* **69**, 315.

KINDLE, C. H. and WHITTINGTON, H. B. (1959). Some stratigraphic problems of the Cow Head area, in western Newfoundland. *Trans. N.Y. Acad. Sci.* **22**, 7.

KING, P. B. (1950). Geology of the Elkton area, Virginia. *Profess. Paper U.S. Geol. Surv.* **230**.

KING, P. B. and FERGUSON, H. W. (1960). Geology of northeasternmost Tennessee. *Profess. Paper U.S. Geol. Surv.* **311**.

KNOPF, E. B. (1962). Stratigraphy and structure of the Stissing Mountain area, Duchess County, New York. *Stanford Univ. Pub. Geol. Sci.* **7**, 1.

LAURENCE, R. A. and PALMER A. R. (1963). Age of the Murray shale and Hesse quartzite on Chilhowee Mountain, Blount County, Tennessee. *Profess. Paper U.S. Geol. Surv.* **475-C**, 53.

LOCHMAN, CHRISTINA (1946). Two upper Cambrian (Trempealeau) trilobites from Duchess County, New York. *Am. J. Sci.* **244**, 547.

LOCHMAN, CHRISTINA (1956). Stratigraphy, paleontology and paleogeography of the Elliptocephala asaphoides strata in Cambridge and Hoosick quadrangles, New York. *Bull. Geol. Soc. Am.* **67**, 1331.

LOCHAMN-BALK, CHRISTINA and WILSON, J. L. (1958). Cambrian biostratigraphy in North America. *J. Paleont.* **32**, 312.

McLAUGHLIN, D. B. and ROOT, S. I. (1966). Comparative tectonics and stratigraphy of the Cumberland and Lebanon valleys. *Geol. Surv. Pa. Guidebook, 31st. Ann. Field Conf. Pa. Geol.*

McLAUGHLIN, D. B. (1967). Structure and stratigraphy of the limestones and dolomites of Dauphin County, Pennsylvania. *Bull. Geol. Surv. Pa. 4th Ser.* **G-44**.

McCARTNEY, W. D. (1967). Whitbourne Map-area, Newfoundland. *Mem. Geol. Surv. Can.* **341**.

MEISLER, H. and BECHER, A. R. (1968). Carbonate rocks of Cambrian and Ordovician age in the Lancaster quadrangle, Pennsylvania. *Bull. U.S. Geol. Surv.* **1254-G**.

NEUMAN, R. B. and NELSON, W. H. (1965). Geology of the western Great Smoky Mountains, Tennessee. *Profess. Paper U.S. Geol. Surv.* **349-D**.

NEUMAN, R. B. (1967). Bedrock geology of the Shin Pond and Stacyville quadrangles, Penobscot County, Maine. *Profess. Paper U.S. Geol. Surv.* **524-I**.

OSBERG, P. H. (1965). Structural Geology of the Knowlton–Richmond area, Quebec. *Bull. Geol. Soc. Am.* **76**, 223.

PALMER, A. R. (1962). *Glyptagnostus* and associated trilobites in the United States. *Profess. Paper U.S. Geol. Surv.* **374-F**.

PALMER, A. R. (1969). Cambrian trilobite distributions in North America and their bearing on the Cambrian paleogeography of Newfoundland, in *Mem. American Assoc. Pet. Geol.*, **12**, 139

RASETTI, FRANCO (1944). Upper Cambrian trilobites from the Lévis conglomerate. *J. Paleont.* **18**, 229.

RASETTI, FRANCO (1959). Trempealeauan trilobites from the Conococheague, Frederick, and Grove limestones of the central Appalachians. *J. Paleont.* **33**, 375.

RASETTI, FRANCO (1961). Dresbachian and Franconian trilobites of the Conococheague and Frederick limestones of the central Appalachians. *J. Paleont.* **35**, 104.

RASETTI, FRANCO (1965a). Middle Cambrian trilobites of the Pleasant Hill formation in central Pennsylvania. *J. Paleont.* **39**, 1007.

RASETTI, FRANCO (1965b). Upper Cambrian trilobite faunas of northeastern Tennessee. *Smiths. Misc. Colln.* **148**, no. 5.

RASETTI, FRANCO (1966). New Lower Cambrian trilobite faunule from the Taconic sequence of New York. *Smiths. Misc. Colln.* **148**, no. 9.

RASETTI, FRANCO (1967). Lower and Middle Cambrian trilobite faunas from the Taconic sequence of New York. *Smiths. Misc. Colln.* **152**, no. 4.

RASETTI, FRANCO and THEOKRITOFF, GEORGE (1967). Lower Cambrian agnostid trilobites of North America. *J. Paleont.* **41**, 189.

RAYMOND, P. E. (1924). New Upper Cambrian and Lower Ordovician trilobites from Vermont. *Proc. Boston Soc. Nat. Hist.* **37**, 389.

RAYMOND, P. E. (1937). Upper Cambrian and Lower Ordovician Trilobita and Ostracoda from Vermont. *Bull. Geol. Soc. Am.* **48**, 1079.

RESSER, C. E. (1938). Cambrian system (Restricted) of the southern Appalachians. *Sp. Paper Geol. Soc. Am.* **15**.

RESSER, C. E. and HOWELL, B. F. (1938). Lower Cambrian *Olenellus* zone of the Appalachians. *Bull. Geol. Soc. Am.* **49**, 195.

ROBISON, R. A. (1964). Late Middle Cambrian faunas from western Utah. *J. Paleont.* **38**, 510.

RODGERS, JOHN (1953). Geologic map of East Tennessee with explanatory text. *Bull. Tenn. Dept. Conserv. Div. Geol.* **58**.

RODGERS, JOHN (1956a). The known Cambrian deposits of the southern and central Appalachian Mountains, *in* Rodgers J., ed., El Sistema Cambrico, su paleogeographia y el problema de su base—pt. 2. *Symp. XX Int. Geol. Cong.*, 353.

RODGERS, JOHN (1956b). The clastic sequence basal to the Cambrian system in the central and southern Appalachians, *in* Rodgers, J., ed., El Sistema Cambrico, su paleogeografia y el problema de su base—Pt. 2. *Symp. XX Int. Geol. Cong.*, 385.

RODGERS, JOHN (1968). The eastern edge of the North American continent during the Cambrian and early Ordovician, *in* ZEN, E-AN and others (1968). Studies of Appalachian Geology, Northern and Maritime. *Wiley Interscience*, 141.

RUSHTON, A. W. A. (1966). The Cambrian trilobites from the Purley Shales of Warwickshire. *Paleontogr. Soc. Mon.* **120**, pub. 511.

SANDO, W. J. (1957). Beekmantown group (Lower Ordovician) of Maryland. *Mem. Geol. Soc. Am.* **68**.

SHALER, N. S. and FOERSTE, A. F. (1888). Preliminary description of North Attleboro fossils, *in* SHALER, N. S., On the geology of the Cambrian district of Bristol County, Massachusetts. *Havard Coll. Mus. Comp. Zool. Bull.* **16**, 13.

SHAW, A. B. (1950). A revision of several Early Cambrian trilobites from eastern Massachusetts. *J. Paleont.* **24**, 577.

SHAW, A. B. (1951). The Paleontology of northwestern Vermont, I. New Late Cambrian trilobites. *J. Paleont.* **25**, 97.

SHAW, A. B. (1952). The Paleontology of northwestern Vermont, II. Fauna of the Upper Cambrian Rockledge conglomerate near St. Albans. *J. Paleont.* **26**, 458.

SHAW, A. B. (1953). The Paleontology of northwestern Vermont, III. Miscellaneous Cambrian fossils. *J. Paleont.* **27**, 137.

SHAW, A. B. (1955). The Paleontology of northwestern Vermont, V. The Lower Cambrian fauna. *J. Paleont.* **29**, 775.

SHAW, A. B. (1957). Paleontology of northwestern Vermont, VI. The early Middle Cambrian fauna. *J. Paleont.*, **31**, 785.

SHAW, A. B. (1958). Stratigraphy and structure of the St. Albans area, northwestern Vermont. *Bull. Geol. Soc. Am.* **69**, 519.

SHAW A. B. (1961). Cambrian of southeastern and northwestern New England. *in* El Sistema Cambrico, su paleogeografia y el problema de su base—Pt. 3, *Symp. XX, Int. Geol. Cong.*, 433.

SHAW, A. B. (1962a). Paleontology of northwestern Vermont, VIII. Fauna of the Hungerford slate. *J. Paleont.* **36**, 314.

SHAW, A. B. (1962b). Paleontology of northwestern Vermont, IX. Fauna of the Monkton quartzite. *J. Paleont.* **36**, 322.

SHAW, A. B. (1966a). Paleontology of northwestern Vermont, X. Fossils from the (Cambrian) Skeels Corners formation. *J. Paleont.* **40**, 269.

SHAW, A. B. (1966b). Paleontology of northwestern Vermont, XI. Fossils from the Middle Cambrian St. Albans shale. *J. Paleont.* **40**, 843.

SHIMER, H. W. (1907a). An almost complete specimen of *Strenuella strenua* (Billings). *Am. J. Sci.* **23**, 199.

SHIMER, H. W. (1907b). A lower-middle Cambrian transition fauna from Braintree, Mass. *Am. J. Sci.* **24**, 176.

STONE, S. W. and DENNIS, J. G. (1964). The geology of the Milton quadrangle, Vermont. *Bull. Vt. Geol. Surv.* **26**.

STOSE, G. W. and JONAS, A. I. (1938). A southeastern limestone facies of Lower Cambrian dolomite in Wythe and Carroll Counties, Va. *Bull. Geol. Surv. Va.* **51-A**.

STOSE, G. W. and JONAS, A. I. (1939). Geology and mineral resources of York County, Pa. *Bull. Pa. Geol. Surv. 4th ser.* **C-67**.

TASCH, PAUL (1951). Fauna and paleoecology of the Upper Cambrian Warrior formation of central Pennsylvania. *J. Paleont.* **25**, 275.

THEOKRITOFF, GEORGE (1964). Taconic stratigraphy in northern Washington County, New York. *Bull. Geol. Soc. Am.* **75**, 171.

THEOKRITOFF, GEORGE (1968). Cambrian biogeography and biostratigraphy in New England, in ZEN, E-AN and others (1968). Studies of Appalachian Geology, Northern and Maritime. *Wiley Interscience*, 9.

WALCOTT, C. D. (1884). On the Cambrian faunas of North America; preliminary studies. *Bull. U.S. Geol. Surv.* **10**.

WALCOTT, C. D. (1886). Second contribution to the studies on the Cambrian faunas of North America. *Bull. U.S. Geol. Surv.* **30**.

WALCOTT, C. D. (1891). Correlation papers—Cambrian. *Bull. U.S. Geol. Surv.* **81**.

WALCOTT, C. D. (1910). *Olenellus* and other genera of the Mesonacidae. *Smiths. Misc. Colln.* **53**, 233.

WALCOTT, C. D. (1912). Cambrian brachiopoda. *Mon. U.S. Geol. Surv.* **51**.

WALCOTT, C. D. (1914). *Dikelocephalus* and other genera of the Dikelocephalinae. *Smiths. Misc. Colln.* **57**, 345.

WALCOTT, C. D. (1916a). Cambrian trilobites. Cambrian Geology and Paleontology III. *Smiths. Misc. Colln.* **64**, 157.

WALCOTT, C. D. (1916b). Cambrian trilobites. Cambrian Geology and Paleontology III. *Smiths. Misc. Colln.* **64**, 303.

WHITAKER, J. C. (1955a). Geology of Catoctin Mountain, Maryland and Virginia. *Bull. Geol. Soc. Am.* **66**, 435.

WHITAKER, J. C. (1955b). Direction of current flow in some Lower Cambrian clastics in Maryland. *Bull. Geol. Soc. Am.* **66**, 763.

WILLARD, BRADFORD (1955). Cambrian contacts in eastern Pennsylvanina. *Bull. Geol. Soc. Am.* **66**, 819.

WILLARD, BRADFORD (1961). Stratigraphy of the Cambrian sedimentary rocks of eastern Pennsylvania. *Bull. Geol. Soc. Am.* **72**, 1765.

WILSON, J. L. (1951). Franconian trilobites of the central Appalachians. *J. Paleont.* **25**, 617.

WILSON, J. L. (1952). Upper Cambrian stratigraphy in the central Appalachians. *Bull. Geol. Soc. Am.* **63**, 275.

WILSON, J. T. (1966). Did the Atlantic close and then re-open? *Nature* **211**, 676.

ZEN, E-AN (1961). Stratigaphy and structure at the north end of the Taconic Range in west-central Vermont. *Bull. Geol. Soc. Am.* **72**, 293.

ZEN, E-AN, and HARTSHORN, J. H. (1966). Geologic map of the Bashbish Falls quadrangle, Massachusetts, Connecticut, and New York. *U.S. Geol. Surv. Geol. Quad. Map* **GQ-507**.

ZEN, E-AN (1967). Time and space relationships of the Taconic allochthon and autochthon. *Spec. Paper, Geol. Soc. Am.* **97**.

THE CAMBRIAN OF CANADA AND ALASKA

F. K. North

Carleton University, Ottawa

Contents

1. General characteristics of the Cambrian in Canada and Alaska

1.1. Distribution of Cambrian rocks

The Cambrian rocks of North America are marginal to and roughly circumferential around the Canadian shield, which for this purpose includes Greenland. Mainland Canada displays the eastern and western sides of this Cambrian girdle, but the junctions across the north and south lie outside the Canadian mainland and outside the scope of this part.

In eastern Canada, the Cambrian is well developed in Canadian Appalachia. By far the best development is in Newfoundland, but the system also crops out extensively along the south side of the St. Lawrence River in Quebec. Rocks proven to be Cambrian in age are exposed in only very small areas in Nova Scotia and New Brunswick, though the areas are classic. Much larger areas expose thick successions of unfossiliferous rocks which may, in part at least, be of Cambrian age.

In the undisturbed area on the foreland of Appalachia, Cambrian rocks are trivially represented in outcrop in Labrador, Quebec, and Ontario. In the subsurface, a slightly larger area of Cambrian sedimentary rocks underlies parts of the Ontario peninsula and the Ottawa–St. Lawrence lowlands.

In western Canada, Cambrian rocks are magnificently exposed in the Rocky Mountains, and only a little less well in the ranges west of the Rocky Mountain trench. Thick sections are also preserved in several of the shorter ranges in the Northwest Territories and the Yukon, though these do not reach the spectacular scenic level of the more southerly ranges. Thinner but still important sections of Cambrian sedimentary rocks are known to occupy an area in excess of half a million square kilometres in the subsurface of Saskatchewan, Alberta, and the Northwest Territories.

1.2. Distribution of facies

At any one time during the Cambrian Period, the depositional régime involved two quite distinct facies, and from four to six sub-facies:

A. *Cratonic or foreland facies*

   i. An inner detrital sub-facies, deposited relatively close to shore, and consisting of light-coloured sands and silts, commonly glauconitic or feldspathic; some muds, and thin beds of limestone or intraformational limestone conglomerate. Over a wide area in the Northwest Territories thick evaporites are developed. Sections in this sub-facies exhibit numerous diastems.

ii. An offshore belt of relatively pure carbonate sediments, with trilobite faunas typical of the carbonate sub-facies. This sub-facies results from pulsatory subsidence in very shallow water, causing rapid cyclic deposition of a wide variety of calcareous rock types. The expansion, contraction, and lateral migration of this sub-facies are responsible for most of the variety in the Cambrian succession.

iii. A fanglomeratic sub-facies of very coarse boulder and sharpstone conglomerate, developed locally adjacent to fault scarps active during the Cambrian on the foreland.

B. *Extracratonic or basinal facies*

i. An outer detrital sub-facies, 'geosynclinal' but not eugeosynclinal, typified by thick, monotonous, dark-coloured muds and silts, commonly laminated, with sands of grey-wacke type probably introduced from outside the basin. Prominently associated with the dark-coloured sediments are reddish or greenish muds and silts. These represent deposition in deeper water, periodically interrupted by the incoming of spectacular limestone conglomerates, presumably from the outer edge of the carbonate sub-facies (Aii). In most areas of sub-facies B(i), the Cambrian rocks, themselves poorly fossiliferous to quite barren, pass upwards into the graptolitic shale facies of the Ordovician.

ii. A condensed, inter-basinal sub-facies consisting of black muds with red or green muds and silts, suggesting an episodically euxinic environment of monotonous stability. Characteristic of this sub-facies are thin calcareous or concretionary bands crowded with fossils. Also characteristic, at least in the early Middle Cambrian, are minor volcanic rocks, chiefly tuffs. The sub-facies invariably passes up into Lower Ordovician rocks in graptolitic shale facies.

iii. A truly eugeosynclinal sub-facies, including submarine volcanic rocks. This sub-facies has not with certainty been identified because, in the absence of fossils, most of its undoubted representatives could plausibly be Ordovician rather than Cambrian (2.3.2). The rare volcanic rocks in proven Cambrian sequences are, with very minor exceptions, associated with non-ophiolitic sediments such as silty dolomites. However, there are clear signs of approach to the sub-facies, in undoubtedly Cambrian rocks, going away from the craton on both sides (in south-eastern Quebec, east-central British Columbia, and the Yukon).

The first facies (A) has been variously called 'cratonic', 'foreland', 'platform', 'stable shelf', 'carbonate terrane', or 'carbonate–quartzite facies'. In this part it will be referred to as the *cratonic facies*. The second facies (B) has been called 'extracratonic', 'basinal', 'geosynclinal', 'miogeosynclinal', 'euxinic', 'outer detrital', 'clastic terrane', or 'grey-wacke–shale facies'. It will be referred to as the *basinal facies*, commonly with modification in recognition of its distinctive sub-facies. The inner and outer detrital sub-facies (Ai and Bi) and the carbonate sub-facies (Aii) are essentially those described by PALMER (1960) for the western United States.

1.3. Contrasts between eastern and western Canada in Cambrian time

The contrasts between the eastern and western developments in Canada's Cambrian System are remarkable. In the west, the Cambrian attains a maximum, present, continuous width of at least 1,150 kilometres, across the strike, in the subsurface and at the surface. The cratonic facies occupies the greater part of this width, the basinal facies being met only close to the western margin and then being almost entirely in the outer detrital sub-facies (Bi). Thus carbonates are very prominent in the western Cambrian, and true shales are relatively restricted. In the east, the known width of the Cambrian development, across the strike, is only about 700 kilometres, and the continuous presence of Cambrian rocks across this width, at present, is very doubtful. The cratonic facies is here extremely narrow, in places either wholly hidden or non-existent. The outcrop areas are dominated by the basinal facies, and the condensed sub-facies (Bii) is superbly developed.

Though the outer margins of both Cambrian troughs are associated with thick Proterozoic sedimentary rocks, there is no uniform association of either of the principal facies with crystalline or non-crystalline Pre-Cambrian. Though the condensed basinal sub-facies (Bii), for example, and the eugeosynclinal sub-facies (Biii) if it exists, lie wholly outside the area in which crystalline basement immediately underlies the Cambrian, the cratonic facies may lie on either crystalline or non-crystalline Pre-Cambrian. At no place in the western Cordillera is the Lower Cambrian known to overlap the Proterozoic sedimentary rocks and transgress over the crystalline foreland. Much of the great width of the western cratonic facies, therefore, lacks the Lower Cambrian. In the east, on the other hand, the Lower Cambrian lapped farther on to the craton than any succeeding Cambrian sediments until the late Cambrian onset of the Ordovician transgression. The cratonic facies in the east, consequently, includes more of the Lower Cambrian than of anything else.

The very unequal distributions, both of time-units and of facies, on the two sides of the Canadian shield, are consequences of the shape and configuration of the shield itself (Fig. 1). The two sides differed widely in their Middle and Late Proterozoic histories. In the east, the late Proterozoic–early Palaeozoic trough developed essentially parallel to an earlier Grenville trough, which had itself not been completely cratonized until about 950 m.y. (STOCKWELL, 1965). The Grenville margin of the craton was abrupt and unstable, apparently sharply downbent and with an exceedingly narrow shelf. Structures inherited from earlier, Pre-Cambrian times, or arising within the new trough during Palaeozoic sedimentation, were parallel to the trough itself, as shown by all Palaeozoic trends in Canadian Appalachia (NEALE and others, 1961). Lateral variations in thickness and lithology were abrupt, and there is no certainty that the Cambrian was deposited continuously throughout the basin. Indeed, it seems likely that it was not. Furthermore, there is no simple, upward progression, through the Cambrian succession, from a cratonic facies to a more and more basinal facies. Those parts of the depositional area which were cratonic at the beginning of Cambrian time remained cratonic throughout the period, and those of basinal character had that character from the beginning.

In the west, the basement is very much older. The Cordilleran geosyncline lay across the outer edges of three shield provinces, the Churchill, Slave, and Bear provinces. The Canadian Rocky Mountains, which expose more Cambrian rocks than any other part of North America, were literally built of the sediments of a Palaeozoic shelf associated with the Churchill province. This province, and the Bear province north of the Arctic Circle, were cratonized by the Hudsonian orogeny, about 1,750 m.y. (STOCKWELL, 1965). The Hudsonian structures were dominated by south-westerly trends, which were overprinted almost at right angles, probably before 1,200 m.y., by the trend of the new Purcell–Windermere trough. These later Pre-Cambrian rocks are little metamorphosed except where they were invaded by *Mesozoic* plutons.

FIG. 1. Distribution of Pre-Cambrian provinces underlying the Cambrian in Canada and Alaska.

The Slave province, which lies between the Churchill and Bear provinces, is still older, being cratonized about 2,500 m.y. Both it and the Bear province, whilst traversed by important faults of south-westerly trend, exhibit fundamental structures almost meridional in trend (STOCKWELL, 1965). It is unlikely that the Slave province any longer underlies any Cambrian rocks (Fig. 1), but the influence on Palaeozoic sedimentation of the meridionally-structured Bear province was quite different from that of the transversely-structured Churchill province. Cambrian and Ordovician rocks, and Lower Cambrian rocks in particular, extend much farther east where they rest on Bear basement than they do to the south; so do the orogenic structures which were imposed upon sediments of those ages at a much later date.

In consequence, the principal inherited or basement-controlled structures in the Palaeozoic Cordilleran trough were transverse to the trough, not parallel to it. Whereas the Cambrian of eastern Canada can be readily understood in terms of facies belts essentially parallel to the depositional and tectonic strike, later telescoped by thrusting, that of western Canada (and Alaska) has to be considered in a succession of blocks controlled by Pre-Cambrian structures effectively perpendicular to both the depositional strike and the strike of the structures later imposed upon the region (Figs 1 and 9).

The margin of the Churchill province of the shield must have tilted downwards, towards the west, in a series of pulsations, acquiring a very wide, gently sloping shelf which permitted a fairly uniform and gradual westward increase in the thickness of its sedimentary content. Lateral variations in thickness or facies take place gradually over wide areas, except at the edges of transverse basement structures. An upward progression from the cratonic facies to one more and more basinal, through the Cambrian succession, is typical, especially in the most westerly belts.

The Bear sector must have assumed a more depressed structural position, and have acquired discrete basins within itself which led to much greater diversity of rock type and stratigraphical thickness than is found in any other part of the Cambrian realm in Canada.

1.4. Faunal contrasts

Long before these lithic and distributional contrasts between the western and eastern Cambrian régimes had become apparent, a startling difference had been recognized between the faunas of the two regions. The Cambrian beds of south-eastern Newfoundland, Cape Breton Island, and southern New Brunswick (and the Boston area of Massachusetts) contained trilobite faunas exactly comparable with those of north-western Europe (especially in species of *Paradoxides*). In contrast, the Cambrian of western Newfoundland, Quebec, and Vermont had few species or genera in common with those faunas, if cosmopolitan agnostids are excluded. Instead, it contained an abundant fauna comparable with that of the rest of North America (including the Cordillera).

Central Newfoundland, central New Brunswick, and central Maine were not known to contain any Cambrian rocks at all. They had not until 1967 been known to contain any fossiliferous Cambrian rocks in the conventional sense. However, small inliers in central Maine had yielded the trace fossil *Oldhamia*, believed to be an index for the Cambrian (NEUMAN, 1962). As the oldest exposed rocks over most of the rest of this central belt are Ordovician, and commonly very early Ordovician, the probability that Cambrian rocks underlie them was obviously strong (2.5.3). It was only a matter of time before they were identified (2.5.6). Nonetheless, prior to the discovery of the *Oldhamia* beds it was possible to postulate a 'New Brunswick geanticline' separating the two faunally distinct realms: a St. Lawrence geosyncline to the north-west, and an Acadian geosyncline to the south-east (Fig. 2). There was a good deal to support this postulate, besides the remarkable faunal contrast:

a. The Cambrian rocks of the inner (St. Lawrence) geosyncline were deposited on a crystalline (Grenville) basement. Those of the outer (Acadian) geosyncline were deposited on a younger, non-crystalline basement.

b. The facies contrast is very obvious. Carbonate rocks are very rare in the Acadian belt, very prominent in the St. Lawrence belt.

c. The Lower Cambrian appears to be greatly dominant over the Middle or Upper Cambrian in the St. Lawrence belt, not so in the Acadian.

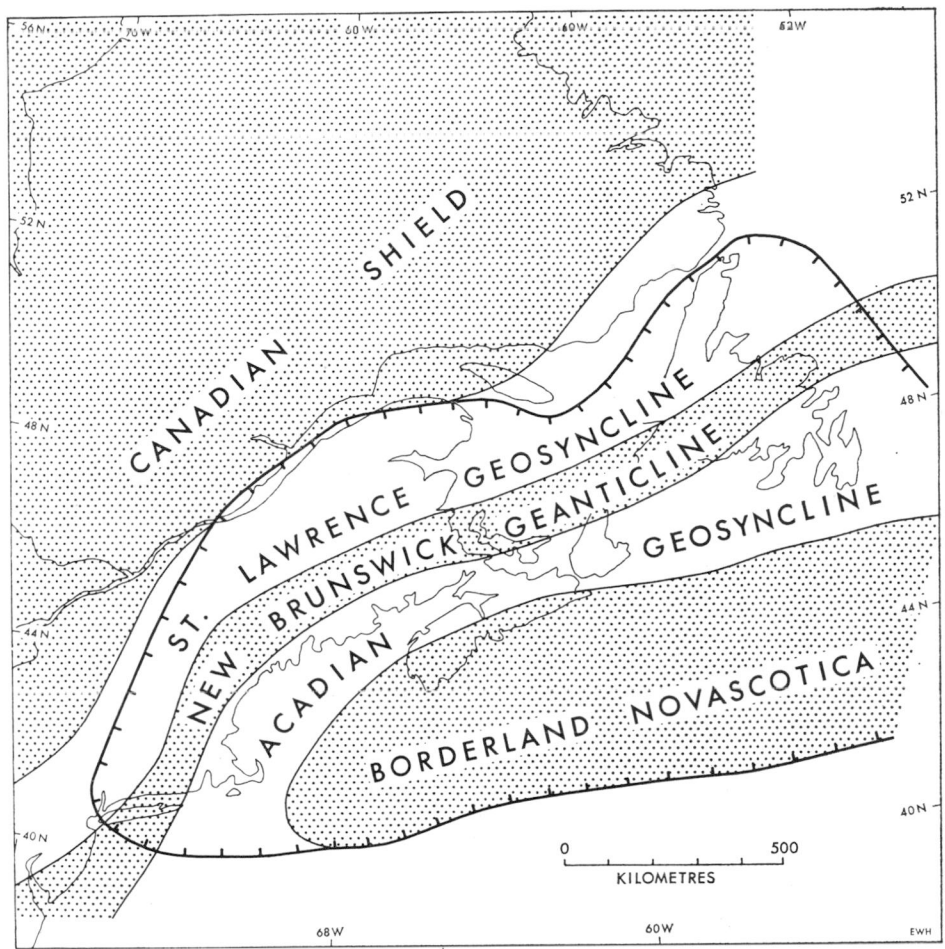

Fig. 2. Cambrian palaeogeography in eastern Canada, according to Schuchert and Dunbar (1934).

d. The structural environment for the Cambrian rocks differs between the two belts. With one relatively trivial exception (2.2.2), all the proven Cambrian rocks of the 'Atlantic' province occur in narrow synclinal remnants within older rocks. It is by no means certain that they were all part of a single continuous depositional area throughout Cambrian time. From Gaspé to Vermont, the Cambrian appears in upfolded or up-faulted belts or patches, surrounded by Ordovician rocks.

Nonetheless, there are serious objections to the idea of decisive structural separation. LOCHMAN-BALK and WILSON (1958) considered the Atlantic and Pacific 'realms' to represent the geosynclinal and cratonic régimes, respectively, and both to extend circumferentially around the Canadian shield. The 'Atlantic' realm is not now apparent in the west because the cratonic environment occupied a very great width there, and the deposits of the bordering geosyncline are largely buried beneath younger rocks, metamorphosed, or obliterated by Mesozoic intrusions. Representatives of the Atlantic fauna do occur among the Cambrian faunas of British Columbia and Texas. The principal components of the Atlantic fauna, however, are literally unrepresented elsewhere in North America. This is not true of the 'Pacific' faunas of the carbonate terrane in the

St. Lawrence belt. The boulders in the Lévis conglomerate and in the Cow Head breccias (2.3.2; 2.3.4) contain distinctly mixed faunas; several of the agnostids, in particular, have Atlantic affinities. The same is true of bedded limestones near Quebec City (2.3.2).

The structural and physical separation of the two realms by an emergent land-ridge was still called for until very recently by some palaeontologists (HUTCHINSON, 1962), but the idea has been discarded by most geologists. WILLIAMS (1964) concisely presented the view that Newfoundland exposes a complete cross-section of a two-sided, symmetrical geosyncline, and that the central part of that geosyncline lacks much of the Cambrian succession because it was deep enough, and far enough from its Pre-Cambrian borderlands, to remain starved of sediment until Ordovician time.

Quite apart from the observation of geological phenomena favourable or unfavourable to the idea of physical separation between the two faunal realms, a philosophical caveat might be entered against it. The New Brunswick geanticline hypothesis requires narrow seaways, with flanking and separating landmasses, receiving only fine-grained dark muds with no coarser marginal clastics except a transgressive basal sandstone. This might be acceptable if there were literally no such clastic rocks available to fill the role, and if the depositional margins of the shale basins could be unambiguously identified on the ground. These conditions are not met in the Atlantic provinces. In Newfoundland, Nova Scotia, New Brunswick, and Quebec, there are thick successions of clastic rocks known to be early Ordovician *or older*. For some of these successions (2.5), a Cambrian age is the most obvious and logical assignment; but the rocks are not fossiliferous.

The reconstruction of the geology of an ancient period not marked by any great igneous activity is dependent on an interpretation of its stratigraphy, and hence of its palaeontology. Much of our traditional view of such matters has come from the interpretations of palaeontologists which, understandably, were designed to explain the distribution of fossils, not the distribution of rocks. Unfossiliferous rocks were commonly misinterpreted, more commonly simply left out of account.

The Cambrian of Canada has been peculiarly susceptible to this treatment. Both the Atlantic provinces and the Western Cordillera have been known since the late nineteenth century to contain very thick successions of sedimentary rocks of unquestionable Pre-Cambrian age. Unfossiliferous Cambrian rocks were therefore all too readily relegated to the Pre-Cambrian—much more readily than unfossiliferous rocks of a younger period would have been. The temptation must have been greatly increased by the circumstance that both regions happened also to contain Cambrian beds so marvellously fossiliferous that they became objects of pilgrimage for palaeontologists from all over the world: the sections in and near the city of Saint John, New Brunswick, and on Manuels River and Smith Sound in Newfoundland; the Burgess and *Ogygopsis* shales on the two sides of the Kicking Horse Pass in British Columbia; and in their different but equally compelling way, the boulders in the Cow Head breccia and the Quebec conglomerates.

The Palaeozoic geosynclines came to be looked upon as linear troughs in which trilobites were deposited; rocks containing no trilobites could not be Cambrian. As recently as 1960, geological maps of Canada showed virtually no Cambrian west of the Rocky Mountain trench, and the areas allotted to the system in the Atlantic provinces were so small that only their palaeontological prestige justified their being coloured separately.

It is now acknowledged that there is enormously more Cambrian exposed at the surface in Canada than was recognized twenty years ago. It will be demonstrated in following chapters that there may yet be much more than is widely acknowledged even now.

1.5. Lower and upper boundaries of the Cambrian in Canada

1.5.1. Lower boundary

The problem of the Pre-Cambrian/Cambrian boundary in general is too big a one for extended discussion by me. The essential aspects of it have been discussed many times, as at a symposium of the International Geological Congress in Mexico (HUTCHINSON, 1956; OKULITCH, 1956), and, most recently, by ROZANOV (1967). It is reviewed by HARLAND elsewhere in the present Series.

In the few areas of eastern Canada (there are none in western Canada or Alaska) in which Lower Cambrian strata rest on crystalline basement, there is of course no difficulty in deciding where to draw their base, even though we know this base is not at the same time-level in all places. Where the Pre-Cambrian rocks are not crystalline, the boundary is conventionally drawn at the first major lithic change below the lowest occurrence of Lower Cambrian fossils.

In eastern Canada, this change is commonly no more than a formational boundary; in few places is it a demonstrable unconformity, and throughout the Acadian belt proper the rocks below the change yield isotopic ages which could as easily be Cambrian as Pre-Cambrian (2.6). Nonetheless, the only active arguments about assignment to the Cambrian or the Pre-Cambrian are those concerning the thin Random Formation in south-eastern Newfoundland (2.2.1) and the much thicker Morrison River Formation on Cape Breton Island (2.2.3).

In western Canada, the quartzites below the lowest Cambrian fossils have conventionally been included in the Lower Cambrian even though a lithic break separates them from the fossiliferous formation. This is because the quartzites rest unconformably on undoubted Pre-Cambrian, Windermere rocks in the Park Ranges of the Rocky Mountains (Fig. 9). In the westernmost outcrop belt, west of the Rocky Mountain trench, this convenient unconformity is not present, or at least is not obvious; the Lower Cambrian is an integral continuation of a thick succession of Proterozoic sedimentary rocks. The quartzite, or its equivalent formation, underlying the lowest Lower Cambrian fossils is still accepted as belonging to the Lower Cambrian, even though we have good evidence that at least some of it is older than the Lower Cambrian quartzite in the Park Ranges (OKULITCH, 1949). It is unlikely that this assignment needs any challenge because, except in the areas close to the 49th parallel, the lowest fossils found are archaeocyathids and olenellids belonging rather high in the Lower Cambrian (OKULITCH and GREGGS, 1958).

1.5.2. Upper boundary

The upper boundary of the Cambrian in Canada is another matter altogether. It is a geological byword that, if the Lower Palaeozoic systems had been defined in eastern North America instead of in Europe, the Lower Ordovician of American usage (and not just the Tremadoc) would have been part of the system below it and not of the system above it. The widespread Knox unconformity follows well-known Lower Ordovician carbonate rocks, and is itself followed by the rocks of the great Middle Ordovician transgression, over the entire central and eastern parts of the continent. The Cambrian and the Lower Ordovician are combined in what is now called the Sauk Sequence in North America (SLOSS, 1963). The Tremadoc Series, which represents only the lower part of the sub-unconformity Ordovician, therefore belongs with the Cambrian here just as it does in Wales, in the Soviet Union, and in the Baltic area (CALVERT, 1964). In the usage of many American geologists of the first half of this century, in fact, an Ozarkian Series was interpolated between the Cambrian and the Ordovician, its upper part being approximately equivalent to the Tremadoc.

Though the verdict of the palaeontologists that the Tremadoc is Ordovician is now in little dispute, European geologists may not be aware that the palaeontological evidence so decisive in Europe is much less impressive in North America. The first asaphid trilobite, *Asaphellus*, occurs here as in Britain, but neither the first trinucleid (*Orometopus*) nor the first cheirurid (*Anacheirurus*) is significant among Canadian faunas (the Canadian being the earliest Ordovician series in North American terminology). Thus North American acceptance of an Ordovician Tremadoc is heavily dependent on the graptolites.

In the 'Atlantic' province, the condensed basinal sub-facies (Bii) of the Cambrian continues uninterruptedly into the Tremadoc, and the succession does not extend beyond the Arenig. As there is no lithic change whatever, the Cambro–Ordovician boundary in that region is wholly palaeontological. In the cratonic realm in western Newfoundland (2.4.1), the Cambrian and Ordovician are continuous in carbonate facies. In the subsurface of Saskatchewan and Alberta, the pre-Devonian erosion surface is at or below the Knox unconformity, so that the Tremadoc (and possibly part of the Arenig as well) is again mappable only with the Cambrian (3.3.5).

In the Cordillera, the Cambro–Ordovician boundary is placed below the zone containing *Symphysurina*. Formationally it is not a difficult boundary to draw where the maximum interfingering of carbonate and shaly half-cycles is developed (3.3.3). It is drawn between the dark, massive, algal carbonate rocks of the Mistaya Formation and the putty-coloured shales of the overlying Chushina Formation. Outside the carbonate facies, however, the boundary occurs within a single group, or even within a single formation. In the south, the McKay and Goodsir Groups extend from the *Elvinia* Zone of the Franconian to Zone D of the Lower Ordovician (ROSS, 1951). In the north, the Kechika Group probably extends from Upper Cambrian to Middle Ordovician (or even to Upper Ordovician), though there are wide areas over which it is all Ordovician and rests unconformably on rocks older than Late Cambrian. Throughout the Mackenzie and Franklin Mountains, similarly, the boundary falls within a single formation. In the complex of ranges in the Yukon, the base of the Ordovician is more distinct, as Upper Cambrian strata are commonly missing, but on the Yukon-Alaska boundary there seems again to be no clear separation, the relatively thin Hillard Limestone containing fossils ranging in age from Middle Cambrian, or older, to very early Ordovician (BRABB 1967).

1.6. Previous Investigations

Cambrian rocks in eastern Canada have been under investigation for about 125 years. Those in the west were first described more than 100 years ago. The total volume of contribution is now so large that only those investigators whose work is of specific significance to students of the Cambrian System can be mentioned in a work like the present one.

1.6.1. Eastern Canada

The earliest professional work on the Cambrian rocks of eastern Canada was that of SIR WILLIAM LOGAN, who named the Quebec Group and recognized its early Palaeozoic age. Wonderfully detailed studies of the Cambrian strata of the other eastern provinces, and of Newfoundland, were made by G. F. MATTHEW, whose long list of contributions was referred to by HUTCHINSON (1952b, p. 4, pp. 16–17; 1956, pp. 312–3).

Towards the end of the 19th century, C. D. WALCOTT began his long association with Canada's Cambrian rocks in these same areas, establishing the significance of the *Olenellus* fauna as MATTHEW had established that of *Protolenus*.

Between the wars, the Cambrian geology of Newfoundland (indeed nearly all the geology of Newfoundland), and that of coastal New Brunswick, became the especial interest of a distinguished group of geologists from Princeton University, notably of B. F. HOWELL. Recent summaries of the Cambrian geology of the Acadian provinces are those of R. D. HUTCHINSON (1956, 1962).

In Quebec, there are no highly fossiliferous black shales carrying Cambrian trilobites and the early work of LOGAN led to no systematic study of Cambrian rocks comparable with that of MATTHEW in the Atlantic provinces. R. W. ELLS had forecast as early as 1889 that the Quebec Group must include Cambrian strata, and the presence of Cambrian fossils in the limestone boulders within the group was recognized early. It was not until the 1930's, however, that the areal extent of (admittedly poorly fossiliferous) Cambrian strata came to be recognized (CLARK, 1934; JONES, 1935; KINDLE, 1942). The only man, other than our own contemporaries, whose work was specifically directed towards Cambrian problems, was FRANCO RASETTI.

Since 1923 reconstructions of the Cambrian palaeogeography of eastern Canada have been profoundly influenced by that proposed by CHARLES SCHUCHERT (1923). It was the wholesale faunal difference between the Cambrian rocks of far eastern Canada and the Cambrian rocks of all the rest of North America that led SCHUCHERT to the postulation of four distinct physiographic provinces in the narrow eastern region during Cambrian time (Fig. 2).

1.6.2. Western Canada and Alaska

The nineteenth century exploration of the Cambrian rocks of the Western Cordillera is peculiarly associated with the names of four men. Only one of them worked specifically on Cambrian rocks and fossils, and this reflects another obvious contrast offered by the system between Canadian Appalachia and the Canadian Cordillera. It is almost impossible to work extensively in the Rocky Mountains without becoming involved with Cambrian rocks even if one's principal interests lie in some other aspect of their geology.

JAMES HECTOR, geologist in the Palliser expedition in the middle of the century, explored the southern ranges in which the Cambrian dominates both the stratigraphy and the scenery. His most lasting contribution was to the topographic nomenclature of the region. In each rib of the Cambrian part of the range, HECTOR gave to the highest peak the name of one of his colleagues in KNOX's anatomy classes in Edinburgh. Unwittingly, he was also responsible for one of the most famous single names in Cordilleran Canada when he suffered serious injury from a kick by his horse.

Early government exploration was totally dominated by two of Canada's greatest pioneer geologists. G. M. DAWSON's position as the father of western Canadian geology is unchallenged. R. G. McCONNELL, who began as DAWSON's assistant, made the first formal stratigraphical subdivision of the Cambrian rocks in the Rocky Mountains, and constructed the first good cross-sections through them.

For the forty years, 1887 to 1927, which followed the work of these two men, the Rocky Mountain Cambrian became effectively the private preserve of C. D. WALCOTT, discoverer of the Burgess Shale fauna. He established formational subdivisions in several parts of the range, and many of the formations he erected (especially those for the Middle Cambrian) are still in use. During WALCOTT's lifetime, the only Canadian geologist working specifically on the Cordilleran Cambrian was LANCASTER BURLING (1914, 1923). The Geological Survey's work on the system was otherwise incidental to regional mapping, by J. A. ALLAN, C. S. EVANS, H. M. A. RICE, and J. F. WALKER. The famous boundary survey by R. A. DALY, in the early years of this century, chanced to be directed along what must be the only strip of territory over twenty degrees of latitude in which no rocks could at that time be certainly assigned to the Cambrian (though many rocks which were not Cambrian were so assigned). In the far north, inevitably, even less was done, the only descriptions of proven Cambrian rocks being made by D. D. CAIRNES and by M. Y. WILLIAMS, for Canada, and BURLING (again, though in this case his work was not published by him) and J. B. MERTIE JR., for Alaska.

During and immediately after the second world war, studies in the mountains remained restricted to a handful of palaeontologists working essentially on their own account— C. F. DEISS, V. J. OKULITCH, and FRANCO RASETTI. A special salute is due to DR. RASETTI, formerly Professor of Physics at the Johns Hopkins University in Baltimore and a remarkable amateur palaeontologist, the only living individual to have worked extensively on Cambrian rocks and faunas in both eastern and western Canada.

In the early 1950's study of the Rocky Mountains was revived as a result of the operations of oil-company geologists, originally seeking surface data on the Devonian System following the discovery of oil in Alberta's subsurface reefs of that age. Enormous strides have been made since the late 1950's and will continue to be made for years ahead, with the provision of modern methods of travel and transport to a brilliant group of mountaineer-geologists in the federal and provincial geological surveys.

1.7. Acknowledgements

The preparation of this summary would have been impossible without the advice of those geologists who are active in work involving Cambrian rocks and fossils in Canada, as well as some working in the United States and in Alaska. This is especially the case for the chapters dealing with eastern Canada; I myself have done no original work whatever on the Cambrian rocks of that part of the country.

For the generous provision of unpublished information, I am indebted to J. D. AITKEN, S. L. BLUSSON, R. B. CAMPBELL, W. H. FRITZ, H. GABRIELSE, L. H. GREEN, G. B. LEECH, E. W. MOUNTJOY, P. H. OSBERG, A. R. PALMER, W. H. POOLE, G. O. RAASCH, PIERRE ST.-JULIEN, P. S. SIMONY, P. J. STREET, G. C. TAYLOR, and H. VAN HEES. For critical reading of parts of the text, and for many helpful comments on it, I am especially indebted to Drs. FRITZ, GABRIELSE, LEECH, and POOLE. The directors of Chevron Standard Company, Calgary, kindly gave me permission to extract information from the company's files, to the great benefit of the sections dealing with the Northwest Territories. For the drafting, by Mr. EDWARD HEARN, I am grateful to him and to the University of Ottawa.

1.8. Works referred to in Chapter 1

BRABB, EARL E. (1967). Stratigraphy of the Cambrian and Ordovician rocks of east-central Alaska· U.S. Geol. Surv. Prof. Paper, **559**-A.

BURLING, L. D. (1914). Early Cambrian stratigraphy in the North American Cordillera, with discussion of *Albertella* and related faunas. *Geol. Surv. Canada Mus. Bull.* **2**, 93.

BURLING L. D. (1923). Cambro–Ordovician section near Mount Robson, British Columbia. *Geol. Soc. America Bull.* **34**, 721.

CALVERT, WARREN L. (1964). Pre-Trenton sedimentation and dolomitization, Cincinnati arch province: theoretical considerations. *Amer. Assoc. Petroleum Geol. Bull.* **48**, 166.

CLARK, T. H. (1934). Structure and stratigraphy of southern Quebec. *Geol. Soc. America Bull.* **45**, 1.

HOWELL, B. F. (chairman) (1944). Correlation of the Cambrian formations of North America. *Geol. Soc. America Bull.* **55**, 993.

HUTCHINSON, R. D. (1952b). The stratigraphy and trilobite faunas of the Cambrian sedimentary rocks of Cape Breton Island, Nova Scotia. *Geol. Surv. Canada Mem.* **263**.

HUTCHINSON, R. D. (1956). Cambrian stratigraphy, correlation, and paleogeography of eastern Canada. *El Sistema Cambrico, Symposium, XX Int. Geol. Cong., Mexico*, 289.

HUTCHINSON, R. D. (1962). Cambrian stratigraphy and trilobite faunas of southeastern Newfoundland. *Geol. Surv. Canada Bull.* **88**.

JONES, I. W. (1935). Dartmouth River map-area, Gaspé Peninsula. *Quebec Bureau of Mines Ann. Rept. 1934*, 3.

KINDLE, C. H. (1942). A Lower (?) Cambrian fauna from eastern Gaspé, Quebec. *Amer. Jour. Sci.* **240**, 633.

LOCHMAN-BALK, CHRISTINA and WILSON, J. L. (1958). Cambrian biostratigraphy in North America. *Jour. Paleont.* **32**, 312.

NEALE, E. R. W., BÉLAND, J., POTTER, R. R. and POOLE, W. H. (1961). A preliminary tectonic map of the Canadian Appalachian region based on age of folding. *Can. Inst. Min. Met. Bull.* **54**, 687.

NEUMAN, ROBERT B. (1962). The Grand Pitch Formation: new name for the Grand Falls Formation (Cambrian?) in northeastern Maine. *Amer. Jour. Sci.* **260**, 794.

OKULITCH, V. J. (1949). Geology of part of the Selkirk Mountains in the vicinity of the main line of the Canadian Pacific Railway, British Columbia. *Geol. Surv. Canada Bull.* **14**.

OKULITCH, V. J. (1956). The Lower Cambrian of western Canada and Alaska. *El Sistema Cambrico, Symposium, XX Int. Geol. Cong., Mexico*, 701.

OKULITCH, V. J., and GREGGS, R. G. (1958). Archaeocyathid localities in Washington, British Columbia, and the Yukon Territory. *Jour. Paleont.* **32**, 617.

PALMER, A. R. (1960). Some aspects of the Upper Cambrian stratigraphy of White Pine County, Nevada, and vicinity. *Intermountain Assoc. Petroleum Geol., E. Nevada Geol. Soc., Joint Field Conf. Guidebook*, 43.

ROSS, R. J. JR. (1951). Stratigraphy of the Garden City Formation in northeastern Utah and its trilobite faunas. *Yale. Univ. Peabody Mus. Nat. Hist. Bull.* **6**.

ROZANOV, A. Yu. (1967). The Cambrian lower boundary problem. *Geol. Mag.* **104**, 415.

SCHUCHERT, CHARLES (1923). Sites and nature of the North American geosynclines. *Geol. Soc. America Bull.* **34**, 151.

SLOSS, L. L. (1963). Sequences in the cratonic interior of North America. *Geol. Soc. America Bull.* **74**, 93.

STOCKWELL, C. H. (1965). Structural trends in Canadian Shield. *Amer. Assoc. Petroleum Geol. Bull.* **49**, 887.

WILLIAMS, H. (1964). The Appalachians in northeastern Newfoundland—a two-sided symmetrical system. *Amer. Jour. Sci.* **262**, 1137.

2. The Cambrian in eastern Canada

2.1. Introduction

Cambrian rocks crop out widely in Quebec, south and south-east of the St. Lawrence River. They are also well exposed on the main eastern and western peninsulas of Newfoundland, and much less well on Cape Breton Island and in southern New Brunswick. Except for small areas in the western limits of outcrop in Quebec and Newfoundland, these regions of Cambrian rock are parts of Canadian Appalachia, an orogenic belt which was deformed in at least two major Palaeozoic orogenies. The relatively trivial areas of Cambrian outcrop lying outside the Appalachian region form part of the covered Canadian shield. To them may be added even smaller outcrops in extreme southern Ontario, and a sizeable area of subsurface Cambrian to the south of them again. The areas concerned represent all the Cambrian sub-facies known in Canada except the fanglomeratic sub-facies (Aiii).

The true cratonic facies of the Cambrian is found in place in eastern Canada in only three small areas:

a. On the two sides of the Long Range of western Newfoundland.

b. In a tiny area at the eastern end of the Gaspé peninsula in Quebec, so small as to be insignificant except for its structural implications.

c. In the immediate hanging wall of the Champlain (Logan) thrust fault in southern-most Quebec, and westwards from there, across the St. Lawrence valley and into the subsurface of the Ontario peninsula.

This distribution is shown in Figs 5 and 7.

From about the latitude of Montreal, northwards and north-eastwards all the way to the Gulf of St. Lawrence, Cambrian rocks of the basinal facies have been transferred tectonically across the foreland and have reached the north-western margin of the original depositional area. Whether a foreland terrane lies continuously beneath them is not known (2.4.2).

Where the present relationships between the cratonic and basinal facies are deci-pherable, they are of three completely contrasting types: relationship by interfingering, as on the west side of the Sutton Mountains in southernmost Quebec (2.4.3); the incorpo-ration of masses of cratonic rocks in limestone conglomerates with basinal rocks as matrix (2.3.2; 2.3.4); and superposition by allochthony (2.3.4). In one extraordinary area on the west coast of Newfoundland, the second and third of these relationships seem to be combined.

All the proven Cambrian rocks of the true cratonic facies (Ai and Aii), and of the true basinal facies (Bi), are found within the so-called St. Lawrence or inner geosyncline (Fig. 2). So many of the fossiliferous representatives in this belt belong to the Lower Cambrian that SCHUCHERT (1923, p. 178) considered that no Middle Cambrian strata were ever deposited in that part of the trough. Though this is now known to be untrue, it remains true that Lower Cambrian fossils are more easily found in it than are fossils of either Middle or Late Cambrian age.

Along the eastern margin of Canada, in south coastal New Brunswick, Cape Breton Island, and south-eastern Newfoundland, neither the true cratonic nor the true basinal facies is developed, at least in fossiliferous form. Instead, the condensed basinal sub-facies (Bii) occurs, dominated by black mudstones with red and green mudstones below, and almost without either carbonates or coarse clastic rocks. These three lithological associations, the condensed, the true basinal, and the cratonic, will be described separ-ately.

2.2. The Condensed Basinal Sub-facies in the Acadian Province

This sub-facies occurs in individually small areas in south-eastern Newfoundland, on Cape Breton Island, and in southern New Brunswick, as well as near Boston in Massa-chusetts. It yields the fauna of the so-called Atlantic realm, dominated by trilobites and brachiopods and completely without archaeocyathids. The Lower and Middle Cambrian are particularly well developed; between the Middle and Upper Cambrian at least one faunal zone is everywhere missing; but the top of the Cambrian, where it is exposed, passes up conformably and without change of lithology into the Tremadoc.

2.2.1. Avalon and Burin Peninsulas, and adjoining coasts, south-eastern Newfoundland

Cambrian rocks occur in a number of narrow, tightly compressed synclinal remnants within a region otherwise nearly all Pre-Cambrian. The beds are for the most part sharply folded, with prominent fracture cleavage. They display an obvious association with the great strike bays which give this part of Newfoundland its peculiar shape; in fact, the axis of maximum Cambrian deposition, from Placentia Bay to Trinity Bay, crosses the narrow Isthmus of Avalon (Fig. 3).

The present distribution of Cambrian rocks suggests that the sea entered the Acadian geosyncline from the north-east (HUTCHINSON, 1962), but thicknesses indicate that it may have entered from both the north-east and the south-west. It left a succession of Cambrian sediments between 300 and a possible 1,200 metres thick, and dominantly shaly. There is a characteristic paucity of sandstones or other coarse clastic rocks, except for a basal conglomerate which is in most places very thin. On the other hand, thin beds of dark limestone, or bands of limestone concretions, are common, and fossils are much commoner in the calcareous members than in the shales (except in the Upper Cambrian Elliott Cove Formation). The sequence has been variously attributed to an 'extracratonic' facies (LOCHMAN-BALK and WILSON, 1958), a 'marine shelf' facies (JENNESS, 1963), and a 'miogeosynclinal' facies (HUTCHINSON, 1962). Though most modern workers seem agreed that the succession represents deposition in very shallow water, neither lithic nor faunal evidence for this belief is apparent in the sections preserved, except in their very lowest and very highest parts.

Two outcrop sections have long been famous among Cambrian palaeontologists: that along the north-west shore of Smith Sound and on adjacent Random Island, on the west side of Trinity Bay, first described by MATTHEW (1899); and that along Manuels River, which flows into Conception Bay 25 kilometres west of St. John's. This second section was described by HOWELL (1925). The most recent stratigraphical descriptions of these two sections are those by JENNESS (1963) and ROSE (1952), respectively.

The huge thickness of late Proterozoic sedimentary and volcanic rocks, which characterizes the Avalon Peninsula, ended with red arkoses, conglomerates and siltstones of the Musgravetown and Hodgewater Groups. Overlying this succession, and still older ones, with varying degrees of discordance is the controversial Random Formation. It consists of one to four whitish quartzite members, with interbedded arkose and greenish siltstone. Giant ripple marks are common. The grains in the quartzite vary from well sorted and rounded (MCCARTNEY, 1967) to angular (JENNESS, 1963). The formation is from zero to 150 metres thick, commonly less than 30 metres. Some of its coloured interbeds resemble the overlying Cambrian sediments, but not all. JENNESS (1963) considered the Random beds to be more like those of the underlying Musgravetown Group than those of the overlying Lower Cambrian; he further regarded the formation's lower contact as conformable and its upper contact as disconformable.

HUTCHINSON (1962) maintained that the pre-Random rocks had been folded prior to the deposition of the Random and the Cambrian, but that the upper surface of the Random had become indurated, and channelled by erosion, before Lower Cambrian sedimentation began. For HUTCHINSON, therefore, as for JENNESS, the Random Formation is Pre-Cambrian.

MCCARTNEY (1958, 1967) considered that, on a regional scale, no angular unconformity exists either below or above the Random. A slightly irregular surface of the formation is in most places disconformably overlain by a thin, red, rounded-pebble conglomerate. This suggests the diachronic shoreline deposit of a transgressing sea. The

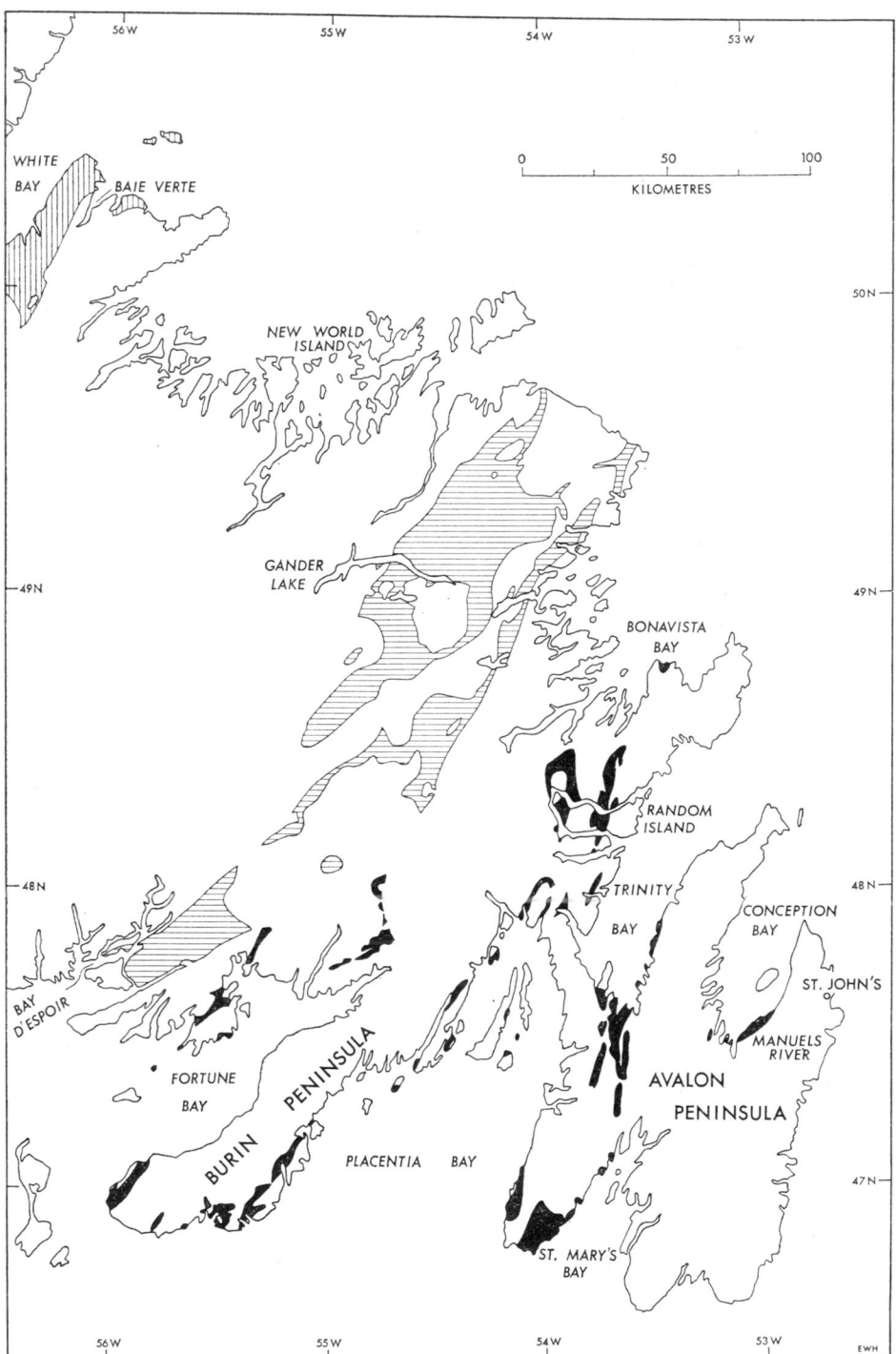

Fig. 3. Outcrop areas of Cambrian and possible Cambrian rocks in eastern Newfoundland. Proven Cambrian in solid black; possible Cambrian (Chapter 2.5) in horizontal pattern; for vertical pattern, see Figure 7.

Random Formation underlies all the Cambrian of south-eastern Newfoundland except that bordering Conception Bay; south-east of the bay, the normally overlying Bonavista and Smith Point Formations are also missing. The Random is therefore more logically mapped with the Palaeozoic than with the Proterozoic, and the present writer agrees with CHRISTIE (1950) and MCCARTNEY (1967) in regarding it as the basal formation of the Cambrian succession (Table 1).

The Lower Cambrian proper was apparently concentrated along the central axis of the peninsulas (HUTCHINSON, 1962). There is no evidence that it was ever deposited along the east shore of the Avalon Peninsula, and it is absent north of Fortune Bay (2.2.2). Elsewhere it consists of bright red and green mudstones or slates, with thin beds of pink or green nodular limestone. At the base is a granule sandstone consisting of pea-sized quartz grains in a white, calcareous, sandstone matrix. This subtle basal grit bed bears to the Random Formation the controversial relation just described. Around the shores of Conception Bay (Fig. 3), the Random Formation is missing, apparently because of the presence of a horst-like platform underlain by the late Pre-Cambrian Holyrood granite (2.6). In this area, a striking conglomerate is developed at the base of the Cambrian. The Pre-Cambrian surface must have presented small-scale angular inequalities like those on a shoreline eroded in rocks of varying hardness. Rectangular blocks of bedrock, whether of the granite or of its neighbouring sedimentary rocks, lay about the surface, especially in hollows and channelways. These blocks were incorporated in the basal deposits of the advancing Early Cambrian sea, forming an irregular layer of block-breccia which passes upwards into a reddish grit. The famous Manuels River section is unusual in having a basal conglomerate which is much thicker, as well as being a true conglomerate. It rests directly on the rhyolite which forms the roof of the Holyrood granite, and the rounded cobbles and pebbles are more of rhyolite than of granite.

Where fully developed, the Lower Cambrian is more than 300 metres thick and is separated into lower (Bonavista) and upper (Brigus) formations by an intervening pink limestone which forms the best marker horizon in the succession. The Smith Point Limestone, 10 metres thick or less, is massive, with some thin interbeds of red shale and pink limestone nodules; it contains prominent algal structures. The limestone seems to have developed about the same time as the arrival of trilobites into the area. Fossils are rare in the underlying Bonavista Formation, and they do not include trilobites; the enigmatic genus *Coleoloides*, originally considered a primitive hyolithid, is found in thin limestone interbeds, with brachiopods and gastropods. The top of the limestone itself, and the overlying Brigus Formation, carry *Callavia*; the Brigus also has *Strenuella* in its lower part and a probable *Protolenus* above. The maximum thickness of the Bonavista Formation is about 200 metres; the Brigus exceeds this thickness slightly at the head of Trinity Bay.

The Middle Cambrian, Upper Cambrian, and Lower Ordovician were divided by HOWELL (1925) and others into formations which are really only faunal zones. Except for the coloured strata representing the base of the Middle Cambrian, all the remaining Cambrian beds, and the Tremadoc, are grey to black mudstones (locally slates) with thin bands of grey to black limestone.

The Chamberlain's Brook Formation, normally about 100 metres thick, comprises greenish shales or slates with interbedded red shales (or slates), and minor pink or grey limestone. The principal fossil is *Paradoxides bennetti* SALTER, representing the Acadian equivalent of the *P. oelandicus* (*P. groomi*) Zone of Britain and southern Sweden. The formation rests without discernible disconformity on the Lower Cambrian, with a promi-

	Zone	Avalon–Burin Peninsulas	North Shore Fortune Bay
Upper Cambrian	Peltura	Elliott Cove	Salmonier Cove
Upper Cambrian	Leptoplastus	Elliott Cove	Salmonier Cove
Upper Cambrian	Parabolina spinulosa	Elliott Cove	Salmonier Cove
Upper Cambrian	Olenus	Elliott Cove	Salmonier Cove
Upper Cambrian	Agnostus pisiformis	Elliott Cove	Salmonier Cove
Middle Cambrian	Lejopyge		
Middle Cambrian	Paradoxides forchhammeri	Manuels River	
Middle Cambrian	Paradoxides davidis	Manuels River	Young's Cove
Middle Cambrian	Paradoxides hicksi	Manuels River	Young's Cove
Middle Cambrian	Paradoxides bennetti	Chamberlains Brook	Young's Cove
Lower Cambrian	Protolenus	Brigus	///
Lower Cambrian	Callavia	Smith Point	///
Lower Cambrian	pre-Callavia	Bonavista Random	Blue Pinion

TABLE 1

Correlation of Cambrian Formations, South-eastern Newfoundland.
(after HUTCHINSON, 1956).

nent manganiferous band at its base. Best developed around the heads of Conception and Trinity Bays (Fig. 3), the manganiferous band consists of beds of nodular black or brown slate and psilomelane (from manganese carbonate), interbedded with green slate. The ore is low grade, syngenetic, associated with barite, and quite unmetamorphosed (MCCARTNEY, 1967). In all essential chemical and stratigraphical respects, the manganiferous interval is comparable with that which in North Wales occurs between the Barmouth Grits, of Solva age, and the underlying Rhinog Grits (MOHR and ALLEN, 1965).

Conformably above the Chamberlain's Brook Formation is the dark-coloured Manuels River Formation, its maximum thickness being about 130 metres but its common thickness no more than 30 metres. Much the most fossiliferous formation in the succession, it contains the zones of *Paradoxides hicksi* and *P. davidis*, and, in places, other trilobites characteristic of the zone of *P. forchhammeri*, though that species itself has not been reported. Along the north-south axis between Trinity Bay and St. Mary's Bay (Fig. 3), the *P. davidis* Zone is replaced by two flows of amygdaloidal, andesitic pillow-lava and breccia (the Chapel Arm Member of MCCARTNEY, 1967). Laterally, the flows give way to a tuffaceous facies within the *P. davidis* Zone. No other flows are known in the Cambrian rocks of the Avalon Peninsula. Similar flows are known in the Cambrian of Cape Breton Island (2.2.3), and of southernmost New Brunswick (2.2.4.), but they appear to be older than the Chapel Arm Member. Probably related to the Chapel Arm Member is a group of small necks of dark green gabbro (the Spread Eagle Gabbro), which intrude rocks up to the lower part of the Chamberlain's Brook Formation.

The entire Upper Cambrian is included in the Elliott Cove Formation, which reaches a maximum thickness of nearly 200 metres. Dark shales with pyrite nodules become progressively more silty and micaceous upwards; features characteristic of very shallow-water deposition (mud cracks, ripple marks, cross bedding, cone-in-cone concretions of grey limestone) continue upwards into the Lower Ordovician Clarenville Formation, which is inseparable from the Elliott Cove except by fossils. For this reason, JENNESS (1963) included all the dark shaly beds of the Clarenville, the Elliott Cove, and the Manuels River in a single group, the Harcourt Group; the conformably underlying green and red beds, down to the top of the Random quartzite, he called the Adeyton Group.

The Elliott Cove Formation is less fossiliferous than the beds below, but it has yielded representatives of four Upper Cambrian zones—those of *Agnostus pisiformis*, *Olenus*, *Parabolina spinulosa*, and *Peltura* (HUTCHINSON, 1962). On palaeontological grounds, HUTCHINSON drew a mild disconformity at the base of the formation.

2.2.2. Area north of Fortune Bay, Newfoundland

A much disturbed Cambrian section appears in a number of small areas north of Fortune Bay, on the south coast of Newfoundland (Fig. 3). It differs from the outcrops on the Avalon Peninsula not only in being conspicuously more sandy, but in being surrounded by younger rocks, not by older rocks (HUTCHINSON, 1962; ANDERSON, 1965).

Resting with probable disconformity on typical red and green clastic rocks of late Proterozoic age is a thick (300 metres) band of white, cross-bedded sandstone and conglomerate, the Blue Pinion Formation. It has not yielded any fossils, but it occupies the stratigraphical position of the Random Formation, and has the same lithology (though much greater thickness).

Overlying this formation, however, is a succession of green to grey, micaceous shale and siltstone, metamorphosed in places to dark slate and quartzite but otherwise containing Middle Cambrian trilobites (*Paradoxides eteminicus* MATTHEW, *P. bennetti* SALTER, and others). The three Lower Cambrian formations of the Avalon Peninsula have therefore wedged out westwards and been overlapped by the basal Middle Cambrian (Table 1). The thickness of this formation, the Young's Cove, is hard to estimate, but it may be about 200 metres.

No contact is seen between the Young's Cove Formation and the younger Salmonier Cove Formation, but it is probably disconformable. The Salmonier Cove consists of soft, thinly-bedded black shale containing black, fossiliferous limestone concretions. Overlain unconformably by volcanic rocks which are probably Ordovician, the formation is only about 60 metres thick. This thickness, however, is sufficient to accommodate, apparently, the whole of the Upper Cambrian, because the trilobites in the concretions include species of *Agnostus*, *Olenus*, *Peltura*, and *Ctenopyge*.

Fifty kilometres to the north-east, a much thicker clastic section is caught up as thermally metamorphosed roof pendants in a large Devonian batholith. Arkosic quartzite and greywacke-conglomerate, with some basalt and argillite (now hornfels), about 600 metres in total thickness, are underlain by a greenish slate of nearly equal thickness (BRADLEY, 1962). No lower beds are seen, but the slate has yielded *Protopeltura?solitaria* (WESTERGARD); the probable age of this trilobite is very late Cambrian.

2.2.3. Cape Breton Island

Cambrian rocks crop out on Cape Breton Island in three narrow strips (Fig. 4). The largest extends south-westwards from Mira Bay, on the north-east corner of the island, for 40 kilometres along the Mira River valley. Two very small strips lie along the southeast side of St. Andrew Channel of Bras d'Or Lake, south-west of the coal town of Sydney. The general structure is a broad syncline striking north-eastwards and broken by cross faults. The Cambrian rocks in the largest, easterly belt are very similar to those on the Avalon Peninsula; the sections in the others are lithically different and much less complete. For reference purposes, the areas will be called Cape Breton East and Cape Breton West respectively. The intervals represented are compared in Table 2.

Proven Lower Cambrian, poorly represented, is known only in the eastern outcrop belt, but there it provides an example of apparently continuous accumulation across the Pre-Cambrian/Cambrian boundary. The Fourchu Group, of volcanic and sedimentary rocks acknowledged to be of late Pre-Cambrian age, is overlain by the Morrison River Formation without any discernible discordance, though there is an abrupt change in lithology. The Morrison River Formation consists of reddish, coarsely clastic sediments which become progressively more fine-grained and micaceous towards the top, and greyish instead of red. The formation exceeds 600 metres in maximum thickness, but in most outcrops it is less than 300 metres thick. At its top is an interval, about 25 metres in average thickness, of whitish quartzite with some interbeds of red quartzite and sandy, micaceous shale. This member is so like the Random Formation of south-eastern Newfoundland, and bears such a similar relation to the Lower Cambrian shales above, that it might be looked upon as the same formation, whatever diachronicity has done to its age with respect to the Random Formation proper. HUTCHINSON (1952b) did in fact regard the Morrison River Formation as the lowest member of the Cambrian succession on Cape Breton Island, though he also regarded the Random Formation in

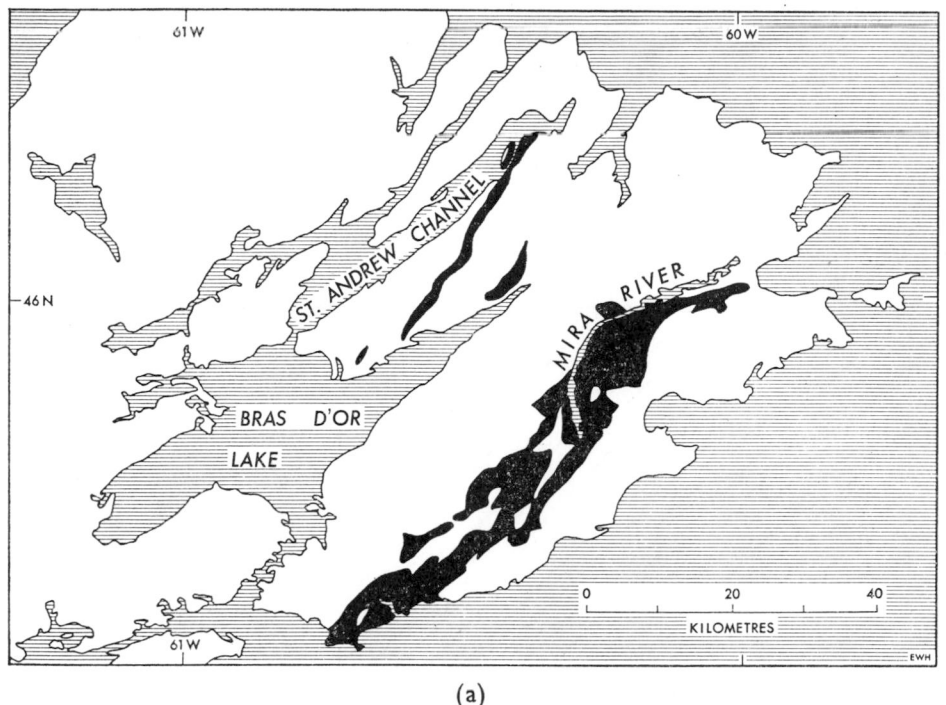

(a)

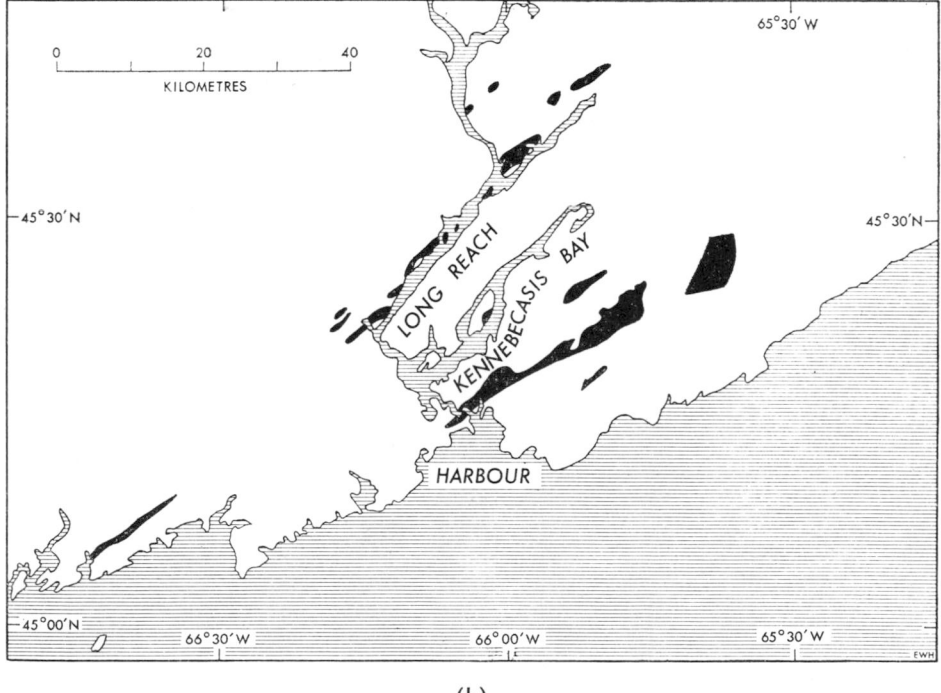

(b)

FIG. 4. Distribution of Cambrian rocks (a) on Cape Breton Island and (b) in the vicinity of Saint John, New Brunswick.

240 F. K. NORTH

	Zone	Saint John New Brunswick	Cape Breton West	Cape Breton East
Upper Cambrian	Peltura	Narrows	MacNeil	MacNeil
	Leptoplastus			
	Parabolina spinulosa			
	Olenus	Black Shale Brook		
	Agnostus pisiformis	Agnostus Cove		
Middle Cambrian	Lejopyge			
	Paradoxides forchhammeri			MacLean Brook
	Paradoxides davidis	Hastings Cove		
	Paradoxides hicksi	Porter Road	MacMullin	Trout Brook
	Paradoxides bennetti	Fossil Brook	Bourinot Group (Gregwa, Dugald, Eskasoni)	
Lower Cambrian	Protolenus	Hanford Brook / Glen Falls	?	Canoe Brook / MacCodrum
	Callavia or pre-Callavia	Ratcliffe Brook		Morrison River

TABLE 2

Correlation of Cambrian Formations, New Brunswick and Nova Scotia (after Hutchinson, 1956).

Newfoundland as Pre-Cambrian. The relationships were considered in detail by WEEKS (1954).

The overlying MacCodrum and Canoe Brook Formations are the Cape Breton equivalent, in nearly all respects, of the Brigus Formation of the Avalon Peninsula— green micaceous shale and shaly sandstone, grading upwards into reddish and mottled micaceous claystones with a few metres of fine red sandstone. The two formations combined are about 300 metres thick, and their lower part carries the *Strenuella* fauna.

As in south-eastern Newfoundland, but here at a lower stratigraphical level, the green and red shaly beds give way abruptly, upwards, to soft, grey to black shales with thin beds of quartzose sandstone. The Trout Brook Formation, about 150 metres thick, contains three of the European *Paradoxides* zones; those of *P. groomi* LAPWORTH (represented here by *P. eteminicus* MATTHEW), *P. hicksi* SALTER, and *P. davidis* SALTER. Above this level, represented by continuous deposition of dark muds in south-eastern Newfoundland, the Cape Breton East section reverts upwards, through a gradational zone marked by increasing sand content, to the greenish, micaceous and quartzose silt- stone elsewhere characteristic of the Lower Cambrian. This formation, the MacLean Brook, 300 metres thick, contains the *Paradoxides forchhammeri* fauna. Overlying it, with an abrupt change in lithology and after the apparent absence of three fossil zones, is the MacNeil Formation, about 300 metres of dark grey to black shales with a few concretions and irregular beds of black, crystalline limestone. It contains representatives of the three youngest Cambrian zones, those of *Parabolina spinulosa*, *Leptoplastus*, and *Peltura*.

The Cape Breton West sections, centred on McLeod Brook and Indian River, are much less complete (Table 2). The Lower Cambrian has not been identified, at least as fossiliferous sedimentary rock. Instead, a succession of amygdaloidal lava flows (in- cluding spilites) and pyroclastic rocks, with interbeds of greywacke and shale, was laid down on top of the Pre-Cambrian George River limestone. This volcanic succession, called the Bourinot Group, is divisible into three formations (Table 2) because of the presence of a tongue of maroon and green tuffaceous shales and siltstones with inter- bedded lavas. The shales and siltstones contain enormous numbers of small inarticulate brachiopods, exhaustively described by MATTHEW (HUTCHINSON, 1952b), and the trilo- bite *Andrarina linnarssoni* (BRÖGGER), indicative of the lowest of the Middle Cambrian *Paradoxides* zones.

The succeeding MacMullin Formation consists of about 300 metres of grey to green, micaceous sandstone and siltstone, with a basal conglomerate. Though it looks some- what like the MacLean Brook Formation of Cape Breton East, and is about the same thickness, the MacMullin Formation is nonetheless older, and its counterpart in Cape Breton East lies within the black shale of the Trout Brook Formation, underlain and overlain by green formations. Between the westernmost outcrop belt, in which the Mac- Mullin Formation is seen, and the easternmost, with the black shale facies, is a narrow zone in which the equivalent beds appear to be a red pebbly arkose and sandstone grading upwards into red and green mudstones. These beds, the Kelvin Glen Group, have yielded only a few brachiopods, but these are of Middle Cambrian age. In the MacMullin Formation, on the other hand, the fauna includes *Paradoxides abenacus*, *Holasaphus centropyge*, and other trilobites discovered and described by MATTHEW. They represent the *Paradoxides hicksi* Zone of south-eastern Newfoundland and the *P. tessini* Zone of Sweden.

The MacMullin Formation is succeeded by a major palaeontological gap. The Mac-Neil Formation of Cape Breton East, the only formation from that area to extend into the Cape Breton West area less than 15 kilometres distant, overlaps the MacMullin Formation but appears to be represented only by its upper beds, in black limestone facies. The only faunal zone identified is the *Peltura* Zone, represented by that genus and by *Lotagnostus trisectus* (SALTER), *Ctenopyge pecten* (SALTER), and *Sphaerophthalmus alatus* (BOECK). Where the volcanic build-up represented by the Bourinot Group was at its maximum, the faunal gap is even greater; the MacNeil Formation there rests directly on the volcanic rocks. As on the Avalon Peninsula and in the Saint John area of New Brunswick, the uppermost Cambrian of Cape Breton West passes up conformably into the Tremadoc, and the boundary between the Cambrian and the Ordovician is drawn palaeontologically.

2.2.4. Saint John area, New Brunswick

The Cambrian beds in the city of Saint John and its environs must have been as carefully picked over as any Cambrian beds in the world. It was here that MATTHEW (1888) appropriated the name of a primitive New Brunswick tribe for a new geological system, the Etcheminian, to apply to Palaeozoic rocks which lie below the Cambrian but contain no trilobites.

Unfortunately, it must be said that the higher beds also provide a deplorable example of improper stratigraphical nomenclature, due to the acceptance of a biostratigraphical basis for formational sub-division. Five Middle and Upper Cambrian formations are based on discontinuous outcrops of a very similar and monotonous lithology. No fewer than fifteen formal stratigraphical units have been erected, ranging in rank from formations to a system, for a cluster of Cambrian outcrops totalling less than 1,000 metres in stratigraphical thickness, of which more than 600 metres are assigned to a single un-fossiliferous formation at the base. The section as pieced together from these outcrops was illustrated by HOWELL (HAYES and HOWELL, 1937, p. 56), who also set out extended fossil lists for correlation with other areas bearing the 'Atlantic' fauna (*idem.* pp. 84–91). The entire Cambrian section was originally referred to the Saint John Group (Table 2).

There are two principal belts of discontinuous Cambrian outcrop. The better known belt extends north-eastwards from Saint John harbour for about 50 kilometres. Structurally it is a compound syncline, narrowing to the south-west where it is steeply overturned towards the north-west. The second belt lies along a narrow estuary of the Saint John River to the north-west (Fig. 4). This estuary is controlled by a fault-zone which extends towards the south-west and strikes into the Bay of Fundy at Beaver Harbour, 25 kilometres short of the Canada–United States boundary which here lies in the estuary of the St. Croix River. For most of its course, the fault-zone separates Pre-Cambrian Coldbrook volcanic rocks, on the south-east, from younger rocks on the north-west. Caught up in it, however, are several small slivers of Cambrian rocks. These differ considerably, in lithic characteristics, from those in and around Saint John, enhancing the suspicion that the section put together by HOWELL is by no means complete.

As on the Avalon Peninsula of Newfoundland and on Cape Breton Island, there is no clear evidence that all the Cambrian formations represented originally extended over the whole area within which the present outcrops appear, small though that area is. The Cambrian seaway may have been so shallow that at times it split up into a number of tortuously-connected sub-basins, controlled by features inherited from Pre-Cambrian erosion.

The Ratcliffe Brook Formation, MATTHEW's type Etcheminian, rests unconformably on the Pre-Cambrian Coldbrook volcanic rocks and appears to have been a direct product of their erosion. The purplish and greenish clastic sediments recall those of the presumably equivalent beds on the Avalon Peninsula, the Bonavista and Brigus Formations. It is conceivable that they were not completely marine. The formation ranges in thickness from 6 metres, in Saint John, to at least 600 metres. It grades upwards to a thin sandstone formation, the Glen Falls, with an easily-recognizable white band, and this in turn is gradationally overlain by the first fossiliferous formation, the Hanford Brook. The lower member of this formation is a grey sandstone, 20 metres thick, which has yielded a peculiar fauna possibly of brackish habitat—brachiopods, pteropods, and the small bivalved crustacean *Beyrichona*. The upper member, a thin shale, has yielded rare specimens of the trilobite *Protolenus*, of which this is the type locality. The *Callavia* beds are therefore unrepresented here. As the 'Etcheminian' lies gradationally below the Hanford Brook Formation, however, it is reasonable to suppose that it is the time-equivalent of at least a part of the Lower Cambrian of the Avalon Peninsula.

A gradational succession of grey to black shales, about 35 metres in total thickness, represents the zones of *Paradoxides eteminicus* and *P. abenacus*, equivalents of the *groomi* and *hicksi* Zones respectively. The zones have been given the formational names Fossil Brook and Porter Road, respectively (Table 2). The early *Paradoxides* fauna of the Fossil Brook Formation has been found again, near Beaver Harbour to the south-west, in a band of grey, siliceous limestone interbedded with slates. Also interbedded, however, are green and purple pyroclastic rocks, no counterparts of which have ever been reported from the Saint John section (HERWART HELMSTAEDT, personal communication, 1968). In yet other downfaulted slivers along the same fault zone there occur red and grey clastic sedimentary rocks, including conglomerates (HAY, 1968). These may be a further facies of the lowest Middle Cambrian, they may represent part of the Ratcliffe Brook Formation, or they may be a quite new part of the succession.

Above the zone of *Paradoxides abenacus* come five further zones, those of *Paradoxides matthewi* (representing *P. davidis*), *Agnostus pisiformis*, *Olenus*, *Parabolina spinulosa*, and *Peltura*. The five have been given the four formational names shown in Table 2, though no definite contacts are known between them and all are of black shale with thin beds, lenses, or concretions of limestone. As in south-east Newfoundland, there is a notable faunal gap between the Middle and Upper Cambrian, but the section continues above the Upper Cambrian into the Tremadoc without change of lithology.

2.3. The Basinal (Geosynclinal) Facies in Quebec and Newfoundland

2.3.1. Introduction

We have seen that the carbonate–quartzite facies typical of the foreland extends eastwards in southern Quebec to the frontal folds of the Champlain (Logan) fault zone, and again in Newfoundland from the west coast to the fault zone east of the Long Range peninsula. Between these two areas, over a linear distance of some 1,200 kilometres, only one tiny group of outcrops of undoubted foreland Cambrian rocks is seen at the surface in place (Fig. 5).

The basinal assemblage, typified by greywackes and shales, lapped both gradationally and unconformably north-westwards on to the Quebec foreland through Cambrian and Ordovician time (CADY, 1960). In addition, the basinal sediments were carried towards

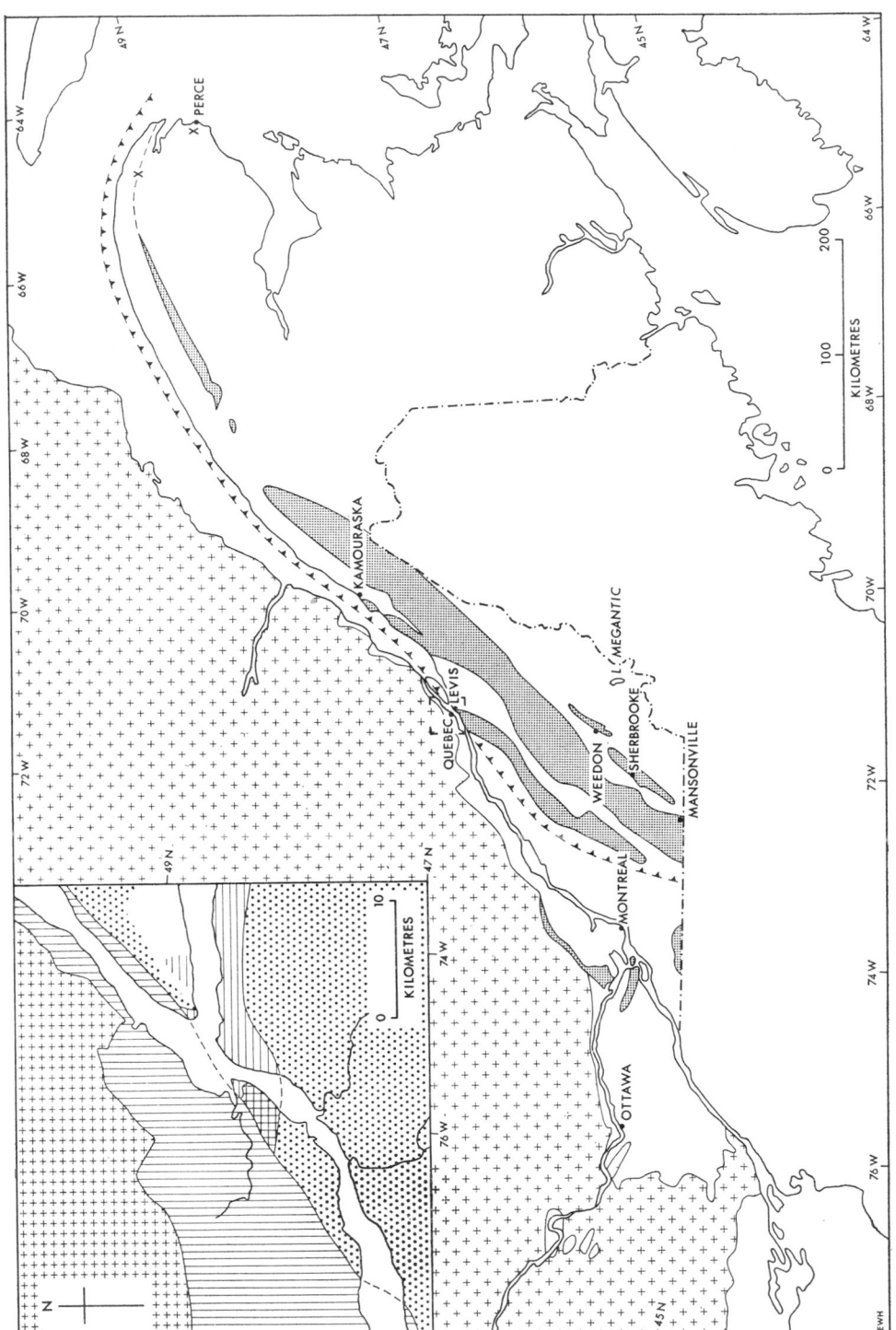

Fig. 5. Outcrop areas of Cambrian and possible Cambrian rocks in Quebec. Localities at the eastern end of the outcrop belt, marked X, are small isolated outcrops. Numerous small outcrops of the Nepean Sandstone, flanking the Pre-Cambrian Frontenac axis in the south-western corner of the map, are not shown. Toothed line is Logan fault.

Inset: Quebec City—Lévis area, showing thrust-slices of Lower Palaeozoic rocks (after Osborne, 1956; St.-Julien, unpublished). Crosses, Pre-Cambrian; vertical ruling, Middle and Upper Ordovician rocks of foreland; dots, Lower Cambrian Charny Formation; horizontal ruling, Lower Ordovician Lévis Formation; cross-hatching, Middle Ordovician Flysch.

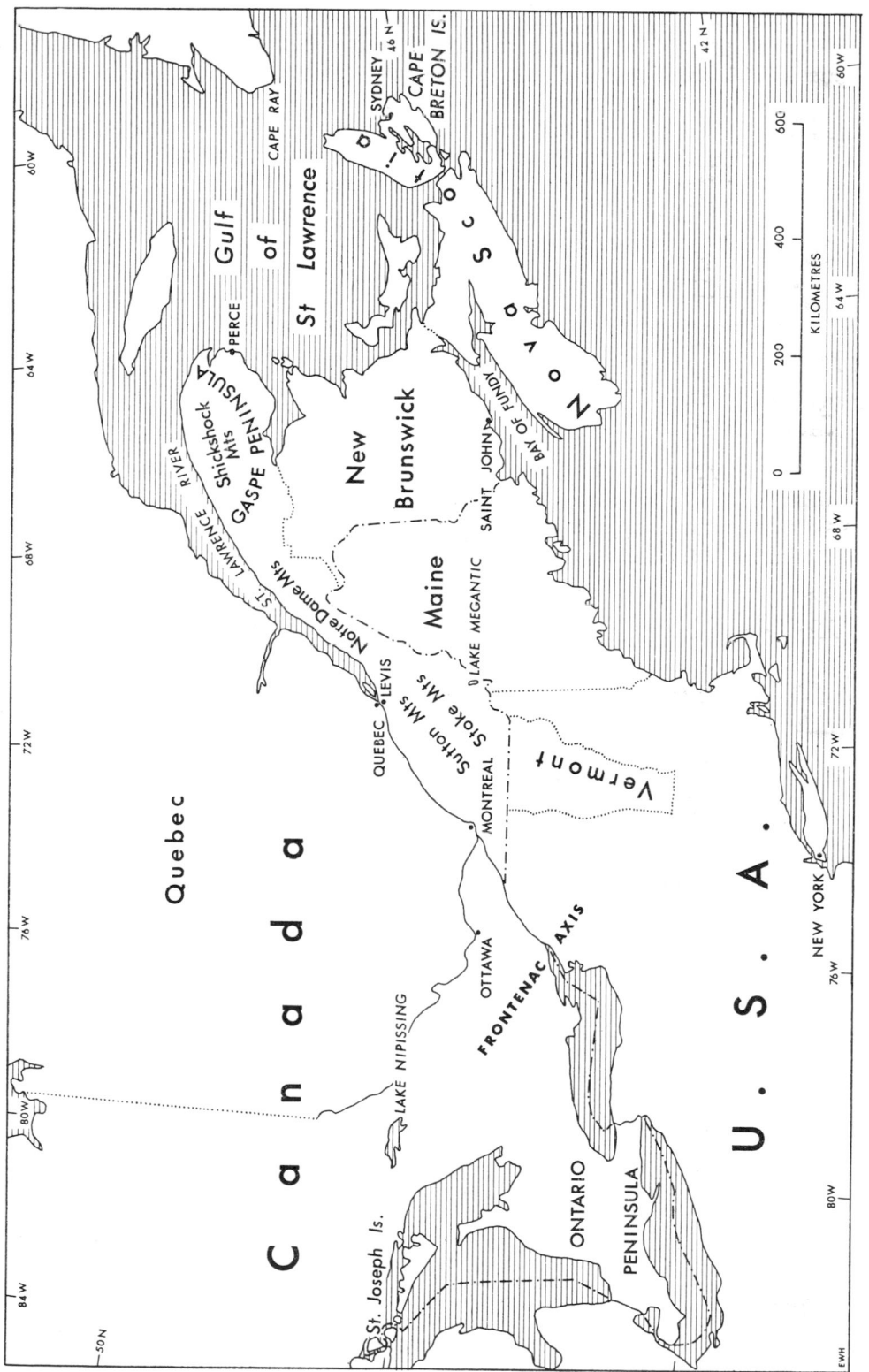

Fig. 6. Eastern Canada, excluding Newfoundland, showing places referred to in the text.

the north-west tectonically during the Taconian orogeny in mid-Ordovician time, over-riding or incorporating any foreland rocks that may have existed there (HENDERSON, 1958). The entire 'Quebec complex', along the south shore of the St. Lawrence River, is in some degree allochthonous, and the axial part of the trough, consisting now of more or less metamorphosed basinal rocks, has been brought to within 40 kilometres of the river's edge. Distinguishing between the Cambrian and the Ordovician in this thick pile of metamorphosed, and unmetamorphosed but sheared and faulted, sedimentary and volcanic rocks is a difficult task, as the views on the age of the Quebec Group testify (OLLERENSHAW, 1967). Formational nomenclature has become exceedingly confused.

2.3.2. South-eastern Quebec

The least understood and defined region, as far as Cambrian stratigraphy is concerned, is that lying east and south of the axis of the Sutton and Notre Dame Mountains (Fig. 6), and extending eastwards to the international boundary between Quebec and Maine.

The belt is one of a great thickness of the greywacke–shale facies, with volcanic rocks in varying degrees of abundance, widespread low-grade metamorphism, contortion of the argillaceous rocks and brecciation of the coarser and more competent ones, lack of marker horizons, and paucity of fossils. According to CADY (1960), no diagnostic fossils have been found in the belt older than Middle Ordovician, and a good deal of the belt exposes Silurian and Devonian strata of less 'geosynclinal' aspect.

Most modern work assigns the lowest exposed rocks of the belt to the 'Cambro-Ordovician'. Because of uncertainty over age and correlation, field geologists in the region have adopted formational and group names of purely local significance rather than imply inter-areal relationships which they cannot demonstrate. Among the names used for formations, or groups, presumed at one time or another to be 'Cambro-Ordovician', are Armagh, Ascot, Bennett, Bonsecours, Brompton, Bunker, Caldwell, Charny, Frontenac, Gagné Brook, Kamouraska, Lauzon, L'Islet, Mansonville, Mansville, Matane River, Ottauquechee, Rosaire, Sherbrooke, Shickshock, Sillery, Sutton, Sweetsburg, and Weedon.

In a very general way, the belt exposes three groups of strata (COOKE, 1937; BÉLAND, 1957, 1962; CADY, 1960; ST.-JULIEN, in press) below the widely-recognized Beauceville (Magog) slates, which are established as being of Middle Ordovician age. The lowest group consists of quartz-, sericite-, or chlorite-schists and schistose quartzites, grading into slates. Metamorphism to chlorite grade is almost ubiquitous; staurolite grade is not uncommon. The schists were originally largely greywackes. This is the group called Sutton when properly schistose, otherwise Bonsecours or 'Lower Sillery'. It at least includes the Lower Cambrian, but beyond that its age is conjectural.

The succeeding unit comprises massive, grey, carbonaceous or feldspathic quartzites with thinly-bedded siltstones or even silty limestones, and black, grey or green slates. The rocks give the impression of having been shelf sediments rather than the deposits of a geosyncline, and only for geographical convenience can they be considered with the basinal facies. They take over the axis of the range farther north, east of Quebec City, where they are variously called the L'Islet or Rosaire Group (BÉLAND, 1962). This unit is also equivalent to the middle part of the Caldwell Group as used by COOKE (1937, 1950), but not as used by BÉLAND (1962); in CADY's opinion (1960), the term Caldwell no longer has any practical value and should be discarded. The unit presumably includes some or all of the Middle and Upper Cambrian, and it equates chronologically with part or all of the Sweetsburg and Scottsmore Formations west of the Sutton Mountains (2.4.3).

The third group, underlying the Beauceville, consists of green, red, purple, and grey slates, phyllites, and argillaceous or feldspathic sandstones, with similarly coloured pillow lavas and tuffaceous beds in many places. This comprises the Mansonville and Brompton Formations along the eastern flank of the Sutton Mountains (ST.-JULIEN, in press); it is also the Caldwell Group as used both by CADY and by BÉLAND. Its equivalent west of the metamorphic axis must be the upper part of the Quebec Group, described below, and its age is approximately Late Cambrian to early Middle Ordovician. In places it may be entirely of Early Ordovician age.

In the Stoke Mountain anticline, midway between the Sutton Mountains and the Maine border, the Ascot Formation of dark phyllites and acid pyroclastic rocks (ST.-JULIEN and LAMARCHE, 1965) seems to be the equivalent of the third of these groups, or to part of it. So also do the Weedon 'schists' of BURTON (1931). The Frontenac 'series' of the Lake Megantic area, in the extreme south-eastern corner of the province, was originally assigned a Cambro–Ordovician age, also, and may include some Cambrian rocks, but most of it is probably Lower Devonian (MARLEAU, 1959).

The thicknesses of the three groups are known only very approximately. The two lower groups show no evidence of very great thickness; they may total only 2,000 metres or thereabouts. The upper unit is more probably 5,000 to 10,000 metres thick. In many parts of the belt there are really only two groups present, the second of the three described being the little-metamorphosed equivalent of the upper part of the first.

The 'schists' of the Sutton Mountain antiform itself belong to the oldest of the three groups, now called Bonsecours. In the far south, close to the United States border, many of the rocks are actually gneisses, granulites, feldspathic quartzites, or slates. Originally they were greywackes, quartzites, siltstones, and grey, green or red shales. The lowest part of the succession can be shown to be the equivalent of the Lower and Middle Cambrian Oak Hill Group of the Quebec foreland (2.4.3). Individual formations of the group, particularly the Sweetsburg slate and the Scottsmore quartzite, can be traced into the 'schist' belt, and even beyond it to the east. They remain quite thin, and are not in any commonly-accepted sense 'geosynclinal'. They are augmented, however, by some thousands of metres of younger rocks, particularly of the Late Cambrian and Early Ordovician.

North-east of Quebec City, where the structural axis of the Notre Dame Mountains coincides approximately with the border between Quebec and Maine, evidence for a south-easterly source of clastic sediment has been presented by HUBERT (1965). Immediately north-west of the axis, which is here in the Rosaire Group, the rocks of the Armagh Group are massive, feldspathic sandstones in a variety of colours, indicating a near-shore environment; the sands contain fragments of Grenville-type basement rock. On the shore of the St. Lawrence River, farther to the north-west and beyond a syncline exposing Lower Ordovician rocks (Fig. 5), similarly coarse sediments are included in the St.-Damase Formation, with polymictic limestone conglomerates in addition. These are underlain, however, by a succession chiefly composed of mudstones and siltstones with only minor coarser beds. This formation, the St.-Roch, furthermore includes some limestone members believed to be of somewhat deeper water origin. The St.-Roch Formation contains *Botsfordia* and is of Early Cambrian age; HUBERT believed that it is also equivalent to the lower part of the Armagh Group to the south-east, and, as no such fine-grained or deeper-water sediments are prominent in that group the implication is that the source of the sediment, at least in Early Cambrian time, lay to the south-east.

The Notre Dame Mountains are flanked on the north-west by a low-lying area yielding sparse outcrops of dark shales with bands of block-bearing Wildflysch. Numerous graptolite localities show that these rocks are of Middle Ordovician (Norman-skill) age, and apparently overlie the rocks of the Notre Dame Mountains in normal stratigraphical succession, just as the presumably equivalent Beauceville slates do on the south-east flank. Proceeding northwards towards the St. Lawrence River, however, the dominant rock is suddenly seen to be a thickly-bedded red or green mudstone or slate, at least 400 metres thick. This is overlain first by massive, dirty grey or green feldspathic grits and further mudstones, and then by green, grey or black shales in which the sandstone intercalations are of a relatively clean, quartzose variety (ST.-JULIEN, personal communication, 1967). These rocks constitute the old Sillery Formation of LOGAN's Quebec Group, and are now grouped together as the Charny Formation (RASETTI, 1946b).

OSBORNE (1954) has shown that the red beds of the lower member were almost certainly red when they were deposited. He has also suggested that the grits and grey-wackes of the middle member, which contain masses of black shale, are adventitious, being introduced by slumping into a sedimentary basin of the order of 100 fathoms deep. RASETTI (1946b), on the other hand, had proposed that all the sediments of the Quebec Group were of very shallow water origin.

In the area across the river from Quebec City (Fig. 5), greenish glauconitic siltstones in the middle member of the Charny Formation contain the brachiopod *Botsfordia pretiosa* (BILLINGS), an index for the Lower Cambrian. This part of the formation is therefore to be equated with the St.-Roch Formation to the north-east. At an unknown stratigraphical distance above the *Botsfordia* beds, RASETTI found trilobites in place in thin, dark, irregular limestone layers within a sequence of grey shales in the upper member of the Charny Formation. The genera represented include *Austinvillia*, *Bonnia*, and *Pagetides*, an association characteristic of the uppermost Lower Cambrian. As this horizon appears to be closely overlain, at that locality, by the famous Lévis limestone-conglomerate, RASETTI concluded that neither the Middle nor the Upper Cambrian occurs in place in the Quebec–Lévis area. In this conclusion, RASETTI was in agreement with SCHUCHERT (1923, p. 178).

The structure, however, is exceedingly complex. As the Charny Formation, which seems to be entirely Lower Cambrian, is both underlain and overlain by Ordovician rocks, some form of allochthony is inevitable. Overturning makes the stratigraphy still more difficult to decipher, and more recent isolated fossil finds suggest that most Cambrian levels may be represented. The Lévis Formation is known to be of earliest Ordovician (Tremadoc-Arenig) age from fossils in its thinly-bedded siltstones. The poorly-defined Lauzon Formation has been considered to lie between the Charny and the Lévis Formations. It lacks the greywackes which typify the Charny; its massive sandstones are richly glauconitic. It also lacks the limestone conglomerates of the Lévis. If the formation really is older than the Lévis, it would have to include the Upper Cambrian, and possibly some of the Middle Cambrian as well, but it may be only a facies of the Lévis Formation. Nonetheless, a *Dendrograptus* from east of Lévis is possibly of Trempealeauan age (OSBORNE and BERRY, 1966). Both the Charny and the Lauzon Formations are in urgent need of stratigraphical revision.

Also east of Lévis, an outcrop of thinly-bedded, fine-grained limestone has yielded a Middle Cambrian fauna described by LAVERDIÈRE (1949). Genera represented include *Peronopsis* and *Ptychagnostus* (both abundant), *Bathyuriscus*, *Olenoides*, and phosphatic brachiopods, indicating a horizon in the middle or late Middle Cambrian. The limestone appears to be in the Lévis Formation, and it is likely that both it and its associated limestone-conglomerate are part of a huge exotic block, about 12 metres thick, in that formation (OSBORNE, 1956).

The presence of both Middle and Upper Cambrian faunas (as well as Lower Cambrian and Lower Ordovician) in the boulders and blocks of the Lévis conglomerate has been recognized for a long time. The faunas have been exhaustively treated by RASETTI (1944–1948, 1963). The Lower Cambrian assemblages contain olenellids, abundant *Bonnia laevigata* RASETTI, and common eodiscids. Relationship to the fauna of the Forteau Formation of the foreland of western Newfoundland (2.4.1) is apparent. The common Middle Cambrian genera are *Bathyuriscidella*, *Hypagnostus*, *Kootenia*, and *Olenoides*. Ptychoparid trilobites occur in both the Lower and the Middle Cambrian assemblages. The commonest Upper Cambrian genera are those of the *Cedaria* and *Crepicephalus* Zones. The less abundantly represented uppermost Upper Cambrian trilobites are dominated by small, smooth species. RASETTI noted that several of the agnostids represented have close relatives in the Atlantic realm, a relationship almost totally unrecognized among other trilobites. LOCHMAN-BALK and WILSON (1958) further noted that the Upper Cambrian assemblage was not one typical of the cratonic régime; about half the genera represented could more properly be regarded as representing a non-cratonic facies (Table 3).

The extraordinary thing is that, although almost all the blocks in the Lévis conglomerates are of a characteristically fine-grained, whitish, massive limestone, that limestone is not known in place anywhere in the region. Yet the fossiliferous boulders continue along the south shore of the St. Lawrence, in the same formation, for at least 250 kilometres north-eastwards from Lévis. The evidence of LAVERDIÈRE's fossil locality led OSBORNE (1956) to the suggestion that the original, autochthonous carbonate bank must have included massive limestones in the Lower and Upper Cambrian, and more thinly-bedded ones in the Middle Cambrian, and that Middle Cambrian fossils are rarer in the boulders than fossils of Early or Late Cambrian age because the Middle Cambrian limestones contributed much smaller pieces to the conglomerates. OSBORNE further preferred a southerly source for the limestones, from a welt in the trough or a southern shelf.

Cambrian rocks are exposed, however, in many areas to the south, especially along the axis of the Notre Dame Mountains, and there is no sign of a limestone suitable for the provenance of the boulders. Moreover, the adventitious material in the Charny Formation appears to be rather clearly of shield derivation. The trilobites themselves, in their mixed aspect, reveal an origin on the outer edge of the bank facies, the matrix of the boulders being basinal. As the entire mass of the Quebec Group gives every sign of being allochthonous, the simplest explanation may be that the boulders came from south of their present position but north of their original position; that their history is comparable with that proposed for the boulders in the Cow Head breccia (2.3.4).

2.3.3. Shickshock Mountains

The axis of the Notre Dame Mountains enters a plunge depression in the area in which Quebec, Maine, and New Brunswick come together. Possible Cambrian rocks are there covered by Siluro–Devonian rocks. However, the main fault zone along the north-western face of the Notre Dame Mountains is probably associated with, or even continuous with, that along the south side of the Shickshock Mountains in Gaspé (Figs 5,

System	Zone	Faunas in Lévis Conglomerates	Faunas in Cow Head Breccias
Upper Cambrian	Saukia / Ptychaspis–Prosaukia	Hungaia, Onchonotus, Plethometopus, Rasettia, Stenopilus, Theodenisia	Keithiella, Loganopeltoides, Plethometopus, Theodenisia
Upper Cambrian	Conaspis / Elvinia		Bienvillia, Ctenopyge, Irvingella, Loganellus, Oligometopus, Parabolinella, Protopeltura, Taenicephalus.
Upper Cambrian	Dunderbergia / Aphelaspis		Aphelaspis, Blountia, Cheilocephalus, Dunderbergia, Onchopeltis, Pterocephalops
Upper Cambrian	Crepicephalus / Cedaria	Blountia, Bynumia, Cedaria, Coosella, Kingstonia, Meteoraspis	Cedaria, Coosella, Crepicephalus, Kingstonia, Meteoraspis
Middle Cambrian	Bolaspidella	Bathyuriscidella, Hemirhodon, Hypagnostus	Centropleura, Kootenia, Phalacroma, Solenopleura, Zacanthoides
Middle Cambrian	Bathyuriscus–Elrathina	Olenoides, Peronopsis	Bathyuriscus, Ogygopsis, Olenoides, Pagetia, Peronopsis
Middle Cambrian	Glossopleura	Kootenia, Pagetia, Zacanthoides	
Middle Cambrian	Albertella / Plagiura–Poliella		
LC	Olenellus–Bonnia	Bonnia, Pagetides, Zacanthopsis	

TABLE 3

Representative Cambrian Faunas from Boulder-bearing Formations Quebec and Newfoundland.

6). In the Shickshock Mountain anticlinorium, 'Cambro–Ordovician' rocks are again exposed. Stratigraphical and structural relationships are both obscure, but the Shickshock Group is older than the Taconian deformation, and it can scarcely be younger than the rocks of the Quebec Group (largely Tremadoc) along the river to the north.

The Shickshock Group (MATTINSON, 1964; OLLERENSHAW, 1967) consists of two main facies, one volcanic and one arkosic. Either facies may dominate in any one part of the range. The volcanic rocks, originally largely basalts and including pillow-basalts, are now for the most part amphibole-schists. Associated with them are minor metasedimentary rocks—mica- and chlorite-schists and meta-arkoses. The arkosic facies consists of arkose with some red slate and schist, and minor chert and tuff. It appears to be both contemporaneous with and older than the volcanic rocks. The lithological association recalls that in the Mansonville Formation to the south (the last of the three facies complexes described for the belt east of the Sutton Mountains); if the two are approximately equivalent, a Late Cambrian age is probable for at least part of the Shickshock Group. This is further implied by a potassium–argon date of 530 ± 35 m.y., based on a muscovite sample, for the group's metamorphism (OLLERENSHAW, 1967).

According to OLLERENSHAW, the arkosic facies of the Shickshock Group had a local source to the south-east in the form of a horst of sodic granite. The arkoses interfinger towards the north-east and north-west with grey slates of the Quebec Group, here called the Matane River Group. These slates, which contain the usual minor beds of coarser clastic rocks, and some red and green slates, have yielded one trilobite. It appears to be a ptychopariid genus, suggesting Early or Middle Cambrian age. Above the slates are tightly folded bands of quartzite in close association with limestone conglomerate. The quartzite-conglomerate unit is the Kamouraska Formation, known from its development in the Notre Dame Mountains to be of Tremadoc age (HUBERT, 1965). The limestones of the conglomerates are less fossiliferous than those of the Quebec and Cow Head conglomerates and breccias, but they have yielded the Lower Cambrian *Salterella*, the Lower or Middle Cambrian *Kootenia*, several Upper Cambrian trilobite genera, and unidentified algae.

From the eastern end of the Shickshock Mountains, the grey slates of the Quebec Group continue in a narrowing band to the tip of the Gaspé peninsula (Fig. 5). Ordovician rocks lie to the north, along the shore, and younger rocks to the south; the narrow intervening belt may include Cambrian strata. The only outcrop along the leading edge of the disturbed belt which is known to be of that age, however, occurs near the northeastern corner of Gaspé (JONES, 1935; HUTCHINSON, 1952a). A black shale bed within the Quebec Group yielded, in addition to *Asaphiscus* and a probable *Ptychagnostus*, the trilobite *Centropleura belli* HUTCHINSON. This is an 'Atlantic' fossil of latest Middle Cambrian age, closely related to *Paradoxides*; its implication is that, if the New Brunswick geanticline existed, it was an imperfect barrier for part of Middle Cambrian time.

2.3.4. Western Newfoundland

As set forth in the next section (2.4.1), a well-known succession of the cratonic facies in western Newfoundland wraps almost completely around the Long Range peninsula. This succession ranges in age from early Cambrian to early Middle Ordovician. Almost wherever rocks are seen overlying this cratonic succession, they are utterly different in facies, and orthodox interpretation of the relationships required the postulation of a most abrupt change in depositional environment.

It is now established that the structurally higher clastic terrane is in fact of the same age as the carbonate (cratonic) terrane, so some explanation other than stratigraphical superposition is required. Some degree of allochthony for the clastic terrane is essential.

Two quite distinct types of 'clastic' terrane are involved:

a. The Cow Head Group of thinly-bedded limestones and shales, with spectacular limestone conglomerates, Middle Cambrian to Middle Ordovician in age and overlain by:

b. The Humber Arm Group: dark green sandstones or greywackes; grey, black, green, red, and purple shales or slates; some limestones and cherts; and mafic volcanic rocks, including pillow basalts. This thick succession is apparently of rather deep water origin; it is very much deformed. Bands of limestone conglomerate are rare, but such as have been described are like those of the Cow Head Group. The Humber Arm Group is known to range in age at least from Late Cambrian to Early Ordovician (KINDLE and WHITTINGTON, 1965).

RODGERS and NEALE (1963) proposed that the allochthony of both these groups of rocks is total; that they are the remains of two large Klippen, one extending along the west coast of the island from Port au Port Peninsula to north of Cow Head, and the other east of the northern end of the Long Range, south of Pistolet Bay (Fig. 7).

If this hypothesis is correct (and it enjoys wide acceptance: BRÜCKNER, 1966; CUMMING, 1967b; POOLE, 1967; WILLIAMS, 1967), the rock masses must have originated in some area to the east of the autochthonous sequence, where large bodies of Lower Palaeozoic clastic and volcanic strata occur as remnants in a terrane now largely granitic. These strata, in place, extend, sporadically, diagonally across the island immediately east of the Long Range fault zone, and several elongate ultramafic intrusions mark their frontal edge. During Cambrian time, sedimentation must have been more or less similar across the whole region, with detritus coming essentially from the Canadian shield to the west and north (BRÜCKNER, 1966). The Cambrian part of the Humber Arm Group includes a good deal of quartzite and shale, and is not totally unlike that in the autochthonous Cambrian of the Long Range. (2.4.1). Strong environmental contrasts developed in the early Ordovician, as the wide, flat, carbonate banks of the St. George and Table Head Groups cut off the supply of detritus from the western shield to the eastern (Baie Verte) depositional area.

As the Taconian compressional régime took effect, new sources of clastic material arose within the geosyncline; a belt of the clastic and volcanic succession was squeezed up and out of its original depositional basin and thrust against the block of the Long Range inlier, to slide by gravity down the inlier's western slope and into a basin in which shale was being deposited on top of the autochthonous carbonate beds. The time at which all this took place is very closely pinpointed: the youngest strata overridden by the allochthon are Middle Ordovician, and so are the oldest neo-autochthonous strata lapping against the allochthon. The more westerly Klippe, at least, was emplaced in Caradoc time.

The Klippe hypothesis was objected to by the late HUGH LILLY (1964). He pointed out that the Humber Arm and Cow Head Groups are quite unlike one another, and that each could be a facies of the autochthonous sequence, deposited before the Ordovician carbonate banks started to form. LILLY divided the Humber Arm Group into five formations. The lowest, the Summerside Formation, begins with reddish and greenish quartzites and siltstones, their base cut off by the thrust- or glide-plane, and ends with red and green shales. The succeeding Irishtown Formation, 300 to 500 metres thick, is

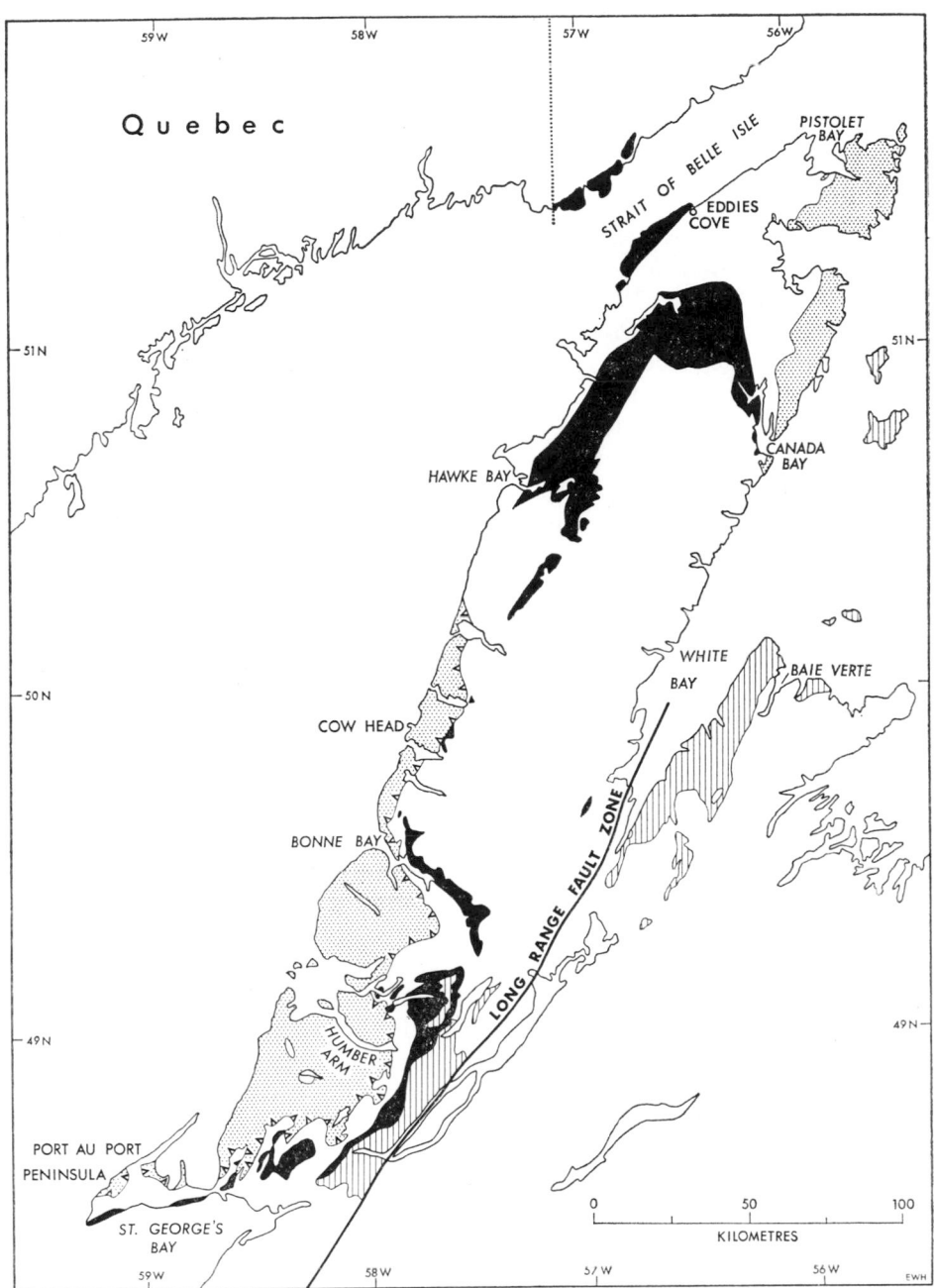

FIG. 7. Outcrop areas of Cambrian and possible Cambrian rocks in western Newfoundland. Cambrian in cratonic facies, solid black; basinal facies of the two Klippen, dotted; possible but unproven Cambrian, vertically ruled.

more shaly, but contains horizons of coarser material looking like fluxoturbidite or slumped units. At least one layer of conglomerate includes limestone pebbles carrying Lower Cambrian fossils. These two formations are sufficiently like those of the cratonic Labrador Group (2.4.1) to justify the belief that they are the basinward equivalents of that group.

The Irishtown Formation is succeeded by the Cooks Brook Formation, about 300 metres of dark shales and limestone breccias (again including some fragments with Lower Cambrian fossils). Some part of this formation may be Cambrian, but its upper part contains Tremadoc graptolites. The two overlying formations of the Humber Arm Group are independently identifiable as Ordovician. The only Upper Cambrian stage definitely established by fossils in other layers of limestone breccia is the Franconian (KINDLE and WHITTINGTON, 1965). The trilobite genera *Buttsia*, *Dunderbergia*, *Oligometopus*, and *Pseudagnostus* represent a horizon younger than any other Cambrian zone identified in place in the region.

If the Klippen exist, LILLY preferred that the zone of slippage beneath the larger one should lie above the lime breccias of the Cooks Brook Formation, and not at the base of the Humber Arm Group as a whole. This interpretation would leave the whole of the group's Cambrian content autochthonous, and intermediate in facies between the Labrador Group of the coast and the Grand Lake Brook Group along the Long Range fault zone (2.4.1). This correlation was illustrated diagrammatically by LILLY (1967).

If, however, as most Newfoundland geologists now believe, the allochthony is genuine, the famous Cow Head breccias become one of the most extraordinary of geological phenomena. Not only do the blocks in the breccias represent a superb example of allochthony by submarine slumping, but the entire mass, blocks and matrix and all, would be highly exotic.

The breccias have been regarded as Middle Ordovician submarine landslide deposits, all formed at the same time (SCHUCHERT and DUNBAR, 1934). The boulders sloughed off the front of an advancing Taconian thrust sheet, and ploughed into the underlying shales. KINDLE and WHITTINGTON (1958) believed that there is represented a normal succession of thinly-bedded limestone and shales, about 300 metres thick, with rare sandstones, and that the spectacular limestone conglomerates occur at intervals throughout it. Some of of the boulders are fossiliferous, and the fossils in any one boulder-layer are all about the same age. The normal beds below any one conglomerate carry fossils showing them to be about the same age as the blocks in the overlying conglomerate. In other words, the conglomerate layers are intraformational and episodic. All the boulders and chips seem to be of limestone; no igneous, metamorphic, or quartzitic material has been seen.

Middle Cambrian trilobites have been found both in bedded limestones and in boulders, but Upper Cambrian trilobites have been found only in boulders in younger conglomerate layers. The shales in the upper part of the succession contain abundant graptolites, and the Cow Head Group must range in age at least from Middle Cambrian to Middle Ordovician, inclusively. The chief Cambrian faunas represented are indicated in Table 3. The assemblage is unusual in its mixture of 'Atlantic' and 'Pacific' faunas; the Ordovician part is even more unusual in its alternation of graptolite and trilobite faunas.

The limestones of the boulders, with their particular fossil assemblages, are unknown in place in western Newfoundland, or in any other part of the Atlantic provinces. The boulders are of all sizes, from small flat chips to blocks large enough to be mapped as formations; they vary in colour and texture; many are quite unfossiliferous; some may have come from bioherms. Like the limestones in the Lévis conglomerates, these limestones in fact represent another facies, in addition to those seen in autochthonous positions. Where the limestone banks lay, and how they were related to the established facies, is still unknown. They need not necessarily have lain in parallel sequence between the carbonate-quartzite facies of the foreland and the shale facies of the basin. They could have formed sporadically on rising submarine ridges, as did many of the rudistid reefs and banks in the Cretaceous of the Caribbean region. They could have been introduced into the basin from either the east or the west (KINDLE and WHITTINGTON, 1958).

If, however, the entire mass is allochthonous, a new possibility is introduced. There is no sign at all of a true carbonate terrane in central Newfoundland, and the Baie Verte and Humber Arms Groups (respectively the autochthonous and allochthonous representatives of the clastic–volcanic facies) are in part eugeosynclinal in aspect. Though reefs or banks could have developed on rising welts within such a depositional area, as indicated above, it would be difficult to regard the resulting terrane as providing exotic material all of one type. The Cow Head breccia is in composition quite unlike the Argille Scagliose. If, however, the breccias were originally formed *east* of the Long Range, the boulders could have come from *east* of their present position but *west* of their original position; that is, from a carbonate bank on part of the Long Range itself. If the now-exposed crystalline core had been a peneplaned platform in latest Pre-Cambrian time, it might have been partly covered by a very shallow-water carbonate bank, of Bahamas type, to the exclusion of other deposits (except perhaps evaporites). If it received little or no other sediment after uplift and eastward tilting during the Taconian movements, the elimination by subsequent erosion of all remaining traces of the banks is not difficult to accept.

2.4. The Cratonic Facies in eastern Canada

2.4.1. Western Newfoundland

The true cratonic or foreland facies of the Cambrian crops out in a discontinuous apron around the flanks of the Long Range, from its northernmost tip to St. George's Bay in the south-west (Fig. 7). The apron is interrupted on the east side by Carboniferous rocks and the waters of White Bay; it is partly interrupted on the west side, also, by being covered up tectonically by the larger of the two Klippen (2.3.4).

There is no doubt that the succession is autochthonous. Lower and Middle Cambrian sandstones, with some carbonates (including reefs), rest with unconformity directly on Pre-Cambrian crystalline rocks, and are succeeded by Middle Cambrian to early Middle Ordovician carbonate rocks, also including reefs. The carbonates are of very shallow water origin, and carry the shelly fauna characteristic of the west side of the Appalachians from this area to Alabama.

Formational sub-divisions (Table 4) are due principally to SCHUCHERT and DUNBAR (1934), BETZ (1939), and WALTHIER (1949). The basal arkosic sandstones, red, purple, and white, with minor conglomerate, are unfossiliferous and may be only incompletely marine (Bradore Formation of the two sides of the Strait of Belle Isle, about 100 metres thick and containing one or more flows of black basalt; Cloud Mountains Formation of Canada Bay, up to 180 metres thick). The succeeding Forteau Formation consists of variegated shales and grey limestones, some of them oolitic, and thin interbeds of white calcareous quartzite. It is exceedingly fossiliferous, carrying *Paedeumias* in dark

	Zone	West flank of Long Range	East flank of Long Range
Upper Cambrian	Saukia	Not fossiliferous	
	Ptychaspis–Prosaukia		
	Conaspis		
	Elvinia		
	Dunderbergia		
	Aphelaspis		
	Crepicephalus	Petit Jardin	
	Cedaria		
Middle Cambrian	Bolaspidella	March Point	Treytown Pond
	Bathyuriscus–Elrathina	Unnamed beds at Eddies Cove	Cloud Rapids
	Glossopleura	Not identified	
	Albertella		
	Plagiura–Poliella		
Lower Cambrian	Olenellus–Bonnia	Forteau	Forteau
	Olenellus		Devil's Cove
	Pre-Olenellus	Bradore	Cloud Mountains

TABLE 4
Correlation of Cambrian Formations, Western Newfoundland.

shales near the base, *Olenellus, Bonnia*; brachiopods (*Kutorgina, Micromitra, Obolella*); *Salterella, Hyolithes*, and large lenses of archaeocyathid 'reef'.

These two formations constitute the Labrador Group. Their equivalent at Bonne Bay, halfway down the west coast, is fossiliferous almost throughout an exposed range of nearly 750 metres; its trilobite and brachiopod fauna recalls that of Vermont. At St. George's Bay, the quartzites and reefs are missing, dark limestones and limy shales making up most of the 400-metre-thick Kippens Formation (WALTHIER, 1949; RILEY, 1962). Its fauna includes the cephalopod-like *Volborthella concavi* (WALTHIER) and button-like cryptozoons.

Intervening between the Bradore and Forteau Formations at Canada Bay is a thin (12 metre) bench of pink limestone with greenish shale interbeds, called the Devil's Cove Formation. Such a limestone is unknown elsewhere in western Newfoundland, but it at once recalls the Smith Point Limestone of the Avalon Peninsula, a comparison given extra significance by the reported presence in the Devil's Cove of *Callavia*, an 'Atlantic' trilobite (BETZ, 1939). The '*Callavia*', however, is in fact a species of *Wanneria*, allied to East Greenland and New York forms. It is associated with *Calodiscus lobatus* (HALL) (A. R. PALMER, personal communication, 1967). There is no 'Atlantic' aspect to the Devil's Cove fauna; nor is the formation as similar to the Smith Point as the original descriptions might suggest.

The Hawke Bay Formation on the west coast represents a reversion to an extremely shallow-water alternation of white and pink quartzites, with some dark shale and greenish sandstone, carbonate beds, and intra-formational flat-pebble conglomerate. The formation as defined by SCHUCHERT and DUNBAR (1934) overlaps both above and below with other formations named elsewhere along the west coast of the island. The beds with *Olenellus* and *Hyolithes* should be assigned to the Forteau Formation. The quartzite above seems to be that included in the Middle Cambrian March Point Formation on the Port au Port Peninsula; formational discrimination would be improved if it were given formational status of its own (WALTHIER, 1949). The quartzite, at least 100 metres thick, is overlain by nearly 200 metres of silty, shaly, and dolomitic beds, the proportion of red and green shales increasing eastwards towards what was originally an outer detrital zone. The lower part of the shaly and calcareous quota of the formation has yielded the late Middle Cambrian trilobites *Marjumia* and *Eldoradia* (LOCHMAN, 1938); the upper part contains only cryptozoons. Only on faunal grounds is the March Point Formation separated from the overlying Petit Jardin Formation; the boundary is drawn artificially below a horizon containing trilobites of the *Cedaria* Zone (and possibly of the *Crepicephalus* Zone also): *Arapahoia raymondi* LOCHMAN, *Coosella, Kingstonia, Maryvillia*, and *Welleraspis*. In Bonne Bay, separated from the Port au Port Peninsula by the largest of the Klippen, the equivalent strata have been called the East Arm Formation (TROELSEN, 1947). No Franconian or Trempealeauan fossils have been found in the Petit Jardin Formation, though Franconian forms, at least, occur in limestone breccias in the Humber Arm Group not far away (2.3.4). The formation ends with an unfossiliferous, cliff-forming dolomite, which again is artificially separated from the overlying Ordovician. Indeed, some of the type section of the Petit Jardin Formation, which is fossiliferous only at the base, may be Ordovician (WALTHIER, 1949).

An essentially similar situation exists wherever the Middle and Upper Cambrian are present in western Newfoundland. On the Strait of Belle Isle, the Forteau Formation passes upwards into an unnamed carbonate unit bearing the Middle Cambrian trilobites *Ehmania, Elrathina*, and *Spencella*? (WHITTINGTON and KINDLE, 1966), and this in turn passes into unchanged carbonates of the Ordovician St. George Group without any Upper Cambrian fossils having been identified in between. In the Canada Bay section, dark limestones, shaly limestones, and thin quartzites of the Cloud Rapids Formation follow disconformably on the Forteau, with a pebble conglomerate at the base (BETZ, 1939). The formation, nearly 100 metres thick, has yielded *Ehmania* and *Kootenia*, in addition to non-diagnostic brachiopods. It is overlain in turn by fine-grained limestones and chert of the Treytown Pond Formation, 200 metres thick, which has not been shown to carry anything beyond the non-diagnostic brachiopods, but which is presumed to be Middle Cambrian as it lies conformably above Middle Cambrian strata which are not latest Middle Cambrian. This section cannot be shown to continue into the Upper Cambrian, a marked faunal break separating the brachiopod-bearing beds from the Lower Ordovician dolomite which is the most prominent formation in the whole Palaeozoic succession. It is also the formation which controls the structural style of much of this strongly-deformed region. The much less competent rocks of the underlying Cambrian formations (especially the shales of the Forteau Formation) have been squeezed diapirically through the breached crests of anticlines in the thickly-bedded St. George Group (LILLY, 1967).

About 125 kilometres south of the Canada Bay section, the Cambrian crops out narrowly at the base of the Palaeozoic west of White Bay (NEALE and NASH, 1963; Fig. 7). More than 200 metres of grey, unfossiliferous shales, calcareous and micaceous and in places metamorphosed to mica-schists, grade downward into micaceous quartzite or quartz-conglomerate. This formation, called the Beaver Brook, appears to be equivalent to part of the Cambrian section north or west of the Long Range—perhaps to the upper part of the Forteau Formation, but more likely to the Cloud Rapids and Treytown Pond Formations because the succeeding marble and limestone (Doucers Formation) are probably Ordovician.

At the south-eastern corner of the Long Range's Cambrian jacket, the rocks pass into an easterly region of metamorphism, as well as giving signs of passing into the outer detrital zone of their original depositional basin. The Labrador Group seems to be represented by the Grand Lake Brook Group (WALTHIER, 1949). Its lower and larger part, the Mount Musgrave Formation, comprises phyllites, schists, and gneisses (BRÜCKNER, 1966; LILLY, 1967) which were originally clastic sediments, arkosic below, and probably as much as 1,500 metres thick. Beds younger than Early Cambrian may be included, because the overlying Reluctant Head Formation passes up conformably into the St. George Group and belongs at least in part to the Upper Cambrian. It consists of thinly-bedded limestones and shales, with some limestone breccias in the lower part, up to 1,000 metres thick in the east (where metamorphism has produced marble) but thinning out drastically westwards. The formation is no doubt equivalent to the March Point and Petit Jardin Formations of the west coast (WALTHIER, 1949).

2.4.2. Eastern Gaspé

The whole of Gaspé lies within the disturbed belt south of the Logan fault. The Lower Palaeozoic rocks exposed in it are all, or nearly all, part of the old 'Quebec Group' of clearly basinal or geosynclinal facies. At the extreme eastern tip of the peninsula, however, near the village of Percé (Fig. 6), tiny outcrops occur of fossiliferous Cambrian limestones and shales. The writer has not seen the outcrops, but their existence in this structural environment is sufficiently unexpected to require speculation on their orgin.

Two faunas are represented, and sufficient distinction exists between their host-rocks for the erection of two formations (KINDLE, 1942; MCGERRIGLE, 1950). The Corner-of-the-Beach Formation consists of a few limestone and shale beds containing the 'Pacific' trilobites *Alokistocare*?, *Bolaspidella*?, *Kingstonoides*, *Modocia*, *Olenoides*, *Peronopsis*, *Spencella*?, and *Zacanthoides*, as well as the brachiopods *Micromitra* and *Prototreta* (W. H. FRITZ, personal communications, 1967–8). This is a late Middle Cambrian fauna. The Murphy Creek Formation is represented by thinly-bedded grey and shaly limestone and fissile aluminous shale with some flat-pebble conglomerate. It has yielded a fauna originally compared with that of the latest Middle Cambrian Maryville Limestone of the southern Appalachians, but now regarded as representing the *Cedaria* and *Crepicephalus* Zones of the basal Upper Cambrian. The formation is overlain by the fossiliferous basal beds of the upper Ordovician to lower Silurian Matapédia Group.

Three preliminary comments on these outcrops seem appropriate. First, the two formations are faunally and lithically directly comparable with the March Point and Petit Jardin Formations on the Port au Port Peninsula of western Newfoundland, the nearest fossiliferous Cambrian strata to the east of this area (2.4.1). Second, the faunas of the two formations are the equivalents, respectively, of the late Middle Cambrian fauna in the bedded limestone block in the Lévis Formation and of the Dresbachian fauna in boulders in that formation (2.3.2). Third, the Corner-of-the-Beach Formation must be about equivalent, in age, to the *Centropleura* beds in eastern Gaspé (2.3.3); these are the nearest fossiliferous Cambrian strata to the west of this area, but they are of basinal facies, not cratonic.

The two formations at Percé, then, are representatives of the cratonic facies (or at the very least of the 'intermediate facies' in the sense used by LOCHMAN-BALK and WILSON, 1958), in a region in which the extracratonic or basinal facies would be confidently expected. Several explanations present themselves; all are structurally intriguing. The first is that the whole outcrop is an exotic block in the Ordovician rocks, as the similar but smaller outcrop near Lévis is believed to be (OSBORNE, 1956). The second is that the Logan fault is here following the base of the Ordovician strata, and that the Cambrian outcrops are a window through it. This would imply a large forward translation for the thrust; it reaches the surface under the St. Lawrence River in a zone usually placed some 50 kilometres or more north of Percé (see, for example, NEALE and others, 1961, Fig. 2). This explanation is easily acceptable to those, including the present writer, who regard all or most of the Quebec Group as allochthonous (HENDERSON, 1958).

A third explanation is that the outcrop area is in fact a window and is very much closer to the thrust-front than it appears to be because the thrust-front turns abruptly southwards off the east end of the Gaspé Peninsula and follows around the Gulf of St. Lawrence to join the great fault zone at Cape Ray in the south-western corner of Newfoundland. A fourth explanation is that the thrust-front diverges widely from the facies boundaries in eastern Gaspé, the boundary between the cratonic and basinal facies in the Middle Cambrian cutting transversely across the Gaspé Peninsula between the Shickshock Mountains and Percé, and then extending across the Gulf of St. Lawrence to Cape Ray. This problem will be returned to in the discussion of the Maquereau Group, one of the successions for which a Cambrian age is merely a possibility (2.5.4).

2.4.3. Southernmost Ontario and Quebec

The Ontario Peninsula, between Lakes Huron and Erie, contains both the north-eastern quadrant of the Michigan basin and the northernmost end of the frontal Appalachian basin (Fig. 6). The axis of the peninsula, the so-called Algonquin arch, separates these two basin edges; along it, Pre-Cambrian crystalline rocks are overlain in the subsurface by the limestones of the Middle Ordovician transgression. Off the flanks of the axial arch, however, three sandy formations of Cambrian age lie in the subsurface between the Middle Ordovician and the basement. How these three formations are correlated depends on whether correlation is made with the Michigan and northern Ohio section to the west, or with the Appalachian foreland section in eastern Ohio and Kentucky to the south.

By correlation with the Michigan basin the three formations reaching into Ontario (ROLIFF, 1954) are called, from bottom to top, the Jacobsville Formation (coloured sandstone with a basal conglomerate), the Mount Simon Formation (more or less white sandstone, up to 45 metres thick), and the Eau Claire Formation (pale sandstone and sandy dolomite, with abundant glauconite). The formations continue to the north-west, beneath the islands which separate Lake Huron from Georgian Bay and which reflect the extension of the Niagara escarpment around the north side of the Michigan basin. On St. Joseph Island, one of those belonging to Canada, the Jacobsville Sandstone appears at the surface, and the sandstones above it are grouped together as the Munising Formation (LIBERTY, 1967). Still younger Upper Cambrian beds of Michigan and Ohio are not known to extend into Canada, according to this correlation.

A correlation with the Appalachian basin led POUNDER (1964) to suggest that the Cambrian units represented are much older. The lowest unit was correlated with the topmost Lower Cambrian Rome Formation, of sandstone and dolomite; the second with the Conasauga shale, which is largely Middle Cambrian and intertongues westward with carbonates; and the uppermost with the Maynardsville Formation, a dolomite-sandstone unit of latest Dresbachian age.

SANFORD and QUILLIAN (1959) compromised by partitioning the peninsula between the two basins, with an arbitrary separation along the 81st meridian. They considered, however, that the two successions were equivalents in all essential respects. In the western section, they did not recognize the Jacobsville Sandstone at the base, but they added an extra formation above the Eau Claire. This extra formation is not a Franconia equivalent, which SANFORD and QUILLIAN believed to be missing, but the Trempealeau, a buff to reddish dolomite with sand at the top. In Ontario, it varies from zero to 60 metres thick. The equivalents on the New York side of the peninsula are the Potsdam Sandstone (like the Mount Simon, but with a basal arkosic conglomerate); the Theresa Formation, (like the Eau Claire but reaching 125 metres in thickness and probably occupying a longer stratigraphical interval); and the Little Falls, a Trempealeauan dolomite at the top.

Each of these formations overlaps the ones below it, and the top of the Cambrian is part of the subcrop of the regional Knox unconformity below the Middle Ordovician. A feature of the sands in the succession is that rounded and frosted grains are common, and within the dolomites the sand grains 'float'. The sandy member at the top of the Trempealeau on the Michigan side of the peninsula is a good example. CALVERT (1964) believed that the sand deposition of this whole region was largely aeolian, due to lack of vegetation cover in Cambrian and Ordovician time.

East of the Frontenac axis, which crosses the St. Lawrence at the eastern end of Lake Ontario, the basal Palaeozoic sandstone is called the Nepean Formation. It conformably underlies the Ordovician of the Ottawa–St. Lawrence valley, and could be either earliest

Ordovician or latest Cambrian in age. It includes talus material from the Frontenac axis itself, fluvial sands and conglomerates, and laminated shoreline sands showing desiccation features. Going east, these basal sandstones thicken considerably and their base becomes progressively older. They have for many years been assigned to a generalized Potsdam Formation, and the history of the usage of this name has been examined by CLARK (1966). CLARK proposed that the basal clastic units south of Montreal should be divided into two formations, each with two members, and that these two formations be assigned to the Potsdam Group, of Trempealeauan age. The lower formation, the Covey Hill, approaches 600 metres in thickness and is virtually all sandstone (with conglomerate below). The red, brown, and buff colours suggest reworked alluvial fan deposits, spread towards the south-east according to the cross-bedding. The upper formation (Châteauguay) is a thinly-bedded orthoquartzite, nearly white, with dolomitic interbeds increasing upwards in its upper member. The lower member contains abundant *Lingulepis acuminata* (CONRAD); the upper member may be equivalent to the March Formation of the Ottawa area, which overlies the Nepean sandstone and is usually assigned to the Lower Ordovician. The Châteauguay Formation is about 200 metres thick.

Eastwards from Montreal, the Potsdam Group thins out, and beyond a point about 100 kilometres down-river from the city it has not been recognized.

In the frontal thrust of the Sutton Mountains anticlinorium (the Champlain or Logan thrust), the Rock River dolomite lies below the Ordovician section and is probably of Late Cambrian age. Whether clastic rocks like the Potsdam lie below the dolomite is not known, because its base is cut off by the thrust fault. Proceeding eastwards, however, the Cambrian becomes almost entirely dolomite. The Middle and Upper Cambrian Saxe Brook dolomite lies directly on the Dunham Formation, also a dolomite but with limestone members carrying Lower Cambrian trilobites. The reason for the formational division is that, across the International Boundary in Vermont, the well-known Parker slate intervenes between the two dolomite units. The base of the Dunham is again cut off by a thrust, eliminating the thick quartzite unit elsewhere known to underlie it.

Between this thrust and the axis of the Sutton Mountains (2.3.2), a complete interfingering is observed between the carbonate–quartzite rocks characteristic of the foreland and the greywacke–shale units of the geosyncline. Some 1,250 metres of Cambrian beds are exposed across a west-facing escarpment; all are gathered into the Oak Hill Group of CLARK (1934). In general, the carbonate–quartzite units thin towards the east and disappear; the quartzites especially disappear eastwards, commonly between beds of dolomite. The greywacke–shale units thin towards the west, and disappear. The carbonates closest to the facies boundary typically include carbonate breccias. This association between the two principal facies, well described by CADY (1960), is illustrated nowhere else in the Cambrian of eastern Canada.

The three lowest formations in the Oak Hill Group all represent the basinal facies: the basal Tibbitt Hill chlorite–schist, derived from sheared volcanic rocks and possibly Pre-Cambrian; the Call Mill slate, no more than 30 metres thick at this position; and the Pinnacle greywacke, with phyllite and arkose. This last formation, about 125 metres thick, illustrates an apparently unusual situation in that its place in the succession farther north is taken by a white quartzite presumably derived from the craton. Overlying the Pinnacle Formation is the first of the foreland fingers, the White Brook dolomite, less than ten to more than twenty metres thick; then the West Sutton slate, thickening rapidly eastwards to over 75 metres and becoming reddish to the north. Next comes the thickest of the foreland formations, the Gilman quartzite, which becomes a siliceous shale to the north and east. It is about 600 metres thick and carries the lowest of the known Cambrian fossils of this section (brachiopods, including *Kutorgina*). It is overlain by the Dunham dolomite, already referred to, and that in turn is overlain by the thin Oak Hill slate, the only sliver of the Parker slate which extends into Canada from Vermont. The Parker carries latest Lower Cambrian trilobites. The section ends with the thin Scottsmore quartzite and conglomerate, and the thick Sweetsburg slate, a prominent black unit with thin, white, siliceous interbeds. These two formations grade westwards into the Saxe Brook dolomite, already mentioned. There is no evidence of any significant breaks in this Cambrian section.

The entire Oak Hill Group grades north-eastwards into the greywacke-shale assemblage, but it retains interbeds of quartzite, dolomite, or limestone. The Gilman and Scottsmore quartzites, in particular, both extend into the region dominated by the shale facies, probably indicating brief episodes of stillstand (CADY, 1960). In the Sutton Mountains, the whole succession is represented within the Sutton 'schists' (2.3.2).

2.5. Rocks of possible but unproven Cambrian age

We have described three principal rock associations in the Cambrian of eastern Canada: a typical foreland association of carbonates and quartzites; a typical basinal association of thick greywackes and coloured mudstones, now considerably metamorphosed in places; and a reduced or condensed mudstone facies. The first two appear, from the description so far, to be restricted to what SCHUCHERT (1923) called the St. Lawrence geosyncline; they probably both lie on a foundation of Grenville metamorphic rocks. The third association appears to be restricted to the outer or Acadian geosyncline of SCHUCHERT; it is known to have been deposited on a post-Grenville succession of Pre-Cambrian sedimentary and volcanic rocks, and there is at present no evidence of Grenville basement below these.

If a 'New Brunswick geanticline' existed during the Cambrian, it is difficult to understand why there is no cratonic facies associated with the condensed shale facies south-east of it. If no 'New Brunswick geanticline' existed, it is equally difficult to understand why there is no coarse, unfossiliferous, clastic facies on one or other side of the condensed shale facies. Such coarse facies do in fact exist, in abundance, but they are normally referred either to the Lower Ordovician or to the Pre-Cambrian. The reasons for these assignments are that the oldest diagnostically fossiliferous rocks in a number of the areas are Middle Ordovician, allowing an Ordovician assignment to unfossiliferous rocks below them; that the region is also one of classically-thick Pre-Cambrian sedimentary and volcanic rocks, so that assignment to unfossiliferous sediments of a Pre-Cambrian age causes no demur; and, most persuasive of all, that the Cambrian sedimentary rocks of Maritime Canada are very fossiliferous, so it is obvious that unfossiliferous rocks in the region cannot be Cambrian.

The present writer believes that some of the old, unfossiliferous successions of uncertain age could just as easily include Cambrian rocks as Pre-Cambrian or Ordovician rocks. Some, indeed, may much more logically be regarded as Cambrian than as either Pre-Cambrian or Ordovician. Seven such groups of strata will be described, with the clear proviso that in no case is Cambrian age more than a reasonable possibility. The areas occupied by the seven groups are shown in Figs 3, 7, and 8.

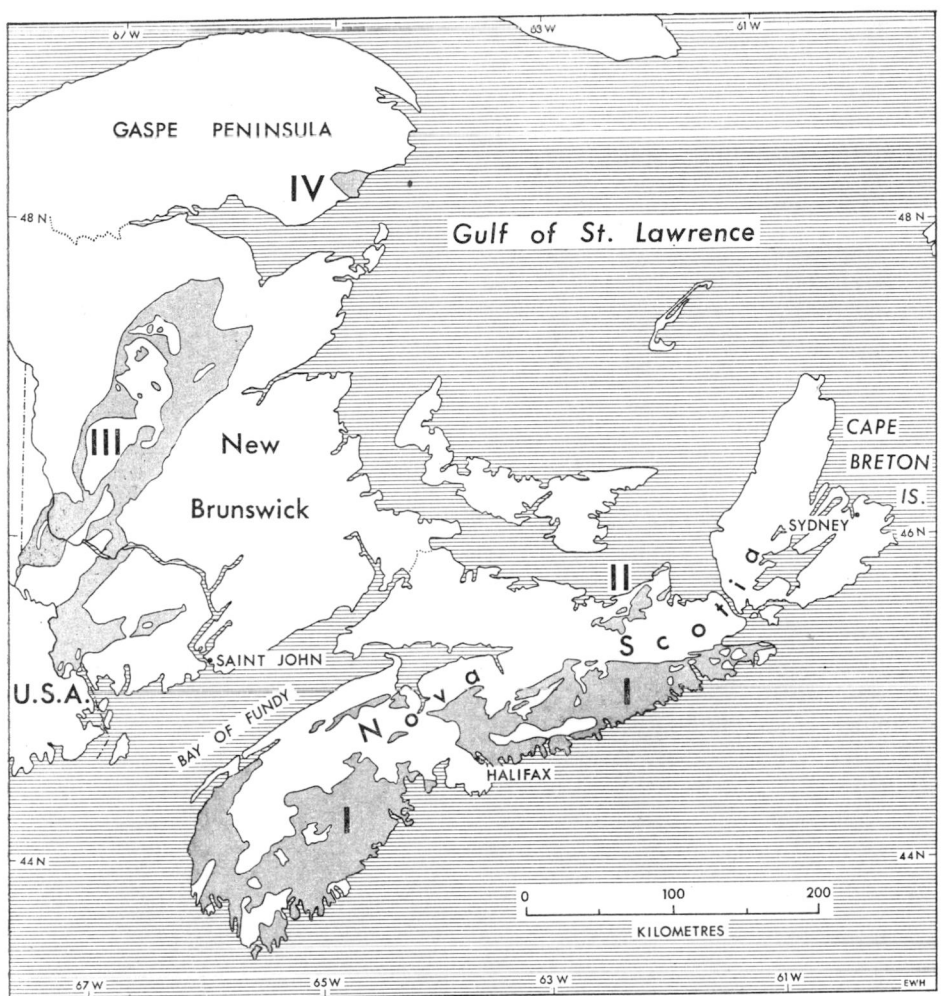

FIG. 8. Possible Cambrian rocks in Maritime Canada, exclusive of Newfoundland.
I, Meguma Group, including Goldenville Formation; II, Browns Mountain Group;
III, variously named rocks of central New Brunswick; IV, Maquereau Group.

2.5.1. Goldenville Formation, Nova Scotia

The Goldenville Formation is the older of the two formations which make up the famous Meguma Group, occupying much of the main peninsula of Nova Scotia. It has been described in many publications, most recently by TAYLOR and SCHILLER (1966) and by TAYLOR alone (1967).

The Goldenville is Tremadoc in age or older, because it conformably and gradationally underlies a slate formation with a few Tremadoc graptolites in it. As there are nearly 6,000 metres of the formation exposed, with the base not seen, it can almost be taken as certain that much of the Cambrian, and perhaps all of it, is represented.

The formation consists of greenish-grey lithic greywackes and feldspathic quartzites, with small amounts of interbedded slate or argillite which increase towards the top. The clastic sediments appear to have been deposited by north-east-flowing turbidity currents, in fairly deep water (CAMPBELL and SCHENK, 1967), and to have been derived from a source to the south-west. Graded bedding is common, and in those areas in which metamorphism has affected the formation one result is that mica-schists form the tops of many graded beds. The rocks are rich in both biotite and pyrite.

The Meguma Group apparently continues under a cover of Carboniferous rocks for at least 150 kilometres, and possibly much further, off the south coast of Nova Scotia (measured at right angles to the coast). This is inferred from magnetic data; the group reveals a characteristic pattern of linear magnetic contours parallel to the coast-line (HOOD, 1967).

According to NEALE and others (1961), the rocks of the Meguma Group were not folded until the Acadian deformation during the Devonian. Tectonically, therefore, mainland Nova Scotia behaved much as did the Avalon Peninsula, different though the two areas were in depositional history. The area occupied by the Meguma Group as a whole was included in a Cambrian landmass, Novascotica, by adherents of the New Brunswick geanticline hypothesis (Fig. 2).

2.5.2. Browns Mountain Group, Nova Scotia

Occupying a small area in the Antigonish highlands of north-central Nova Scotia, the Browns Mountain Group consists of mudstones, red and grey slates, and greywacke, with some interbedded volcanic flows and tuffs. The clastic rocks above them are believed to be Ordovician. The group itself has yielded a few fossil linguloids which could be Early Ordovician or Cambrian in age. The Ordovician age is preferred, because some oolitic hematite is associated and the analogy with the Wabana iron ore of Newfoundland suggests an Early Ordovician assignment. However, the base of the group is not seen. It may continue downwards into the Cambrian; as the area lies within the Acadian geosyncline even on the SCHUCHERT reconstruction, it is perfectly possible that Cambrian rocks of the condensed, Atlantic type, as exposed on Cape Breton Island (2.2.3), underlie the Browns Mountain Group.

2.5.3. Unnamed beds in central New Brunswick

From the Baie de Chaleur to the Maine border, New Brunswick is cut in half by a band of old rock which separates the Gaspé synclinorium from the Carboniferous basin of the central part of the province. The band, which constitutes the province's mineral belt, may be called the Miramichi geanticline (POOLE, 1967).

The band exposes quartzose argillite, phyllite, schist, and quartzite of unknown thickness, which grade into gneisses to the south-west. The rocks appear to pass up conformally into a eugeosynclinal succession of Middle or Late Ordovician age. The upper part of the older group carries a shelly fauna referable to either the Lower Ordovician or the early Middle Ordovician (W. H. POOLE, personal communication, 1967). The shaly Charlotte Group in the southern part of the province has yielded Arenig graptolites (CUMMING, 1967a). The sedimentary rocks are intruded and metamorphosed by granitic bodies which have always been assigned a Devonian age but some of which are pre-Devonian, and probably late Ordovician, according to isotopic data (POOLE, 1963).

The area concerned may be called the type locality for the New Brunswick geanticline in Cambrian time. Yet, in north-eastern Maine, at least one formation of probable Cambrian age crops out in an inlier striking directly towards western New Brunswick (NEUMAN, 1962; EKREN and FRISCHKNECHT, 1967). It is not exposed at the surface in New Brunswick, but the probability that it underlies the rocks seen there is clear.

2.5.4. Maquereau Group, Gaspé Peninsula

The Maquereau Group crops out between Port-Daniel and Chandler, on the south-east corner of Gaspé. It consists of more than 8,000 metres of clastic sedimentary and volcanic rocks, completely surrounded by faults except where they form the sea coast. The whole succession is pre-Middle Ordovician because it contributed fragments to a thick conglomerate in the adjacent Mictaw Group, which contains Middle Ordovician graptolites (2.6).

The group has been divided into formations by AYRTON (1967). The Newport Formation comprises purple and green, poorly stratified quartzose greywacke with oriented purple slate fragments, and red-banded siltstones. The formation is about 2,500 metres thick, and rests with apparent unconformity (though the actual contact is not seen) on the Port-Daniel Formation. This is a succession, more than 5,000 metres thick, of green, red, and purple clastic rocks and andesitic volcanic rocks (including pillow lavas).

The Port-Daniel Formation in turn is in fault contact with a third formation, the Chandler, which is not unlike the Port-Daniel Formation lithologically but appears to lack conglomerates and volcanic rocks. Its thickness and its original relation to the other two formations are unknown.

ALCOCK (1935) likened the rocks of the Maquereau Group to those lying north-east of Lake Matapédia, in the Shickshock Mountains; those rocks are now referred to the Matane River and Shickshock Groups (2.3.3). ALCOCK also believed that the Maquereau Group was cut by intrusions which did not penetrate the younger Ordovician rocks. This has been taken as evidence of a pre-Trenton, and probably Cambrian, deformation (2.6). Later work has not substantiated a pre-Trenton intrusive episode, but there is no doubt that the Maquereau rocks had been folded, somewhat metamorphosed, and eroded before Trenton time.

The rocks of the Maquereau Group could not be more unlike the well-bedded limestones and shales of the two cratonic formations cropping out near Percé, less than 40 kilometres to the north-east. This means that one of the odd structural circumstances set out in connection with those formations (2.4.2) must apply, unless both the Maquereau Group and the Matane River–Shickshock Groups are Pre-Cambrian. OLLERENSHAW's trilobite seems to eliminate that possibility for the Matane River Group, but the Maquereau Group could admittedly as easily be Pre-Cambrian as Cambrian.

2.5.5. Fleur de Lys Group, western Newfoundland

Rocks of this group form the east shore of White Bay, a fault-controlled arm of the sea in north-western Newfoundland (Fig. 7). They occupy a north-north-east trending anticlinorium, the core of which is formed of micaceous schists and gneisses, some of them quartzitic and some, which might be called eclogites, garnetiferous. On the flanks are bands of white marble, amphibolite, and conglomerates with stretched quartz pebbles. Thin sills and dykes of meta-gabbro are common in the group, the total thickness of which may exceed 6,000 metres.

NEALE and NASH (1963) believed that at least the flanking sedimentary rocks of the group are likely to belong to the Lower Palaeozoic. They appear to be older than the Baie Verte Group, which post-dates the initial phases of Fleur de Lys deformation. The Fleur de Lys Group may possibly, though not certainly, in that event be pre-Arenig, and the first phase of its deformation and metamorphism may be of an age comparable with the Sardic episode in the Dalradian region of Scotland (NEALE and KENNEDY, 1967). WILLIAMS (1964) agreed that the group as a whole is no younger than Middle Ordovician and no older than late Pre-Cambrian, and considered it to be probably the basal clastic sequence and overlying carbonates of the cratonic Cambrian succession of western Newfoundland.

If the Humber Arm Group of the west coast of the island is allochthonous (2.3.4), it presumably originated as a basin-margin facies immediately east of the present Fleur de Lys anticlinorium, incorporated some material from it, and then overrode it, all subsequent to the initial ('Sardic') phase of Fleur de Lys deformation but prior to the final (Acadian) phase of its metamorphism. By the time the latter took place, the Humber Arm Group was safely transferred too far to the west to be affected by the intrusions.

2.5.6. Gander Lake Group, eastern Newfoundland

The Gander Lake Group occupies a structural position in north-eastern Newfoundland comparable with that of the Fleur de Lys Group in the north-west (WILLIAMS, 1964). It consists of three unnamed units (JENNESS, 1963). The lowest rests on the Pre-Cambrian and represents the greywacke–slate assemblage by a monotonous and thinly-bedded sequence of micaceous and quartzose sandstones, greywackes, siltstones, and slates. The thickness cannot be accurately assessed, but it may be over 3,000 metres. Volcanic rocks are almost absent. The clastic rocks are all more or less metamorphosed, the grade of regional metamorphism increasing from the chlorite zone in the north-west to the biotite zone in the south-east. In places, the rocks have also been metamorphosed to coarse gneisses and schists during the Devonian deformation.

This lowest unit is overlain by the second or 'middle unit', of pyroclastic rocks and slates containing Middle Ordovician graptolites. One locality believed to represent nearly the top of the lowest unit (JENNESS, 1963) yielded species of *Climacograptus*.

Whether any part of the lowest unit reaches stratigraphically downwards into the Cambrian cannot be determined at present. On New World Island to the north, however, a limestone interbedded or incorporated within pillow lavas has yielded ellesmeroceratids (KAY, 1967) and pliomerids, suggesting an age between Tremadoc and Arenig. As no late Cambrian deformation can be detected in the area, it is as likely as it is in New Brunswick (3.5.3) that Cambrian beds lie below. Indeed, Middle Cambrian trilobites have now been reported (HORNE, 1968; KAY and ELDREDGE, 1968) from the boulders in a Wildflysch-type of deposit which covers a large area in south-eastern Notre Dame Bay. Below Caradoc graptolite-bearing shales, on New World and Dunnage Islands (Dunnage Island being actually a pair of small islands between New World Island and the main island of Newfoundland), lies a very complex boulder- and pebble-bearing argillite succession called the Dildo Sequence (KAY, 1967). It has been likened to the Argille Scagliose of the Apennines (HORNE, 1968). Lying still farther down in the sequence is a lava-greywacke assemblage containing bedded limestone members.

The limestone has yielded trilobites identified as *Bailiaspis* and *Kootenia* (KAY and ELDREDGE, 1968). These fossils are not only the oldest ever found along the 'New Brunswick geanticline', and the first of undoubted Cambrian age. They are also, if correctly identified, the first clear association of a typical Atlantic and a typical Pacific form. *Bailiaspis*, a blind conocoryphid, is a Swedish genus common in the condensed basinal facies of the Atlantic seaboard (HUTCHINSON, 1952b, 1962); *Kootenia* is a Quebec and Cordilleran genus (Table 3). Further comment on this new discovery will have to await more extensive collecting and confirmation of generic identification.

The Gander Lake Group represents a succession very similar to that in the Meguma Group of Nova Scotia (2.5.1), even though inevitable diachronicity has caused variation in the age of the lower parts of the slate component. In SCHUCHERT's reconstruction, the Gander Lake Group lies on the axis of the New Brunswick geanticline.

2.5.7. Baie d'Espoir Group, eastern Newfoundland

The Baie d'Espoir Group appears to be the structural and stratigraphical continuation of the Gander Lake Group to the south-west. It is now separated from that group by the large Devonian intrusive which carries as roof pendants the Cambrian rocks north-east of Fortune Bay (2.2.2). It differs from the Gander Lake Group only in that it passes downwards and eastwards into granitic gneiss and is not known to have Pre-Cambrian rocks adjacent to it. Otherwise, the same comments apply to its lower member as to the lower member of the Gander Lake Group.

2.6. Diastrophism and isotopic dates, Cambrian of eastern Canada

No significant episode of folding can be invoked during Cambrian time in eastern Canada. At least two significant episodes of *uplift* have, however, been persistently referred to in the literature.

Both these episodes have been based on SCHUCHERT's contention that either the Middle or the Upper Cambrian was absent by non-deposition in all or parts of the so-called St. Lawrence geosyncline (1923, p. 178). Uplift was invoked for most of this trough at the close of Early Cambrian time, followed by erosion for most of the Middle Cambrian (WEEKS, 1957, p. 143). At or about the close of the Cambrian, a further episode of uplift was claimed to have affected north-western Vermont and adjacent parts of Quebec, accompanied by tilting and even possibly some folding. This episode, described as an 'arch-making movement' by SCHUCHERT (1930, p. 708), was called upon to divide the St. Lawrence geosyncline into two seaways by elevation of the Green Mountain–Sutton Mountain 'anticlinorium'. It has been called the Vermont or Green Mountain disturbance.

The Middle Cambrian is now established to be present over much of the St. Lawrence geosyncline (2.3; 2.4), though there are at the same time both cratonic and basinal segments of the geosyncline in which it appears to be absent, as at Lévis (SCHUCHERT and DUNBAR, 1934; RASETTI, 1946b). This absence is less likely to have been due to an isolated, identifiable episode of uplift, however, than to persistent minor instability through much of Cambrian time. This is demonstrated by the apparent absence of different parts of the succession in different regions (Tables 1–4). Examples are the persistent lack of early Middle Cambrian faunas from the 'Pacific' realm, in Quebec and western Newfoundland, and the absence of the *Lejopyge* Zone from the 'Atlantic' realm. Even these are only apparent palaeontological gaps; no regional disconformities can be identified in the field. The minor instability is further demonstrated by the occurrence, almost throughout the succession, of formations containing boulder beds and exotic material. The adventitious greywackes in the Lower Cambrian Charny shales, the blocks in the Cow Head breccias, and limestone conglomerates in the Lévis and Kamouraska Formations are merely the best-known examples.

The second uplift, at or close to the end of the Cambrian, can be postulated upon quite different criteria, but it would no longer be appropriate to refer to it as the Green Mountain disturbance. ALCOCK (1935) believed that the Maquereau Group (2.5.4) was intruded by serpentinite, amphibolite, and granite, and metamorphosed, sufficiently early to supply material to the basal conglomerate of the Mictaw Group, which is dated as Middle Ordovician by graptolites. ALCOCK concluded that this intrusive episode was an aspect of a deformation during Cambrian time (a deformation referred to by others as the Gaspesian orogeny).

The conglomerate referred to is a major formation at least 300 metres thick, composed almost entirely of Maquereau material. AYRTON (1967) doubted that it is a basal conglomerate, however. It seems rather to be a sedimentary breccia within the Mictaw Group, representing a talus deposit from a fault scarp of Maquereau rocks. If this is in fact what it is, the deformation responsible for it would be Middle Ordovician, not Cambrian, and it would post-date the regional Knox unconformity (v.i.).

In a similar way, COOKE (1955) and RIORDON (1957) deduced a Cambrian folding and thrusting episode in south-eastern Quebec. The succession for which COOKE and RIORDON used the group name 'Caldwell' was folded and metamorphosed before providing quartzitic debris to a breccia at the base of the Ordovician Beauceville Group. There is no evidence, however, that the Beauceville Group is any older than Middle Ordovician. The 'Caldwell Group' as used by COOKE probably extends into the Ordovician (2.3.2), and the unconformity below the Beauceville is again the Knox unconformity, which is post-Beekmantown and definitely Ordovician.

The most convincing evidence for Cambrian diastrophism anywhere in the St. Lawrence geosyncline is to be found in the Shickshock Mountains (2.3.3). OLLERENSHAW (1967) called for a late Cambrian or very early Ordovician deformation to be responsible for the metamorphism of the Shickshock Group, which is probably of Cambrian age. Tremadoc shales with abundant perfectly preserved graptolites occur along the river shore to the north and north-west (including the famous Matane localities of LAPWORTH and others). They seem quite unaffected by any metamorphism, but the loss of metamorphism northwards from the Shickshock Mountains seems to be gradational, and there is no clear evidence that the event producing it was pre-Tremadoc. The event has, however, been dated radiometrically as taking place 530 ± 35 m.y.; it may, of course, have been quite local.

Few of the occurrences of Cambrian igneous rocks offer any support to a proposal of serious Cambrian diastrophism. The basalts in the Lower Cambrian Bradore Formation at the Strait of Belle Isle (CHRISTIE, 1951; CLIFFORD, 1965) were probably subaerial. The lavas and pyroclastic rocks in the Middle Cambrian shales of eastern Newfoundland (2.2.1), Cape Breton Island (2.2.3), and southern New Brunswick (2.2.4) are parts of fossiliferous marine successions and have no large tectonic significance.

The lavas of the Shickshock Group (2.3.3) seem originally to have been basalts, probably of taphrogenic origin (OLLERENSHAW, 1967). However, they include pillow lavas, and some chert and tuff are associated. This assemblage reinforces the evidence noted above as suggesting some local Cambrian diastrophism in this part of Gaspé. The volcanic rocks in the Port-Daniel Formation of the Maquereau Group (2.5.4) would do the same if they could be proved to be Cambrian.

The widespread volcanic successions in the 'Cambro-Ordovician' of south-eastern Quebec may all be Ordovician. Greenstones, chloritic schists, and pillow lavas occur in the so-called Caldwell Group (COOKE, 1937; RIORDON, 1957); at the visible base of the Ascot Formation at Sherbrooke (ST.-JULIEN and LAMARCHE, 1965); in the upper part of the Brompton Formation to the west (ST.-JULIEN, 1961b); and in the Weedon schists to the north (BURTON, 1931). The trachytes, rhyolites, basalts, and tuffs below the Beauceville Group in the Lake Memphremagog area (COOKE, 1950; RIORDON, 1957) are almost certainly Ordovician.

A totally different type of igneous activity characterizes a south-eastern part of the Canadian shield. It is represented by a series of alkaline ring complexes with associated carbonatites, bearing a general spatial relationship to the Ottawa–Bonnechère graben system. Potassium-argon dates on biotites show that the complexes are of a great variety of ages, but at least one complex is isotopically Cambrian and stratigraphy does not contradict this assignment. The Manitou Island complex is within Lake Nipissing, east of the mining town of Sudbury and north-east of Lake Huron (Fig. 6). It is a ring complex of syenitic soda-pyroxene rocks, with associated coarse white carbonatite carrying large crystals of apatite, biotite, and sodic amphiboles and pyroxenes (LOWDON and others, 1963). The complex is intrusive into Grenville metamorphic rocks, and is partly covered unconformably by Middle Ordovician limestones. Biotite from the carbonatite has yielded an age of 560 m.y.

If this age is reliable (and it has been independently corroborated by a 570 m.y. age: GITTINS, MACINTYRE, and YORK, 1967), two possibilities are brought to light. Obviously other alkalic bodies within the shield may prove to be of possible Cambrian age.* In addition, however, the possibility arises that some of the numerous minor intrusions of other types, cutting rocks of the Grenville 'Series' in many places, may be Cambrian: dykes of diabase and lamprophyre, and small stocks of mafic rock. The immediate Appalachian foreland may therefore not have been as quiescent in latest Proterozoic and Cambrian times as has been imagined, but this does not alter the record of calm in the depositional basin.

The lack of clear evidence for Cambrian diastrophism is reflected in the survey, by NEALE and others (1961), of folding episodes in Canadian Appalachia. For those authors, the rocks now lying between the Champlain–Logan thrust front and the Notre Dame–Shickshock Mountains axis were folded only during the Ordovician (Taconian orogeny). Most of the rocks east and south of this deformational axis were folded only during the Devonian (Acadian orogeny). The rocks of the axial belt itself were folded in both orogenies.

This obscures a critical point. The Appalachian fold belt in Canada includes areas classically 'geosynclinal' during the Ordovician, the Silurian, and the Devonian. It also includes a narrow belt, from the Bay of Fundy to the Avalon Peninsula, which shows no evidence whatever of having presented a 'geosynclinal' régime at any time since the Pre-Cambrian. This belt is the so-called Acadian geosyncline (2.2). A non-orogenic Cambrian Period leading into a violently orogenic Ordovician would be expected to show closely related distributions of Cambrian and Ordovician successions. Alternatively, a non-orogenic Cambrian finally folded in the Devonian might be expected to show Cambrian rocks related to a thick Palaeozoic succession. The Acadian 'geosyncline' does neither; it was a hang-over from the Proterozoic, not a precursor of the Palaeozoic. As far as present evidence goes, it died during the Ordovician *without being folded*. During the Ordovician, it was a band of minimum deposition between two geosynclines of maximum deposition. If there had been no Cambrian diastrophism, the band was probably the same during the Cambrian, and not itself a geosyncline between two emergent mountain arcs, as required by SCHUCHERT. Its Cambrian history was a reflection of Proterozoic activity, not a preparation for Ordovician activity.

* Mr. M. SHAFIQULLAH of the author's own department informs him that at least one such body has already done so. The carbonatite from the Brent crater, on the northern edge of Algonquin Park about 90 kilometres east-south-east of the Manitou Islands complex, has yielded a potassium-argon date of 576 m.y. Like the Manitou Islands complex, the Brent body was known to be younger than the Grenville 'series' and older than Middle Ordovician.

The late Proterozoic Musgravetown and Hodgewater Groups of the Avalon Peninsula of Newfoundland consist of typical molasse sediments—red and green conglomerates, arkoses, and siltstones—with a basal volcanic formation, the Bull Arm. The molasse is overlain without significant discordance by the Random Formation (2.2.1). Below, the molasse passes down without significant discordance into an underlying group of Proterozoic sedimentary rocks, but below that group is a controversial unconformity above the Harbour Main Group.

The Harbour Main was deformed and intruded in a late Pre-Cambrian deformation (McCARTNEY, 1967), to yield the molasse material. This is the Avalonian orogeny of POOLE (1967). A granite emplaced during the deformation has been dated as 574 ± 11 m.y. by the rubidium–strontium isochron method (McCARTNEY and others, 1966). The granite is unconformably overlain by fossiliferous Lower Cambrian rocks along the south shore of Conception Bay. The Avalonian orogeny and its associated granite are thus exactly comparable with the Neponset orogeny in Massachusetts, and its associated Hoppin Hill granite (FAIRBAIRN and others, 1965).

The Bull Arm felsite, at the base of the molasse section, has yielded an isotopic date very close to 500 m.y. by the rubidium–strontium whole-rock method. So have the Coldbrook Group of the Saint John region and the Fourchu Group of Cape Breton Island (FAIRBAIRN and others, 1966). These three volcanic groups underlie the fossiliferous Lower Cambrian rocks of the three principal areas of the Acadian facies of the Cambrian; the contact in each case is essentially conformable. The age of 500 million years is too low even for the Lower Cambrian on any standard time scale; the M.I.T. geologists responsible for it consider this to be due to diffusion consequent upon further diastrophism since the rocks were deposited. The probable relationships between all these groups, and between them and the Cambrian, were well set out by WEEKS (1957, p. 138).

The conclusion is a startling one. The classical Acadian geosyncline of the Cambrian, with its Atlantic fauna, had ceased to be a geosyncline in Proterozoic time. During the Cambrian, it had the form of a rather stable submarine bank, platform or horst (POOLE, 1967), receiving a thin, condensed shale section, whilst its flanking basins (not flanking landmasses) began the subsidence that was to be carried on until the Devonian. This condition for the bank lasted at least until the end of Early Ordovician time (though on easternmost Cape Breton Island there is no Ordovician).

2.6.1. Later deformations affecting Cambrian rocks

Major overthrusting towards the north and west took place during the Taconian orogeny, which over the region as a whole occupied much of mid- and late-Ordovician time. The allochthonous masses in western Newfoundland are a reflection of these movements, as are several of the limestone breccias and conglomerates. The entire 'Quebec complex', along the south shore of the St. Lawrence, might be regarded as allochthonous (HENDERSON, 1958).

The principal episode of folding and intrusion affecting the Cambrian rocks took place during Devonian time (Acadian orogeny). The original shape of the Avalonian platform, carrying the condensed, black shale sub-facies (Bii) of the 'Atlantic' province, was still further disrupted by large-scale faulting in Westphalian (Cumberlandian) time. The faulting extends from Massachusetts to north coastal Newfoundland, and includes the ancestral faults of the Bay of Fundy graben system. Strike-slip movement, inevitably postulated, is by no means proven, but if it took place it must have been right-handed.

2.7. Economic Geology, Cambrian of eastern Canada

In rocks of Cambrian age in eastern Canada, metallic deposits are far more important than non-metallic deposits. None of the large or famous deposits of the region, however, are found in Cambrian rocks.

The best known district is the gold-producing area of Nova Scotia. The Goldenville quartzite, which must include rocks of Cambrian age even if it also includes older ones (2.5.1), was widely referred to for many years as the Gold-bearing Series. Mining began in 1862. The gold occurs in quartz veins which form saddle reefs in large anticlines; abundant pyrite and some arsenopyrite are associated. The deposits were assigned to the mesothermal category by LINDGREN. Elsewhere in the Goldenville Formation, scheelite is an important constituent in the quartz veins, accompanied by tourmaline and a manganese garnet.

The principal base metal mined is copper, and its host rock can nowhere be proven to be Cambrian. The deposits in the New Brunswick mineral belt (2.5.3) lie in a succession of metamorphic and semi-metamorphic rocks known to be Middle Ordovician or older; quite possibly all are Ordovician. Similarly the numerous small mines in the copper belt east of the Sutton Mountains, in southern Quebec, are almost all in rocks determinable only as 'Cambro-Ordovician' (2.3.2). The ore-bodies are typically concentrated along contacts between metavolcanic rocks, of the Ascot and Mansonville Formations, and either metasediments or Ordovician ultramafic rocks (ST.-JULIEN, 1965 and in press). The principal deposits were in the Sherbrooke area. They were pyritic, with chalcopyrite the principal ore mineral. In belts closer to the Sutton Mountains, pyrrhotite or bornite took the place of pyrite as the chief sulphide mineral.

The relative absence of lead–zinc mineralization, which is in marked contrast to its abundance in rocks of Cambrian age in western Canada (3.8), is no doubt due to the relative absence of carbonate rocks from the Cambrian of the deformed parts of Canadian Appalachia. The only deposit of any significance is at the Mindamar mine on Cape Breton Island, now out of operation. The Bourinot Group of volcanic rocks, where crossed by a major fault, is cut by numerous dykes. They caused alteration to talcose schists, which were later partly replaced by sulphides. Some copper sulphides occurred in addition to galena and sphalerite.

Also on Cape Breton Island there is a local development of iron formation in the Middle Cambrian slates. Thin bands of hematite, with some magnetite, are separated by slate; the ore appears to be syngenetic. In a similar stratigraphical position is the manganese band at the base of the Middle Cambrian shales around Conception Bay in eastern Newfoundland (2.2.1).

A less familiar type of deposit produces niobium, the carbon-combining metal used in steelmaking, from alkalic igneous complexes. The Newman deposit is in the Manitou Islands complex, which is apparently Cambrian in age (2.6). Uranium pyrochlore, averaging more than 0.5% of Cb_2O_5, occurs in a large fracture zone cutting across alkaline silicate and carbonatite rocks, which are also invaded by lamprophyre dykes.

There is very small production of oil and gas from Cambrian sandstones in two fields on the Ontario Peninsula. The Clearville oilfield and the Gobles oil and gas field are both in that part of the peninsula which represents the northern subsurface end of the Appalachian basin (2.4.3). The equivalent Potsdam sandstone in the Montreal area is under investigation as a possible underground gas-storage reservoir.

272 F. K. NORTH

2.8. Bibliography: eastern Canada

ALCOCK, F. J. (1935). Geology of Chaleur Bay region. *Geol. Surv. Canada Mem.* **183**.

ALCOCK, F. J. (1938). Geology of Saint John region, New Brunswick. *Geol. Surv. Canada Mem.* **216**.

ANDERSON, F. D. (1965). Belleoram, Newfoundland. *Geol. Surv. Canada* Map 8-1965.

AYRTON, W. G. (1967). Chandler–Port Daniel area. *Quebec Dept. Nat. Res. Geol. Rept.* **120**.

BAIRD, D. M. (1960). Observations on the nature and origin of the Cowhead breccias of Newfoundland. *Geol. Surv. Canada Paper* **60–3**.

BÉLAND, J. R. (1957). St. Magloire and Rosaire–St. Pamphile areas. *Quebec Dept. Mines, Geol. Surv. Branch, Geol. Rept.* **76**.

BÉLAND, J. R. (1962). Ste. Perpétue area, Kamouraska and L'Islet Counties. *Quebec Dept. Nat. Res. Geol. Rept.* **98**.

BETZ, FREDERICK, JR. (1939). Geology and mineral deposits of the Canada Bay area, northern Newfoundland. *Newfoundland Geol. Surv. Bull.* **16**.

BRADLEY, D. A. (1962). Gisborne Lake and Terrenceville map-areas, Newfoundland. *Geol. Surv. Canada Mem.* **321**.

BRÜCKNER, W. D. (1966). Stratigraphy and structure of west-central Newfoundland. *Geol. Assoc. Canada Guidebook*, Geology of parts of Atlantic provinces, Halifax, 137.

BURTON, F. R. (1931). Vicinity of Lake Aylmer, eastern Townships. *Ann. Rept. Quebec Bureau of Mines* for 1930, pt. D, 99.

CADY, WALLACE M. (1960). Stratigraphic and geotectonic relationships in northern Vermont and southern Quebec. *Geol. Soc. America Bull.* **71**, 531.

CALVERT, WARREN L. (1964). Pre-Trenton sedimentation and dolomitization, Cincinnati arch province: theoretical considerations. *Amer. Assoc. Petroleum Geol. Bull.* **48**, 166.

CAMPBELL, F. H. A. and SCHENK, P. E. (1967). Paleocurrent and basin analysis of the Meguma Group, Nova Scotia. *Maritime Sediments* **3**, 27.

CHRISTIE, A. M. (1950). Geology of Bonavista map area, Newfoundland. *Geol. Surv. Canada Paper* **50–7**.

CHRISTIE, A. M. (1951). Geology of the southern coast of Labrador from Forteau Bay to Cape Porcupine, Newfoundland. *Geol. Surv. Canada Paper* **51–13**.

CLARK, T. H. (1934). Structure and stratigraphy of southern Quebec. *Geol. Soc. America Bull.* **45**, 1.

CLARK, T. H. (1966) Châteauguay area. *Quebec Dept. Nat. Res. Geol. Rept.* **122**.

CLIFFORD, P. M. (1965). Palaeozoic flood basalts in northern Newfoundland and Labrador. *Can. Jour. Earth Sci.* **2**, 183.

COOKE, H. C. (1937). Thetford, Disraeli, and eastern half of Warwick map-areas, Quebec. *Geol. Surv. Canada Mem.* **211**.

COOKE, H. C. (1950). Geology of a southwestern part of the Eastern Townships of Quebec. *Geol. Surv. Canada Mem.* **257**.

COOKE, H. C. (1954). The Green Mountain anticlinorium in Quebec. *Geol. Assoc. Canada Proc.* **6**, 37.

COOKE, H. C. (1955). An early Palaeozoic orogeny in the eastern townships of Quebec. *Geol. Assoc. Canada Proc.* **7**, 113.

CUMMING, L. M. (1967a). Geology of the Passamaquoddy Bay region, Charlotte County, New Brunswick. *Geol. Surv. Canada Paper* **65–29**.

CUMMING, L. M. (1967b). Platform and Klippe tectonics of western Newfoundland: a review. *Roy. Soc. Canada Spec. Pub.* **10**, 10.

EKREN, E. B. and FRISCHKNECHT, F. C. (1967). Geological–geophysical investigations of bedrock in the Island Falls quadrangle, Aroostook and Penobscot Counties, Maine. *U.S. Geol. Surv. Prof. Paper* **527**.

FAIRBAIRN, H. W., BOTTINO, M. L., PINSON, W. H., and HURLEY, P. M. (1966). Whole-rock age and initial 87Sr/86Sr of volcanics underlying fossiliferous Lower Cambrian in the Atlantic Provinces of Canada. *Can. Jour. Earth Sci.* **3**, 509.

FAIRBAIRN, H. W., MOORBATH, S., RAMO, A. O., PINSON, W. H., and HURLEY, P. M. (1965). Rb-Sr age of granitic rocks of southeastern Massachusetts and the age of the Cambrian at Hoppin Hill. *M.I.T. 13th Ann. Prog. Rept.* 3.

GITTINS, J., MACINTYRE, R. M., and YORK, D., (1967). The ages of carbonatite complexes in eastern Canada. *Can. Jour. Earth Sci.* **4**, 651.

HAY, P. W. (1968). Geology of the St. George–Seven Mile Lake area, southwestern New Brunswick. *New Brunswick Dept. Nat. Res. Map Series* **68–1**.

HAYES, A. O. and HOWELL, B. F. (1937). Geology of Saint John, New Brunswick. *Geol. Soc. America Spec. Paper* **5**.

HENDERSON, W. R. S. (1958). 'Blountian' allochthone in Appalachians of Quebec. *Alberta Soc. Petrol. Geol. Jour.* **6**, 120.

HOOD, P. J. (1967). Geophysical surveys of the continental shelf south of Nova Scotia. *Maritime Sediments* **3**, 6.

HORNE, G. S. (1968). Chaotic clastic deposits of Early-Medial Ordovician age in the central mobile belt of northeastern Newfoundland. *Geol. Soc. America, Northeastern Section*, 1968 *Annual Meeting Program*; Washington, D. C., abstracts, 36.

HOWELL, B. F. (1925). The faunas of the Cambrian Paradoxides beds at Manuels, Newfoundland. *Bull. American Paleont.* **11**, no. 43.

HOWELL, B. F. (1926). The Cambrian-Ordovician stratigraphic column of southeastern Newfoundland. *Can. Field-Naturalist*, **40**, 52.

HOWELL, B. F. (1943), Faunas of the Cambrian Cloud Rapids and Treytown Pond Formations of northern Newfoundland. *Jour. Paleont.* **17**, 236.

HUBERT, CLAUDE (1965). Stratigraphy of the Quebec complex in the L'Islet-Kamouraska area, Quebec. *Maritime Sediments* **1**, 13.

HUBERT, CLAUDE (1967). Tectonics of part of the Sillery Formation in the Chaudière-Matapédia segment of the Quebec Appalachians. *Roy. Soc. Canada Spec. Pub.* **10**, 33.

HUTCHINSON, R. D. (1952a). Middle Cambrian of the Atlantic realm in eastern Gaspé. *Amer. Jour. Sci.* **250**, 275.

HUTCHINSON, R. D. (1952b). The stratigraphy and trilobite faunas of the Cambrian sedimentary rocks of Cape Breton Island, Nova Scotia. *Geol. Surv. Canada Mem.* **263**.

HUTCHINSON, R. D. (1956). Cambrian stratigraphy, correlation, and paleogeography of eastern Canada. *El Sistema Cambrico, Symposium, XX Int. Geol. Cong., Mexico*, 289.

HUTCHINSON, R. D. (1962). Cambrian stratigraphy and trilobite faunas of southeastern Newfoundland. *Geol. Surv. Canada Bull.* **88**.

JENNESS, S. E. (1963). Terra Nova and Bonavista map-areas, Newfoundland. *Geol. Surv. Canada Mem.* **327**.

JONES, I. W. (1935). Dartmouth River map-area, Gaspé Peninsula. *Quebec Bureau of Mines, Ann. Rept.* 1934, 3.

KAY, MARSHALL (1967). Stratigraphy and structure of northeastern Newfoundland bearing on drift in North Atlantic. *Amer. Assoc. Petroleum Geol. Bull.* **51**, 579.

KAY, MARSHALL and ELDREDGE, N. (1968). Cambrian trilobites in central Newfoundland volcanic belt. *Geol. Mag.* **105**, 372.

KINDLE, C. H. (1942). A Lower (?) Cambrian fauna from eastern Gaspé, Quebec. *Amer. Jour. Sci.* **240**, 633.

KINDLE, C. H. and WHITTINGTON, H. B. (1958). Stratigraphy of the Cow Head region, western Newfoundland. *Geol. Soc. America Bull.* **69**, 315.

KINDLE, C. H. and WHITTINGTON, H. B. (1959). Some stratigraphic problems of the Cow Head area in western Newfoundland. *New York Acad. Sci. Trans.* ser 2, **22**, 7.

KINDLE, C. H. and WHITTINGTON, H. B. (1965). New Cambrian and Ordovician fossil localities in western Newfoundland. *Geol. Soc. America Bull.* **76**, 683.

LAVERDIÈRE, J. W. (1949). Bedded limestones in the Lévis Formation. *Roy. Soc. Canada Trans.* ser. 3, **43**, 71.

LIBERTY, B. A. (1967). Stratigraphic studies of Middle Ordovician and Cambrian strata in the St. Joseph Island–Sault Ste. Marie area. *Geol. Surv. Canada Paper* 67-1, pt. A, 154.

LILLY, H. D. (1964). Possible "Taconic" Klippen in western Newfoundland. *Amer. Jour. Sci.* **262**, 1130.

LILLY, H. D. (1967). Some notes on stratigraphy and structural style in central west Newfoundland. *Geol. Assoc. Canada Spec. Paper* **4**, 201.

LOCHMAN, CHRISTINA. (1938). Middle-Upper Cambrian faunas from western Newfoundland. *Jour. Paleont.* **12**, 461.

LOCHMAN-BALK, CHRISTINA and WILSON, J. L. (1958). Cambrian biostratigraphy in North America. *Jour. Paleont.* **32**, 312.

LOWDON, J. A., STOCKWELL, C. H., TIPPER, H. W., and WANLESS, R. K. (1963). Age determinations and geological studies. *Geol. Surv. Canada Paper* 62-17, 93.

MARLEAU, RAYMOND A. (1959). Age relations in the Lake Megantic Range, southern Quebec. *Geol. Assoc. Canada Proc.* **11**, 129.

MATTHEW, G. F. (1888). On a basal series of Cambrian rocks in Acadia. *Can. Record of Science* **3**, 21.

MATTHEW, G. F. (1888–9). On the classification of the Cambrian rocks in Acadia. *Can. Record of Science* **3**, 71, 303.

MATTHEW, G. F. (1899). The Etcheminian fauna of Smith Sound, Newfoundland. *Prel. Trans. Roy. Soc. Canada* **5**, 97.

MATTINSON, C. R. (1964). Mount Logan area, Matane and Gaspé-North Counties. *Quebec Dept. Nat. Res. Geol. Rept.* **118**.

McCARTNEY, W. D. (1958). Geology of Sunnyside map-area, Newfoundland. *Geol. Surv. Canada Paper* **58-8**.

McCARTNEY, W. D. (1967). Whitbourne map-area, Newfoundland. *Geol. Surv. Canada Mem.* **341**.

McCARTNEY, W. D., POOLE, W. H., WANLESS, R. K., WILLIAMS, H., and LOVERIDGE, W. D. (1966). Rb/Sr age and geological setting of the Holyrood granite, southeast Newfoundland. *Can. Jour. Earth Sci.* **3**, 947.

McGERRIGLE, H. W. (1950). The geology of eastern Gaspé. *Quebec Dept. Mines Geol. Rept.* **35**.

MOHR, P. A. and ALLEN, R. (1965). Further considerations on the deposition of the Middle Cambrian manganese carbonate beds of Wales and Newfoundland. *Geol. Mag.* **102**, 328.

NEALE, E. R. W., BÉLAND, J., POTTER, R. R., and POOLE, W. H. (1961). A preliminary tectonic map of the Canadian Appalachian region based on age of folding. *Can. Inst. Min. Met. Bull.* **54**, 687.

NEALE, E. R. W. and KENNEDY, M. J. (1967). Relationship of the Fleur de Lys Group to younger groups of the Burlington Peninsula, Newfoundland. *Geol. Assoc. Canada Spec. Paper* **4**, 139.

NEALE, E. R. W. and NASH, W. A. (1963). Sandy Lake (east half), Newfoundland. *Geol. Surv. Canada Paper* **62-28**.

NEUMAN, ROBERT B. (1962). The Grand Pitch Formation: new name for the Grand Falls Formation (Cambrian?) in northeastern Maine. *Amer. Jour. Sci.* **260**, 794.

OLLERENSHAW, N. C. (1967). Cuoq-Langis area, Matane and Matapédia Counties. *Quebec Dept. Nat. Res. Geol. Rept.* **121**.

OSBORNE, F. FITZ (1954). The petrology of the Charny Formation. *Geol. Assoc. Canada Proc.* **6**, 111.

OSBORNE, F. FITZ (1956). Geology near Quebec City. *Le Naturaliste Canadien* **83**, 157.

OSBORNE, F. FITZ and BERRY, W. B. N. (1966). Tremadoc rocks at Lévis and Lauzon. *Le Naturaliste Canadien* **93**, 133.

POOLE, W. H. (1963). Hayesville, New Brunswick. *Geol. Surv. Canada* Map 6-1963.

POOLE, W. H. (1967). Tectonic evolution of Appalachian region of Canada. *Geol. Assoc. Canada Spec. Paper* **4**, 9.

POUNDER, J. A. (1964). Cambrian of Ontario. *Ontario Petroleum Inst.* 3rd Ann. Conf., 1.

RASETTI, FRANCO (1944). Upper Cambrian trilobites from the Lévis conglomerate. *Jour. Paleont.* **18**, 229.

RASETTI, FRANCO (1946a). Early Upper Cambrian trilobites from western Gaspé. *Jour. Paleont.* **20**, 442.

RASETTI, FRANCO (1946b). Cambrian and early Ordovician stratigraphy of the lower St. Lawrence valley. *Geol. Soc. America Bull.* **57**, 687.

RASETTI, FRANCO (1948a). Lower Cambrian trilobites from the conglomerates of Quebec (exclusive of the Ptychopariidea). *Jour. Paleont.* **22**, 1.

RASETTI, FRANCO (1948b). Middle Cambrian trilobites from the conglomerates of Quebec (exclusive of the Ptychopariidea). *Jour. Paleont.* **22**, 315.

RASETTI, FRANCO (1963). Middle Cambrian ptychoparioid trilobites from the conglomerates of Quebec. *Jour. Paleont.* **37**, 575.

RESSER, C. E. and HOWELL, B. F. (1938). Lower Cambrian *Olenellus* Zone of the Appalachians. *Geol. Soc. America Bull.* **49**, 105.

RILEY, G. C. (1962). Stephenville map-area, Newfoundland. *Geol. Surv. Canada Mem.* **323**.

RIORDON, P. H. (1957). Evidence of a pre-Taconic orogeny in southeastern Quebec. *Geol. Soc. America Bull.* **68**, 389.

RODGERS, J. and NEALE, E. R. W. (1963). Possible "Taconic" Klippen in western Newfoundland. *Amer. Jour. Sci.* **261**, 713.

ROLIFF, W. A. (1954). The pre-Middle Ordovician rocks of southwestern Ontario. *Geol. Assoc. Canada Proc.* **6**, 103.

ROSE, E. R. (1952). Torbay map-area, Newfoundland. *Geol. Surv. Canada Mem.* **265**.

SANFORD, B. V. and QUILLIAN, R. G. (1959). Subsurface stratigraphy of Upper Cambrian rocks in southwestern Ontario. *Geol. Surv. Canada Paper* **58-12**.

SCHUCHERT, CHARLES (1923). Sites and nature of the North American geosynclines. *Geol. Soc. America Bull.* **34**, 151.

SCHUCHERT, CHARLES (1930). Orogenic times of the northern Appalachians. *Geol. Soc. America Bull.* **41**, 701.

SCHUCHERT, CHARLES and DUNBAR, CARL O. (1934). Stratigraphy of western Newfoundland. *Geol. Soc. America Mem.* **1**.

ST.-JULIEN, PIERRE (1961a). Fraser Lake area, Shefford and Stanstead Counties. *Quebec Dept. Mines, Prelim. Rept.* no. **439**.

ST.-JULIEN, PIERRE (1961b). Lac Montjoie area, Sherbrooke, Richmond, and Stanstead Counties. *Quebec Dept. Nat. Res. Prelim. Rept.* no. **464**.

ST.-JULIEN, PIERRE (in press). Orford–Sherbrooke area. *Quebec Dept. Nat. Res.* map 1619.

ST.-JULIEN, PIERRE and LAMARCHE, R. -Y. (1965). Geology of Sherbrooke area. *Quebec Dept. Nat. Res., Prelim. Rept.* no. **530**.

TAYLOR, F. C. (1967). Reconnaissance geology of Shelburne map-area, Queens, Shelburne and Yarmouth counties, Nova Scotia. *Geol. Surv. Canada Mem.* **349**.

TAYLOR, F. C. and SCHILLER, E. A. (1966). Metamorphism of the Meguma Group of Nova Scotia. *Can. Jour. Earth Sci.* **3**, 959.

TROELSEN, J. C. (1947). A new Cambrian fauna from western Newfoundland. *Amer. Jour. Sci.* **245**, 537.

WALTHIER, T. N. (1949). Part I: Geology and mineral deposits of the area between Corner Brook and Stephenville, western Newfoundland. Part II: Geology and mineral deposits of the area between Lewis Hills and Bay St. George, western Newfoundland. *Newfoundland Geol. Surv. Bull.* **35**.

WEEKS, L. J. (1954). Southeast Cape Breton Island, Nova Scotia. *Geol. Surv. Canada Mem.* **277**.

WEEKS, L. J. (1957). The Appalachian region. *Geol. Surv. Canada Econ. Geol. Ser.* **1**, ed. 4, 123.

WHITTINGTON, H. B. and KINDLE, C. H. (1966). Middle Cambrian strata at the Strait of Belle Isle, Newfoundland, Canada. *Geol. Soc. America*, N. E. Section, Annual Meeting, *abstracts*, 46.

WILLIAMS, H. (1964). The Appalachians in northeastern Newfoundland—a two-sided symmetrical system. *Amer. Jour. Sci.* **262**, 1137.

WILLIAMS, H. (1967). Geology, Island of Newfoundland. *Geol. Surv. Canada* Map 1231 A.

3. The Cambrian in western Canada and Alaska

3.1. Introduction

As explained in the preamble (1.3), the principal Palaeozoic trough in western Canada succeeded a Proterozoic trough formed across the edges of crystalline basement provinces. The nature and distribution of Cambrian rocks were consequently profoundly controlled by structures transverse to the depositional basin. There were three principal transverse elements:

i. The linear element Montania, and its eastern extension the Swift Current platform. This feature partially closes the western Canadian depositional basin of Cambrian time on the south. Over it, Windermere and Lower Cambrian rocks are absent; the Upper Cambrian is absent over most of the crest of the structure, and the Middle Cambrian over some of it (3.2).

ii. The Peace River arch, a much wider structure extending approximately from 55° to 60° north latitude. In the north it is referred to as the Macdonald or Tathlina 'high'. Pre-Middle Devonian erosion caused the Cambrian to become restricted to the western edge of this element, so that the width of preserved Cambrian, across the strike, is here at its minimum for western Canada. The Lower Cambrian is thick over the arch's western flank; the Middle and Upper Cambrian are much reduced (3.4).

iii. An enigmatic and unnamed element which impinges on the basin from the east-north-east, north of the Arctic Circle. This element may have been, or may have become, associated with the Franklinian geosyncline, which gave rise in later Palaeozoic time to the mountain belt which fringes the Canadian Arctic islands on the north-west (3.6).

In the two sectors lying between these three transverse structures, the Cambrian has its greatest stratigraphical development and very wide areal extension. The Rocky Mountain trough of Cambrian time extended from the north side of Montania to the *north* side of the Peace River arch. The so-called Park Ranges of the Rockies (HOLLAND, 1964), which include most of the famous peaks and which are carved almost entirely out of Cambrian rocks, extend only from Montania to the *south* side of the Peace River arch. The Mackenzie Mountain trough extended from the north side of the arch to the enigmatic third element.

The Canadian Rocky Mountains are, by definition, bounded along their western side by the Rocky Mountain trench (Fig. 9). The trench lies within a band of Proterozoic, Cambrian, and Lower Ordovician rocks, and bears such a geographical relation to the present position of important boundaries within these rocks that it provides an appropriate line of division in any consideration of longitudinal facies belts.

West of the trench, in the Interior Ranges, Cambrian rocks are almost continuously exposed from south of the 49th parallel to beyond the 60th. Most of the rocks of the western Cambrian that can be assigned to an extracratonic facies are found within this far western belt; they follow conformably on thick Proterozoic clastic rocks of the Windermere System and associated 'groups', and are lost westwards, either through metamorphism or by burial beneath younger rocks, with no sign of approach to a western depositional limit.

The trench, like the Rocky Mountains which it bounds, ends south of the 60th parallel. Some 250 kilometres to the north-west of its disappearance into the Liard Plain, another trench begins, having a more westerly strike and bearing to the Cambrian rocks a much less obvious geographical relation than does the Rocky Mountain trench. We thus have, in addition to the convenient longitudinal division at the trench, an equally convenient latitudinal division along the 60th parallel, representing a map approximation to the northern edge of the Peace River arch (Figs 10–12).

3.1.1. Principal facies of the western Cambrian

Each of the five sectors thus established has not only its own characteristic successions of Cambrian formations, vertically, but also its own characteristic succession of facies belts circumferential to the craton. Lateral shifts of the principal facies affected much greater areas than they did in eastern Canada. Though at any one time during the Cambrian there was a simple passage from inner to outer detrital zones via an intervening carbonate bank, all decipherable areas of Cambrian rock reveal some prolonged episode of cratonic sedimentation.

In each of the areas of great Cambrian development between the transverse elements, five longitudinal belts of differing depositional histories can be distinguished, but they are not the same in the two sectors. In the sector between the Peace River arch and Montania, the belts are these, from the craton outwards:

i. An inner detrital facies (Ai) of Middle and Upper Cambrian only, resting on crystalline Pre-Cambrian without any Lower Cambrian. This facies is now entirely in the subsurface.

ii. A belt of complete section in cratonic facies, including great carbonate development (sub-facies Aii) in the Middle Cambrian, and lesser development in the Upper Cambrian. The Lower Cambrian is present below, commonly as a quartzite, but rests on Proterozoic sedimentary rocks with major unconformity.

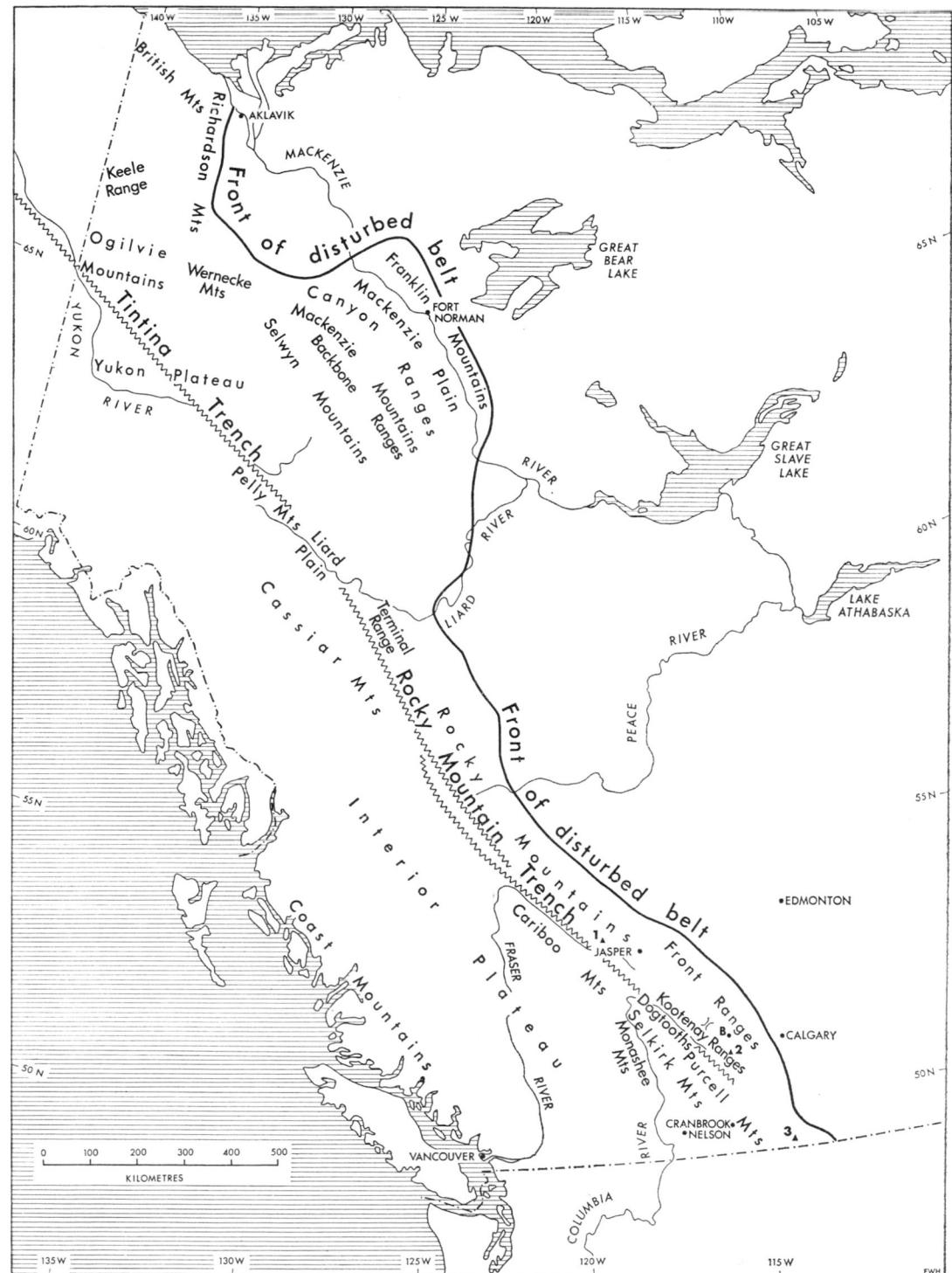

Fig. 9. Western Canada and easternmost Alaska, showing principal ranges, peaks, rivers, and place names referred to in the text. Towns (dots) (*B*, Banff). Peaks (triangles): 1, Mount Robson; 2, Mount Assiniboine; 3, Windsor Mountain. Kicking Horse Pass indicated by the symbol ⊐⊏.

iii. A belt in which the section may be complete, but in which the Middle Cambrian has changed to a poorly fossiliferous argillaceous facies (Bi); the only impressive carbonate development is here in the Upper Cambrian. Belts (ii) and (iii) gave rise to the Park Ranges or main axis of the Rocky Mountains.

iv. A narrow belt characterized by very incomplete sections. Either the Lower or the Middle Cambrian is invariably missing or very much reduced. Such section as is developed (or preserved) is dominantly clastic; the only considerable carbonate is in the Upper Cambrian. This narrow belt displays an obvious linear association with the much younger Rocky Mountain trench.

v. An outer, western, detrital belt, which yet includes at least one major carbonate unit in the Lower Cambrian of clearly cratonic association. The Cambrian in this belt follows conformably on Proterozoic sedimentary rocks. The belt lies wholly west of the Rocky Mountain trench, and extends westward into a metamorphic zone or under the cover rocks of the Interior Plateau. Volcanic rocks occur in the Cambrian, but are rare.

In contrast, the belts distinguishable in the Northwest Territories and Yukon, again beginning at the craton, are as follows:

i. An inner detrital and evaporitic facies belt (Ai), becoming platform carbonate at the top. This belt lacks the Lower Cambrian or has it considerably reduced.

ii. A belt of great subsidence in the early Cambrian, following on thick Proterozoic rocks but followed by incomplete Middle and Upper Cambrian sections. This belt was structurally separated from the first one by a positive strip. Its contents appear all to be of shallow water origin, but their fossils are those of the outer detrital sub-facies (Bi).

iii. A wide belt having little or no Cambrian in it now. This belt is cut off on the south-west by the Tintina trench.

iv. A further cratonic belt, showing an upward transgressive progression through the Cambrian and being related to a succession which farther south lies wholly east of the Rocky Mountain trench.

v. An outer, basinal facies, if one may presume that it must have existed, must now be obliterated by the younger batholiths to the west. It is nowhere recognizable.

In comparing these two successions of facies belts, it is important to bear in mind that such similarities as exist between them obscure one absolutely fundamental distinction. The Pre-Cambrian and Palaeozoic sections in the far north, though folded and faulted, offer no evidence of being allochthonous. The amount of crustal shortening has been moderate. The sections in the south, on the other hand, have undergone crustal shortening of 50% or more, and eastward translation by overthrusting of scores of kilometres. Though no adequate palinspastic reconstruction for a period as old as the Cambrian has yet proved possible for that part of the Cordillera, all the exposed Cambrian rocks there are allochthonous and the degree of allochthony increases westward (SHAW, 1963; BALLY, GORDY, and STEWART, 1966).

In spite of this, there is no evidence that the facies belts described for the southern Cordillera are out of their original order. No belt of the original Cambrian depositional basin offers signs of having in its entirety achieved greater allochthony than its neighbouring belts, as the basinal facies has done in Newfoundland and Quebec (2.3.4; 2.3.2). Original relationships between the belts are, however, widely obscured by the thrusting or by other types of faulting, especially along the Rocky Mountain trench.

The distribution of the Cambrian, by series and by facies, is shown in Figs 10–12. The ranges of the Cordillera significant to a study of the Cambrian in the west are named on Fig. 9.

3.2. The Cambrian in the Montania sector

The core of the high area called Montania lies in the extreme north-western corner of the state of Montana and the south-eastern corner of British Columbia. It is bounded on the north-west by an important zone of faulting, passing approximately through the town of Cranbrook (Fig. 9). The high area originated at least as early as the close of Purcell time, because it lacks Windermere and Lower Cambrian rocks completely. Over its small crest, it lacks all Lower Palaeozoic rocks.

North-eastwards along Montania from this small crest, about 300 metres of Cambrian strata overlie eroded rocks of the Purcell System, which here is underlain by the Lewis overthrust. All the Cambrian strata are Middle Cambrian (NORRIS and PRICE, 1966). At the base, the transgressive basal clastic unit, commonly a quartzite, is called the Flathead Formation. Its maximum thickness is about 40 metres. A gradational succession above begins with the Gordon Formation, it and the underlying Flathead Formation being the only stratigraphical units in the western Canadian Cambrian to have American type localities (in Montana). The Gordon is a distinctive, greenish grey shale, fissile and micaceous, 50 to 90 metres thick; near its base, glauconitic sandstones mark its passage into the underlying sandstone, and near its top are nodular and argillaceous limestones marking a similar gradation into the overlying carbonate unit.

The Gordon Formation is exceedingly fossiliferous, especially in the limestone beds. *Albertella*, *Glossopleura*, and *Zacanthoides* are all found in it. The rest of the succession is all carbonate; the Elko Formation, 60 to 160 metres thick, is unfossiliferous, but the Windsor Mountain Formation above it has yielded *Ehmaniella*, *Kootenia*, and other trilobites.

On the north side of Montania, immediately south of the bounding fault or flexure, the Cambrian section is still further reduced. Only 120 metres of Middle Cambrian dolomite (Elko) and shale (Gordon) remain between the underlying Purcell and the overlying Devonian. There is no Flathead Formation below, though the Gordon Formation has a thin basal conglomerate of Pre-Cambrian pebbles. Trilobite genera in the Gordon Formation here include *Glossopleura*, *Kochaspis*, and *Plagiura*.

Eastwards, in the subsurface over the Swift Current platform, the Middle Cambrian section thickens only slightly and becomes somewhat more calcareous, before thinning out again relatively abruptly to a depositional zero edge about on the 108th meridian (Fig. 11). The Upper Cambrian, absent over the crest of the 'high', is unaffected in the subsurface to the east, its nature and distribution being exactly as they are in the rest of southern Alberta and Saskatchewan. There was, however, considerable local relief on the Pre-Cambrian surface along the Swift Current platform. Several wells have been drilled on anomalously 'high' spots, reaching the crystalline basement with either no Middle Cambrian above it, or with only partial sections preserved in the Middle Cambrian, the Upper Cambrian, or both.

3.3. The Cambrian between Montania and the Peace River arch

3.3.1. The Cambrian in the Ranges west of the Rocky Mountain trench

The complex strip of Cambrian outcrops lying west of the Rocky Mountain trench at these latitudes is an excellent example of continuous sedimentation across the Pre-Cambrian/Cambrian boundary. Nowhere along the western edge of the outcrop belt is there known to be a gap representing the Lipalian interval. As the base of the Cambrian must therefore be arbitrary, not all the rocks currently assigned to the system are

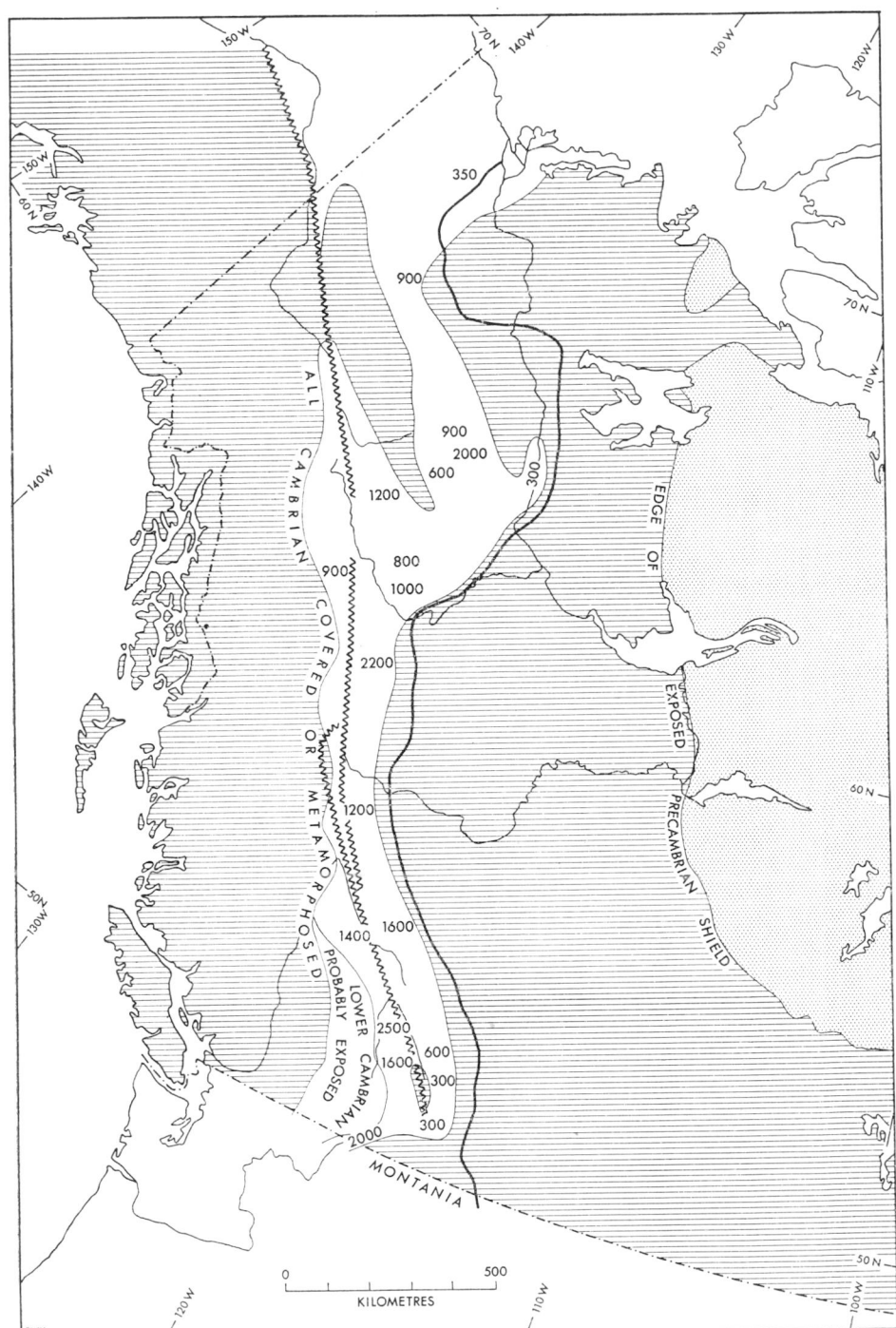

FIG. 10. Lower Cambrian rocks in western Canada and easternmost Alaska. Spot thicknesses in metres (map not palinspastic). Heavy line is eastern front of disturbed belt.

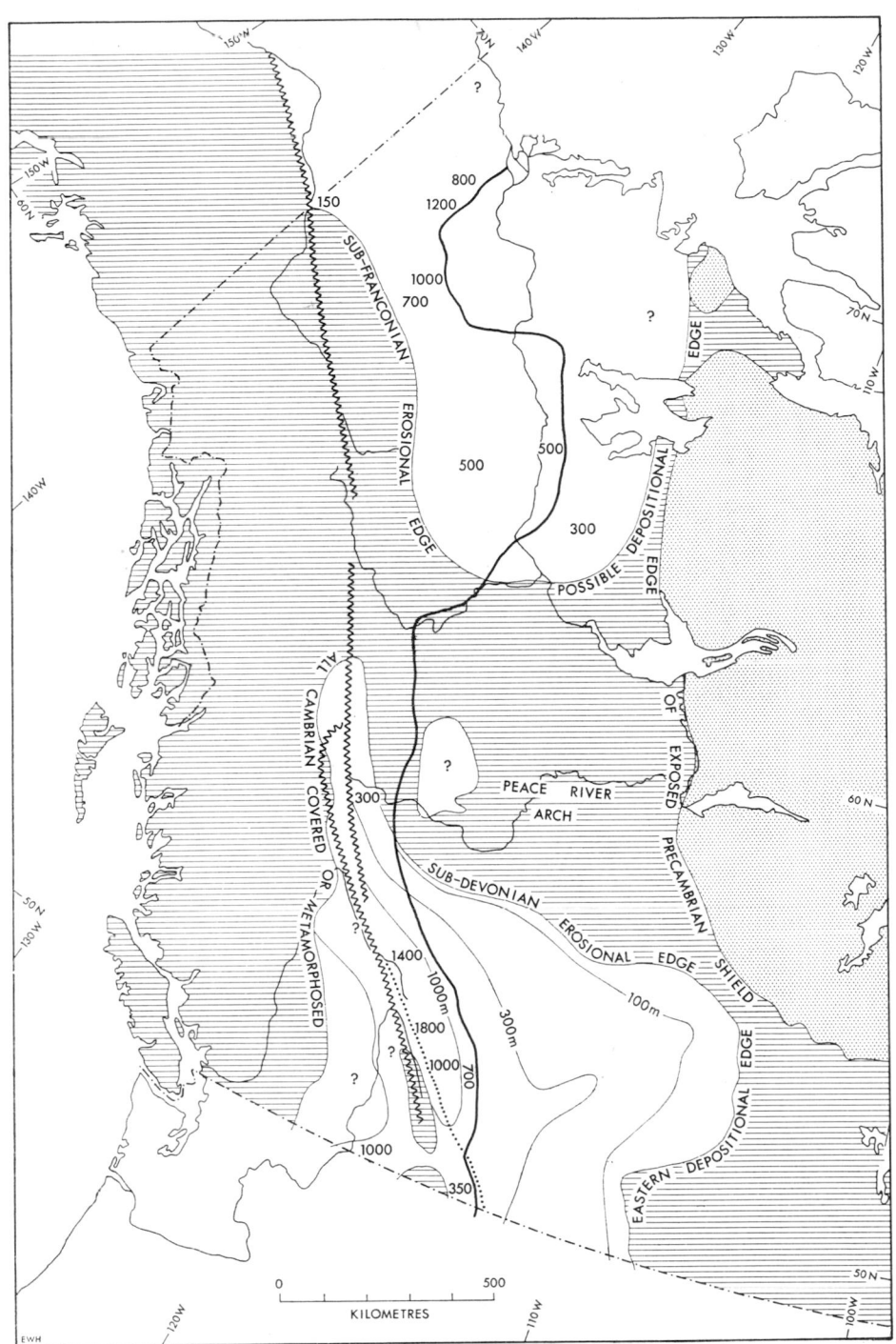

FIG. 11. Middle Cambrian rocks in western Canada and easternmost Alaska. Spot thicknesses in metres (map not palinspastic); generalized isopachs in south, where subsurface rocks are unaffected by crustal shortening. Heavy line is eastern front of disturbed belt. Dotted line separates calcareous (eastern) from argillaceous (western) facies in southern sector.

10CAM

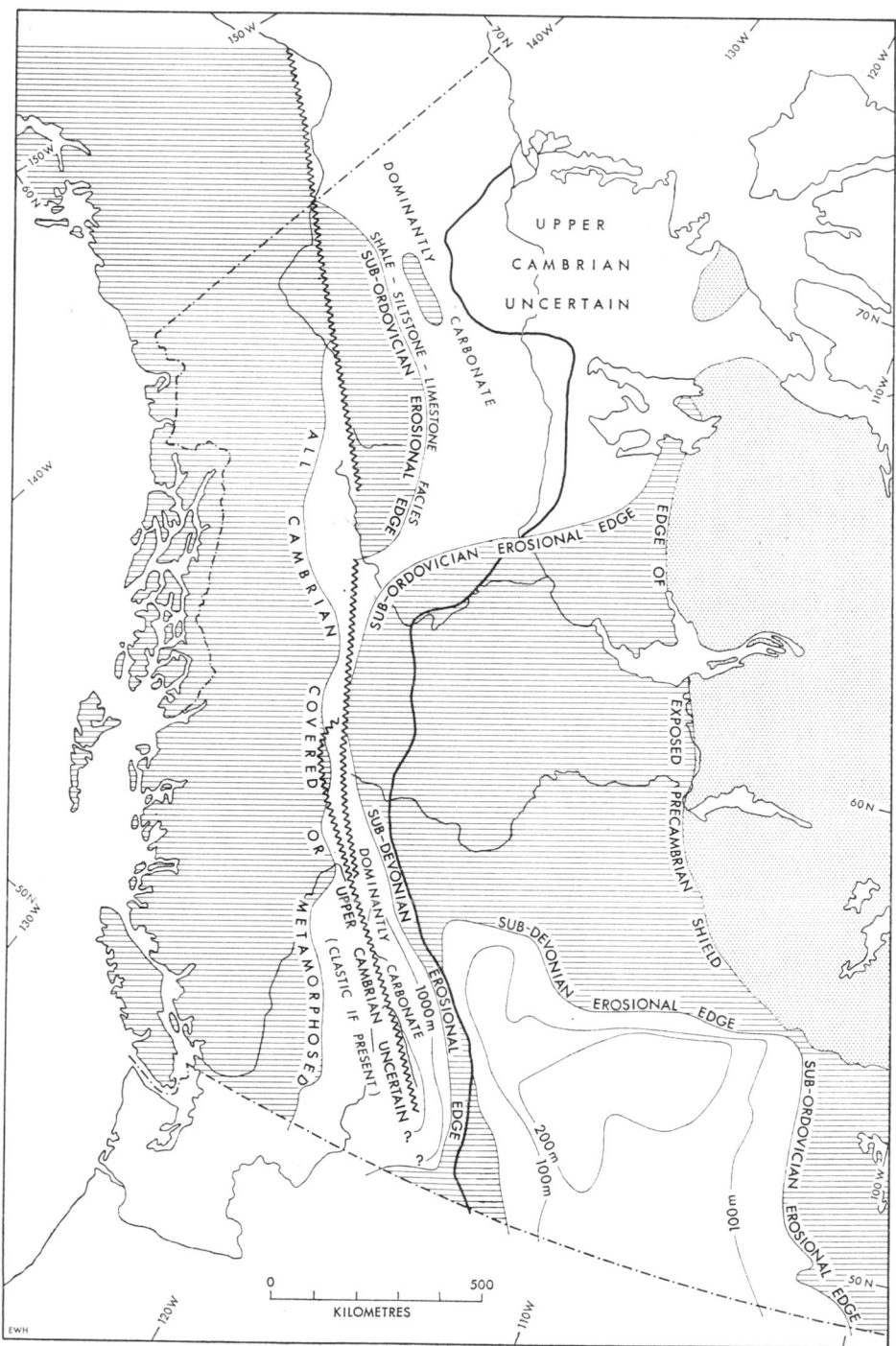

FIG. 12. Upper Cambrian rocks in western Canada and easternmost Alaska. Generalized isopachs, in metres, in southern subsurface. Heavy line is eastern front of disturbed belt.

conclusively proved to belong to it, but those which are so proved are integral continuations of thick Proterozoic sequences.

From the Metaline Falls area of north-east Washington State (PARK and CANNON, 1943), the Monk Formation and the lower part of the Gypsy quartzite can be traced into Canada as the Horsethief Creek Group of the Windermere System, which is of late Proterozoic age. The upper part of this Proterozoic succession is also in all lithic characteristics a duplicate of the Miette Group of the central Rocky Mountains, a group also overlain by early Cambrian sediments.

Along the entire length of the westernmost outcrop belt, Lower Cambrian quartzites and associated argillaceous and calcareous rocks follow conformably on rocks of this type. Between the town of Nelson and the International Boundary (Fig. 9), the quartzite succession includes two formations (LITTLE, 1960). Below is the Quartzite Range Formation, of white, green, or pinkish quartzites with some sericite-schists interbedded. The conformably overlying Reno Formation consists largely of greenish argillites and quartzites, with some phyllites and schists, representing a transition zone between the underlying quartzites and overlying limestones. The two formations have a combined maximum thickness of more than 1,500 metres, but this thickness is halved in a short distance both northwards and westwards, and in these directions the formations also become more fine-grained and more argillaceous. Even in these farthest westerly sections, all observations bearing upon the provenance of the quartzites indicate transport of the detritus from the shield source in the east.

Northwards, the quartzites are almost continuously exposed to the Big Bend of the Columbia River and beyond. In general, this belt comprises a great anticlinorium of Purcell rocks, the Purcell Mountains, on the east, and the synclinorium of the western Selkirk Mountains, exposing Windermere and Cambrian rocks, on the west (Fig. 9).

In the Selkirk Mountains, the Hamill Group lies conformably on typical Windermere grits and again consists of quartzites, white to pinkish and massive below but with thin interbeds of phyllite appearing in the upper 600 metres. The group attains a maximum thickness of some 2,500 metres, and is responsible for some of the highest peaks in the range. The greatest development of volcanic rocks in the Hamill Group occurs in the Big Bend region itself (WHEELER, 1965); the complex includes bedded greenstones, tuffs, and breccias, and thickens towards the west, perhaps testifying to an original westward approach towards a truly eugeosynclinal régime. In the same area, the quartzites of the Hamill Group become somewhat thinner and more shaly towards the north-west.

Where the quartzites are next seen, in the northern Cariboo Mountains (Fig. 9), they certainly have nothing like the thickness of the Hamill Group. Their equivalents are to be found in the upper part of the Cariboo Group (SUTHERLAND BROWN, 1963). At least the two lower formations of that group, including a thick limestone, can be assigned to the Pre-Cambrian (R. B. CAMPBELL, personal communication, 1967). Overlying them are brown to green shales and siltstones, with brown quartzites and minor limestones, called the Yankee Belle Formation. It may represent the lower part of the Hamill Group, but if it does so its much shalier character is apparent. The succeeding Yanks Peak Formation, however, is a true quartzite, a very pure, dense rock with well-rounded grains which suggest the reworking of a pre-existing sandstone. These two formations, which are now in various stages of metamorphism, have a combined thickness of about 1,000 metres, but this is irregularly reduced towards the north-west. The quartzite portion, less than half the total even at its thickest, wedges out relatively abruptly in that direction between two siltstone–shale formations.

The quartzites are everywhere succeeded, with apparently perfect conformity, by a calcareous group or formation carrying archaeocyathids. In the south, this is represented by the Laib Formation, which follows the Reno Formation. It comprises a basal limestone, massive, white, and siliceous, with a good deal of quartzite above it, followed by a thick mass of grey-green argillite and phyllite with thin interbeds of limestone. The formation is about 1,200 metres thick, and the proportion of limestone increases westward. Archaeocyathids are abundant in the basal limestone, the commonest being *Ethmophyllum whitneyi* MEEK and several species of *Ajacicyathus*. The Laib Formation as a whole, therefore, is probably equivalent to the Maitlen phyllite of Washington, or to all except the basal part of that formation (LITTLE, 1960).

In the Selkirk Mountains, the comparable Badshot limestone is divided into a lower, shaly Mohican Formation and an upper, thickly bedded Lade Peak Formation. It thickens very abruptly towards the north-west, from less than 80 metres in the Purcell Mountains to more than 300 metres and possibly to more than 600. Archaeocyathids occur in the upper 100 metres. Though the formations are partly argillaceous or phyllitic, the thick bedding of the upper limestone makes it a feature former. Where it stands almost vertically, between the western synclinorium and the eastern anticlinorium, it was referred to by the mining men as the Lime Dyke.

In the Cariboo Mountains, the shaly lower formation is called the Midas, a dark carbonaceous phyllite 150 to 200 metres thick, with some siltstone and limestone. Its bedding surfaces exhibit great numbers of 'trace fossils' (F. G. YOUNG, personal communication, 1967). The overlying archaeocyathid limestone, with associated shale and dolomite, reaches a thickness of about 250 metres. It contains algal pisolites, which may be *Girvanella*. Because at this latitude there happens also to be an archaeocyathid limestone in this stratigraphical position in the Rocky Mountains, on the other side of the Rocky Mountain trench, its formational name, Mural, is tentatively used for the limestone in the Cariboo Mountains.

Strata overlying the archaeocyathid limestones are dominantly shaly throughout the western belt, and their age is in most cases in doubt. In the south, the Laib Formation is gradationally overlain by the Nelway Formation, restricted to small areas near the International Boundary. Its lower parts contain a good deal of grey to black phyllite and calcareous schist; in the centre is a prominent cream-weathering dolomite; and the top is dense, grey limestone with abundant chert nodules. The true formational top is in fact not seen in Canada, but the amount of the formation exposed exceeds 1,200 metres. Its immediate equivalent south of the border, the Metaline limestone, contains numerous trilobites, concentrated in the lower beds of the formation (PARK and CANNON, 1943). Genera include *Elrathia*, *Elrathina*, and *Ogygopsis*, showing that the fossiliferous part is the direct correlative of the Pagoda Formation of Montana and the Stephen Formation of the Canadian Rocky Mountains, both medial Middle Cambrian (3.3.3). This correlation raises the unprovable speculation that the big dolomite above is about the equivalent of the Eldon Formation in the Park Ranges of the Rocky Mountains (though shaly equivalents would have to lie between the two), and that the upper parts of the Nelway Formation equate with the Ottertail limestone of the western Rockies (3.3.3). The latter is Upper Cambrian.

In the Selkirk and Purcell Mountains, the Badshot limestone is overlain by the Lardeau Group. This has been divided into five formations (FYLES and EASTWOOD, 1962), but none has ever yielded any fossils and whether all five are Cambrian is doubtful. The oldest, the Index Formation, 500 to 750 metres thick, consists of dark grey and green phyllites and argillites with very minor limestones and volcanic rocks. It is overlain by a highly variable succession of grey to black, siliceous or carbonaceous argillites and slates, divisible in places into three formations because of the appearance of a lenticular, though massive, quartzite unit in the middle. The Lardeau Group ends in this area with a volcanic formation, the Jowett, comprising pillow lavas, green agglomerates and breccias and minor sedimentary rocks. The thickness of the Jowett Formation ranges up to at least 1,200 metres, but it is normally less than this because of truncation at the top by an overthrust which carries lower Windermere strata over the Lardeau Group. The Jowett is very probably Ordovician, or even younger. The three clastic formations below it may be either Upper Cambrian or Ordovician. The evidence of incomplete sections to the east (3.3.2) suggests that the whole of the Middle Cambrian may be absent here.

In the Cariboo Mountains, the unnamed section above the Mural limestone is dominated by black shales with minor dark limestones and greenish siltstones. The lower parts carry olenellid trilobites like those of the Eager Formation (3.3.2); genera represented include *Bonnia*, *Olenellus*, *Kootenia*, and *Wanneria*. The presence of the Middle Cambrian above is indicated by finds of *Glossopleura* and *Ogygopsis*, unfortunately from obscure stratigraphical positions (SUTHERLAND BROWN, 1963). Whether the higher beds are Middle Cambrian, Upper Cambrian or Ordovician is not known. There is less reason here than in the Selkirk Mountains, however, to believe that the Cambrian section is incomplete; the nearest rocks east of the trench include one of the thickest and most complete Cambrian sections in Canada. The lithological difference between the two sections, however, could scarcely be more complete. That in the Cariboo Mountains is nearly all of dark coloured, thinly-bedded clastic rocks, abundantly bored and bearing ample evidence of deposition in shallow water. That in the Rocky Mountains is more than two-thirds carbonate, most of it dolomite, and the major carbonate units show a tendency to thicken towards the trench (3.3.3).

The Lardeau and Cariboo Groups are everywhere highly deformed. The original folding of the Cariboo Group may have come shortly after the group's deposition, during the so-called Cariboo orogeny (WHITE, 1959), approximately the Cordilleran equivalent of the Caledonian. The principal tectonic transport of the Cariboo rocks is towards the south-west. The principal episode of deformation for both groups, however, occurred during the Mesozoic, when they were heavily intruded by granitic rocks which caused the development of garnet, staurolite, or kyanite in the schistose rocks and marmorization of the limestones. To the south-west, both the Lardeau and the Cariboo Groups, with underlying Proterozoic and overlying Palaeozoic rocks, are involved in the Shuswap Complex.

This extensive metamorphic terrane, with unique structural style, has been the subject of controversy since the days of G. M. DAWSON. Within it, metamorphosed sedimentary rocks give way abruptly, westwards, to a complex of granitic gneisses, saturated with pegmatites, but preserving recognizable bands of quartzite and marble. Unpropitious for stratigraphical subdivision, this high-grade metamorphic complex is lumped together as the Monashee Group (JONES, 1959). The Shuswap Complex as a whole is now known to involve the entire stratigraphical succession from the Windermere System to the Upper Palaeozoic, its latest and most intense phase of metamorphism coming apparently in the Jurassic (Nevadan or Coast Range deformation).

The areas of Cambrian rocks in the Selkirk and Cariboo Mountains are separated by a wide band of granitic hornblende-gneisses and gneissic granite. These rocks are unlike any others in this part of the Cordillera, and the episode of metamorphism which they represent is unique in that it extends across the Rocky Mountain trench south-west of Mount Robson. The fact that Cambrian and latest Pre-Cambrian rocks strike directly towards the metamorphic rocks from both ends might be taken to imply the presence of the first among the second, exactly as in the Shuswap terrane. On the north-east side of the trench, in fact, the Miette Group and the Lower and Middle Cambrian can all be shown to be involved in the metamorphism, with stretching and flattening of pebbles and the development of gneissic structures (PRICE, 1967). The main triangle of gneisses and gneissic granite, however, has not been shown to be structurally or stratigraphically continuous with these rocks, and it has been suggested that it is a much older terrane, resulting from a remobilization of crystalline Hudsonian basement (R. B. CAMPBELL, personal communication, 1967).

The westernmost boundary of the Cambrian is everywhere in doubt. There is no sign of approach to a western shoreline; the Cambrian rocks disappear under younger rocks, or into the thick metamorphosed and granitized successions just described, or are abruptly cut off by faults whilst still in full thickness. At latitude 53° N., a tiny inlier of archaeocyathid limestone appears from beneath Mesozoic rocks at the extreme eastern boundary of the Interior Plateau, east of the Fraser River. Elsewhere, the presence or absence of Cambrian rocks in the metamorphic belt has to be guessed at on lithological grounds.

An example is provided by a minor fault block of Shuswap-style metamorphic rocks on the International Boundary at about 118° 30' W. (PRETO, 1967). A group of paragneisses and calcareous schists is followed in turn by a massive quartzite, a marble, and further schists and gneisses. This is so suggestive of the Horsethief Creek–Hamill–Badshot–Lardeau succession that equivalence is confidently postulated. Further fault blocks of similar rocks occur at the same latitude as far west as 119° 30' W. If these still include Cambrian rocks, the present width of preserved Cambrian amounts to 1,400 kilometres measured along the International Boundary. Nearly 500 kilometres of this present width would lie in the foreshortened belt of the Cordillera, and the original depositional limit would not be seen in either east or west.

3.3.2. The Cambrian in and adjacent to the Rocky Mountain trench

The region described in the preceding section is separated from the classic Cambrian areas of the southern Rocky Mountains by the Rocky Mountain trench. This enormous structural lineament (Fig. 9) follows Pre-Cambrian or Cambro–Ordovician rocks for its entire length. Between the latitudes of our present concern (50° to 55° North), the trench approximately coincides with the line along which the Lower Cambrian quartzites begin to be underlain by the unconformity which progressively cuts down-section eastwards into older and older Proterozoic rocks. Except in the immediate western wall, there is no locality west of the trench at which the Cambrian is significantly unconformable on the Pre-Cambrian. In no locality east of the trench is the Cambrian known to be conformable on the Pre-Cambrian.

South of latitude 52° N., at least, the trench also appears to lie within a zone along which the Cambrian section is everywhere reduced by the loss of either the Lower Cambrian or the Middle Cambrian, or both. This same interval of latitude is also the only one over which the Rocky Mountains include distinct ranges lying to the west of their principal structural axis. These ranges, collectively called the Kootenay Ranges (Fig. 9), are different in structural style from the rest of the Rocky Mountains and they form the east wall of the trench between latitudes 49° 45′ and 51°45′ N. (HOLLAND, 1964).

The abrupt northern flank of Montania crosses the trench at about 49° 30′ N. Within a very short distance northward, the Lower Cambrian and the Upper Cambrian appear in considerable thickness, whilst the Middle Cambrian, which is all that represents the system on Montania, disappears altogether or becomes thin and unfossiliferous (LEECH, 1954). The quartzite part of the Lower Cambrian is here called the Cranbrook Formation. It is in pale pastel colours dominated by green, and is much interbedded with sandstone, grit, and quartz–pebble conglomerate. *Nevadella* has been found near the top. Also near the top of the formation, bodies of rock magnesite occur locally in the Purcell Mountains, in association with dolomite (RICE, 1937). The quartzite lies unconformably on Horsethief Creek strata, older Windermere beds, and eroded Upper Purcell, illustrating the southward and eastward stratigraphical downcutting of the oldest Cambrian clastic unit on to Montania and towards the crystalline shield. The formation thins rapidly northwards, from a maximum of nearly 300 metres, to zero within the Kootenay Ranges.

The Cranbrook Formation is conformably overlain, in the south, by the Eager Formation (RICE, 1937; LEECH, 1954, 1958). This is a thick clastic unit of very shallow water origin, largely reddish or greenish shales and sandstones becoming progressively more calcareous upwards and giving way westwards to rusty-weathering argillites with coloured streaks. In rather brightly coloured beds near its base are found olenellid trilobites and the brachiopod *Kutorgina*, making this part of the formation the same age as, or very slightly older than, the Peyto limestone in the Park Ranges (3.3.3). The Eager Formation as a whole appears to be the same lithic unit as the Chancellor Group in those ranges, though whether it originally had the same time-stratigraphical boundaries as the Chancellor cannot be proven. The base of the Chancellor Group is not seen at these latitudes; the top of the Eager Formation is seen only as an erosion surface. The Eager is, however, disconformably overlain by the formation that conformably overlies the Chancellor. If the correlation is correct, much of the unfossiliferous part of the Eager Formation west of the trench must be Middle Cambrian.

Northwards along the Kootenay Ranges, the Eager Formation thins even more abruptly than does the Cranbrook Formation. The thinning takes place at the expense of the upper part of the formation (HENDERSON, 1954; LEECH, 1954), and is principally due to pre-late Cambrian erosion. There are a number of localities within the southern sector of the trench, and in both walls opposite them, at which the entire Middle Cambrian is known to be absent (NORTH and HENDERSON, 1954). Whether intervals of deficient or no deposition also contributed to this lack is not certain, but the area over which the lack is observed is one in which the Lower Cambrian is also missing (or reduced to insignificant proportions) and the Upper Cambrian very well and typically developed.

The Upper Cambrian succession in the Rocky Mountains as a whole is dominated by cliff-forming carbonate units. A very well developed representative occurs throughout the Kootenay Ranges; it also dominates three intra-valley ridges within the trench (EVANS, 1933; REESOR, 1957). It is here called the Jubilee Formation. As it contains no fossils, it has long been of uncertain age, but it is now known to be younger than the *Crepicephalus* Zone of the Upper Cambrian and older than the *Elvinia* Zone (G. B. LEECH, personal communication, 1967). At the southern end of the Kootenay Ranges, it is 1,200 metres thick and almost entirely of dolomite (LEECH, 1954). It retains a thickness of 700 metres or more towards the north, as the earlier Cambrian formations wedge out through erosion. In places, the base of the formation carries a 5 to 10 metre

	Zones	Montania Sector	Kootenay Ranges
	Overlain by:	Devonian	L. Ordovician
Upper Cambrian	Saukia		Lower McKay Group
	Ptychaspis- Prosaukia Conaspis Elvinia		
	Dunderbergia Aphelaspis		Jubilee
	Crepicephalus Cedaria		
Middle Cambrian	Bolaspidella		
	Bathyuriscus- Elrathina		
	Glossopleura	Windsor Mountain Elko	
	Albertella	Gordon	
	Plagiura- Poliella	Flathead	
L Cambrian	Olenellus		Eager
	Pre-Olenellus		Cranbrook
	Underlain by:	Purcell	Purcell-Windermere

TABLE 5
Correlation of Cambrian Formations, Southern Canadian Rocky Mountains

Southern Park Ranges (West Flank)	Southern Park Ranges (East Flank)	Northern Park Ranges	Front Ranges
L. Ordovician	L. Ordovician	L. Ordovician	Devonian
Lower McKay Group	Mistaya	Lynx Group	/////
	Bison Creek		Bison Creek
Ottertail	Lyell		Lyell
Chancellor Group	Sullivan Waterfowl Arctomys ———	——— Arctomys ———	Sullivan Waterfowl Arctomys ———
	Pika Eldon	Pika Titkana	Pika Eldon
	Stephen	Tatei	Stephen
	Cathedral	Chetang	Cathedral
	Mount Whyte	Hota-Adolphus	/////
Gog Group	Gog Group	Mahto Mural McNaughton	/////
Windermere	Windermere	Miette	Pre-Cambrian

TABLE 5 Continued

conglomerate, largely of Purcell boulders and pebbles, resting on the Cranbrook Formation or on the Pre-Cambrian. The most northerly appearance of the Jubilee Formation in the trench is at 51° 30′ N., resting disconformably on Lower Cambrian and older rocks. At that latitude, the Kootenay Ranges as a unit are wedging out between the Park Ranges and the trench (WHEELER, 1963).

The Jubilee Formation is overlain, throughout the Kootenay and Park Ranges, by the Cambro–Ordovician McKay Group. Where it is not sheared, the McKay Group consists of shales, bluish limestones, a little dolomite, and a good deal of intraformational limestone conglomerate. It is the thickest single map-unit in the Kootenay Ranges, as well as being the most widespread in outcrop. No complete section is known in the vicinity of the trench, but the group is locally at least 1,500 metres thick (EVANS, 1933; LEECH, 1960). On the east face of the trench, there are also places at which it is absent altogether, through excision by thrusting or strike-slip faulting. Though pre-late Ordovician erosion reduced the thickness of the group in parts of the ranges, it nowhere removed it altogether.

No figure for the average thickness of the McKay Group in the Kootenay Ranges would therefore have much meaning, but the amount of the bottom of the group which belongs in the Cambrian may not vary very much. The base of the group is certainly Franconian; its trilobites include species of *Briscoia*, *Irvingella*, and *Taenicephalus*. The bulk of the group is equally certainly Lower Ordovician. In the Kootenay Ranges generally, the Cambrian quota may be no more than 350 or 400 metres. Eastwards, in the Park Ranges, the whole group is very much thicker, and the Cambrian portion of it is thicker accordingly (3.3.3).

The McKay Group, converted to slates, extends across the trench into the Purcell Mountains in one spot (WALKER, 1926; REESOR, 1957). Reduced to inconsequential thickness, it rests on Jubilee dolomite, itself only 100 metres thick, and this in turn rests unconformably on Horsethief Creek strata with no Middle or Lower Cambrian. As the McKay Group is followed upwards by the younger formations that normally follow it farther east, and as these also are here very thin, the attenuation of every unit cannot be due to erosion. We must here be very close to the western depositional edge of the Upper Cambrian carbonate facies, as well as being in a zone in which pre-late Cambrian erosion removed much of the Middle and Lower Cambrian. Farther to the west, in the Selkirk Mountains, the eastwardly-overthrust equivalents of the Upper Cambrian carbonate unit must have been in more argillaceous facies, and either have been eroded away or are represented by some part of the Lardeau Group. The possibility is thus reinforced that the Middle Cambrian is essentially unrepresented in the Selkirk Mountains (3.3.1).

Less than 100 kilometres to the north of the locality with the very thin section, and still in the immediate west wall of the trench, a totally different Cambrian succession occurs in the Dogtooth Mountains (Fig. 9). The Lower Cambrian Hamill Group, about 1,200 metres thick, contains not only an unusually high quota of shaly material but also the most easterly volcanic rocks known in the Lower Cambrian of western Canada (P. S. SIMONY, personal communication, 1967). The volcanic rocks are pillow-flow greenstones and explosive breccias. The overlying formation, called the Donald, consists of limestones, sandstones, and limy slates, with a central band of fossiliferous limestone carrying Lower Cambrian trilobites, *Salterella*, and archaeocyathids. The latter are not like those in the Laib Formation near the International Boundary, however. *Ethmophyllum*, the principal archaeocyathid genus in the Laib, has not been reported from the

Donald; its principal archaeocyathid is *Ajacicyathus purcellensis* OKULITCH a younger form apparently belonging to the late Lower Cambrian (OKULITCH and GREGGS, 1958). The Donald Formation is about 600 metres thick.

In the trench itself, the Donald appears to be overlain by limy slates with thin interbeds of limestone (Canyon Creek Formation). These do not closely resemble any known Middle or Upper Cambrian rocks in the southern Rocky Mountains or the Purcell Range. They have yielded a single trilobite, probably a *Proceratopyge*, indicating a Late Cambrian age. (W. H. FRITZ, P. S. SIMONY, personal communications, 1967.) The slates are overlain by other slates of the McKay Group. As the latter, at all localities to the east, overlies the Jubilee Formation or its equivalent, the absence of this major feature-maker only forty kilometres along the strike from its type locality is puzzling even if the contact between the Canyon Creek and Donald Formations is a fault. If the Canyon Creek is stratigraphically equivalent to the Jubilee, the westward shale-out of the latter must be as abrupt as that of the Middle Cambrian carbonates in the Park Ranges (3.3.3).

The exposures of Cambrian rocks in and adjacent to the Rocky Mountain trench, between Montania and the Big Bend of the Columbia River, are thus anomalous in two respects: in lying between a western Cambrian zone (3.3.1) in which the only big carbonate is in the Lower Cambrian, and a central Cambrian zone (3.3.2) in which the only big carbonate is in the Upper Cambrian; and in exhibiting an unusual amount of intra-Cambrian erosion. Much of this southern part of the trench itself (indeed much of the whole trench) was eroded into highly faulted, shattered and mashed strata of the McKay Group, including the Cambrian McKay.

3.3.3. The Cambrian in the Park Ranges of the Rocky Mountains

The Park Ranges of the southern Canadian Rocky Mountains are loosely defined as the axial zone of rugged peaks carved largely from Cambrian sedimentary rocks. As so defined, they extend from 49° 45′ N., approximately, to 54° N., and over all but the southern extremity of this extent their axis forms the continental divide and hence the boundary between the provinces of British Columbia and Alberta (HOLLAND, 1964). Beyond the two extremities of the Park Ranges, we approach the transverse axes of Montania and the Peace River Arch (3.1), and the loss of the ranges is a consequence of the loss, against these axes, of the great Lower Palaeozoic succession from which the ranges were developed.

Southwards from 51° 45′ N., the Park Ranges are separated from the Rocky Mountain trench by the series of smaller, en echelon ranges collectively called the Kootenay Ranges of the Rocky Mountains (3.3.2). Northwards from 51° 45′ N., the Kootenay Ranges have wedged out into the trench and the western boundary of the Park Ranges is then at the trench. The eastern boundary of the Park Ranges is marked by a set of great thrust faults, which carry the Pre-Cambrian and Lower Palaeozoic rocks eastwards over the Upper Palaeozoic rocks which dominate the Front Ranges at all latitudes (Fig. 9).

The Cambrian rocks of the Park Ranges have been the subject of at least five distinct formational nomenclatures. Through the work of AITKEN (1966), GREGGS (1963), and MOUNTJOY (1962), these have now hopefully been stabilized as a two-fold nomenclature, the different nomenclatures for the northern and southern sectors of the range reflecting in large part the different distributions of carbonate units in the Middle and Upper Cambrian (Table 5). In the areas of the great mountains at the north and south ends of the range, the section goes down into the Proterozoic; the Lower and Middle Cambrian formations make up the bulk of the range there, with the Upper Cambrian dolomites and limestones forming some of the higher peaks. About latitude 52° N., in the central sector of the range, Mississippian, Devonian, and Ordovician beds form most of the high peaks, and the base of the exposed section lies in the Middle Cambrian.

We have seen that, in the ranges west of the Rocky Mountain trench, the rocks assigned to the Lower Cambrian (and especially the characteristic quartzites) thin progressively from south to north. The quartzites, at least, are more than twice as thick in the Selkirk Mountains as in the Cariboo Mountains (3.3.1). In the Park Ranges of the Rocky Mountains, exactly the opposite situation prevails. The quartzites are unconformable, throughout the ranges, on eroded Windermere rocks which are themselves largely grits and argillites. A rather conspicuous colour change characterizes the break. The beds below it are commonly greenish; the overlying quartzites tend to be either grey or whitish, purplish or pink.

Near the southern end of the range, in the massif of Mount Assiniboine (Fig. 9), the quartzite attains a thickness of about 400 metres and contains no significant shaly inter-beds (DEISS, 1940). Both thickness and degree of monotony are so untypical that it is unfortunate that this locality provides the type for the Gog Group, the name now in use for this part of the Lower Cambrian throughout the Park Ranges. Northwards, the group thickens and thins in an undulatory manner, indicating lenticular deposition. It dominates the cluster of mountains familiar to visitors to Lake Louise, and, in this part of the range, one or more thin shale members within the group carry such fossils as *Hyolithes* and small brachiopods, as well as the trails of annelids or trilobites.

Dropping below valley level in the central sector of the range, the group reappears to the north as a great lens of massive, well sorted quartzose sandstone, with interbeds of shale or siltstone which in general increase in number and thickness westwards. The lower parts are heavily cross-bedded, the beds indicating transport from the east and north (SLIND and PERKINS, 1966). The group is here of the order of 2,500 metres thick, resting on the Miette Group with a basal quartz-pebble conglomerate 10 to 30 metres thick. At a stratigraphical distance below the top of the Gog Group varying between about 50 and about 300 metres, the quartzites are interrupted by some 200 metres of massive, light grey limestone with thin interbeds of pale-weathering sandstone and argillite. Where it is as well developed as this, the limestone permits the division of the group into three formations—the McNaughton below, the Mural limestone near the top, and the Mahto Formation at the top. The limestone carries archaeocyathids (otherwise rare east of the trench), and olenellids. It also carries the rare edelsteinaspid trilobite *Polliaxis*, essentially a Siberian genus (PALMER, 1968). The quartzites both above and below the limestone are in places riddled with *Scolithus* tubes. Even where the Mural limestone is quite thin, it is clear that there is very much more Lower Cambrian in the Jasper region, and especially very much more Lower Cambrian quartzite, than there is in the Cariboo Mountains (3.3.1), 35 kilometres across the trench to the north-west.

Shortly before the close of Early Cambrian time, coarse sand deposition of shield provenance gave way to the phase of more fine-grained and calcareous sedimentation that was to continue for much of the remainder of Palaeozoic time. The basin from which the Park Ranges were eventually to arise became differentiated into two contrasting longitudinal facies belts. The western half received grey to brown muds and limestones and thin bluish limestones, with more brightly coloured and shalier rocks following later. This basinal facies interfingered over a startlingly short distance, along what is now almost the central axis of the range, with a bank or platform succession of massive carbonate rocks (RASETTI, 1951; COOK, 1967).

The basinal facies, in which the thin calcareous members are now irregularly dolomitized and the thicker argillaceous members highly cleaved, is called the Chancellor Group (NORTH and HENDERSON, 1954). Only in its most north-westerly outcrops, close to the trench, is it seen to rest on the quartzites (WHEELER, 1963). As it is of the order of 2,500 metres thick and overlain by unfossiliferous carbonate rocks, there has been some controversy about the true stratigraphical interval it occupies. Largely unfossiliferous itself, it has recently yielded a possible *Ptychoparella* from the lower part (WHEELER, 1963); the unexpected Atlantic trilobite *Lejopyge* from its middle section, and a group of agnostids of the *Cedaria* Zone from near the top. Also near the top, a variety of trilobite genera from the *Cedaria* and *Crepicephalus* Zones permits correlation of the uppermost Chancellor with the much more fossiliferous carbonate-shale facies to the east: *Arapahoia*, *Bynumia*, *Cedarina*, *Coosella*, and *Modocia* (W. H. FRITZ in COOK, 1967).

The Chancellor Group as a whole is probably equivalent to the Eager Formation farther south in the trench, though the top of the Eager is everywhere an erosion surface whereas in the case of the Chancellor it is the base which is poorly understood. Moreover, the present outcrop area of the Chancellor Group is separated from that of the Eager Formation by the belt of incomplete Cambrian sections (3.3.2) and by younger rocks. The stratigraphical interval represented by the Chancellor is probably that from latest Lower Cambrian to early Upper Cambrian. Its relationship to the great carbonate formations to the east proves that at least the lower half of the group belongs largely in the Middle Cambrian; this part retains the highest proportion of carbonate beds and blocks within the cleaved argillaceous mass (COOK, 1967). The evidence of the trilobites assigns the upper half of the group, or thereabouts, to the early Upper Cambrian. This part consists almost entirely of cleaved calcareous slates.

In no part of the more easterly, carbonate facies is the equivalent range of formations anything like 2,500 metres thick. The westward thickening may be due at least in part to penetrative flow and the development of intense cleavage, transmitted upwards through the section (COOK, 1967). Shaly formations within the carbonate facies become very much thicker as they enter the western facies, and the structural style there is quite different from that controlled by the competent formations in the eastern facies.

The carbonate facies of the Middle Cambrian in fact represents a subtle interplay between three sub-facies, those of the true cratonic carbonate shelf (sub-facies Aii) and the immediately inner and outer detrital zones (Ai and Bi). Minor lateral shifts of these three sub-facies, all deposited in very shallow water, result in a protracted succession of paired half-cycles which began with the uppermost Lower Cambrian and extended throughout the Middle and Upper Cambrian and on into the Ordovician (AITKEN, 1966).

Above an abrupt lower contact, which may approximate to a time plane, each lower, shaly half-cycle is gradationally overlain by an upper, carbonate half-cycle. The upper half-cycles commonly end in coarse, algal biostromes. Biocalcarenites, littered with trilobite remains, occur as interbeds in the shaly half-cycles but are rare in the carbonate half-cycles, in which the limestones tend to be fine-grained, dense, and unfossiliferous. They are also commonly dolomitized, but the dolomites have no obvious stratigraphical value. There are also lesser rhythms distinguishable within the cycles. The cyclic character of the succession becomes indistinguishable both towards the craton and towards the outer basin. Formational boundaries are consequently dependent on the lateral shifts of the three sub-facies, and those separating the two parts of a single cycle, in particular, may easily be drawn differently by different geologists.

The succession of zonal trilobites represented in the shaly half-cycles is remarkably constant no matter what changes of thickness affect the intervening carbonate half-cycles. The thick, unfossiliferous carbonates must therefore have accumulated extremely rapidly wherever and whenever limestone bank conditions were established.

Five complete cycles are readily recognizable, along the belt of maximum inter-fingering, between the top of the Lower Cambrian quartzites and the top of the Cambrian. Not all westward shifts of the inner detrital belt were equally extensive, of course, and there are parts of the range in which some half-cycles are lost and larger intervals in the succession become essentially all of carbonate.

The lowest shaly half-cycle constitutes the Mount Whyte Formation. It consists of thinly bedded limestones, some of them oolitic, and greenish siliceous shales, but with enough thin sandstone beds to weather to a reddish colour and form an easily recognizable unit. Several levels in it are persistently fossiliferous. The lowest zone yields *Bonnia fieldensis* (WALCOTT), *Fremontia canadensis* (WALCOTT), as well as species of *Onchocephalus*, *Syspacephalus*, and other trilobites, and algal 'buns'. This is an uppermost Lower Cambrian faunule, and, as all fossils in the rest of the formation are Middle Cambrian, this lower member was separated from the Mount Whyte Formation by RASETTI (1951) and included in the underlying quartzite formation as the Peyto Limestone Member. It should be restored to the Mount Whyte Formation. The numerous trilobite species in the upper part of that formation include *Amecephalus cleora*, *Plagiura cercops*, and *Poliella prima*, all WALCOTT species. In the southern part of the Park Ranges, the thickness of the formation varies from zero to nearly 200 metres, averaging less than 100 metres. The Peyto Limestone Member at the base is nowhere more than 80 metres thick, commonly less than 10 metres. This variability led to a suspicion that the formation either contains an erosional disconformity, or begins or ends at one, but there is not much evidence for one on the ground. The equivalent formation in the northern part of the range is called the Hota, which in the Mount Robson region thickens westward to nearly 300 metres (MOUNTJOY, 1962). Still farther north, towards the Peace River arch, the formation is difficult to recognize as a separate unit from the overlying Middle Cambrian rocks.

The Mount Whyte Formation is overlain in the southern part of the range by the mountain-forming Cathedral Formation, the oldest of the carbonate half-cycles. The Cathedral is typically a bedded dolomite with a good deal of limestone below. However, there are also numerous lens-like bodies of either of these lithologies within the other, and dolomite–limestone relationships within the formation are so complex that no stratigraphical significance can be attributed to them (RASETTI, 1951). The lower part of the formation locally carries a thin shaly member, the Ross Lake shale, which yields an early Middle Cambrian fauna with *Albertella bosworthi* WALCOTT. The average thickness of the Cathedral Formation, well exposed on many mountains north and south of Kicking Horse Pass, is 350 metres. It is through the dolomite of this formation, in Mount Ogden, that the lower of the two famous spiral tunnels of the Canadian Pacific Railway line was driven; the upper tunnel, on the other side of the Kicking Horse River, passes through Gog quartzites.

North-westward, the Cathedral becomes a dark grey limestone with a much lower degree of dolomitization (WHEELER, 1963), and this character is retained into the Mount Robson area in the northern sector of the range. The formation is there called Chetang. It is, however, much reduced in thickness by comparison with the Cathedral, being little more than 200 metres thick at maximum. This is because it is replaced, in its upper and lower parts, by the facies of the shaly half-cycles—greenish calcareous shales or mudstones cyclically interbedded with fragmental limestones and limestone conglomerates. Eastwards across the range, this facies takes over the whole formation, which then bears the name Tatei. The Tatei Formation in full development, as in the Jasper region, is 600 metres thick, but it thins towards the north and becomes a yellow-weathering, thinly bedded, sandy dolomite with red shale and sandstone interbeds (SLIND and PERKINS, 1966). Thus the depositional strike of the carbonate bank of this age was more westerly than the strike of present structures; the eastward change to shaly facies, which in this northern sector takes place in the middle of the Park Ranges, in the south takes place far out under the plains (3.3.5).

The Tatei Formation contains trilobites, such as *Glossopleura*, which in the southern sector are characteristic of the second, higher, shaly half-cycle. This is represented by the Stephen Formation, which consists chiefly of dark grey, fine-grained limestones with interbedded greenish shales and argillaceous limestones and occasional limestone breccia or conglomerate. The thickness, necessarily variable, is characteristically 100 to 200 metres. This figure has meaning, of course, only where the formation is preceded and succeeded by fully-developed carbonate half-cycles. A very short distance west of the formation's type locality, both these half-cycles pass abruptly into the western shaly facies already described (Chancellor Group), and the Stephen then loses all identity.

The consequence of this facies boundary for the fauna of the Stephen Formation is extraordinary. The formation, as a shaly half-cycle, is in any event highly fossiliferous, its numerous trilobite species including several of *Amecephalus*, *Glossopleura*, and *Zacanthoides*, and, at the top, abundant *Bathyuriscus adaeus* WALCOTT. Near the western boundary of the carbonate facies, however, the formation acquires a remarkable variety of extra faunules within thin, localized lenses of black shale. Two of these occurrences have become world famous. Unfortunately, both lie now in a zone of complex faulting, and their exact position within the Stephen Formation is difficult to determine (RASETTI, 1951). COOK (1967) appears to have demonstrated that the better-known of the two localities lies in an eastwardly-transported Klippe.

The first lens to be discovered was that of the *Ogygopsis* shale, its fauna dominated by great numbers of *Ogygopsis klotzi* (ROMINGER). It was this fauna which led to WALCOTT's interest in the area surrounding the Kicking Horse Pass. This led in turn to the discovery of the second fauna, that of the Burgess Shale. A lens about two metres thick contained a remarkable assortment of fossils not represented in the normal succession, nor in any other known succession, including bivalved crustaceans, holothurians, merostomata, medusae, and annelids. They were described by WALCOTT in a series of papers published in the Smithsonian Miscellaneous Collections, especially in the years 1911 and 1912. More recently, RASETTI (1951) has described the fauna in the context of three separate, subsidiary faunules. The lowest is typified by *Olenoides serratus* (ROMINGER); the second, the famous 'Phyllopod Bed' in WALCOTT's main quarry, by *Pagetia bootes* WALCOTT; and the uppermost by *Ehmaniella burgessensis* RASETTI. WALCOTT's quarry, long ago picked over, is now being re-excavated, and large numbers of new specimens are being unearthed.

The carbonate half of the second cycle is represented by the Eldon Formation, typically a great mass of dolomite which forms the peaks of many of the most familiar mountains. In the southern sector of the range, the dolomite displays some reefoid aspects: pale

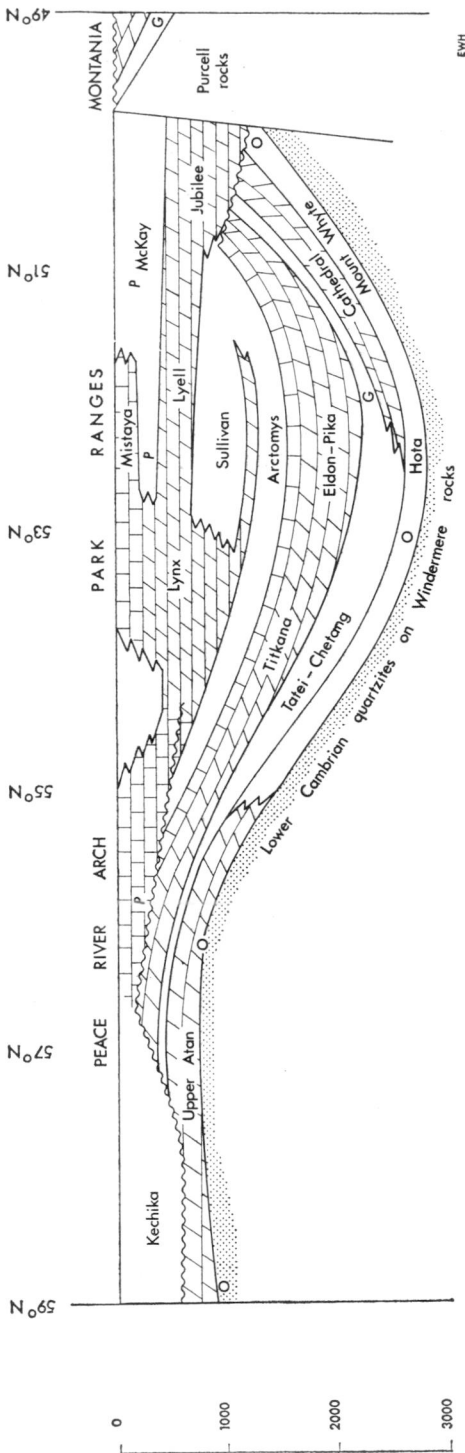

FIG. 13. Longitudinal section along the axis of maximum carbonate deposition in the Middle and Upper Cambrian, Canadian Rocky Mountains. Length of section about 1500 kilometres. P = *Ptychaspis-Prosaukia* Zone; G = *Glossopleura* Zone; O = *Olenellus* Zone. Note sub-Franconian unconformity in northern third of section.

colours, coarse and vuggy texture, lack of obvious bedding, and total absence of fossils except for a few algae. Towards the north, as well as across the strike, this dolomitic facies gives way to dark coloured limestones with only irregular blotches of dolomitization (and those concentrated towards the top). This second facies is still cliff-forming and apparently massive, but in detail it can be seen to be thinly bedded, almost laminated (WHEELER, 1963). In the northern sector of the range, as at Mount Robson, the formation bears the name Titkana. Under either name, the formation is about 350 metres thick, but general westward thickening brings it to a maximum of nearly 600 metres. Still farther north, the lower, non-dolomitic portion becomes largely shale, the effect being to increase the thickness of the lower, shaly half-cycle (Tatei Formation) at the expense of the upper, carbonate half-cycle (Titkana Formation).

The Eldon–Titkana Formation is not the end of the second carbonate half-cycle. Overlying it is a very persistent unit of thinly bedded limestones with interbeds of grey, calcareous shales, some intraformational limestone conglomerate, and very little dolomite. This is the Pika Formation, 100 to 200 metres thick, but thickening towards the trench to as much as 500 metres. It is in general very sparsely fossiliferous, though locally abundantly so. Its few species belong to such late Middle Cambrian genera as *Glyphaspis*, *Modocia*, and *Rowia*?, variously attributed to a *Bolaspidella* Zone or to a *Thompsonaspis* Zone.

The third shaly half-cycle is represented by the Arctomys Formation, so persistent and characteristic that it bears the same name from one end of the range to the other. It consists largely of coloured shales—maroon, red, buff, or green—dolomitic or siliceous, with thin bands of dolomite or siltstone. Mud cracks and casts of salt crystals are common, pointing to an extreme shallowing of a depositional basin which must have remained very shallow indeed throughout. Orange and red weathering colours make the formation the best single marker horizon in the whole Cambrian succession. It thickens rather consistently from south to north, from less than 100 metres to an average of about 130 metres and a maximum, near Mount Robson, of over 300 metres. The Arctomys is essentially unfossiliferous, though a '*Deissella*' fauna has been reported from it (SLIND and PERKINS, 1966). As early Upper Cambrian fossils are found shortly above it in a number of areas, the formation can be taken as being of latest Middle Cambrian age, or as straddling the Middle-Upper Cambrian boundary.

The carbonate half-cycle following the Arctomys is actually part of the type-section of the original Arctomys itself. It is now separated as the Waterfowl Formation. Varying in thickness from less than 50 metres to nearly 200, it consists of mottled limestone and dolomite, with some limestone breccia and a little olive-green shale. Its fossils are those of the lower *Cedaria* Zone.

In the southern sector of the Park Ranges, the fourth cycle begins with the thickest of the recessive units, the Sullivan Formation. It consists of green shales with thin beds of oolitic or bioclastic limestone concentrated near the top and the base. The formation thickens more or less regularly across the range, from 200 metres in the east to more than 400 metres in the west, and in its full development it embraces parts of three Upper Cambrian faunal zones—the upper part of the *Cedaria* Zone, the whole of the *Crepicephalus* Zone, and the lower part of the *Aphelaspis* Zone. Genera represented by several species include *Arapahoia*, *Blountia*, and *Bynumia*; in several beds, brachiopods greatly outnumber trilobites.

The carbonate companion of the Sullivan is the Lyell Formation, another cliff-former comprising cyclically bedded limestone and dolomite between 250 and 350 metres thick. In the central part of the range, the formation extends from the upper part of the *Aphelaspis* Zone to include the base of the *Elvinia* Zone.

The fifth and last of the completely Cambrian cycles begins with the Bison Creek Formation, a succession of olive-grey shales or mudstones with thin interbeds of lime-stone, many of them oolitic or calcarenitic. The thickness of 150 to 200 metres embraces the upper *Elvinia* Zone and the zones of *Conaspis* and *Ptychaspis-Prosaukia*. The succeeding carbonate half-cycle, the Mistaya Formation, comprises massive, clastic limestone, with some oolitic intervals, and equally massive stromatolitic biostromes typified by the form-genus *Collenia*. The clastic limestones have yielded a few trilobites (*Eureka, Stenopilus*) of *Saukia* Zone affinity (RAASCH and BRUCE, in the press). The formation thickens westwards, across the Park Ranges, from nearly 200 to more than 300 metres. In places, uppermost Cambrian trilobites of the *Saukia* Zone can also be found in the lower parts of an overlying, sixth half-cycle, represented by the Survey Peak Formation which is otherwise all Ordovician.

In this most perfectly developed form, the cyclicity of sedimentation in the Middle and Upper Cambrian is characteristic only of the eastern and central parts of the southern sector of the Park Ranges. In the western parts of the ranges, as we have seen, at least the three lowest cycles, and the fourth shaly half-cycle, are merged in the argillaceous facies of the Chancellor Group, in which no equivalent cyclicity has yet been dis-tinguished. For the fourth, Upper Cambrian carbonate half-cycle, however, exactly the opposite change takes place.

Along the western side of the southern sector, the argillaceous Chancellor Group is followed by a huge formation of essentially uninterrupted carbonate, the Ottertail Formation. This is everywhere at least 600 metres thick, in places as much as 1,000 metres. The upper half or more of the formation is of almost unbedded, bluish limestone, forming steep cliffs more than 500 metres high. This part is certainly the Lyell Formation of the eastern part of the range. The lower half of the Ottertail is well bedded, even laminated, but still essentially all of limestone. This part, quite unfossiliferous, was thought to be equivalent to the Waterfowl and Sullivan Formations farther east, but it is now established that the whole of the Ottertail is simply a greatly thickened equivalent of the Lyell. That the Waterfowl and Sullivan Formations are represented along the west side of the range by the upper part of the Chancellor Group is indicated by the discovery of trilobites of the *Cedaria* and *Crepicephalus* Zones near the top of that group (COOK, 1967). Except that it is typically a limestone and not a dolomite, the Ottertail is also equatable with the Jubilee Formation of the Kootenay Ranges (3.3.2). The area of development of the two formations, along the eastern side of the trench from south of the 50th parallel to north of the 52nd, is thus one in which there is only one major carbonate unit in the whole of the Cambrian, and it is of Late Cambrian age.

In the northern sector of the Park Ranges, the whole of the Upper Cambrian changes to the carbonate facies, and this change can be followed in detail because it takes place along the present structural strike. Northwards from the 52nd parallel, first the fifth shaly half-cycle, (the Bison Creek Formation), and then the fourth, (the Sullivan Formation), grades into massive carbonates and so loses its identity (AITKEN and GREGGS 1965). The whole of the Upper Cambrian, above the Arctomys Formation, is then a single carbonate unit, the Lynx Group, its type section being near Mount Robson.

In the region dominated by that mountain (Frontispiece), the Lynx Group consists of well-bedded, silty dolomite and limestone, with bands of intraformational limestone-conglomerate and very minor shale. Fossils have been found only in the lower part; they include species of *Arapahoia* and *Coosia*. Thickness of the group varies from 800 metres, in the east, to 1,200 in the west (MOUNTJOY, 1962).

The Lynx is overlain by another shaly half-cycle, the Chushina Formation, which is definitely Lower Ordovician. Proceeding still farther northwards along the Park Ranges, however, this shaly facies appears progressively lower in the section. The typical Lynx dolomite is steadily reduced, and more and more of the upper part of the group becomes indistinguishable from the argillaceous limestone and shale of the Chushina Formation. Trilobites show, however, that the succession represents all three of the North American Upper Cambrian stages (SLIND and PERKINS, 1966).

The continuation of carbonate deposition until the very end of Cambrian time is therefore characteristic of only a restricted belt within what are now the Park Ranges. This is further illustrated in the western part of those ranges, where the Ottertail limestone is overlain, conformably, by another pair of half-cycles still of Cambrian age. These are the lower part of the Goodsir Group, which continues through the Lower Ordovician in a facies typified by limestone, some of it very cherty, calcareous shale, and limestone-conglomerate. Thin stripy bedding renders the group easily recognizable. The group as a whole is exactly equivalent to the McKay Group of the Kootenay Ranges (3.3.2), and that name is preferred to the name Goodsir. The Cambrian quota of the group is equivalent to the Bison Creek and Mistaya Formations, the last pair of Cambrian half-cycles along the axis of the Park Ranges.

The lower, shaly half-cycle, of light green shale and argillaceous limestone, is easily phyllitized by shearing, but it has yielded a trilobite fauna sufficiently abundant to permit biostratigraphical zonation by RAASCH and BRUCE (in the press). From the base of the group, the *Elvinia* Zone of the lower Franconian Stage is represented by *Elvinia roemeri* (SHUMARD), *Housia canadensis* (WALCOTT), and *Irvingella major* ULRICH and RESSER. These are followed by *Bernia obtusa* FREDERICKSON and *Parabolinoides hebe* FREDERICKSON, fossils of the lower part of the *Conaspis* Zone. The *Ptychaspis-Prosaukia* Zone is represented by *Ptychaspis striata* WHITFIELD, *Prosaukia longicornis* ULRICH and RESSER, and species of *Briscoia*, *Drumaspis* and *Ellipsocephaloides*. These faunas are widespread in the interior of North America.

The upper, carbonate half-cycle is a cliff-forming limestone about 160 metres thick; no fossils have been found in it, but the beds above it contain a basal Ordovician fauna and equivalence of the limestone to the Mistaya Formation may confidently be postulated.

The axis of the southern Canadian Rocky Mountains, called the Park Ranges, though young in structural age, is thus essentially Cambrian in both content and control. The ranges were raised along the narrow belt of greatest development of the cratonic carbonate sub-facies (Aii) of the Middle Cambrian, and especially along the boundary between this sub-facies and the basinal sub-facies (Bi) to the west. This boundary was itself controlled by features along the eastern shelf of an earlier, Proterozoic basin. Near the close of Middle Cambrian time, the very shallow-water beds of the Arctomys Formation were spread with remarkable continuity over carbonate bank and shale basin alike, and the carbonate bank then expanded westwards, for the Upper Cambrian to overlie what had been the shale basin during the Middle Cambrian. Eastwards from the Park Ranges, no complete Cambrian section is found again in western Canada.

3.3.4. The Cambrian in the Front Ranges of the Canadian Rocky Mountains

The front Ranges of the Rocky Mountains in Canada are a series of eastwardly directed thrust slices, pushed aside by the great overthrust mass of the Park Ranges.

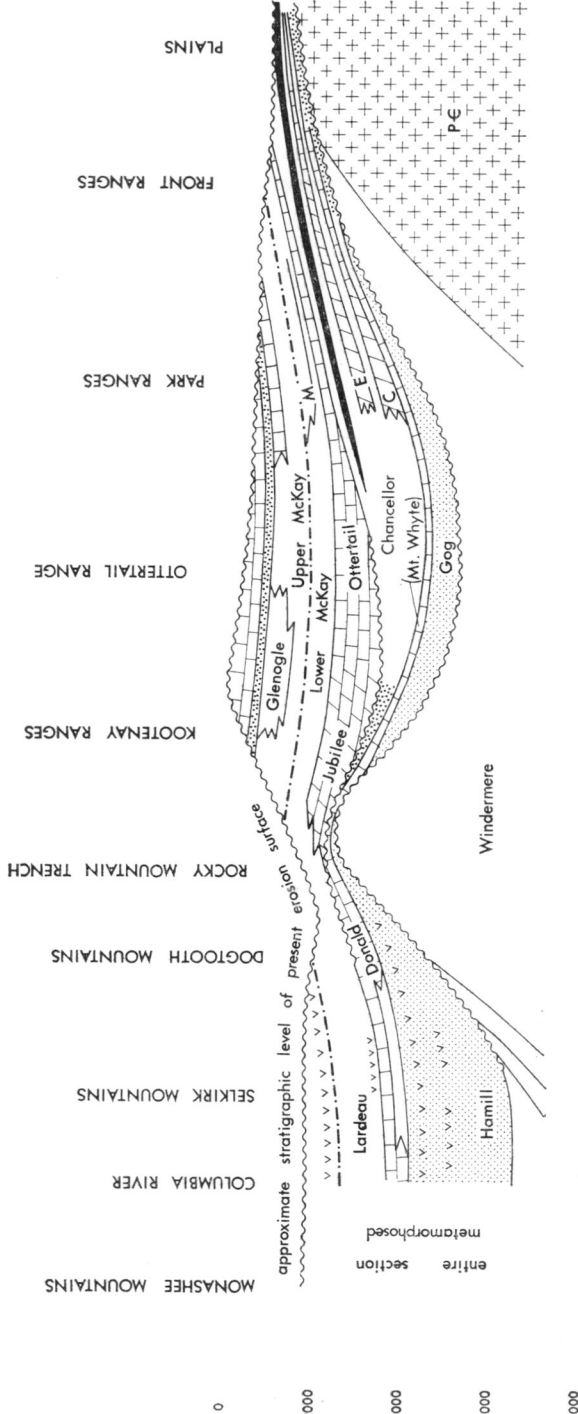

FIG. 14. Cross-section approximately along 51° 30'N. latitude, showing disposition of Cambrian (and Lower Ordovician) rocks before subsequent deformation. Length of section at least 400 kilometres, now reduced by shortening to less than 250 kilometres. Dominant lithologies only represented; absence of indication means dominantly argillaceous strata.

Dash-dot line represents base of Ordovician; solid black is Arctomys Formation. M = Mistaya Formation; E = Eldon Formation; C = Cathedral Formation.

The Front Ranges are totally dominated by Upper Palaeozoic rocks, and Cambrian rocks crop out only in narrow strips in the hanging walls of some of the thrusts. There is no continuous Cambrian outcrop belt as there is in the Park Ranges.

Lower Cambrian rocks are not seen, and probably do not extend to the eastern edge of the thrust belt. Both the Middle and the Upper Cambrian are well represented, and, because of their progressive eastward truncation below the sub-Devonian unconformity, the oldest rocks are exposed in the easternmost ranges.

In the southern sector, east of the Kicking Horse Pass and centred on the resort town of Banff (Fig. 9), there are four front range thrust blocks. In the most westerly block, closest to the Park Ranges, the oldest beds seen belong high in the Middle Cambrian. The section from the Arctomys Formation to the top of the Cambrian, and on into the Ordovician, is like that already described for the eastern part of the Park Ranges, except that the formations are individually thinner. The zones of *Cedaria*, *Elvinia*, and *Ptychaspis-Prosaukia* are well represented faunally by such species as *Bynumia elegans* and *B. rangerensis*, *Drumaspis alberta*, *Ellipsocephaloides sawbackensis*, and *Irvingella alberta*, all RESSER species.

In the second range from the front, the exposed Cambrian succession both begins and ends at lower stratigraphical levels. The Eldon Formation, here a sufficiently pure limestone to be quarried for lime, is the oldest horizon seen; the basal Upper Cambrian, carrying fossils of the *Cedaria* Zone, is the youngest, being unconformably overlain by the Devonian.

In the easternmost range, the frontal thrust is overlain by an almost complete Middle Cambrian succession, but has no Upper Cambrian at all (FITZGERALD, 1962). At the base is a sandstone-shale unit about 100 metres thick, representing the basal transgressive formation and carrying Middle Cambrian trilobites of the *Albertella* Zone. This formation, which would be called the Flathead sandstone in the Montania sector (3.2), is therefore the same age as the lower part of the Cathedral Formation in the Park Ranges to the west, and it becomes progressively younger eastwards across the prairie sub-surface (3.3.5).

Above it come successively the rest of the Cathedral Formation, reduced to about 200 metres of dolomitic limestone; the Stephen, in its normal recessive-weathering facies and with its usual fossils; the Eldon, about 260 metres thick and composed of cliff-forming dolomite and limestone; and the lower part of the Pika Formation. This succession is overthrust on to the Cretaceous along the mountain front, and is itself unconformably overlain by the Devonian.

In the northern sector, east of Jasper, the Front Ranges differ in style from those in the south, but the exposed Cambrian rocks exhibit an exactly similar pattern. In the more westerly front ranges, a little Upper Cambrian is all that is seen between the Devonian strata and the thrusts; in places, no Cambrian is visible at all, and the oldest rocks above the thrust are earliest Ordovician in age. In the more easterly front ranges, the exposed section represents the upper half of the Middle Cambrian. The oldest horizon is a thinly-bedded argillaceous limestone with partings of green shale, yielding fossils of the Stephen Formation but here more properly called Tatei. It is in fact the same unit as that seen all along the Front Ranges and even over Montania (3.2). It is overlain by massive, dolomitic Eldon, with a limestone-conglomerate at the top. The Eldon here is only about 120 metres thick, and, as there appears to be a disconformity between it and the overlying Arctomys, there may have been a brief erosional episode in Pika time. The

Arctomys, however, is in full thickness of nearly 200 metres, and of completely character-istic lithology (3.3.3).

3.3.5. The Cambrian beneath the plains of southern Alberta and Saskatchewan

Over the interval of latitude occupied by the Park Ranges of the southern Rocky Mountains, Cambrian rocks extend far to the east beneath the plains of southern Alberta and Saskatchewan. They have been penetrated, wholly or in part, by about 1,000 wells. They everywhere rest directly on crystalline Pre-Cambrian rocks of the Churchill province of the Canadian shield; the Proterozoic sedimentary rocks, which underlie the Cambrian throughout the Cordillera, are entirely absent. Lower Cambrian rocks are also absent. The Cambrian rocks in turn are covered by younger rocks, mostly Devonian and Cretaceous, ranging in thickness from about 1,000 metres, in the east, to more than 3,000 metres close to the disturbed belt. The maximum known thickness of Cambrian rocks now preserved in the subsurface is a little over 500 metres, in east–central Alberta (FULLER and PORTER, 1962; VAN HEES, 1964).

The base of the Cambrian section is a transgressive sandstone 30 to 50 metres thick. This unit is in fact seen at the surface on the Montania 'high', where it is represented by the Flathead Formation (3.2). At the western limit of the plains, this basal clastic unit is about in the stratigraphical position of the lower part of the Cathedral Formation. It becomes progressively younger eastwards, cutting out more and more of the Middle Cambrian as the transgression proceeded. Over the Swift Current platform, the sub-surface extension of Montania, the sea had extended only as far eastward as the 108th meridian by the end of Middle Cambrian time. Off the northern edge of the platform, the sea reached the 106th meridian. Eastwards from these positions, all the Cambrian ever deposited was Upper Cambrian (Figs 11, 12).

The limestone of the Cathedral Formation, with some dolomite, lapped only a short distance on to the shield before being taken over by the basal clastic unit. The overlying Stephen Formation is already in detrital facies in the mountains, but its extension in the subsurface can be recognized by the recovery, from a number of wells, of fossils repre-senting the Glossopleura and Bathyuriscus-Elrathina Zones. The genus Glossopleura itself is especially persistent. Electric log data indicate a thickness for the subsurface Stephen of between 30 and 60 metres.

The carbonate of the Eldon Formation extends much farther eastwards than that of the Cathedral, because it represented a later stage of a continuous transgression. Typically 60 to 80 metres thick, it passes first into shale, which acquires tongues of sandstone, and then into the basal clastic unit. Either the upper Eldon or the lower Pika is in shaly facies throughout; this unit averages 50 metres in thickness. The Pika, or its upper part, is a very persistent limestone, though less than 30 metres thick. Its top provides an excellent electric log marker across much of the region, and, as no distinctive Arctomys can be recognized in the wells, this marker is taken as the boundary between the Middle and the Upper Cambrian in the subsurface.

The maximum subsurface thickness known for the Middle Cambrian is 470 metres, near Calgary. Over the area of its distribution, the parallelism between the carbonate facies and the thicknesses is so close that it is evidence for control of that facies by some critical water depth (VAN HEES, 1964).

The transgression continued during the Late Cambrian. The Upper Cambrian is from 150 to 250 metres thick, and is overlain conformably by the Tremadoc. The lowest of the Upper Cambrian stages, the Dresbachian, transgressed beyond the margin of the

Middle Cambrian, and in the eastern part of its subsurface extent it lies directly on the shield. Glauconitic shales and siltstones are characteristic, with only minor limestone and that chiefly at the top. The most commonly recovered fauna is of the Dresbachian *Crepicephalus* Zone; as it occurs high up in the succession, there can be very little deposit of the succeeding Franconian and Trempealeauan Stages. The zones of these stages have not in fact been identified with certainty in the Canadian subsurface, but electric log correlations with wells in the north-central United States show that the *Saukia* Zone, at least, is represented (VAN HEES, 1964).

As in so much of the interior of North America, the first major unconformity within the Phanerozoic is the Knox unconformity (1.5.2), ending the Sauk Sequence (SLOSS, 1963). This unconformity is younger than the Tremadoc, and strata of Tremadoc age are preserved below it over much of the southern prairies. The age is determined by electric log correlation with wells in North Dakota and Montana, where the equivalent formation, the Deadwood, contains at its top a typical earliest Ordovician fauna (FULLER and PORTER, 1962).

One result of the movements causing the unconformity was the creation of the Williston Basin, a large cratonic sag the northern sector of which underlies south-western Manitoba and south-eastern Saskatchewan. There is no sign of the basin before the end of Sauk time, and Cambrian rocks are absent from its eastern half. Within the basin, Devonian rocks rest on Silurian and Ordovician; outside it, they rest on Cambrian or Tremadoc. The edges of the Upper Cambrian in the subsurface are now all erosional. The eastern limit is a consequence of the Knox unconformity, which cuts down eastwards on to the Pre-Cambrian. The northern limit is at the sub-Devonian Meadow Lake escarpment, forming the southern, subsurface boundary of the Peace River arch (3.4).

The Cambrian shoreline at any one time is marked by clastic rocks, chiefly deltaic and littoral sands. On the south side of the Meadow Lake escarpment, in west-central Saskatchewan, coarse sand dominates both the Middle and the Upper Cambrian (VAN HEES, 1964). This sand was derived from the ancestral Peace River arch. Both the clastic rocks and the carbonates are persistently glauconitic. Samples of glauconite from them have yielded Mississippian and Pennsylvanian radiometric dates (STEVENS, 1965), probably because of partial loss of radiogenic argon consequent upon burial and later erosional unloading.

3.4. The Cambrian over the Peace River arch

The prairie region of north-western Alberta and north-eastern British Columbia lies over a structurally high block called the Peace River arch. The arch possesses an east-north-easterly trend, almost exactly at right angles to the Cordillera; it is much the largest of the transverse structures controlling the nature and distribution of Palaeozoic sediments.

The arch is sharply bounded on both north and south. Its southern margin lies along the northern foot of a prominent, north-facing escarpment, which was eroded in Lower Palaeozoic rocks in pre-Middle Devonian time. This feature, commonly referred to as the Meadow Lake escarpment, now lies entirely in the subsurface, approximately along 55° north latitude (Figs 11 and 12). Its crest is formed of Ordovician limestone, with Upper and Middle Cambrian below. The northern boundary of the arch coincides with the westward, subsurface extension of a major zone of faulting along Great Slave Lake; it crosses the front of the Cordillera about at the 60th parallel.

Northwards from the Meadow Lake escarpment, the Upper and Middle Cambrian thin rapidly to zero by sub-Devonian truncation. Over 300 metres of Cambrian rocks were removed in central Alberta (VAN HEES, 1964). Along the abrupt southern border of the arch, Upper Palaeozoic rocks lie directly on the buried crystalline shield. Cambrian rocks are found at this latitude only in the mountains, which are here at their narrowest. Northwards, one effect of the arch is to cause important divergences between Palaeozoic depositional and erosional trends, on the one hand, and subsequent structural trends, on the other. A system of almost northerly-trending fault blocks, active during Cambrian time, appears within the front ranges north of 57° N. latitude, and this pattern becomes increasingly prominent northward to beyond 60° N. A consequence is that facies boundaries and erosional edges there follow more northerly trends than do present structures, and the amount of section preserved beneath the Devonian differs sharply from one fault block to another. Farther west, in the main range of the northern Rocky Mountains, the reverse relation is displayed. Cambrian stratigraphical trends cross the Rocky Mountain trench in a north-westerly direction on the east flank of the Cassiar Mountains (Fig. 9), and facies boundaries and erosional edges there exhibit more westerly trends than do present structures.

The block faulting began in Pre-Cambrian time, and continued to be active during the Early Cambrian. The principal uplift of the arch must be post-Early Cambrian; the Lower Cambrian is not reduced in thickness in the mountains within its latitudes. Most of the uplift is post-Ordovician, because the limestone of that system is in full development south of the erosional Meadow Lake escarpment; both Ordovician and Silurian are similarly in full thickness north of the arch. The main uplift was very probably Siluro-Devonian, synchronous with important Caledonian movements elsewhere.

Where the Peace and Pine Rivers flow out of the trench and across the mountains, the westernmost recognizable Cambrian rocks form part of the metamorphic Wolverine Complex (ROOTS, 1954). It consists of clastic sedimentary rocks, with some carbonates and lavas, all intruded and granitized to form gneisses, schists, quartzites, marbles, and amphibolites. The metamorphism antedated the folding and appears to be unrelated to the great Mesozoic batholith to the west. Rather it seems to have been a consequence of deep-seated intrusion during the Cariboo orogeny, which in a broad sense was the Cordilleran equivalent of the Caledonian.

Where metamorphism is less intense, the Lower Palaeozoic part of the complex is called the Ingenika Group, which can be shown to include equivalents of the Cariboo Group, from the south, and the Atan and Kechika Groups, from the north (3.3.1; 3.5). It is dominated by finely grained and thinly bedded schistose and phyllitic rocks with abundant quartz and chlorite, and green or brown feldspathic quartzite. Large lenses of limestone yield archaeocyathids, the commonest being *Ajacicyathus purcellensis* OKULITCH. The rocks were originally argillaceous sandstones, shales, and algal limestones, with some conglomerates, and even in their present condition they display ample evidence of origin in shallow water (ROOTS, 1954). This evidence, combined with what we know of the structural history of the Peace River arch, suggests that we are dealing with an essentially cratonic facies, and that any basinal facies developed at this latitude must have been deposited still farther west, where no Cambrian rocks can now be recognized.

On the east side of the trench, the base of the Cambrian is still ill-defined. A succession of schists, slates, and greywackes, the Misinchinka Group, is probably assignable to the Proterozoic. However, it passes up gradationally into a quartzite which is in turn overlain by 450 metres of dolomite with interbeds of red shale and sandstone near the base. The lower part of the dolomite is exceedingly fossiliferous. Two archaeocyathid horizons have been identified (STREET, 1966), and the trilobites include species of *Bonnia*, *Kootenia*, *Olenellus*, *Olenoides*, *Paedeumias*, *Wanneria*, and *Zacanthopsis*. This fauna is characteristic of both the eastern and the western Cambrian of this age, being of similar aspect to the faunas in the Lévis conglomerates of Quebec and the Parker slate of Vermont (2.3.2). In the east, some of the genera are regarded as typifying an intermediate facies realm (LOCHMAN-BALK and WILSON, 1958) rather than a truly cratonic one, indicating that the Lower Cambrian dolomite in the Pine Pass district lay close to the western edge of the carbonate bank. In the more easterly thrust blocks, the quartzite is considerably thicker, exceeding 700 metres, and the overlying dolomite correspondingly both thinner and younger. No olenellids have been found in it there, but *Ogygopsis* and *Syspacephalus* suggest a very early Middle Cambrian age (IRISH, 1964).

Thicknesses of the order given here make it clear that the Peace River arch was not a significantly positive element in Early Cambrian time. The Middle Cambrian, however, is greatly reduced. The maximum thickness demonstrable for it is about 260 metres, and the reduction from the more normal thickness seems to be at the expense of the lower half of the series, which has gone over to 100 metres or less of yellow-weathering, thinly bedded, dolomitic siltstone with red shale interbeds. The rest of the series is an unnamed formation almost entirely of carbonate. Fossils from the middle of it include *Kootenia*, *Ogygopsis*, *Olenoides*, *Pagetia*, and *Pachyaspis*, a distinctly western assemblage representing the outer detrital realm and a stratigraphical interval which includes the *Bathyuriscus-Elrathina* trilobite zone (STREET, 1966).

The Upper Cambrian is equally reduced, and again the 250-metre thickness identified by fossils lacks the lower half of the series. Rather thinly bedded, nodular limestones, with shaly or sandy interbeds, pass directly upwards into typical Lower Ordovician like the Chushina Formation of the Park Ranges farther south. *Pseudagnostus*, *Saratogia*, and *Wilbernia* represent the uppermost *Conaspis* Zone and that of *Ptychaspis*. Above come *Geragnostus*, the brachiopod *Westonia*, and other fossils of the *Saukia* Zone.

These reduced sections of the Middle and Upper Cambrian increase considerably towards the east, not towards the west as would be the case south of the arch. But the great cliff-forming carbonates of these ages in the Park Ranges (3.3.3) are entirely absent here.

In the subsurface to the east, the whole of the Lower Palaeozoic is absent, through a combination of reduced deposition and pre-Middle Devonian erosion. Where any Cambrian rocks remain, we would expect them to be Lower Cambrian rather than Middle or Upper Cambrian. Within the area bounded by 56° and 58° N. latitude and 119° and 122° W. longitude (Fig. 11), wells completely penetrating the proven Devonian enter a section of glauconitic shales, sands, and siltstones with a persistent dolomite member in the middle. This section, which rests directly on crystalline Pre-Cambrian rocks, could be of almost any age from early Devonian to Proterozoic. However, the lithology and the electric log characteristics resemble those of the Middle Cambrian south of the Peace River arch; in one well, some fossil forms resembling *Acrotreta* were recovered. It is therefore possible that, in this small area, an isolated patch is preserved of the lower part of the Middle Cambrian, in inner cratonic facies (VAN HEES, 1964).

Farther north in the Cordillera, in the drainage area of the Liard River, the divergences between depositional and structural strike become evident. The Cambrian everywhere rests on Proterozoic sedimentary rocks, not on crystalline basement, and these rocks, though immensely variable in detail, become progressively older eastwards. Structures inherited from them exercised profound control over Lower Palaeozoic sedimentation. The most westerly Cambrian is clearly divisible into two groups of sedimentary rocks, which continue without any very obvious changes for a considerable distance northward and beyond the 60th parallel.

Beds assignable to the Lower Cambrian are divisible into two formations. The lower, 500 to 700 metres thick, comprises feldspathic quartzite with minor pebble-conglomerate, and greenish- or maroon-weathering finer clastic rocks. This unit yields fossils in a few localities; they include *Olenellus gilberti* MEEK. Above comes a unit, 200 to 500 metres thick, essentially of carbonate rocks—well-bedded to massive limestones, orange-weathering sandy dolomites, limestone conglomerates, and, again, minor shale or slate. This formation yields archaeocyathids, especially *Coscinocyathus dentocanis* OKULITCH. The two formations, individually unnamed, make up the Atan Group in the central part of the belt (GABRIELSE, 1963b). The overall resemblance of the group to the Labrador Group, on the foreland of western Newfoundland (2.4.1), is apparent. The similarity reinforces the basic difference in the nature of the Cambrian transgressions on the two sides of the continent (1.3). In the east, the archaeocyathid limestones typify the sections closest to the outcropping shield; in the west, they are found only in the sections farthest from it.

The remainder of the Cambrian, fossils from which still await detailed study, consists of more thinly bedded strata—grey, nodular, shaly limestone, commonly now phyllitic; calcareous or dolomitic siltstone; limestone cobble-conglomerate; and minor clastic rocks. Towards the craton there appears a good deal of orange-weathering sandy dolomite, a rock very characteristic of the Cambrian sediments of the whole northern region. In places, some volcanic rocks are associated, but these may be younger. Their proportion increases northwards into the Pelly Mountains (3.5).

In the area immediately south of the 60th parallel, where the true Rocky Mountains end, this formation constitutes the lower division of the Kechika Group (GABRIELSE, 1963b). Uppermost Cambrian is certainly represented in it; the Middle Cambrian cannot be thick, and may again be absent. The upper division of the Kechika Group, which is dominated by black, thinly bedded, pyritic and carbonaceous shales and slates, and argillaceous limestones, carries graptolites in places and may be all Ordovician.

Thus the rocks, mainly Cambrian, responsible for the axial ranges of the Rocky Mountains are here transferred to the west side of the Rocky Mountain trench. The true Rocky Mountains therefore die out into the Liard Plain, where exposures are too poor for any detailed stratigraphy. Rocks which may be Cambrian are almost entirely clastic, and most outcrops are of the Kechika Group which may here be wholly Ordovician. As we have seen, an incomplete Cambrian section, but with a well-developed, highly incompetent uppermost Cambrian and Lower Ordovician formation at the top of it, is characteristic of the immediate vicinity of the trench throughout much of its length (3.3.2).

Eastwards, in the appropriately named Terminal Range, the Lower Cambrian Atan Group reappears in full thickness (nearly 1,000 metres), with *Olenellus* in the clastic members and archaeocyathids in the limestones. The Middle Cambrian has not been identified palaeontologically, but where the range crosses the 58th parallel an unnamed limestone formation 700 metres thick carries near its top a fauna of the lower part of the Upper Cambrian—*Baltagnostus*, *Clavagnostus*, *Blountia*, and *Cedaria*, with some brachiopods in addition. Both northwards and eastwards, however, this limestone wedges out between two unconformities. The Kechika Group, above, though thick, is

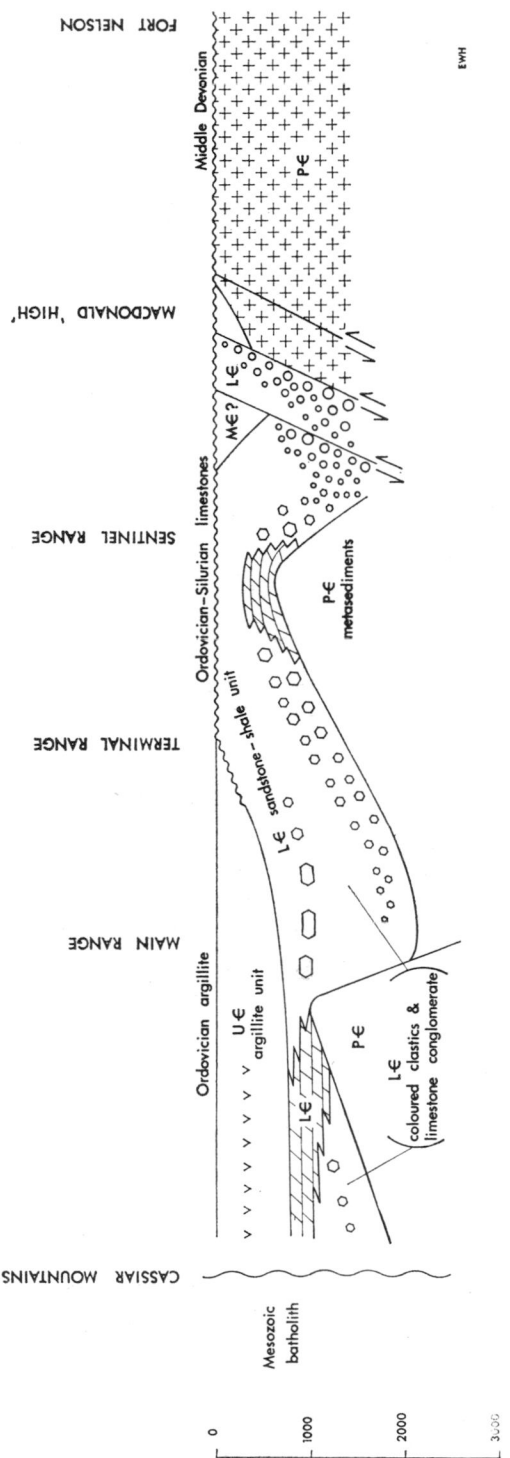

FIG. 15. Cross-section approximately along 59°N. latitude, showing disposition of Cambrian rocks before subsequent deformation. Length of section about 400 kilometres. Dominant lithologies indicated; absence of indication means dominantly argillaceous strata.

then all Ordovician, lying unconformably on the Atan Group with both Middle and Upper Cambrian missing. We are here in the area of the northerly-trending fault blocks, and lateral changes in thickness and facies become very pronounced.

The Kechika Group thins out to zero eastwards. The Atan Group, below, changes to an almost entirely clastic facies—dark, laminated argillite, siltstone, sandstone, and sandy limestone, reddish-weathering sandy dolomite and cross-bedded dolomitic sandstone or quartzite, and a wedge of coarse, polymictic pebble and cobble conglomerate 200 metres thick. In the next fault block to the east, this conglomerate takes over the whole section, becoming a 2,000-metre pile of coarse fanglomerate where it lies in downdropped fault blocks of Pre-Cambrian rocks (G. C. TAYLOR, personal communication, 1967). The principal fault was downthrown to the west. The fanglomerate, with boulders up to almost two metres across, therefore wedges out abruptly westwards, becoming first a conglomerate with smaller, rounded boulders and pebbles of quartzite and basalt, then successively a coloured feldspathic sandstone, a siltstone-argillite unit with patch reefs in it, and, finally, ordinary carbonates of the Atan Group. The boulder beds are therefore the stratigraphical equivalent of part of the Lower Cambrian. They are overlain either by the Lower Ordovician part of the Kechika Group, or by the Silurian. There is no Middle or Upper Cambrian, and commonly no Upper Ordovician. On the upthrown side of at least one fault block, the Silurian rests directly on Pre-Cambrian carbonate rocks.

These conditions continue northwards, for a short distance, across the 60th parallel, before the structural and stratigraphical patterns flare out eastwards and westwards beyond the northern edge of the arch. The northern extension of the Liard Plain yields only a little archaeocyathid limestone with some unfossiliferous siltstones and shales which seem to represent a more thinly-bedded equivalent of the Atan Group (GABRIELSE, 1967b). In the first range to the east, the Lower Cambrian assumes a mixed facies, with very fossiliferous limestone, siltstone, and reddish-weathering fine-grained sandstone totalling about 700 metres in thickness. Archaeocyathids and trilobites occur together, and the development may be regarded as transitional between the Atan facies, to the south, and the Sekwi facies which typifies the Lower Cambrian in the Mackenzie Mountains to the north (3.5).

Beds both above and below this unnamed formation may also belong to the Cambrian. Underlying it are more than 1,000 metres of drab siltstones with some conglomerate which are probably a less coarse version of the Lower Cambrian fanglomerate to the south-east. They rest on a thick pile of amygdaloidal greenstones. Overlying the Atan-Sekwi unit is the normal, argillaceous Kechika Group.

Eastwards again, the Lower Cambrian is represented by sandy dolomite with amygdaloidal flows and breccias, overlain by Upper Cambrian and Lower Ordovician Kechika Group in more calcareous facies. Rocks below the Lower Cambrian dolomite are not seen, however. This brings us, eastwards, to the region of the northerly-trending fault blocks. The Lower Cambrian sandy dolomite and volcanic rocks are underlain by the big conglomerate-sandstone unit, but here this includes in its middle part a thick band of limestone and dolomite carrying olenellids. Further flows and breccias occur below, but whether these also belong to the Lower Cambrian, or are Pre-Cambrian, is not known. The upper part of the succession is clear, however. The Lower Cambrian carbonate-volcanic unit is unconformably overlain by the Middle Ordovician; the Middle Cambrian, Upper Cambrian, and Lower Ordovician are all missing. Farther east still, on the Yukon–Northwest Territories boundary and immediately north of 60° N. latitude, all that is left of the Cambrian is 100 or 200 metres of sandstone and conglomerate, leading to a feather edge of the fault-fanglomerates.

3.5. The Cambrian between the 60th and 65th parallels

We have seen that the Cambrian depositional area underwent major modifications northward across the Macdonald 'high'. The Atan-Kechika succession, making up the Rocky Mountains along the western side of the 'high', crosses the Rocky Mountain trench, diagonally, at about 59° N. latitude. Northwards from that latitude, this facies is restricted to a relatively narrow zone south-west of the Tintina trench, and the true Rocky Mountains have ended. The more easterly front ranges were controlled on the 'high' by a series of fault blocks already in existence in Cambrian time (3.4), and the resulting diversification of Cambrian facies continues northwards across the 60th parallel. Passing into the region in which the Slave and Bear provinces form the basement, the diversification abruptly becomes much more pronounced. A complete cross-section through the Cambrian rocks of this region shows five distinct components (Fig. 16).

The Atan-Kechika succession extends, in effect, from the Peace River to the north-western end of the Pelly Mountains, where it ends against the Tintina trench (Fig. 9). Pre-Cambrian sedimentary rocks in varying degrees of metamorphism are overlain by the usual quartzite, much of it green, with argillite and other clastic rocks. It is between 500 and 700 metres thick, and has yielded *Olenellus*, *Pagetides*, and *Wanneria*. The quartzite is again succeeded by the Atan limestone and dolomite, which commonly includes the distinctive orange- or reddish-weathering silty dolomite at its base. The formation is typically about 300 metres thick, but it thickens distinctly to the south-west, being swallowed up by the Cassiar batholith without revealing any sign of approach to a western shoreline. The limestone contains the usual archaeocyathids, the commonest again being *Coscinocyathus dentocanis* OKULITCH.

The Atan Group is overlain, as in the south, by the Kechika Group, thinly and wavily banded argillaceous limestone, calcareous shale, and some quartzite and limestone cobble-conglomerate. The beds are typically strongly cleaved and deformed; much of the group is in fact lustrous phyllite (WHEELER, 1960a, b). It may include some of the Middle Cambrian, though there is no fossil evidence for it; Upper Cambrian strata are very probably present, and the upper part of the group is certainly Ordovician, in grapto-litic facies. The whole Cambrian succession is strongly metamorphosed where it it is close to the batholith.

In the north-west, between Keno Hill and Whitehorse, the Harvey Group (CAMPBELL, 1954, 1967b) may be equivalent to the combined Atan and Kechika Groups. It originally consisted of impure quartz-sandstone and siltstone below, at least several thousand metres thick and with minor shale and limestone; thinly-bedded limestone above, about 400 to 500 metres thick; and shaly rocks at the top, perhaps as much as 1,500 metres thick. The whole forms a conformable succession, resting on the Pre-Cambrian Yukon Group without obvious unconformity and being apparently overlain by Siluro–Devonian rocks. Unfortunately, the only evidence for the age of the group is the similarity of its succession to that of the Atan and Kechika Groups; the Harvey Group is totally isolated by large faults and widely metamorphosed, especially in its lower parts, by Mesozoic granite intrusions. The original sediments are now quartzites, quartz–biotite schists, lime–silicate gneisses, spotted slates, and hornfels.

The second component of our cross-section at this latitude is a band on the north-east side of the Tintina trench in which no Cambrian is known with certainty. For a width

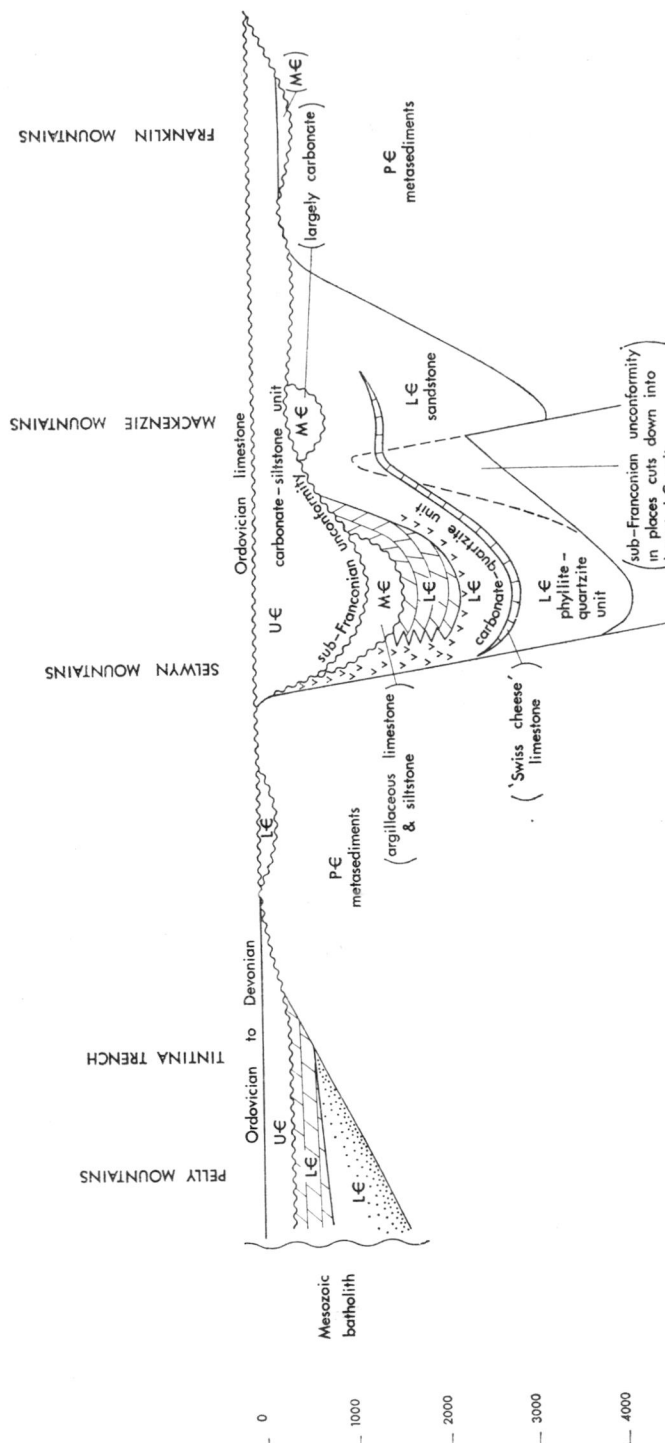

FIG. 16. Cross-section approximately along 62°N. latitude, showing disposition of Cambrian rocks before subsequent deformation. Length of section about 600 kilometres. Dominant lithologies indicated; absence of indication means dominantly argillaceous strata.

across the strike varying between 100 and more than 200 kilometres, Upper Palaeozoic rocks rest on a Pre-Cambrian unit resembling the Miette Group. Over some eastern part of this blank zone, the absence of Cambrian rocks is certainly due to a combination of pre-Franconian erosion and failure of late Cambrian deposition. Only at the extreme north-western end of this trend, at the western end of the Ogilvie Mountains on the Yukon–Alaska boundary, do proven Cambrian rocks appear (3.6).

East and north-east of the blank zone comes the great Lower Cambrian basin of the Mackenzie Mountains. In the Backbone Ranges (Fig. 9), sandstones and feldspathic quartzites almost 2,000 metres thick yield from the top a fauna of the upper *Olenellus* Zone (*Fremontia*, *Paedeumias*). Westwards, this upper part of the coarse clastic unit is replaced by a unit of markedly different facies, the Sekwi Formation, itself up to 1,000 metres thick in the axis of the trough. The Sekwi Formation comprises thinly bedded siltstones, weathering yellow, green, or maroon; sandy, fine-grained carbonate rocks, some with algal structures; quartz-pebble conglomerate; and a peculiar basal limestone in which pods and lenses of limestone have suffered preferential solution within irregularly bedded siltstone and dolomite. This very characteristic basal unit is called the Swiss Cheese limestone by the field geologists. The formation yields *Nevadella addyensis* (OKULITCH), *Olenellus gilberti* MEEK, and other Lower Cambrian trilobites, and several archaeocyathid genera. Basic flows and tuffs occur near the top of the formation in places.

Both these Lower Cambrian formations thin out very abruptly to both east and west. On the Yukon–Northwest Territories boundary at 64° N., west of the trough, the Lower Cambrian is only about 30 metres thick, and in the Canyon Ranges to the east of the trough it is absent. The margins of the basin were uplifted and eroded at the beginning of Cambrian time, removing thousands of metres of Proterozoic sedimentary rocks from the Canyon Ranges, and again uplifted at the close of the Early Cambrian, establishing a new Palaeozoic depositional basin. In it, the Middle Cambrian to Middle Ordovician succession is in basinal facies.

The Middle Cambrian in the basin consists of about 500 metres of dark grey, platy, nodular, argillaceous limestone and calcareous siltstone, resting on an unconformity with a basal evaporitic siltstone. The siltstone contains casts of salt crystals. The series becomes more calcareous upwards, as well as laterally, and is exceedingly fossiliferous. Where the unconformity is in its lowest position, the series yields trilobites of four Middle Cambrian zones: that of *Plagiura-Poliella* (*Fieldaspis*, *Wenckchemnia*); that of *Albertella* (*Albertella*, *Chancia*, *Yohoaspis*); that of *Glossopleura* (*Glossopleura*, *Kootenia*, *Spencia*); and that of *Bathyuriscus-Elrathina* (*Bathyuriscus*, *Elrathina*, *Olenoides*, *Pagetia*, *Zacanthoides*). Elsewhere, the unconformity below the Middle Cambrian cuts out the lower part of the series in varying amounts; the lowest part represented may locally be as high as the *Bolaspidella* Zone.

Further uplift followed at the close of the Middle Cambrian. The earliest Upper Cambrian is apparently everywhere missing, and the sub-Franconian unconformity bevels everything below it to the south-east and south-west. To the south-west, this unconformity is responsible for at least part of the area which lacks Cambrian and which intervenes between the Mackenzie and Pelly Mountains. The Cambrian ends with well-bedded dolomite and limestone, crossing all previous structures and passing up into the Ordovician (Franklin Mountain Formation). Its fauna represents a complete sequence from the *Elvinia* Zone to the top of the Cambrian and beyond: *Briscoia*, *Hungaia*, *Prosaukia*, *Pseudagnostus*, *Ptychaspis*, and the Trempealeauan brachiopod

Finkelnburgia. About 500 metres of the formation belong to the Upper Cambrian. The series contains several small reefs of massive pale dolomite.

The eastern boundary of the Mackenzie Mountains basin remained positive throughout the Cambrian, especially during the Middle Cambrian, restricting the area of the Canyon Ranges and the Franklin Mountains from access by the open sea. In the Canyon Ranges proper, a relatively thin (400 metres) succession of fine-grained, buff and yellow weathering dolomite, siltstone, and mudstone carries all the hallmarks of very shallow water deposition—mud cracks, salt casts, purple- and green-weathering laminations. Near its base occur fossils of the *Plagiura-Poliella* Zone (*Fieldaspis* and *Kochiella*). The beds rest unconformably on the Pre-Cambrian, and are themselves overlain unconformably by Ordovician and Silurian rocks. This positive element represents the fourth component in our cross-section.

The fifth is the restricted basin underlying the Franklin Mountains and the Mackenzie plain (Figs 9, 16). A basal transgressive sandstone or quartzite is Early Cambrian in the mountains—the Mount Clark Formation, 250 metres thick, with *Salterella*, resting on the Pre-Cambrian Katherine Group. The sandstone becomes progressively younger eastwards, and its equivalent on the north arm of Great Slave Lake may be Ordovician (Old Fort Island Formation). The basin was thereafter filled by the sediments of the Macdougal Group, unique in the Cambrian of Canada. The rocks are shaly and evaporitic, black and bituminous below, but otherwise green and pink, with siltstone, dolomite, and a good deal of gypsum. Maroon or reddish weathering colours are common. Sand increases in amount and becomes coarser towards the east, indicating a shield source. Olenellids are found at the bottom, and Middle Cambrian trilobites (*Bathyuriscus*, *Glossopleura*) higher up. The gypsiferous part of the group, about 300 metres thick in outcrop but displaying abrupt lateral variations, was called the Saline River Formation by WILLIAMS (1923). A well in the Mackenzie River valley penetrated 700 metres of halite, with minor anhydrite, dolomite, and shale, between layers of Middle Cambrian green shale. Even if thickened by flowage or diapirism, this concentration of salt is unique in the Cambrian of North America.

Like the sandstone below it, the Macdougal Group becomes younger eastwards. Its equivalent on the edge of the shield is Middle Ordovician in age. The group is present throughout the Mackenzie plain and valley area, but thins north-westwards from Norman Wells. As in the Mackenzie Mountains, the Cambrian ends with a major dolomite formation (Franklin Mountain Formation), which continues into the Ordovician. The lower part is a blocky dolomite with some interbeds of red and green shale. Floating quartz grains are common, as well as algal structures and intraformational breccias. Upper Cambrian trilobites, such as *Welleraspis*, have been recorded, but are rare.

3.6. The Cambrian north of 65° North latitude

The far northern Cordillera are, inevitably, the least well known. For the Cambrian, however, we know enough to be sure that no ready generalization about parallel facies belts is adequate.

A short distance north of the 65th parallel, the frontal Mackenzie ranges swing sharply around to the west, and the north-trending Franklin Mountains die out (Fig. 9). Thence northwards, individual ranges show more conspicuous variation of trend and style than they do anywhere else in the frontal Canadian Cordillera. This variation is a consequence of the development of a number of discrete basins, which came into existence before the beginning of Cambrian time and were apparently accentuated during it. They persisted, *mutatis mutandis*, until late in the Devonian, when they were more or less uniformly buried by clastic rocks. The existence of the basins may itself have been a consequence of the interaction in this region of two geosynclinal trends, the Cordilleran from the south and the Franklinian from the north-east (Fig. 1).

Evidence for the nature and behaviour of this most northerly transverse element, the Franklinian geosyncline, is confusingly contradictory (3.1). We have already seen that, between the 60th and 65th parallels, the disturbed belt swings abruptly eastwards in conformity with the great eastward extension of Lower Palaeozoic rocks north of the Peace River arch (3.5). In this still more northerly area, north of the Arctic Circle, though the disturbed belt swings back to the west, the area of preserved Lower Palaeozoic rocks undergoes a vast extension to the east and continues across Victoria Island towards the Boothia Peninsula. Though at latitude 62° N. the formational equivalents of the Franklin Mountains succession appear to become wholly Ordovician eastwards towards Great Slave Lake (3.5), they may at latitude 68° N. remain Cambrian for a much greater eastward extent. North of Great Bear Lake, a succession just like that in the Franklin Mountains—basal sandstone, coloured shales and minor dolomite of the Macdougal Group (though without exposed evaporites), and Franklin Mountain dolomite—rests on Proterozoic dolomite of the Hornby Bay Group. The Macdougal equivalent has been assigned an age of 470 m.y. by the potassium–argon method on glauconite. This suggests that the rocks are Ordovician, but the fossils appear to resemble Middle or Upper Cambrian forms. Lower Cambrian strata are known on Victoria Island, farther to the north-east.

Thus the eastern shelf succession, typified by that in the Franklin Mountains and the subsurface of the middle Mackenzie valley (3.5), here becomes a *northern* shelf succession. It probably continues in Macdougal facies as far as the delta area at Aklavik. There massive, pulverized gypsum forms the cores of at least two diapiric structures (KENT and RUSSELL, 1961). The cores contain xenolithic blocks and fragments of red, maroon, and green shale, and are associated with isolated masses of diabase and gabbro. The association is obviously reminiscent of the Macdougal Group (3.5), and the correlation is reinforced by the appearance of Lower Cambrian rocks forced up along faults. East of the delta at about the same latitude (68°10′ N.) there occurs the most northerly known outcrop of probable Cambrian rocks in the Mackenzie River Basin—a carbonate–shale succession, so far without diagnostic fossils (NORRIS, PRICE and MOUNTJOY, 1963).

What appears to be a representative succession of the cratonic facies at this latitude crops out in the White Mountains, a north-east trending horst-like mass at the northern end of the Richardson Mountains, west of the delta. At the base is the usual quartzite, about 300 metres thick and carrying the usual interbeds of red and green shale and sandy limestone with *Olenellus*. Following a higher limestone-shale interval about 550 metres thick, with the Middle Cambrian trilobites *Acrothele* and *Alokistocare*, comes an unknown thickness of dolomite. No fossils are known from it, but some layers within it are severely brecciated, as if by the solution of evaporites like those in the Macdougal Group. The succession ends with 350 metres of grey algal and pelletal limestone which seems to represent the Upper Cambrian part of the Franklin Mountain Formation (3.5).

No similarly brecciated or evaporitic units have been identified in the Barn and British Mountains, extending north-westerly across the International Boundary into Alaska. However, such beds may reappear in the Keele Range, a south-west-trending block circumscribed by the Porcupine River at 67° N. latitude and the Alaska boundary

I I CAM

Fig. 9). Carbonates with many brecciated beds in their lower part are again overlain by a dolomite like the Franklin Mountain Formation.

These data suggest that the cratonic facies of the Cambrian in the far north makes a wide circular swing from its northerly trend in the Franklin Mountains, to westerly in the White Mountains and south-westerly in the Keele Range. This possibility is enhanced by the behaviour of the positive Cambrian ridge which separated the cratonic-evaporitic facies from the trough of the Mackenzie Mountains (3.5). The zero edge of the Lower Cambrian trends north-westerly along the northern Mackenzie Mountains and then westerly into the Werneckes (Fig. 10). North of it, thin Upper Cambrian shelf carbonates overlie Middle Cambrian sandstones and conglomerates, which in turn are unconformable on the Pre-Cambrian. This thin succession appears to pass southwards into a dolomite, with reefs fringing the old Lower Cambrian basin. Farther to the north-west, in the north-western Wernecke Mountains, the thin (450 metres) Upper Cambrian shelf carbonates are again present, with Dresbachian fossils of cratonic aspect (*Densonella, Komaspidella, Meteoraspis*). So are the Middle Cambrian sandstones and conglomerates, about 600 metres thick and interbedded with amygdaloidal andesite and basalt flows (GREEN and RODDICK, 1962). Again, these clastic rocks lie with angular unconformity on the Tindir Group. They thin out westwards and disappear in the Ogilvie Mountains.

This attractively simple picture is interfered with by a narrow but conspicuous shale basin which crosses the arcuate facies pattern almost at right angles. The greater part of the Richardson Mountains, and the central Wernecke Mountains, developed from this persistent basin, which lasted into Devonian time. The present width of the clastic facies is only about 50 kilometres; carbonate banks flanked it on both east and west sides. The basin's chief consequence was that the narrow mountain belt evolved from it differs in structure, trend, and Cambrian stratigraphy from the wide east–west belts on its two sides, the Ogilvie and Mackenzie Mountains (Fig. 9). In particular, the ejective style of folding in the Richardson Mountains contrasts markedly with the dejective style (wide anticlines and narrow synclines) in the Mackenzie Mountains.

The Cambrian section in the Richardson Mountains is more than 3,000 metres thick, and may be nearly 5,000 metres. Most of it is Middle and Upper Cambrian. The Lower Cambrian, about 300 metres thick, is a limestone in reef or shelly facies, with a typical cratonic fauna—olenellids, *Paedeumias, Wanneria*, and brachiopods (including *Yorkia*). This represents the carbonate bank which later existed only on the western side of the trough. The Middle and Upper Cambrian consist of brown, platy siltstones, shales, and thinly bedded limestones, at least 3,000 metres thick, carrying *Olenoides, Scenella*, asaphids, and abundant spicules of *Protospongia*. Marginally to both the east and the west, this basinal facies gives way to limestone and limestone breccia.

Sections at the south end of the trough, in the valleys of the Wind and Bonnet Plume Rivers, differ in that they lack the Upper Cambrian, which is nearly 2,000 metres thick in the north. The Lower Cambrian comprises limestone and dolomite, both more or less silty or sandy, and coloured siltstone; where the base is seen, it is a 50-metre bench of quartzite or fine-grained sandstone resting unconformably on Pre-Cambrian Tindir sedimentary rocks. Fossils include *Olenellus truemani* WALCOTT and other olenellids, *Bonnia, Strenuella, Pagetiellus*, brachiopods, and the European gastropod *Latouchella*. The Middle Cambrian consists entirely of coloured, calcareous or dolomitic siltstone, shale, and silty dolomite, about 750 metres thick. It carries a typical Cordilleran fauna

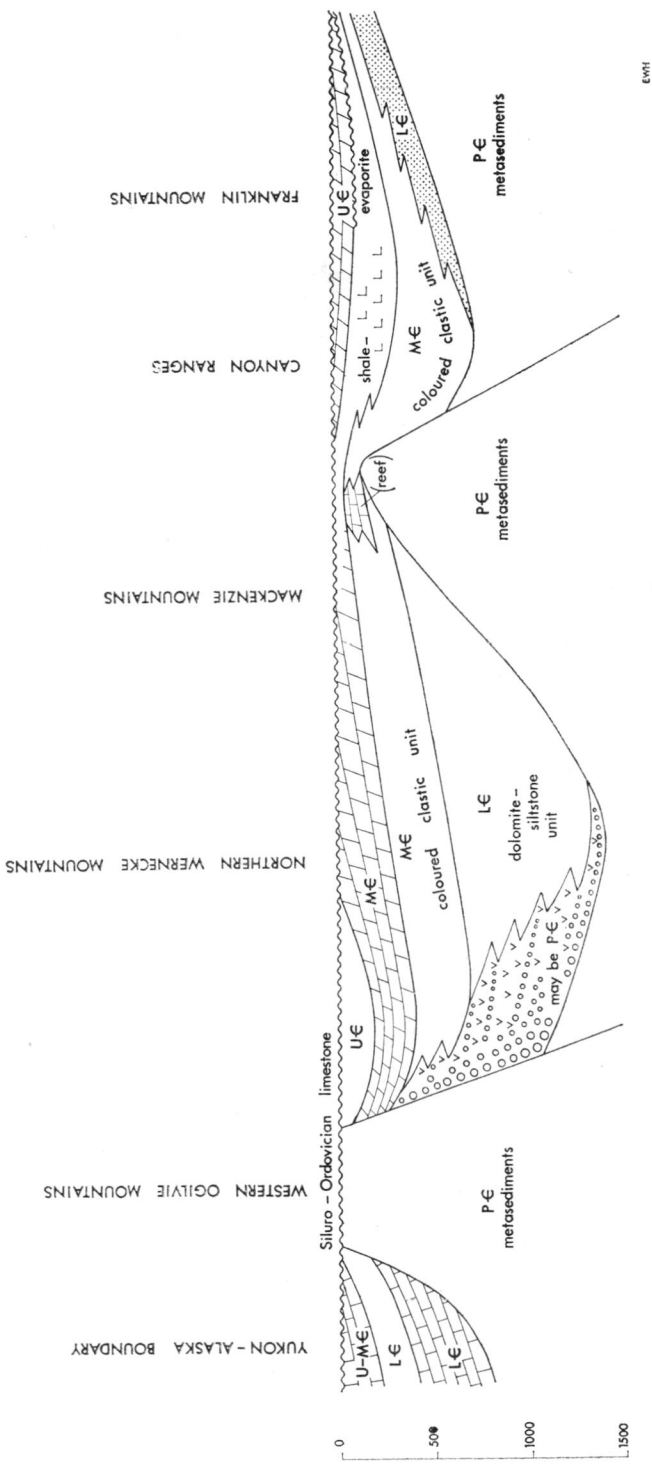

FIG. 17. Cross-section approximately along 65°N. latitude, showing disposition of Cambrian rocks before subsequent deformation. Length of section about 1000 kilometres. Dominant lithologies indicated; absence of indication means dominantly argillaceous strata.

including *Amecephalus, Chancia, Fieldaspis, Glossopleura,* and *Wimanella.* Unconformably above comes the ubiquitous Ordovician dolomite and limestone.

The narrow basin was persistently unstable. At various localities within it, unconformities are of pre-Early Cambrian, pre-Middle Cambrian, and pre-Ordovician ages. The Lower Cambrian is rapidly reduced to zero going from west to east, on to the post-Lower Cambrian uplift which is continuous with that separating the Backbone and Canyon ranges of the Mackenzie Mountains (3.5). The angular unconformity at the base of the Upper Cambrian increases in effect to the east, where the Middle Cambrian is sharply truncated, and to the south; in each direction, the Cambrian is eventually cut out completely. On the edge of the carbonate bank in front of the Wernecke Mountains, the Upper Cambrian is in places reduced to zero by pre-Ordovician erosion. This sub-Ordovician unconformity rises in the section northwards, exposing several hundred metres of further Middle Cambrian clastic rocks which become more fine-grained to the north. They were therefore derived from the south. North–south faulting must have been episodically active, and the Middle Cambrian conglomerates are among the results of it.

The most southerly sections of the clastic facies occur in the valleys of the Wind and Bear rivers where they trench through the Wernecke Mountains (about at latitude 65° N.). The great Lower Cambrian trough of the Mackenzie Mountains (3.5) narrows abruptly from the south-east, between its flanking ridges, at about the same latitude. As the two extremities are separated by no more than 100 kilometres in an east-west direction, it seems likely that the two basins were in Proterozoic and early Palaeozoic time part of one structural trend, which controlled the shape of the present Mackenzie, Wernecke, and Richardson Mountains.

The extreme north-west corner of mainland Canada is occupied in part by the Barn and British Mountains (Fig. 9). No Cambrian has so far been proven to crop out in the Barn Mountains, though Ordovician and Silurian rocks are known. In the British Mountains, the Neruokpuk Group of Alaska consists of mildly metamorphosed, varicoloured slates, quartzites, greywackes, and carbonate rocks. Formerly believed to be Pre-Cambrian, this group of rocks is now believed to include strata as young as Devonian (W. P. BROSGÉ, personal communication, 1967). It may include some Cambrian rocks also. There are also volcanic rocks, associated with carbonates, which may be Cambrian in age.

The only proven Cambrian rocks in Alaska occur immediately west of the Yukon–Alaska boundary, north-east of the crossing of that boundary by the Yukon River at 65° N. latitude. The succession, first investigated by BURLING and CAIRNES, and later by MERTIE, has been fully described by BRABB (1967), the fauna by PALMER (1968). Parts of the succession can be traced into the Yukon at the western extremity of the Ogilvie Mountains (GREEN and RODDICK, 1962).

The exposures appear to straddle the western edge of a carbonate bank, with a much more variable, boulder-bearing succession lying off its edge (Figs 10–12). The area is similarly transected by the boundary between predominantly carbonate and predominantly siliceous rocks in the Ordovician succession.

The carbonate facies is represented by the lower member of the Jones Ridge Limestone, which rests without obvious discordance on the Pre-Cambrian Tindir Group. The limestone is fine-grained, partly oolitic, and interbedded with some dolomite. Archaeocyathids occur about 150 metres above the base. No Middle Cambrian fossils have been reported, and the Upper Cambrian forms first appear at a level only 75 metres below

Lower Ordovician forms. The *Dunderbergia* Zone of the Dresbachian Stage is represented by a large number of trilobite species, but no Franconian forms have been found succeeding them. The uppermost 60 metres of the Cambrian quota of the formation have yielded faunas of two Trempealeauan zones: *Briscoia*, *Rasettia*, and *Tatonaspis* below, and *Bayfieldia* and *Yukonaspis* above. Though the thickness of the Cambrian part of the Jones Ridge Limestone approaches 1,000 metres, the formation has not been shown to extend into the Yukon.

South-west of the facies boundary, a much more varied succession is found, and it is this succession which can be traced into the Yukon. The Funnel Creek Limestone at the base of the Palaeozoic section is about 400 metres thick and apparently conformable on the Tindir Group. It is massive, with some oolitic and dolomitic horizons, but it also contains thin interbeds of grey chert and layers of edgewise limestone conglomerate, indicating depositional conditions beyond the edge of the bank. The formation is essentially unfossiliferous, though at least one archaeocyathid horizon may prove to belong to it.

Conformably overlying the Funnel Creek Limestone is the Adams Argillite, 100 to 200 metres of green and red shales with beds of siltstone or limestone and a basal quartzite. Some thin greenstone members, perhaps altered andesite flows, occur in places. At least one thin limestone unit carries the archaeocyathid *Ethmophyllum*. Trilobites, of general Siberian aspect, include *Neocobboldia*, *Pagetides*, and *Serrodiscus*. Worm burrows are common, and BRABB (1967) has also reported the fan-shaped trace fossil *Oldhamia* (1.4).

The youngest Cambrian formation in the area is the Hillard Limestone, surprisingly cliff-forming in spite of its apparently representing a basinal facies. Good lenticular bedding is interrupted by layers of edgewise limestone conglomerate, especially in the upper part of the formation. Shale, siltstone, chert, and the common orange-weathering dolomite are interbedded; some horizons are highly organic and phosphatic. In spite of this variety, and in spite of the presence of at least nine faunal zones (PALMER, 1968), the formation is no more than 150 metres thick.

A basal conglomerate of limestone boulders yields *Bonnia* and olenellids like those in the Quebec conglomerates (Table 3). The lower Cambrian fauna in place includes *Kootenia*, *Pagetides*, and *Zacanthoides*. Middle Cambrian fossils appear all to represent the *Bolaspidella* Zone—*Dorypyge*, *Lejopyge*, *Marjumia*, *Modocia*, *Semisphaerocephalus*, *Spencella*, and other genera. On the Canadian side of the border, the equivalent beds have yielded *Agnostus*, *Anomocare*, and *Solenopleura* (GREEN and RODDICK, 1962). The apparent absence of all Middle Cambrian faunas except those of the latest Middle Cambrian may indicate a disconformity between the Hillard Limestone and the Adams Argillite; it recalls the similar situation in parts of the Mackenzie Mountains (3.5) and on the flanks of the unstable clastic basin to the north of them (3.6).

The Dresbachian Stage is represented by the *Cedaria* and *Dunderbergia* Zones; the Franconian by the *Elvinia* Zone (*Peratagnostus*, *Proceratopyge*) and the *Ptychaspis-Prosaukia* Zone (*Drumaspis*, *Hungaia*, *Onchonotus*). The topmost stage, the Trempealeauan, yields *Bayfieldia* and *Yukonaspis*. Thus at least five Upper Cambrian faunal zones are represented in an interval of about 50 metres, above which the formation is unconformably overlain by fossiliferous Lower Ordovician.

In the Alaska Range, south-west of the Yukon River and east of the city of Fairbanks, pre-Devonian rocks underwent metamorphism to the amphibolite facies. The rocks may have been of Pre-Cambrian age, but it is equally possible that they were Lower Palaeozoics involved in a Caledonian metamorphic event (RAGAN and HAWKINS, 1966). There may therefore be more Cambrian rocks in Alaska than have been identified so far.

3.7. Cambrian diastrophism in western Canada and Alaska

No episode of folding can be demonstrated to have taken place during Cambrian time in western Canada. At least two episodes of uplift, tilting, and erosion took place, however, and at least one episode of block faulting in addition to faulting which may have accompanied the uplifts.

Block faulting of approximately meridional trend was active in the north-western sector of the Peace River arch in the early Cambrian. It gave rise to the large wedges of coarse fanglomerate of Atan age there (3.4). Similar faulting may have continued during the Middle Cambrian in the Northwest Territories and the Yukon, producing thinner but otherwise comparable conglomerate wedges and lenses.

At the close of Early Cambrian time, there was widespread uplift across the Yukon and Northwest Territories, and the base of the Middle Cambrian is an unconformity in much of the Richardson, Wernecke, Ogilvie and Mackenzie Mountains (3.5; 3.6). A graben-like structure along the eastern side of the Wernecke Mountains was active at this time. A still more widespread episode of uplift, tilting, and erosion took place in either late Middle Cambrian or earliest Late Cambrian time; the most accurate placement of it that can be made is that it was pre-Franconian (GABRIELSE, 1967a). An angular unconformity marks the base of the remaining Upper Cambrian rocks in the Richardson and Wernecke Mountains, the southern Mackenzie Mountains, and, 1,300 kilometres farther south, in the westernmost Rocky Mountains and easternmost Purcells, now separated from one another by the Rocky Mountain trench (3.3.2). In all these ranges, Upper Cambrian rocks which are not earliest Upper Cambrian rest on a variety of older, eroded rocks. Limestone cobble-conglomerates are widespread in the northernmost Rocky Mountains, in both the Atan and the Kechika Groups; some are of Early Cambrian age; others may be Early Ordovician. Unconformities between the Cambrian and the Ordovician, as in the Wernecke Mountains (where no Upper Cambrian remains), probably represent the Knox unconformity and are a consequence of uplift in the Early Ordovician.

Cambrian vulcanicity was negligible except in the far west and the far north. Even there it was only minor, but Cambrian volcanic rocks occur, nearly all in the Lower Cambrian, in the Selkirk and Dogtooth Mountains, in the Selwyn and Wernecke Mountains, off the northern edge of the Peace River arch, and along the Yukon-Alaska boundary. Any truly eugeosynclinal realm which may have existed during Cambrian time in western Canada must have lain still farther west and been obliterated during later Palaeozoic and Mesozoic intrusion and metamorphism. Early Palaeozoic history in the coastal region of the Mesozoic batholiths is a complete blank, except in south coastal Alaska. Between the arcuate Alaska Range and the Pacific, more than a thousand metres of Lower and Middle Ordovician geosynclinal sedimentary and volcanic rocks appear to rest on Pre-Cambrian metamorphic basement. Though no Cambrian sedimentary rocks have been distinguished, Cambrian time may have been occupied here by episodes of subaerial vulcanicity (THOMPSON, 1960). No unquestionably Cambrian isotopic date has been established from anywhere in western Canada, or Alaska, so far as the present writer can discover.

3.7.1. Later deformations affecting Cambrian rocks

The first important deformation to affect the Cambrian rocks of western Canada, the so-called Cariboo orogeny (WHITE, 1959), cannot be pinpointed more closely than that it was post-Ordovician, pre-Mississippian. It affected wide areas of the Cambrian rocks now lying west of the Rocky Mountain trench, from the Shuswap terrane in the south (3.3.1), through the Wolverine Complex west of the Peace River arch (3.4), and into the Yukon. However, it is highly likely that this orogeny was responsible for the ubiquitous sub-Devonian unconformity (HARKER, HUTCHINSON and McLAREN, 1954; VAN HEES, 1964), and for the elevation of the Peace River arch. In this event, the orogeny was younger than the Middle Silurian and older than the Middle Devonian, and was effectively the Cordilleran equivalent of the Caledonian. The present structures in which the Cambrian rocks are involved are of Mesozoic and early Tertiary age.

3.8. Economic geology, Cambrian of western Canada and Alaska

Both metallic and non-metallic mineral deposits have been exploited commercially from Cambrian rocks in western Canada. They differ conspicuously from those in rocks of the same age in the east.

The principal metals won have been gold, lead and zinc with or without silver, and tungsten. Commercial quantities of copper are insignificant; the important copper camps in British Columbia lie far to the west of the westernmost known Cambrian rocks, within the Mesozoic eugeosyncline. Non-metallic minerals of commercial importance include barite, magnesite, and lime. Mineral production from Lower Palaeozoic rocks in the west has, however, been much below that from Proterozoic and Mesozoic rocks (McKECHNIE, 1966).

Metallic deposits sufficiently large to be noted here occur in two principal environments. The first is in clastic sedimentary rocks, especially quartzites, fractured as a consequence of brittleness induced by metamorphism. All such occurrences are close to the eastern edge of the Mesozoic batholiths, and so in the far western succession in which the Cambrian is a continuation of the Proterozoic. All the gold occurs in rocks of this succession.

The two largest gold camps are those of Sheep Creek, south of Nelson in south-eastern British Columbia, and the Cariboo, within the loop of the Fraser River in the east-central part of the province (Fig. 9). In the Sheep Creek mines, gold and silver were recovered from fissure-filled veins in the Quartzite Range and Reno Formations (3.3.1). In the Cariboo, gold–quartz veins are associated with the quartzites of the Midas Formation. They also contain pyrite, arsenopyrite, galena, and sphalerite, as well as tungsten and bismuth minerals.

Some lead–zinc–silver mines have been worked in rocks of this type. An example is the Silver Cup Mine in the Lardeau district, in which mineralization occurred in slates of the Cambrian or Ordovician Triune Formation. Galena and sphalerite accompanied a variety of other sulphides, most of the silver occurring in tetrahedrite.

The second principal type of metallic deposit is the lead-zinc sulphide deposit in carbonate rocks. Such bodies are formed by replacement, closely controlled by zones of dolomitization and by formational contacts (especially those with overlying argillaceous rocks). They are also, in a number of cases, distinctly controlled by lateral facies changes from carbonate rocks (in the east) to more argillaceous equivalents (in the west). Many of the deposits are consequently much farther from the main metamorphic belt than are the deposits of the first type.

In the Nelson area, the lead–zinc–silver mineralization is in dolomitized limestone of the Lower Cambrian Laib Formation (3.3.1). In the southern Rocky Mountain trench, the Silver Giant Mine is in the Upper Cambrian Jubilee dolomite, with a barite gangue (3.3.2). Much farther east, the Monarch and Kicking Horse mines were in cylindrical orebodies in the lower part of the Middle Cambrian Cathedral Formation (3.3.3), their precarious entries on opposing mountainsides being familiar for many years to travellers through the Kicking Horse Pass. In the far north, a similar example is the Tintina Silver Mine in the Pelly Mountains, Yukon Territory, the host rock of which is a carbonate of the Lower Cambrian Atan Group (3.5). On the opposite side of the Tintina trench (Fig. 9), the Anvil Mine is in a very massive galena-sphalerite body concordantly replacing calcareous phyllites of the lower part of the Kechika Group or the upper part of the Atan. Numerous smaller deposits are known, some of historical interest. One such was Silver City on old Castle Mountain, now Mount Eisenhower, opened up about 1880 by miners on their way to the Cariboo.

Tungsten minerals are also widely distributed in Cambrian limestones. Several mines near Nelson recover the metal from ores in the Badshot limestone and associated argillaceous rocks. A much larger deposit is that at the Canada Tungsten Mine in the Selwyn Mountains, on the Yukon–Northwest Territories boundary (3.5), in which scheelite, pyrrhotite, and chalcopyrite occur in Lower Cambrian carbonates close to a small granitic stock.

A very large sedimentary iron ore deposit, also near the Yukon–Northwest Territories boundary and north of the 65th parallel, has been thought to be Cambrian in age. It is overlain with angular unconformity by a massive carbonate assigned to the Ordovician. The ferruginous, siliceous, and conglomeratic mudstone, called the Rapitan Group, is now confidently assigned to the Pre-Cambrian (GABRIELSE, 1967a).

Barite dominates the non-metallic minerals of economic significance. It is a common gangue mineral in the lead–zinc deposits in the limestones; in places, chiefly in the southern part of the Rocky Mountain trench, it occurs in sufficiently large bodies to be mined itself. Magnesite also occurs in large bodies in several places near the southern end of the trench. The largest are in the Purcell Mountains, in the Cranbrook Formation which otherwise consists chiefly of quartzite (3.3.2). At least one occurs in the Rocky Mountains, shortly south of Mount Assiniboine, in metamorphosed Cathedral dolomite close to its westward change to an argillaceous facies (3.3.3).

The Middle Cambrian Eldon Formation is in places a sufficiently pure limestone to be quarried for lime. One quarry, at Kananaskis in the Front Range, east of Banff, has been in operation for a number of years.

No commercial oil or gas has been found in Cambrian rocks in Alberta or Saskatchewan, but helium is produced from them from several wells in extreme southern Saskatchewan (VAN HEES, 1964).

3.9. Bibliography: Western Canada and Alaska

AITKEN, J. D. (1966). Middle Cambrian to Middle Ordovician cyclic sedimentation, southern Rocky Mountains of Alberta. *Canadian Petrol. Geol. Bull.* **14**, 405.

AITKEN, J. D. (1967). Classification and environmental significance of cryptalgal limestones and dolomites, with illustrations from the Cambrian and Ordovician of southwestern Alberta. *Jour. Sed. Petrol.* **37**, 1163.

AITKEN, J. D. (1968a). Cambrian sections in the easternmost southern Rocky Mountains and the adjacent subsurface, Alberta. *Geol. Surv. Canada Paper* **66–23**.

AITKEN, J. D. (1968b). Pre-Devonian history of the Southern Rocky Mountains. *Alberta Soc. Petrol. Geol. 16th Ann. Field Conf. Guidebook*, 15.

AITKEN, J. D. and GREGGS, R. G. (1967). Upper Cambrian formations, southern Rocky Mountains of Alberta, an interim report. *Geol. Surv. Canada Paper* 66–49.

ALLAN, J. A. (1914). Geology of Field map-area, B. C. and Alberta. *Geol. Surv. Canada Mem.* 55.

BALLY, A. W., GORDY, P. L., and STEWART, G. A. (1966). Structure, seismic data, and orogenic evolution of southern Canadian Rocky Mountains. *Canadian Petrol. Geol. Bull.* 14, 337.

BELL, W. A. (1959). Stratigraphy and sedimentation of Middle Ordovician and older sediments in the Wrigley–Fort Norman area, Mackenzie District, N.W.T. *Canadian Min. Met. Bull.* no. 561, 3.

BELYEA, H. R. and NORRIS, A. W. (1962). Middle Devonian and older Palaeozoic formations of southern District of Mackenzie and adjacent areas. *Geol. Surv. Canada Paper* 62–15.

BLUSSON, S. L. (1966). Frances Lake, Yukon Territory and District of Mackenzie. *Geol. Surv. Canada* Map 6-1966.

BOSTOCK, H. S. and LEES, E. J. (1938). Laberge map-area, Yukon. *Geol. Surv. Canada Mem.* 217.

BRABB, EARL E. (1967). Stratigraphy of the Cambrian and Ordovician rocks of east-central Alaska. *U.S. Geol. Surv. Prof. Paper* 559-A.

BURLING, L. D. (1914). Early Cambrian stratigraphy in the North American Cordillera, with discussion of *Albertella* and related faunas. *Geol. Surv. Canada Mus. Bull.* 2, 93.

BURLING, L. D. (1923). Cambro–Ordovician section near Mount Robson, British Columbia. *Geol. Soc. America Bull.* 34, 721.

BURLING, L. D. (1955). Annotated index to the Cambro–Ordovician of the Jasper Park and Mount Robson region. *Alberta Soc. Petrol. Geol. 5th Ann. Field Conf. Guidebook*, 15.

CAIRNES, D. D. (1914). The Yukon–Alaska International boundary between Porcupine and Yukon Rivers. *Geol. Surv. Canada Mem.* 67.

CAMPBELL, R. B. (1954). Glenlyon, Yukon. *Geol. Surv. Canada Paper* 54–12 (map only).

CAMPBELL, R. B. (1961). Quesnel Lake (west half), British Columbia. *Geol. Surv. Canada* Map 3-1961.

CAMPBELL, R. B. (1963). Quesnel Lake (east half), British Columbia. *Geol. Surv. Canada* Map 1-1963.

CAMPBELL, R. B. (1967a). McBride map-area. *Geol. Surv. Canada Paper* 67-1, pt. A, 53.

CAMPBELL, R. B. (1967b). Geology of Glenlyon map-area, Yukon Territory. *Geol. Surv. Canada Mem.* 352.

CAMPBELL, R. B. (1968). Canoe River, British Columbia. *Geol. Surv. Canada* Map 15-1967.

COOK, DONALD G. (1967). Structural style influenced by a Cambrian regional facies change in the Mount Stephen–Mount Dennis area, Alberta–British Columbia. Unpublished Ph.D. thesis, Queen's University, Kingston, Ontario.

DALY, R. A. (1912). North American Cordillera, Forty-ninth Parallel. *Geol. Surv. Canada Mem.* 38.

DAWSON, G. M. (1886). Preliminary report on the physical and geological features of that portion of the Rocky Mountains between latitudes 49° and 51° 30′. *Geol. Surv. Canada Ann. Rept.* 1.

DEISS, C. F. (1939). Cambrian formations of southwestern Alberta and southeastern British Columbia. *Geol. Soc. America Bull.* 50, 951.

DEISS, C. F. (1940). Lower and Middle Cambrian stratigraphy of southwestern Alberta and southeastern British Columbia. *Geol. Soc. America Bull.* 51, 731.

DOUGLAS, R. J. W. and NORRIS, D. K. (1963). Dahadinni and Wrigley map-areas, District of Mackenzie, Northwest Territories. *Geol. Surv. Canada Paper* 62-33.

EVANS, C. S. (1933). Brisco-Dogtooth map-area, British Columbia. *Geol. Surv. Canada Summ. Rept.* 1932a, II, 106.

FITZGERALD, E. L. (1962). Early Middle Cambrian formations in the Front Range near Ghost River, Alberta. *Alberta Soc. Petrol. Geol. Jour.* 10, 501.

FRITZ, W. H. and NORRIS, D. K. (1966). Lower Middle Cambrian correlations in the east-central Cordillera. *Geol. Surv. Canada Paper* 66-1, 105.

FULLER, J. G. C. M. and PORTER, J. W. (1962). Cambrian, Ordovician, and Silurian formations of the northern Great Plains, and their regional connection. *Alberta Soc. Petrol. Geol. Jour.* 10, 455.

FYLES, J. T. (1966). Lead–zinc deposits in British Columbia. *Canadian Inst. Min. Met. Spec. Vol.* 8, 231.

FYLES, J. T. and EASTWOOD, G. E. P. (1962). Geology of the Ferguson area, Lardeau District, British Columbia. *British Columbia Dept. Mines Bull.* 45.

GABRIELSE, H. (1962a). Cry Lake, British Columbia. *Geol Surv. Canada* Map 29-1962.

GABRIELSE, H. (1962b). Kechika, British Columbia. *Geol. Surv. Canada* Map 42-1962.

GABRIELSE, H. (1963a). Rabbit River, British Columbia. *Geol. Surv. Canada* Map 46-1962.

GABRIELSE, H. (1963b). McDame map-area, Cassiar District, British Columbia. *Geol. Surv. Canada Mem.* **319**.

GABRIELSE, H. (1966a). Jennings River map-area. *Geol. Surv. Canada Paper* **66-1**, 42.

GABRIELSE, H. (1966b). Operation Selwyn. *Geol. Surv. Canada Paper* **66-1**, 42.

GABRIELSE, H. (1967a). Tectonic evolution of the northern Canadian Cordillera. *Can. Jour. Earth Sci.* **4**, 271.

GABRIELSE, H. (1967b). Watson Lake, Yukon Territory. *Geol. Surv. Canada* Map 19-1966.

GABRIELSE, H., RODDICK, J. A., and BLUSSON, S. L. (1965). Flat River, Glacier Lake, and Wrigley Lake, District of Mackenzie and Yukon Territory. *Geol. Surv. Canada Paper* **64-52**.

GREEN, L. H. and RODDICK, J. A. (1961). Nahanni, Yukon Territory and District of Mackenzie. *Geol. Surv. Canada* Map 14-1961.

GREEN, L. H. and RODDICK, J. A. (1962). Dawson, Larsen Creek, and Nash Creek map-areas, Yukon Territory. *Geol. Surv. Canada Paper* **62-7**.

GREEN, L. H., RODDICK, J A., and BLUSSON, S. L. (1968). Nahanni, District of Mackenzie and Yukon Territory. *Geol. Surv. Canada* Map 8-1967.

GREGGS, R. G. (1962). Upper Cambrian biostratigraphy of the southern Rocky Mountains, Alberta. Unpublished Ph.D. thesis, University of British Columbia.

GREGGS, R. G. (1963). Upper Cambrian–Lower Ordovician rock nomenclature in the southern Rocky Mountains. *Edmonton Geol. Soc. 5th Ann. Field Trip Guidebook*, 1.

GUSSOW, W. C. (1957). Cambrian and Precambrian geology of southern Alberta. *Alberta Soc. Petrol. Geol. 7th Field Conf. Guidebook*, 3.

HANDFIELD, R. C. (1967). A new Lower Cambrian archaeocyatha? *Jour. Paleont.* **41**, 209.

HARKER, P., HUTCHINSON, R. D., and McLAREN, D. J. (1954). The sub-Devonian unconformity in the eastern Rocky Mountains of Canada. *Amer. Assoc. Petroleum Geol. Symposium, Western Canada Sedimentary Basin*, 48.

HECTOR, JAMES (1863). *Journals of the exploration of British North America*.

HENDERSON, G. G. L. (1954). Geology of the Stanford Range of the Rocky Mountains. *British Columbia Dept. Mines Bull.* **35**.

HOLLAND, STUART S. (1964). Landforms of British Columbia, a physiographic outline. *British Columbia Dept. Mines Pet. Res. Bull.* **48**.

HUGHES, R. D. (1955). Geology of portions of Sunwapta and Southesk map-areas, Jasper National Park, Alberta, Canada. *Alberta Soc. Petrol. Geol. 5th Ann. Field Conf. Guidebook*, 69.

HUME, G. S. (1954). The Lower Mackenzie River area, Northwest Territories and Yukon. *Geol. Surv. Canada Mem.* **273**.

IRISH, E. J. W. (1963). Halfway River, British Columbia. *Geol. Surv. Canada* Map 22-1963.

IRISH, E. J. W. (1964). Preliminary account of the Lower Palaeozoic strata of a part of northeastern British Columbia. *Canadian Petrol. Geol. Bull.* **12**, 808.

JONES, A. G. (1959). Vernon map-area, British Columbia. *Geol. Surv. Canada Mem.* **296**.

KENT, P. E. and RUSSELL, W. A. C. (1961). Evaporite piercement structures in the northern Richardson Mountains. *Geology of the Arctic*, ed. G. O. RAASCH, **1**, 584.

KOBAYASHI, T. (1938). Upper Cambrian fossils from British Columbia with a discussion on the isolated occurrence of the so-called 'Olenus' beds at Mt. Jubilee. *Jap. Jour. Geol. and Geog.* **15**, nos. 3-4, 149.

LEECH, G. B. (1954). Canal Flats, British Columbia. *Geol. Surv. Canada Paper* **54-7**.

LEECH, G. B. (1958). Fernie map-area, west half, British Columbia. *Geol. Surv. Canada Paper* **58-10**.

LEECH, G. B. (1959). Canal Flats, Kootenay District, British Columbia. *Geol. Surv. Canada* Map 24-1958.

LEECH, G. B. (1960). Fernie map-area (west half), Kootenay District, British Columbia. *Geol. Surv Canada* Map 11-1960.

LEECH, G. B. (1966). Kananaskis Lakes, west half, area. *Geol. Surv. Canada Paper* **66-1**, 65.

LITTLE, H. W. (1960). Nelson map-area, west half, British Columbia. *Geol. Surv. Canada Mem.* **308**.

LOCHMAN-BALK, CHRISTINA and WILSON, J. L. (1958). Cambrian biostratigraphy in North America. *Jour. Paleont.* **32**, 312.

MARTIN, L. J. (1959). Stratigraphy and depositional tectonics of north Yukon–lower Mackenzie area, Canada. *Amer. Assoc. Petroleum Geol. Bull.* **43**, 2399.

McCONNELL, R. G. (1887). Report on the geological structure of a portion of the Rocky Mountains. *Geol. Surv. Canada Ann. Rept.* for 1886, **2**, D.

McKechnie, N. D. (1966). Distribution of productive mineral deposits related to time-stratigraphic sequences in British Columbia. *Canadian Inst. Min. Met. Spec. Vol.* no. **8**, 193.

Mertie, J. B., jr. (1930). Geology of the Eagle–Circle District, Alaska. *U.S. Geol. Surv. Bull.* **816**.

Mertie, J. B., jr. (1933). The Tatonduk–Nation District, Alaska. *U.S. Geol. Surv. Bull.* **836-E**.

Mertie, J. B. jr. (1937). The Yukon–Tanana region, Alaska. *U.S. Geol. Surv. Bull.* **872**.

Mountjoy, E. W. (1960). Miette, west of Fifth Meridian, Alberta. *Geol. Surv. Canada* Map 40-1959.

Mountjoy, E. W. (1961). Rocky Mountain Front Ranges along the Athabasca valley, Jasper National Park, Alberta. *Edmonton Geol. Soc. 3rd Ann. Field Trip Guidebook*, 14.

Mountjoy, E. W. (1962). Mount Robson (southeast) map-area, Rocky Mountains of Alberta and British Columbia. *Geol. Surv. Canada Paper* **61-31**.

Mountjoy, E. W. (1964). Mount Robson (southeast quarter), Alberta–British Columbia. *Geol. Surv. Canada* Map 47-1963.

Mountjoy, E. W. and Aitken, J. D. (1963). Early Cambrian and late Precambrian paleocurrents, Banff and Jasper National Parks. *Canadian Petrol. Geol. Bull.* **11**, 161.

Muller, J. E. (1961). Pine Pass, British Columbia. *Geol. Surv. Canada* Map 11-1961.

Muller, J. E. and Tipper, H. W. (1962). McLeod Lake, British Columbia. *Geol. Surv. Canada* Map 2-1962.

Norford, B. S. (1962). Illustrations of Canadian fossils: Cambrian, Ordovician, and Silurian of the Western Cordillera. *Geol. Surv. Canada Paper* **62-14**.

Norford, B. S. (1968). A Middle Cambrian *Plagiura—Poliella* faunule from southwest District of Mackenzie. *Geol. Surv. Canada Bull.* **163**, 29.

Norris, D. K. and Price, R. A. (1966). Middle Cambrian lithostratigraphy of southeastern Canadian Cordillera. *Canadian Petrol. Geol. Bull.* **14**, 385.

Norris, D. K., Price, R. A., and Mountjoy, E. W. (1963). Northern Yukon Territory and north-western District of Mackenzie. *Geol. Surv. Canada* Map 10-1963.

North, F. K. (1953). Cambrian and Ordovician of southwestern Alberta. *Alberta Soc. Petrol. Geol. 3rd Ann. Field Conf. Guidebook*, 108.

North, F. K. (1964). Cambrian of the Cordillera. *Geological History of Western Canada; Alberta Soc. Petrol. Geol.*, 28.

North, F. K. and Henderson, G. G. L. (1954). Summary of the geology of the southern Rocky Mountains of Canada. *Alberta Soc. Petrol. Geol. 4th Ann. Field Conf. Guidebook*, 15.

Okulitch, V. J. (1949). Geology of part of the Selkirk Mountains in the vicinity of the main line of the Canadian Pacific Railway, British Columbia. *Geol. Surv. Canada Bull.* **14**.

Okulitch, V. J. (1956). The Lower Cambrian of western Canada and Alaska. *El Sistema Cambrico, Symposium, XX Int. Geol. Cong., Mexico*, 701.

Okulitch, V. J. and Greggs, R. G. (1958). Archaeocyathid localities in Washington, British Columbia, and the Yukon Territory. *Jour. Paleont.* **32**, 617.

Palmer, Allison R. (1968). Cambrian trilobites of east-central Alaska. *U.S. Geol. Surv. Prof. Paper* **559-B**.

Park, C. F. and Cannon, R. S. (1943). Geology and ore deposits of the Metaline Quadrangle, Washington. *U.S. Geol. Surv. Prof. Paper* **202**.

Poole, W. H., Roddick, J. A., and Green, L. H. (1960). Wolf Lake, Yukon Territory. *Geol. Surv. Canada* Map 10-1960.

Preto, V. A. (1967). Grand Forks (west half) map-area. *Geol. Surv. Canada Paper* **67-1**, pt. A, 84.

Price, R. A. (1962). Fernie map-area, east half, Alberta and British Columbia. *Geol. Surv. Canada Paper* **61-24**.

Price, R. A. (1965). Flathead map-area, British Columbia and Alberta. *Geol. Surv. Canada Mem.* **336**.

Price, R. A. (1967). Operation Bow-Athabasca, Alberta and British Columbia. *Geol. Surv. Canada Paper* **67-1**, pt. A, 106.

Raasch, G. O. and Bruce, C. J. (in press). Canadian–Chazyan succession, White River area, British Columbia. *Canadian Petrol. Geol. Bull.*

Raasch, G. O. and Campau, D. E. (1957). Cambrian biostratigraphy of California Standard Parkland no. 4-12. *Alberta Soc. Petrol. Geol. Jour.* **5**, 140.

Ragan, Donal M. and Hawkins, James W. (1966). A polymetamorphic complex in the eastern Alaska Range. *Geol. Soc. America Bull.* **77**, 597.

Rasetti, Franco (1951). Middle Cambrian stratigraphy and faunas of Canadian Rocky Mountains. *Smithsonian Misc. Coll.* **116**, no. 5.

RASETTI, FRANCO (1956). The Middle and Upper Cambrian of Western Canada. *El Sistema Cambrico, Symposium, XX Int. Geol. Cong., Mexico,* 735.

REESOR, J. E. (1957). Lardeau (east half), Kootenay District, British Columbia. *Geol. Surv. Canada* Map 12-1957.

RICE, H. M. A. (1937). Cranbrook map-area, British Columbia *Geol. Surv. Canada Mem.* **207**.

RICE, H. M. A. (1941). Nelson map-area, east half, British Columbia. *Geol. Surv. Canada Mem.* **228**.

ROOTS, E. F. (1954). Geology and mineral deposits of Aiken Lake map-area, British Columbia. *Geol. Surv. Canada Mem.* **274**.

SHAW, E. W. (1963). Canadian Rockies—orientation in time and space. *Amer. Assoc. Petroleum Geol. Mem.* **2**, 231.

SLIND, O. L. and PERKINS, G. D. (1966). Lower Paleozoic and Proterozoic sediments of the Rocky Mountains between Jasper, Alberta, and Pine River, British Columbia. *Canadian Petrol. Geol. Bull.* **14**, 442.

SLOSS, L. L. (1963). Sequences in the cratonic interior of North America. *Geol. Soc. America Bull.* **74**, 93.

SMITH, PHILIP S. (1939). Areal geology of Alaska. *U.S. Geol. Surv. Prof. Paper* **192**.

STEVENS, R. D. (1965). K-Ar age of Cambrian glauconite from Alberta. *Geol. Surv. Canada Paper* **65-2**, 32.

STREET, P. J. (1966). Trilobite zones in the Murray Range, Pine Pass map-area, British Columbia. Unpublished M.Sc. thesis, University of British Columbia.

SUTHERLAND BROWN, A. (1963). Geology of the Cariboo River area, British Columbia. *British Columbia Dept. Mines Bull.* **47**.

THOMPSON, RAYMOND M. (1960). Geology and Petroleum possibilities of Alaska. *XXI Int. Geol. Cong., Norden,* **XI**, 27.

VAN HEES, HENDRIK (1964). Cambrian of the Plains. *Geological History of Western Canada, Alberta Soc. Petrol. Geol.,* 20.

WALCOTT, C. D. (1911a). Middle Cambrian Merostomata. *Smithsonian Misc. Coll.* **57**, no. 2, 17.

WALCOTT, C. D. (1911b). Middle Cambrian holothurians and Medusae. *Smithsonian Misc. Coll.* **57**, no. 3, 41.

WALCOTT, C. D. (1911c). Middle Cambrian annelids. *Smithsonian Misc. Coll.* **57**, no. 5, 109.

WALCOTT, C. D. (1912). Middle Cambrian Branchiopoda, Malacostraca, Trilobita, and Merostomata. *Smithsonian Misc. Coll.* **57**, no. 6, 145.

WALCOTT, C. D. (1919). Middle Cambrian algae. *Smithsonian Misc. Coll.* **67**, no. 5, 217.

WALCOTT, C. D. (1920). Middle Cambrian Spongiae. *Smithsonian Misc. Coll.* **67**, no. 6, 261.

WALCOTT, C. D. (1928). Pre-Devonian Paleozoic formations of the Cordilleran provinces of Canada. *Smithsonian Misc. Coll.* **75**, no. 5, 185.

WALKER, J. F. (1926). Geology and mineral deposits of Windermere map-area, British Columbia. *Geol. Surv. Canada Mem.* **148**.

WEBB, J. B. (1965). Cratonic depositional sequences and Cordilleran orogenies: Palaeozoic Era—Western Canada. *Geol. Assoc. Canada Proc.* **16**, 11.

WHEELER, J. O. (1960a). Quiet Lake, Yukon Territory. *Geol. Surv. Canada* Map 7-1960.

WHEELER, J. O. (1960b). Finlayson Lake, Yukon Territory. *Geol. Surv. Canada* Map 8-1960.

WHEELER, J. O. (1963). Rogers Pass map-area, British Columbia and Alberta. *Geol. Surv. Canada Paper* **62-32**.

WHEELER, J. O. (1965). Big Bend map-area, British Columbia. *Geol. Surv. Canada Paper* **64-32**.

WHEELER, J. O. (1966a). Lardeau (west half) map-area. *Geol. Surv. Canada Paper* **66-1**, 102.

WHEELER, J. O. (1966b). Eastern tectonic belt of Western Cordillera in British Columbia. *Canadian Inst. Min. Met. Spec. Vol.* **8**, 27.

WHITE, W. H. (1959). Cordilleran tectonics in British Columbia. *Amer. Assoc. Petroleum Geol. Bull.* **43**, 60.

WILLIAMS, M. Y. (1923). Reconnaissance across northeastern British Columbia and the geology of the northern extension of the Franklin Mountains, N.W.T. *Geol. Surv. Canada Summ. Rept.* 1922, pt. B, 65.

THE CAMBRIAN OF THE NORTH AMERICAN ARCTIC REGIONS

John Watson Cowie

Department of Geology, University of Bristol, England

Contents

1. Introduction

This review of the Cambrian geology of the North American arctic regions includes the whole of the Canadian arctic archipelago, a small section of the mainland of Canada in Boothia Peninsula, and the island of Greenland. The Cambrian of the District of Mackenzie and the Yukon in Canada, and of Alaska in the United States of America, are dealt with elsewhere in this volume.

The climate of the whole region, which is shown with neighbouring regions in Fig. 1, is arctic. Seas freeze annually and the vegetation is arctic or high arctic with the tree-line

FIG. 1. The North American shores of the Arctic Ocean with selected Cambrian localities approximately indicated in Greenland, Canada, and Alaska (unnumbered western Canadian localities in part taken from F. K. North, this volume). The Cambrian of Spitsbergen is described in another volume of this series.

Plate 3. North coast of Devon Island in arctic Canada looking westwards from south of Cape Newman Smith, (see Fig. 5 and page 351), towards Bear Bay. Flat lying sedimentary strata of Cambrian age unconformably overlie the Pre-Cambrian metamorphic shield complex which is intensely foliated and cut by faults and igneous intrusions. (*This photograph T 435L-39 is reproduced by permission of the National Air Photo Library, Surveys and Mapping Branch, Department of Energy, Mines and Resources, Ottawa, Canada*).

(To face page 326)

far to the south of the region. Precipitation varies considerably but is low in general and some areas are arid. Relief and land forms are of great variety ranging from alpine topography in parts of Ellesmere Island and north-east and east Greenland to low-lying featureless plateaus in parts of north Greenland and the Queen Elizabeth Islands. Rough crystalline shield areas with disordered drainage and low, glacially-eroded, relief are of great extent. Ice sheets, caps, and glaciers play an important role in constraining geological field research in some of the terrains. Glacial deposits locally may be an obscuring agent and weathered mantles may be a serious difficulty due to the intensity of frost-shattering in most areas. The impression left with geologists who have worked in many parts of the high arctic lands of Canada and Greenland is of the vast expanses of well-exposed bedrock (Plate 3). The fjords and inlets slice sections through the rocks, exposing varied aspects of the geological story in continuous exposures which are open to view over many kilometres of steep, bare cliff profiles. Much of the area is only accessible at great expense and a few hours field work may be all that was possible so that only relatively cursory reconnaissance field observations have been published. The studies also range over a considerable span of the twentieth and late nineteenth centuries so that sedimentological and palaeoecological aspects are either not well known or unknown. Because of the widely separated localities and the unevenness of knowledge it is thought that palaeogeographical speculations in the form of maps can hardly be justified.

The major localities are shown in a generalized way in Fig. 1: numbered when referred to below, unnumbered when dealt with by other contributors to this volume or in another volume of this series. The largest geological unit is the Canada–Greenland Shield, a major crustal crystalline metamorphic basement complex which is Pre-Cambrian in age. It outcrops widely in the eastern and southern parts of the Canadian sector and is also extensively present in south, west, and north-west Greenland; it does not seem to outcrop in the ice-free lands of north Greenland and is only seen in a few limited areas in east Greenland north of 70° N. The usual interpretation is that it underlies much of the inland ice-cap of Greenland. There are three major settings for known or presumed Cambrian rocks in Greenland (see Figs 7 and 8):

1. In cratonic or marginal facies of unconformable cover rocks near to the crystalline basement (numbers in brackets refer to Fig. 1) in the Danmark Fjord area (2), Inglefield Land (5), and probably also in Prudhoe Land in north-west Greenland.

2. The miogeosynclinal area of north Greenland which flanks on the south the Palaeozoic mountain belt, in Peary Land (3) and Nyeboe Land (4).

3. The suprastrata of the fold belt of east Greenland which were folded in the Pre-Cambrian Carolinidian and the Palaeozoic Caledonian orogenies in the fjord zone of east Greenland (1) and Kronprins Christian Land at the northern end of the fold belt, to the east of Danmark Fjord.

The similarity of the geology of north-west Greenland and Ellesmere Island is now well known; the separation line, which is relatively straight, figures in continental drift speculation and the search for significant major crustal lineaments (Fig. 2).

The stratigraphical and structural relationships of late Pre-Cambrian to Devonian sediments on opposite sides of the Nares Strait lineament (Smith Sound–Kane Basin–Kennedy Channel–Hall Basin–Robeson Channel strait between Canada and Greenland) suggest that it is a submarine rift valley—an extensional structure which formed between north Greenland and Ellesmere Island allowing rotation. Prior to late Cretaceous time there was a single land mass and geological provinces including the Pre-Cambrian

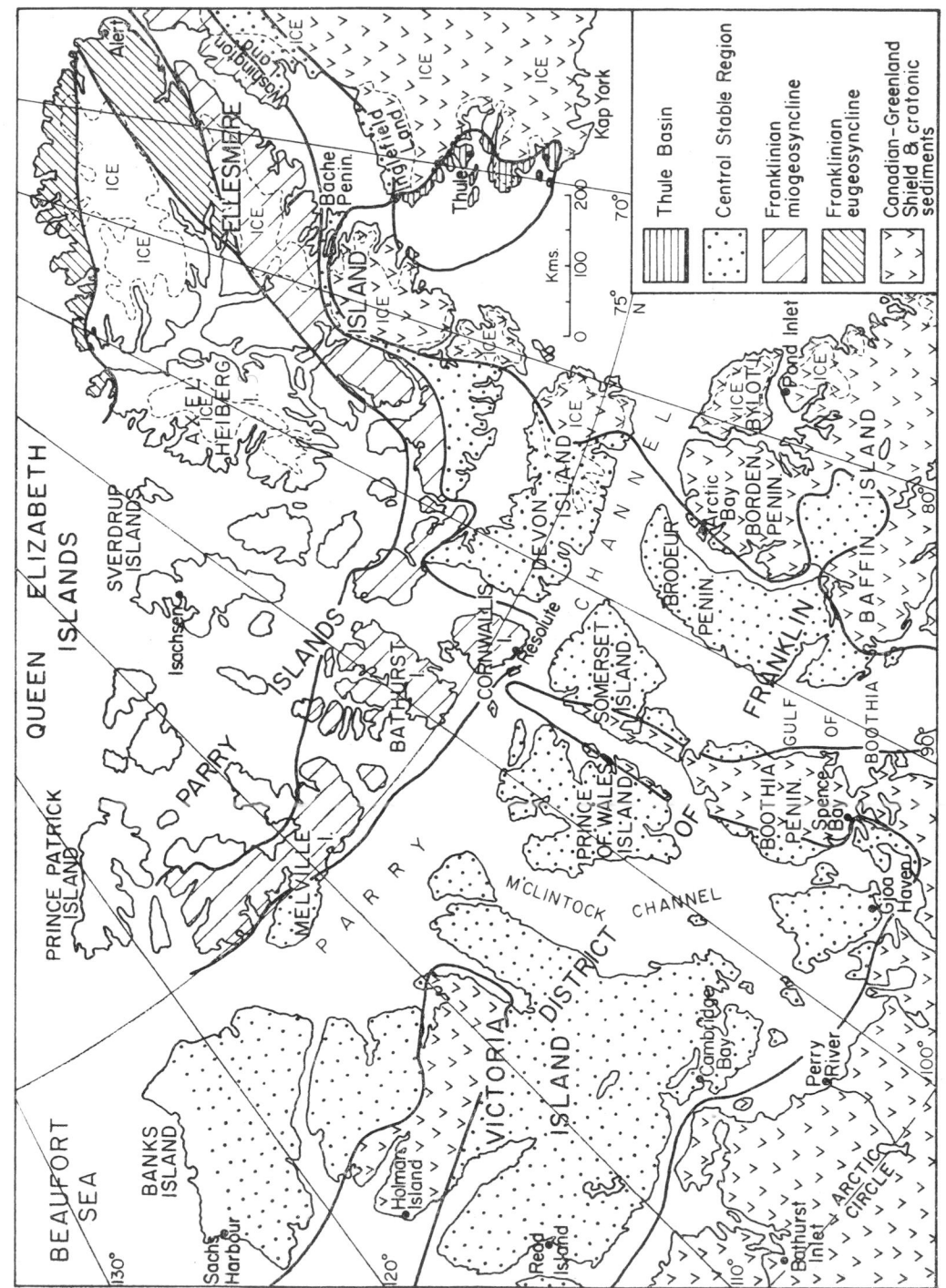

Fig. 2. Selected parts of stratigraphical–structural provinces of arctic Canada and north-west Greenland.

crystalline basement were continuous between the two sides (KERR, 1967b). The structure and stratigraphy of the Pre-Cambrian strata support this interpretation.

Stratigraphical–structural provinces in the Canadian Arctic Archipelago were convincingly defined in 1960 by THORSTEINSSON and TOZER, following earlier work by FORTIER, McNAIR, and THORSTEINSSON (1954). With continuing field and laboratory work, however, their original ideas have been extended and elaborated through the work of a number of authors including BLACKADAR, CHRISTIE, and KERR. The Canadian Shield exposed in the south and east is essentially continuous with the Greenland Shield and both are covered in the north and west, in the Thule area, Inglefield Land, Washington Land, Ellesmere Island, Devon Island, and other islands to the south and west by thin Pre-Cambrian and Palaeozoic sedimentary rocks. These cover rocks are partly cratonic sediments and are also made up to a greater extent of slightly thicker, little folded strata of the Central Stable Region (KING, 1959); they are shown in Figs 2, 7, and 8 which follow.

The Franklinian geosyncline, trending north-easterly through the Canadian Arctic islands can be followed through Washington Land and along the north coast of Greenland. It is built of about 12,000 metres of Pre-Cambrian and Palaeozoic strata. It can be divided into a south-easterly miogesynclinal belt and a north-westerly eugeosynclinal belt and was deformed by earth movements during the Palaeozoic.

The junction between the Franklinian miogeosyncline and eugeosyncline is covered in places by up to 12,000 metres of late Palaeozoic to mid-Tertiary sediments of the Sverdrup Basin, which were in their turn deformed by a late Mesozoic–Tertiary orogeny.

In discussing the possible downward extension of the Cambrian part of the succession it has been necessary to consider late Pre-Cambrian formations and also formations which might be Cambrian or late Pre-Cambrian in age. Some transfers of age assignment have been suggested. No general discussion of the problem of the base of the Cambrian System has been attempted because it is dealt with in a separate contribution to these volumes; it is the subject of international and national debate. The placing of the base of the Cambrian in the various sections which follow is partly pragmatic but the guiding principle is the position of any fauna which can be taken as clearly of a Cambrian character. Reliance is placed mainly on biostratigraphical methods but speculative lithostratigraphical correlations have been thought worthwhile. In these arctic regions there is little current discussion regarding the Cambro-Ordovician boundary, probably because there is a lack of late Cambrian, or putative late Cambrian, faunal evidence to justify such a discussion.

Place-names in Greenland (politically a part of Denmark) are given in the Danish form in accordance with international convention and as shown on official Danish maps. In accordance with stratigraphical priority, however, the geographical components of formation names are retained in the form given by the original author, even though it may be an anglicised or translated version of the present Danish language place-name.

2. The Cambrian around the southern part of Nares Strait, in Greenland and Canada

2.1. Inglefield Land and Washington Land, north-west Greenland

In most parts of Inglefield and Washington Lands (Fig. 3) the igneous and metamorphic rocks of the basement complex are overlain unconformably by a comparatively thin

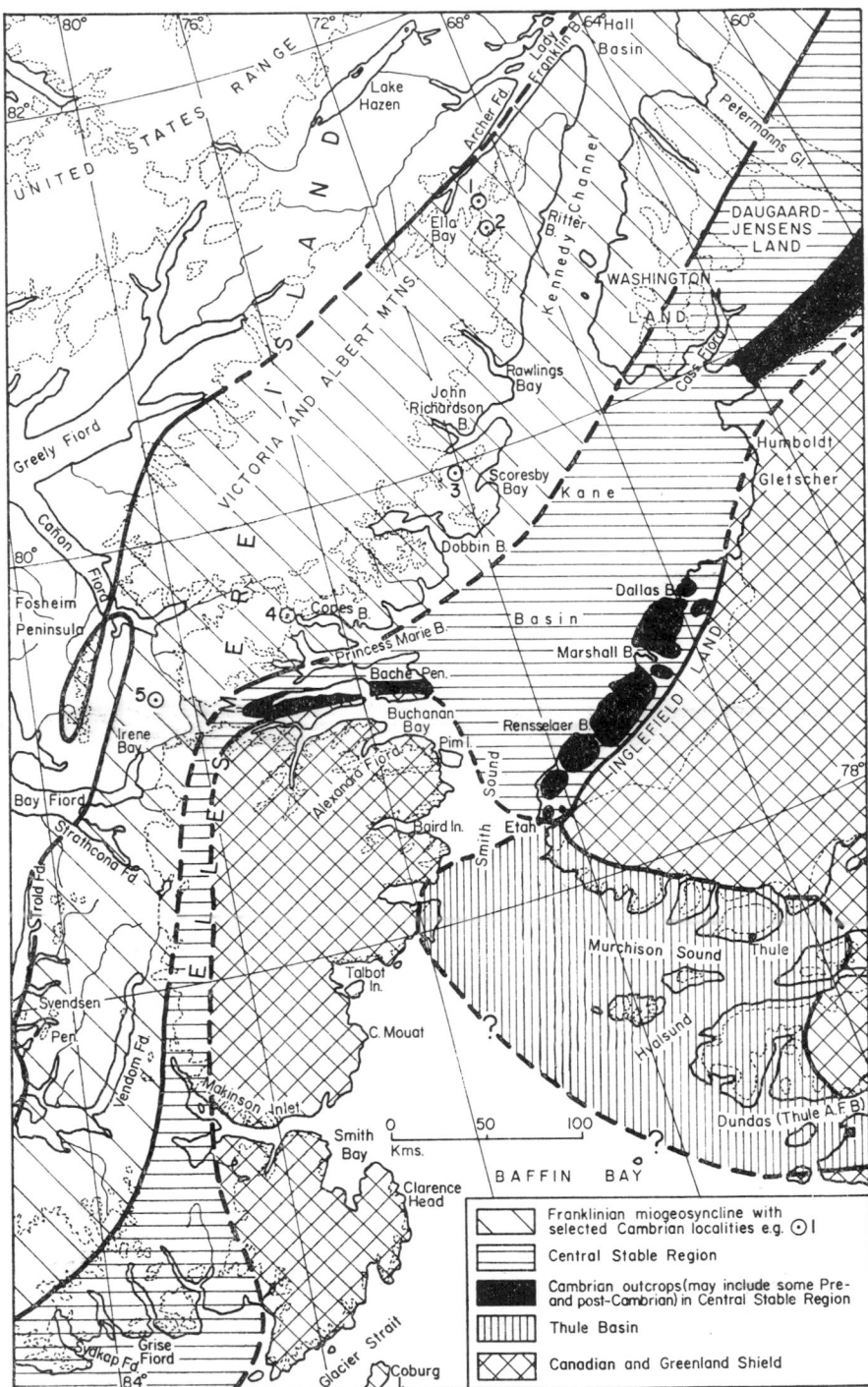

FIG. 3. Parts of Ellesmere Island, arctic Canada, and north-west Greenland around northern Baffin Bay and the southern end of Nares Strait. The edge of ice-caps is shown by a dashed line. Cambrian outcrops occur in southern Ellesmere Island which are not shown here because details are not yet published (see page 353).

succession of sedimentary rocks which have suffered practically no flexuring and only minor faulting. By correlations with other areas, discussed later, it seems that the whole of this sedimentary sequence up to the base of the Cass Fjord Formation (see below) should be assigned to the Cambrian. Prior to the recent Ellesmere Island surveys by the Geological Survey of Canada, the older unfossiliferous beds lying with very marked unconformity on the basement were referred to as late Pre-Cambrian, 'Eo-Cambrian', or Lower Cambrian; only at a higher level were Lower and Middle Cambrian fossils found. The Middle Cambrian is followed with regional disconformity, or unconformity in the broad sense of the term, by Lower Ordovician strata; the late Middle Cambrian and Upper Cambrian, if present at all, are unfossiliferous. The main, best known, Cambrian outcrops are found in Inglefield Land which is the type area for the Cambrian formations of the following succession:

Age	Formation and maximum thickness in metres	Lithology
Lower Ordovician	Cass Fjord, 400	Limestone, conglomerate, shale.
Middle Cambrian	Cape Wood, 90	Limestone, dolomite, sandstone, conglomerate.
Lower Cambrian	Cape Kent, 20	Limestone, dolomite, sandstone, shale.
Lower Cambrian	Wulff River, 35	Limestone, sandstone, conglomerate.
———— ? slight erosional disconformity ————		
? Lower Cambrian	Cape Ingersoll, 10	Dolomite, limestone.
? Lower Cambrian	Cape Leiper, 40	Dolomite.
? Lower Cambrian	Rensselaer Bay, 150	Sandstone, conglomerate, dolomite.
———— MARKED ANGULAR UNCONFORMITY ————		
Pre-Cambrian		Schists, gneisses, granite, diorite, gabbro.

Rensselaer Bay Formation

Conglomeratic at its base where it rests with major unconformity on the basement complex, this formation can be divided into two members which can be correlated with the three members of the same formation described from the Bache Peninsula area

(CHRISTIE, 1967, pp. 14–20; KERR, 1967a, pp. 30–31). It is therefore proposed here that the following names should be used:

Inglefield Land	Bache Peninsula area
Sverdrup Member, 50 m	Sverdrup Member, 50–90 m
Hatherton Member, 100 m	{ Bache Peninsula Member, 0–27 m { Camperdown Member, 0–80 m

———— MARKED ANGULAR UNCONFORMITY ————

PRE-CAMBRIAN BASEMENT COMPLEX

The beds of the Hatherton Member (named after HATHERTON BUGT 10 kilometres north of Etah in south-west Inglefield Land, Fig. 3) are distinguishable in early descriptions by KOCH (1929a, p. 220), BENTHAM (1936, p. 429), and TROELSEN (1950b, pp. 35–37). COWIE (1961, p. 12) visited the type locality of the member. They are red (possibly maroon to purple), ferruginous sandstones which are feldspathic, conglomeratic, cross-bedded and ripple-marked; also present is dolomite with stromatolites (*Collenia*) at a number of horizons. There is a basal conglomerate up to 10 metres thick. This member is represented by the Lower Beds subdivision at Hatherton Bugt (COWIE, 1961, p. 12) and Bed 43 at Kap Ingersoll (COWIE, 1961).

This basal member of the Rensselaer Bay Formation suggests shallow water by the presence of ripple marks, with cross-bedding indicating variation in the direction of depositing current, and intermittent drying out or extreme shallowing shown by fossil mud-cracks and penecontemporaneous erosion. Stromatolites are taken to represent algal reefs, lagoonal or littoral growths in the early Cambrian seas.

The Sverdrup Member is a cream, yellow, or buff sandstone or siltstone with ripple-marks and cross-bedding in places, interbedded with buff, fine-grained siltstone. A change in climate is indicated from Hatherton Member times in the change of colour from red to cream, yellow, or buff and deeper water by the finer grade but the ripple marking and coarser horizons suggest continued instability.

Sills and dykes

Mafic intrusives mainly of quartz–dolerite cut the crystalline basement complex and the Rensselaer Bay Formation in south-western Inglefield Land. The relationship with the overlying carbonate rocks is not clearly known and interpretation should await further investigation (COWIE, 1961, p. 21); it is possible that the basic intrusions into the sedimentary series of north-west Greenland are of more than one age.

Cape Leiper Formation

A stylolitic, arenaceous, fine- to medium-grained dolomite, which is grey or buff when fresh and which weathers dark buff to orange, is a cliff-forming series. Quartz grains in the dolomite are locally macroscopic. The formation is probably conformable at base and top, but TROELSEN claims induration of the upper surface (1950b, p. 36).

This formation indicates by its finer grade and much reduced percentage of psammitic material, coupled with important carbonate content, that chemical deposition had assumed major importance and land was probably farther away. Further work may support the idea that the dolomite is of primary origin.

Cape Ingersoll Formation

At Kap Ingersoll a medium-grey, fine-grained limestone is found but according to TROELSEN (1950b, p. 36; 1956a, p. 85; 1956b, p. 76) farther north the formation is represented by a grey, medium-grained, hard crystalline dolomite which weathers to a rusty-red. In all areas it is described as showing large irregular-shaped blebs of white or buff, medium-grained crystalline dolomite which branch and form connecting channels; these blebs at Kap Ingersoll associated with limestone give an aggregate petrology of a dolomitic limestone. Thus again the dolomite may be primary while the calcareous content may be secondary. TROELSEN (1950b, p. 36) suggests a diastem at the base of the formation and an erosional disconformity at the top. The uppermost part is strongly brecciated (BERTHELSEN and NOE-NYGAARD, 1965, p. 216). On general grounds (COWIE, 1961, pp. 22–23) only a short time gap can be suggested before the deposition of the Wulff River Formation commenced. The change of carbonate character at this level in the Cambrian succession probably also indicates palaeogeographical changes.

The older sedimentary formations in Inglefield Land suggest a gradual change in sedimentation from coarser to finer clastic types, with an accompanying introduction and eventual predominance of chemical deposition which is mainly magnesium rich earlier and lime rich later. During a short time gap between the deposition of the Ingersoll and Wulff River Formations slight erosion and brecciation of underlying rocks with an erosion surface of minimal relief occurred, accompanied by regression of the sea and a pause in deposition which may now be shown by a slight erosional disconformity.

Wulff River Formation

Above the slight unconformity, with no angular discordance, at the top of the underlying Cape Ingersoll Formation there is a marked change of lithology to a variable series of beds which is best known near the type area in north-east Inglefield Land, where it is a grey hard limestone with grey calcareous sandstone and thin conglomerates (TROELSEN, 1956a, p. 78). Earlier reports by KOCH described the formation as green glauconitic sandstone with beds of limestone and conglomerates containing pebbles of fossiliferous dark brown sandstone, quartz, and diabase.

Outcrops in south-west Inglefield Land differ in petrology and at Kap Ingersoll the beds are entirely glauconitic siltstones and sandstones with some phosphatic horizons; bedding surfaces are irregular with some contortion. The clastic deposits contain glauconite and phosphate suggesting renewed submergence in a shallow sea and coinciding with the migration of organisms: the first metazoan fossils in Inglefield Land occur in this formation. Unstable conditions are shown by lateral variation in deposits, in the laying down of conglomerates, and in irregular bedding surfaces.

LAUGE KOCH made fossil collections between 1916 and 1923 from thin arenaceous limestone bands and sandstone pebbles from the conglomerates. Many of KOCH's collections were made from scree (*pers. comm.*); they were described by C. POULSEN (1927) and include: *Paterina lata* (POULSEN), *Obolus?* sp., *Botsfordia caelata* (HALL), *Acrothele? pulchra* POULSEN, *Salterella expansa* POULSEN, *Strenuaeva groenlandica* (POULSEN), *Paedeumias? breviloba* (POULSEN), and *Wanneria arcticus* (POULSEN).

TROELSEN's collections made between 1939 and 1941 from fairly pure limestone were also described by C. POULSEN (1958); it seems they were at least partly obtained from talus (TROELSEN, 1950b, p. 40), they include: *Kutorgina reticulata* POULSEN, *Salterella expansa*, *Olenellus carinatus* POULSEN, *O. laevis* POULSEN, *O. troelseni* POULSEN, *Paedeumias groenlandicus* POULSEN, *Holmia mirabilis* POULSEN, *Wanneria abnormis* POULSEN, *W. inermis* POULSEN, *W. mediocris* POULSEN. *W. ruginosa* POULSEN, *W. subglabra* POULSEN, *W. troelseni* POULSEN, and *Bonnia arctica* POULSEN.

TROELSEN's collections have only *Salterella expansa* in common with KOCH's collections so that it seems likely that two different faunal horizons have been sampled by the two field workers; this explanation fits the difference in lithology already noted.

The faunal assemblages, including as they do olenellid and non-olenellid trilobites, perhaps suggest a younger Lower Cambrian age. The inclusion of scree specimens, understandable as it is when the difficult exploratory conditions of their collection are taken into account, nevertheless causes some doubt in the biostratigraphical relationships. The fauna as a whole shows some affinity to the Atlantic Province.

Cape Kent Formation

In north-east Inglefield Land this formation is a yellowish white or cream, almost pure, oolitic limestone. TROELSEN (1950b, p. 41) states that the lithologic character of the formation is the same throughout Inglefield Land; but the writer (1961, pp. 24–25) correlated a thin sequence of about 8 metres of unfossiliferous dolomite, sandstone, and shale found at Kap Ingersoll with the Cape Kent Formation because of its position in the succession and the high proportion of carbonate rocks. TROELSEN (1950b, p. 41) noted that fossils are only visible on weathered surfaces and that his collections, like those of KOCH (described by C. POULSEN, 1927), were nearly all from talus blocks and may represent more than one faunal horizon. Recent work by V. POULSEN (1964) now gives a faunal list which includes: *Bristolia groenlandica* (POULSEN), *B. kentensis* (POULSEN), *Dolichometopsis resseri* POULSEN, *D. septentrionalis* POULSEN, *Hyolithes poulseni* RESSER, *Inglefieldia affinis* POULSEN, *I. inconspicua* POULSEN, *I. porosa* POULSEN, *I. venulosa* (POULSEN), *Kochiella arcana* POULSEN, *K. gracilis* POULSEN, *K. tuberculata* POULSEN, and *Poulsenia groenwalli* (POULSEN). This is a late Lower Cambrian assemblage.

KURTZ, MCNAIR, and WALES (1952, p. 651) speculated as to whether the fossils described from the Cape Kent Formation may represent two faunas which may be stratigraphically separated. *Dolichometopsis* is found at Dundas Harbour, Devon Island (p. 66) in the Bear Point Limestone which was assigned by these authors to the lower Middle Cambrian on the grounds of the reported presence of unnamed representatives of the *Albertella* zone. *Olenellus* occurs in the underlying Rabbit Point Sandstone. KURTZ and others suggested by their correlations between Devon Island and north-west Greenland that the Cape Kent Formation ranges in age from late Lower to early Middle Cambrian. Arguments presented in some detail by the author in 1961 (pp. 24–25) are probably still valid; V. POULSEN agreed in 1964 (p. 60) that the entire Cape Kent Formation should be considered late Lower Cambrian. NORFORD in 1968 described *Kochiella* and *Inglefieldia* from unnamed beds near the South Nahanni River, south-west District of Mackenzie, northern Canada, which also contained *Fieldaspis* and brachiopods. NORFORD discounted a Lower Cambrian age for the faunule on the grounds of the absence of olenellids (found widely in the District of Mackenzie and the Yukon Territory) and the lower Middle Cambrian associations of *Fieldaspis*. He assigned the faunule to the *Plagiura-Poliella* Zone, the lowest Middle Cambrian zone in western North America. It seems entirely reasonable to postulate, as NORFORD does (1968, p. 33), that the presence of *Kochiella* and *Inglefieldia* in both the Cape Kent Formation and the Middle Cambrian beds from south-west Mackenzie District suggests an age range of the genera and/or the Cape Kent Formation from Lower to Middle Cambrian times. It seems clear that only further, more detailed, field work will satisfactorily settle this question of the age range of these Cambrian formations.

The Cape Kent Formation, which is found widely in this region, marks the predominance of calcareous, true limestone, deposition and the presence of oolites and the lithological uniformity may agree with deposition in a relatively shallow sea but not in a littoral site. Rapid variations in the sedimentary environment were still prevailing in the south where intraformational conglomerates and glauconite occur with a proportion of pelitic material.

Cape Wood Formation

This formation has been divided by TROELSEN (1950b, pp. 42–46) into two members: the Blomsterbaek Limestone Member, 2 to 5 metres thick, and the Cape Russell Member, 45 to 90 metres thick.

The older Cape Russell Member in the type area is variable vertically through the sequence and also laterally; it includes glauconitic sandstone, arenaceous limestone, limestone, dolomite, and conglomerates containing fragments of limestone and sandstone. This member may be of shallow water origin and rests with a basal conglomerate on the Cape Kent Formation. At Kap Ingersoll the member is a series of alternating limestones and dolomites which are often argillaceous and occasionally glauconitic and contains an abundance of intraformational conglomerates and breccias (COWIE, 1961, pp. 25–26).

The Blomsterbaek Limestone Member in north-east Inglefield Land is a thin-bedded, grey, fine-grained limestone which weathers yellow and has a thin basal conglomerate with fragments of limestone (TROELSEN, 1956a, p. 24). At Kap Ingersoll this member is represented by a fine-grained, thinly-bedded grey dolomite which is green and easily shattered when weathered. The overlying formation is the widespread and characteristic Lower Ordovician (Canadian) Cass Fjord Formation of thinly stratified, nodular, muddy limestones and shales with many intraformational conglomerates. The Upper Cambrian is, so far, unrepresented by fossils in Inglefield Land and Washington Land.

Unstable sedimentary conditions during Cape Wood Formation times are indicated by both clastic and chemical deposits being accompanied at frequent intervals by penecontemporaneous brecciation and the formation of conglomerates. This is clear also from faunal correlations as zones or parts of zones are missing according to V. POULSEN (1964, fig. 9), who suggested the tabulation shown on page 336.

The north-west Greenland faunas which belonged to the cratonic realm are most closely related to the Cordilleran faunas from North America. The fact that many of the fossils were collected from scree is a difficulty in the assessment of faunal asssociations; this is fully discussed by V. POULSEN and his conclusions are adopted here.

The Cape Wood Formation basal conglomerate contains elements of the *Glossopleura* assemblage but species of *Fieldaspis*, *Amecephalus*, and ? *Kochaspis* occur. Although the last two genera are only represented in loose material it is considered by V. POULSEN that they come from pebbles of the basal conglomerate of the Cape Russell Member. At Blomsterbaekken, between Dallas Bugt and Marshall Bugt in north-east Inglefield Land (Fig. 3), unfossiliferous calcareous sandstones, which may be early Middle Cambrian in age, intervene between the Cape Kent Formation and the typical Cape Russell Member limestone. Elsewhere, as noted, there are no such beds. It is possible that *Fieldaspis*, *Amecephalus*, and *Kochaspis*, found only as remanié fossils in the Cape Russell

Standard Zones Pacific Faunal Province	Biostratigraphy	Lithostratigraphy
Bathyuriscus- *Elrathina*	*Blainiopsis* Faunule	Blomsterbaek Limestone Member
Bathyuriscus- *Elrathina*	present but unnamed ⎱	
Glossopleura	*Clavaspidella* Faunule *Glossopleura* Faunule ⎰	⎬ Cape Russell Member
Albertella	not present	—
Plagiura-Poliella	present but unnamed and remanié.	not identified.

Member basal conglomerate, represent the *Plagiura-Poliella* Zone derived from beds which now, in the main at least, have been eroded away and which probably represented only a short duration of sedimentation. If the Cape Kent Formation did extend its deposition into Middle Cambrian times (p. 335) the pebbles containing the fossils could have been derived from this formation. (NORFORD, 1968, p. 33). The ensuing period of erosion and break in sedimentation persisted for the duration of the whole of the *Albertella* Zone which is unrepresented in north-west Greenland.

The *Glossopleura* Zone has an upper boundary within the Cape Russell member. The older, *Glossopleura* faunule, contains predominantly *Glossopleura* and *Polypleuraspis; Kootenia, Poulseniella, Ptychoparella,* and *Solenopleurella* are present but extremely scarce. The younger *Clavaspidella* faunule, has the characteristic genera *Clavaspidella* and ? *Kootenia.*

The *Bathyuriscus-Elrathina* Zone includes as dominant genera *Glyphaspis* and *Acrocephalops* which seem to be uniformly distributed throughout the sequence. *Elrathiella* tends to be concentrated in the younger part of the zone and *Blainiopsis* may represent an upper faunule and occurs only in the Blomsterbaek Limestone Member. Other genera in this zone in Inglefield Land are *Acrocephalops, Kootenia,* and *Zacanthoides.* Miogeosynclinal conditions are not represented by the character of the faunas.

V. POULSEN (1964, p. 75) is of the opinion that towards the end of Lower Cambrian times, represented by the Cape Kent Limestone, a regression set in and the nearby coastlines moved to the west. Following a certain lapse of time a new temporary transgression occurred at the time of the *Plagiura-Poliella* Zone, but the sea receded during the *Albertella* Zone when erosion occurred. At the time of the *Glossopleura* Zone a prolonged submergence supervened with only a brief break until the end of the zone. During much of *Bathyuriscus-Elrathina* Zone time there was no sedimentation in the area but by the end of this zone maximal transgression is suggested by faunal association with both Cordilleran and eastern parts of North America. Regression set in again, however, and persisted throughout the time of the *Bolaspidella* Zone until the next submergence in the Ordovician.

Near Washington Land KOCH (1929a, pp. 7–9) described sedimentary rocks belonging to his 'Thule Formation' and the Cape Wood Formation at localities close to the northern margin of the Humboldt Gletscher, south-east of Cass Fjord (Fig. 3). On the island of Putlersuak was found yellow saccharoidal crystalline dolomite with? annelid trails and? algal stromatolites which appeared to belong to the Cape Leiper Formation. Loose blocks of red sandstone and diabase occur which may have come from outcrops of the Rensselaer Bay Formation. Thick hard limestone resembling the upper part of the Cape Wood Formation and a reddish-yellow limestone form part of a putative Cambrian stratified series of about 200 metres thickness stretching towards Cass Fjord where it is covered by the Cass Fjord Formation of Lower Ordovician age. As shown in Fig. 3 an outcrop of Cambrian rocks is suggested from south-east of Cass Fjord across Daugaard-Jensen Land to the Petermann Gletscher.

2.2. The Thule Basin, Prudhoe Land to Kap York, north-west Greenland

The Thule Basin (Figs 2 and 3) is a term used by KERR (1967b, p. 487) for the down-faulted sedimentary basin which is known to stretch from near Dundas (Thule Air Force Base) in the Kap York district, through ice-free areas on the shores of Hvalsund and Murchison Sound, to Etah. The northern outcrops occur in Prudhoe Land: the name given to the peninsulas north of Murchison Sound between Thule (originally Kanak) and Etah. An extension of this basin is inferred in Ellesmere Island south of Baird Inlet (Fig. 3) (CHRISTIE, 1962a) and was suspected in southern Ellesmere Island (WORDIE, 1938; BENTHAM, 1941) but is probably absent (CHRISTIE, 1962b). In the extreme east of Devon Island near Philpots Island and in the Borden and Brodeur Peninsulas of Baffin Island there may be remnants of the same basin but it is quite possible that some of these localities may have belonged to separate and discrete basins.

The sedimentary series found in the Thule Basin have usually been considered together and have, over the years, been assigned various ages, either late Pre-Cambrian, Eocambrian, or Lower Cambrian. LAUGE KOCH's original Thule Formation which he set up in 1916 was dated (1929c) as Late Algonkian and had its type locality at 'Wolstenhome Fjord, around Thule'. BERTHELSEN and NOE-NYGAARD summarized many of the arguments in 1965 and suggested that the Pre-Cambrian suprastrata in their 'Cape York district' were Eocambrian (as part of the Pre-Cambrian). More recent work, reviewed elsewhere in this chapter, shows a need for a reassessment of the evidence regarding the correlations of this sedimentary succession. This is attempted in section 2.5. and the evidence which is available as a basis is summarized below.

Many of the stratigraphical details are due to LAUGE KOCH's early work between 1916 and 1923 when he travelled over much of the Thule Basin area in Greenland from Kap York to Etah (1925, 1926) and recognized units (in progressively younger order) of red sandstone with *Cryptozoon* reefs, yellow sandstone, dolomite, and shale. MUNCK (1941) measured 500 metres of sandstones and conglomerate near Robertson Bay in Prudhoe Land in a summer visit in 1936. TROELSEN (1950b, p. 19) measured 1,000 metres of sandstones and conglomerates, with no observed base or top to his succession, at a locality about 60 kilometres south-east of Etah. KURTZ and WALES visited the Dundas area and published details in 1950; the latest work known to the author is by DAVIES, KRINSLEY, and NICOL (1963).

It is implicit in most geological accounts that the whole Thule Basin and also Inglefield Land (not to mention other areas further afield) could be dealt with as a unit and the sedimentary successions observed in the various localities lumped together in the litho-stratigraphical unit, the Thule Formation or Group. The Thule Group gradually assumed a mantle of chronostratigraphical significance which was perhaps never merited, or intended by early workers. In order to avoid confusing the issues involved it is advisable to follow BERTHELSEN and NOE-NYGAARD in avoiding the generalised use of the term Thule Group or Formation and to resort to other, local, terms until much more evidence has been obtained from the widely scattered outcrops where systematic mapping and section measuring has so far only sporadically been carried out over a long period by different geologists. In this contribution the term Thule Group is confined to the Dundas area.

Only two areas within the Thule Basin have adequate published results—the Dundas area near Wolstenholme Fjord and Prudhoe Land.

2.2.1. The Dundas area

The area around Dundas (Thule Air Force Base) in the vicinity of North Star Bugt has been studied in some detail (DAVIES, KRINSLEY, and NICOL, 1963). It should be noted that these are the southernmost exposures of sedimentary rocks in north-west Greenland, at the south-east end of the Thule Basin and over 200 kilometres in a direct line south-south-west from Etah at the north-west end of the basin.

The Thule Group is divided into three formations which can be briefly categorized as, quartzite (below), black shale, and red beds with dolomite (above). The black shale is an unusual sedimentary type for the Pre-Cambrian of Greenland which according to DAVIES and others (1963, p. 36), appears to be found only in the Dundas area as far north as Inglefield Bredning (the inner part of Murchison Sound).

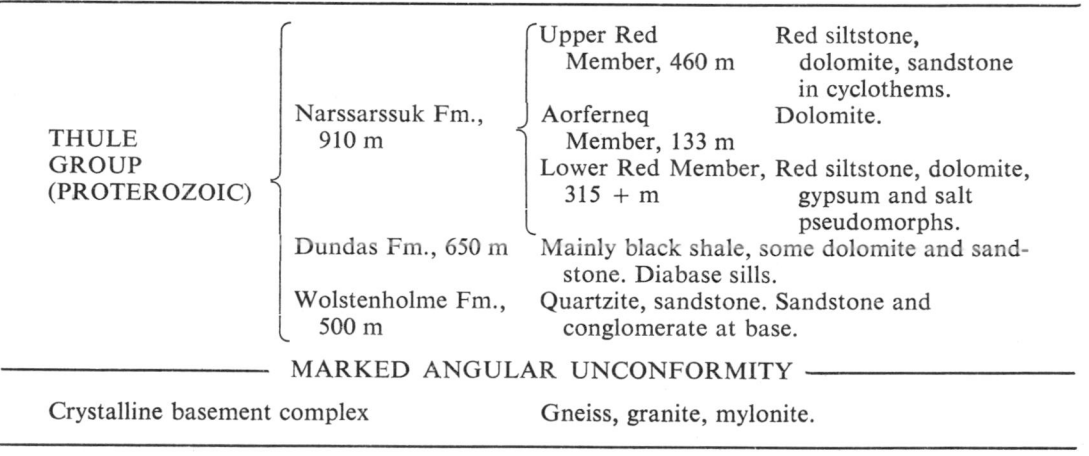

THULE GROUP (PROTEROZOIC)	Narssarssuk Fm., 910 m	Upper Red Member, 460 m	Red siltstone, dolomite, sandstone in cyclothems.
		Aorferneq Member, 133 m	Dolomite.
		Lower Red Member, 315 + m	Red siltstone, dolomite, gypsum and salt pseudomorphs.
	Dundas Fm., 650 m	Mainly black shale, some dolomite and sandstone. Diabase sills.	
	Wolstenholme Fm., 500 m	Quartzite, sandstone. Sandstone and conglomerate at base.	

———————————— MARKED ANGULAR UNCONFORMITY ————————————

Crystalline basement complex   Gneiss, granite, mylonite.

No fossils have been found in these strata (or in loose blocks) and no radiometric dates have been published for the succession or for the mafic dykes and sills which cut both the crystalline basement complex and the overlying sediments and occur mainly in the fissile shales of the Dundas Formation. The igneous rocks are petrographically similar to the dykes and sills of Inglefield Land.

2.2.2. Prudhoe Land

In a paper by LAUGE KOCH published in 1926 many details are given in maps and sections concerning the whole Thule Basin. His publications provide details of the ice-free strip of land along the inner parts of the fjords and the outer parts of peninsulas of the northern shores of Murchison Sound from Thule to Etah (Fig. 3)—Prudhoe Land.

In this latter area the sequence is given as red sandstone, yellow sandstone, dolomite and shale from older to younger subdivisions of the sedimentary sequence. The north-west end of LAUGE KOCH's maps and sections includes Kap Alexander which is only 17 kilometres from Etah. Since these visits by LAUGE KOCH in the early years of this century few geological results have been published concerning Prudhoe Land.

In 1957, the author did not land in the area to the south of Foulke Fjord (Etah lies on its northern shore) but the sequence was thought without doubt to belong to the Rensselaer Bay Formation and was provisionally mapped accordingly (COWIE, 1961, p. 5, fig. 1). The return motorboat journey from Thule (Kanak) (Fig. 2) to Etah by the author in 1957 was made close to the headlands and outer coast so that lithologies and succession could be made out. There seemed to be no reason to postulate a radical change in the general character of the exposed sedimentary successions proceeding along the northern shores of Murchison Sound from Thule (Kanak) to Etah and around to Kap Ingersoll on Inglefield Land.

It can be reasonably assumed in LAUGE KOCH's early work in the Thule Basin that his references (in his written descriptions, maps and sections) to 'red sandstone', 'yellow sandstone', and 'dolomite and shale' were primarily concerned with lithological descriptions at an early stage of reconnaissance. Later, he and others came to believe that the strata were of the same age and that these descriptive units could be applied throughout the Thule Basin; recent work suggests that this may not be true (see page 349).

CHRISTIE (1962a) reports a series of sandstone, shale, tuff and volcanics which is 620 metres thick, intruded by diabase dykes and resting unconformably on crystalline Pre-Cambrian rocks on eastern Ellesmere Island south of Baird Inlet between latitudes 78° N. and 78° 15' N. These sedimentary rocks are overlain by 1,240 metres of white and pale brown sandstone, red shale, and thick lava beds. From their lithology and stratigraphical position they seem to be of Proterozoic type and CHRISTIE assigns them a late Pre-Cambrian age and correlates them with the Dundas area succession. In Fig. 3 this occurrence is included in the Thule Basin while the other possible areas in Ellesmere, Devon, and Baffin Islands mentioned on page 337 are excluded. ? Symbols and a dashed margin in Fig. 3 suggest a possible further extension of the basin to the south-west.

2.3. Bache Peninsula area, Ellesmere Island

The sedimentary formations exposed on Bache Peninsula, the neighbouring areas of Knud Peninsula, and the shores of Flagler Bay have for a number of reasons received a good deal of attention from geologists compared with other parts of the Kane Basin area. Ellesmere Island to the south and the north of the peninsula is heavily glaciated. This, in conjunction with its position near Smith Sound at the entrance to Kane Basin and its proximity to Greenland, has made Bache Peninsula a focal point for expeditionary travel, and a gateway for the overland route through the Sverdrup Pass westwards to the ice-free land of Ellesmere Island and Axel Heiberg Island. During the present century the area has been visited by many interested geologists: SCHEI, BENTHAM, WORDIE, DREVER, TROELSEN, CHRISTIE, KERR, and COWIE.

The Bache Peninsula Arch is part of the Central Stable Region which can be followed from Devon Island through south-eastern Ellesmere Island to Bache Peninsula and Flagler Bay and indeed across Smith Sound into Inglefield Land. The Bache Peninsula Arch is a 'broad, mildly positive basement arch that extends across the Central Stable Region from Greenland to Ellesmere Island.' (KERR, 1967a, p. 6). During much of later Pre-Cambrian time this arch was probably a site for erosion whilst in early

Phanerozoic time deposits were thinner along its crest than in flanking regions; westwards the arch can be followed into the Franklinian miogeosyncline by variations in
facies and thicknesses.

The Pre-Cambrian crystalline basement rocks consisting of gneiss, pegmatite,
granite, and crystalline limestone are overlain unconformably by sedimentary formations with gentle dips and only minor faulting. By correlations with other areas, discussed
later, it can be suggested that the lowest sedimentary strata may be Cambrian; fossils
of Lower and Middle Cambrian age are found higher in the succession. The Middle
Cambrian is followed without visible break by Lower Ordovician strata. The most
complete account is given by CHRISTIE and the following composite succession is taken
from CHRISTIE (1967) and KERR (1967a).

Age	Formation and thicknesses in metres	Lithology
Lower Ordovician	Cass Fjord, 360–480	Carbonate rocks and gypsum.
Middle Cambrian	Cape Wood, 40	Dolomite, limestone, sandstone. Intraforma-tional conglomerate.
Lower Cambrian	Cape Kent, 15	Dolomite, dolomitic limestone.
Lower Cambrian	Police Post, 4·5	Limestone, sandy limestone, sandstone.
? Lower Cambrian	Cape Ingersoll, 74	Vuggy dolomite.
? Lower Cambrian	Cape Leiper, 30–45	Dolomite.
? Lower Cambrian	Rensselaer Bay, 50–200	Sandstone, arkose, conglomerate, shale dolomite.

———————— MARKED ANGULAR UNCONFORMITY ————————

Pre-Cambrian		Gneiss, pegmatite, granite, crystalline limestone.

Rensselaer Bay Formation

This formation consists of unfossiliferous and mainly arenaceous beds of clastic
origin which are restricted to the area around Bache Peninsula and Inglefield Land.
It was deposited only on the Central Stable Region and rests unconformably on the
crystalline basement complex. It is divided into three members, shown on page 341.

The two older members, 1 and 2, are restricted in outcrop and in most parts of the
area outcrops of the Sverdrup Member appear as the basal stratum resting on the
Pre-Cambrian basement complex. CHRISTIE (1967, p. 14) suggests that the Sverdrup
Member is transgressive; in that case, as shown in section in his figure 2, there may be
simple overlap of the lower members by the upper. Lateral variation is nevertheless
possible. KERR (1967a, p. 29) suggests attenuation, pinching out, and overlap of the
Sverdrup Member as outcrops are traced westwards from the easternmost exposures

3. Sverdrup Member, 50–90 m	Sandstone.
2. Bache Peninsula Member, 0–27 m	Arkose, conglomerate, sandstone.
1. Camperdown Member, 0–80 m	Sandstone, shale, carbonate rocks with algal traces.

on Ellesmere Island, near Cape Camperdown. The possibility of a gap between the times of deposition of the lowest beds of the Sverdrup Member and of the highest beds of the Bache Peninsula Member is perhaps also worth noting; conformable relationships are recorded by CHRISTIE and also in places a 'rather abrupt contact' between the two members, but with some apparent gradation or interfingering on weathered surfaces which may reflect sedimentary gradation (1967, pp. 17–21). An important feature of the Sverdrup and Bache Peninsula Members is the presence in the sandstone of scolithid-like tubes or pipes which in so many parts of the world seem to be associated with the advent of the Cambrian System.

In the major part of its outcrop the Rensselaer Bay Formation is represented by the Sverdrup Member only, resting unconformably on a gneissic basement complex; and outcrops in this way extend outside the Bache Peninsula–Flagler Bay area, and are found in Sverdrup Valley east of Irene Bay and southwards from there near the ice-cap.

Lower Cambrian carbonates

The Cape Leiper, Cape Ingersoll, Police Post and Cape Kent Formations are thin but persistent, predominantly carbonate rocks which are conformable to one another. KERR (1967a, p. 36), suggests that some erosion or a period of non-deposition may have occurred between the Cape Ingersoll Formation and the Police Post Formation. In addition to the Bache Peninsula–Flagler Bay area they occur with slightly greater thickness south of Irene Bay and to the west of the ice-cap between the rivers leading into Strathcona and Vendom Fjords.

Cape Leiper Formation

This is a yellow or grey, fine-grained dolomite with little lateral variation, resting conformably on the Rensselaer Bay Formation. It is a competent cliff-forming series which is fairly thinly-bedded (1–3 metres), is locally petroliferous, and has strongly developed stylolites and beds of intraformational conglomerate or breccia. No fossils have been found.

Cape Ingersoll Formation

Of reddish-brown weathering, medium to coarsely crystalline, grey-brown dolomite; the formation rests with sharp conformable contact on the Cape Leiper Formation. It is characteristically vuggy to cavernous with mottled weathering, some breccia, and white carbonate blebs which CHRISTIE (1967, p. 25) ascribes to remnant or pseudofossil forms: otherwise the formation shows no indications of fossils. KERR (1967a, p. 36) notes that the vugs are filled with white crystalline dolomite aligned along bedding planes. There seems to be little lateral variation within the formation.

Police Post Formation

This thinly-bedded, glauconitic, arenaceous limestone is dark grey in the middle, becoming green near the top and the bottom of the formation. It has abundant worm trails and thin pyrite stringers. This thin unit has yielded the first undoubted Lower Cambrian fauna of this area and may rest with slight unconformity on the Cape Ingersoll Formation (either due to erosion or non-deposition). The nature of the beds in this part of the Cambrian succession indicates extremely shallow water conditions with periodic emergent movements. The Lower Cambrian carbonates become more calcareous and less dolomitic above the Cape Ingersoll Formation. Collections over the last seventy years or so were made by SCHEI, BENTHAM, DREVER, TROELSEN, and CHRISTIE. POULSEN described and illustrated in 1946 some of the specimens collected by SCHEI and BENTHAM. The latter explorer's fossils were obtained from scree and were preserved in a conglomeratic limestone thought to have come from a 'sandstone, grit, conglomerate series' near the Royal Canadian Mounted Police post on Bache Peninsula (POULSEN, 1946, p. 301). This was termed the *Bonniopsis* horizon: the exact stratigraphical provenance is far from certain but seems likely to be the Police Post Formation as set up by TROELSEN (1950b, p. 38). The fauna is: *Paedeumias? borealis* POULSEN; olenellid indet.; *Bonniopsis nasuta* POULSEN; *B. rostrata* POULSEN; *Acrothele? pulchra* POULSEN; and *Hyolithes* (*Hyolithes*) sp. The collections made by CHRISTIE have been described (COWIE, 1968) and are partly listed by CHRISTIE (1967, p. 30) and KERR (1967a, p. 37). They comprise *Bonniopsis* sp.; *Paedeumias turmalis* COWIE; olenellid indet.; *Hyolithes* sp.; and *Circotheca* sp. They came from the basal 3 metres of the Police Post Formation and were *in situ*. This reliably located fauna gives a Lower Cambrian age which is probably not early in that subdivision; it seems to be contemporaneous with the collections described by POULSEN.

Cape Kent Formation

Medium-grained, thick-bedded, pale brown dolomitic limestone or dolomite which weathers brown (light, reddish, or orange) makes up the formation. It is sometimes oolitic, has mottled weathering, and is arenaceous. It is conformable to the Police Post Formation below and the Cape Wood Formation above. In Inglefield Land the Cape Kent Formation is fossiliferous and can be assigned to the later part of the Lower Cambrian but no fossils are found in the beds in the Bache Peninsula–Flagler Bay area which are correlated across Smith Sound on the basis of lithology and stratigraphical position in a closely comparable succession. The Scoresby Bay Formation found in neighbouring parts of Ellesmere Island (see p. 345) contains fossils of Cape Kent Formation age and is correlated by CHRISTIE and KERR with the Cape Kent Formation of Bache Peninsula and there seems little doubt that the latter is of late Lower Cambrian age.

Cape Wood Formation

Medium-grained, grey-brown, brown-weathering, petroliferous dolomite and limestone can in this formation be massive, mottled, and cavernous. The lithologies are variable and locally sandy and a grey, calcareous sandstone (1·5 metres thick) marks the base. The lower boundary of the formation is conformable, but the upper contact with the overlying Lower Ordovician Cass Fjord Formation is subject to different interpretations: CHRISTIE comments (1967, p. 33) that it is conformable and abrupt on Bache Peninsula while KERR (1967a, pp. 42–3), apparently referring to the Police Post outcrops on Bache Peninsula, also suggests a probable disconformity which may be coextensive with a widespread disconformity eliminating Upper Cambrian strata. Fossils collected from talus by BENTHAM were described by POULSEN (1946). A lower horizon, the *Glossopleura*, gave: *Clavaspidella sp.; Glossopleura* cf. *expansa* POULSEN; *G. longifrons* POULSEN; and *G. walcotti* POULSEN; whereas an upper, the *Blainiopsis* horizon, gave: *Blainiopsis benthami* POULSEN; *B. holtedahli* POULSEN; *B. scheii* POULSEN; *B.* ? sp.; *Elrathia* ? sp.; *Kochina arctica* POULSEN; and *Olenoides* ? *fallax* POULSEN. These both indicate a Middle Cambrian age and were separated into the two assemblages because the fossils 'originate from two lithologically and palaeontologically different horizons' (POULSEN, 1946, p. 301). The *Elrathiella* horizon which is found in north-west Greenland between the *Glossopleura* and *Blainiopsis* horizons has not been found on Bache Peninsula. TROELSEN (1950b, pp. 43–46) apparently collected both the *Glossopleura* and *Blainiopsis* faunas and distinguished two members. An upper one, the Blomsterbaek Member, has a thin conglomerate at its base and contains fossils of the *Glossopleura* assemblage; the lower is the Cape Russell Member which TROELSEN gives (*op. cit.*, p. 43) as the equivalent of the *Blainiopsis*-horizon. In the summer of 1968 a field party of the Geological Survey of Canada (led by R. L. CHRISTIE and accompanied by J. W. COWIE) made new collections from the upper part of the formation; these are being studied by the author (CHRISTIE, 1969, p. 234). The Cape Wood Formation is a littoral facies of the Parrish Glacier Formation (p. 48), the latter by its shaly character suggesting a deeper water, geosynclinal, association.

2.4. Franklinian Miogeosynclinal area of central and eastern Ellesmere Island

In the Franklinian Miogeosyncline, which lies north and west of the Central Stable Region and trends east and north-east through the Queen Elizabeth Islands, (Fig. 2) the Cambrian has a different facies from that shown by cratonic sediments of the Central Stable Region in north-west Greenland and Bache Peninsula. The Franklinian miogeosyncline is separated from the Central Stable Region by a flexure. This large area has recently been described by KERR (1967a and b), and on the basis of sections from localities 1–5 (Fig. 3) the following stratigraphical scheme can now be suggested:

Age	Formation and range of thicknesses in metres	Lithology
Lower Ordovician	Copes Bay, 500–1,450	Limestone, dolomite, conglomerate.
REGIONAL DISCONFORMITY		
Middle Cambrian	Parrish Glacier, 260–640	Limestone, dolomite, sandstone.
Lower Cambrian	Scoresby Bay, 260–860	Dolomite, limestone, shale.
Lower Cambrian & ? Lower Cambrian	*Ellesmere Group* — Kane Basin, 85–400	Siltstone, sandstone, phyllite.
	Rawlings Bay, 280–920	Sandstone, conglomerate.
	Ritter Bay, 330–420	Phyllite, slate.
	Archer Fjord, 90	Sandstone.
ANGULAR UNCONFORMITY		
Pre-Cambrian	Ella Bay, 150–760	Dolomite.

Ella Bay Formation

This formation is a resistant predominantly dolomitic formation of presumed late Pre-Cambrian age which occurs widely in central north Ellesmere Island. It is overlain unconformably by the Ellesmere Group and frequently rests on a thrust plane so that the full thickness is not seen.

Ellesmere Group

A clastic sequence widely exposed on the shelf and miogeosynclinal areas of central and eastern north Ellesmere Island which includes (KERR, 1967a, p. 18) not only the formations shown on page 343 but also the Rensselaer Bay Formation (p. 331) of the Central Stable Region. The group is almost entirely of marine clastics and rests unconformably on older rocks with angular discordance, and a basal conglomerate or sandstone; the miogeosynclinal development is thickest in the north near Ella Bay at localities 1 and 2 and decreases south-westwards through localities 3 and 4 to locality 5 at Irene Bay (Fig. 3). A gradation from pure quartzose sandstones of the shelf (Bache Peninsula) to shaly sands and black phyllites deep within the geosynclinal belt, as seen in the north (Ella Bay), probably reflects increasing water depth and a greater distance from the source of quartz sand, which seems to have been to the south-east. Deposition was more rapid in the geosyncline and probably began earlier. Emergence of the Central Stable Region along the Bache Peninsula Arch commenced in pre-Ellesmere Group times; this is suggested by the angular unconformity between the group and the crystalline basement complex and overlap relationships seen at Bache Peninsula and illustrated by CHRISTIE (1967, fig. 4). This emergence may have persisted for some part of Ellesmere Group time and caused differences in thickness between the shelf and the miogeosyncline reflecting either non-deposition at times or slower deposition at others; however, it seems that widespread deposition prevailed in later Ellesmere Group time but was slower near and on the shelf. The upper contact of the group is apparently conformable with overlying formations on the shelf and in most of the geosynclinal exposures except in localities 1 and 2 near Ella Bay where there is evidence of local erosion.

Archer Fjord Formation

This is a quartz sandstone with pebbly sandstone at the base. An angular unconformity is seen in the type area to which the formation is confined and it may be a rather local, coarse clastic deposit: a tongue of sediments derived from the shelf to the east-south-east. No fossils or trace fossils have been found in the formation.

Ritter Bay Formation

The formation interfingers between the Archer Fjord Formation and the Rawlings Bay Formation at locality 2. In north-east Ellesmere Island it includes phyllites and minor slates which are dark grey to black and weather medium grey with a white bloom which is an aid to distant observation. Metamorphism is of a low grade. No fossil remains or trace fossils have been found.

Rawlings Bay Formation

Mainly quartz sandstone and pebble conglomerate. Near the shelf, for example at Copes Bay, it contains coarse pebble conglomerates and rests with angular unconformity on older rocks; followed northwards its thickness increases and the pebbles become smaller and less numerous. Near the Kennedy Channel its lower parts grade into the Archer Fjord Formation and the Ritter Bay Formation. The formation is widely exposed in the miogeosynclinal areas of eastern and central Ellesmere Island. It is conformable to both overlying and underlying formations.

Indications of organisms are found at many levels throughout the formation and include algal stromatolites, worm trails, burrows and tubes such as the abundant *Scolithus linearis*, and probable trilobite trails. *Cruziana jenningsi*, for example, occurs about 90 metres from the top of the formation at locality 1 (Fig. 3) and is usually considered to be with little doubt a trilobite trail (or egg burrow?).

Kane Basin Formation

This formation of fine grained, quartzose sandstone, which is thinner nearer the shelf, thickens and grades to black shale and phyllite to the north-west as it is followed into the deeper geosyncline. It is conformable on the Rawlings Bay Formation, and with the overlying Scoresby Bay Formation except in the north where there is a local angular unconformity at its top.

At locality 1 (Fig. 3) *Olenellus* sp. has been collected from about 325 metres above the base of the formation which is there 400 metres thick. At locality 3, *Cruziana jenningsi*, probable trilobite trails, worm trails and burrows, and other possible organic markings occur at about 10 metres above the base of the 160 metres thick formation. At the same locality at 130 metres above base occur well-preserved specimens of *Olenellus praenuntius* COWIE (COWIE, 1968). No other trilobites have been found in the formation and the Kane Basin Formation can be placed in the early part of the Lower Cambrian.

The Ellesmere Group stratigraphical relationships show that clastic strata of south-easterly provenance thicken and become shaly to the north-west and onlap the shelf region. The Archer Fjord Formation is a basal conglomeratic tongue restricted to the deeper parts of the geosyncline. Shelfward it merges with the rudaceous and arenaceous Rawlings Bay Formation. In deeper parts of the basin lower parts of the Rawlings Bay Formation grade to phyllite of the Ritter Bay Formation and upper parts grade to phyllite of the Kane Basin Formation. The Kane Basin Formation is a thin quartz sandstone in the south-east but thickens and becomes more shaly in the miogeosyncline where its age span is probably greater (KERR, 1967a and b).

Scoresby Bay Formation

Mainly consisting of cliff-forming dolomite which may be vuggy, this formation increases in thickness to the north-west. Stromatolites occur near the base at locality 5. It is conformable on the Ellesmere Group except in northerly exposures where it is locally unconformable. The Middle Cambrian Parrish Glacier Formation succeeds conformably. Sections at the different localities 1–5 indicate a progressive thickening from the region adjacent to the shelf at Irene Bay (locality 5) into the Franklinian miogeosyncline east of Ella Bay (locality 1) but the formation remains fairly uniform in lithology although towards the north bedding is thinner, colour is darker, shale and limestone intercalations slightly increase, and vugs decrease in number.

Fossils are confined, so far, to locality 5 where at 106 metres from the base of a 260 metre thick succession of the formation are *Bonniopsis*? sp., *Fremontia* sp., *Paterina* sp., *Hyolithes* sp. indet., and ptychoparioid gen et sp. indet. The association of olenellids with other trilobites suggests an assignment of the fauna to the later, rather than earlier, part of the Lower Cambrian.

Parrish Glacier Formation

Varied rock types are included in the formation: alternating limestone, shale, and silt-stone, with flat-pebble conglomerates, arenaceous dolomite, and sandstone; the sequence is characteristically brightly coloured. In addition to the outcrops in central and eastern parts of Ellesmere Island, which have been under discussion in relation to older formations, it is found south of Bay Fjord, to the east of Trold Fjord (78° 30′ N. approx., Fig. 3). It is conformable on the Scoresby Bay Formation and is overlain with presumed regional disconformity by the Copes Bay Formation, which is probably Lower Ordovician although the contact is not seen at the type locality.

Specimens of a new genus of bathyuriscid trilobites which may be related to *Clavaspidella*, found 15 to 24 metres above the base at locality 4, where the formation has a total thickness of 265 metres, suggest an early Middle Cambrian age. It is assumed that Upper Cambrian rocks are absent here as in other arctic and boreal regions (KERR, 1967a, p. 41). The Copes Bay Formation seems to be clearly correlatable with the Cass Fjord and Cape Clay Formation of Bache Peninsula, north-west Greenland and elsewhere (Fig. 4).

2.5. Correlations within the region

The discovery of Lower Cambrian fossils in the Franklinian miogeosynclinal area of Ellesmere Island affects interpretations of the possible downward extension of the Cambrian in arctic regions of North America and, perhaps, elsewhere. KERR has discussed this point at some length in two recent papers (1967a, 1967b). The youngest formation of the Ellesmere Group (Fig. 4)—the Kane Basin Formation—with its olenellid fauna is clearly Lower Cambrian and the underlying formations of the group rest unconformably on the Ella Bay Formation in central and south-eastern parts of the miogeosyncline. It seems justifiable at present to provisionally assign the whole of the Ellesmere Group to the Lower Cambrian and to take the unconformity at its base as the base of the Cambrian because:

1. Trilobite faunas found in the Kane Basin Formation establish the age of the upper part of the Group as Lower Cambrian.

2. *Scolithus linearis* is found in abundance in the underlying Rawlings Bay Formation and this can be considered as evidence of Lower Cambrian age (POULSEN, 1960, p. 11; MOORE, 1962, p. 215).

3. Trilobite trails (*Cruziana*) and other trace fossils ascribed to the activities of trilobites are found in the Rawlings Bay Formation and Kane Basin Formation. It is usually assumed that trilobites are only found in Phanerozoic series.

4. No important depositional breaks occur in, or between, the formations of the Ellesmere Group; it has been, and continues to be, common policy to place the Pre-Cambrian–Phanerozoic boundary at the first significant unconformity beneath the oldest occurrence of diagnostic Lower Cambrian fossils. CLOUD and NELSON's paper (1966) on Phanerozoic–Cryptozoic and related transitions is a recent example of this approach. This point is put forward here not as an argument on grounds of general principle but as a precedent which is apparently not yet outmoded in recent practice. It may well be discarded in the course of current discussions prior to the proposal for an international standard for the base of the Cambrian System by the Cambrian Subcommission of the International Union of Geological Sciences.

5. It is claimed by many stratigraphers that the lowest fossiliferous subdivision of the Lower Cambrian contains a skeletal fauna which includes archaeocyathans and

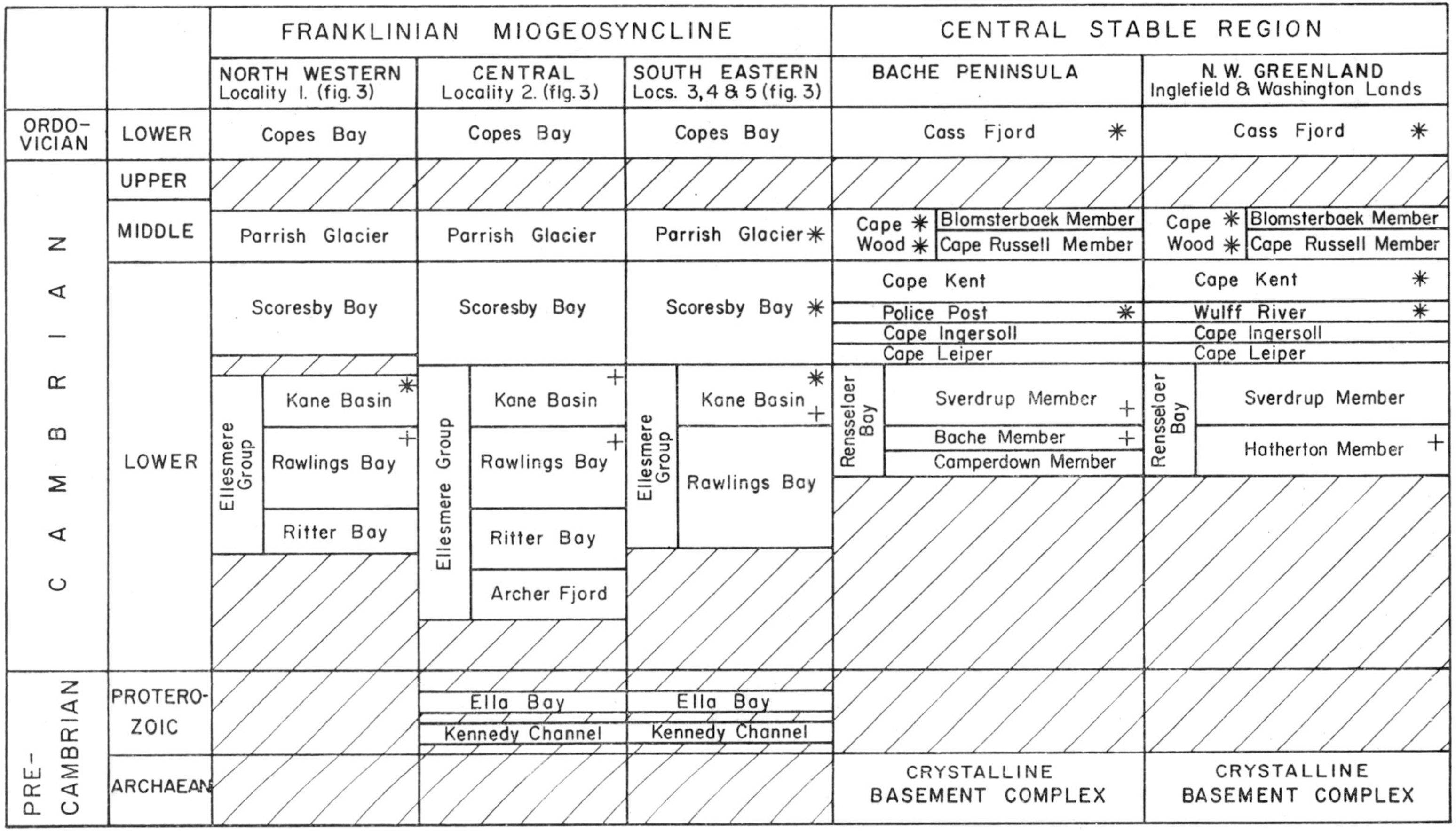

		FRANKLINIAN MIOGEOSYNCLINE			CENTRAL STABLE REGION	
		NORTH WESTERN Locality 1. (fig.3)	CENTRAL Locality 2. (fig.3)	SOUTH EASTERN Locs. 3,4 & 5 (fig.3)	BACHE PENINSULA	N.W. GREENLAND Inglefield & Washington Lands
ORDOVICIAN	LOWER	Copes Bay *	Copes Bay	Copes Bay	Cass Fjord *	Cass Fjord *
CAMBRIAN	UPPER					
CAMBRIAN	MIDDLE	Parrish Glacier	Parrish Glacier	Parrish Glacier *	Cape Wood * Blomsterbaek Member / Cape Russell Member *	Cape Wood * Blomsterbaek Member / Cape Russell Member *
CAMBRIAN	LOWER	Scoresby Bay / Kane Basin *+ / Rawlings Bay + / Ritter Bay / Ellesmere Group	Scoresby Bay / Kane Basin + / Rawlings Bay + / Ritter Bay / Archer Fjord / Ellesmere Group	Scoresby Bay * / Kane Basin *+ / Rawlings Bay / Ellesmere Group	Cape Kent / Police Post * / Cape Ingersoll / Cape Leiper / Rensselaer Bay / Sverdrup Member + / Bache Member + / Comperdown Member	Cape Kent * / Wulff River * / Cape Ingersoll / Cape Leiper / Rensselaer Bay / Sverdrup Member / Hatherton Member +
PRE-CAMBRIAN	PROTEROZOIC		Ella Bay / Kennedy Channel	Ella Bay / Kennedy Channel	CRYSTALLINE BASEMENT COMPLEX	CRYSTALLINE BASEMENT COMPLEX
PRE-CAMBRIAN	ARCHAEAN				CRYSTALLINE BASEMENT COMPLEX	CRYSTALLINE BASEMENT COMPLEX

FIG. 4. Correlation of Cambrian formations around the southern part of Nares Strait, in north-west Greenland and in Ellesmere Island. The table is diagrammatic and not to scale on a spatial or temporal basis and the thicknesses of formations are not represented. Horizontal lines at the same level are meant to be isochronous, however. No attempt is made to indicate degree of certainty of correlation: this is dealt with in the text.

* Fossils. + Trace fossils. The positions of these two symbols are only approximately located with relation to the timescale of the left-hand margin. Archaean and Proterozoic are not used here with a precise time-significance.

brachiopods but no trilobites (ROZANOV, 1967, p. 430; SOKOLOV, 1968. p 32). If this pre-trilobite zone is universally valid then it may be represented in the older formations of the Ellesmere Group. The Rawlings Bay Formation yields trilobite trace fossils; the Ritter Bay and Archer Fjord Formations could represent a pre-trilobite Lower Cambrian zone even though they have not yet been found to contain trace or skeletal fossils. From the point of view of correlations with other areas the important point established is that the Rawlings Bay Formation seems to be clearly Lower Cambrian in age on the basis of trilobite trace fossils such as *Cruziana*.

6. The Ellesmere Group is overlain by the Lower Cambrian Scoresby Bay Formation which yields diagnostic fossils at Copes Bay. This shows that the top of the Ellesmere Group cannot range into the Middle Cambrian.

The general problem of the base of the Cambrian System and the beginning of the Cambrian Period is not discussed here. It is dealt with for these volumes by W. B. HARLAND.

The Scoresby Bay Formation overlying the Ellesmere Group was first described by THORSTEINSSON (1963, p. 386) and referred to the 'Thule Group'. Later work by the Geological Survey of Canda has led to the abandonment of the term 'Thule Group' in arctic Canada and the renaming of the series. The Scoresby Bay Formation which is found in the Franklinian miogeosyncline is now considered to be equivalent in age to the Cape Leiper, Cape Ingersoll, Wulff River, Police Post, and Cape Kent Formations found in the Central Stable Region of Bache Peninsula, Canada, and Inglefield Land, Greenland (Fig. 4). The older formations of this latter region were considered part of the 'Thule Group' by earlier workers including TROELSEN (1950) and THORSTEINSSON (1963) and it is this correlation between the Scoresby Bay Formation at Copes Bay and the Cape Leiper to Cape Kent series of formations at Bache Penninsula which is the crucial correlation affecting the Cambrian of arctic Canada and Greenland, not only in the neighbourhood of Nares Strait (the Smith Sound–Kane Basin–Kennedy Channel–Hall Basin–Robeson Channel strait between Canada and Greenland) but also further afield. Fossil evidence is, so far, lacking in the Cape Leiper and Cape Ingersoll Formations but the correlation with the Scoresby Bay Formation on lithological and successional grounds is firmly established by the mapping of the Geological Survey of Canada. Faunal correlation as well as lithological-stratigraphical correlation is, without much doubt, fully achieved between the upper part of the Scoresby Bay Formation of the miogeosyncline and the Cape Kent Formation of the Central Stable Region by the discovery by KERR in 1962 of Lower Cambrian trilobites in the Scoresby Bay Formation at Irene Bay (locality 5 on Fig. 3). This fossiliferous dolomite unit of the Scoresby Bay Formation at Irene Bay can be mapped in continuous outcrop ('walked into'—KERR, 1967a, p. 8) with the dolomite of the Scoresby Bay Formation at Copes Bay; the dolomites of THORSTEINSSON (1963) are therefore closely correlated and the conformable clastics below are also correlated with the Bache Peninsula succession (Fig. 4).

The successions in the Lower Cambrian of Bache Peninsula in Ellesmere Island have long been closely compared with those of Inglefield Land in north-west Greenland by SCHEI, BENTHAM, WORDIE, TROELSEN, POULSEN, and COWIE. The formational names are, with one exception, shared. KERR has recently shown (1967b, p. 491) that the isopachs of Cambrian sediments show a continuity and congruity when traced across Nares Strait.

In Fig. 3 it can be seen that the relationship between the sedimentary rocks of the Thule Basin and the Central Stable Region is a critical one for the correlation of the older sedimentary sequences of north-west Greenland. In section 2.2.2. (page 339) reasons are given why it seems reasonable to correlate the strata of Prudhoe Land with the lower part of the succession in Inglefield Land (Fig. 10 p. 337). Earlier workers such as SCHEI, KOCH, and TROELSEN always assumed this correlation to be true. It is debatable whether any major change of facies takes place as exposures are followed from Inglefield Land into Prudhoe Land south of the fjord at Etah although solid facts and detailed stratigraphy are not available.

In past studies, however, it has been assumed that the whole of the Thule Basin area shows successions of strata of roughly similar age; the most recent example is the review by BERTHELSEN and NOE-NYGAARD (1965).

KOCH (1929c) grouped together all putative Pre-Cambrian sedimentary rocks which unconformably overlie the peneplained crystalline basement complex under the term Thule Formation (named after the old 'Thule' trading station at North Star Bugt, Wolstenholme Fjord which is now renamed Dundas). This was a lithostratigraphical term without any attempt at chronostratigraphical precision and was applied to the whole region from Kap York to Etah and beyond. It was not until 1922 that LAUGE KOCH referred the Thule district sedimentary series to the Pre-Cambrian; other ages up to Tertiary had been considered, including Cambrian on the basis of correlations with Ellesmere Island influenced by HOLTEDAHL's opinion regarding SCHEI's material from Bache Peninsula. KOCH (1925, p. 304) commented—'The age of the series of strata . . . remained uncertain until July 1922 when I found the series of sandstone in Inglefield Land overlain by Lower Cambrian conglomerates and limestone containing a well-preserved *Olenellus* fauna and separated from the sandstone by a distinct unconformity. Thus it was proven that the sandstone series in the Cape York district is Pre-Cambrian.'

Nowadays it is clearer that an unconformity below beds with an *Olenellus* fauna does not necessarily signify the base of the Cambrian, even though this may be a convenient level to select as part of a working hypothesis for a time (p. 346). Unfossiliferous beds below such an unconformity may be expected to be correlated at a later date with beds elsewhere with newly found Lower Cambrian fossils. Even if an unconformity at the level postulated by KOCH in Inglefield Land is accepted, and there is doubt, it is reasonable to accept a slight unconformity within the Lower Cambrian sequence if evidence requires it. It is now realised, moreover, that the Lower Cambrian may span 30–60 m.y. of the 70–100 m.y. of Cambrian time.

As stated above, there is considerable similarity between the strata in Prudhoe Land and those in Inglefield Land but the sequence in the Dundas area varies from both these as described in sections 2.1 and 2.2. There are more lithological variations, far greater thicknesses, and—compared with Inglefield Land—no overlying fossiliferous Palaeozoic beds. The presence of igneous sills and dykes in all three regions is almost certainly of no correlative value. One of the pertinent facts is that the great thickness of black shales of the Dundas Formation (650 metres in the Dundas area) is not found north of Inglefield Bredning (the gulf at the inner end of Murchison Sound) in Prudhoe Land where the succession is basically different. There seems to be at the present every reason for denying any age equivalence between the sediments of the Dundas area and Prudhoe Land although facies change will need consideration in any new field research.

Another point which may be significant is that neither SOLE MUNCK (1941) or DAVIES and others (1963) mention stromatolites in the Dundas area; these trace fossils are known from the Rensselaer Bay Formation in Inglefield Land and near Etah and seem from old records to occur also in the red sandstone beds of Prudhoe Land; but location tends to be vague. BERTHELSEN and NOE-NYGAARD acknowledge (1965, p. 223) that not much is known about the stratigraphy between Inglefield Bredning and Etah and that in this stretch there is a lithological difference from the Dundas area.

The discovery of Lower Cambrian fossils in the Franklinian miogeosyncline of central and eastern Ellesmere Island, seems to reinforce the inclination, shown by COWIE (1961, p.16), CHRISTIE (1967, pp. 26–29) and KERR (1967a, p. 8, p. 20) to cast doubt on correlation between the earliest sedimentary strata in Inglefield Land (and similarly, therefore, Bache Peninsula) and the succession in the Dundas sector of the Thule Basin. KERR (1967b, p. 487), while retaining an age connotation of Upper Proterozoic for the Thule Group in the Thule Basin and firmly maintaining that the base of the Cambrian lies below the Rensselaer Bay Formation in Inglefield Land and Bache Peninsula, states, after some discussion: 'There is little evidence to suggest a choice between a Late Proterozoic and Cambrian age for the Thule Group of the Thule basin, although their nearness to similar Cambrian rocks on Inglefield Land might suggest they are Cambrian. However because these rocks have been generally regarded in the past as Precambrian they will be referred to in this paper as Upper Proterozoic.'

The evidence regarding occurrence or non-occurence of sills and dykes in various early sedimentary successions in arctic Canada and Greenland is set aside for the present because the data and mapping are inadequte in many areas. The author agrees with BERTHELSEN and NOE-NYGAARD (1965, p. 217) that the cutting of the various sedimentary series by dykes and sills is of 'no great stratigraphic value' because intrusives of widely differing ages occur in the same area or in different sequences in different areas, which may lead to mistaken correlations on these grounds.

It is suggested here as a working hypothesis, until further field research is carried out in the Thule Basin, that there must be two sedimentary series found there which could satisfy the quotation from KERR on more than one count:

1. Sandstones, dolomites, and shales in Prudhoe Land, which by correlation with the contiguous area of Inglefield Land are considered to be of Lower Cambrian age.

2. The Thule Group of the Dundas area which is considered to be of Proterozoic (late Pre-Cambrian) age. The two series are possibly geographically separated by the wide fjord of Inglefield Bredning.

The overall correlations affecting Cambrian and late Pre-Cambrian (Proterozoic) formations are considered further in Chapter 6 where the Proterozoic age of the Thule Group is dealt with in a wider context. The important possibility of a correlation between the Thule Group and the Upper Proterozoic sedimentary series of north-west Baffin Island (see page 358) (BLACKADAR, 1957) was substantiated by LEMON and BLACKADAR in 1963 (table II). The close similarity in lithology between the two regions, which lie slightly over 500 kilometres apart (Fig. 2) is most striking and convincing:

North-west Greenland	North-west Baffin Island		
Narssarssuk Fm., 910 m, siltstone, dolomite, sandstone.	Sandstone, siltstone, mudstone, shale. } upper	} Uluksan Group 2,100 m	
Dundas Fm., 650 m, shale, dolomite, sandstone.	Dolomite, shale—lower		
Wolstenholme Fm., 500 m, quartzite, sandstone.	Equalulik Group, 1,500 m, quartzite, volcanics.		

The Equalulik and Uluksan Groups are dated on radiometric grounds as Upper Proterozoic and Pre-Cambrian, on any acceptable time-scale, (p. 359). Other Proterozoic sedimentary series which can be correlated with the Thule Group are the Kennedy Channel and Ella Bay Formations (KERR, 1967b)—see Figs 4 and 10.

If the Thule Group near Dundas is Proterozoic and if the sedimentary series of Prudhoe Land are Lower Cambrian like the earliest strata in Inglefield Land then perhaps the Bache Peninsula Arch (page 339) was influencing sedimentation as suggested by KERR (1967b, p. 489) but in the following manner:

1. During Proterozoic time the Franklinian miogeosyncline was subsiding (the Kennedy Channel and Ella Bay Formations were then deposited) and so was the Dundas area of the Thule Basin (the Thule Group was then deposited), but the Bache Peninsula Arch (Bache Peninsula–Inglefield Land–Prudhoe Land) was emergent and received no sediments.

2. In Lower Cambrian time the Dundas area of the Thule Basin ceased to receive sediments but deposition continued in the Franklinian miogeosyncline, the Central Stable Region, and the Prudhoe Land part of the Thule Basin. Alternative theories are possible in line with KERR's suggestions (*op. cit.*).

The Lower Cambrian rocks are succeeded in arctic Canada and Ellesmere Island by fossiliferous Middle Cambrian rocks but sedimentation seems to have been terminated in the later part of Middle Cambrian times before the *Bolaspidella* Zone. The succeeding strata are Lower Ordovician, also suggesting a period of non-deposition or possible erosion during late Middle Cambrian and Upper Cambrian times. The Middle Cambrian Parrish Glacier Formation in the miogeosyncline has a new genus and species of bathyuriscid trilobite related to *Clavaspidella* which can be correlated with the fauna, including *Clavaspidella*, which is assigned to the *Glossopleura* Zone of the Pacific Faunal Province and is represented in the Cape Russell Member of the Cape Wood Formation in the Central Stable Region.

The Cape Wood and Parrish Glacier Formations are not lithologically closely similar. The variability and sandy character of the Cape Wood Formation of Inglefield Land and Bache Peninsula suggests that it is a littoral or shallow-water facies of the shaly, miogeosynclinal Parrish Glacier Formation.

3. The Cambrian of southern Ellesmere Island and the rest of the Canadian Arctic Archipelago

The islands flanking Jones Sound (Fig. 5) and the north side of the Parry Channel (Fig. 2) have been explored since the early days of arctic travel in the seventeenth century when Baffin sailed north in 1616. Prior to that it is possible that this area was traversed by Norse from Greenland and, of course, Eskimo tribes have hunted in the area before that and since. Early geological work was, not surprisingly, relatively sparse but mention should be made in the context of Cambrian formations of SCHEI (Norwegian *Fram* Expeditions of 1898–1902) at the beginning of the century: BENTHAM, WORDIE, and TROELSEN followed in the late thirties and early forties of this century; and PREST, KURTZ, and WALES, the geologists of 'Operation Franklin' of the Geological Survey of Canada, CHRISTIE, and COWIE in the last two decades.

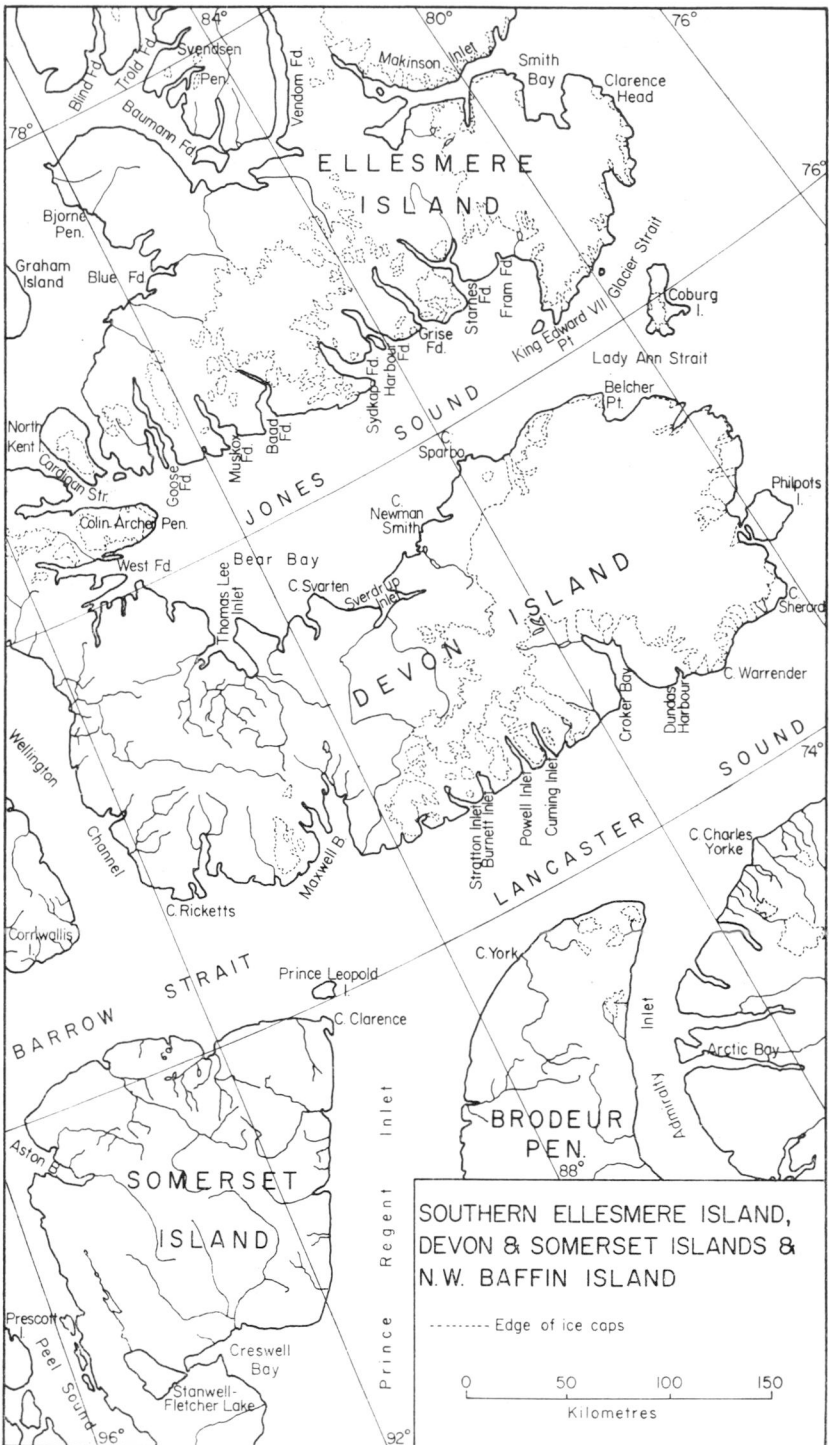

F IG. 5. Southern Ellesmere Island, Devon and Somerset Islands, and north-west Baffin Island.

3.1. Southern Ellesmere Island

From published data it seems probable that Cambrian strata of the Central Stable Region and also, perhaps, of the Franklinian miogeosyncline (Fig. 2) occur below the western parts of the icecaps of south-eastern Ellesmere Island. These strata are known to outcrop in ice-free areas and one locality is described west of the eastern icecap between the Bache Peninsula–Sverdrup Pass–Bay Fjord line and the south coast of Ellesmere Island.

KERR (1967a, p. 40 and fig. 2) mentions the occurrence of beds assigned to the Parrish Glacier Formation east of Trold Fjord about 20 kilometres north of Starfish Bay. The formation is here divisible into a lower unit, which is mainly sandstone with minor shaly limestone, and an upper unit of sandstone and siltstone with minor beds of dolomite and flat-pebble conglomerate. At this locality the Copes Bay Formation overlies, probably disconformably, the Parrish Glacier Formation.

Lower Ordovician outcrops of Copes Bay Formation are recorded from Starfish Bay near Trold Fjord (Fig. 3) on a map (1100A) produced by Fortier and others in 1963 to accompany Memoir 320 of the Geological Survey of Canada on 'Operation Franklin.' Areas elsewhere in Svendsen Peninsula between Vendom and Trold Fjords are given as probably mainly Lower Palaeozoic.

Near the south coast of Ellesmere Island between Sydkap Fjord and Baad Fjord an area of Cambrian and/or Ordovician outcrop is given on Map 1100A of the Geological Survey of Canada; the strata here include limestone and limestone conglomerate, may include some Proterozoic, and unconformably overlie the Pre-Cambrian metamorphics (NORRIS, 1963, pp. 29–31). CHRISTIE in 1962 showed undifferentiated outcrops of 'Cambrian, Ordovician, Silurian' sedimentary rocks in a preliminary map with descriptive notes of south-east Ellesmere Island, based on field work in 1960. This was followed by further investigations in the area in 1968 by CHRISTIE and COWIE referred to in a report by CHRISTIE (1969, p. 234), when further fossil collections from Cambrian beds were made in southern Ellesmere Island and Devon Island. Descriptions of Cambrian formations and fossil faunas from this area are planned for future publications of the Geological Survey of Canada. Earlier workers in this south coast region, including SCHEI who spent a considerable time in the area at the beginning of this century, have suggested that the rocks overlying unconformably the Pre-Cambrian crystalline basement resemble the sequence on Bache Peninsula which they also had seen, and that a part of the sedimentary succession is of Cambrian age. In the writer's experience in 1968 the resemblance is not close. Lithological correlations were made by WORDIE (1938, p. 399) who suggested the presence of 'Thule' sandstone, possibly Rensselaer Bay Formation, above Craig Harbour in the area of King Edward VII Point (Fig. 5). Also BENTHAM (1941, p. 37) suggested the presence of 'Thule beds' and Cambrian sediments near Fram, Grise, and Harbour Fjords (Fig. 5). BENTHAM and WORDIE seem to have seen no evidence of unconformity within the sedimentary sequence immediately overlying the crystalline basement complex. Discussion of these earlier results, which was perforce pursued without palaeontological evidence, was undertaken by FORTIER (1963, pp. 303–309).

3.2. Devon Island

The Cambrian in Devon Island can be considered part of the Central Stable Region (Fig. 2) but putative Cambrian in the Parry Islands to the west belongs to the Franklinian miogeosynclinal belt.

Published results from Devon Island indicate fossiliferous Cambrian beds from only one comparatively small area around Dundas Harbour described by KURTZ, MCNAIR, and WALES in 1952. Further field work (COWIE, 1962) in this and other areas of Devon Island suggested that it is possible to correlate the formations set up by KURTZ and others for the Dundas Harbour area with successions found over wide tracts near the north and south coasts of central Devon Island. Fossils collected in 1961 from Dundas Harbour and other parts of the island will be described in a forthcoming paper. It is considered (COWIE, 1962, p. 255) that no Pre-Cambrian sedimentary rocks are present on Devon Island west of the main icecap. Philpots Island (Fig. 5), a peninsula or tidal island of eastern Devon Island, and possibly other neighbouring areas as well, may include sedimentary rocks (page 337). TAYLOR (1956, pp. 35–36) records earlier observations: (a) at a distance, from the sea, by Inglefield, of stratified rocks; (b) by PHILPOTS, who landed, that the centre of the island is a swampy plain with scattered boulders of hard red stone; and (c) his own conclusion from aerial photographs that isolated uplands look like remnant tablelands and may be sedimentary residuals. These features may possibly be formed in Thule Group or other sedimentary series. Cambrian rocks occur in the north of Devon Island (Plate 3) in considerable outcrops from near Cape Sparbo south-westwards to the area drained by the rivers running into Sverdrup Inlet. In ice-free areas along the south coast many of the lower valley slopes are formed in Cambrian strata from Dundas Harbour to the area between Stratton Inlet and Maxwell Bay (Fig. 5). Away from the Dundas Harbour–Croker Bay area the major part of the Cambrian succession in the rest of Devon Island becomes much more dolomitic and less fossiliferous so that from the point of view of palaeontology and biostratigraphy the south-eastern Devon Island outcrops seem to be the most valuable. In the following succession from Dundas Harbour the thicknesses of the formations set up by KURTZ, MCNAIR, and WALES have been revised to accord better with observations in 1961. The original estimates from 1952 are shown in brackets, and the age of the Bear Point Limestone is extended into the Lower Cambrian, otherwise the information is taken from the paper by KURTZ and others in 1952.

Age	Formation and thicknesses in metres according to COWIE, unpublished m.s. (in brackets MCNAIR, KURTZ, and WALES, 1952)	Lithology
Lower Ordovician	Mingo River 60 (80)	Limestone with minor dolomitic limestone and shale.
Middle Cambrian	Ooyahgah 110 (170)	Dolomitic limestone, shale, sandstone, and limestone.
Lower and Middle Cambrian	Bear Point 150 (240)	Limestone and dolomitic limestone.
Lower Cambrian	Rabbit Point 6 (26)	Sandstone.

———————————————— MARKED ANGULAR UNCONFORMITY ————————————————

| Pre-Cambrian | | Granites, gneisses, and quartzites with diabase dykes. |

The Pre-Cambrian crystalline basement complex is cut by several diabase dykes which are truncated by the marked angular unconformity underlying the Palaeozoic sedimentary rocks.

Rabbit Point Formation

Calcareous and glauconitic sandstone of this formation rests unconformably on the basement complex and is fossiliferous at certain horizons, with *Olenellus* and linguloid brachiopods. The beds also contain abundant vertical tubes of *Scolithus*. The author has identified *Bristolia* from collections made in 1961. This Lower Cambrian clastic formation can be regarded as the deposit of advancing seas that reached Devon Island later than other areas to the north which are considered elsewhere in this contribution. The earlier part of the Lower Cambrian which is found in Ellesmere Island and north-west Greenland is probably unrepresented by deposits.

Bristolia is found in the Cape Kent Limestone and may only occur in younger Lower Cambrian strata, so that a partial correlation between the Rabbit Point Sandstone and the Cape Kent Limestone may be reasonably hazarded at this stage in continuing research.

Bear Point Formation

This more varied succession of limestone and dolomitic limestone with some minor shale and sandstone rests conformably on the Rabbit Point Sandstone. The carbonate rocks contain numerous thin, flat-pebble conglomerates and clastic beds which suggest a shallow water depositional environment with interruptions of sedimentation. Fossils occur at a number of horizons; the basal beds contain *Dolichometopsis* and unlisted representatives of the *Albertella* Zone according to KURTZ, MCNAIR, and WALES (1952, p. 651) but their faunas have not yet been described. *Dolichometopsis* has been found associated with olenellids (COWIE, 1961, p. 24) and V. POULSEN in 1964 (p. 12) stated that the genus is restricted to the Lower Cambrian. Further publication of results must be awaited before age relationships for these basal beds can be considered firmly established; they are here assumed to be late Lower Cambrian on the basis of the published occurrence of *Dolichometopsis*. A reassessment may be necessary if descriptions of the representatives of the *Albertella* zone are forthcoming. Beds higher in the sequence are provisionally accepted as Middle Cambrian.

Ooyahgah Formation

Subdivided (KURTZ and others, 1952, p. 651) into five lithological units, this is a formation of varying lithology with cyclical sedimentation in part and gypsum near the middle. The gypsum suggests the existence of an enclosed or partially enclosed basin during the deposition of these evaporites. The only fossils noted (KURTZ and others, 1952, p. 646) are brachiopods including *Paterina*, a genus which ranges through the Cambrian. (The paterinids range from Lower Cambrian to Middle Ordovician.) The Middle Cambrian age of this formation has not yet, therefore, in published work, been clearly established. The author would agree, however, with KURTZ, MCNAIR, and WALES that it is most likely that Upper Cambrian rocks are not present in the Dundas Harbour area or in the rest of Devon Island. The lower beds of the overlying Mingo River Limestone, which is considered on faunal evidence to be Lower Ordovician in age, are predominantly irregularly bedded limestone with flat-pebble conglomerates which are correlated (*op. cit.*, p. 653) with the uppermost beds of the Cass Fjord Formation of Greenland. Both the Lower Ordovician and the Lower Cambrian are perhaps unconformable in the broad sense of the word on older rocks and the basal strata in each case are younger than those found in Ellesmere Island and north-west Greenland.

Parties of the Geological Survey of Canada visited the central (longitudinally) part of Devon Island during 'Operation Franklin' (FORTIER and others, 1963, pp. 156–256) and outcrops near Burnett Inlet on the south coast (Fig. 5) and near Sverdrup Inlet on the north coast may be Cambrian.

GLENISTER reported 900 metres of marine strata above the Pre-Cambrian gneisses along the shores of Burnett Inlet and referred most of this to the Ordovician System. The 'Lower Ordovician and (?) Earlier' beds include a sequence resting unconformably on the Pre-Cambrian gneisses: partly dolomitic limestone with beds of sandstone and of limestone edgewise conglomerate. A 4·5 metres thick stratum of quartzose sandstone, coarse-grained at the base and medium-grained in the upper part, was observed (GLENISTER in FORTIER and others, 1963, p. 181) resting unconformably on the Pre-Cambrian crystalline basement. This stratum is probably equivalent to the Rabbit Point Sandstone and higher beds above suggest equivalence to the Bear Point Limestone. No fossils are recorded from these beds but Ordovician fossils are found higher in the succession. GLENISTER found that the junction of the sedimentary rocks with the metamorphic basement is commonly concealed by talus in the Sverdrup Inlet area but it is likely that here also he was dealing with a different facies in rocks which are equivalent in age to the Cambrian formations of the Dundas Harbour area.

CHRISTIE and COWIE in 1968 visited the south coast of Devon Island (CHRISTIE, 1969, p. 234) between Croker Bay and the west side of Cuming Inlet. Sections of Palaeozoic rocks (including Cambrian) were measured and fossils collected.

3.3. Cornwallis, Bathurst, Melville, and Victoria Islands

Cornwallis and Little Cornwallis Islands lie within the Franklinian miogeosynclinal region of folded sedimentary strata. They are adjacent to the relatively undisturbed Central Stable Region beds of western Devon Island, which lies eastwards across Wellington Channel, and Somerset and Prince of Wales Islands to the south across Barrow Strait (part of the Parry Channel—Fig. 2). Cornwallis and Little Cornwallis Islands have been surveyed in detail by THORSTEINSSON (1958) and it is almost certain that Cambrian strata do not outcrop. The oldest formation exposed is the Eleanor River Formation which consists of 150 metres of unfossiliferous limestone overlain by the Cornwallis Formation, which is Middle Ordovician as shown by fossils in the middle part and above. THORSTEINSSON tentatively assigns the Eleanor River Formation an Ordovician age.

The Bathurst Island group of islands was investigated during 'Operation Franklin' by parties of the Geological Survey of Canada and most of the area lies within the fold belt of the Franklinian miogeosyncline (Fig. 2). The structure is mainly one of closely spaced anticlines and synclines, locally complex and steeply plunging. In the eroded anticlines the oldest rocks exposed are Ordovician and from fossil evidence there seems little chance of finding Cambrian beds on these islands (FORTIER and others, 1963).

Moving westwards to Melville Island and the neighbouring islands (grouped together as the Western Queen Elizabeth Islands), these have been described by TOZER and THORSTEINSSON (1964). The Canrobert Formation, which includes about 300 metres of limestone, dolomite, and shale is the oldest stratigraphical subdivision in these islands and is exposed as part of the Franklinian miogeosynclinal area (Parry Islands Fold Belt) in north-west Melville Island. In the upper beds of this formation Lower Ordovician Arenig graptolites occur and the underlying unfossiliferous beds could be pre-Ordovician. On published evidence, however, there seems little grounds for suspecting the presence of Cambrian outcrops.

Crossing the Parry Channel from the Western Queen Elizabeth Islands sedimentary strata of the Central Stable Region are encountered again over much of Victoria Island associated with a broad uplift of the Canadian Shield (the Minto Arch) (Figs. 1 and 2). This structure causes inliers of interbedded Proterozoic sedimentary strata, sills, and lava flows. On the flanks of these older rocks in northern Victoria Island Cambrian strata appear to occur and also near to isolated outcrops of Proterozoic in the south-east of the island and along the south coast. The Cambrian of the mainland of Canada to the south of Victoria Island is dealt with by F. K. NORTH in the previous part of this volume. FORTIER in STOCKWELL, 1957 (p. 418) reported Upper Cambrian or Lower Ordovician fossils in float south of the island and an Upper Cambrian fossil from float on the north-east coast of the island itself. At that time the strata from which these loose fossils had come were not located.

The Shaler Group, exposed in the Minto Arch area in western Victoria Island, was assigned to the later Pre-Cambrian by THORSTEINSSON and TOZER (1962); it is made up of five formations which are overlain by basalt flows and intruded by gabbroic dykes and sills. The sedimentary group and the associated igneous rocks are unconformably overlain by sandstone with minor shale and siltstone of map unit 10 (*op. cit.*, pp. 19 and 22) which is up to about 120 metres thick. This unit is structurally conformable with overlying rocks and occupies hollows in the substrata. It was provisionally dated as Upper Cambrian on the basis of chitinoid brachiopods first collected by WASHBURN (1947, p. 33). The faunal list comprises *Iphidella* ? sp., *Lingulella* ? sp., *Dicellomus* ? sp., and *Scolithus* sp. *Iphidella* could be Middle or Upper Cambrian but the possible presence of *Dicellomus* suggests an Upper Cambrian age. The fossil localities are in north-west and south Victoria Island. Map unit 10a appears to be absent around the small inliers of older rocks that are scattered north of Coronation Gulf (THORSTEINSSON and TOZER, 1962, p. 39). The next younger map unit (10b) is mainly dolomite which is much thicker (900 metres) and may range in age from Upper Cambrian (although there is no fossil evidence) to Middle Silurian. Fossils are scarce and difficult to locate in the succession owing to the character of the terrain.

Recent work, including oil exploration, has yielded additional information which is unpublished but F. K. NORTH (this volume, page 313) states that Lower Cambrian strata are now known on Victoria Island.

In 1965 A. H. McNAIR reported abundant brachiopods and other fossils and trace-fossils from western Victoria Island. McNAIR (*pers. comm.*, 1968) remarks 'the beds containing the brachiopods and worm trails occur in Lower Cambrian sediments that were down-faulted into the Precambrian Shaler Group'.

3.4. Boothia Peninsula, Prince of Wales and Somerset Islands

To the east of Victoria Island another uplift of the Canadian Shield, known as the Boothia Arch, affects Boothia Peninsula and Prince of Wales and Somerset Islands (Fig. 2), so that large tracts of Archaean schists and gneisses outcrop and older sedimentary series are found along its flanks. Parts of Somerset and Prince of Wales Islands were investigated during 'Operation Franklin' in 1955 (FORTIER and others, 1963). In north-western Somerset Island two new formations were described for the first time and considered ? Proterozoic: the Aston Formation with 2,100 metres of quartzite and red sandstone and the conformably overlying Hunting Formation with 680 metres of dolomite. Both are unfossiliferous except for possible protozoans in the former and are intruded by basic dykes and sills. The Aston and Hunting Formations and the dykes and sills are overlain by Ordovician and Silurian dolomites of the Allen Bay Formation. Work from 1964 onwards by DINELEY and others has presented further details of the two formations, in particular the fact that the arenaceous, fluvial, or shallow-marine Aston Formation grades upward into the lagoonal Hunting Formation and the presence of stromatolites and worm burrows in both formations. DINELEY (1966, p. 272) remarked on the similarity of the Hunting to the Cornwallis Formation (Middle Ordovician) and that the absence of any break in the Hunting-Palaeozoic succession suggests that the Hunting may be Palaeozoic in age. Dykes and sills are common in the Aston but only a few thin sills are present in the Hunting and no intrusives were found in the local Palaeozoic beds. The evidence is interpreted by DINELEY as giving no certain indication that the intrusives are of pre-Palaeozoic date.

Near the line of latitude 70° N. to the north of Spence Bay on Boothia Peninsula (Fig. 2) Cambrian fossils were collected during 'Operation Prince of Wales' of the Geological Survey of Canada in 1962 (BLACKADAR and CHRISTIE, 1963). They came from the lower beds of a series of weakly cemented sandstone, dolomite, and intraformational breccias and conglomerates (map unit 8) which totals about 90 metres in thickness. Near Kangikjuke Lake the following fossils have been identified (*op. cit.*, p. 9): ? *Hyolithes* sp., cf. *Glossopleura* sp., and cf. *Elrathia* sp. which suggest a Middle Cambrian age and a correlation with the main, higher, part of the Bear Point Limestone of Dundas Harbour on Devon Island. These Middle Cambrian beds were placed above the Aston and Hunting Formations by BLACKADAR and CHRISTIE although the older formations apparently do not outcrop near Spence Bay and were not mapped anywhere on Boothia Peninsula. Map unit 8 is overlain conformably by map unit 9 of dolomite and sandstone containing Ordovician, and probable Silurian, fossils.

The results obtained so far, taking Somerset Island and Boothia Peninsula together, are therefore interpreted (TUKE, DINELEY, and RUST, 1967) as suggesting that the Aston and Hunting Formations comprise a Palaeozoic succession continuous with overlying fossiliferous Ordovician rocks and which includes fossiliferous Middle Cambrian beds in the Spence Bay area of Boothia Peninsula. Clearly the presence of stromatolites in the two oldest sedimentary formations does not necessarily indicate a Proterozoic age. The ages can now be suggested (*op. cit.*, pp. 709–710) as Middle Cambrian (and ? older) for the Aston Formation and Middle Cambrian to Ordovician for the Hunting Formation.

3.5. North-west Baffin Island

In north-west Baffin Island, on Borden and Brodeur Peninsula (Figs. 2 and 5), are sedimentary rocks which are of Palaeozoic age. This area was visited by the Fifth Thule Expedition, 1921–1924, under the leadership of KNUD RASMUSSEN and geological results were subsequently published. Otherwise geological exploration has been carried out mainly by staff of the Geological Survey of Canada, particularly by R. G. BLACKADAR. In Borden and Brodeur Peninsulas the Pre-Cambrian igneous, metamorphic, sedimentary, and volcanic series is unconformably overlain by sedimentary strata which may be Cambrian near the base; higher beds are Ordovician and Silurian in age (TRETTIN, 1965). Research on the Pre-Cambrian rocks of this area has been supported by radiometric dates (BLACKADAR, 1965) and these clearly indicate that an Upper Proterozoic period of sedimentation and vulcanicity is represented there by the Eqalulik and Uluksan Groups. These groups are found in north-east as well as north-west Baffin Island and also near Fury and Hecla Strait in west-central parts of the island. Dykes which cut these Proterozoic rocks and not the unconformably overlying Gallery Formation give K-Ar radiometric dates of 915 and 1,140 m.y. (*op. cit.*, p. 22). The possible correlation of the Eqalulik and Uluksan Groups with the Thule Group of Greenland is discussed on page 351.

The Gallery Formation rests with very low angular discordance on the Pre-Cambrian basement complex and its contact with the overlying Turner Cliffs Formation is structurally conformable but marked by an abrupt change from sandy to dolomitic, shaly, and silty rocks with abundant worm markings. It seems likely that the two formations originated in a similar environment and are not separated by a major disconformity.

No diagnostic fossils have been found in the Gallery Formation but the lack of a significant break at the junction with the overlying Turner Cliffs Formation suggests a Cambrian and/or Lower Ordovician age, while lithology and stratigraphical position

Age	Formation and thicknesses in metres	Lithology
Lower Ordovician and early Middle Ordovician	Ship Point, 45–270	Dolomite; minor conglomerate, siltstone, sandstone, shale.
———————— contact relationships uncertain ? major hiatus ————————		
Lower Ordovician and/or Cambrian	Admiralty Group { Turner Cliffs, 0–300	Dolomite, sandstone, conglomerate; minor siltstone, shale, iron ore.
	Gallery, 0–340	Sandstone; minor siltstone, shale, conglomerate.
—————————— ANGULAR UNCONFORMITY ——————————		
Upper Proterozoic	Uluksan Group, 7,000	Sandstone, siltstone, dolomite. mudstone, shale.
Upper Proterozoic	Eqalulik Group, 1,500–5,000	Quartzite, volcanics.
Middle Proterozoic		Gneiss and migmatite.

prompt a correlation with the Rabbit Point Sandstone (late Lower Cambrian) of Dundas Harbour in Devon Island which lies almost directly across Lancaster Sound (TRETTIN, 1965, p. 6).

The Turner Cliffs Formation, in addition to the worm markings noted above, shows stromatolites. Linguloid brachiopods (*Lingulella* s.s.) have been collected from the lowest strata so that there is a clear case for assigning it a Palaeozoic age. The upper contact with the fossiliferous Lower and Middle Ordovician Ship Point Formation is problematical but regional relationships suggest the presence of a major hiatus. TRETTIN (1965, p. 10) suggests a Cambrian and/or Lower Ordovician age because of the age range of *Lingulella* s.s. (Lower Cambrian to Middle Ordovician) coupled with the age of the overlying

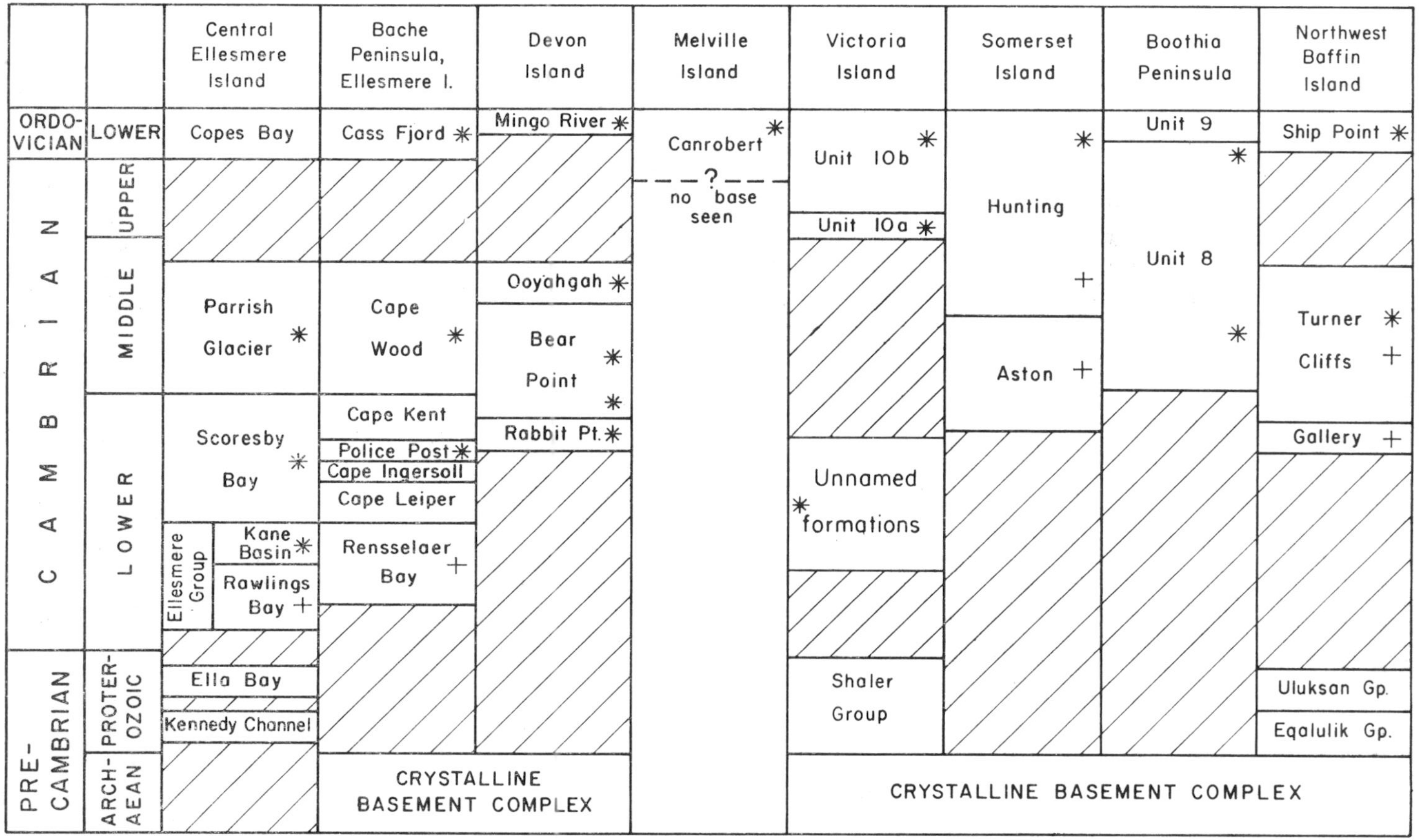

FIG. 6. Correlation of Cambrian formations of the Canadian Arctic Archipelago. The table is diagrammatic and not to scale on a spatial or temporal basis and the thicknesses of the formations are not represented. Horizontal lines at the same level are meant to be isochronous, however. No attempt is made to indicate degree of certainty of correlation: this is dealt with in the text.

∗ Fossils. +Trace fossils. The positions of these two symbols are only approximately located with relation to the timescale of the left-hand margin. Archaean and Proterozoic are not used here with a precise time-significance.

Ship Point Formation. He correlated the Turner Cliffs Formation with the Middle Cambrian beds on Boothia Peninsula and the combined Bear Point and Ooyahgah Formations of Dundas Harbour.

3.6. Correlation of Cambrian formations of the Franklinian miogeosyncline and Central Stable Region of north-eastern arctic Canada

These are summarized in Fig. 6 which is based on the evidence presented in the various sections of the text.

4. The Cambrian of northern Ellesmere Island and northern Greenland

The structural elements of Ellesmere Island discussed earlier extend into Greenland as shown on Fig. 7. The Central Stable Region can be traced across north Greenland to the Independence Fjord–Hagen Fjord–Danmark Fjord region of north-east Greenland. It is flanked to the south-west by strata which were folded during the Pre-Cambrian Carolinidian orogeny (HALLER, 1961a) while the younger strata of the Central Stable Region were laid down after the earth-movements. The miogeosynclinal belt of arctic Canada–north Greenland can possibly be followed around the margin of the Wandel Sea into the land bordering the east coast of Danmark Fjord and southwards from there. The miogeosynclinal belt in the north and the Pre-Carolinidian strata to the south are flanked to the east by the East Greenland fold belt, which was subject to episodes of folding, faulting, and metamorphism during both the Carolinidian and Caledonian orogenies. The Greenland Shield, which is well-exposed in north-west Greenland, is not exposed in north Greenland (although previously reported at the head of Victoria Fjord) or north-east Greenland and the tentative lines indicated are quite hypothetical as no geophysical evidence is known to the author. Presumably the crystalline basement complex is not deeply buried as 'upthrust wedges of Basement rocks' occur in the northern end of the East Greenland fold belt (Fig. 7) in north-east Greenland (BERTHELSEN and NOE-NYGAARD, 1965, p. 224).

4.1. Northern Ellesmere Island and north Peary Land

In these two regions of Greenland and Canada are found folded and metamorphic belts which may represent dissected parts of the same eugeosynclinal belt (Fig. 7).

The eugeosynclinal belt of north Ellesmere Island is centrally covered by younger sediments of the Sverdrup Basin so that the orogenic system comprises three tectonic provinces: (1) a coastal belt of metamorphic rocks, (2) tightly folded early and middle Palaeozoic rocks, (3) the Sverdrup Basin of late Palaeozoic, Mesozoic, and Cainozoic rocks. A miogeosynclinal belt occurs further to the south-east (Fig. 7) in central Ellesmere Island (discussed earlier on page 343).

The best known part of this extensive region is north-eastern Ellesmere Island; from here CHRISTIE (1964), continuing earlier work by BLACKADAR and himself, described the Cape Columbia Group and suggested a pre-Middle Ordovician age. The group consists mainly of gneisses and schists. A specimen of biotite gneiss from this group gave a K-Ar radiometric age of 545 m.y. which suggests a Lower Cambrian age (on both the Geological Society of London's Phanerozoic time-scale, 1964, and KULP's scale of 1960) for the latest metamorphism undergone by the rocks. The rock itself may be older. The Mount Disraeli Group and the M'Clintock Group, which are probably younger than the Cape Columbia Group, are considered Ordovician or earlier on the basis of fossil evidence,

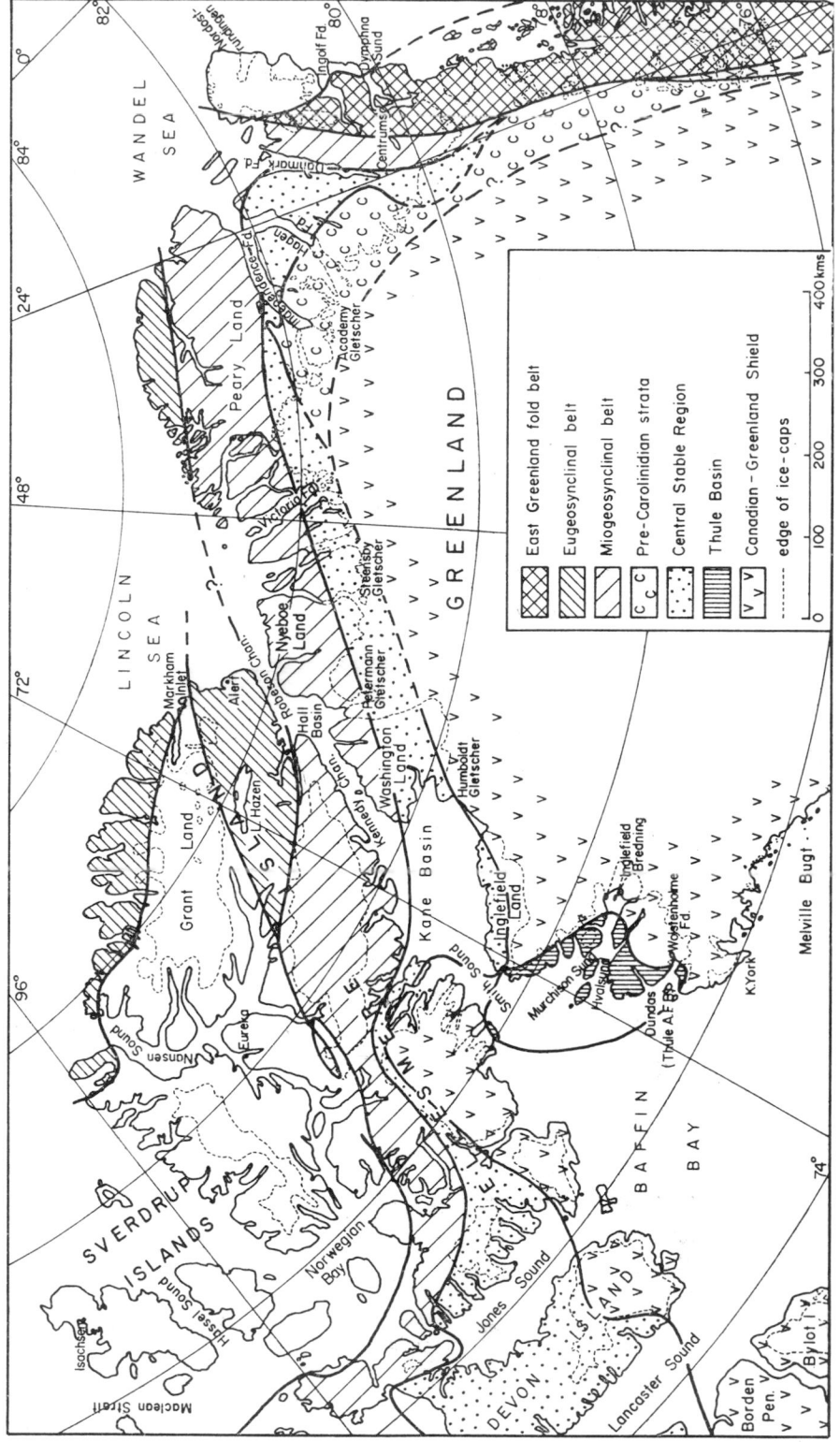

FIG. 7. Structural-stratigraphical provinces in north Greenland and northern arctic Canada. Abbreviations: Thule A.F.B.—Thule Air Force Base; K. York—Kap York; Fd.—Fjord. The east Greenland fold belt includes pre-Caledonian strata which may belong to the western foreland.

volcanic associations, and tectonic grade; they are sediments and volcanics which are much less metamorphosed. The Daly River Terrane of limestones and quartzites with Ordovician fossil-bearing beds are a sedimentary series which is clearly dated and correlative with similar strata in Washington and Hall Lands.

TRETTIN (1964) investigated parts of northernmost Axel Heiberg Island and north-westernmost Ellesmere Island (Figs 2 and 6) in 1961 and 1962. His working area was centred in the Nansen Sound area. Earlier work had been done by SCHEI in 1902, CHRISTIE in 1954, ROOTS in 1955, THORSTEINSSON in 1957, and FRICKER in 1960 and 1961. In Axel Heiberg Island the Rens Fjord Complex is a thick assemblage of clastic sediments, carbonates, chert, and volcanic rocks. It differs markedly in lithology and tectonic grade from Silurian units (dated by fossil evidence) of the Nansen Sound area and must be pre-Silurian in age. A K-Ar radiometric age of 535 m.y. from muscovite out of a phyllitic sandstone suggests that the schistosity developed in Cambrian times, the accuracy is ± 35 m.y. Parts of the Rens Fjord Complex are comparable in lithology and meta-morphic grade to parts of the Mount Disraeli Group of Ellesmere Island and units 2 and 3 mentioned in the following paragraph.

In north-westernmost Ellesmere Island the Cape Columbia Group (named from north-eastern Ellesmere Island) is found and TRETTIN suggests a Lower Cambrian (?) or older age for the group; it comprises a great variety of sedimentary and plutonic rocks showing a wide range, but distinctly higher grade, of metamorphism than other units of the region. Unit 2 and Unit 3 (TRETTIN, 1964, p. 4) can be compared to the Rens Fjord Complex and the Mount Disraeli Group respectively; with the younger Bourne Group they are assigned a pre-Silurian (?) age.

North Peary Land was visited by LAUGE KOCH in 1921; he was the first to traverse the whole of the coast bordering the Arctic Ocean. FRÄNKL and MULLER in 1953 made a return journey on foot across northern Peary Land from Frigg Fjord (83° 09′ N.) to Kap Morris Jesup (83° 39′ N.) which is the northernmost point of land in the world. FRÄNKL (1955b) named a large number of new formations from the sedimentary and metasedimentary sequences which display increasing metamorphism from south to north. A Lower Palaeozoic age was suggested for some of the formations but no specific assignment to the Cambrian was put forward. The oldest fossil trails, in the Grønnemark Shales, were taken as indicating a Phanerozoic age. It is perhaps a permissible speculation in dealing with such remote and little-known areas to suggest until better evidence is available that the Grønnemark Shales may be correlated with the Schley Fjord Shale (page 365) and that the underlying Grønnemark Sandstones correlate with unit 5 of sandstones and shales in southern Peary Land (page 364), both of which are assigned to the Lower Cambrian in this contribution.

From the published accounts it seems probable that Cambrian rocks occur in the eugeosynclinal belts of northern Canada and Greenland but that diagnostic fossils, as yet not forthcoming, may not be easily found because of the high tectonic and meta-morphic grade of the rocks. Radiometric ages will be increasingly useful but may only give the age of metamorphic events and not the age of formation of the rock.

4.2. Nyeboe Land, Wulff Land, and south Peary Land

In the summers of 1965 and 1966 the Geological Survey of Canada and the Greenland Geological Survey jointly investigated parts of the north coast of Greenland. Operations were mounted inland towards the icecap from Polaris Promontory on the Greenland side of the Robeson Channel and eastwards as far as Victoria Fjord to the south-east of

Peary Land. This area had been traversed by sledge-parties a number of times at the turn of this century and earlier but geological observations have been scarce and scattered. LAUGE KOCH made some observations in the earlier part of this century but he had to spend a high proportion of his time on topographical surveying and was hampered by snow conditions and the threat of starvation, so that geological collections had in some cases to be abandoned and have not since been recovered.

DAWES found (1966, p. 11) that the oldest rocks exposed were unfossiliferous quartzites and conglomerates on the north-west coast of Wulff Land and suggested a possible Cambrian age. Cambrian fossils were found in a dolomite sequence in northern Nyeboe Land which passes upward into Ordovician limestone and dolomitic limestone. According to V. POULSEN (*pers. comm.* in 1968) a Cambrian trilobite fauna dominated by agnostids was found by DAWES in northern Nyeboe Land in dark bitumous limestone. The agnostid species allow a correlation with the Atlantic fauna from the *Paradoxides davidis* Zone in south-eastern Newfoundland; a few additional species indicate the presence also of the *Paradoxides forchhammeri* Zone, possibly equivalent to the upper part of the St. Albans Shale in Vermont. In the Pacific standard zonation the fauna corresponds to the upper part of the *Bolaspidella* Zone (of the Middle Cambrian) and is thus younger than the well-known Middle Cambrian fauna from Inglefield Land to the south.

Southern Peary Land is part of the miogeosynclinal belt and its southern margins form part of the Central Stable Region (Fig. 6) flanking the Greenland Shield. Earlier geological investigations by PEARY, FREUCHEN, and LAUGE KOCH were followed by the observations of TROELSEN, ELLITSGAARD-RASMUSSEN, and FRÄNKL.

The following succession in southern Peary Land was established by TROELSEN (1949, 1956b) but the age assignments are new:

Age	Formation, thicknesses in metres	Lithology
Lower Ordovician	Wandel Valley, 350	Limesone, dolomite
— ? Stratigraphical break —		
Lower Cambrian	Brønlund Fjord, 156	Dolomite
— erosional disconformity —		
Lower Cambrian	unnamed unit (5)	
	approx. 40	Sandstone, shale
	150	Scree
	245	Quartzite, sandstone
? Lower Cambrian	unnamed unit (4), approx. 250	Dolomite with stromatolites.
? Lower Cambrian	unnamed unit (3), approx. 70	Sandstone, shale.
Pre-Cambrian	unnamed unit (2), 1–100	Tillite
— ? unconformity —		
Pre-Cambrian	unnamed unit (1), approx. 1,000	Sandstone

West and north of Independence Fjord around Midsommersø and Jørgen Brønlund Fjord gently dipping sandstones occur (unit 1), intruded by dykes and sills of dolerite, with a thickness of approximately 1,000 metres. According to TROELSEN (1950a, 1956b) these sandstone are overlain by a tillite (unit 2) which is 1 to 100 metres thick. The pebbles and boulders of the tillite are scratched but no striated pavement below was found and other associated glacial features have not been recorded up to the present. From regional considerations, discussed later, there is probably an unconformity below the tillite: it oversteps across a thick pre-tillite sequence in Kronprins Christian Land on to the Pre-Cambrian sandstones of the Danmark Fjord area (where the tillite has not yet been found—see Fig. 10, p. 377) and south Peary Land (BERTHELSEN and NOE-NYGAARD, 1965, fig. 26). Unit 3 of sandstone and shale follows and units 4 and 5 complete the succession below an erosional disconformity. The Brønlund Fjord Formation occurs above this disconformity in south Peary Land. In north-east Peary Land the Lower Cambrian Schley Fjord Formation, which is less than 100 metres thick, is regarded as the age equivalent of the Brønlund Fjord Formation.

Within the whole region of North Greenland there are a number of dykes and sills which cut the various parts of the sedimentary sequences and seven generations have been distinguished in southern Peary Land by ELLITSGAARD-RASMUSSEN (1950 and 1955). It seems best at this stage in field research to retain the policy suggested by BERTHELSEN and NOE-NYGAARD (1965, p. 229) and continue an earlier line by COWIE (1961, p. 22) . . . 'the presence or absence of intrusives cannot be used for chronostratigraphic correlations.' The Lower Cambrian age of the Brønlund Fjord Formation is based on fragments of olenellids and *Salterella* which come from the basal stratum of the formation near Midsommersø. Similarly, the shales of the Schley Fjord Formation yielded olenellids at its type locality, which is an island in the north-eastern part of G.B. Schley Fjord in north-east Peary Land.

4.3. Danmark Fjord area and Kronprins Christian Land

FRÄNKL in 1954 described a major stratigraphical break within the sedimentary succession in eastern Kronprins Christian Land (the land area between the Dijmphna Sund–Nordostrundingen coast and Danmark Fjord, Fig. 7). He termed the earth movements and prolonged associated emergence which caused this break the 'Hekla Sund phase' (1956, p. 29). The tectonics of the region are complex and further field research on the ground, to follow up FRÄNKL's work, is needed. PEACOCK and WYLLIE during geological work in western Dronning Louise Land (the extensive group of nunataks around 77° N. in East Greenland) in the early fifties of this century also recognised a period of folding which had taken place before the deposition of a younger group of sediments. The age of these widely separated movements seemed to the field workers to be Pre-Cambrian. HALLER (1961a) from his field work in the East Greenland Fold Belt, and studies from the air and from aerial photographs in north Greenland as far west as southern Peary Land, concluded that a Pre-Cambrian orogeny had affected eastern and northern parts of Greenland to the north of latitude 76° N. This orogeny, the Carolinidian, was claimed to have manifestations in the East Greenland Fold Belt particularly in the north. Older rocks were formed before the orogeny but show no structural effects to the west of the miogeosynclinal area. Their approximate extent is shown as 'Pre-Carolinidian strata' in Figs 7 and 8. The structural evolution of north-eastern Greenland is illustrated by an admirable series of sections in a diagram by BERTHELSEN and NOE-NYGAARD (1965, fig. 26).

ADAMS amd COWIE examined the Lower Palaeozoic and older strata in the region of western Kronprins Christian Land and around Danmark Fjord. The following tabulation is based on details published in 1953 modified by later work including field work in 1954; the Danmark Fjord and Amdrup Formations are the basal strata of the previously undivided Centrum Limestone:

Age	Formation, thicknesses in metres	Lithology
Lower Ordovician	Amdrup, 240	Limestone and dolomitic limestone.
	stratigraphical break	
Lower Cambrian	Danmark Fjord, 10	Limestone, dolomite, conglomerate.
	erosional disconformity	
Lower Cambrian	Kap Holbaek, 135	Sandstone, quartzite, shale.
	? stratigraphical break	
? Lower Cambrian	Fyn Sø, 324	Dolomite, minor shale.
? Lower Cambrian	Campanuladal, 250	Sandstone, limestone.
	unconformity	
Pre-Cambrian	Norsemandal, 300 + base not seen	Sandstone, quartzite, with basic dykes.

Norsemandal Formation

The sandstones which weather a predominantly red colour and associated quartzites weathering yellow and grey of this formation are traversed by dolerite and porphyrite dykes which were not found in overlying strata, suggesting an unconformity although no angular discordance was observed.

Campanuladal Formation

At the base of the formation 150 metres of sandstone and shale are succeeded by 100 metres of thinly-bedded limestone with shale and intraformational conglomerates; stromatolites occur near the top.

Fyn Sø Formation

Stromatolites are abundant in the dolomites of this formation and often the whole rock is built up of concentric structures through great thicknesses out of the lower 300 metres. Near the top a 20 metres thick band made up almost entirely of nested cones are assigned to the stromatolite *Conophyton* (REZAK, 1957; COWIE, 1961, pp. 30–31); it is overlain by 4 metres of shale.

Kap Holbaek Formation

Three sandstones and two quartzite subdivisions make up the formation which shows imperfectly developed examples of *Scolithus* in the lower quartzite subdivision; considered to suggest a possible Lower Cambrian age. (POULSEN, 1960; MOORE, 1962; KERR, 1967a, p. 25).

In the Danmark Fjord region the Campanuladal, Fyn Sø, and Kap Holbaek Formations are an apparently continuous sequence with only a possible slight break between the upper two formations; but in eastern Kronprins Christian Land there is an unconformity between the Fyn Sø and Kap Holbaek Formations. In the Danmark Fjord region and in eastern Kronprins Christian Land there is a marked disconformity or unconformity between the Kap Holbaek and Danmark Fjord Formations; the latter is a highly siliceous contorted limestone with chert and flint and many large-scale intraformational conglomerates. The Danmark Fjord Formation and the Brønlund Fjord Formation have a close resemblance in their lowest beds, the former is unfossiliferous but shows numerous relict fossil structures whilst the latter has an olenellid fauna which gives a Lower Cambrian age: they are correlated on these grounds and that of stratigraphical position (COWIE, 1961, pp. 29–30). No Cambrian fossils have been found in the Danmark Fjord–Kronprins Christian Land region except for *Scolithus* noted above. The Lower Ordovician is unequivocally represented by faunas in the Amdrup Formation which succeeds after a stratigraphical break.

In the east coast regions of Kronprins Christian Land the above sequence can be correlated with similar rocks from the Campanuladal Formation upwards but the lower part of the column exhibits considerable variation, a composite succession can be suggested, based on FRÄNKL (1954, 1955a, 1956):—

Lower Ordovician	Centrum Limestone (as marbles)
Lower Cambrian and ? Lower Cambrian	Danmark Fjord Formation, 10–30 m ————————disconformity———————— Kap Holbaek Formation, 2–5 m ————————unconformity———————— Fyn Sø Formation, 250 m Campanuladal Formation, 150 m
Pre-Cambrian	Ulveberg Sandstone and Tillite, 20–35 m Rivieradal Sandstone, 1,000–2,000 m Taagefjeldene Greywacke, 700–1,000 m Sydvejdal Marble, 100–400 m Stenørken Phyllite, 1,000 m — unconformity—Carolinidian orogeny — unnamed sandstone and arkose (Unit A)

Detailed consideration of this succession is out of context here (see BERTHELSEN and NOE-NYGAARD, 1965). FRÄNKL correlated the Campanuladal Formation of this region with the Campanuladal Formation of the Danmark Fjord region and also correlated his lowest unit of sandstone and arkose (Unit A) with the Norsemandal Formation of the Danmark Fjord region (FRÄNKL, 1955a, fig. 11). The Ulvebjerg Tillite which is only 1 to 2 metres thick exhibits rounded and polished porphyries and quartzites (1–4 mm in diameter) in a matrix of well-rounded quartz grains and tiny carbonate crystals. Features commonly displayed by tillites or associated with them are not described in this occurrence but FRÄNKL (1954) correlated the Ulvebjerg Tillite with the upper tillite of East Greenland (Fig. 10) and also with the tillite described by TROELSEN from Peary Land (p. 364) with apparent confidence.

The Kap Holbaek Formation may be represented by fissure fillings of sandstone and in some sections it is probably found as a bed resting with a slight discordance on an eroded surface of the Fyn Sø Formation and overlain by the Danmark Fjord Formation, surmounted by marbles of the Centrum Limestone which may include the Amdrup Formation. The unconformity at the base of the Kap Holbaek Formation was not observed in the Danmark Fjord area but is clearly of importance in regional considerations affecting the whole of north-east Greenland.

4.4. Correlations within the region

These are summarized in Fig. 10. The pre-tillite sandstones in south Peary Land and eastern Kronprins Christian Land are correlated on grounds of lithological type and position within the stratigraphical succession, which is below the tillites in the two areas. The character and association of these two tillites in north-east Greenland are not well authenticated by detailed work but they have gained acceptance. Although there is still a danger that tillites may be too easily accepted as synchronous (COWIE, 1961, p. 20) it is a useful working hypothesis to assume that they were formed in the same glacial episode—the Varangian. They could be of Cambrian age but are here placed near the top of the later Pre-Cambrian (Proterozoic). The Carolinidian orogeny affected the older strata in eastern Kronprins Christian Land; the Stenørken Phyllite to Rivieradal Sandstone sequence which lies conformably below the tillite is unconformable on the older sandstones due to this phase of movement, which may also be partly the cause of the unconformity between the Campanuladal Formation and the Norsemandal Formation. The same gap in the succession at this level may be present also below the tillite in south Peary Land (BERTHELSEN and NOE-NYGAARD, 1965, figs. 26 and 27).

As BERTHELSEN and NOE-NYGAARD have commented (1965, p. 231), HALLER's term Hagen Fjord Group (1961a, p. 157) is too broad in scope, covering both pre- and post-tillite formations which are here considered to span Pre-Cambrian and Cambrian ages; it is not used here. The Campanuladal Formation seems to be correctly assigned to the lowest post-tillite part of the succession in Kronprins Christian Land as does the Fyn Sø Formation above and thus a correlation is achieved with the type areas of these two formations in the Danmark Fjord region. The post-tillite sandstone of south Peary Land seems to be contemporaneous with the Campanuladal Formation on grounds of general lithology and stratigraphical position and this same argument applies to the dolomite with stromatolites which appears to be a correlative of the Fyn Sø Formation with its abundant stromatolites. From consideration of the details given in previous parts of this chapter it seems most reasonable to correlate sandstone, quartzite, and shale of the Kap Holbaek Formation with *Scolithus* of the Danmark Fjord–Kronprins Christian Land region with similar strata of sandstone, quartzite, and shale in south Peary Land. Although there is a considerable difference in thickness the Danmark Fjord Formation, which shows only relict fossil markings, is correlated in age with the fossiliferous Brønlund Fjord Formation and the different facies development of the Schley Fjord Formation. The Amdrup Formation, named in this paper for the lower part of the thick Centrum Limestone of ADAMS and COWIE (1953) follows after a stratigraphical break and correlates satisfactorily with the Wandel Valley Formation of south Peary Land, which follows a probable stratigraphical break at a similar level in the geological column.

The base of the Cambrian System was previously placed at the base of the Kap Holbaek Formation but from correlations with north-west Greenland and Ellesmere Island discussed on page 378 it is now placed above the tillites and below the Campanuladal Formation and equivalent beds.

5. The Cambrian of east Greenland

The complexities of the East Greenland fold belt (Fig. 8) are beyond the scope of this contribution and have been the subject of prolonged investigations, particularly by Danish expeditions under LAUGE KOCH.

Recent generalized comments on the fold belt of East Greenland, which stretches from Scoresby Sound near 70° N. to near the Wandel Sea at more than 81° N. (about 1,400 kilometres in length), have been given by FRÄNKL (1956) and HALLER (1961b). In the inner parts of Scoresby Sund near Gaase Land a small outcrop of crystalline basement complex is seen at the western border of the belt. This has been dated as 1,900 m.y. (HALLER and KULP, 1962). The basement is again seen in Dronning Louise Land (a series of nunataks between 70° and 77° N.), otherwise the Greenland Shield is un-exposed in north-east and east Greenland north of Knud Rasumussen Land where basic igneous rocks of the Brito-Arctic Province make up the majority of the outcrops stretching away southwards.

In Dronning Louise Land the western margin of the East Greenland fold belt is seen overlying the 'Archaean' metamorphic basement. On the western margin of these 'Archaean' outcrops the quartzites of the Trekant Series can be correlated, on the most general grounds, with the Thule Group of the Dundas area in north-west Greenland or the Norsemandal Formation and strata of equivalent age in north-east Greenland. The Trekant Series may therefore be pre-Carolinidian and Proterozoic in age.

The Zebra Series overlying the Trekant Series is correlated by PEACOCK (1956; 1958, p. 132) with hypothetical beds below the Stenørken Formation of FRÄNKL (1955a) of post-Carolinidian Proterozoic age. If these hypotheses are correct there seems to be little likelihood that Cambrian rocks (or indeed fossils) will be found in this area.

Rocks formed before the Carolinidian orogeny, and involved in it, probably form part of the east Greenland fold belt between 76° and 78° N. and were 'reworked' in the Caledonian orogeny (BERTHELSEN and NOE-NYGAARD, 1965, p. 241). The Caledonian orogeny, a term which is used in a broad sense for mountain building movements which commenced in Palaeozoic times, was the main orogeny in east Greenland.

The Cambro–Ordovician strata, which are underlain by a great thickness of Pre-Cambrian sediments, are found in the fjord zone where they are fossiliferous. Lower Palaeozoic rocks may possibly occur also in the nunataks of the inner, ice-cap, region where fossils have not been found. Both areas lie between latitudes 70° and 75° N.

5.1. The fjord zone

The Cambrian outcrops in east Greenland between latitudes 72° and 75° N. are shown in Fig. 9. The pre-Devonian sedimentary series show large-scale disharmonic folding which is part of an upper nonmetamorphic or very low grade metamorphic suprastructure which reflects deformation of the crystalline, migmatitic infrastructure of the Caledonides. The Cambrian succession participates with older and younger strata in folding and faulting into relatively broad anticlines and synclines. These folds have varying plunge and form a series of axial culminations and depressions. The Lower Palaeozoic strata are usually found only in synclinal depressions and this accounts to a great extent, in conjunction with faulting, for the scattered nature of the outcrops shown in Fig. 9.

Although there are local variations the Cambrian of this north–south zone is remarkably uniform and belongs to one facies, so that the whole fjord zone can be summarized in one succession and the faunas dealt with together.

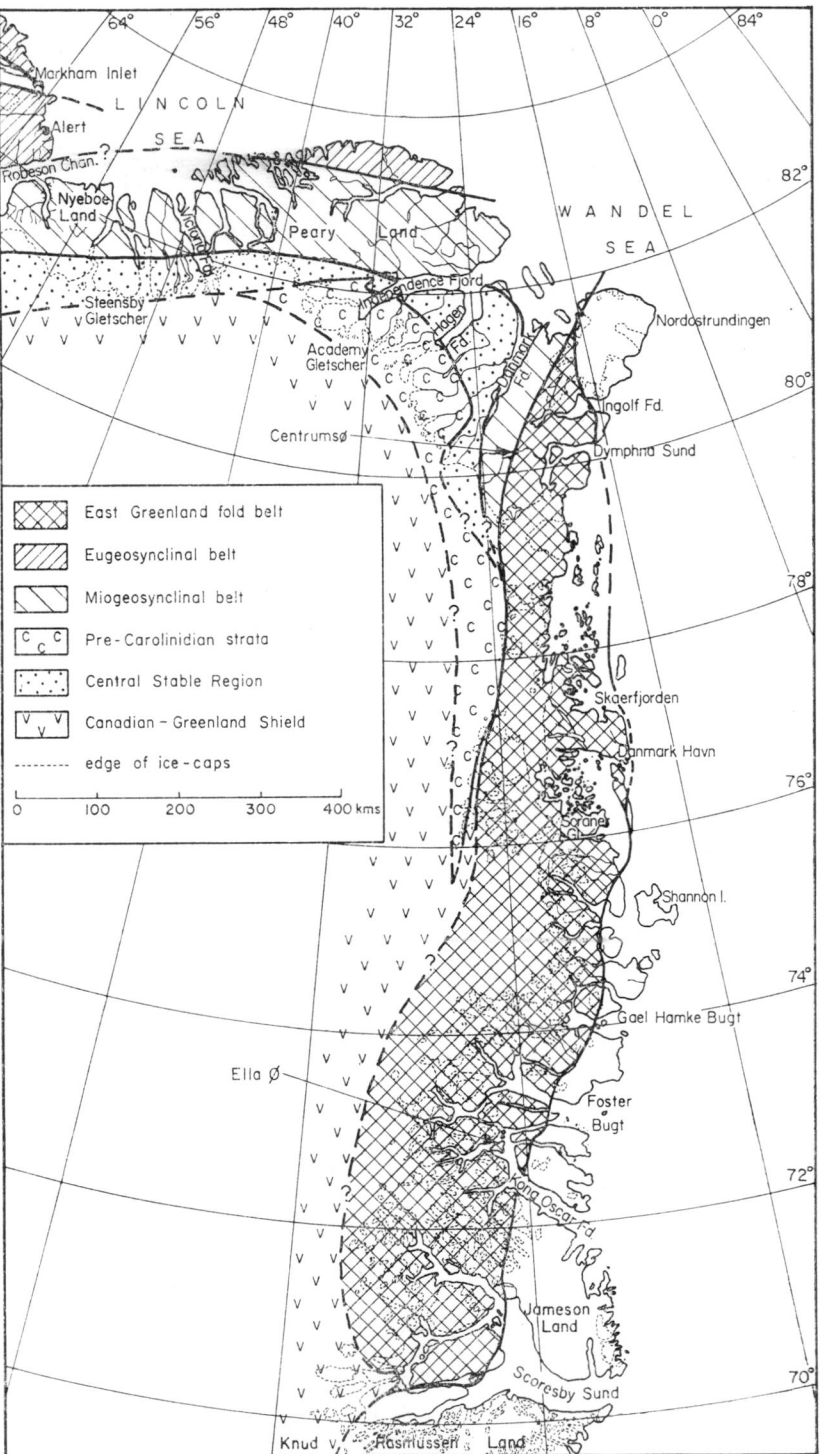

Fig. 8. Structural-stratigraphical provinces in east and north-east Greenland. The east Greenland fold belt includes pre-Caledonian strata which may belong to the western foreland. Fd: Fjord, ø: Island, Chan.: Channel.

German and Swedish nineteenth-century explorers were the first geological investigators in the fjord zone under consideration: KOLDEWEY in 1870 and NATHORST in 1899. WORDIE led a series of expeditions from 1926 to 1929 and quartzite erratics with *Scolithus* were found and compared with Scottish Cambrian quartzite with *Scolithus*. LAUGE KOCH commenced his great series of expeditions in 1926 which were to continue with little interruption until 1958 when the leader was in his seventh decade! He found the first Cambrian fossils *in situ* on Ella Ø in 1927. C. POULSEN in 1929 established the main stratigraphical framework for the Lower Palaeozoic rocks and also investigated the tillites. Many geologists on these Danish expeditions made random observations on the fossiliferous Cambrian rocks but systematic work has mainly been done by T. HEINRICHSON, C. POULSEN, P. J. ADAMS, and the author.

The generalized succession for the fjord zone is as follows:

Age	Formation and maximum thicknesses in metres	Lithology
Lower Ordovician	Cass Fjord, 230	Limestone, limestone-shale, conglomerates.
————————————————stratigraphical break————————————————		
? Middle Cambrian	Dolomite Point, 400	Dolomite, minor calcareous dolomite.
? Middle Cambrian and Lower Cambrian	Hyolithus Creek, 210	Dolomite
Lower Cambrian	Ella Island, 90	Limestone, minor shale.
Lower Cambrian	Bastion, upper part, 100	Shale, minor shell-limestone.
————————————————stratigraphical break————————————————		
Lower Cambrian	Bastion, lower part, 100	Shale, sandstone.
————————————————stratigraphical break————————————————		
Lower Cambrian	Kløftelv, 70	Quartzite, minor sandstone.
————————————————regional unconformity————————————————		
? Lower Cambrian	Spiral Creek, 25	Dolomite, siltstone.
? Lower Cambrian	Canyon, 300	Dolomite, shale, silstone.
Pre-Cambrian	Tillite Group	

The Tillite Group, consisting of two tillite complexes of regional extent separated by a sequence of varied sedimentary strata including shales and sandstones, was discovered in 1929. The group seems to be well-authenticated as a series of true glacial deposits representing a major glaciation, part of the Varangian glaciation which affects many regions around the North Atlantic.

By correlation of the Tillite Group with the tillites described earlier in north-east Greenland it can be suggested that the Canyon Formation is ? Lower Cambrian in age (Fig. 10).

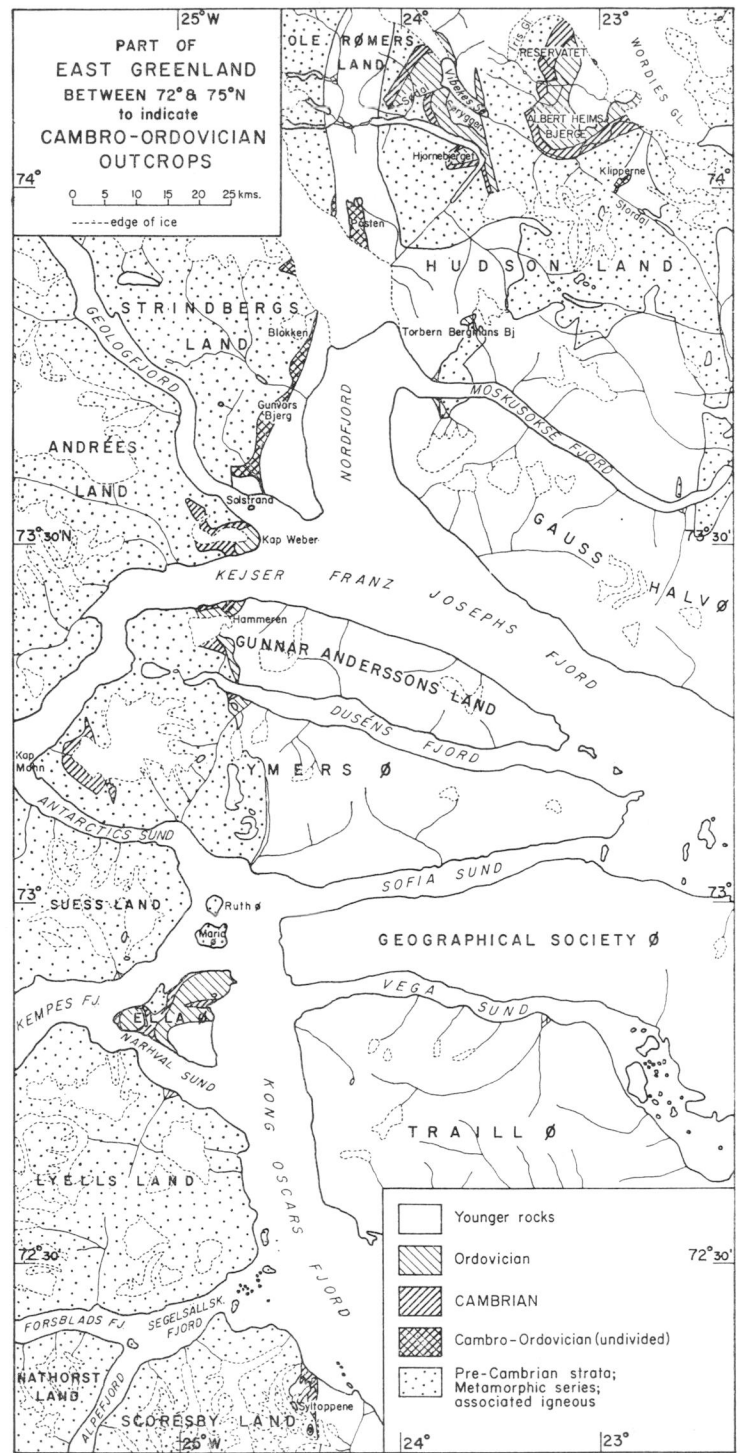

Fig. 9. Part of east Greenland to indicate Cambrian outcrops in the fjord zone between 72° and 75° N. The map is oriented approximately north–south. Outcrops of Cambrian (and Ordovician) rocks occur just to the north of this map in C. H. Ostenfelds Nunatak (ca. 23° W., 74° 20′ N.).

Canyon Formation

The beds are divided into three units:

3. Dolomitic limestone and limestone with some shale and dolomite.
2. Dolomite with dark shale.
1. Shale interbedded with limestone.

Cyclic deposits in the lowest unit 1 have been interpreted as varves but perhaps they are cyclothems of a non-glacial character. The formation is conformable to both the underlying Tillite Group and the overlying Spiral Creek Formation. SCHAUB (1950, 1955) discussed Pre-Cambrian to Cambrian sedimentation in east Greenland and regarded the lamination of unit 1 as due to the annual melting of glaciers with the sea as the site of deposition. Higher in the succession the aqueous site may have been brackish or even fresh water. Dessication of laminae in higher beds of units 2 and 3 and intraformational breccias suggest shallow water with periodic drying out—perhaps the setting was an area of extended shoals in front of a land mass. Sedimentation seems to have been proceeding either at a faster rate than any subsidence or at an equal rate. Stromatolites (*Collenia*) are found in the upper beds of the formation; oolites and tubes which may be algal are also present.

Spiral Creek Formation

The basal dolomites of this thin but varied and complex series contain many intra-formational conglomerates and are succeeded by ripple-marked dolomitic siltstones with well-preserved salt (NaCl) impressions in the form of hopper cubes. The upper beds are of dolomite with chert in lenticles, bands, blebs, streaks, pisolites, and oolites; in some beds the siliceous material is a more important constituent than the dolomite and a 20 centimetres thick band of chert comes at the top of the formation.

The salt casts seem to be proof that shallowing of the sea sufficient to allow desiccation occurred. The Spiral Creek Formation as a whole indicates changes in climate and in depth of the sea that were a contrast to the conditions prevailing in underlying formations; presumably the Canyon Formation and Tillite Group were deposited in cold seas. Dessication, ripple-marking, cross-bedding, intraformational breccias, salt casts, and the presence of feldspar all suggest shallowing, intermittent drying-out, and penecontemporaneous erosion in a warm, humid climate (SCHAUB, 1950). Markings of possible organic origin have been found by A. M. SPENCER, in 1968 (*pers. comm.*).

Kløftelv Formation

This series of quartzites and sandstones rests with slight disconformity on the Spiral Creek Formation in Ella Ø. In north Scoresby Land (Fig. 9) the Canyon and the Spiral Creek Formations are absent from the succession and the Kløftelv Formation was deposited on the Upper Tillite. In Hudson Land and Ole Rømers Land (near 74° 15′ N.) the characteristic lithology of the Spiral Creek Formation is absent and the Kløftelv Formation rests on characteristic Canyon Formation beds. It seems unlikely that lateral variation or diachronism is possible and therefore a major break of sedimentation occurred so that the Kløftelv Formation transgressed across strata below with a regional unconformity (COWIE and ADAMS, 1957, p. 148; COWIE, 1961, fig. 5). The main lithology of the formation is a massive pure quartzite which shows cross-bedding, ripple-marks, and 'swash' marks; this and the uniformity of the thickness and its consistent development in five similar subdivisions suggests rapid marine transgression over a peneplained land area followed by quiet stable conditions while a blanket sand was laid down in shallow waters. Numerous tracks and trails are the only traces of fossils.

Bastion Formation

This formation is divided into two parts by a stratigraphical break and there is also a stratigraphical break at its base.

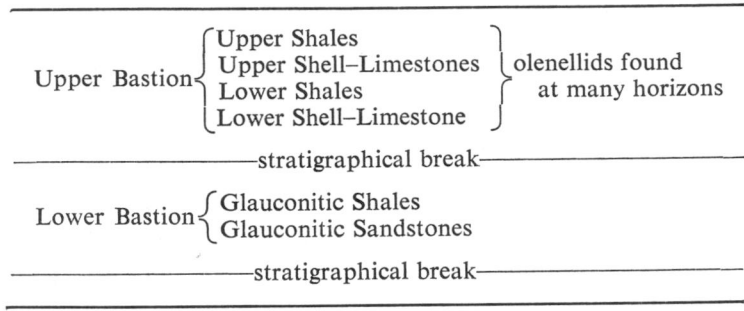

The Glauconitic Sandstones have a glauconitic siliceous conglomerate at the base with phosphate bodies composed of collophane and this mineral could be organic in origin; the overlying sandstones and the succeeding Glauconitic Shales are unfossiliferous. The upper Bastion Formation commences with the Lower Shell–Limestone which is glauconitic, ferruginous, and locally conglomeratic and phosphatic. In places it is almost entirely composed of fossil fragments. Fossils are found at a number of horizons in the Lower and Upper Shales and this lithology of glauconitic, arenaceous shale is intercalated with occasional fine-grained massive bands which vary from limestone and arenaceous limestone to siltstone. The Upper Shell–Limestone consists of bands rich in brachiopod fragments. The predominance of shale may suggest a deepening of the sea although quartz is still present as a rock component. The older beds, with olenellids only, probably belong to the earlier Lower Cambrian but the upper beds include eodiscids as well as olenellids so the later part of the Lower Cambrian is suggested. The fauna includes: *Fordilla troyensis* WALCOTT, *Botsfordia caelata* (HALL), *Discinella braastadi* POULSEN, *D. micans* (BILLINGS), ? *Dicellomus* sp., *Kutorgina* sp., *Lingulella* (*Lingulepis*) *prisca* POULSEN, *Micromitra* sp., ? *Mickwitzia* sp., *Obelella congesta* POULSEN, *Paterina* sp., *Rustella* sp., ? *Helcionella cingulata* COBBOLD, ? *Dentalium* sp., *Helenia bella* WALCOTT, *Hyolithellus micans* (BILLINGS), *Hyolithes* (*Orthotheca*) *bayonet* MATTHEW, *H.* (*Hyolithes*) *americanus* BILLINGS, *H.* (*H.*) *billingsi* WALCOTT, *H.* (*H.*) *mutatus* POULSEN, *H.* (*H.*) *poulseni* RESSER, *H.* (*H.*) *similis* WALCOTT, *Calodiscus lobatus* (HALL), *Olenellus* cf. *arcticus* (HALL), *O.* cf. *thompsoni* (HALL), *O.* cf. *truemani* WALCOTT, ? *Weymouthia* sp., and *Bradoria* sp.

Ella Island Formation

This formation consists almost entirely of fairly massive, hard, fine-grained limestone much of which is arenaceous in character but there is a thin group of shaly beds in the middle of the sueccession. The massive strata are cross-bedded and also show slumping features and intraformational conglomerates. A characteristic feature is the presence of abundant fragments of trilobite tests at many horizons. The junctions at base and top are conformable.

It seems likely that the sea may not have deepened very much during the limestone deposition of this formation but a strong carbonate chemical factor in the sedimentation was introduced. Presumably the seas were warmer and the reduced clastic percentage may have been due to increased distance from land and reduction of its relief. The trilobite faunas are varied and olenellids persist throughout the formation. *Proliostracus* and other genera in the highest beds suggest a late Lower Cambrian faunal association. The fauna includes: *Archaeocyathus atlanticus* BILLINGS, *Scolithus linearis* HALDEMANN, *Acrothele* sp., *Discinella braastadi*, *D. micans*, *Kutorgina* aff. *cingulata* BILLINGS, *K. reticulata* POULSEN, *Micromitra* sp., *Paterina mediocris* POULSEN, *Hyolithellus micans* (BILLINGS), *Salterella rugosa* BILLINGS, *Stenothecoides poulseni* RESSER, alokistocarid ind. (cf. *Kochiella*), *Bonnia groenlandica* POULSEN, *Calodiscus agnostoides* (KOBAYASHI), *Kootenia* sp., *Olenellus curvicornis* POULSEN, *O. simplex* POULSEN, *Paedeumias hanseni* POULSEN, *P. tricarinatus* (POULSEN), *Proliostracus strenuelliformis* POULSEN, ? *Protypus* sp., *Wanneria ellae* POULSEN, and *W. nathorsti* POULSEN.

Hyolithus Creek Formation

This formation of massive, hard, resistant, irregularly-bedded, dark-grey or black dolomite frequently shows stylolites and intraformational conglomerates at many horizons. *Salterella*, which is the only fossil known apart from numerous algal stromatolites, is found in the lower part and taken to indicate a Lower Cambrian age whereas the higher beds could be Middle Cambrian. The environment of deposition of these dolomites may have been warm seas with increased depth, but intermittent shallowing occurred indicated by the intraformational conglomerates. No quartz grains have been found in the strata so perhaps land was distant and of low relief. Organisms were perhaps absent as there seems to be no relict fossil evidence; secondary dolomitization, if it occurred, may alternatively have destroyed any fossils.

Dolomite Point Formation

Extended and intensive search in this formation for fossils has so far been unavailing. It is presumed that the age is Cambrian and pre-Ordovician. From the ubiquitous absence of Upper Cambrian fossils in Greenland, Scotland, and Spitsbergen a questionable Middle Cambrian age assignment seems most reasonable. The strata are massive, fine-grained dolomite but thinly bedded silty dolomite and dolomitic shales are present. Chert and flint are common especially in the lowest and highest beds, occurring as ovoid bodies, lenticles, and blebs of hard, black, splintery chert or, more rarely, as grey to white flint. Stromatolite 'mushrooms' have been silicified and at irregular intervals thin and laterally impersistent intraformational breccias and conglomerates show that penecontemporaneous erosion was operative.

The stratigraphical break between the Dolomite Point Formation and the overlying Cass Fjord Formation may represent a considerable interval of time, as suggested above. The Cass Fjord Formation is typically thinly-stratified, nodular, muddy limestones and shaly partings, alternating with bands of a purer and more homogeneous limestone. The age of the basal beds is not established by trilobites but the brachiopod-gastropod fauna seems to strongly suggest a Lower Canadian age of the early Lower Ordovician. Higher beds of the formation contain diagnostic trilobites and are Middle Canadian.

5.2. The nunatak zone

The metasediments of the nunatak zone were discovered in 1929 by a British expedition under the leadership of WORDIE and named the Petermann Series. Later work, including mapping by KATZ (1952) and WENK and HALLER (1953), established a more complete description of the stratigraphy of the nunataks. This strongly suggested that the

Petermann Series represents the western facies of the Eleonore Bay Group which is characteristically developed in the fjord zone.

It is not known if the western nunatak zone sedimentary series, or strata now covered by ice even further to the west, range up into the Cambrian Period: no firm evidence is available. HALLER and KULP (1962, p. 30) mention under 'nunatak region' in a diagram comparing stratigraphical sections of Pre-Cambrian and Lower Palaeozoic formations in the nunatak and fjord regions that 'fossiliferous boulders indicate presence of Palaeozoic strata west of the exposed area.' Boulders so far examined by the author which have been considered to be fossiliferous have been unconvincing but other collections perhaps exist. Exceptions, however, are erratic quartzite blocks with *Scolithus* which have been found near the east Greenland coast and inland. They are associated with heterogeneous erratics which include igneous and metamorphic types. These blocks are found in a number of places where iceflow could in the past have brought them from the western nunatak zone.

6. The Cambrian of eastern arctic North America, including Greenland, and its relationship to the Cambrian of Spitsbergen and Scotland

The correlations between the widely scattered regions of this study, some of which are numbered in Fig. 1, are illustrated in Fig. 10: only brief comments are needed to explain certain points and supplement earlier notes. Two main critical horizons can be used for correlation but both may be diachronic and unfortunately radiometric dates are as yet of no detailed assistance. These two horizons are:

1. The tillite group, which is thought to be of the same Varangian age in east Greenland, Kronprins Christian Land, and south Peary Land.

2. The first horizon with olenellid trilobites which occurs in

    (*a*) the Bastion Formation in east Greenland,

    (*b*) the Brønlund Fjord and Schley Fjord Formations in south Peary Land,

    (*c*) the Wulff River and Police Post Formations in Inglefield Land and Bache Peninsula, and

    (*d*) the Kane Basin Formation in central Ellesmere Island.

Other factors which are also useful are:

    (i) the presence of trace-fossils such as *Scolithus*, *Cruziana* and, to a limited extent, stromatolites,

    (ii) unconformities which may be associated with widespread earth movements such as the Carolinidian,

    (iii) lithological correlations which may indicate palaeogeographical conditions, and

    (iv) stratigraphical associations and palaeontological criteria relating to later Lower, Middle, and Upper Cambrian faunas.

The east Greenland Cambrian sequence dealt with in the last section can be correlated with north-east Greenland on the basis of the possible contemporaneity of the tillites in the two areas and the Pre-Cambrian formations below may be of similar age. In east Greenland the Eleonore Bay Group may represent a long almost continuous period of diverse sedimentation; in the south it perhaps spans the Carolinidian orogeny. The Eleonore Bay Group could perhaps be equivalent to the pre-tillite formations in north-east Greenland. The Kløftelv Formation and the Kap Holbaek Formation both follow a period of regional unconformity and are correlated on their lithology and stratigraphical position below the fossiliferous Bastion-Ella Island and Brønlund Fjord Formations

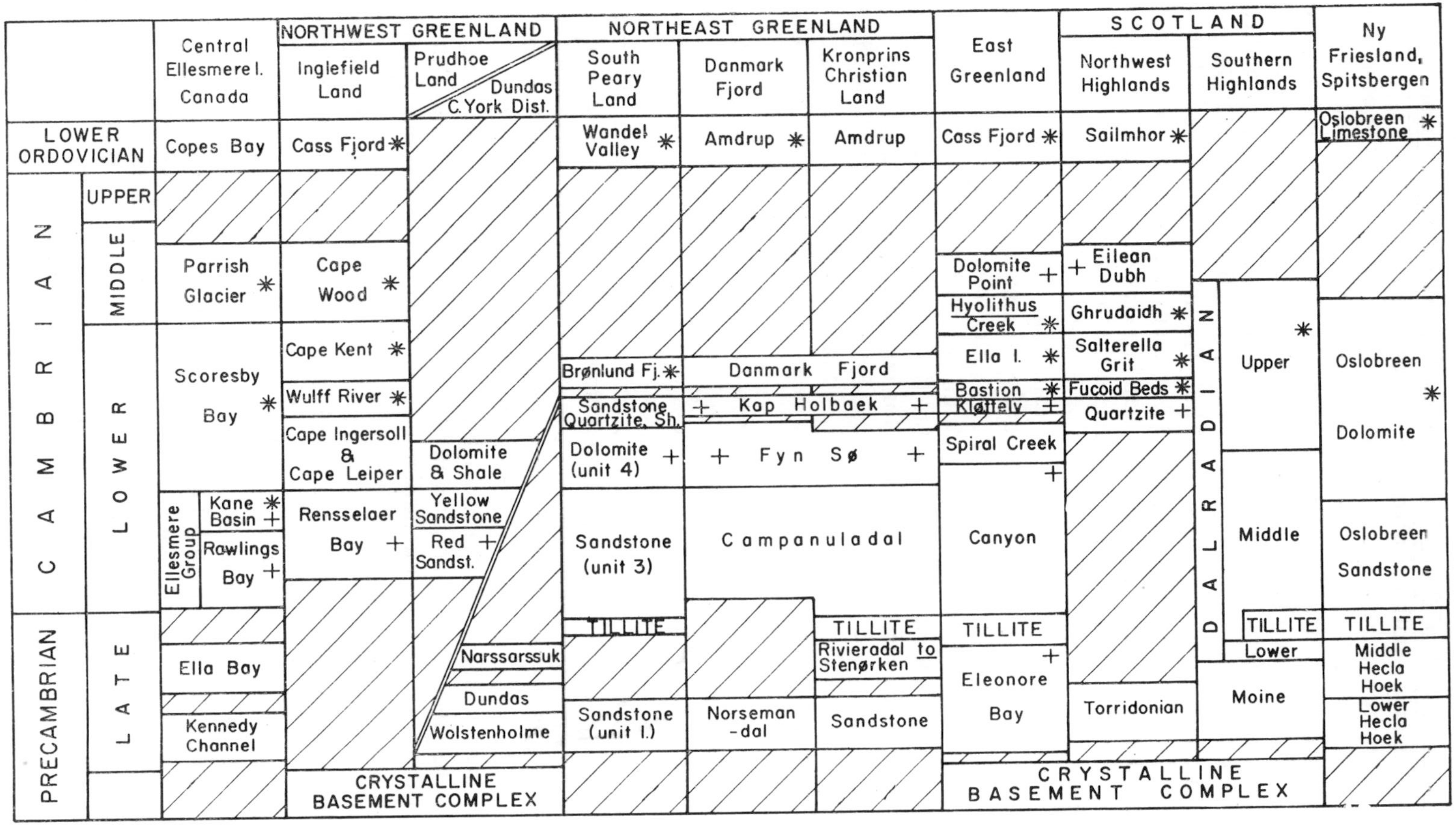

Time scale (left-hand margin):
- LOWER ORDOVICIAN — UPPER
- CAMBRIAN — UPPER, MIDDLE, LOWER
- PRECAMBRIAN — LATE

NORTHWEST GREENLAND

Central Ellesmere I. Canada: Copes Bay *, Parrish Glacier *, Scoresby Bay *, Kane Basin *+, Rowlings Bay +, Ella Bay, Kennedy Channel; Ellesmere Group

Inglefield Land: Cass Fjord *, Cape Wood *, Cape Kent *, Wulff River *, Cape Ingersoll & Cape Leiper, Rensselaer Bay +

Prudhoe Land / Dundas C. York Dist.: Dolomite & Shale, Yellow Sandstone, Red Sandst. +, Narssarssuk, Dundas, Wolstenholme; CRYSTALLINE BASEMENT COMPLEX

NORTHEAST GREENLAND

South Peary Land: Wandel Valley *, Brønlund Fj. *, Sandstone Quartzite Sh., Dolomite (unit 4) +, TILLITE, Sandstone (unit 3), Sandstone (unit 1.)

Danmark Fjord: Amdrup *, Danmark Fjord, Kap Holbaek +, Fyn Sø +, Campanuladal, Norseman-dal, Sandstone

Kronprins Christian Land: Amdrup, TILLITE Rivieradal to Stengærken, Sandstone

East Greenland: Cass Fjord *, Dolomite Point +, Hyolithus Creek *, Ella I. *, Bastion *+, Kløftelv +, Spiral Creek +, Canyon, TILLITE, Eleonore Bay; BASEMENT COMPLEX

SCOTLAND

Northwest Highlands: Sailmhor *, Eilean Dubh +, Ghrudaidh *, Salterella Grit *, Fucoid Beds *+, Quartzite +, Torridonian; CRYSTALLINE BASEMENT COMPLEX

Southern Highlands: Upper *, Middle, Lower, Moine, TILLITE; DALRADIAN

Ny Friesland, Spitsbergen: Oslobreen Limestone *, Oslobreen *, Dolomite, Oslobreen Sandstone, TILLITE, Middle Hecla Hoek, Lower Hecla Hoek

FIG. 10. Correlations between selected areas in Greenland, arctic Canada, Scotland, and Spitsbergen. The table is diagrammatic and not to scale on a spatial or temporal basis and the thicknesses of formations are not represented. Horizontal lines at the same level are meant to be isochronous, however. No attempt is made to indicate degree of certainty of correlation; this is dealt with in the text.

* Fossils. + Trace fossils. The positions of these two symbols are only approximately located with relation to the timescale of the left-hand margin.

respectively. The Danmark Fjord Formation is considered to be the age equivalent of the Brønlund Fjord Formation—both are carbonate successions.

Until further details are available, particularly from north Greenland, the correlation between north-east Greenland and north-west Greenland is a difficult one. A tillite horizon is not known in north-west Greenland but the pre-tillite Proterozoic is probably represented by the Thule Group in the Dundas area which can be correlated with the Equalulik and Uluksan Groups of north-west Baffin Island (which are undoubtedly Proterozoic) and the Kennedy Channel and Ella Bay Formations of central Ellesmere Island. DAVIES and others. (1963) note that the 'Thule Group' at the head of Danmark Fjord (presumably including the Norsemandal Formation) seems to be more closely equivalent to the Dundas area than to Inglefield Land (presumably including the Rensselaer Bay Formation) and they specifically correlate the Wolstenholme Formation of the Thule Group with the Norsemandal Formation. The Pre-Carolinidian, Proterozoic, Norsemandal Formation and other equivalent sandstone formations in north-east Greenland are here considered to be of approximately the same age as the Thule Group of the Dundas area and the Proterozoic sedimentary series of arctic Canada.

The Ellesmere Group, with older Lower Cambrian fossils, correlates with the Rensselaer Bay Formation and may be equivalent to the sequence in Prudhoe Land and, moreover, there seem to be reasonable grounds for correlating these sandstones with unit 3 in south Peary Land and the Campanuladal Formation. This correlation is reinforced by the fact that the Fyn Sø Formation is similar to the dolomite of unit 4 in south Peary Land and both are lithologically similar to the Cape Leiper and Cape Ingersoll Formations.

Pre-Cambrian–Cambrian–Ordovician sequences in Scotland fall into two categories: (a) the geosynclinal belt, the southern Highlands belonging to this region, and (b) the foreland region of north-west Scotland. The tillite known from the southern Highlands correlates to a high degree of probability with the Spitsbergen and east Greenland tillites. A. M. SPENCER, who has worked on the Dalradian Portaskaig Tillite and who is now commencing studies of the East Greenland Tillite Group, is of the opinion (*pers. comm.*) that the two formations are of the same age and exhibit similar successions and lithologies. The fossil horizon of the Leny Limestone in the Upper Dalradian is probably indicative of a late Lower Cambrian age and may be of similar date to the Cape Kent Formation in north-west Greenland, the Ella Island Formation in east Greenland, and the upper beds of the Oslobreen Dolomites in Ny Friesland and their equivalents in other parts of Spitsbergen The Lower and Middle Hecla Hoek and the Moine Series are both older than tillite horizons and probably of similar age.

The foreland region of Scotland correlates with both Spitsbergen and east Greenland, mainly on the basis of the younger strata. There is a close similarity between these regions in lithology and faunas. The Scottish quartzites are characterized in their upper parts by the trace-fossil *Scolithus* and can therefore be considered to be Lower Cambrian and to correlate on lithological similarity with quartzites of the Kløftelv Formation of east Greenland. *Scolithus* has not been found *in situ* in east Greenland but, as already noted, numerous erratic blocks of quartzite with *Scolithus* have been found there, probably derived from the nunatak zone of the ice-cap to the west. The main ground for suggesting age equivalence between the Fucoid Beds of Scotland and the Bastion Formation of east Greenland is the faunal content which includes closely comparable species of *Olenellus*, *Paedeumias*, and *Wanneria* which come from similar grey, highly fissile, arenaceous shales, weathering rusty-brown, interbedded with bands of a hard grey arenaceous mudstone with bands and lenticles of more arenaceous material. The *Salterella* Grit of Scotland, with a great variety of rock types, bears little lithological or palaeontological correlation with the arenaceous limestones of the Ella Island Formation in east Greenland. The Ghrudaidh Group and the *Hyolithus* Creek Formation both have *Salterella* in their lower beds and also have a very close resemblance lithologically, being made up of dark leaden-grey, medium-grained, hard dolomite which weathers in a carious manner to a pale buff colour. Irregular bedding is common and stylolites paralleling the bedding have jet-black, greasy-sheened partings. There is a transition at the top of these formations in both Scotland and Greenland to a paler-grey, more flaggy, dolomite. *Salterella* is probably confined to the Lower Cambrian and is found also in the Oslobreen Dolomite in Ny Friesland, Spitsbergen. The Eilean Dubh Group and the Dolomite Point Formation seem to be unfossiliferous except for stromatolites and the lithological similarity is remarkable; both are predominantly fine-grained to porcellaneous, pale-grey dolomites with disseminations of siliceous bodies, mainly chert, weathering to a pale-buff, carious surface. Stromatolites are often silicified and similar forms are seen in Scotland and east Greenland. There seems to be a strong possibility that Upper Cambrian times are unrepresented by deposits in Spitsbergen, Scotland, and Greenland; in any case no fossils of that age are known from these boreal regions.

The Torridonian, which is unconformably overlain with strong angular discordance by the Cambrian quartzites in the foreland region of Scotland, is considered to be pre-tillite and late Pre-Cambrian in age. Age relationships in the Pre-Cambrian of Scotland have been subject to controversy but the Moine metamorphic assemblage is here considered to be partly equivalent in age to the Torridonian. Pre-Cambrian age relationships and correlations are not, however, relevant to the present contribution and are only considered at all, as mentioned in the introduction, because they have an indirect bearing on Cambrian stratigraphy.

The faunal provinces of the Cambrian of arctic North America (including Greenland), Spitsbergen, and Scotland have been the subject of a good deal of comment. The concept of major faunal realms and provinces (COWIE, 1960, p. 60) may still be used, however, to some advantage. If the Lower Cambrian of the world can be considered to show two major realms—the olenellid and the redlichiid—then the olenellid realm may be divided into: (1) the Acado–Baltic or Atlantic Province with characteristic trilobites which include *Callavia, Holmia, Kjerulfia, Protolenus, Strenuaeva,* and *Strenuella;* and (2) the Pacific Province with characteristic trilobites with include *Bathynotus, Bonnia, Bonniella, Olenellus, Paedeumias,* and *Protypus.* Although there is some admixture of provinces the details of faunas from the regions described in this contribution seem to belong in the main to the Pacific Province. Atlantic Province elements may be suggested by the presence in east Greenland of one specimen of *Holmia;* of *Calodiscus* in east Greenland and Spitsbergen; of *Strenuaeva* and *Holmia* in north-west Greenland. Thus in Lower Cambrian times there appear to have been incursions of Atlantic Province faunal elements into the Pacific Province in east Greenland and in north-west Greenland. It must be admitted, however, that faunas are sparse and poorly preserved in many of the localities so that evidence which is sufficient to establish provincial identity without doubt is lacking so far.

Middle Cambrian faunas are known from these regions only in north and north-west Greenland and Ellesmere Island. They seem to be most closely related to the Cordilleran faunas of North America which are part of the Pacific Province (V. POULSEN, 1964). The recent discovery of trilobites in Nyeboe Land (page 364), which include agnostid species allowing a correlation with the Atlantic Province fauna in south-east Newfoundland, shows that an incursion may have taken place just before Upper Cambrian times. VALDEMAR POULSEN has extensively discussed Middle Cambrian Pacific–Atlantic correlation (1964, p. 68).

Acknowledgments

I wish to record my sincere thanks to Professor D. L. DINELEY and Sir JAMES STUBBLEFIELD for reading the manuscript and making many helpful suggestions; cooperation was also a pleasure with A. W. A. RUSHTON, A. M. SPENCER, and F. K. NORTH.

Selected Bibliography

ADAMS, P. J. and COWIE, J. W. (1953). A geological reconnaissance of the region around the inner part of Danmark Fjord, Northeast Greenland. *Medd. Grønland*, **111**, No. 7.

BENTHAM, R. (1936) *in* HUMPHREYS, N. Oxford University Ellesmere Land Expedition. Appendix I: Geology. *Geog. J.* **87**, 427.

BENTHAM, R. (1941). Structure and glaciers of southern Ellesmere Island. *Geog. J.* **97**, 36.

BERTHELSEN, A. and NOE-NYGAARD, A. (1965). The Precambrian of Greenland, in *The Precambrian* **2**, ed. K. RANKAMA. New York.

BLACKADAR, R. G. (1957). The Proterozoic stratigraphy of the Canadian Arctic Archipelago and Northwestern Greenland in *The Proterozoic in Canada* ed. J. E. GILL. *Roy. Soc. Can.* Spec. Pub. 2., 93.

BLACKADAR, R. G. and CHRISTIE, R. L. (1963). Geological reconnaissance, Boothia Peninsula, and Somerset, King William, and Prince of Wales Islands, District of Franklin. *Geol. Surv. Can.* Paper 63–19.

BLACKADAR, R. G. (1965). Geological reconnaissance of the Precambrian of northwestern Baffin Island, Northwest Territories. *Geol. Surv. Can.* Paper 64–42.

CHRISTIE, R. L. (1962a). Geology, Alexandra Fjord, Ellesmere Island, District of Franklin. *Geol. Surv. Can.* Map 9–1962.

CHRISTIE, R. L. (1962b). Geology, Southeast Ellesmere Island, District of Franklin. *Geol. Surv. Can.* Map 12–1962

CHRISTIE, R. L. (1964). Geological reconnaissance of northeastern Ellesmere Island, District of Franklin. *Geol. Surv. Can.* Mem. 331.

CHRISTIE, R. L. (1967). Bache Peninsula, Ellesmere Island, Arctic Archipelago. *Geol. Surv. Can.* Mem. 347.

CHRISTIE, R. L. (1969). Eastern Devon Island and Southeast Ellesemere Island, District of Franklin. *Geol. Surv. Can.* Paper 69–1A, 231.

CLOUD, P. E. and NELSON, C. A. 1966. Phanerozoic–Cryptozoic and related transitions: new evidence. *Science* **154**, 166.

COWIE, J. W. and ADAMS, P. J. (1957). The geology of the Cambro–Ordovician rocks of Central East Greenland. Part I: Stratigraphy and structure. *Medd. Grønland* **153**, (1).

COWIE, J. W. (1960). Notes on Lower Cambrian stratigraphy in the boreal regions. *Intern. Geol. Congr., Norden* **8**, 57.

COWIE, J. W. (1961). Contributions to the geology of North Greenland. *Medd. Grønland* **164**, (3).

COWIE, J. W. (1962). Geology, p. 255 *in* The Devon Island Expedition of the Arctic Institute of North America. *Arctic* **14**, 252.

COWIE, J. W. (1968). Contributions to Canadian Palaeontology. Lower Cambrian faunas from Ellesmere Island, District of Franklin. *Geol. Surv. Can.* Bull. **163**, 1.

DAVIES, W. E., KRINSLEY, D. B., and NICOL, A. H. (1963). Geology of the North Star Bugt area, Northwest Greenland. *Medd. Grønland* **162**, (12).

DAWES, P. R. (1966). Lower Palaeozoic geology of the western part of the North Greenland fold belt. *Grønlands Geol. Undersøgelse*, Nr. 11.

DINELEY, D. L. (1966). Geological studies in Somerset Island, University of Ottawa Expedition, 1965. *Arctic* **19**, 270.

EHA, S. (1953). The pre-Devonian sediments on Ymers ø, Suess Land and Ella ø (East Greenland) and their tectonics. *Medd. Grønland* **111**, (2).

ELLITSGAARD-RASMUSSEN, K. (1950). Preliminary report of the geological field work carried out by the Danish Peary Land Expedition in the year 1949–1950. *Med. Dansk. Geol. Foren.* **11**, 589.

ELLITSGAARD-RASMUSSEN, K. (1955). Features of the geology of the folding range of Peary Land, North Greenland. *Medd. Grønland* **127**, (7).

ETHERIDGE, R. (1878). Palaeontology of the Coasts of the Arctic Lands visited by the late British Expedition under Captain Sir GEORGE NARES, R.N., K.C.B., F.R.S. *Q. Jl. geol. Soc. Lond.* **34**, 556.

FEILDEN, H. W. and DE RANCE, C.E. (1878). Geology of the coasts of the arctic lands visited by the late British Expedition under Capt. Sir GEORGE S. NARES. *Geog. J.* **34**, 556.

FORTIER, Y. O., McNAIR, A. H., and THORSTEINSSON, R. (1954). Geology and petroleum possibilities in Canadian arctic islands. *Bull. Am. Assoc. Petrol. Geol.* **38**, 2075.

FORTIER, Y. O. *et al.* (1963). Geology of the north-central part of the arctic archipelago Northwest Territories (Operation Franklin). *Geol. Surv. Can.* Mem. 320.

FRÄNKL, E. (1953). Die geologische Karte von Nord–Scoresby Land (NE-Grönland). *Medd. Grønland* **113**, (6).

FRÄNKL, E. (1954). Vorläufige Mitteilung über die Geologie von Kronprins Christians Land. *Medd. Grønland* **116**, 52.

FRÄNKL, E. (1955a). Weitere beiträge zur Geologie von Kronprins Christians Land. *Ibid* **103**, (7).

FRÄNKL, E. (1955b). Rapport über die Durchquerung von Nord Peary Land (Nordgrönland) im Sommer 1953. *Medd. Grønland* **103**, (8).

FRÄNKL, E. (1956). Some general remarks on the Caledonian mountain chain of East Greenland. *Medd. Grønland* **103**, (11).

HALLER, J. (1956). Die Strukturelemente Ostgrönlands zwischen 74° and 78° N. *Medd. Grønland* **154**, (2).

HALLER, J. (1961a). The Carolinides: An orogenic belt of late pre-Cambrian age in North-east Greenland. *Geol. Arctic* **1**, 155.

HALLER, J. (1961b). Account of the Caledonian orogeny in Greenland. *Geol. Arctic* **1**, 170.

HALLER, J. and KULP, J. L. (1962) Absolute age determination in East Greenland. *Medd. Grønland* **171**, (1).

HOLTEDAHL, O. (1913a). The Cambro–Ordovician Beds of Bache Peninsula and the neighbouring regions of Ellesmere Land. *Report of the Second Norwegian Arctic Expedition in the 'Fram'* 1898–1902, 3, (28), 1–14. Videnskabs–Selskabets i Kristiania. *Kristiania.*

HOLTEDAHL, O. (1913b). On the Fossil Faunas from Per Schei's series B in Southwestern Ellesmere Land. *Ibid.* 4, (32), 1–48.

HOLTEDAHL, O. (1917). Summary of geological results. *Ibid.* No. 36.

HUMPHREYS, *et al.* (1936). Oxford University Ellesmere Land Expedition. *Geog. J.* **87**, 385.

KATZ, H. R. (1952). Ein Querschnitt durch die Nunatakzone Ostgrönlands (ca. 74° n.B.). Erbegnisse einer Reise vom Inlandeis (in Zusammenarbeit mit den Expéditions Polaires Françaises von P.–E. Victor) ostwärts bis in die Fjordregion, ausgeführt in Sommer 1951. *Medd. Grønland* **144**, (8).

KATZ, H. R. (1961). Late Precambrian to Cambrian stratigraphy in East Greenland. *Geol. Arctic* **1**, 299.

KERR, J. W. (1967a). Stratigraphy of central and eastern Ellesmere Island, arctic Canada. Pt. 1. Proterozoic and Cambrian. *Geol. Surv. Can.* Paper 67–27 Pt. I.

KERR, J. W. (1967b). Nares submarine rift valley and the relative rotation of North Greenland. *Bull. Can. Pet. Geol.* **15**, 483.

KING, P. B. (1959). *The evolution of North America.* Princeton.

KOCH, L. (1920). Stratigraphy of Northwest Greenland. *Medd. Dansk. Geol. Foren.* **5**, No. 17.

KOCH, L. (1923). Preliminary report upon the geology of Peary Land, arctic Greenland. *Am. J. Sci.* **5**, 189.

KOCH, L. (1925). The geology of North Greenland. *Am. J. Sci.* **9**, 271.

KOCH, L. (1926). A new fault zone in Northwest Greenland. *Am. J. Sci.* **12**, 301.

KOCH, L. (1929a). The geology of East Greenland. *Medd. Grønland* **73**, Pt. 2, No. 1.

KOCH, L. (1929b). The geology of the south coast of Washington Land. *Medd. Grønland* **73**, Pt. 1, No. 1.

KOCH, L. (1929c). Stratigraphy of Greenland. *Medd. Grønland* **73**, (2) afd., Nr. 2.

KOCH, L. (1933). The geology of Inglefield Land. *Ibid.* **73**, (2).

KOCH, L. (1935). *Geologie von Grönland.* (Geologie der Erde.) Berlin.

KULLING, OSKAR (1930). Stratigraphic studies on the geology of North-east Greenland, *Medd. Grønland* **74**, (13).

KURTZ, V. E. and WALES, D. B. (1950). Geology of the Thule Area, Greenland. *Proc. Oklahoma Acad. Sci.* **31**, 83.

KURTZ, V. E. McNAIR, A. H., and WALES, D. B. (1952). Stratigraphy of the Dundas Harbour area, Devon Island, Arctic Archipelago. *Amer. Journ. Sci.* **250**, 636.

LEMON, R. R. H. and BLACKADAR, R. G. (1963). Admiralty Inlet area, Baffin Island, district of Franklin *Geol. Surv. Can.* Mem. 328.

McNAIR, A. H. (1965). Pre-Cambrian metazoan fossils from the Shaler Group, Victoria Island Canadian Archipelago. *Geol. Soc. Amer.* Program 1965 Annual Meetings, Abstracts, 106.

MOORE, R. C. (1962). Treatise on Invertebrate Paleontology, Part W, Miscellanea, Univ. of Kansas Press.

MUNCK, SOLE (1941). Geological observations from the Thule district in the summer of 1936. *Medd. Grønland* **124**, (4).

NATHORST, A. G. (1901). Bidrag til Nordösta Grönlands geologi. *Geol. Fören. i. Stockholm Förh.* **23**, (4).

NORFORD, B. S. (1968). Contributions to Canadian Paleontology. A Middle Cambrian *Plagiura-Poliella* faunule from Southwest District of Mackenzie. *Geol. Surv. Can.* Bull. 163, 29.

NORRIS, A. W. (1963). Cambrian stratigraphy, pp. 29–31 *in* FORTIER *et al.*, Geology of the north-central part of the arctic archipelago Northwest Territories (Operation Franklin). *Geol. Surv. Can.* Mem. 320.

PEACOCK, J. D. (1956). The geology of Dronning Louise Land, N.E. Greenland. *Medd. Grønland*, **137**, (7).

PEACOCK, J. D. (1958). Some investigations into the geology and petrography of Dronning Louise Land, N.E. Greenland. *Medd. Grønland* **157**, (4).

POULSEN, CHR. (1927). The Cambrian, Ozarkian and Canadian faunas of Northwest Greenland. *Medd. Grønland* **70**, 233.

POULSEN, CHR. (1930). Contributions to the stratigraphy of the Cambro–Ordovician of East Greenland. *Medd. Grønland* **74**, 297.

POULSEN, CHR. (1932). The Lower Cambrian faunas of East Greenland. *Ibid.* **87**, 1.

POULSEN, CHR. (1946). Notes on Cambro–Ordovician fossils collected by the Oxford University Ellesmere Land Expedition 1934–1935. *Q. Jl. geol. Soc. Lond.* **102**, 299.

POULSEN, CHR. (1956). The Cambrian of the East Greenland geosyncline. *Int. Geol. Cong., Mex.* **1**, 59.

POULSEN, CHR. (1958). Contribution to the palaeontology of the Lower Cambrian Wulff River Formation. *Medd. Grønland* **162**, 1.

POULSEN, CHR. (1960). Notes on some Lower Cambrian fossils from French West Africa; *Mat. Fys. Medd. Dan. Vid. Selsk.* **32**, 1.

POULSEN, V. (1964). Contribution to the Lower and Middle Cambrian palaeontology and stratigraphy of Northwest Greenland. *Medd. Grønland* **164**, (6).

REZAK, R. (1957). Stromatolites of the Belt Series in Glacier National Park and vicinity, Montana. *U.S.G.S. Prof. Paper* 294-D, 124.

ROZANOV, A. YU. (1967). The Cambrian Lower Boundary Problem. *Geol. Mag.* **104**, 415.

SCHAUB, H. P. (1950). On the Pre-Cambrian to Cambrian sedimentation in NE-Greenland. *Medd. Grønland* **114**, (10).

SCHAUB, H. P. (1955). Tectonics and morphology of Kap Oswald (NE-Greenland). *Medd. Grønland* **103**, (10).

SCHEI, P. (1903a). Summary of Geological results. *Geog. J.* **22**, 56.

SCHEI, P. (1903b). Preliminary Report on the Geological Observations made during the Second Norwegian Polar Expedition of the 'Fram'. Royal Geog. Soc. *London*.

SCHEI, P. (1904). Preliminary account of the geological investigations made during the Second Norwegian Polar Expedition in the *Fram*, 1898–1902 *in* OTTO SVERDRUP, *New Land* **2**, 455. London.

SOKOLOV, B. S. (1968). Stratigraphic Boundaries of Lower Paleozoic Systems. *Int. Geol. Cong., Czech.* **9**, 31.

SOMMER, M. (1957). Geologie von Lyells Land (NE-Grönland). *Medd. Grønland* **155**, (2).

STOCKWELL, C. H. (ed.) (1957). Geology and Economic Minerals of Canada. *Geol. Surv. Can.* Econom. Geol. Ser. No. 1 (4th. ed.).

STUBBLEFIELD, C. J. (1956). Cambrian palaeogeography in Britain. *Int. Geol. Cong., Mex.* **1**, 29.

TAYLOR, A. (1956). Physical geography of the Queen Elizabeth Islands, Canada. American Geographical Society, New York.

TEICHERT, C. (1939). Geology of Greenland. *In* Geology of North America, **1**. *Geologie der Erde.* Berlin.

THORSTEINSSON, R. (1958). Cornwallis and Little Cornwallis Islands, District of Franklin, Northwest Territories. *Geol. Surv. Can.* Mem. 294.

THORSTEINSSON, R. and TOZER, E. T. (1960). Summary account of structural history of the Canadian Arctic Archipelago since Pre-Cambrian Time. *Geol. Surv. Can.* Paper 60–7.

THORSTEINSSON, R. and TOZER, E. T. (1962). Banks, Victoria and Stefansson Islands, Arctic Archipelago. *Geol. Surv. Can.* Mem. 330.

THORSTEINSSON, R. (1963). Copes Bay, Ellesmere Island *in* Operation Franklin, Fortier *ed., Geol. Surv. Canada* Mem. 320, 386.

TOZER, E. T. and THORSTEINSSON, R. (1964). Western Queen Elizabeth Islands, Arctic Archipelago. *Geol. Surv. Can.* Mem. 332.

TRETTIN, H. P. (1964). Pre-Mississippian rocks of Nansen Sound area, District of Franklin. *Geol. Surv. Can.* Paper 64–26.

TRETTIN, H. P. (1965). Lower Palaeozoic sediments of northwestern Baffin Island, District of Franklin. *Geol. Surv. Can.* Paper 64–47.

TROELSEN, J. C. (1949). Contributions to the geology of the area around Jørgen Brønlunds Fjord, Peary Land, North Greenland. *Medd. Grønland* **149**, (2).

TROELSEN, J. C. (1950a). Section on geology *in* WINTHER, P. C. A preliminary account of the Danish Peary Land Expedition 1948–1949. *Arctic* **3**, (1).

TROELSEN, J. C. (1950b). Contributions to the geology of Northwest Greenland, Ellesmere Island and Axel Heiberg Island. *Medd. Grønland* **149**, (7).

TROELSEN, J. C. (1956a). The Cambrian of North Greenland and Ellesmere Island. *Int. Geol. Cong., Mex.* **1**, Pt. 1, 71.

TROELSEN, J. C. (1956b). Greenland *in* Lexique Stratigraphique International. **1**, (1a). *Paris.*

TUKE, M. F., DINELEY, D. L., and RUST, B. R. (1966). The basal sedimentary rocks in Somerset Island, N.W.T. *Can. Journ. Earth Sci.* **3**, 697.

WASHBURN, A. L. (1947). Reconnaissance geology of portions of Victoria Island and adjacent regions Arctic Canada. *Geol. Soc. Amer.* Mem. 22.

WENK, E. (1961). On the crystalline basement and the basal part of the pre-Cambrian Eleonore Bay Group in the south-western part of Scoresby Sound. *Medd. Grøland* **168**, (1).

WENK, E. and HALLER, J. (1953). Geological explorations in the Petermann region, western part of Fraenkels Land, East Greenland. *Medd. Grønland* **111**, (3).

WORDIE, J. M. (1938). An expedition to North-west Greenland and the Canadian Arctic in 1937. *Geog. J.* **92**, 398–399.

THE CAMBRIAN OF SOUTH AMERICA

A. V. Borrello

Department of Geology, Faculty of Natural Sciences, National University of La Plata,
Argentina

1. Historical review

Argentina and Bolivia are the countries which, by reason of the important extent of their outcrops of Lower Palaeozoic strata, provide a firm basis for the development of investigations into the problems of the South American Cambrian.

We owe the first reference to the Cambrian rocks to KAYSER (1876), who described the Cambrian trilobite fauna collected by LORENTZ and HYERONIMUS in the Cordillera Oriental of northern Argentina and southern Bolivia. KAYSER attributed an Early Cambrian age to this fauna without providing further geological data. These fossils have subsequently been considered by HARRINGTON and LEANZA (1957) to be of Ordovician age. KAYSER (1897) later classified another association, collected by VALENTÍN in northern Argentina (on the border of Salta–Jujuy), as indicating the presence of the 'Middle Cambrian', into which he incorporated his previous determinations (1876). The Ordovician age of these fossils is also now accepted. In the decade between 1880 and 1890, northern Argentina was extensively surveyed by BRACKEBUSCH (1891); the results of his work were recorded on an exploratory geological map which comprised the area up to the southern border of Bolivia. In this map the reference to the Cambrian was very generalized, and included other sedimentary series (e.g. Pre-Cambrian and Ordovician rocks), which have later been discriminated as the result of more systematic and detailed work.

The first investigator to recognize the Cambrian rocks was KEIDEL (1907) during his investigations in the Cordillera Oriental of Argentina. In the Sierra de Zenta, (Jujuy), he observed and separated a group of arenites from the folded and metamorphosed phyllites, schists, and quartzites. The arenites were attributed to the Cambrian, being covered by other Lower Palaeozoic sediments (lower Silurian) and characterized by a simple structural style. In his subsequent works, KEIDEL (1912, 1914), among other data concerning the Cambrian, sheds light on the primary tectonic relationships between the Proterozoic basement and its Cambrian cover. In one of his last publications on the Lower Palaeozoic of northern Argentina (1943), KEIDEL brought up to date the relationships with the Ordovician deposits and made further observations on the Cambrian sediments. KEIDEL also introduced in 1937 a stratigraphical subdivision of the Cambrian of the Cordillera Oriental (*in* HARRINGTON, 1937).

In the southern zone of Bolivia, STEINMANN and HOEK (1912) assigned to the Cambrian a fossiliferous section of the Lower Ordovician. This section is exposed to the west of Tarija, nearly on the border of Argentina. The real local Cambrian, which is a northern prolongation of that found in the Cordillera Oriental in northern Argentina, has been named the Sama Formation by AHLFELD and BRANIŠA, who also describe in the area fossiliferous Cambrian rocks which diverge from the correlations established in Argentina (HARRINGTON and LEANZA, 1957). The so-called Cambrian associations (Cambro–Ordovician), described by KOBAYASHI (1937) from Argentina and Bolivia, are also held to be of Ordovician age.

HAUSEN (1925) recognized as Cambro–Silurian, whitish and reddish quartzites which he descriptively designated 'high mountain quartzites', indicating how they compared with the strata which KEIDEL (1907) among others described in northern Argentina.

Mention must be made of the stratigraphical contributions to the Cambrian in northern Argentina furnished by DANIEL (1940), DE FERRARIIS (1940), and HERRERO DUCLOUX (1940), whose doctorates were supervised by KEIDEL.

During the last few years TURNER (1959, 1960, 1963a, 1964b) and VILELA (1956, 1961) have carried out various geological surveys of areas with Cambrian deposits in the Cordillera Oriental. The former introduced a regional stratigraphical terminology for the Cambrian (Mesón Group), and prepared an up-to-date summary of the Cambrian of northern Argentina (TURNER, 1963a).

The extensive Cambrian area of the Precordillera of San Juan and Mendoza, in the western centre of Argentina, has its own history of geological investigations.

In 1873 STELZNER collected remains of trilobites in the Precordillera of San Juan (Sierra de Zonda) which were studied by KAYSER who described them as belonging to an association of 'primordial and Infra-silurian' fossils (KAYSER, 1876). HARRINGTON and LEANZA (1943) determined that these fossils, and others collected by KEIDEL and HARRINGTON in the Sierra de Zonda between 1937 and 1942, in reality belong to the marine facies of the Middle Cambrian. This was the first confirmation of the presence of fossiliferous Cambrian rocks in South America.

On the basis of some finds by RUSCONI (1945a), LEANZA (1947) showed that the fossiliferous limestones of San Isidro in the extreme south of the Precordillera of Mendoza belong to the Middle Cambrian and contain remains of a fauna of *Kootenia*. Between 1945 and 1962 RUSCONI collected and described a large amount of fossiliferous material which indicated the presence of horizons containing Middle and Upper Cambrian faunas. The largest palaeontological collections obtained came from the north of Mendoza.

Between 1954 and 1955 CHRISTIAN and VALDEMAR POULSEN examined part of the extensive outcrops of the Cambrian beds which in the Precordillera extend over the southern part of San Juan and the northern part of Mendoza. V. POULSEN (1958) produced a brief and detailed study of the Middle Cambrian, determining the biostratigraphical characteristics of the *Glossopleura* Zone. He indicated the palaeontological similarities of this area with those of the Arrojos Formation of Mexico. CHR. POULSEN (1960) made a specific study of the Cerro Solitario area (Cerro Blanco) in northern Mendoza, establishing the presence of *Bolaspidella*, *Baltagnostus*, and acado-baltic agnostids, which he correlated with horizons containing *Paradoxides forchhammeri* in the northern hemisphere. The determinations of the trilobites were revised in both cases, introducing forms which are figured in their papers.

The discovery of the olenellidian Lower Cambrian was made by BORRELLO (1963a, 1964), firstly in the Sierra de Villicúm and then in the Sierra de Zonda to the north and south of the San Juan River, in the Precordillera of San Juan. The same author described the first Lower Cambrian fossils in these areas. Detailed studies of this assemblage of Lower Cambrian fossils were later initiated in Paris by HUPÉ and BORRELLO in 1965; work which is still in progress. A previous classification of the biostratigraphical zones of the Cambrian of the Precordillera of western Argentina has recently been presented by BORRELLO (1965), indicating the palaeogeographical relationships which are equivalent to those established in the United States.

HARRINGTON and KAY (1951) described remains of trilobites obtained from the Cordillera Oriental of Colombia (Rio Duda) which they identified as species of the genus *Ehmania* of the Middle Cambrian. The mention of acado-baltic trilobites in this sector is more recent, remains of *Paradoxides* having been described by RUSHTON (1963).

Outside of the Andean Cordillera efforts to prove the existence of Cambrian deposits have been hampered by the lack of palaeontological evidence. HARRINGTON (1962) has made a reappraisal of observations made to this purpose in Brazil, occurring in a number of studies extending from that of DERBY (1906) to that of MORALES (1959), in order to improve knowledge of the Cambrian history of this area. OLIVEIRA (1956) and GUIMARAES (1964) have contributed to this discussion with their condensed papers on the stratigraphy of the Lower Palaeozoic rocks in the Brazilian cratonic area. HARRINGTON (*op. cit.*) also refers to a similar possibility of a Cambrian age for the formations of clastic rocks described by CECIONI (1956) in the southern archipelago of Chile. This is a view shared in principle by the present author as the result of reviewing the extent of the geosynclinal evolution of southern Patagonia. Also of uncertain age are beds outcropping on the basement east of Corumbá, in eastern Bolivia, where the probable presence of Cambrian rocks has been proposed (AHLFELD and BRANIŠA, 1960) as the result of the finding of algae (BEURLEN and SOMMER, 1957).

The geosynclinal evolution of the Cambrian in Argentina is among the problems which are being studied by the present author in connection with geotectonic aspects of Argentina.

2. Regional distribution

The proven extent of Cambrian deposits in South America includes the Andean belt where outcrops and fossiliferous localities have been cited in Colombia and Argentina. In the sub-Andean area (Bolivia), and cratonic region (Brazil), the Cambrian age of beds attributed to the Lower Palaeozoic is generally uncertain, though it will be shown later that more definite conclusions can be reached concerning the time of accumulation of certain of these deposits.

2.1. Proven Cambrian deposits

In the South American continent there are three regions which contain sediments of proven Cambrian age: the Cordillera Oriental of Colombia; the Cordillera Oriental of northern Argentina, including the southern border of Bolivia; and the Precordillera of western Argentina (Fig. 1).

The information available on the Cambrian rocks of Colombia is very limited. HARRINGTON and KAY (1951) refer only to the area of the Río Duda (Quebrada de Agua Bonita) in the region of the Cordillera Oriental, which lies between the Río Magdalena and the headwaters of the Río Guaviare, approximately latitude 3° 20′ N., north-east of Uribe. The locality is somewhat imprecisely defined and the identification is solely based on the finding of loose blocks derived from outcrops not seen. RUSHTON (1963) enlarged somewhat on this information as he had new palaeontological material available for study, giving the locality, according to the collectors, as being the Uribe Trail to the west of Bogotá.

In the Cordillera Oriental of northern Argentina (Jujuy and Salta) and southern Bolivia the Cambrian outcrops extend between latitude 21° 25′ S. and latitude 25° S. The area common to both countries coincides with the mountainous massif of the Sierra and Cordillera of Santa Victoria, which is cut by the international border approximately on the latitude 22° 05′ S. In Bolivian territory, the zone of the desert plain and the hills of Tacsara in the extreme north, and the continuation of the Sierra of Santa Victoria have a Cambrian extension and clearly terminate the expanded outcrops of the northern

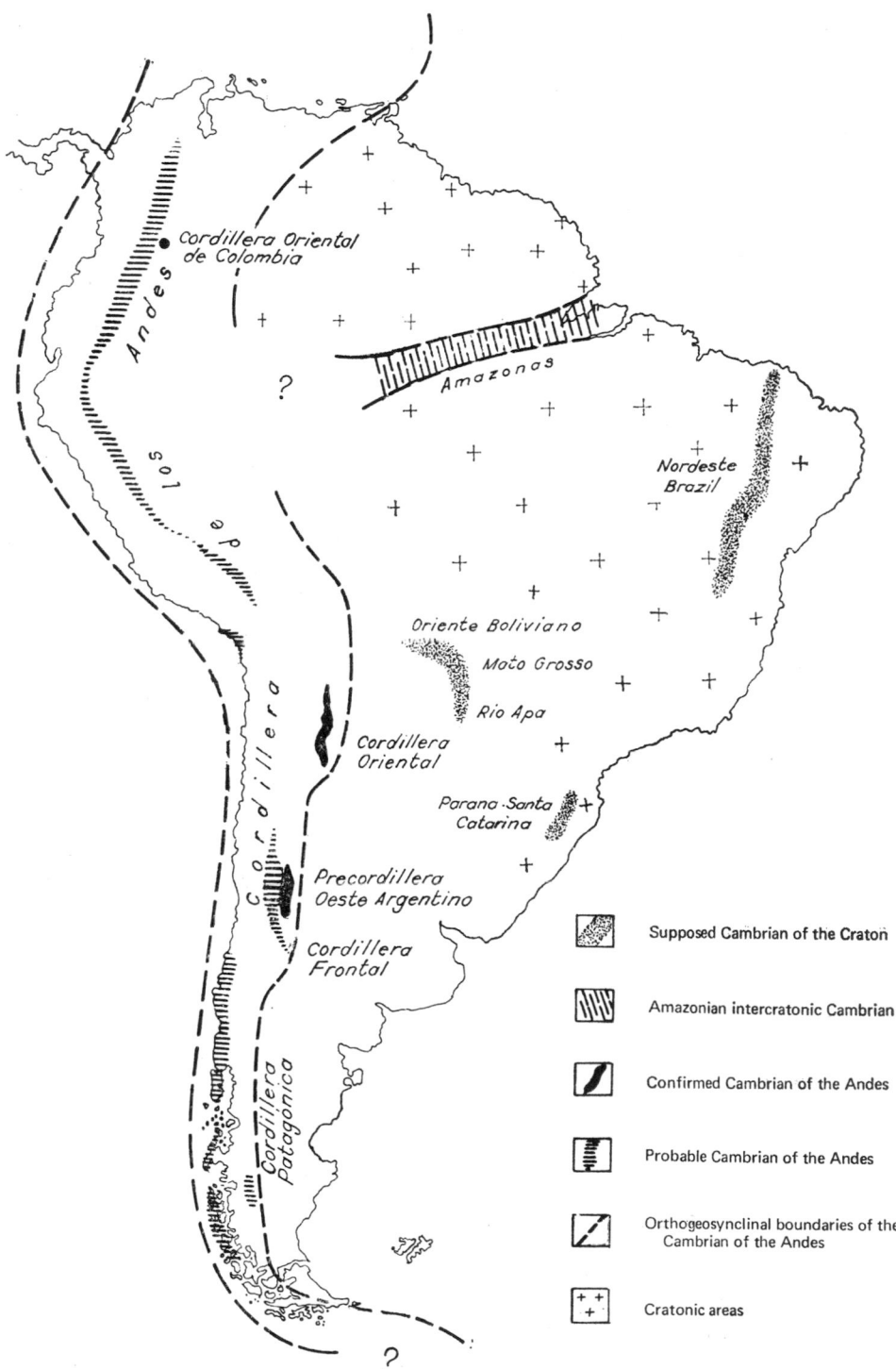

Fig. 1. Sketch schematizing the generalized distribution of the Cambrian in South
America (according to the author, and based on Harrington, 1962).

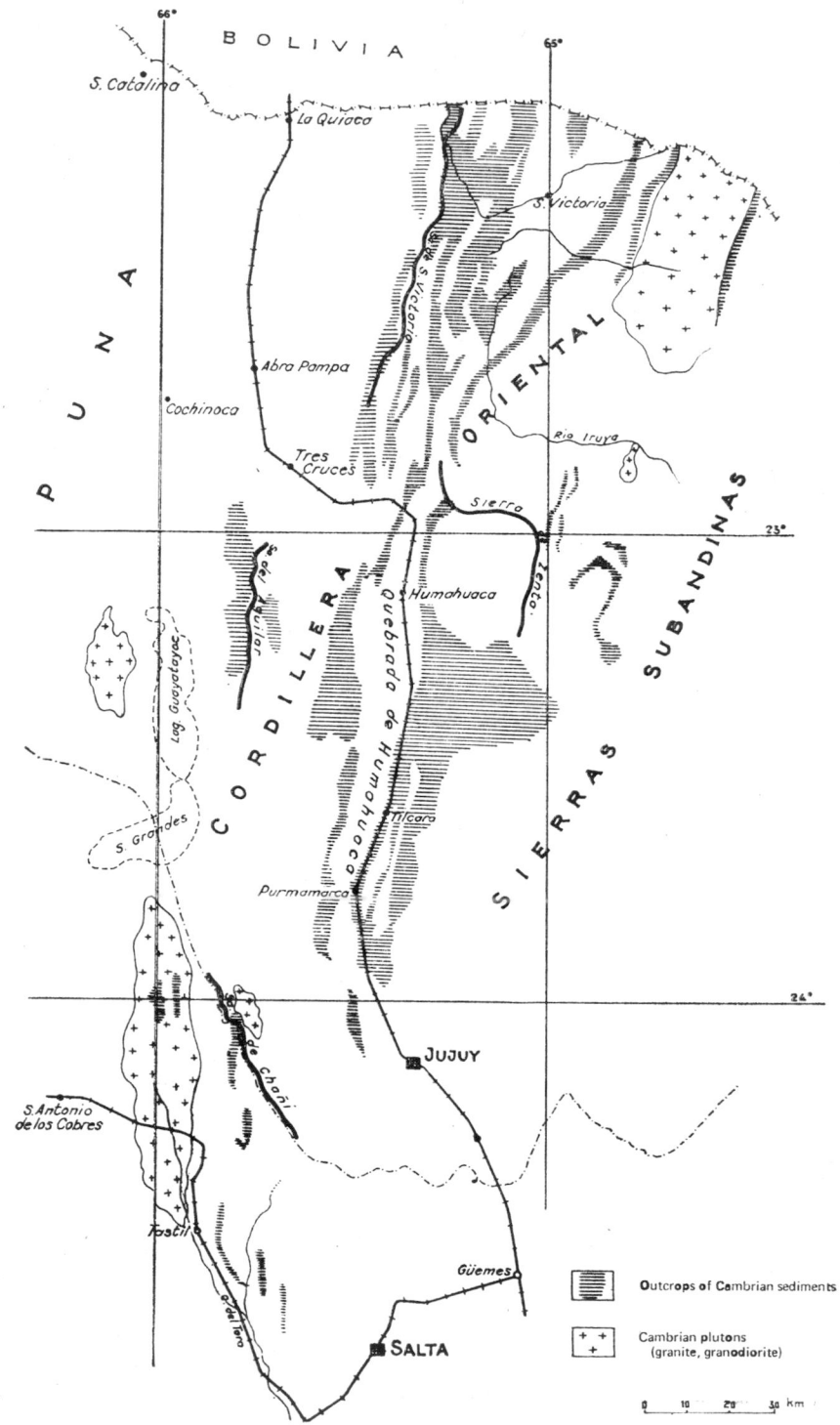

FIG. 2. Distribution of Cambrian rocks in northern Argentina (based on Turner, 1963a).

part of the Cordillera Oriental of Argentina. The limits of this area are approximately defined by meridians 64° 30' W. and 66° W. In this area the hills of Santa Victoria, Zenta, Aguilar, and Chañi (Fig. 2), among others, contain such rocks beneath the surface. It is further thought that these may have a considerable extension towards the east of the western belt of the Sierras Subandinas, as they have recently been found west of San Ramón of the Nueva Orán, near the Tropic of Capricorn, east of the Sierra de Zenta.

The Sierras Subandinas to the east and the Puna to the west, border the Cambrian terrain of the Cordillera Oriental.

The extent of the Precordillera of western Argentina (San Juan and Mendoza) with fossiliferous Cambrian outcrops is limited by latitude 31° S. and approximately latitude 33° S. The lateral limits correspond with meridians 68° 30'W. and 69°10' W. The mountains of Villicúm and Zonda in San Juan and those to the east of Uspallata, from the Cerro Pelado rising on the eastern flanks of the Precordillera from San Isidro to the Cerro Solitario, contain Cambrian rocks (Fig. 3). The northern limits are somewhat indefinite and require further studies before they are finally defined. The southern limit is clear, the structural relationships being observed with the neighbouring terrain, which constitutes the so-called Cordillera Frontal, north of the Mendoza River. Towards the south-west, west, and north-west the Cambrian deposits (Cambro–Ordovician) of the Precordillera seemingly pass in transition into the Cordillera Frontal, outcropping in Uspallata and in the contiguous meridional sectors. Large folds mark the eastern limits and, in San Juan, the Cambrian sediments occur totally separated from the neighbouring cratonic structure which extends into the Sierra Pié de Palo and the valley of Tulúm.

2.2. Outcrops of rocks of uncertain or probable Cambrian age

A Cambrian age* has been discarded for the Lower Palaeozoic beds of Venezuela, which now have been classified as Ordovician on the basis of their trilobite fauna (ROD, 1955). HARRINGTON (1962) adhered to the idea that a clastic succession of conglomerates and pelites which outcrops in Cochabamba in the Cordillera Oriental of Bolivia (AHLFELD and BRANIŠA, 1960) could be of Cambrian age. He added that diverse pre-Devonian non-metamorphic rocks and metamorphic granulites, which in Brazil occur mainly in the form of clastics in the eastern, southern, and Mato Grosso regions of the Amazon River and partly pass on into Bolivia, might be of Cambrian age. In this same area similar sedimentary deposits on an ancient basement were assigned by AHLFELD and BRANIŠA to the Cambrian, following the result of work by various Brazilian authors.

In the middle west of Mendoza outcrops of submetamorphic to metamorphic rocks which appear on the southern flanks of the Cordillera Frontal could be of Cambrian age. In stretches of the Pacific coast of Peru and Chile it is perhaps possible to assign a Cambrian age to the outcrops of metamorphic rocks which have been foliated and subjected to plutonic activity of Palaeozoic age.

The Patagonian Cordillera between latitude 46° 20' and latitude 49° S. has revealed the presence of folded limestones in the sector of Lake Buenos Aires (HEIM, 1940) and quartzites in the Sierra de Sangra, north of Lake San Martin (REYES, 1966), which suggest by reason of their position underneath beds of Palaeozoic 'flysch', firstly a Cambro–

* Stratigraphical Lexicon of Venezuela. Minist. Min., Div. Geol. Sp. Publ., Caracas. 1956.

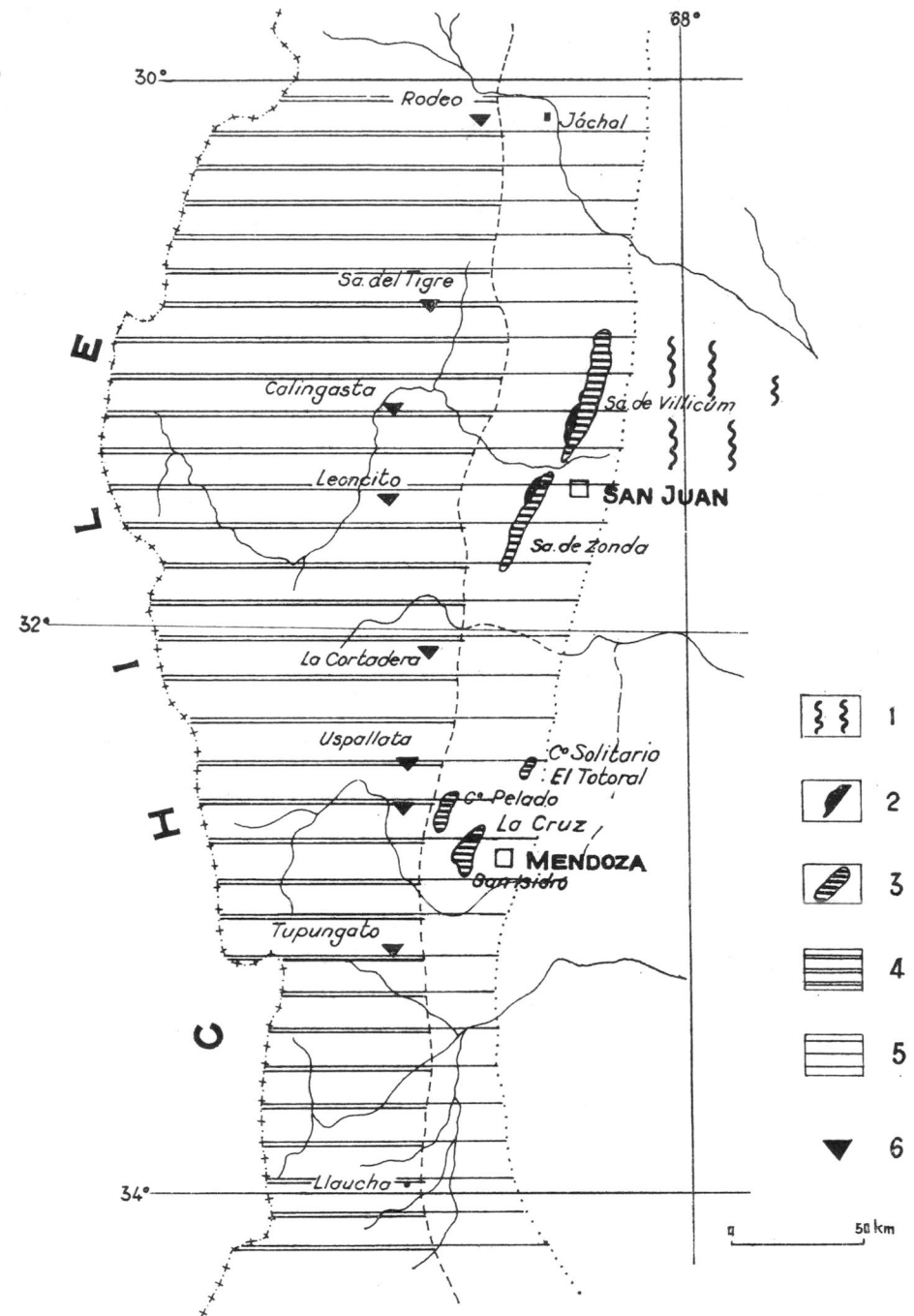

FIG. 3. Orthogeosynclinal character of the Cambrian in western Argentina. *1*, Pre-
Cambrian of the Sierras Pampeanas; *2*, Lower Cambrian and *3*, Middle and Upper
Cambrian of the Precordillera; *4*, Eugeosynclinal area, phyllitic-euxinic (Cambro-
Ordovician) of the Precordillera and Cordillera Frontal; *5*, The calcareous miogeo-
synclinal area; *6*, The Lower Palaeozoic ophiolites (according to the author, 1967).

Ordovician sequence, and, secondly, an authentic miogeosynclinal régime in certain ways comparable to that of the Precordillera of western Argentina. HARRINGTON (1962) equally indicated the possibility of the development of Cambrian rocks in the southern archipelago of Chile, conforming to the data published by CECIONI (1956).

3. Stratigraphy. Typical sections

3.1. Proven stratigraphical sections

Knowledge of the proven Cambrian of the Cordillera Oriental and Precordillera, which is synthetized in the following pages, helps to explain the corresponding stratigraphical development in Bolivia and Argentina.

3.1.1. Cordillera Oriental, area south of Bolivia

AHLFELD and BRANIŠA (1960, p. 32) have indicated by the name of *Sama Formation* an assemblage of Cambrian rocks which in the Cordillera of Santa Victoria lies directly beneath fossiliferous rocks of Lower Ordovician age and is composed of compact quartzose arenites, greyish or reddish in colour, with no outcropping basement. It was indicated that these deposits pass into arenites and lutites containing Cambrian fossils (*sic*) which were in fact assigned to the Ordovician by HARRINGTON and LEANZA (1957). The type section occurs in the Abra de Sama on the highway Villazón-Tarija in the axis of the mountain range of Tacsara. The original geological cross-section provided by STEINMANN and HOEK (1912) (Fig. 4), is of this area. It is possible that the basal Cambrian of Sama is developed very thickly in the subsurface, despite the stratigraphical implications. Only tubes of *Skolithos* sp. have been found in the Cambrian sediments of Sama (AHLFELD and BRANIŠA, *op. cit.*, p. 43).

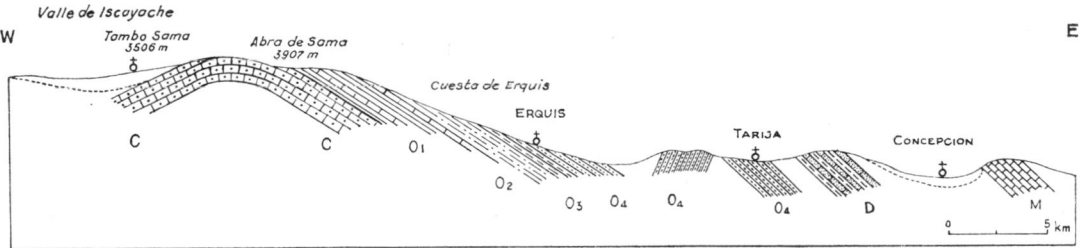

FIG. 4. The Cambrian in the Palaeozoic sequence of the Cordillera Oriental in southern Bolivia. C, Cambrian (Sama Formation); O1–O4, fossiliferous Ordovician; D, Devonian; M, Mesozoic (adapted from Steinmann and Hoek, 1912).

3.1.2. Cordillera Oriental, area of northern Argentina

The development of the Cambrian in this area consists of psammitic sequences which are typical of the whole of the Cordillera Oriental (Fig. 2).

KEIDEL (*in* HARRINGTON, 1937, p. 101) divided the Cambrian of Salta and Jujuy into three lithological units: lower quartzitic arenites; arenites with 'Scolithus' (*sic*), and upper quartzitic arenites. He later (KEIDEL, 1943, p. 108) systematized the classification denoting the units as K_1, K_2, and K_3, a scheme which has been used subsequently by many geologists.

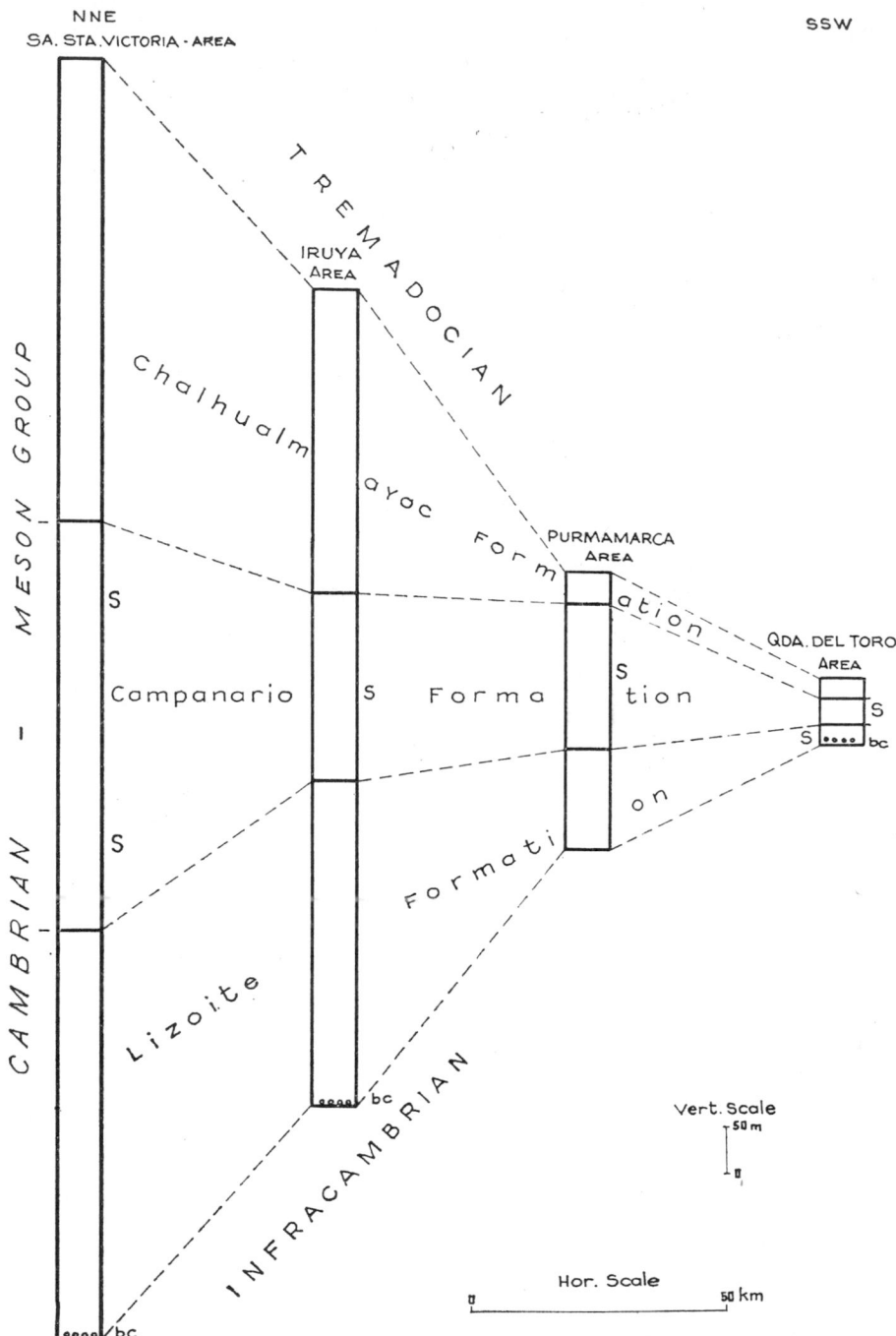

FIG. 5. Correlation scheme of the Cambrian formations of the Cordillera Oriental (Salta and Jujuy), northern Argentina, showing the wedging of the sequence in south-south-western direction. *bc*, basal conglomerate; *S*, *Skolithos* (according to the author, 1967, based on the writings of Keidel, Turner, and Vilela).

TURNER (1960) has proposed revised stratigraphical terms to replace the former and has described for the type section of the Sierra de Santa Victoria (1963a, pp. 197–198) the following ascending sequence (Fig. 5):

> Pre-Cambrian.
> *angular discordance.*

Meson Group—Cambrian

A. *Lizoite Formation*: Basal conglomerate, 10 metres thick, with phenoclasts 3 to 20 centimetres in diameter derived from quartzitic rocks, chloritic schists, and granodiorites from the neighbouring basement. In conformity with the above occur greyish-white silicified arenites showing cross-lamination, and dark green quartzitic schists. Thickness: 1,000 metres.

B. *Campanario Formation*: Reddish and purplish arenites with tubes of *Skolithos* intercalated in the basal and middle parts of the sequence with micaceous schists. The colours of the rocks vary from pink to greyish-white, or reddish to dark red. The sediments are in part lenticular. Thickness: 1,000 metres.

C. *Chalhualmayoc Formation*: Silicified arenites similar to those of the Lizoite Formation in massive clearly defined, beds white or yellowish-white in colour, with intercalations of dark green schists and with cross-lamination. Thickness: 1,100 metres.

> *angular discordance.*
> Lower Ordovician.

The areal distribution of the Chalhualmayoc Formation is greater than that of either of the other two Cambrian formations which have been described. The discordance between the Cambrian and Ordovician rocks of this area can be observed to the west of Cerro Blanco, Jujuy, on the eastern slopes of the Sierra de Santa Victoria (Fig. 6), among other places.

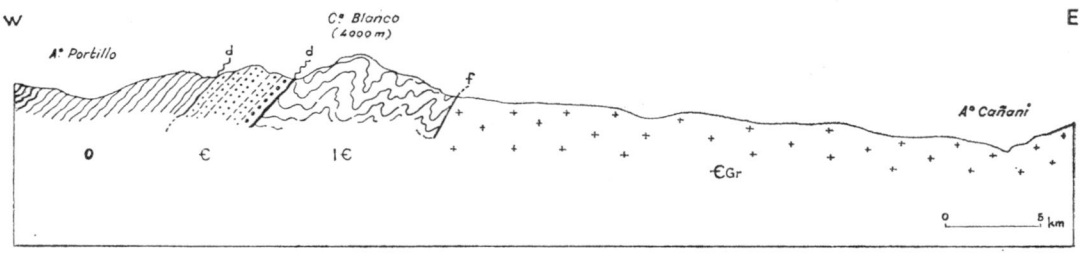

FIG. 6. Partial geological cross-section of the eastern flanks of the Sierra de Santa Victoria in the Cerro Blanco sector. *I*€, Pre-Cambrian; €, Cambrian; *O*, Ordovician; €*Gr*, Cambrian granodiorites; *d*, discordances; *f*, fault (according to Turner, 1960).

The development of the Meson Group in the section of Iruya, studied by VILELA (1961), reveals a thickness of only 2,000 metres for the three formations of the group. The basal conglomerate in this area is 27 metres thick and the maximum recorded development is in northern Argentina. Beneath this, with marked unconformity, is the Pre-Cambrian basement ('Infracambrian'), and the Cambrian is followed, also with marked unconformity, by the overlying fossiliferous Lower Ordovician strata.

The minor stratigraphical development of the Cambrian rocks in northern Argentina (Fig. 5) is shown in the section of the Quebrada del Toro (Quebrada de Incamayo). The

Meson Group, present in aligned outcrops (Fig. 7) according to the studies of KEIDEL (1943, pp 194–198) and VILELA (1956, pp 27–30), is in the following sequence:

Pre-Cambrian.
angular discordance.

Meson Group—Cambrian.
  A. *Lizoite Formation* (o K₁): Discontinuous basal conglomerate up to 2 metres in thickness composed of rounded, pinkish grains of quartzites and Pre-Cambrian quartz; reddish and violet silicified arenites, sometimes with cross-lamination and ripple marks on the bedding planes. Thickness: 40 metres.
  B. *Campanario Formation* (o K₂): Red, reddish, or reddish-brown arenites with intercalations of whitish arenites, which contain abundant tubes of *Skolithos* in thin bands. In parts there are greenish arenites which pass into psammites with similar colour to those just described. Various members in this formation can be distinguished. Thickness: 65–70 metres.
  C. *Chalhualmayoc Formation* (o K₃): Fine and medium-grained compact arenites similar to the psammites of the Lizoite Formation. The colour varies from red to whitish-greyish, yellow, or pink, in thick bands, sometimes with cross-lamination and ripple-marks on the bedding surfaces. Thickness: 45–50 metres.

angular discordance
Lower Ordovician.

The relations between the Cambrian and its Pre-Cambrian basement can be clearly seen in the Quebrada de Humahuaca, Jujuy, particularly in their stratigraphical details; and also immediately to the east of the Quebrada del Toro, Salta (Fig. 8, p. 399) where the rocks are more accessible than in other parts of northern Argentina.

The preceding descriptions reveal a gradual diminishing of the thickness of the Cambrian from north-north-east to south-south-west (Fig. 5). The reduction in thickness is also evident towards both flanks of the Cordillera Oriental.

In Purmamarca, Jujuy, on the right hand side of the Quebrada de Humahuaca, the Campanario Formation (which contains numerous tubes of *Skolithos* which are deep red in colour) shows sedimentological characters of the rhythmical 'flysch' (*orthoflysch*) type. The same Cambrian beds can be observed on the left-hand side of the Quebrada de Purmamarca where they have a vertical disposition (Plate 4). *Skolithos* tubes are predominant in the Campanario Formation of Salta and Jujuy, as is indicated by TURNER's descriptions (1963a, pp 197–202). In the area of the Quebrada del Toro, Salta, VILELA (1956, pp 29–30) indicated the presence of tubes of *Skolithos* in the three divisions of the outcropping Cambrian. The present author found some remnants of markings of the problematical *Corophioides* (traces of annelida?) in fragments of pinkish quartzitic rock which cover the floor of the Quebrada de Humahuaca between Tilcara and Humahuaca. Five kilometres to the north of Tilcara, in the Puerta de Juella above the Quebrada de Humahuaca, HERRERO DUCLOUX (1940, p. 10) has also found remains of poorly preserved brachiopods in quartzitic arenites.

Outcrops of quartzites and calcareous quartzites, grey to reddish in colour, which are approximately 2,000 metres thick form the base of a considerable succession in the area of the Aguilar Mine, Jujuy, from whence SPENCER (1950, pp 405 and 410) described them as 'Aguilar Quartzites', attributing them to the Cambrian. TURNER (1963a, fig. 2)

Plate 4. Cambrian bedded flysch, Campanario Formation, on the left slope of the Quebrada de Purmamarca, to the west of the Quebrada de Humahuaca, Cordillera Oriental, Jujuy, Argentina, partly obscured by drift (on the right). (See page 396).
(Photograph by the author, 1967).

Plate 5. Limestone of the Upper Cambrian in the northern Aguja (3,300 meters o.d.) of the Cerro Pelado, Precordillera of Mendoza, Argentina, bordered by faults. In the foreground is seen the folded structure of the Palaeozoic and Triassic rocks of the Sierra de Uspallata. In the background are seen the rugged crests of the high Cordillera of the Andes (Mesozoic).
(Photograph by the author, 1965).   (To face page 396)

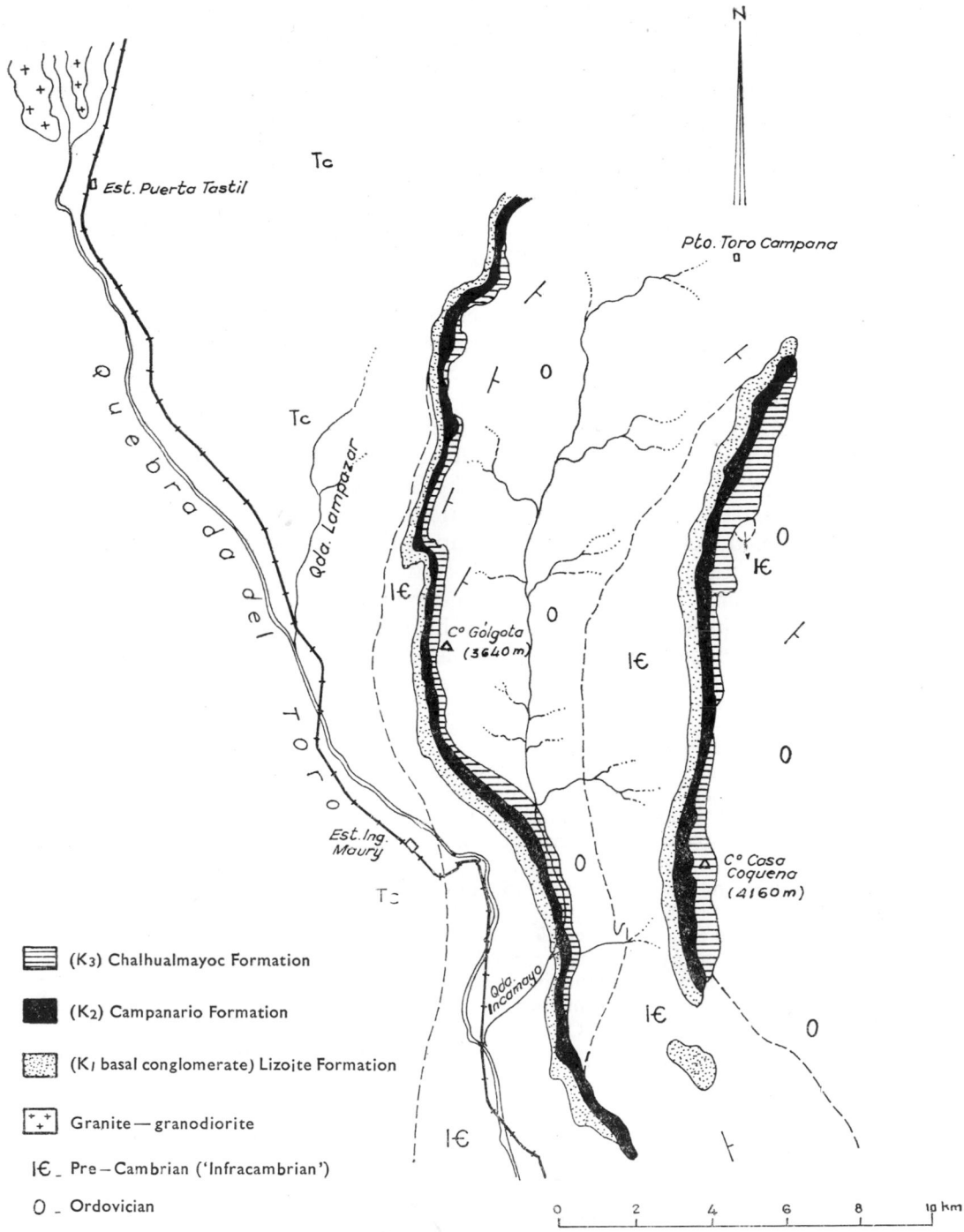

Fig. 7. Distribution of the typical Cambrian of the Quebrada del Toro, Cordillera Oriental, Salta (according to Vilela, 1956).

indicated other similar localities in the extreme south of the Cambrian outcrops of the Cordillera Oriental, Salta, which he considered less certain and requiring further study to prove their correct age.

The precise Cambrian age of the sediments throughout the Cordillera Oriental cannot be established because of lack of adequate palaeontological evidence.

The presence of supposed Cambrian beds resting on the metamorphic basement in northern Tucumán, indicated by BONARELLI and PASTORE (1918–1919, p. 34, Lámina I) in the southern extension of the outcrops of the zone of the Cordillera Oriental of Salta, has not been confirmed.

3.1.3. Precordillera of western Argentina

Groups of calcareous Cambrian rocks predominate in the eastern half of the Precordillera of San Juan and Mendoza. In the western half, a phyllitic and pelitic sequence which is accepted as Cambro–Ordovician extends as parallel outcrops from southern La Rioja to north-eastern Mendoza (Fig. 3).

BORRELLO (1962) introduced the term La Laja Limestone to describe the series of Cambrian fossiliferous carbonate rocks which outcrop typically in the section of the Sierra de Zonda (Valley of Zonda) between the San Juan River to the north and the Quebrada de La Laja to the south, on the national route 20, which runs through the area to the west of the city of San Juan. An up-to-date adaptation of the formational names is proposed, in part, here in order to describe the stratigraphical entities studied within the requirements of current American nomenclature. Correlatory columns of similar calcareous sequences are illustrated to this effect. (Fig. 9).

The Cambrian formation occurring in the area of the Precordillera of San Juan, which corresponds to the already mentioned Sierra de Zonda, is described as follows:

> Pre-Cambrian?
> discordance? (fault zones)
> *La Laja Limestone*: Limestones, dolomitic limestones, and dolomites in medium and thick regular beds. The limestones and dolomitic limestones are grey, dark grey, and bluish in colour: the dolomites are greyish yellow and whitish. In the basal part (Sierra de Zonda, north-western sector) the calcareous sediments in corrugated bands have intercalations of fine-grained, argillaceous limestones, somewhat sandy in parts (Lower Cambrian). The same section contains thin bands of yellowish and brownish arenites (Sierra de Villicúm, south-western flank). The middle section contains well stratified calcareous horizons. In the upper parts can be distinguished within the limestone irregular plates and intercalations of black or brownish black, compact cherts, which are highly developed in the Ordovician part of the formation in the Precordillera of San Juan*.
> *Fossils*: the trilobites correspond to the zones of *Olenellus* (Upper), *Glossopleura*, and *Bathyuriscus-Elrathina*, (Fig. 9). *Visible thickness*: in the western flanks of the Sierra de Zonda, San Juan, 600 to 700 metres. *Age*: The whole of the Cambrian is represented, though the fossils (trilobites) have only been determined in the Lower and Middle Cambrian sections. The succession is continuous into the:
>
> *Ordovician* (San Juan Limestone).

The only outcrops of the fossiliferous Lower Cambrian calcareous rocks, in the first mountainous belt to the east of the Precordillera of San Juan (Fig. 10), are found to the north and west of the city of San Juan.

Cambrian formations from the eastern flanks of the Precordillera of Mendoza (Fig. 3) have been described. These are extensively developed between the city of Mendoza

---

* HARRINGTON and LEANZA (1957, pp 18 and 20) have described the Llanvirn limestones and shales of San Juan as the *San Juan Limestone*; the limestones contain abundant chert.

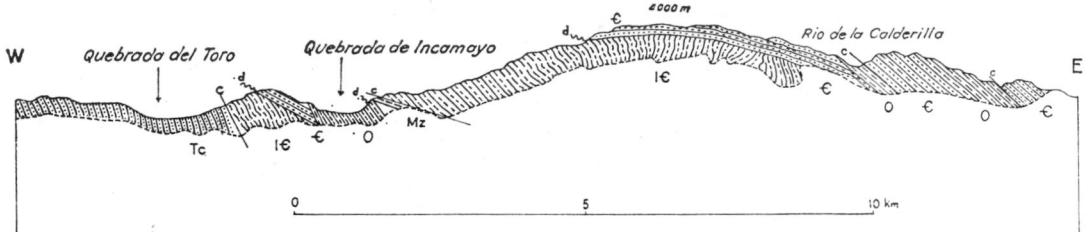

FIG. 8. Structural disposition of the Cambrian beds in the southern part of the Cordillera Oriental, Salta, above the Quebrada del Toro. $I\text{-}\mathrm{C}$, Pre-Cambrian; C, Cambrian; O, Ordovician; Mz, Mesozoic; Tc, Tertiary; c, thrusts; d, discordances (according to Keidel, 1947).

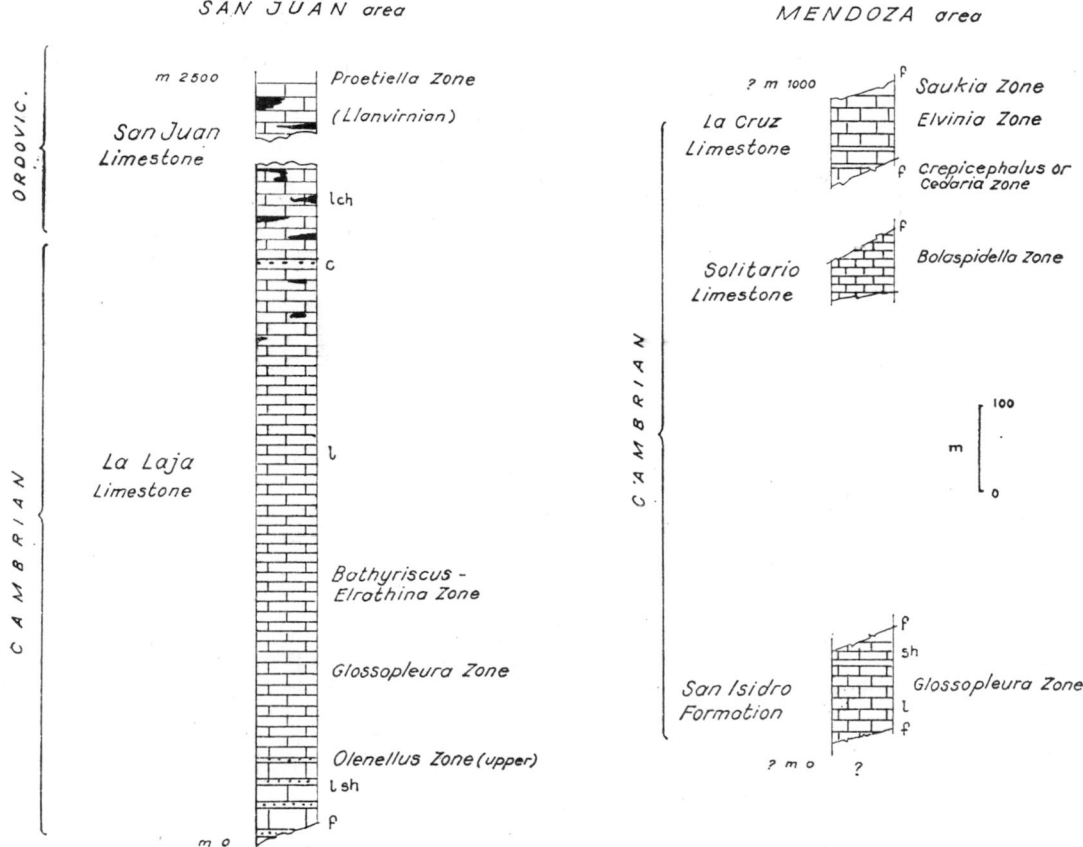

FIG. 9. Correlation of the Cambrian of the Precordillera of western Argentina; f, faults; c, quartzite; l, limestone; lch, limestone with cherts; lsh, limestone and lutites; sh, lutites (according to the author, 1967).

(west) and Canota (approximately 50 kilometres from south to north, from San Isidro to Cerro Solitario.

From the San Isidro sector proper, the following description of the homonymous formation is presented:

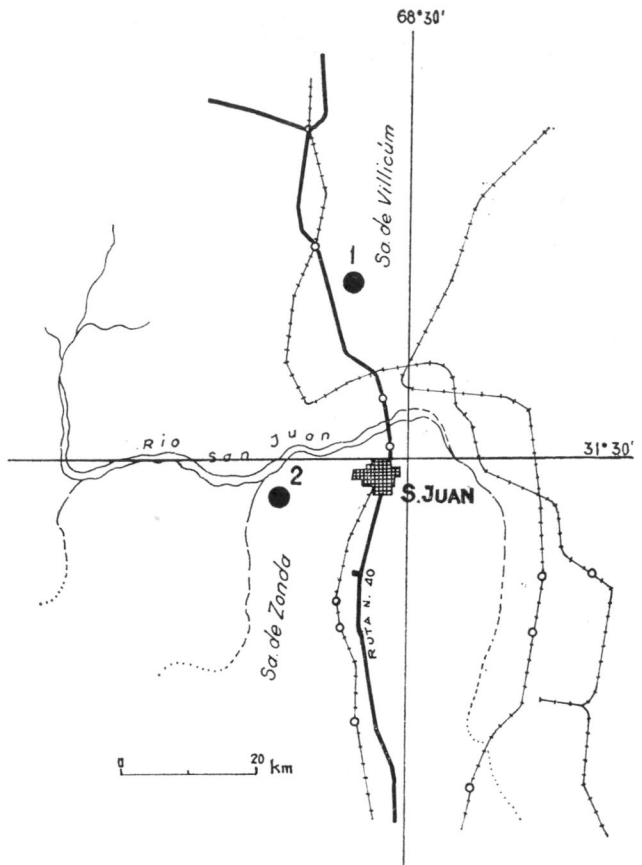

FIG. 10. Map showing detailed position of the only two known Lower Cambrian localities with the *Olenellus* Zone in the southern hemisphere, in the Precordillera of San Juan, Argentina. *1*, *Olenellus* and *Fremontella* (Villicúm); *2*, *Fremontella* (Zonda) (according to the author, 1965).

base: fault

San Isidro Formation: Dark grey to pale grey limestones with intercalated thin bands of greenish grey shales. The shales are well preserved farther to the north in the profile of Cerro Martillo–Quebrada Oblicua. The dark limestones are cut by very thin veins of calcite which in part have a corrugated appearance. Also included are some lenticular masses, and comminuted and brecciated lenses. Toward the north (Rio Seco of the Quebrada Empozada) arenaceous limestones and clear greenish-grey limestones appear to form the base of the calcareous section of the local Cambrian of San Isidro. *Fossils*: the trilobites belong to the *Glossopleura* zone. *Visible thickness*: in the sector of San Isidro-Empozada-Quebrada Oblicua, approximately 100 metres. *Age*: Middle Cambrian. top: faulted contact with upper Cambrian.

In northern Mendoza, a short distance away from and directly to the west of the international route to Chile, opposite the Estancia Canota, the following Cambrian formation occurs:

base: fault

Solitario Limestone: Dark grey limestones in thin, compact bands which are irregularly fractured, sometimes in the form of sharp splinters; outwardly very discoloured by the action of weathering.

They are cross-cut by very white, thin veins of calcite. The outcrops present an appearance of marked lithological regularity. *Fossils*: trilobites of the *Bolaspidella* Zone and associated shelly fossils. *Visible thickness*: some 70 metres. *Age*: late Middle Cambrian.
top: fault.

In the belt of Cambrian beds which extends from the Quebrada de San Isidro to the Quebrada de La Cruz (Cerro Martillo), there is a calcareous section to the west of the city of Mendoza from whence the following formation has been described:

base: faulted contact with Middle Cambrian.
La Cruz Limestone: Thinly stratified fine-grained limestones, sometimes with poorly defined bedding planes, ranging in colour from black, brownish black, and greyish black, to grey, whitish grey, and ash grey. The darker limestones tend to have intercalations of dark chert in irregular bands and lenses. The fracture of the calcareous sediments is even, irregular, or conchoidal. The pale limestones are in part richly fossiliferous sediments (Quebrada Oblicua, north of San Isidro), *Fossils*: the trilobites found in this formation belong to the zones of *Cedaria* or *Crepicephalus*, *Elvinia*, and *Saukia*. *Visible thickness*: in the Quebrada de la Cruz (Cerro Martillo–Quebrada Oblicua) about 100 metres. *Age*: Upper Cambrian
top: faulted contact with fossiliferous Upper Ordovician 'flysch'.

The *Saukia* Zone appears in the base of a thick calcareous sequence (approximately 1,000 metres) which without doubt passes up into the Ordovician and forms part of the structure of the Precordillera between Mendoza and Uspallata (Cerro Pelado, 3,452 metres). Part of this formation is also represented in the section of El Totoral to the south of the Cerro Solitario in northern Mendoza (Fig. 3).

There are insufficient data available to determine the detailed stratigraphy of the Cambrian formations developed in Colombia from the brief references available (HARRINGTON and KAY, 1951; RUSHTON, 1963).

3.2. Stratigraphical character of probable Cambrian formations

The cratonic areas of Brazil, eastern Bolivia, and the Andean Cordillera contain a group of stratigraphical units which are indicated as being of questionable Cambrian age on account of the scant amount of information available.

3.2.1. Cratonic region of Brazil and eastern Bolivia

The stratigraphical groups and formations which are briefly described in this work have been selected as those most representative of a probable Cambrian age on the South American Cratonic basement, in some cases inevitably shedding a certain amount of doubt on the accuracy of previous theories. The groups of formations are described according to American stratigraphical criteria current in the literature, conserving as far as possible the hierarchy accorded to them by the original names of 'series' and 'formations'.

Amazonas Area

Uatumá Formation: On both sides of the Amazon geosyncline OLIVEIRA (1956, p. 19) and GUIMARAES (1964, pp 248–253) describe as Cambrian rocks, this unit of approximately 100 metres thickness, made up of compact reddish and greenish arenites, with fragments of quartz, biotite, and feldspar in an argillaceous cement. The formation is unconformable upon the underlying Pre-Cambrian and is itself covered by Silurian beds (Trombetas Formation).

In the Amazon geosyncline HARRINGTON (1962, p. 1781) recognized, according to data from MORALES (1959), that in the subsurface in the middle and upper river basin of the Amazon River, another formation called the Acarí-Jaú, can provisionally be included within the Cambrian.

North-eastern area of Brazil

Bambuí Formation: This formation is a succession of slates, phyllites, dark or pinkish crystalline limestones, and reddish arenites with sponge spicules and annelid tubes of *Arthraria riachaoensis* Ruedemann. The thickness is 1,000 metres in Mina Gerais, Ceará, and Sergipe, where OLIVEIRA (1956, p. 24) described it as being of Silurian age. In Bahia it apparently transgresses onto the sediments of the Jaibara Group and is composed of some 100 metres of bluish calcareous rocks, to which KEGEL and his associates (1958, pp 13–15) ascribed a Lower Palaeozoic age, following the criteria of WILLIAMS (1926). It is stratigraphically connected with the San Francisco Group up to Parnaíba and Tocantins (OLIVEIRA, *op. cit.*). GUIMARAES (1964, pp 272–278) placed it within the Ordovician System. In the boundary zone of Bahia–Minas Gerais its base is unconformable upon the Lavras Group (Fig. 11). Its radiometric age is approximately 600 m.y.*

Jaibara Group: From the north-east of Ceará, KEGEL and his associates (1958) have described this group as being a formation composed of the following (*sic*): *Aprazivel*—100 metres of basal conglomerate with pebbles of granite, porphyritic granite, felsite, and melaphyre, and *Trapiá*—600 metres of arenites, lutites, and dark grey limestones (Fig. 12) which lie below the sediments of the Bambuí Formation (KEGEL and others, *op. cit.*, p. 11). OLIVEIRA (1956, pp 23–24) considered that on the contrary the Jaibara Group only represents the northern outcrops of the Bambuí or San Francisco Formations, and attributed a Silurian age to it, whereas GUIMARAES, (1964, pp 266–271), partly following KEGEL (*op. cit.*), referred to this group as being of Ordovician age. KEGEL and his associates attributed it generically to the Lower Palaeozoic, having described it as the base of the Bambuí Formation (1958, pp 13–16). HARRINGTON (1962, p. 1781), who bases his writings on their work, estimated that the group can provisionally be indicated as having a Cambrian age.

Lavras Group: Along the length of the Cordillera del Espinazo from the centre of Bahia to the centre of Minas Gerais, the Cambrian sequence outcrops principally as 200 metres of conglomerates, arenites, tillites, lutites, and limestones, whose basement is the complex of rocks of the upper Pre-Cambrian and whose cover is the sediments of the Bambuí Group (OLIVEIRA, 1956, pp 19–20). The sequence was known as a clastic unit and a glacial origin was assigned to it; thus it is represented in the Minas Gerais as the Sopa Formation (fluvioglacial) and Macaúbas Formation (glacial). The deposits represent the secondary matrix of the diamonds which have been found in Bahia and in the northern part of the Minas Gerais (OLIVEIRA, *op. cit.*). The original information given by DERBY (1906) and WILLIAMS (1930) was useful to HARRINGTON (1962, p. 1781) in attributing a Cambrian age to the group, though it has been referred to by GUIMARAES (1964, p. 221) as of Pre-Cambrian age. The group lacks fossils and is in part folded.

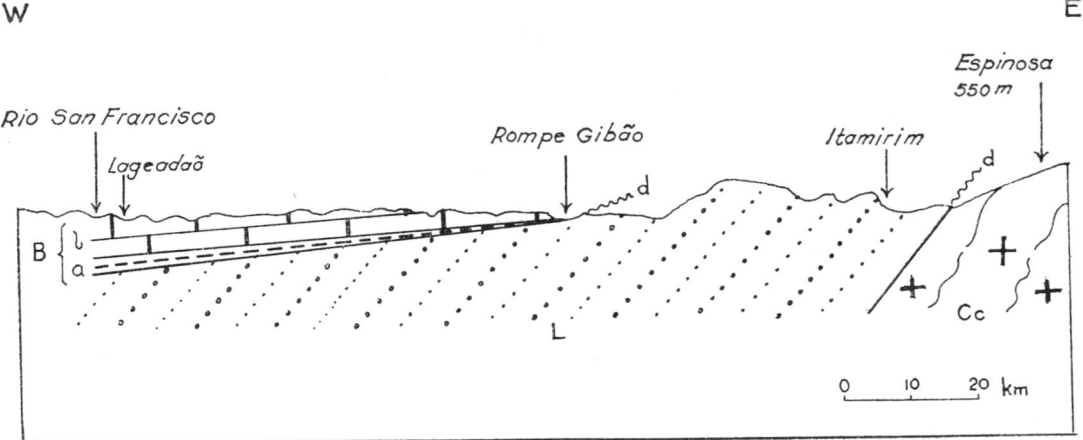

FIG. 11. Geological cross-section to the east of the San Francisco River, in north-east Brazil, showing the relations between the Pre-Cambrian basement and the metamorphic-sedimentary cover. *Cc*, Crystalline complex, in parts granitic; *L*, Lavras Group, diamandiferous conglomerate; *B*, Bambuí Formation (*a*, arkose; *l*, limestone); *d*, discordances (according to Moraes, 1937).

Personal communication of U. Cordani, Geocronological Laboratory, University of Sao Paulo, Brazil, 1967.

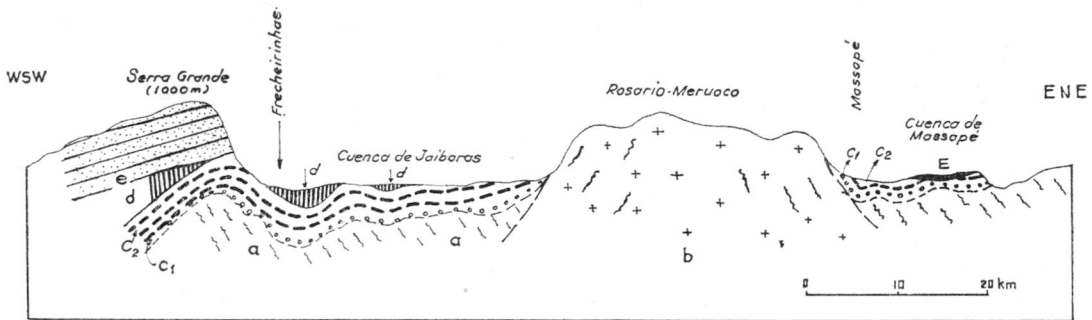

FIG. 12. Geological cross-section between the basins of Jaibaras and Massapé, north-east of Ceará, Brazil. *a*, gneiss; *b*, porphyritic granite; *c*, Jaibara Group (c_1, Aprazivel Formation, c_2, Trapiá Formation); *d*, Bambuí Formation, *e*, Serra Grande Formation; *E*, spilite (according to Kegel and associates, 1958).

Area of Paraná–Santa Catarina

Gaspar Formation: This unit of conglomerates, lutites, and violet and reddish arenites between 380 and 530 metres thick, is superposed, according to MAACK (1947, p. 107), in Santa Catarina on lutites, conglomerates, slates, and quartzites with calcareous lenses of the Ibirima Formation (1,000 metres), beneath which is the Pre-Cambrian granitic or gneissic basement. Lacking fossils it is provisionally believed by HARRINGTON (1962, p. 1781) to be within the Cambrian, which accords with information from FREITAS (1945) and MAACK (*op. cit.*). The last author was quoted in this same context by OLIVEIRA (1956, pp 20–21), when referring to the group called Itajaí Series, in what is known as the Cambrian of Santa Catarina.

Ribeira and Itajaí Groups: These belong to the geosyncline of the Paraná. The first, which was correlated with the Lavras Group by OLIVEIRA (1956, p. 20) in Sao Paulo and Paraná, includes tilloidal conglomerates, phyllites, and arenites which have been affected by epizonal metamorphism. GUIMARAES has indicated (1964, pp 244–246) the age of these groups as being Proterozoic. The second, north-east of Santa Catarina was correlated with the Ribeira Group and comprises 1,000 metres of conglomerate, purplish arenites, grey leptometamorphic phyllites, and lutites of fluvial origin, cut by porphyries, granite, pegmatites, and diorites.

Area of Eastern Bolivia–Mato Grosso–Apa River

Corumbá Group: Perhaps this is the most doubtful of all the groups which are here assigned to the Cambrian; it comprises 420 metres of pisolitic, brecciated, sandy dolomites (Bocaina Formation, 300 m.), and dark, lenticular, oolitic, graphitic limestones with ripple marks, intercalated by greenish and reddish lutites, silstones, and grey micaceous arenites (Tamengo Formation, 120 m.). Its basement is the Pre-Cambrian with which it has a discordant relationship. It occurs in the territory of the Paraguay River extending over the Mato Grosso and eastern Bolivia, and a part of Paraguay, south of the Apa River. GUIMARAES (1964, pp 254–262) has assigned a Cambrian age to this group, while OLIVEIRA (1956, p. 21) attributed it to the Ordovician, and CHAMOT (1962, pp 14–16) ascribed the group to the Pre-Cambrian in that he compared it with the 'Bodoquena Series' of Brazil. AHLFELD and BRANIŠA (1960, pp 21–22) preferred to place these rocks within the Cambrian, accepting as valid the disputable argument that the remains of algae which are found in these beds, and which have been described as *Aulophycus* by BEURLEN and SOMMER (1957) in Mato Grosso (Ladario) Brazil, correspond to the fossil forms of that name studied in the United States of North America by FENTON and FENTON. If one accepts its correlation with the said 'Bodoquena Series', which was originally ascribed to the Cambrian (ARROJADO LISBOA, 1909), one could further extend the stratigraphical equivalence ot the 'Itapucumí Series', (HARRINGTON, 1950) in eastern Paraguay. This latter having earlier been considered to belong to the Lower Palaeozoic, has recently been attributed to the Proterozoic (HARRINGTON, 1962, p. 1782). ECKEL (1959) also included this last series in the Lower Palaeozoic.

Jacadigo Group: This sequence follows upon the Corumbá Group and occurs to the west of Corumbá, in the Mato Grosso, and in eastern Bolivia. It comprises the Urucúm Formation, 400 to 500 metres in thickness, of sandstones and conglomerates, buff to brown in colour; the

Corrego das Pedras Formation, 95 metres thick of arenites and hematitic jaspers with various iron minerals; and the Banda Alta Formation, 300 metres thick of arenites and silstones with iron and manganese ores (DORR, 1945, pp 9–15). AHLFELD and BRANIŠA (1960, pp 22–25) placed the group within the Ordovician when giving a detailed account of the geology of Bolivia. HARRINGTON (1962, pp 1780–1781) and GUIMARAES (1964, pp 262–265) assigned it to the Cambrian System, though this last author also used it in a descriptive reference to the Proterozoic (GUIMARAES, op. cit., pp 240–243). OLIVEIRA (1956, p. 18), however, expressed the opinion that the beds of the Paraguay River have a Silurian age.

In the north-east of Brazil, then, the Cambrian sequences are represented by the Lavras Group and Bambuı Formation; while in the Brazilian region of the Mato Grosso and eastern Bolivia, the equivalents are found in the Corumbá Group and Jacadigo Group in this order, though the stratigraphical equivalence of the former is more certain.

3.2.2. The Andean Cordillera

The outcrops of rocks, in great part metamorphic, which are visible on the Pacific coast of the South American Andes in Peru, between Ocoña and Mollendo, on both sides of latitude 17° S.; and in Chile, from Pichilemú, latitude 34° 20′ S., and south of the peninsula of Taitao on latitude 45° S. (Fig. 1); can tentatively be referred to as Cambrian or in general as Lower Palaeozoic. These outcrops, on account of the presence of ophiolites (Valdivia, Chile) in their midst, and because of their relative structural position within the orthogeosynclinal Palaeozoic belt, would coincide with the Cambro–Ordovician eugeosynclinal zone.

In the Cordillera Oriental of Bolivia, FRAENKL (1959) described as the Limbo Formation a succession which is over 2,600 metres in thickness and is composed of anhydrite, lutites, conglomerates, and quartzites. As the result of faulting it appears to be below the Ordovician, which rises up to 3,500 metres in the Limbo sector, between the 107 and 146 kilometre points on the Cochabamba–Tunari highway. HARRINGTON (1962, p. 1780) indicates the almost certainty of its Cambrian age and AHLFELD and BRANIŠA (1960, pp 36–37) in principle express a similar opinion.

In the eastern flanks of the Andes in Mendoza, the section of the Frontal Cordillera exposed between Llaucha and the Diamante River contains in the base of the very marked Palaeozoic sequence (KITTL, 1934, pp 35–36) a pale grey gneiss and tectonically deformed reddish porphyrites, which BORRELLO (1963, pp 5–8) associates with the eugeosynclinal pliomagmatic zone of the Lower Palaeozoic. According to the same author (op. cit.) the following also belong to this zone: the masses of quartzitic and calcareous rocks and schists which are equated with a metamorphic complex with green schists and amphibolites in central and northern Mendoza (POLANSKI, 1958, p. 169); and the non-metamorphic and metamorphic rocks of the Farallones and Bonilla area of the Precordillera in the Sierra de Uspallata, among which there are: garnetiferous quartzites (metaquartzites), schists, chloritic and sericitic schists, and crystalline limestones, densely penetrated by basic and ultrabasic rocks of the ophiolitic Lower Palaeozoic association (KEIDEL, 1939, pp 28–43).

The quartzitic (orthoquartzitic) strata of the Sierra de Sangra in the extreme north of the San Martín Lake in the Patagonian Cordillera, which south of latitude 46° S. passes from Argentina to Chilean territory (REYES, 1966, pp 324–326) indicate, on account of their position beneath the middle Palaeozoic 'flysch', a miogeosynclinal mass of the Lower Palaeozoic, and are possibly of Cambrian age. This is also suggested by the presence of quartzitic phyllites and limestones metamorphosed into the marble of the Buenos Aires Lake, Chile, on latitude 47° S. (HEIM, 1940, pp 34–38 and plate 1), where thcy appear penetrated by granite. There may be a correlation with the deposits of lutites, quartzites, and limestones which CECIONI (1956, pp 194–195) has described from southern Chile (islands of Contrera and Ramirez) on latitude 52° S., and which drew the attention of HARRINGTON to this, admittedly hypothetical, possibility (1952, p. 1781).

Data from the fossiliferous Cambrian rocks of Colombia, Argentina, and Antarctica can be used to support an extrapolation which permits a Cambrian age to be assigned, at least in part, to these groups of the above-mentioned Andean regions. HARRINGTON (1962, p. 1781) accepted also in this case that at least part of the metamorphic formations of the Andes of Venezuela, Ecuador, and Peru might have a Cambrian age.

4. Palaeontology

In South America the only Cambrian faunas that are sufficiently abundant to permit an important development in the collation of biostratigraphical knowledge are those of the Precordillera of San Juan and Mendoza. The succession and composition of the trilobite faunas which are indicated in the following lists are considered to be generally representative of the Cambrian of western Argentina and are known through the works of RUSCONI (1945a–1962), LEANZA and HARRINGTON (1943), LEANZA (1947), V. POULSEN (1958), CHR. POULSEN (1960), BORRELLO (1963–1967), the contribution of HUPÉ (Sorbonne, Paris) and BORRELLO (La Plata, Argentina) in preparation, and some unpublished results.

4.1. Faunal successions

4.1.1. Argentina

(a) *Lower Cambrian*—upper *Olenellus* Zone

Localities: Cerro del Molle (Sierra de Villicúm) in the north-west of the Sierra de Zonda, San Juan.
Stratigraphy: La Laja Formation (lower).
    Pagetides? dubius Hupé
    Laudonia sp.
    Olenellus aff. *brachycephalus* (Emmons)
    Fremontella inopinata Borrello
    Eoptychoparia acuminata Hupé
    Villicumia borrelloi Hupé
    Zacanthopsis sp.
    ? *Paedeumias* sp.
    Kootenia sp.
Associated shelly fauna: Brachiopoda, *Hyolithes* sp. (Calyptoptomatida), and traces of Annelida.

(b) *Middle Cambrian*—(i) *Glossopleura* Zone

Localities: San Isidro and Cerro Martillo, Mendoza.
Stratigraphy: San Isidro Formation and La Laja Formation (lower, upon the *Olenellus* Zone).
    Dorypyge joanirusconii Rusconi
    Kootenia incerta (Rusconi)
    Olenoides sp.

Ogygopsis sp.
Oryctocephalus asperoensis Rusconi
Athabaskiella parabolica (Rusconi)*
Chilonorria perlottii Rusconi
Mendospidella digesta (Leanza)
Asperocare argentinum V. Poulsen
Alokistocare elongatum V. Poulsen
Alokistocare australe V. Poulsen
Alokistocare mendozanum (Rusconi)*
Alokistocarella mexicana Lochman
Amecephalina argentina Kayer
Glossopleura inexsulcata (Rusconi)
Zacanthoides ferula Leanza

Associated shelly fauna: Acrotreta sp., *Lingulella* sp. (Brachiopoda); *Hyolithes* sp. (Calyptoptomatida); algae; traces of Annelida; Porifera (Hyalospongea): *Protospongia* sp. and *Chancelloria* sp.

(ii) *Bathyuriscus-Elrathina* Zone

Localities: Quebrada de Juan Pobre, Quebrada La Laja, Quebrada de Zonda (on national route 20), and the western flanks of the Sierra de Zonda (Quebrada del Carmen) San Juan.

Stratigraphy: La Laja Formation (lower, upon the *Glossopleura* Zone).

Elrathia carmenensis Hupé
Amecephaline argentina Kayser*
Ehmania (?) *lajensis* (Kayser)
Ehmania hypselogena Harrington and Leanza
Eteraspis orbignyana (Kayser)†‡
Eteraspis prosorysa Harrington and Leanza‡

Associated shelly fauna: Billingsella sp. (Brachiopoda) and *Hyolithes* sp. (Calyptoptomatida).

(iii) *Bolaspidella* Zone

Localities: Cerro Solitario (Cerro Blanco), Canota, Mendoza.

Stratigraphy: Solitario Formation.

Agnostus exsulatus Chr. Poulsen
Baltagnostus hospitus Chr. Poulsen
Baltagnostus mendozensis Chr. Poulsen
Culipagnostus chipiquensis Rusconi
Diplagnostus jarillensis Rusconi
Kormagnostus?propinquus Chr. Poulsen
Oedorhachis australis Chr. Poulsen
Peronopsis ultima Chr. Poulsen
Phoidagnostus solitariensis (Rusconi)
Pseudagnostus parabolicus Rusconi
Stigmagnostus canotensis Rusconi

* Revision by H. J. HARRINGTON; written communication.
† Indicated by V. POULSEN (1958, p. 16) in the Zone of *Glossopleura* of Mendoza.
‡ The genus *Asaphiscus* Meek is unrelated, according to H. J. HARRINGTON (written communication).

Prometeoraspis canotensis Chr. Poulsen
Williamsina cortesi Chr. Poulsen
Williamsina harringtoni Chr. Poulsen
Williamsina mikkelseni Chr. Poulsen
Williamsina ornata Chr. Poulsen
Talbotinella communis Chr. Poulsen
Talbotinella leanzai Chr. Poulsen
Talbotinella rusconii Chr. Poulsen
Bolaspidella luciae Chr. Poulsen
Canotaspis aliena Chr. Poulsen
Goycoia* tellecheai Chr. Poulsen
Goycoia brevicauda Chr. Poulsen
Goycoia pecoralis Chr. Poulsen

Associated shelly fauna: *Dicellomus* sp. (Brachiopoda) and *Orthotheca* sp. (Calyptopto-matida).

(c) *Upper Cambrian*—(i) *Cedaria* or *Crepicephalus* Zone

Localities: North of San Isidro-Quebrada Oblicua (north-north-east of San Isidro) and the El Totoral area, Mendoza. From the latter Upper Cambrian faunas of various zones have been studied.
Stratigraphy: La Cruz Limestone (lower).
  Tricrepicephalus anarusconii Rusconi
  ? *Blountia* sp.
Associated shelly fauna: Brachiopoda; Calyptoptomatida.
This horizon coincides with the Dresbachian zones of North America.

(ii) *Elvinia* Zone

Localities: Quebrada Oblicua (north-north-east of San Isidro).
Stratigraphy: La Cruz Limestone (middle).
  Elvinia obliquoensis Rusconi
  Irvingella jorusconii Rusconi
Associated shelly fauna: Brachiopoda, Mollusca.
This zone corresponds to that of the same name in the Franconian of North America.

(iii) *Saukia* Zone

Localities: Quebrada La Cruz, south of the Cerro Aspero and Cerro Pelado, Mendoza.
Stratigraphy: La Cruz Limestone (upper).
  Lotagnostus sp.
  Hemirhodon huentotianum (Rusconi)†
  Bienvillia sp.
  Parabolinella peladoensis n. sp.‡
  Hungaia puelchana Rusconi
  Rasettia crucensis (Rusconi)
  '*Thriarthropsis*' *pampanum* (Rusconi)†
  Cuyanaspis empozadensis Rusconi

* H. J. HARRINGTON and also the present author consider that this genus must correspond to *Asaphiscus* Meek.
† Revision by H. J. HARRINGTON (written communication).
‡ Only in the Cerro Pelado.

Associated shelly fauna: Brachiopoda, Mollusca, Calyptoptomatida. This zone is equivalent to that of the same name in the Trempealeau of North America.

4.1.2. Colombia

Middle Cambrian—Bathyuriscus–Elrathina? Zone

 Locality: Rıo Duda, Uribe trail, west of Bogotá.
 Stratigraphy: imprecise.
  Peronopsis (or *Hypagnostus* sp.)*
  Ehmania akanthophora Harrington and Kay
  ? *Ehmania amphibola* Harrington and Kay
  Paradoxides sp.*
Associated shelly fauna: unknown.
Sufficient data are not available in this case to indicate the possible biostratigraphical equivalence.

4.2. Palaeobiogeographical relationships

The biostratigraphy proposed for Argentina evidently suggests relationships with the faunal classifications provided by LOCHMAN-BALK and WILSON (1958), though as far as the faunal zones of the Precordillera of San Juan and Mendoza are concerned, corresponding studies are still being carried out.

According to these investigators, North American faunal realms characterized by trilobites are controlled in their distribution by well defined geotectonic factors. On this basis the faunal lists provided here show the relations which exist between the cratonic and euxinic (eugeosynclinal) environments during the Cambrian Period. It can be demonstrated that the Cambrian trilobite associations of the Andes of western Argentina have palaeobiogeographical affinities which enable extensive comparisons to be made with certain domains of the North American region.

Details given in Table 1, in which the typical North American Cambrian realms, which can be termed structural realms, are represented in the abscissa, will be considered to this effect. In the ordinate are listed the representative genera of the seven trilobite zones, proven to date, which characterize the development of the Cambrian of San Juan and Mendoza in the Precordillera.

From the data presented in the Table, certain features emerge concerning relationships, though generalized, of the structural permanence of the corresponding biological distributions during the Cambrian Period. Main conclusions can be summarized as follows:

In the Lower Cambrian the trilobites of the upper *Olenellus* Zone of San Juan, whose genera are *Pagetides, Olenellus, Eoptychoparia, Zacanthopsis*, and *Paedeumias*, are all included in the similar biostratigraphical zone of the Intermediate Cratonic realm of eastern United States of America, showing only in part a tendency towards permanence in the Intermediate environment (*Olenellus, Zacanthopsis*) or the External Margin of the Intermediate realm (*Pagetides, Eoptychoparia*) respectively.

From the Middle Cambrian of San Juan and Mendoza in the *Glossopleura* Zone, the genera *Dorypyge, Kootenia, Olenoides, Ogygopsis, Alokistocare, Alokistocarella, Glossopleura*, and *Zacanthoides* are prevalent in the Cratonic and Intermediate realms of the Cordillera, and of the eastern United States. The genera *Kootenia* and *Dorypyge* extend to the Extracratonic and Euxinic realms of the North Atlantic (cosmopolitan). *Orycto-*

* See RUSHTON (1963, pp 254–257).

TABLE 1

Palaeobiogeographical Relations of the principal genera of Cambrian Trilobites of the Precordillera of Western Argentina with genera common to the Cambrian Realms of North America

Trilobite Zones	Common Genera	N. American Realms: Argentine Areas	CCd	CId	CIN	OmI	EEu	CCr	CLu	CMW	INv	ICA	IVG
Sk	Rasettia	MEb						X	X	X	X		X
	Hungaia												X
	Parabolinella										X		X
	Lotagnostus												
El	Irvingella	MGi						X	X	X	X	X	
	Elvinia							X	X	X	X	X	
Tc	Tricrepicephalus	MGi						X	X	X		X	
Bl	Bolaspidella		X	X	X	X				X	X		X
	Peronopsis		X		X		X				X		
BE	Ehmania	MGi	X	X	X	X							
	Elrathia		X	X									
Gl	Zacanthoides	MGb	X	X	X								
	Glossopleura		X		X								
	Amecephalina		X	X	X								
	Alokistocare		X	X	X	X							
	Oryctocephalus		X	X									
	Ogygopsis		X	X	X								
	Olenoides		X	X	X	X				X			
	Kootenia		X	X	X		X						
	Dorypyge		X	X	X		X						
Ol	Paedeumias	MGb			X								
	Zacanthopsis			X	X	X							
	Eoptychoparia				X								
	Olenellus			X	X	X							
	Pagetides					X							

KEY

N. American Realms (1):
CCd, Lower and Middle Cambrian Cratonic Cordilleran
CId,   ,,   ,,   ,,   ,,   Cordilleran Intermediate
CIN,   ,,   ,,   ,,   ,,   Eastern N. America Cratonic and Intermediate
OmI,   ,,   ,,   ,,   ,,   Outer margin of Intermediate
EEu,   ,,   ,,   ,,   ,,   North Atlantic Extracratonic Euxinic
CCr, Upper Cambrian Cratonic in Croixan area
CLu,   ,,   ,,   ,,   in Llano Uplift Texas
CMW, ,,   ,,   ,,   in Montana–North Wyoming
INv,   Intermediate in Nevada
ICA,   ,,   ,,   in Central Appalachians
IVG,   ,,   ,,   in Vermont-Gaspé

Argentine Areas: MGb, Miogeosynclinal border
MGi, Miogeosynclinal interior
MEb, Miogeosynclinal–Eugeosynclinal boundary

Trilobite Zones: Ol, Olenellus (sup.); Gl, Glossopleura;
BE, Bathyuriscus-Elrathina; Bl, Bolaspidella;
Tc, Cedaria or Crepicephalus; El, Elvinia;
Sk, Saukia (only proved zones).

(1) LOCHMAN-BALK and WILSON (1958).

cephalus, however, remains restricted in comparison with other forms of the Intermediate realms of the Cambrian of the North American Cordilleras.

For the higher Middle Cambrian, the affinities of the genera *Elrathia* and *Ehmania* of the *Bathyuriscus–Elrathina* Zone and of the genera *Peronopsis* and *Bolaspidella* of the *Bolaspidella* Zone, with the equivalents of the Cratonic realm of the Cordillera, Intermediate realm of the Cordillera, and Cratonic and Intermediate realms of eastern North America, can be observed. The presence of *Peronopsis* in the Extracratonic–Euxinic realm of the North Atlantic would indicate its cosmopolitan character.

From the Upper Cambrian of Mendoza the *Tricrepicephalus* Zone (*Cedaria* or *Tricrepicephalus*) and the *Elvinia* Zone are coincident with those identified in the Cratonic realm of the Croixan area, Cratonic of the Llano Uplift of Texas, Cratonic of Montana and northern Wyoming, and the Intermediate realm of Nevada and the Central Appalachians.

With the exception of *Rasettia*, which appears in nearly all of the previously cited realms of the Upper Cambrian, the association of *Lotagnostus*, *Parabolinella*, *Hungaia*, and *Rasettia* of the *Saukia* Zone is present in what is known as the Intermediate realm of Vermont-Gaspé in eastern North America, where the Middle Cambrian genus *Bolaspidella* is also represented, sometimes in its ancestral form. The *Hungaia* fauna of Argentina with *H. peladense* Rusconi is a replica of that of *Hungaia magnifica* (Billings) in eastern Canada.

In South America the Cambrian trilobite associations are essentially of the Intermediate Cratonic realm, thus indicating as a group their 'Pacific province' character. This would also be the impression from the Middle Cambrian fauna of Colombia (Río Duda), though it must be admitted that the presence of *Paradoxides* causes one to recognize within this the allocthonous interaction of an Atlantic element.

5. Correlations

In the preceding chapter the order of the biostratigraphical zones, representative of the fossiliferous Cambrian rocks of Argentina, was first presented, and then the palaeobiogeographical relations which can be derived from the palaeontological lists were discussed. These data serve as a basis for the stratigraphical correlation of the fossiliferous Cambrian of Mendoza and San Juan, and further, they are sufficiently valid to indicate generic correspondence with the Cambrian trilobites of Colombia. In Table 2, together with the correlation of the South American Cambrian of San Juan–Mendoza and Colombia, are incorporated the non-fossiliferous series of the western Precordillera of Argentina, assigned to the Cambro-Ordovician, and those similarly assigned in the Cordillera Oriental of Salta-Jujuy and southern Bolivia, in which the only included fossil remains of relative stratigraphical value are of *Skolithos*. This table allows a more exact integration of knowledge of the regional development of the Cambrian of the continent, in the context of which discussion of the most recent data will permit detailed correlations to be made.

The fossiliferous Cambrian strata of the Precordillera of San Juan and Mendoza provide an essential basis for South American regional correlation. The column shows in San Juan the development of a continuous Cambrian succession from Lower Cambrian beds containing *Olenellus* and *Fremontella* to the Llanvirn with *Proetiella* and *Maclurites* (Fig. 9). The contact of the base of these strata with the Pre-Cambrian

(Infracambrian) of the so-called Caucete Group is not seen.* In the western Pre-cordillera the Bonilla–Alcaparrosa Formations, their base again not seen, culminate in a concordant sequence which contains remains of *Climacograptus* and *Amplexograptus* of the Caradoc. Two continuous successions in San Juan, one being relatively rich in fossils (miogeosynclinal) and the other nearly devoid of fossils (eugeosynclinal), have generally in their upper parts fossiliferous Ordovician sequences in common. It is therefore possible, to a great extent, to establish the correlation of similar Lower Palaeozoic beds in western Argentina with a fair degree of certainty. In the Cordillera Oriental of northern Argentina, the Mesón Group which extends over Salta and Jujuy, particularly in the Sierra de Santa Victoria, has as a basement folded sub-metamorphic sediments of the Pre-Cambrian (Infracambrian), and is unconformable on this basement. The first beds of the *Parabolina argentina* Zone of the fossiliferous lower Tremadoc cover it discordantly (HARRINGTON and LEANZA, 1957, pp 8–9). The lower Tremadoc also covers the Sama Formation of southern Bolivia. The same relationship apparently occurs between the Cambrian and the Ordovician in the Cordillera Oriental of Colombia, though obviously fuller geological and palaeontological data are needed to allow corresponding correlations with the extreme north of the South American Andes.

In conclusion, only the Cambrian sequences of western Argentina with their continuous orthogeosynclinal development permit the establishment of a complete Cambro–Ordovician succession. The exact placing of the remaining Cambrian strata is still uncertain. It has not been possible to establish correlations with the supposedly Cambrian deposits of the South American Cratonic area, nor yet with those distributed on the Pacific coastline and in the Patagonian Cordillera of Argentina, and which extend into Chilean territory.

6. Igneous, metamorphic, and sedimentary petrology

A well-balanced picture of the petrological types referred to is presented here to conform with the theme of the present account; the Andean region is, however, better provided with references.

6.1. Sedimentary petrology

Psephitic deposits are frequent in the Lizoite Formation of the lower part of the Cambrian Meson Group of northern Argentina (TURNER, 1963a). They represent either heterogeneous masses of marine origin or the transition of continental to marine conditions in the beginning of the process of transgression. They contain abundant fragments of rocks from the Pre-Cambrian basement. They are, however, unknown in Bolivia and other South American Cambrian areas at the base of the Lower Palaeozoic sequence.

In the marine environment certain isolated banks of clastic masses were also formed, resembling those which in the San Isidro Formation of the Cambrian of northern Mendoza represent types of slide breccia in a tectonic facies. In this latter they are found in succession up to the level of the Ordovician. The conglomerates of the Limbo Formation (FRAENKL, 1959) in general would be more representative of a continental environment, and some, associated with quartzites, certainly suggest an origin like that of molasse. HARRINGTON (1962) stated that the beds attributed to the Cambrian of Brazil, which spread over Ceará, Bahía, Santa Catarina, and Mato Grosso, contain remains of fanglomerates which originated in the uplift of the Pre-Cambrian basement.

* BORRELLO *in* Léxico Estratigráfico Internacional–Argentina, in preparation, Paris.

TABLE 2

Statigraphical Correlation of the Cambrian of South America

Zones	Precordillera–Western Argentina			Cordillera Oriental—N. Argentina		Cord. Oriental
	Precordillera, Western	Precordillera, Eastern, Mendoza	Precordillera, Eastern, San Juan	Salta-Jujuy S. Santa Victoria (S. of Bolivia)	S. Bolivia C. Victoria-Tacsara	Colombia
Ordovician	Climacograptus aff. antiquus, Amplexograptus sp.	?	Sierra of Talacasto, Villicúm, etc. Maclurites Proetiella	T R E M A D O C	T R E M A D O C	?
Saukia	Formations: Alcaparrosa	La Cruz Form. (upper) Lotagnostus Hungaia	continuous succession	Mesón Group — Chalhuamayoc Form. Campanario Form. (Skolithos) Lizoite Form.	Sama Form. (Skolithos)	?
Ptychaspis Prosaukia	Yerba Loca					
Conaspis	Rio Blanco	La Cruz Form. (middle) Irvingella-Elvinia				Rio Duda Form. Ehmania Paradoxides
Elvinia	Farallones					
Dunderbergia	Bonilla					
Aphelaspis	(Cambro-Ordovician formations)					
Crepicephalus		La Cruz Form. (basal) Tricrepicephalus				
Cedaria		Solitario Form. Bolaspidella				
Bolaspidella						
Bathyuriscus Elrathina			La Laja Limestone Ehmania, Elrathia			
Glossopleura		San Isidro Form. Glossopleura, Kootenia	La Laja Limestone Glossopleura, Kootenia			
Albertella						
Poliella-Plagiura		?				
Olenellus Upper			La Laja Limestone (basal) Olenellus Fremontella			
Olenellus Lower			?			
Pre-Cambrian	no outcrops	Infracambrian	Infracambrian	Infracambrian	no outcrops	?

The Lavras Group appears as an important Cratonic entity because of its psephitic facies with diamonds at Lavras Diamantinas Bahía, Brazil, where it was described by DERBY (1906). OLIVEIRA (1956) ascribes to it a glacial origin. It extends to Paraná in southern Brazil, where it is called the Ribeira Group. In the Minas Gerais, the psephitic beds of the Lavras Group are restricted to the Macaúbas Formation, and are evident in the form of tillites, with large, angular blocks and smaller clasts of granites, gneiss, conglomeratic quartzite, and quartz, among others.

Typical psammitic sediments of the marine Cambrian form the outcrops of the miogeosynclinal trench of northern Argentina occupied by the Cordillera Oriental. They belong to the Mesón Group, and are composed of medium to fine-grained arenites, these last including remains of tubular *Skolithos*. In the locality of Purmamarca (Jujuy) there are fine to coarse reddish arenites intercalated with sandy lutites in an evidently alternating pattern of sedimentation of the flysch type, the remains of the already-mentioned *Skolithos* predominating.

Arenites are also the major constituent of the Sama Formation in southern Bolivia (AHLFELD and BRANIŠA, 1960). In the Sierra de Zonda, Precordillera, San Juan, a band of quartzitic arenites intercalated between limestones appears above the beds of the La Laja Formation to the north of national highway 20. Dark yellowish calcareous arenites can be observed in the lower part of the Cambrian in the outlet of the Quebrada Empozada to the north of San Isidro, and beds of the same sediments have been observed in the upper regions of the same section, in the limestones of the San Isidro Formation. HARRINGTON and KAY (1951) described quartzose arenites in the Middle Cambrian of Río Duda, Uribe Trail, Colombia. The supposed Cambrian beds of the Amazonian plain contain arkosic psammites (OLIVEIRA, 1956). The quartzites of the Sierra de Sangra, Santa Cruz, could in Patagonia be another psammitic sequence of the South American Cambrian (Cambro–Ordovician).

In the Andean regions pelitic deposits are primarily represented in the widely developed thick orthogeosynclinal Cambro–Ordovician series. Such are the lutites of the Alcaparrosa Formation, San Juan, in the eugeosynclinal upper part of a sequence which does not clearly reveal, because of faulting, an accessible section of the Cambrian. Pelites are not uncommon in the form of lutites in the Cambrian of northern Argentina, within the Mesón Group. The mouth of the Amazon contains sediments of this character in the local, supposedly Cambrian, sequence.

The eastern half of the Precordillera of western Argentina is of calcareous composition (Fig. 3). Fossiliferous Cambrian sections, with calcareous lithology, have been described from the extreme north of the Sierra de Villicúm, San Juan, up to San Isidro, Mendoza, and these constitute the sediments of the older Middle Cambrian formations which are referred to as La Laja and San Isidro, respectively. They are in general calcareous carbonates and homogeneous dolomites, greyish yellow to dark grey or black, nodular in part, and of regular deposition. STELZNER (1885) noted the presence of oolites and bituminous substances in the Cambrian limestones which outcrop in the Quebrada de La Laja, San Juan. The Upper Cambrian limestones of the La Cruz Limestone or Formation, Mendoza, which are grey or bluish grey in colour, and especially those of the Solitario Limestone or Formation, present a well-defined parallel stratification which is in part schistose. In the Cerro Pelado and its northern Aguja, Mendoza, the very fine-grained Upper Cambrian limestones (Plate 5) are compact and of yellowish and greyish colour.

On the western side of the Sierra de Zonda and in the extreme north of the Sierra de Villicúm, Precordillera of San Juan, the calcareous sediments of the Middle and Upper Cambrian are overlain by carbonates with intercalations of black chert, which are predominant in the upper part, particularly in the Ordovician limestone and dolomite succession.

The limestones of the Precordillera of Mendoza resemble those described by HEIM (1940) from the Chilean side of Lake Buenos Aires, though they show signs of intense recrystallization. HARRINGTON and KAY (1951) indicated that the fragments with Cambrian fossils of the Rio Duda, Colombia were composed of a calcareous psammitic material. The controversial Corumbá Formation contains calcareous and dolomitic rocks (AHLFELD and BRANIŠA, 1960; CHAMOT, 1962), and extends in a similar facies towards the depression of the Paraguay River. All these calcareous sequences are free of chert intercalations.

Sediments of chemical origin include those of the Limbo Formation of Bolivia which contains masses of anhydrite (FRAENKL, 1959).

6.2. Igneous petrology

Basic and ultrabasic igneous rocks are present in the Cambro–Ordovician, and have been known since the last century within the Precordillera–Cordillera Frontal of western San Juan and Mendoza, (AVÉ-LALLEMANT, 1892).

The basic rocks were referred to by KEIDEL (1937) who described the gabbro and the sausuritized diabase in the Farallones Formation of Uspallata. POLANSKI (1958) and ZARDINI (1959, 1960, and 1961) have referred to the character of the ultrabasic rocks in the same region. The first two authors support a Pre-Cambrian age for these rocks.

In the region of La Cortadera–Bonilla, ZARDINI (1960) described serpentinite bodies related to peridotitic intrusions, with serpentine, chlorite, chromiferous magnetite, limonite, and talc. In the Cordón del Portillo, Tunuyán River, GONZÁLES DÍAZ (1958) stated that these rocks form a large ultrabasic mass where anthophyllitic, talcose, tremolitic, serpentinitic, and olivine-bearing schists appear in associated rocks. The bodies are numerous and are concordant. They are parallel to the schistosity. Ultrabasic rocks are predominant in the make-up of the Bonilla Formation and its equivalents in northwestern Mendoza, in which region they would belong exclusively to the Lower Palaeozoic of the Andes, in accordance with the present state of geological knowledge of the continent. Outside of the Andean Cordillera, the basic and ultrabasic rocks of the Cordillera Costanera of Chile could be considered to be contemporaneous with the initial simatic magmatism during the Cambro–Ordovician.

The presence of basic and ultrabasic rocks in the geosynclinal Lower Palaeozoic of Argentina does not constitute an independent event in the eugeosynclinal magmatization. As will be shown later (Chapter 9) these rocks form part of an extended ophiolitic association which is well developed in the Precordillera and Cordillera Frontal of Argentina up to and including the Ordovician Period.

It is not improbable, according to their geological position, that the gneiss remnants of Llaucha have an igneous origin. These are moderately silicic acid rocks for which the possibility of a Hercynian granitization is unlikely.

On the basis of the first concrete geochronological references it is possible to indicate that the masses of granodiorite which serve as a basement to the Mesón Group in the Cordillera Oriental of northern Argentina originated during a post-assyntic process, and thus belong to the Lower Cambrian period. Such is also the case for the plutons of the

Puna-Cordillera Oriental of Salta, northern Argentina, which consist mainly of pinkish grey granites (KEIDEL, 1943, pp 170–175), and also granites and granodiorites (VILELA, 1956, pp 26–37). On the Argentine–Bolivian border the groups of granodioritic rocks have identical relationships. TURNER (1960, p. 168) has described these granodiorites as belonging to the Cañaní Formation. He indicated that they are coarse-grained rocks, as a mass have a clear grey colour with variations in tone, and that they contain abundant quartz. Granitic and tonalitic facies are associated with the granodiorites in the regional plutonism. It is fairly certain that a Cambrian age can be assigned to the group of granitic rocks of microgranular and granular structures, with a trondhjemitic texture, described by the same author (1964a, pp 28–30) as the Cachi Formation, which outcrops in the Nevados of Palermo in the Puna, to the south of San Antonio de los Cobres, Salta (Fig. 2), approximately on the meridian and to the south of the plutonic body of Tastil. Pegmatitic masses which intrude the metamorphic basement in Cordoba in the central area of Argentina, related to the granitic post-assyntic plutonization, have yielded radiometric ages within the Cambrian Period. To these may be added the pegmatites of the other Pre-Cambrian sectors of the country, which are very frequent and are emplaced e.g. in Catamarca, San Juan, and San Luis, into the basement and contain various mineralizations.

6.3. Metamorphic petrology

The deeper part of the orthogeosynclinal Cambro–Ordovician zone of the Andes presents a broad metamorphic picture which extends to the west of the Precordillera and Cordillera Frontal of Argentina.

STAPPENBECK (1917) advanced knowledge of a zone of gneisses in the Cordón del Plata, Cordillera Frontal of Mendoza, linking these with injected mica-schist which controls the granitic plutonization (Upper Palaeozoic?).

Granodioritic schistose gneisses were described by KITTL (1934) in the lower part of a succession of phyllites and mica-schists between Llaucha and the Diamante River (Cruz de Piedra–Carrizalito), and they probably represent the southernmost base of the Cambro–Ordovician structure outcropping in western Argentina. Mica-schists with andalusite and staurolite can be added. In the Precordillera to the north of the Mendoza River, the Uspallata range contains a similar complex, known through the works of AVÉ-LALLEMANT (1892) for its phyllites, mica-schists, calcareous phyllites, and chloritic and garnetiferous schists. KEIDEL (1937) also described green schists and lustrous schists in the Bonilla Formation and quartzitic garnetiferous rocks in the Farallones Formation. To these ZARDINI (1959) added the presence of quartzites and amphibolites in the area of Yalguaraz (Sierra de La Cortadera). It is in part the same series, south of Mendoza, which STAPPENBECK (op. cit.) described as the 'stratified rocks of the Lower Palaeozoic', and GROEBER (1939) as the metamorphosed Lower Palaeozoic. In a recent study DE ROMER (1964) has revised in detail the sector of El Choique, Quebrada de Santa Elena (Uspallata), on the south-western border of the Precordillera of Mendoza, and described in the local metamorphic sections two series: the lower one formed of muscovite and chlorite schists, schistose psammites and green schists, plus amphibolites and altered gabbro; and the upper series containing arenaceous chlorite schists, chloritic and carbonaceous schists, serpentinites, dolomites, and chloritic-dolomitic schists, succeeded by laminated calcareous rocks, phyllites, phyllitic schists, and chloritic and carbonaceous schists of low grade. For the area of the Tunuyán River, Mendoza, POLANSKI (1957) interpreted a similar complex as elements of a metamorphic basement with basic intrusions.

In the west of the Precordillera of San Juan in the zone of Calingasta and its environs, HARRINGTON (1957), on the basis of his own and other observations, described phyllites, sericitic arenites, and greywackes of the Alcaparrosa Formation. Also FURQUE (1963) described quartzites, lutites, sericitic lutites, and greywackes in the complex of the Yerba Loca and Rio Blanco Formations, to the north-west of San Juan, which extends from basal Cambrian to Ordovician.

GONZÁLES DÍAZ (1958) and ZARDINI (1960) studied the problem of the regional meta-morphism of the Cambro–Ordovician structure of the Precordillera–Cordillera Frontal. The latter recognized for the Precordillera of Mendoza a zone of green schists (lower part) in the area of La Cortadera–Bonilla and another of amphibolites in the Cordillera, in the trench of the Cordillera Frontal corresponding to the sector of the Las Tunas River. The relation between the regional metamorphism of the country rocks and the ultrabasic rocks suggested to ZARDINI (1960) that the garnetiferous mica-schists, crystalline lime-stones, and peridotitic ultrabasic rocks are in the amphibolite facies, whilst the phyllites and schists of the green schist facies are directly associated with the serpentinites of the area.

OLIVEIRA (1956) indicated that the Lavras Formation (*sic*), said to be of Cambrian age, has a slight metamorphism, as can be observed in Bahia and Minas Gerais; but certain sectors exist where the rocks indicated are affected (exceptionally?) by a mesozonal metamorphism, e.g., Canaveiras (Bahia) Carandaí and Sierra de Catuni (Minas Gerais). In general, the rocks of the supposed cratonic Cambrian are either metamorphic granu-lites or non-metamorphic types (HARRINGTON, 1962).

7. Sedimentary structures

An examination of the sedimentary structures of the Cambrian strata of South America reveals certain simple features of the sedimentation. The problem has, however, not been treated in detail.

The oolitic limestones of the Corumbá Formation (Bodoquena) in Bolivia–Brazil (Paraguay) contrast with others of the succession. The Cambrian limestones of San Juan, Argentina, are also oolitic (STELZNER, 1885). In the Cordillera Oriental of northern Argentina we may note the cross-laminations of the numerous Cambrian psammite beds of the Mesón Group (TURNER, 1963), adding that ripple-marks and mud-cracks have also been observed on the bedding planes, e.g. in the Chalhualmayoc and Campanario Formations of the geological section of Santa Victoria, Torrential structures have been observed in the arenites and quartzites in the Lizoite Formation in Purmamarca, Jujuy (TURNER, 1963, p. 201).

The calcareous sediments of the Precordillera of western Argentina, present in San Juan, show primary corrugations in the lower levels of the La Laja Limestone (BORRELLO, 1962), which are exposed in the western flanks of the Sierra de Zonda (*Glossopleura* Zone). In this area one can observe structures of organic origin in the limestones, no doubt due to traces of worms ('vermiglyphen').

Rocks resembling the (Upper?) Cambrian of San Juan contain stylolites. The occurrence in some sectors of the western flank of the Sierra de Zonda and to the north of Villicúm, San Juan, of exclusive horizons of dark to black chert, which are prevalent from the Middle to Upper Cambrian, is a spectacular feature. The upper part of the outcropping

Cambrian of the San Isidro Formation, Mendoza, contains similar siliceous masses which are deformed by the folding.

Compact and dense nodular bodies are locally found above and below the fossiliferous Middle Cambrian in the San Isidro Formation, Mendoza. This same formation, can be shown to contain some horizons with massive flanking breccias, no doubt of tectonic origin. Breccias of this type have also been observed in the eastern part of the Sierra de Zonda, west of the city of San Juan, where they are unconformable on the dipping strata of the uppermost parts of the Cambrian or Lower Ordovician.

From the Rio Blanco Formation, in the Cambro–Ordovician to the west of the Precordillera of San Juan, FURQUE (1963) mentioned the presence of symmetrical and octagonal mud-crack structures in sandy lutites along the homonymous river, preserved despite the tectonic dislocation of these horizons during and after the Upper Palaeozoic period.

8. Sedimentation

The processes of Cambrian sedimentation are characterized by a more clearly individualized development, as the result of structural control, in the geosynclinal regions than in the cratonic regions. The uniformity of the respective lithologies is more marked in the former. A relatively larger accumulation of sedimentary deposits can be observed in the orthogeosynclinal regions, a fact which confirms the greater mobility and subsidence of these regions. In the orthogeosynclinal areas the sedimentary pattern of Cambrian times sometimes continued uniformly into the Ordovician, as occurs in the evolution of the Lower Palaeozoic environment of the Precordillera of western Argentina. In the Cratonic areas, where the age of the oldest Cambrian formations which overlie the Pre-Cambrian basement is really uncertain, one can observe an appreciable variability in the known sedimentary types.

8.1. Sedimentation in geosynclinal areas

The Cordillera Oriental of Colombia, the Cordillera Oriental of northern Argentina and southern Bolivia, and the Precordillera of western Argentina reveal sedimentary processes of marine environments, which can be attributed to evolution in an orthogeosynclinal belt. In the Colombian area the scanty information available makes it possible only to guess at the presence of an external geosynclinal complex with miogeosynclinal characteristics. Its relations with the deeper or internal orthogeosyncline are unknown. In principle, it represents the sedimentation of a geosynclinal trough which has evolved prior to the so-called 'flysch' patterns.

The sedimentary environment in which the horizons of the Mesón Group in the Cordillera Oriental of northern Argentina, in Salta and Jujuy, accumulated presents a picture of the marginal geosynclinal areas with true psammitic miogeosynclinal deposits without cratonic relationship, excepting in the basal part where a discontinuous formation of transgressive conglomerates is found. The thickness, which is over 3,000 metres, testifies to the magnitude of subsidence in the area. The basal conglomerates reveal the nature of the tectonic regeneration of the regional Pre-Cambrian structures which were accommodated to receive the infilling of psammitic sediments during the Cambrian Period. The source of sediment for the basin is apparently found in the dorsal zone which separates it from the internal geosynclinal area. This latter has not yet been defined but it would be necessary to locate it contiguously with the Cordillera Oriental

(Puna). The conditions of sedimentation of the Sama Formation in southern Bolivia are similar on account of their equivalence of facies.

In Purmamarca, Jujuy, the reddish rhythmically bedded sediments of the Campanario Formation, rich in tubular remains of *Skolithos*, show the patterns of 'orthoflysch', which are unusual in a miogeosynclinal area. They testify to a premature advent of pre-orogenic processes which no doubt extends also into the Cordillera Oriental of Argentina as a representative of geosynclinal tectofacies.

The calcareous and calcareous-dolomitic sedimentation in the eastern half of the Precordillera of San Juan and Mendoza shows typical evidence of the persistence of the miogeosynclinal trough. The amount of orthoquartzite involved is limited. Progressive deepening of the trough can be deduced from east to west, i.e. from the Craton of the Sierras Pampeanas to the eugeosynclinal region of the Precordillera and Cordillera Frontal. The intermediate zone between the mio- and eugeosynclinal zones in San Juan shows a mixed pattern of sedimentation with the presence of calcareous-dolomitic and lutitic-phyllitic rocks, the latter belonging to the deep orthogeosynclinal zone. The presence of horizons with chert, which have possibly originated as the result of metasomatism in the middle of the geosyncline and which are related to the simatic vulcanicity through the loss of silica during the albitization of the ophiolites, reveals a greater relative depth of the miogeosyncline of the Precordillera during the time of calcareous sedimentation. It is possible that the limestones and dolomites are partly clastic. This would be due to the predominance of calcareous and dolomitic rocks in the Pre-Cambrian of the contiguous region of Sierra Pie de Palo (Los Angacos Formation of the Caucete Group), which is the structural base of the Lower Palaeozoic of San Juan.

The limestones and lutites of the San Isidro Formation on the southern border of the Precordillera of Mendoza belong to a well-defined marginal environment which lacks psammites and basal conglomerates, with the exception of the calcareous arenites of the Quebrada Empozada which further north intercalates with the above-mentioned formation. The Cambrian rocks of this sector show the minimum recorded thickness as a result of fracturing towards their base.

The maximum thicknesses of the Cambrian sediments of the Precordillera are approximately 1,000 metres for the calcareous and calcareous dolomitic deposits, and probably around 3,000 metres for the phyllites and lutites, respectively in the mio- and eugeosynclinal Cambro–Ordovician of western Argentina.

8.2. Sedimentation in the cratonic terrain

The calcareous sedimentation of the Corumbá Formation (or Bodoquena) at the mouth of the Paraguay River, which is extensive in eastern Bolivia and the Mato Grosso area of Brazil, has a tendency because of its pericratonic structural position to a sedimentation of orthogeosynclinal type (KAY, 1951), though in the upper parts it is noted that psammitic and pelitic sediments of epicontinental origin occur above the limestones and dolomites. The total thickness of the sequence is only 300 metres, resting on the rigid Pre-Cambrian basement.

The progression towards an increasingly continental nature of sedimentation could be demonstrated by examination of the depositional features of the supposed Cambrian of the Brazilian platform, to which belong the deposits of the Jaibara of Ceará Group, the Lavras Group of Bahia, the Gaspar Formation of Santa Catarina, and the Jacadigo de Mato Grosso Group, together with masses of fanglomerates and subfluviatile deposits of diminished or variable thickness.

It is not possible at present to attempt a correlation of these facies, because of incomplete knowledge of their relationships and particularly the absence of biostratrigraphical correlations between the cratonic and geosynclinal deposits. Even in the case of the rocks of the Limbo Formation of Bolivia, which are a series of terrigenous or littoral deposits within the Andean structure, and which could represent a series of Cambrian strata, an autonomous or even misleading structural and palaeogeographical pattern would be obtained and be difficult to interpret in its relations to the strictly geosynclinal pattern of sedimentation.

9. Vulcanicity

The vulcanicity which is associated with the Cambro–Ordovician strata of the Precordillera of western Argentina in San Juan and Mendoza is of submarine origin. Basaltic pillow-lavas and spilites are relatively frequent in the San Juan region which extends between Calingasta and Leoncito (BORRELLO, 1965). Other outcrops appear farther to the north between the San Juan River and the Jachal River, intercalated between the eugeosynclinal sedimentary deposits of the Lower Palaeozoic which are at present being studied.

The data derived from AVÉ-LALLEMANT (1892) and KEIDEL (1939), on diabase and sheets of basic rocks intercalated in the metamorphic Lower Palaeozoic rocks of the region of Uspallata, refer in a strict sense to the emplacement of gabbroic rocks. STAPPENBECK (1917) had earlier described similar outcrops on the eastern sides of the Cordillera del Plata, Cordillera Frontal, Mendoza, indicating the diabase as being an example of vulcanicity even in the form of intrusive bodies in the Rio de las Tunas. Phenomena of spilitization in the volcanics of the north-eastern sector of the Cordón del Portillo, in the Cordillera Frontal de Mendoza were mentioned by GONZÁLES DÍAZ (1958).

According to this last author, within the local metamorphic environment rocks of porphyritic structure, varying from acid to moderately silicic in composition and with a high degree of spilitization, are classified as granitoid porphyries and are of Lower Palaeozoic age in this area. The described porphyritic types can be compared with those of more limited development which occur within gneisses and mica-schists in Llaucha, on the road to the Diamante Lake in the extreme south of the Cordillera Frontal of Mendoza.

The Lower Palaeozoic volcanic rocks which to a large extent can be assigned to the Cambrian in the Precordillera and Cordillera Frontal occur in a group which demonstrates the effects of albitization in an association of the spilite-keratophyre type which is related to the processes of the initial magmatism in the eugeosyncline. The relationship of the spilitic rocks to the initial basic magmatism was examined by BORRELLO (1963) and recently included in the ophiolitic association of the simatic magmatism (BORRELLO, 1965) as another example of the various characteristics of the magmatico-structural evolution, during the persistence of the trough in the deeper parts of the geosynclinal area.

Cambrian vulcanicity is not well known in other South American regions. However, KEGEL and his associates (1958; pp 17–18, 34–35, and 44; fig. 3) have alluded to the presence of spilites north of Ceará (Cuenca de Massapé) which discordantly overlie the sediments of the Jaibara Group (Fig. 12). They have also described from the same area a keratophyro-spilitic rock with large phenocrysts of albite (op. cit., p. 17). These albitized rocks are assigned to a phase of subsequent vulcanicity (Stille) without it being possible to establish their relationship with the geomagmatic episodes of the Lower Palaeozoic of the Andean region.

10. Mineral deposits

The mineral deposits of the South American Cambrian are only partly known. From the available information it can be deduced that they belong to diverse mineral assemblages which can only partly be fitted into an over-all picture of the classification of the mineral deposits.

10.1. Metalliferous minerals

WOKITTEL (1960) in referring to the Cambro–Ordovician of Colombia mentioned certain Andean areas—where the rocks are probably largely of Ordovician age—containing deposits of gold, lead, and copper, related to acidic intrusions occurring in the Silurian period.

According to DORR (1945, pp 1, 34–43) the Jacadigo Group of the Morro de Urucúm, Brazil, contains reserves of hematite, totalling 1,310,000,000 metric tons (55% Fe and 20% SiO_2) and manganese ore with 4,420,000 metric tons of measured mineral, 11,750,000 metric tons of indicated mineral, and 17,500,000 metric tons of inferred mineral (45·6% Mn and 11·1% Fe). This author also mentions iron ore of doubtful quality and also detrital minerals (64% Fe and 4% SiO_2) with appreciable reserves. The western part of this iron-bearing region continues into Bolivia, where its importance is also recorded (AHLFELD and BRANIŠA, 1960, p. 23).

The calcareous quartzites of the Aguilar mine, Jujuy, in northern Argentina, contain exploitable deposits of lead, silver, and zinc, which originated through a process of replacement during the Cainozoic period (SPENCER, 1950).

10.2. Non-metalliferous minerals

The Lavras Group, in the Brazilian area of Bahia and Minas Gerais, is noted for the presence of diamondiferous conglomerates (OLIVEIRA, 1956) with one representative locality: Lavra Diamantina, Bahia, conveniently described by DERBY (1906). The gem-bearing conglomerates are believed to represent secondary products derived from primary accumulations formed during the emplacement of pegmatites into the Minas and Itacolomí Series, which in part are made up of graphitic pelites. According to BATEMAN (in 1957, pp 912–913) Brazil is considered to be the only important diamondiferous country apart from Africa. He summarises the various aspects of their occurrence and expresses the view that the diamonds are found in ancient terraces, in ancient conglomerates and igneous breccias, which have been the source of most of the present alluvial deposits. The deposits of Anga Sija, Minas Gerais, have as sedimentary matrix a deposit of angular blocks with a pelitic matrix which could have been derived from the alteration of a pre-existing rock of the kimberlite type. The Brazilian diamond production is of 150,000 carats per annum, including the extraction from the Lavras Group and alluvials of Bahia and other regions of Brazil.

The pegmatites which are geochronologically dated as Cambrian and emplaced in the basement of the Sierra de Cordoba, Argentina, are known, as in the case of the Las Tapias mine, for having provided profitable quantities of beryl and spodumene of good quality.

The upper parts of the Cambrian sequence of San Juan and the Ordovician contain accumulations of calcite, known since STELZNER's voyage (1885). These are irregularly shaped and have given rise to a moderate exploitation of the mineral, which is fairly good to good in quality. The Lower Cambrian in the middle of the north-eastern part of the Sierra de Zonda contains secondary deposits of sulphur which irregularly cover the limestones, and which have up to the present not been found to be exploitable.

The serpentinitic bodies of La Cortadera and environs, Bonilla (Precordillera), and Las Tunas (Cordillera Frontal) are the main talc producing areas, the principle industry being at Bonilla. The ultrabasic rocks contain asbestos, though in general not of industrial quality. It is in the form of crisotile in veinlets or fibrous grains formed from an antigoritic mass (ZARDINI, 1961). The presence of chromiferous magnetite irregularly dispersed in the serpentinite, though sometimes forming thick aggregates (ZARDINI, 1961), has no industrial importance.

10.3. Economic geology of the Cambrian rocks

The Precordillera of San Juan and Mendoza in western Argentina, which has large reserves of Cambro–Ordovician calcareous rocks, has given rise to intense exploitation for the production of cement and lime since 1913. The larger industry has been concentrated between Zonda, San Juan, and Capdevila, Mendoza. To a lesser extent it is spread out over all of the eastern flanks of the Precordillera from Jachal, San Juan up to San Isidro, Mendoza involving mining and quarrying operations of varying importance.

In the main body of the Sierra de Zonda there have been demonstrated levels of Cambrian and Cambro–Ordovician carbonates, noted for the predominance of dolomites, whose development and importance has been locally examined (KITTL, 1960, pp 79–80).

The Corumbá Formation (or Bodoquena Series or Itapucumí Series–eastern Paraguay) in the pericratonic zone of Bolivia, Paraguay, and Brazil, containing beds of limestone, promises to be a productive source for the cement and lime industry. The plant for manufacturing these products, as yet on a limited scale (ECKEL, 1959), is located in the valley of the Paraguay River near to Vallemí, Paraguay. The reserves are enormous and will exceed the demands of exploitation for a long time.

VILELA (1956, p. 52) has indicated that the granodioritic pluton of Tastil is being exploited on a small scale, but that it could have many applications as revetment rock in the construction of buildings. Regarding the sedimentary Cambrian of the Quebrada del Toro (Salta) he indicates the existence of a bed of quartzitic arenites one metre thick in the lower levels of K_1 (*sic*) or the Lizoite Formation, with a 99·18 % content of silica, which is useful in the manufacture of colourless glass.

In southern Bolivia and northern Argentina, the psammitic rocks of the Cambrian have been used in shaped blocks for the building of dwellings for the rural population.

11. Palaeoecology

The palaeoecology of the South American Cambrian which is accessible for study, appears to be nearly wholly restricted to the area of the Precordillera of western Argentina, and the fossiliferous sequences discussed in this work (3.1 and 4.1) were described from here. The respective palaeobiogeographical relationships constitute an important element in the judgement of the palaeoecological aspect of the situation, and to this purpose, were examined in a previous chapter (4.2). The ecological character of the Cambrian of northern Argentina is more difficult to study because of the scarcity of fossils except for traces of worms, and is even more so in the case of the supposed Cambrian of the cratonic region. In this context is it necessary to examine the palaeoecological scene of the Cambrian, and as far as possible in each case to establish its physical and biological characteristics.

11.1. The Cambrian environments of the Precordillera, western Argentina

A sub-littoral zone, of a substantial platform in an epicontinental environment equivalent to a marginal-cratonic and an extra-cratonic marginal area is discernible in the Lower Cambrian of the Precordillera of San Juan, where the olenellid association appears in the Sierra de Villicúm. The equivalent situation in North America has been demonstrated by LOCHMAN-BALK and WILSON (1958, p. 318). Arenaceous, calcareous, and calcareo-dolomitic rocks have produced excellently preserved fossils (e.g., *Olenellus*, *Fremontella*).

These environmental conditions generally appear to have been permanent during the Middle Cambrian. The benthonic environment is indicated by the development of calcareous formations in the area of the basin (RICH), and by the presence of such well preserved genera as, for example, *Kootenia*, *Glossopleura*, and *Zacanthoides*. In the Middle Cambrian of San Juan (Sierra de Zonda) the calcareous sequences are uniform up to the level of the Zone of *Bathyuriscus–Elrathina* (e.g., *Elrathia*, *Ehmania*). This is not the case, however, in northern Mendoza where the Middle Cambrian is partly lutitic. Argillaceous sediments, intercalated between thin and thick bands of limestone, contain remnants of trilobites (e.g. *Oryctocephalus*) and other fossils which are sometimes poorly preserved.

The upper part of the Middle Cambrian with the *Bolaspidella* Zone tends to reveal conditions of major relative depth within the benthonic part of the basin. The trilobites and calyptoptomatides (*Baltagnostus*, *Bolaspidella*, *Orthotheca*) are magnificently represented. They are small and even delicate, preserved in black fine-grained muddy limestones, which characterize the intermediate euxinic cratonic zone of the Lower Palaeozoic geosyncline.

A similar situation occurs in the Upper Cambrian from the *Crepicephalus* (*Cedaria* or *Crepicephalus*) Zone to that of *Saukia* in the Precordillera of Mendoza. Grey and dark limestones frequently contain small trilobites (e.g. *Elvinia*, *Irvingella*, *Parabolinella*) with the exception of *Hungaia* which has a large exoskeleton. In all these cases the state of preservation is extraordinarily good. An accumulation of fragments (*Hungaia puelchana* Rusconi) has been observed forming irregular lenticular bands up to one metre thick.

In the Precordillera of Mendoza remains of algae replaced by black silica have been found in the Upper Cambrian rocks (north of San Isidro). The existence of dark chert suggests the deepening of the Palaeozoic geosyncline in the Upper Cambrian, although in the extreme north of the Sierra of Villicúm, San Juan, these siliceous rocks were found above the Zone of *Glossopleura*. Traces of annelids are found in great profusion in San Juan in the cratonic and intermediate environments; but those which would be expected to occur typically in the deposits of the Lower Cambrian platform are scarce. However, the presence of *Hyolithes* (Calyptoptomatida) seems to indicate an ecological régime of shallower depth in San Juan (Lower Cambrian) and Mendoza (Middle Cambrian).

11.2. The Cambrian environment in the Cordillera Oriental, northern Argentina

The dominant presence of *Skolithos*, which forms part of the original sedimentary fabric, in the rocks of the middle part of the outcropping Cambrian sequence of Salta and Jujuy in this area of northern Argentina, only enables this sequence to be placed in the extension of a basin of alternating sediments of neritic type or 'undaform' (RICH). In such an external miogeosynclinal zone, the ecological significance of these remains is moreover only relative. However, they seem to be associated with a 'flysch' series (Campanario Formation of Santa Victoria to Quebrada del Toro, Cordillera Oriental).

In the same region *Skolithos* tubes are very frequent, together with other vermiformal and problematical remains, in the Ordovician rocks.

11.3. Environmental conditions in the supposed Cambrian of the cratonic areas

In a case similar to the previous one worm tubes of *Arthraria* have been found in the Bambuí Formation of Brazil, occurring in arkosic rocks together with sponge spicules. The regional cratonic environment is the platform for all the parageosynclinal sedimentation, so that the local series must be interpreted as having frequently developed under neritic or epicontinental conditons. The same can be said for the ecological environment of the Corumbá Formation of Brazil (and Bolivia), based upon the presence of algae of genus *Aulophycus*, which in any case is very imprecise. In calcareous sediments of the cratonic formations, (e.g., Bambuí Formation, Corumbá Formation) we can perhaps find evidence of sedimentary environments and sometimes of organic accumulations of greater relative depth (clinoforms, RICH).

12. Palaeogeography

A cursory examination of the expressive South American morphostructural frame with its Pacific orogenic spine—which is a geological framework for the platforms of the continent which extend from the Atlantic Ocean—leads one to recognize the distinction between the geosynclines and cratons of which it is formed. Within the scope of the Pacific geosynclines the greater permanence of the marine systems can be accepted, while in the cratonic regions we can detect the evolution of the terrigino-fluvial basins.

12.1. Cambrian conditions in the geosynclinal zones

WEEKS (1948) offered an earlier palaeogeographical scheme for South America, gathering for the purpose the marine facies of the Cambro–Ordovician into one single unit which he extended from the geosynclinal areas of the Pacific to the Atlantic border of the Brazilian Craton. In a more detailed work, and having at his disposal the data of subsequent geological investigations, HARRINGTON (1962) produced a more complete palaeogeographical map of South America, specifically for the Cambrian Period. In this map he clearly shows: firstly, the considerable extension of the marine facies; secondly, the expansion of these facies from the Andean region to the inter-cratonic areas (Guiana–Brazilian shields); and lastly, the lesser extension of Cambrian deposits indicated as being of continental origin, which are limited to the Brazilian Craton.

In our interpretations of palaeogeography we shall review the data provided by these authors and its bearing on the reassessment of those formations which, outside of the typical documented palaeontological localities, are in any way correlated with the Cambrian sequences of the continent.

The distribution of the Andean Cambrian in the palaeogeographical orthogeosynclinal system, though it is more continuous between the parallels 22° S. and 35° S., in general appears to be connected with the Colombian Cambrian, where the trilobites apparently indicate a biofacial equivalence of the Middle Cambrian of the Cordillera Oriental of Colombia and the section south of the Precordillera of San Juan, Argentina. The limestones and other miogeosynclinal sediments of the Buenos Aires Lake (Chile) and San Martin Lake (Argentina) zones are, in principle, regarded as part of the Andean Lower Palaeozoic orogenic belt. This facilitates understanding of the continuity of the geosynclinal basin including the supposed Cambrian of the archipelago of southern Chile, at least that portion of the Cordillera situated between latitudes 3° N. and 52° S. It is not possible to extend the belt of Cambrian horizons farther south into the Patagonian Cordillera and any connections there might be with the Cambrian of the Antartic region, including that of the region of the Wedell Sea which contains archaeocyathids (Lower or Lower to Middle Cambrian), are not known.

It may perhaps be possible one day to prove that there exists a normal continuity of the South American Cambrian geosynclinal basin towards the subpolar latitudes, in the same way as in the opposite extreme there seem to appear connections between South America and North America. In the latter case, the zone of relatively greater proximity is that which occurs between the Rio Duda, in Colombia, and Sonora, in Mexico, where the intermediate data which would define a real and continuous palaeogeographical link are still missing. This idea, if substantiated, would demonstrate the extensive or Pan-Pacific character of the Cambrian orthogeosyncline (Cambro–Ordovician) on the western border of the Americas.

It is nearly certain that the permanence of the orogenic sequence which commences with the geosynclines of the Lower Palaeozoic in South America did not undergo major modifications during the Cambrian Period. In this sense, as emerges from pertinent studies, there should exist a manifest relationship between the distribution of the Andean marine Cambrian and the site of the actual Andean belt, which is placed in the continental structure as the result of orogeny (HARRINGTON, 1962). With the exception of the Amazonian zone on the eastern border, the orthogeosynclinal zones can be discerned as a whole. This is evident since the respective miogeosynclinal zones (Sama Formation and Mesón Group) of Rio Duda to the Cordillera Oriental of Bolivia and Argentina, and up to the east of the Precordillera of San Juan and Mendoza (La Laja Limestone, San Isidro Formation), on the one hand, and a eugeosynclinal zone in the western half of this Precordillera (Alcaparrosa, Farallones, and Bonilla Formations), on the other, exhibit the same characteristics as those of the vast territory of orthogeosynclinal Cambrian which terminates in the Ordovician of the Argentine Andes (Cordillera Frontal).

The development of the Limbo Formation of the Cordillera Oriental of Bolivia (Cochabamba) does not fit very well into this orthogeosynclinal picture. This is so in relation to its régime of sedimentation, which is in part more of platform and terrigenous type rather than of geosynclinal marine or continental marine type. The presence of anhydrite is only frequent in the mollasic sequences, and is absent in the Cambrian orthogeosynclinal regions which are known in southern Bolivia and in Argentinian territory.

The persistence of the Cambrian sedimentary sequences, including the marine facies, in the geosynclinal intercratonic area of the Amazon in northern Brazil is reasonable, as it would only require the recognition of a type of geosyncline which is not strictly orthogeosynclinal in that it lacks the necessary magmatic activity. The best indication of this could be the presence in the base of the Palaeozoic sequence of the Acarí–Jaú Formation, which has only been proved below the extensive river beds of the middle and upper course of the Amazon River (MORALES, 1959), and consists of continental, subcontinental, and shallow marine facies.

12.2. Cambrian conditions in the cratonic areas

The Andean and Amazon regions are palaeogeographically and structurally zones of subsidence and deposition which deserve the definition of geosynclinal; but the remainder of the South American continent must surely be classified, for the Cambrian Period, as an area of platforms or cratons. The corresponding palaeogeographical pattern is of massive, stable areas with infrequent autonomous basins which lie far from each other. Above the craton the basins appear to contain continental sediments with fanglomerates and fluviatile strata, for example the Jaibara Group of Ceará (KEGEL and others, 1958), the Lavras Group of northern Bahia (DERBY, 1906; WILLIAMS, 1930), the Gaspar Formation of Santa Catarina (FREITAS, 1945; MAACK, 1947), and the Jacadigo Group to the south of Mato Grosso (DORR, 1945) revised by HARRINGTON (1962) in assigning to the Cambrian stratigraphical units of parageosynclinal development. A glacial origin for the deposits of the Lavras Group is not very acceptable. It is not known whether the tectonics of the assyntic cycle have had any important repercussion in the cratonic regions. It appears to be difficult to correlate the details of the palaeogeographical evolution of these basins with their tectonics. In principle, the establishment of the fanglomerate deposits as being of Cambrian age and their further evaluation would permit the assertion that the assyntic movements—which in the Andean regions produced areas of subsidence as a result of tectonic deformation—caused a pattern of doming and fracturing which would have allowed the development of certain types of tafrogeosynclines (KAY, 1951). This would demonstrate to some extent the palaeogeographical development of psephite deposition upon the shield areas. They could also have a greater age and represent molassic types developed in a Pre-Cambrian cycle, as in fact does happen in the top of the Pre-Cambrian basement south of the Puna in Catamarca, Argentina (Colana–Pomán area).

The south-eastern region of the craton of Brazil, up to northern Paraguay and eastern Bolivia, is the site of a sedimentary basin which contains calcareous and other sediments of the Corumbá Formation (or Bodoquena), also considered to be of Cambrian age. It evidently reflects pericratonic conditions and does not maintain any connection with the Andean or cratonic sedimentary sequences of the South American continent. This basin is similar to the orthogeosynclinal types described by KAY (1951) in North America, because of the facies developed in it rather than its palaeogeographical situation. In this case its development as an environment of subsidence and marine sedimentation does not support the argument that the assyntic tectonic movements caused the formation of the corresponding structural depression, though two problems remain *viz.*: the inexplicable lack of relation with the subsiding Andean orthogeosynclinal zones, including the biofacial sequence; and the fact that the contiguous pericratonic basin of the Chaco does not exhibit similar sequences and in fact does not contain any Cambrian strata. The Bambui Formation in north-eastern Brazil contains, among other sediments, beds of limestone in the middle of a cratonic area, but their lithological characteristics do not resemble those of the calcareous rocks previously described. Its sedimentary sequence, which is typical of platform deposits, conforms with that of parageosynclinal accumulation, stages of epirogenesis controlling the sedimentation of diverse pelites. Such calcareous formations provide logical evidence of the accumulation of marine sequences, this having been discussed already in Chapter 11.3

Controls of the palaeogeographical evolution of the Lower Palaeozoic (Cambro–Ordovician) of South America are dependant upon the geotectonic pattern and consequent structures, though future advances in the knowledge of the features discussed above will allow a reassessment of the chronology of the Cambrian and its distributions of land and sea.

13. Structural evolution

Tectonic developments in South America during Cambrian times must be studied in the areas of the Cordillera Oriental of northern Argentina and the Precordillera of western Argentina. The Cambrian geology of other regions of the continent does not at the present offer useful information for similar analyses, as will have been deduced from the various aspects discussed here. In Argentina the top and bottom of the Cambrian sequence is exposed, in some regions marked by discordancies with the Pre-Cambrian and Ordovician respectively.

13.1. Beginning of structural evolution. Establishment of the Cambrian geosyncline

In Argentina, the end of the Pre-Cambrian, Upper Proterozoic, sedimentation also coincided with the assyntic tectonic movements which caused orthotectonic fold structures in northern Argentina (Salta and Jujuy) and another folded or paratectonic structure in western Argentina (San Juan). Leptometamorphic sediments, metamorphic rocks, and granitic plutons occur in both areas. In general these assyntic structures become cratonic with the culmination of a short geotectonic cycle, the *Protoidic*, which represents the oldest recorded in the Andean regions of Argentina, and which must surely have a more considerable extension in the South American Andes.

With the commencement of the Lower Palaeozoic, begins the geosynclinal sedimentation of the following cycle which we can call *Palaeoidic*, and the basins which originated with the advent of the Cambrian are a consequence of the transformation of the cratonic areas to subsiding orthogeosynclinal zones through geotectonic regeneration (STILLE). This geosynclinal rehabilitation led to marine Cambrian sedimentation on an eroded surface of Pre-Cambrian rocks of diverse composition. As a result, the Cambrian has a transgressive contact with its Pre-Cambrian basement and the discordance between them is regional and angular, being particularly well represented in northern Argentina. In the Precordillera of western Argentina, longitudinal folds of the Andean margin separate the Pre-Cambrian of the western Sierras Pampeans from the Cambrian of San Juan. Equivalent conditions can be accepted for Bolivia, where, however, there are no outcrops of the basement of the Cordillera Oriental.

The Cambrian geosynclinal sequence in northern Argentina presents a picture of a trough with an intermediate phase of epirogenesis, represented in the Campanario Formation by a rhythmical sequence of orthoflysch with regular and alternating sediments. When this event is considered in general terms, and extended to other areas of the Cordillera Oriental of northern Argentina, it is found to coincide with the early subsidence of the trough in the middle of the miogeosyncline.

13.2. Persistence of the régime of subsidence in the Cambrian

In the Cordillera Oriental of northern Argentina and southern Bolivia, the miogeosynclinal Cambrian occupies a considerable time span though lack of palaeontological evidence makes it difficult to ascertain its exact duration. The only indication of prolonged regional subsidence is provided by the maximum thickness of some thousands of metres of deposits. This subsidence was initiated after the post-assyntic plutonization of Early Cambrian age. Conversely, in the Precordillera, the persistent subsidence is recorded in the calcareous facies of the miogeosynclinal area, from the upper section of the Lower Cambrian (*Olenellus* Zone) in San Juan, to the top of the Upper Cambrian (*Saukia* Zone) in Mendoza. In northern Argentina the Cambrian sequence ends in an unconformity resulting from tectonic movements, probably Sardic, earlier than the lower Tremadoc (*Parabolina* Zone). Meanwhile, in the Precordillera, Upper Cambrian sedimentation was followed by Ordovician up to and including the Llanvirn. Consequently, the duration of deposition of the calcareous miogeosynclinal sediments was more prolonged and finally culminated in an early phase of Taconic movements. In San Isidro, Mendoza between the beds of the La Cruz Formation (Upper Cambrian) and those of the Empozada Formation (Llandeilo–Caradoc), the angular discordance which is observed can be attributed to a phase in the Sardic or Vermontian orogenies, the latter being a little earlier (Younger Appalachian).

The Cambrian sequences which have been described represent from a structural point of view tectonism of the geosynclinal trough followed by deposition of 'flysch': of Tremadoc age in northern Argentina (and southern Bolivia) and of Llandeilo age in the Precordillera. In both cases the upper limit of the trough and the base of the 'flysch', which are planes of unconformity, are the discordancies mentioned as Sardic and Pretaconic phases respectively.

13.3. Palaeotectonic structure of the Cambrian

Termination of the development of the geosynclinal trough is conditioned by preorogenic movements corresponding to the picture of regional embryotectonics. These movements, prior to the initiation of the 'flysch' régime, defined structures in space and time in the Cordillera Oriental and Precordillera respectively. They represent the first stage in the orogenic evolution of the South American Andes within the Palaeoidic structural sequence (Cambro–Triassic).

The embryotectonics began to develop with the stage of moderate regional tectonic inversion during which the subsiding environments were transformed into sub-positive elements of geanticlinal character. A major fracture with an orientation along the direction of the longitudinal Andean axis would have affected such elements. This can be deduced from the fact that in the Palaeozoic 'flysch' of the Precordillera sizeable blocks of limestone of Cambrian and Ordovician age have been deposited as the result of sliding within the sedimentational trough, so producing features familiar in the Alpine and Appenitic tectonics of Europe.

Processes of early gravity folding prior to the main geosynclinal tectonics, associated with the marginal zone and referred to the embryotectonics, have been proved in the southern sector of the Lower Palaeozoic geosyncline of the Precordillera of San Isidro, Mendoza, although these are only of local development. In this region beds of limestone with bands of chert are highly folded and stacked. It would appear probable that this is not an isolated example.

The age of the embryotectonical processes here described corresponds to that of the discordancies which produce the sharp base for the deposits of Tremadoc 'flysch' in northern Argentina and southern Bolivia. SANFORD and LANGE (1960, p. 1348) indicated that the uplift and gentle folding of the Andean Cordillera culminated in the upper part of the Cambrian.

13.4. Ophiolitization

The Cambro–Ordovician eugeosynclinal sequence of the Precordillera and Cordillera Frontal of San Juan and Mendoza of phyllitic and euxinitic composition was described in Chapter 9 as containing, in the middle of the sedimentary development, inclusions of the corresponding ophiolitic association. This association of phyllite-euxinite and ophiolite is determined by the state of maximum orthogeosynclinal subsidence, to the point where the eugeosynclinal belt had masses of simatic magmatic material incorporated in its middle part during the main period of trough development. One can observe a continuation of the ophiolites into the 'flysch' facies, in the north-east of San Juan (pillow-lavas); as for instance in the case of the Palaeozoic outcrops east of Rodeo. It has been estimated that the main development of the ophiolites begins with the initiation of geosynclinal sedimentation and culminates when the tectonic inversion implants the embryonic tectonic pattern in the Palaeoidic geosyncline of western Argentina, leading into 'flysch' deposition.

13.5. Geosynclinal tectonics

In northern Argentina and southern Bolivia the geosynclinal or orogenic tectonics, *sensu stricto*, are due to Hercynian movements occurring during the Carboniferous. The Cambrian beds of the trough and the Upper Ordovician to Upper Carboniferous flysch-oidal sediments occur within a group which has been affected by earth movements and then covered by post-Ordovician molassic sediments. The Cambrian strata do not show any evidence of an intensive phase of geosynclinal or tectonic deformation because of their position overlying the rigid basement of Pre-Cambrian rocks of the Cordillera Oriental, which represents a series of highly tectonised mixed 'flysch' deposits containing calcareous masses (Quebrada de Humahuaca: León-Volcán).

In the Precordillera the picture is similar but the lack of a structurally high or active basement during tectonic movements permitted the generation of complex structures along the length of the geosynclinal margin to the east, towards the craton of San Juan and Mendoza. Cambrian and Ordovician strata, together with the mid-Palaeozoic 'flysch', outcrop within a zone of deformation containing local folding and thrusting (Fig. 13) which may be classified as an autochtonous tectonic zone.

In the extreme south the Precordillera presents a distinctive structure in the Cambrian base of the La Cruz Formation The dark grey limestones and dense silicic horizons of this formation have been affected to a greater degree than the limestones of the San Isidro Formation by intensive folding. The folding is recumbent in style and the folded rocks are covered with angular discordance by slide breccias and Ordovician flysch (Fig. 14). This represents a markedly marginal tectonic episode prior to the real geosynclinal Palaeoidic tectonics of the region.

The Cambrian beds of Argentina and southern Bolivia contain other dislocations resulting from basement tectonics of Cainozoic age. Consequently they are involved in the superposed tectonic structure of the Andean region.

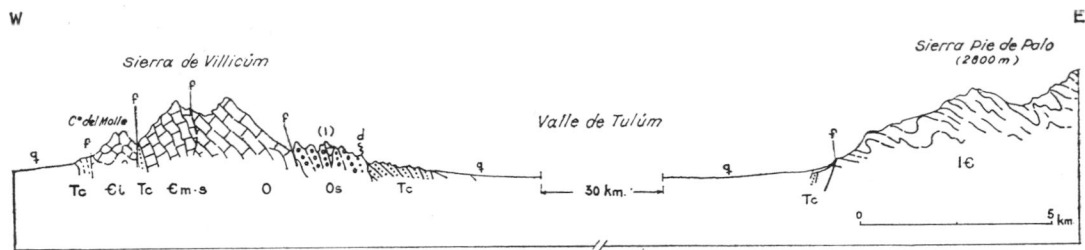

FIG. 13. Geological cross-section of the eastern border of the Precordillera (Sierra de Villicúm) facing the Sierras Pampeanas (Sierra Pié de Palo), San Juan, western Argentina, showing the tectonics of the basement. ⵜ€, Pre-Cambrian; €i, Lower Cambrian (*Olenellus*) and €m-s, Middle and Upper Cambrian (La Laja Limestone); O, Ordovician (San Juan Limestone); Os, Ordovician (Caliza San Juan); Os, Olistostroma (Wildflysch) Upper Ordovician-Devonian, (1) Olistolito; Tc, Tertiary; q, Quaternary alluvium; f, faults (according to the author, 1967).

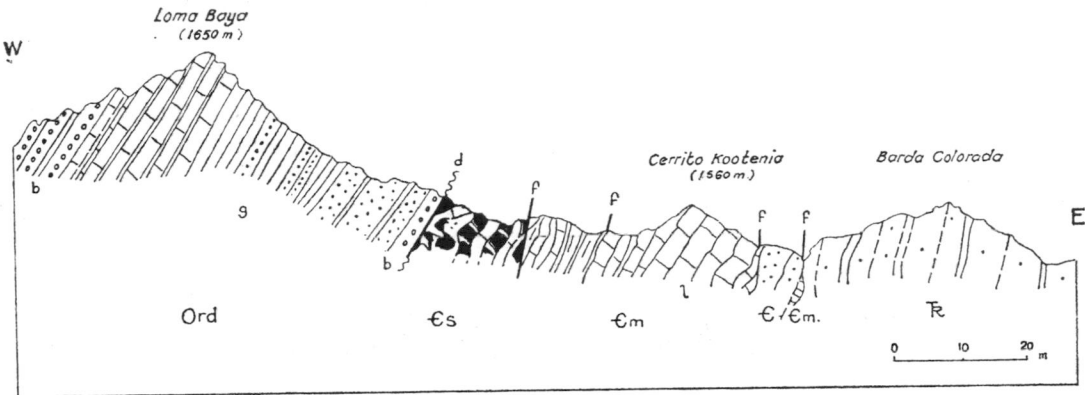

FIG. 14. Geological cross-section of the San Isidro section, Precordillera, Mendoza. €, Undifferentiated Cambrian; O, Ordovician flysch (b, breccias; g, richly graptolitic horizons); Tr, Continental Triassic (or Permo-Triasic) (outline by the author, 1967).

It is probable that the Cambrian beds of Colombia, together with the Ordovician of the Río Duda and the Cordillera Oriental, present a pattern of similar structural evolution of the tectonic area in which they appear in the South American Cordillera.

14. Summary of geological history

A summary of the Cambrian history of South America can be presented on the basis of the previous descriptions of palaeogeographical and structural elements as follows:

In the cratonic regions structures of the Pre-Cambrian basement originated after the assyntic movements with the formation of terrigenous basins of sedimentation in Brazil of reduced extent and autonomous in their geographical distribution.

The Brazilian pericratonic region is assumed to be in part occupied by Cambrian to Upper Palaeozoic deposits, from Bolivia to the upper course of the Amazon River, where the existence of marine sediments has been indicated, though lacking significant fossils. The subsiding belt of the Amazon River, which is intercratonic between Brazil and Guiana, is the site of geosynclinal sedimentation, whose development included the Cambrian Period.

In the South American Andean zone, from Colombia to the centre of western Argentina, the Cambrian sequence which is essentially marine and in part richly fossiliferous, characterises the development of an orthogeosyncline which originated in the Lower Cambrian and in some regions has a continuous development into the Middle Ordovician. The miogeosynclinal zones in the Precordillera of western Argentina are composed of calcareous sediments which, for the Cambrian period, indicate the development of cratonic and intermediate sub-euxinic series with trilobites. In the Cordillera Oriental of Bolivia and Argentina the psammitic Cambrian represents a miogeosynclinal facies, next to neighbouring cratonic platforms. The corresponding eugeosynclinal basins are not known. These only appear clearly defined in the western centre of Argentina in the area of the Cordillera Frontal and western Precordillera of San Juan and Mendoza, where they are characterized by a well-developed ophiolitic association.

The history of the South American Cambrian begins with post-assyntic tectonic regeneration and terminates with Sardic movements in the Cordillera Oriental of Bolivia and Argentina. In the Precordillera of western Argentina movements of this phase are only known in the southern extremity of the Cambrian structure of Mendoza (San Isidro) where they could even be of Vermontian age. However, in this region of the Precordillera and in the contiguous Cordillera Frontal of Mendoza and San Juan, there is a passage from Cambrian to Ordovician up to the Llanvirn, seen in all of the Andean area, and representing subsidence of the Lower Palaeozoic trough within the orthogeosyncline.

In the Cordillera Oriental of Argentina and southern Bolivia, and partly in the Puna of Salta and Jujuy, granodioritic, and in some sectors also granitic, plutons were intruded during Cambrian times and affected by the assyntic tectonics.

The history of the Cratonic platforms in central Argentina, which appear to be related to the Cambrian by virtue of their late cratonic plutonization is little known.

The areas near the border between Buenos Aires Province and La Pampa, to the west of the Sierras Australes, appear to be mainly composed of red granites which, according to their radiometric dating, were emplaced at the end of the late Cambrian or near the beginning of the Early Ordovician.

The extent of continental regions or land areas during the Cambrian was considerable as depositional areas with fluviatile or terrigenous sedimentation must have covered a large part of the cratonic surface of South America which is now below the surface.

The Andean history of South America is in many ways similar to the Pacific history of North America, particularly as it concerns the biofacies of the trilobites. However, the history of the Precordillera of San Juan and Mendoza presents passages in its geotectonic development which are also similar to those which characterize the succession of geological events in the Appalachians of eastern United States.

15. Radiometric dates

The available published information refers to radiometric determinations of minerals from a variety of rocks. It has been compiled from a variety of sources which illustrate examples of South American Cambrian dates, between the limits of 500 and 600 m.y. according with the latest time-scale of KULP (1961).

The information in the following table has been extracted from the work of HERZ and his associates (1961, p. 1112) and relates only to granitic rocks. The second table is taken from a publication by ALDRICH and his associates (1964, p. 329) and provides radiometric ages for a variety of rocks, also from Brazil. In both cases the potassium–argon method was applied; though in the second there are comparative data obtained by the rubidium–strontium method.

Mineral used	Ar^{40}/K^{40} Locality	Ages of Minerals from Granitic Rocks, Minas Gerais, Brazil Quadrangle	Age m.y.
Biotite	Moeda road	Marinho da Serra	600
Muscovite	Barreiro	———	595
Biotite	Casa Braca	Ibirite	550
Muscovite	Serra de Ouro Branco	Dom Bosco	530
Biotite	Moeda	———	514
Biotite	Itabirá	Itabirá	500

Ages of Brazilian Rocks

Rock	Location	Mineral	Rb-Sr Age, m.y.	K-Ar Age, m.y.
Schist	Minas Gerais 20° 17·2′ S 43° 57·6′ W	Biotite	530	610
Pegmatite, amazonite	19° 53·7′ S 43° 10·7′ W	Feldspar	545	
Granite, gneiss	Sergpipe and Bahía 9° 28′ S 38° 15′ W	Biotite	600	570
Granodiorite	10° 15′ S 37° 53′ W	Biotite	610	500
Granodiorite	10° 09′ S 37° 29′ W	Biotite	585	580
Schist	10° 16′ W 37° 27′ W	Biotite	550	600

Geochronological determinations of minerals from pegamtites by the uranium–thorium–lead method and of granite by the potassium–argon method have been presented by GUIMARAES (1964, pp 287–288, Tables X and XI) from whose publication the following data have been extracted: (shown on next page)

CORDANI and his associates (1967) suggested for the Bambuí Formation (cratonic zone of the San Francisco region, Brazil), an age of approximately 600 m.y. They added that the South American orogenic events occurred within the limits of 400–600 m.y., by reference to age determinations of some of the associated rocks which come from the north-east and east of Brazil.

Age determinations of minerals from Brazilian pegmatites by the uranium, thorium, lead method.

Mineral	Source	Ages m.y.	Region
Fergusonite	Fazenda Guandú	547·9	Espiríto Santo
Fergusonite	Fazenda Guandú	569·4	Espiríto Santo
Polycrasite	Faz. Sta. Clara, Pomba, M.G.	546	—
Polycrasite	Faz. Sta. Clara, Pomba, M.G.	569	—
Samarskite	Divino de Ubá, M.G.	523	—
Samarskite	Divino de Ubá, M.G.	557	—

Preliminary radiometric data from Argentina have been provided by LINARES (1961, pp 207–209) through chemical and isotopic methods, and the results of his analyses on rocks of the basement from the middle section of the country are given below:

	Isotopic age determinations					
Source	Pb^{206}/U^{238}	Pb^{207}/U^{235}	Pb^{207}/Pb^{206}	Aver. Age m.y.	Rock	Observations
Las Tapias, Córdoba	514 ± 1	516 ± 11	526 ± 64	520 ± 15	pegmatite	without correction
Las Tapias, Córdoba	513 ± 1	$502 \pm 11·3$	456 ± 64	500 ± 15	pegmatite	with correction

Other radiometric data obtained on the granodiorites of Tastil (Salta, northern Argentina) for which an average age of 530 m.y. is accepted (Shell Capsa Argentina) are based on the following analyses (unpublished information):

	Comparative determinations biotite-orthoclase
Biotite	Rb-Sr Method = 563 m.y. $\pm$ 30 m.y.
	K-Ar Method = 507 m.y. $\pm$ 20 m.y.
Orthoclase	Rb-Sr Method = 540 m.y. $\pm$ 40 m.y.
	K-Ar Method = 440 m.y. $\pm$ 20 m.y.

The granodiorite of the Sierra de Santa Victoria, in the Cordillera Oriental, Jujuy, northern Argentina, is considered to be similar to that previously mentioned in Tastil in the same Andean region.

Acknowledgements

The author has been considerably assisted by written discussions with Professor HORACIO J. HARRINGTON of Texas, U.S.A. concerning the stratigraphical material. Kind assistance has also been given by Professor GERADO BOTERO ARANGO, University of Medellin, Colombia and Professor JEAN-CLAUDE VICENTE of the Institute of Geology of the University of Santiago, Chile, on matters concerning the Bibliography. The attention given by Professor JOHN REYNOLDS of the Department of Physics, University of California at Berkeley, in answering many questions concerning geochronological investigations is greatly esteemed. Dr. NORMAN HERZ of the U.S. Geological Survey is thanked for drawing the author's attention to many of the pertinent references.

The geologist UMBERTO CORDANI of the Laboratory of Geochronology, Sao Paulo, Brazil, provided valuable data and opinions regarding geochronological material on the Lower Palaeozoic of his country; at the same time mention must be made of the most helpful collaboration provided by Dr. MARCELO MESIGOS of Shell Capsa, Argentina, who provided the author with geochronological analyses of rocks from northern Argentina, which had been carried out in Texas, U.S.A.

The faunal lists of the Cambrian of the Precordillera of western Argentina used in this study include forms which have been studied and revised with the valuable cooperation of Professor PIERRE HUPÉ of the Faculty of Science in Paris. Professor CHRISTIAN POULSEN of the Museum of Mineralogy and Geology of the University of Copenhagen is thanked for allowing the author to consult his collections of trilobites from Argentina.

To Professors ARTURO AMOS, ALFREDO J. CUERDA, and JUAN CARLOS M. TURNER of the Faculty of Natural Sciences and Museum of the National University of La Plata, Argentina the author offers sincere gratitude for reading the original manuscript and making helpful suggestions which were taken into account during final revision.

Sincere thanks are offered to the scientific personnel of the Geological Section of that Faculty for their help in the library, in the laboratory, and above all in the field during studies of the revision of the Cambrian, help which has proved most useful in the preparation of the present contribution. The same must be said for the staff of the Photographic Laboratory of the Museum of La Plata whose efficient work has been much appreciated.

Finally the author wishes to express a personal acknowledgement to the Comisión de Investigación Científica de la Provincia de Buenos Aires, La Plata, Argentina for their assistance in the preparation of some of the graphic material presented here.

Bibliography

AHLFELD, FEDERICO (1946). Geología de Bolivia. *Rev. Mus. La Plata (N. Ser.) Secc. Geol.*, **3**.

AHLFELD, FEDERICO Y BRANIŠA, LEONARDO (1960). *Geología de Bolivia*. D. Bosco. La Paz.

ALDRICH, L. T., HART, S. R., TILTON, G. R., DAVIS, G. L., RAMA, S. N. I., STEIGER, R., RICHARDS, J. R., and GERKEN, J. S. (1964). Isotope Geology. In: *Annual Report of the Director of the Department of Territorial Magnetism. Carnegie Inst. Washington Year Book*, **63**, 328.

ALMEIDA, FERNANDO F. M. DE (1945). Geología do Sudoeste Mattogrossense. *Brasil Div. Geol. Min. Bol.*, **116**.

AMOS, ARTURO J., CAMACHO, HORACIO, H., CASTELLARO, HILDEBRANDA A., and MENDENEZ, CARLOS, A. (1963). Guía Paleontologica Argentina. *Publ. Cons. Nac. Inv. Cient. y Técn.* Bs. Aires.

AVÉ LALLEMANT, GERMÁN (1892). Observaciones sobre el Mapa del Departamento de Las Heras. I. Provincia de Mendoza. *An. Mus. La Plata, Secc. Geol. y Miner.* La Plata.

BARBOSA, OCTAVIO (1949). Contribuçao a Geología da região Brasil-Bolivia. *Miner. e Metal.*, **13**, 271.

BARBOSA, OCTAVIO (1957). Sobre a idade de Serie de Corumbá. *An. Acad. Brasil. Cienc.*, **29**, 249.

BATEMAN, ALAN M. (1950). *Economic Mineral Deposits*. Yale Univ. 2nd. Ed. (Spanish Ed. *Yacimientos minerales de rendimiento económico*. Omega. Barcelona. 1957).

BEDER, R. (1928). Los yacimientos de mineral de plomo en el departamento de Yavi, provincia de Jujuy. *Publ. Dir. Gral. Min. y Geol.*, **38**. Bs. Aires.

BEURLEN, KARL and SOMMER, FRIEDRICH W. (1957). Observaçoes estratigráficas e paleontológicas sobre o Calcario Corumbá. *Bol. Dept. Nac. Prod. Min., DGM,* **168.**

BONARELLI, GUIDO (1921). Tercera contribución al concoimiento geológico de las regiones petrolíferas del norte. (Provincias de Salta y Jujuy). *An. Min. Agric. Nac., Secc.,* **XV,** 1. Bs. Aires.

BONARELLI, GUIDO y PASTORE, FRANCO (1918–1919). Bosquejo geológico de la Provincia de Tucumán. *Prim. Reun. Nac. Soc. Arg. Cienc. Naturales.* (Tucumán, 1916), **27.** Bs. Aires.

BORELLO, ANGEL V. (1962). Caliza La Laja (Cámbrico Medio-San Juan). *Not. Com. Invest. Cient. Prov. Bs. Aires,* **I,** (2).

BORELLO, ANGEL V. (1963a). Fremontella inopinata n.sp. del Cámbrico de la Argentina. *Ameghiniana,* **III,** 51.

BORELLO, ANGEL V. (1963b). Elementos del magmatismo simaico en la correlación de la secuencia geosinclinal de la Precordillera. *Mus. Arg. Cienc. Nat. "Bernardino Rivadavia", Inst. Nac. Inv. Cienc. Naturales, Cienc. Geol.,* **I,** (19), Bs. Aires.

BORELLO, ANGEL V. (1964). Sobre la presencia del Cámbrico inferior Olenellidiano en la sierra de Zonda, Precordillera de San Juan. *Ameghiniana,* **III,** 313.

BORELLO, ANGEL V. (1965a). Cámbrico. In: Borrello, Angel V. (Ed.). Indice bibliográfico de Estratigrafía Argentina. *Com. Invest. Cient. Prov. Bs. Aires,* 63.

BORELLO, ANGEL V. (1965b). Sobre dl desarrollo bioestratigráfico del Cámbrico de la Precordillera. *Act. Geol. Lilloana,* **VII,** 39 (Vol. III, *Seg. Jorn. Geol. Argentinas*).

BORRELLO, ANGEL V. (1965c). Sistemática estructural sedimentaria en los procesos de la orogénesis. *An Com. Invest. Cient. Prov. Bs. Aires,* **VII,** 65.

BORRELLO, ANGEL V. (1966.) Geosinclinales. *Centr. Argentino Geol.* (Confer.). Bs. Aires.

BORELLO, ANGEL V. (1967). El Género *Elrathia* (Trilobita) en el Cámbrico de San Juan *Ameghiniana,* V, (4) 158.

BORRELLO, A. V. (1969). Los Geosinclinales de la Argentina. *An. Inst. Nac. Geol. Minería.* XIV. Bs. Aires.

BORRELLO, ANGEL V. y PERNAS, R. D. (1965). Sobre la presencia del género Kootenia en el Cámbrico de San Juan. *Act. Geol. Lilloana,* **VII,** 57 (Vol. III *Seg. Jorn. Geol. Argentinas*).

BRACKEBUSCH, LUIS (1891). MAPA GEOLOGICO DEL INTERIOR DE LA REPUBLICA ARGENTINA. Scale 1 : 1.000.000, color sheets, Gotha.

CECIONI, GIOVANNI (1956). Primeras Noticias sobre la Existencia del Paleozoico superior en el Archipiélago Patagónico entre los Paralelos 50° y 52° S. *Publ. Inst. Geol. Univ. Chile,* **8,** 183.

CHAMOT, GUY A. (1962). Bosquejo geológico de la plataforma del Escudo Brasileño en el oriente Chiquitano-Bolivia. *Bol. Inst. Bol. Petrol.,* **3,** (4), 11.

COMISION DE LA CARTA GEOLOGICA DEL MUNDO. MAPA GEOLOGICO DE AMERICA DEL SUR. (1964). Escala 1 : 5.000.000, 2 sheets. Brasil.

CORDANI, U. G., MELCHER, G. C., and ALMEIDA F. F. M. DE (1967). Outline of Precambrian Geochronology of South America. (Abstract). Geochronology Conference Edmonton. Canadá.

DANIEL, JOAQUIN (1940). Sobre la constitución, disposición transgresiva ya tectónica de los estrator mesozoicos en Alfarcito, departamento de Tilcara (Provincia de Jujuy). *Tesis Mus. La Plata,* **3.**

DE FERRARIIS, CARMELO, I. C. (1940). Corrimientos de bloques de montaña en los alrededores de Purmamarca, departamento de Tumbaya (Provincia de Jujuy). *Tesis Mus. La Plata,* **1.**

DERBY, O. A. (1906). The Serra do Espinhaço, Brazil. *J. Geol.,* **14,** 394.

DE ROMER, HENRY S. (1964). Sobre la geología de la zona de "El Choique" entre el Cordón de los Farallones y el Cordón de Bonilla, Quebrada Santa Elena, Uspallata (Provincia de Mendoza). *Rev. Asoc. Geol. Argentina,* XIX, 9.

DORR, JOHN VAN N. 2d. (1945). Manganese and iron deposits of Morro de Urucúm, Mato Grosso, Brazil. *U.S. Geol. Survey Bull.,* **946-A.**

ECKEL, E. B. (1959). *Geology and mineral resources of Paraguay-A reconnaissance* (with sect. on Igneous and metamorphic rocks, by Ch. Milton and E. B. Eckel; Soils by P. Tirado Sulsona). *U.S. Geol. Survey, Prof. Paper,* **327.**

FERUGLIO, EGIDIO (1931). Observaciones geológicas en las provincias de Salta y Jujuy. YPF, *Contr. Prim. Reun. Nac. Geogr.,* **7,** 5. Bs. Aires.

FRAENKL, E. J. (1959). La Formación Limbo. *Bol. YPF Boliviano,* **2** (5).

FREITAS, RUY OZORIO DE (1945). O conglomerado do Bau (Serie Itajai-Santa Catarina). *Bol. Univ. Sao Paulo, Fac. Fil. Cienc. Letr.,* **50,** Geol. (2), 37.

FURQUE, GUILLERMO (1936). Descripción geológica de la Hoja 17b Guandacol, prov. La Rioja.-prov. San Juan. *Bol. Dir. Nac. Geol. y Min.*, **92**. Bs. Aires.

GARCÍA, ERNESTO (1951). Contribución al conocimiento de la Precordillera Mendocina. *Act. XV Sem. Geogr. GAEA*, 491. República Argentina.

GERTH, HEINRICH (1955). *Der Geologische Bau der Südamerikanischen Kordillere Gebr.* Borntraeger Berlin-Nikolasee.

GONZALEZ DIAZ, EMILIO F. (1958). Estructuras del basamento y del Neopaleozoico en los contrafuertes nordorientales del Cordón del Portillo, provincia de Mendoza. *Rev. Asoc. Geol. Argentina*, **XII**, 65.

GUIMARAES, DJALMA (1964). Geología do Brasil. *Mem. Dept. Prod. Min.*, **1**.

HARRINGTON, HORACIO J. (1937). On some Ordovician fossils from northern Argentina. *Geol. Mag.*, **74**, 97.

HARRINGTON, HORACIO J. (1950). Geología del Paraguay oriental *Contrib. Cient. Fac. Cienc. Ex. Fís. y Nat. Univ.*, **1**, Bs. Aires.

HARRINGTON, HORACIO J. (1956). Argentina. In: Jenks, Williams F. (Ed.). Handbook of South American Geology. *Mem. Geol. Soc. Amer.*, **65**.

HARRINGTON, HORACIO J. (1961). The Cambrian formations of South America. (El Sistema Cámbrico, Symp.) *Congr. Geol. Int. 20th*, México, 3, 504.

HARRINGTON, HORACIO J. (1962). Paleogeographic development of South America. *Bull. Amer. Assoc. Petr. Geol.*, **46**, 1773.

HARRINGTON, HORACIO J. y LEANZA, ARMANDO F. (1943). Paleontología del Paleozoico inferior de la Argentina. I. Las faunas del Cámbrico medio de San Juan. *Rev. Mus. La Plata*, (*N. Ser.*) 2, Secc. Pal., 207.

HARRINGTON, H. J. and LEANZA, A. F. (1957). Ordovician trilobites of Argentina. *Sp. Publ., Dept of Geol. Kansas Univ.*, **1**.

HARRINGTON, HORACIO J. and KAY, MARSHALL (1951). Cambrian and Ordovician faunas of Eastern Colombia. *J. Paleont.*, **25**, 655.

HAUSEN, J. (1925). Sobre un perfil geológico del borde oriental de la Puna de Atacama. *Bol. Acad Nac. Cienc. Córdoba*, **XXVIII**, 1.

HAUSEN, H. (1930). Geologische Beobachtungen in den Hochgebirgen der Provinzen Salta und Jujuy Nordwestargentinien. *Act. Geographica*, 3, (1).

HEIM, ARNOLD (1940). Geological observations in the Patagonian Cordillera (Preliminary report). *Eclog. Geol. Helvet.*, 33, 25.

HERZ, NORMAN, HURLEY, P. M., PINSON, W. H., and FAIRBAIRN, H. W. (1961). Age measurements from a part of the Brazilian shield. *Bull. Geol. Soc. Amer.*, **72**, 1111.

HERRERO DUCLOUX, ABEL (1940). Sobre los fenómenos de corrimiento en ambos lados dela quebrada de Juella, departamento de Tilcara (Provincia de Jujuy). *Tesis Mus. La Plata*, **2**.

HUTCHINSON, R. D. (1956). Cambrian stratigraphy, correlation and paleogeography of Eastern Canada. (El Sist. Cámbrico, Symp.) *Congr. Geol. Intern.* 20th, Mexico, 2, 289.

KAYSER, E. (1876). Ueber Primordiale und Untersilurische Fossilien aus der Argentinischen Republik. Palaeontographica, **Supp. III**, (2).

KAYSER, E. (1897). Beitraege zur Kenntniss einiger Faunen Sudamerikas. *Zeits Deutcsh. Geol. Gesellschaft*, **XLIX**, 274.

KEGEL, WILHELM, PENNA SCORZA, EVARISTO and CHAGAS PINTO COELHO, FRANCISCO DAS (1958). Estudios geológicos no Norte do Ceará. *B:. Brasil Div. Geol. Min.*, **184**.

KEIDEL, H. (1907). Ueber den Bau der Argentinischen Anden. *Sitzungsb. K. Akad. Wiss. Wien, Math.-Nat. Klas.*, **CXVI**, 649.

KEIDEL, J. (1910). Estudio geológico de la quebrada de Humahuaca, en la de Iruya y en la de algunos de sus valles laterales. Provincias de Jujuy y Salta. In: Memoria Div. Minas, Geol. e Hidrog. 1908. *An Minist. Agric. Nac. Secc. Geol.*, V, (2), 76. Bs. Aires.

KEIDEL, J. (1911). Sobre los progresos en la exploración geológica de la Republica Argentina Guía Expos. Turín.

KEIDEL, H. (1912). Die neueren Ergebnisse der staatslichen geologischen Untersuchungen. *Inter. Geol. Congr.* 11th, Stockholm, **II**, 1127.

KEIDEL, J. (1914). Ueber das alter die Verbreitung und die Gegenseitigen Beziehungen der Verschiedenen tecktonischen struktur in den Argentinischen Gebirgen. *Compt. Rend. Congr. Geol. Intern.*, 12th, 671.

KEIDEL, J. (1917). Noticias sobre exploraciones geológicas en la provincia de Jujuy. *Physis, Rev. Asoc. Arg. Cienc. Nat.*, **III**, 112, Bs. Aires.

KEIDEL, J. (1937). La prepuna de Salta y Jujuy. *Rev. Centr. Est. Doct. Cienc. Nat. Bs. Aires*, **1**, 125.

KEIDEL, J. (1939). Las estructuras de corrimientos paleozoicos de la sierra de Uspallata. (Provincia de Mendoza). *Physis, Rev. Asoc. Arg. Cienc. Nat.* **XIV**, (*Geol. y Paleont.*), **3**. Bs. Aires.

KEIDEL, J. (1943). El Ordovícico inferior de los Andes del Norte argentino y sus depósitos marino glaciales. *Bol. Acad. Nac. Cienc. Córdoba*, **XXXVI**, 140.

KEIDEL, J. (1947). El Paleozoico. In: Geografía de la República Argentina. GAEA, **I**, 127, Bs. Aires.

KITTL, ERWIN (1934). Informe preliminar sobre un viaje de estudio en la zona cordillerana del sur de Mendoza. *Rev. Min.*, **6**, 33.

KITTL, ERWIN (1960). Sobre la formación de algunas dolomitas y magnesitas del país. *Rev. Min.*, **25**, 77.

KOBAYASHI, TEIICHI (1937). The Cambro-Ordovician shelly faunas of South America. *J. Fac. Sci. Imp. Univ. Tokio*, **II** (Geol., IV), (4).

LEANZA, ARMANDO F. (1947). El Cambrico medio de Mendoza. *Rev. Mus. La Plata*, (*N. Ser.*), Secc. *Pal.*, **III**, 223.

LEANZA, ARMANDO F. (1958). Geología Regional. In: La Argentinea. Suma de Geografía. **I**. (3), 277. Peuser, Bs. Aires.

LEVORSEN, A. I. (1945). Geological map of South America. Part 2. *Sp. Pap. Geol. Soc. Amer.*, **61**.

LINARES, ENRIQUE (1961). Los métodos geocronológicos y algunas edades de minerales de la Argentina, obtenidos por medio de la relación plomouranio. *Rev. Assoc. Geol. Argentina*, **XIV**, 181.

LISBOA, MIGUEL ARROJADO RIBEIRO (1909). Oeste de Sa Paolo, Sul de Matto Grosso Brasil. Estr. Ferro Noroeste. *Relat. Com. Schnorr.* Río de Janeiro, 172.

LOCHMAN-BALK, CHRISTINA and WILSON, JAMES LEE (1958). Cambrian biostratigraphy in North America. *J. Paleont.*, **32**, 312.

MAACK, REINHARD (1947). Breves noticias sobre a geología das Estados do Paraná e Santa Catarina. *Arg. Biol. Tecnol.*, **II**, 67. Curtiba.

MORAES, LUCIANO JACQUES de y otros (1937). Geología económica de norte de Mina Geraes. *Bol. Serv. Form. Prod. Min.* ,**19**.

MORALES, L. G. (1959). General Geology and oil possibilities of the Amazonas Basin, Brazil. 5th *World Petrol. Congress., Proc. Secc.*, **1**, 51. New York.

OLIVEIRA, A. I., DE y LEONARDOS, O. N. (1943). Geología do Brazil. *Min. Agric. Ser. Did.*, **2**, 2a. ed. Río de Janeiro.

OLIVEIRA, AVELINA IGNACIO DE (1956). Brazil. In: Jenks, Williams F. (Ed.) Handbook of South American Geology. *Mem. Geol. Soc. Amer.*, **65**.

PERNAS, RICARDO DARIO (1963). El perfil de la quebrada de la Cruz, Mendoza. *Prim. Reun. Comun. Cient. Div. Geol. Fac. Cienc. Nat. Mus. La Plata*, in *Rev. Asoc. Geol. Arg.*, **XVIII**, 107. 1962. Bs. Aires.

PERNAS, DARIO A. (1964). El género Protaspongia en el Cámbrico de Mendoza. *Seg. Reun. Com. Cient. Div. Geol. Fac. Cient. Nat. Mus. La Plata, CIC, Resúmenes, Misc.*, **1**.

POLANSKI, J. (1958). El bloque varíscico de la Cordillera Frontal de Mendoza. *Rev. Asoc. Geol. Argentina*, **XII**, 165.

POULSEN, CHR. (1960). Fossils from the late Middle Cambrian Bolasipdella zone of Mendoza, Argentina. *Mat. Fys. Medd. Dan. Vid. Selsk*, **32**, (11).

POULSEN, VALDEMAR (1958). Contributions to the Middle Cambrian paleontology and stratigraphy of Argentina. *Mat. Fys. Medd. Dan. Vid. Selks*, **31**, (8).

REYES, JULIO CESAR (1966). Reconocimiento geológico del brazo nordoriental del Lago San Martín. Sierra de Sangra. *An Univ. d. Salvador*, **2**, 321. Bs. Aires.

ROD, E. (1955). Trilobites in 'Metamorphic' rocks of El Baul, Venezuela, *Bull. Amer. Assoc. Petroleum Geol.* **39**, 1865.

RUSCONI, CARLOS (1945a). Trilobites silúricos de Mendoza. *An. Soc. Cient. Arg.*, **CXXXIX**, 216.

RUSCONI, CARLOS (1945b). Nuevos trilobites del Cámbrico de Mendoza. *Bol. Paleont. Bs. Aires*, **19**.

RUSCONI, CARLOS (1946a). Varias especies de trikobitas y sesterias del Cámbrico de Mendoza. *Rev. Soc. Hist. Geogr. Cuyo.* **1**, 1.

RUSCONI, CARLOS (1946b). Los trilobites del Cámbrico de Mendoza. *Bol. Soc. Geol. Perú*, **XIX**, 45.

RUSCONI, CARLOS (1947). Especie de trilobites del Cámbrico de Mendoza. *An Soc. Cient. Arg.*, **CXLIV**, 560.

RUSCONI, CARLOS (1948). Apuntes sobre el Triásico y el Ordovícico de El Challao, Mendoza. *Rev. Mus. Hist. Nat. Mendoza*, **II**, 165.

RUSCONI, CARLOS (1949). Neuvo género de trilobita del Cámbrico medio de Mendoza. *Rev. Mus. Hist. Nat. Mendoza*, **III**, 212.

RUSCONI, CARLOS (1950a). Diferentes organismos del Ordovícico y del Cámbrico de Mendoza. *Rev. Mus. Hist. Nat. Mendoza*, **IV**, 63.

RUSCONI, CARLOS (1950b). Trilibitas y otros organismos del Cámbrico de Canota. *Rev. Mus. Hist. Nat. Mendoza*, **IV**, 71.

RUSCONI, CARLOS (1950c). Nuevos trilobitas y otros organismos del Cámbrico de Canota. *Rev. Mus. Hist. Nat. Mendoza*, **IV**, 85.

RUSCONI, CARLOS (1951a). Más trilobitas cámbricos de San Isidro, Cerro Pelado y Canota. *Rev. Mus. Hist, Nat. Mendoza*, **V**, 3.

RUSCONI, CARLOS (1951b). Trilobitas cámbricos del Cerro Pelado (Mendoza). *Bol. Paleont. Bs. Aires*, **24**.

RUSCONI, CARLOS (1952a). Varias especies de trilobitas del Cámbrico de Canota. *Rev. Mus. Hist. Nat. Mendoza*, **VI**, 5.

RUSCONI, CARLOS (1952b). Fósiles cámbricos del cerro Adpero, Mendoza. *Rev. Mus. Hist. Nat. Mendoza*, **VI**, 63.

RUSCONI, CARLOS (1953a). Lista de agnóstodis e hiolites del Cámbrico de Mendoza. *An Soc. Cient. Arg.*, **CLVI**, 3.

RUSCONI, CARLOS (1953b). Trilobitas ordovícos y cámbricos de Mendoza. *Bol. Paleont. Bs. Aires*, **25**.

RUSCONI, CARLOS (1953c). Nuevos trilobitas del Cámbrico de la Quebrada de La Cruz. *Bol. Paleont. Bs. Aires*, **27**.

RUSCONI, CARLOS (1953d). Siete especies de trilobitas del Cámbrico de la Quebradita Oblicua, Sud del Cerro Aspero. *Bol. Paleont. Bs. Aires*, **28**.

RUSCONI, CARLOS (1954a). Trilobitas cámbricos de la Quebrada Oblicua, Sud del cerro Aspero. *Rev. Mus. Hist. Nat. Mendoza*, **VII**, 3.

RUSCONI, CARLOS (1954b). Las peizas "tipos" del Museo de Mendoza. *Rev. Mus. Hist. Nat. Mendoza*, **VII**, 81.

RUSCONI, CARLOS (1954c). Neuvas especies cámbricas del cerro Aspero. *Bol. Paleont. Bs. Aires*, **29**.

RUSCONI, CARLOS (1954d). Fósiles cámbricos y ordovícicos de San Isidro. *Bol. Paleont. Bs. Aires*, **30**.

RUSCONI, CARLOS (1955a). Más fósiles cámbricos y ordovícicos de San Isidro, Mendoza. *Bol. Paleont. Bs. Aires*, **31**.

RUSCONI, CARLOS (1955b). Notas previas sobre organismos ordovícos y cámbricos de San Isidro, Mendoza. *Bol. Paleont. Bs. Aires*, **32**.

RUSCONI, CARLOS (1955c). Fósiles cámbricos y ordovícosic al Oeste de San Isidro, Mendoza. *Rev. Mus. Hist. Nat. Mendoza*, **VIII**, 3.

RUSCONI, CARLOS (1956a). Mares y organismos extinguidos de Mendoza. *Rev. Mus. Hist. Nat. Mendoza*, **IX**, 2.

RUSCONI, CARLOS (1956b). Fósiles cámbricos al Sud de El Totoral. *Rev. Mus. Hist. Nat. Mendoza*, **IX**, 115.

RUSCONI, CARLOS (1956c). (Listas de géneros y especies fundadas por el autor). *Rev. Mus. Hist. Nat. Mendoza*, **IX**, 121.

RUSCONI, CARLOS (1956d). Correlaciones cambro-ordovicicas entre Mendoza y Norteamérica. (El Sistema Cámbrico, Symp.). *Congr. Geol. Int. 20th*, México, **2**, 751.

RUSCONI, CARLOS (1958a). Nuevos trilobitas de la Quebrada Oblicua. *Rev. Mus. Hist. Nat. Mendoza*, **XI** 81.

RUSCONI, CARLOS (1958b). Nuevos fósiles cámbricos de El Totoral, Mendoza. *Rev. Mus. Hist. Nat. Mendoza*, **XI**, 93.

RUSCONI, CARLOS (1958c). Trilobites olénidos de Mendoza y de otras regiones del Mundo. *Rev. Mus. Hist. Nat. Mendoza*, **XI**, 109.

RUSCONI, CARLOS (1959). El Cámbrico medio de Mendoza según Pouslen. *Rev. Mus. Hist. Nat. Mendoza*, **XII**, 3.

RUSCONI, CARLOS (1962a). Discusión acerca de alguhos trilobites del Cámbrico de Mendoza, Argentina. *Rev. Mus. Hist. Nat. Mendoza*, **XIV**, 3.

RUSCONI, CARLOS (1962b). Fósiles cámbricos del Oeste de la Quebrada de la Cruz. *Rev. Mus. Hist. Nat. Mendoza*, **XIV**, 55.

RUSCONI, CARLOS (1962c). Correlación de organismos Cambro-ordovícicos de Mendoza. *Rev. Mus. Hist. Nat. Mendoza*, **XIV**, 97.

RUSHTON, A. W. A. (1963). Paradoxides from Colombia. *Geol. Mag.*, **100**, 265.

SANFORD, ROBERT M. and LANGE, F. W. (1960). Basin-study approach to oil evaluation of Parana miogeosyncline of South Brazil. *Bull. Amer. Assoc. Petr. Geol.*, **44**, 1316.

SCHMIEDER, OTTO (1926). The east Bolivian Andes-South of the Rio Grande or Guapay. *Univ. Calif. Publ. Geogr.*, **II**, (5), 85.

SPENCER, FRANK N. (Jr) (1950). The geology of the Aguliar lead—zinc mine. *Econ. Geol.* **45**, 405.

STAPPENBECK, RICHARD (1910). La Precordillera de San Juan y Mendoza. *An. Min. Agric. Nac., Secc. Geol.*, **IV**, (3), Bs. Aires.

STAPPENBECK, R. (1917). Geologia de la falda oriental de la Cordillera del Plata (Prov. de Mendoza). *An Min. Agric. Nac. Secc. Geol.*, **XII**, (1), Bs. Aires.

STEINMANN, G. und HOEK, H. (1912). Das Silur and Cambrian des Hochlandes von Bolivia und ihre Fauna. *N. Jahrb. f. Miner. Geol. u. Pal.*, **XXXIV**, 176.

STELZNER, ALFRED (1885). Beitraege zue Geologie und Paleontologie der Argentinischen Republik. I. Geol. Teil. Cassel (Spanish text. *in: Act. Acad. Cienc. Córdoba*, **VIII**, 1924).

STOSE, GEORGE W. (1950). GEOLOGIC MAP OF SOUTH AMERICA. Approx. scale 1 : 5.000.000 (published by The Geol. Soc. Amer.). 2 sheets.

TURNER, J. C. M. (1959). Estratigrafía del Cordón de Escaya y de la Sierra de Rinconada (Jujuy). *Rev. Asoc. Geol. Argentina*, **XIII**, 15.

TURNER, J. C. M. (1960). Estratigrafía de la Sierra de Santa Victoria y adyacencias. *Bol. Acad. Nac. Cienc. Córdoba*, **XLI**, 163.

TURNER, JUAN CARLOS M. (1963a). The Cambrian of Northern Argentina. (Symp. on Petrol. Geol. South America)., *Tulsa Geol. Soc. Dig.*, **31**, 193.

TURNER, JUAN C. M. (1963b). Perfil transversal de la Puna, latitud 22° 15'S aproximadamente. *An. Seg. Jorn. Geol. Argentinas.* (en prensa).

TURNER, JUAN CARLOS M. (1964a). Descripción geológica de la Hoja 7c–Nevado de Cachi (Provincia de Salta). *Bol. Dir. Nac. Geol. y Min.* **99**, Bs. Aires.

TURNER, JUAN CARLOS M. (1964b). Descripción geológica de la Hoja 2c–Santa Victoria (Provincia de Salta y Jujuy). *Bol. Inst. Nac. Geol. y Min.*, **104**. Bs. Aires.

TERMIER, HENRI and TERMIER, GENEVIEVE (1964). *Les temps fossilifères.* I. Paleozoique inférieur. Masson, París.

VILELA, CESAR R. (1956). Descripción geológica de la Hoja 7d, Rpsario de Lerma (Provincia de Salta). *Bol. Div. Nac. Minería*, **84**. Bs. Aires.

VILELA, CESAR R. (1961). Algunos rasgos particulares de la Geología de Iruya (Salta-Jujuy). *Rev. Asoc. Geol. Argentina*, **XV**, 119.

WEEKS, L. G. (1947). Paleogeography of South America. *Bull. Amer. Assoc. Petr. Geol.*, **31**, 1194. (ibid. (1948) *Bull. Geol. Soc. Amer.*, **59**, 249).

WILLIAMS, HORACIO E. (1926). Notas sobre a geología e recursos mineraes do norte do Ceará. *Bol. Serv. Geol. e Min. do Brasil*, **16**. Río de Janeiro.

WILLIAMS, HORACIO E. (1930). Estudios geológicos na Chapada diamantina Estado da Bahiá. *Bol. Serv. Geol. Min.*, **44**, Río de Janeiro.

WINDHAUSEN, ANSELMO (1931). *Géología Argentina.* Seg. Part., Geología Histórica y Regional del Territorio Argentino. Peuser. Bs. Aires.

WOKITTEL, ROBERTO (1960). Recusros minerales de Colombia. *Serv. Geol. Nac.*, Bogotá.

ZARDINI, RAÚL ALBERTO (1959). Serpentinitas de río de las Tunas, Cuchilla de Yalguaraz, Mendoza. *Rev. Asoc. Geol. Argentina*, **XIII**, 67 (1958).

ZARDINI, RAÚL A. (1961a). Serpentinitas de la mina "La Mendocina", Uspallata (Mendoza). *Rev. Asoc. Geol. Argentina*, **XV**, 43. (1960).

ZARDINI, RAÚL A. (1961b). Esquito talco-actinolítico en la mina "Sol de Mayo", (Mendoza). *Rev. Asoc. Geol. Argentina*, **XV**, 181 (1960).

ZARDINI, RAÚL A. (1962). Significade geológico de las serpentinitas de Mendoza. *An. Prim. Jorn. Geol. Argentinas.* **II**, 437, 1960.

INDEX